Xuan Wang, Sajid Bashir, Jingbo Louise Liu (Eds.)
Nanochemistry

Also of interest

Materials Science.
Volume 2: Phase Transformation and Properties
In cooperation with: Shanghai Jiao Tong University Press
Gengxiang Hu, Xun Cai and Yonghua Rong, 2021
ISBN 978-3-11-049515-7, e-ISBN 978-3-11-049537-9

Nanomaterials.
Volume 1: Electronic Properties
Engg Kamakhya Prasad Ghatak and Madhuchhanda Mitra, 2019
ISBN 978-3-11-060922-6, e-ISBN 978-3-11-061081-9

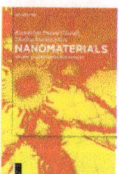

Nanomaterials.
Volume 2: Quantization and Entropy
Engg Kamakhya Prasad Ghatak and Madhuchhanda Mitra, 2020
ISBN 978-3-11-065972-6, e-ISBN 978-3-11-066119-4

Nanoscience and Nanotechnology.
Advances and Developments in Nano-sized Materials
Marcel Van de Voorde (Ed.), 2018
ISBN 978-3-11-054720-7, e-ISBN 978-3-11-054722-1

Nanochemistry

From Theory to Application for In-Depth Understanding
of Nanomaterials

Edited by
Xuan Wang, Sajid Bashir, Jingbo Louise Liu

DE GRUYTER

Editors
Prof. Xuan Wang
Texas A&M University Higher Education Center at McAllen
6200 Tres Lagos Blvd.
McAllen, TX 78504-9703
United States of America
xuan.wang@science.tamu.edu

Prof. Sajid Bashir
Texas A&M University-Kingsville
MSC 161, 700 University Blvd., Kingsville, TX 78363-8202
United States of America
br9@tamuk.edu

Prof. Jingbo Louise Liu
Texas A&M University-Kingsville
MSC 161, 700 University Blvd., Kingsville, TX 78363-8202
United States of America
and
Texas A&M Energy Institute
Frederick E. Giesecke Engineering Research Building
TAMU, TX 77843-3372
United States of America
kfjll00@tamuk.edu

ISBN 978-3-11-073985-5
e-ISBN (PDF) 978-3-11-073987-9
e-ISBN (EPUB) 978-3-11-073999-2

Library of Congress Control Number: 2022939582

Bibliographic information published by the Deutsche Nationalbibliothek
The Deutsche Nationalbibliothek lists this publication in the Deutsche Nationalbibliografie;
detailed bibliographic data are available on the internet at http://dnb.dnb.de.

© 2023 Walter de Gruyter GmbH, Berlin/Boston
Cover image: Olemedia/E+/getty images
Typesetting: Integra Software Services Pvt. Ltd.
Printing and binding: CPI books GmbH, Leck

www.degruyter.com

Contents

Section 4: **Focus on select example applications of nanoscience in energy, environment, and health**

Common abbreviations

δ	Delta
α	Elevation angle
λ	Wavelength
Φ	Polarization potential
θ	Theta
AC	Activated carbon
ACN	Acetonitrile
AFC	Alkaline fuel cell
AOR	Alcohol oxidation reaction
ASU	Air separation unit
ATR	Autothermal reforming of methane
BFB	Bubbling fluidized bed
BP	Black phosphorus
CB	Conduction band
CBM	Conduction band minimum
CCS	Carbon capture and storage
CCS	CO_2 capture and storage
CCT	Clean coal technology
CE	Coulombic efficiency
CFS	Capacitive faradic storage
C_H	Helmholtz layer capacitance
CHP	Combined heat and power
CLG	Chemical looping gasifier
CNF	Carbon nanofibers
CNT	Carbon nanotubes
CO	Carbon monoxide
CO_2	Carbon dioxide
COF	Covalent organic frameworks
COS	Carbonyl sulfide
CP	Conductive polymers
CQD	Carbon quantum dots
CS_2	Carbon disulfide
C_{SC}	Charge layer capacitance
CTMA	Cetyltrimethylammonium chloride
CV	Cyclic voltammetry
CVD	Chemical vapor deposition
des	Desorption
DFT	Density functional theory
dI	Decrement of the light intensity
DL	Double layer
DMF	*N, N*-Dimethylacetamide
DMFC	Direct methanol fuel cell
DOS	Density of states
dx	Film thickness
E	Occupied by the electrons

https://doi.org/10.1515/9783110739879-203

E_C	CB energy
E_F	Fermi level
ESPW	Electrochemical stable potential window
ESR	Equivalent series resistance
E_V	VB energy
FEC	Fluoroethylene carbonate
FOM	Figure of merit
GHG	Greenhouse gases
GO	Graphene oxides
GPL	Gel polymer electrolyte
H_2O	Water
H_2S	Hydrogen sulfide
HCN	Hydrogen cyanide
HER	Hydrogen evolution reaction
HF	Heating fluid
HHV	Higher heating value
HOMO	Highest occupied molecular orbital
HOR	Hydrogen oxidation reaction
HRSG	Heat recovery steam generator
HTMM	High-temperature mixing method
HyPr-RING	Hydrogen production by reaction-integrated novel gasification
$h\nu$	Photon energy
I	Light intensity
IEA	International Energy Agency
IGCC	Integrated coal gasification combined cycle
IGFC	Integrated coal gasification fuel cell
IHP	Inner Helmholtz plane
ISE	Inorganic solid electrolytes
k	Boltzmann constant
LCI	Life cycle impact
LDH	Layered double hydroxides
LSV	Linear sweep voltammetry
LUMO	Lowest unoccupied molecular orbital
MCFC	Molten carbonate fuel cell
MEA	Membrane-electrode assemblies
MeOH	Methanol
MEXT	Ministry of Education, Culture, Sport, Science, and Technology of Japan
MH	Metal hydride
MIEC	Mixed ionic/electronic conductors
MO	Methyl orange
MOF	Metal-organic frameworks
N_C	Effective densities of CB
NFCS	Non-faradaic capacitive storage
NMP	N-Methyl-2-pyrrolidone
N_V	Effective densities of VB
OER	Oxygen evolution reaction
OHP	Outer Helmholtz plane
OOH	Adsorbed (per) hydroxide species
OOR	Oxidation of organic reactant

ORR	Oxygen reduction reaction
PAFC	Phosphoric acid fuel cell
PAH	Polycyclic aromatic hydrocarbons
PANI	Porous nanoflower polyaniline electrode
PCE	Power conversion efficiency
PCM	Phase change material
PCP	Porous coordination polymers
PEMFC	Proton exchange membrane fuel cell
PHES	Pumped hydroenergy storage
PM	Particulate matter
PPy	Polypyrrole
PVA	Polyvinyl alcohol
QFL	Quasi-Fermi level
RDE	Rotating disk electrode
RDS	Rate-determining step
ROS	Reactive oxygen species
SDC	Static dielectric constant
SEI	Solid electrolyte interphase
SHE	Standard hydrogen electrode
SIB	Sodium-ion batteries
SNCR	Selective noncatalytic reduction
SOFC	Solid oxide fuel cell
SPE	Solid polymer electrolyte
SRM	Steam reforming of methane
SSE	Solid-state electrolyte
STEM	Scanning transmission electron microscopy
STP	Standard temperature and pressure
TES	Total energy supply
TFE	Tetrafluoroethylene copolymer
TGA	Thermogravimetric analysis
TiO_2	Titanium dioxide
TMB	Transition metal based
TMO	Transition metal oxides
TOC	Total organic carbon
TXM	Transmission X-ray microscopy
UPD	Underpotential deposition
VB	Valence band
VBM	Valence band minimum
VOC	Volatile organic compounds
WGS	Water-gas shift
WtE	Waste to energy
XAS	X-ray absorption spectroscopy
XRD	X-ray diffraction

Ashraf Abedin, Sai Raghuveer Chava, Jingbo Louise Liu,
Sajid Bashir

Preface

Nanomaterials for catalysis and remediation: a select review on synthesis, structure, stability, and function

Introduction

This is a book that deals with the aspects of nanotechnology related to bottom-up and top-down synthesis, characterization by electron microscopy and spectroscopy, and nanoapplications. These include nanomaterials in sustainable energy, water remediation, biomedicine, and forensic science. The purpose is to give you a brief overview of the textbook and highlight emerging topics and examples that were not included or areas where the editors are more relevant today. The selected highlights and opinions, by definition, will be unique to each editor. The textbook covers theory, how-to practicality, and multiple examples across multiple areas. The first section on synthesis has wet chemistry related to sol–gel, solvothermal, hydrothermal, and a brief comparison to top-down ball milling. The information is of great interest to undergraduate, graduate, and postgraduate students who may know the experimental approach but not the underlying rationale. This book will also be of great value to technicians and material technologists who are focused on applications by giving them insight and justification of the merits of these synthetic techniques and, lastly, to faculty, researchers, laboratory managers, and directors who need a compendium of universal techniques and detailed examples in handy bound volume.

The advantage of nanomaterials where the dimensions are less than 100 nm is that most of the atomic space is available for chemical interactions, catalysis, and

Authors' contributions and acknowledgments: The bulk of the preface was written by AA, and the last section by SB. SRC collated the data, and XW and JLL edited the sections. This work was partially supported by the Welch Foundation (AC 0006).

Ashraf Abedin, The Cain Department of Chemical Engineering at Louisiana State University, 3307 Patrick F. Taylor Hall, Baton Rouge, LA, 70803, e-mail: mabedi1@lsu.edu
Sai Raghuveer Chava, Department of Chemistry, Texas A&M University-Kingsville, MSC 161, 700 University Blvd., Kingsville, TX, 78363-8202, e-mail: br9@tamuk.edu
Jingbo Louise Liu, Sajid Bashir, Department of Chemistry, Texas A&M University-Kingsville, MSC 161, 700 University Blvd., Kingsville, TX, 78363-8202, e-mail: br9@tamuk.edu; Texas A&M Energy Institute, Frederick E. Giesecke Engineering Research Building, 1617 Research Pkwy, 3372 TAMU, College Station, TX, 77843-3372, Tel: 01 361 593 2919, e-mail: jingbo.liu@tamu.edu

https://doi.org/10.1515/9783110739879-001

bonding [1]. Materials that are inert under bulk conditions are reactive when synthesized on the nanoscale. These strategies have led to an exponential increase in design, synthesis, characterization, and diverse applications [2].

The textbook deals with the various wet synthetic methods of generating nanomaterials. Techniques include sol–gel, solvothermal, hydrothermal, and electrochemical in generating metals, ions, metal oxides, carbon nanotubes, both single- and multiple-walled, and graphene-based nanoparticles (NPs), graphene oxides, and reduced graphene oxides for the multiple catalytic applications [3].

The three board areas are synthesis, characterization methods, and applications. The applications of nanomaterials are in the field of energy [4], environment [5], and healthcare [5], so they connect science to societal challenges and showcase potential solutions. Nanomaterials are also used in the area of wastewater remediation [7]. Life cannot exist without water, and as the population increases, the demand for clean water will rise. Therefore, the issues of waste management are critical for life. The Earth has 70 % water, but only 2–3 % of it is freshwater [8]. The problem has been made worse by the disposal of energy materials (batteries), electronics, and plastics. Wastewater has heavy metals, microplastics, and other xenobiotics, which can cause harm if consumed and must be removed [9] (Fig. 1).

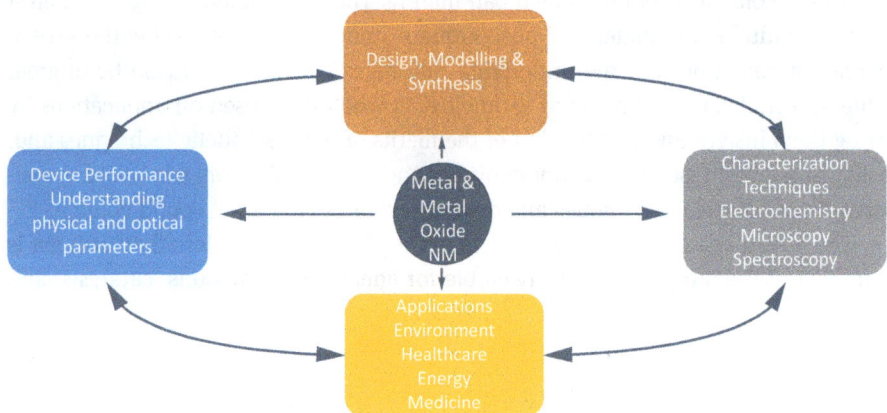

Fig. 1: The interconnection between design, synthesis, and application of nanomaterials.

Usage of nanomaterials in environmental applications

Nanomaterials are nanoadsorbents that can also enact disinfection through electron ejection of the metal oxide to generate reactive oxygen species and promote carbon–carbon bonds for decolorization. Examples include nano-oxides of silver and titanium for both disinfection and photocatalysis [10]. Removal of carbon waste products can be accomplished using carbon-based adsorbents such as activated carbon, graphene, and carbon nanotubes that are hydrophobic, have a high surface area, and have excellent stability at varying pH. Stability can be increased by synthesizing covalent-organic frameworks (COFs) and metal-organic frameworks (MOFs). The metal is the catalytic center, and the organic framework is the skeleton containing the metal or metal oxides [11]. Due to the variability of the metal and organic linkers, the stoichiometry of the MOF can be varied to include gas and heavy metal capture. The recent introduction of graphene is attractive due to its high electrical conductivity and hydrophobicity, which can be tuned, enabling graphene-based materials to behave as ultra-adsorbents [12]. The outer rings can be oxidized or reduced to tailor the degree of hydrophobicity with multiple oxygen sites.

The carbon-to-carbon pie bonds and oxygen functionality can serve as absorbent and catalytic sites for the photodecomposition of dyes and pharmaceuticals. The graphene sheet can be tuned through functionalization to include hydrophobic interactions, covalent bonds (C->C=O→C-OH), and dipolar interactions, including hydrogen bonds, to capture a variety of xenobiotics [13]. The carbon backbone consists of sigma bonds (three electrons) and the weaker pie bond (one delocalized electron). The difference between localized and delocalized electrons enables two to three interactions accessible on the surface [14]. These surface energetics enable heavy metal capture from the graphene oxide/reduced oxide and hydrophilic copolymers such as polyvinyl alcohol and organic and dye capture through weaker pie-to-pie interactions [15]. Carbon nanotubes can also be employed where the graphene sheet is rolled into a tube, and absorption on the surface of the tube would result in removal through sorption [16] (Figs. 2 and 3).

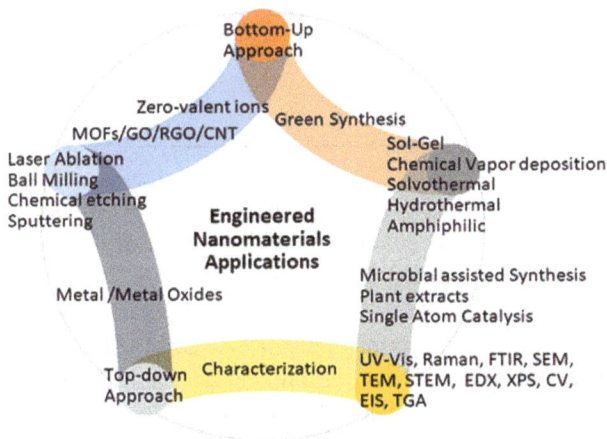

Fig. 2: Relationship between synthesis approach, incorporation of plant-based coatings, characterization, and applications of engineered nanomaterials.

Zero-valent transition metals such as zero-valent iron (ZVI) are excellent heterogeneous catalysts for the degradation of organic and inorganic pollutants, often as NPs, where the magnetic properties of iron NPs can be used during the adsorption process [17].

Nanometal oxides are another class widely used in metal capture, which, when fabricated at the 1–4 nm scale, can have high surface energies and aggregate, which are used in tandem with porous supports such as cyclodextrins, gum arabic, polysaccharides, and diatoms to enable recycling and regeneration of the catalyst [18]. Others have used these to remove other metals such as Cd^{2+}, Pb^{2+}, Ni^{2+}, hematite (alpha-Fe_2O_3), and goethite (alpha-Fe-OOH) [19]. These metal oxides are coated with biopolymers such as chitosan or coated on silica to remove dyes and other organics in removing drugs and organics [20]. For example, Fe_3O_4/GO can remove Pb^{2+}/Cd^{2+}, can inactivate *E. coli*, and can be recovered by adjusting the magnetic field, often exhibiting a pseudo-second-order kinetic profile [21]. Our textbook shows the usefulness of gum arabic and chitosan-coated iron NPs for fingerprint detection and separate removal of As and Cr in oxides such as AsO_3^{3-}, AsO_4^{3-}, CrO_4^{2-}, and $Cr_2O_7^{2-}$.

The last two areas of interest are nanomaterials as photocatalysts and disinfectants [22]. Titania was historically used for disinfection and cleanup of water but only under UV conditions [23]. By coupling transition metals with titania, visible light can generate electron/hole (e^-/h^+) pairs, which promote the formation of reactive species such as OH radicals [24]. The synthesis and applications of nanomaterials are summarized as follows:

Xenobiotic	Catalyst	Synthesis	Reference
Rhodamine	TiO_2/RGO	Hydrothermal	[25]
Methyl orange	Kaolinite/TiO_2	calcination	[25]
Methylene blue	Graphene/TiO_2	Solvothermal	[26]
Rhodamine B	TiO_2/SnO_2	Sol–gel	[27]

Since titania has a bandgap in the UV range, visible light exciton generation can occur by forming a heterojunction [28]. The conduction and valance bands' redox potential can be adjusted by doping other metal or metal oxides [29]. Ternary or quaternary metal–metal oxides such as $ZnFe_2O_4$–$AgTiO_2$ can be introduced for disinfection, decolorization, and recycling through magnetic field sweep [30].

A new approach is required to remove heavy radionucleotide, and COF/MOF must be thermally and chemically stable. Here COF offers opportunities, when complexed with bidentate complexing agents, can be used to detect and capture divalent ions such as Cu^{2+}, Ni^{2+}, Pd^{2+}, and even Ag^+, Fe^{3+}, and Cr^{3+} depending on the interaction with the complexing ligands within the COF [31]. Lastly, nanomaterial toxicity must also be assessed since these materials, comparable to viruses, can enter the body through inhalation, absorption, or ingestion and generate reactive oxygen species causing disease [32]. Certain materials like CdSe are more toxic than others, such as carbon black, but their risk is important in examining their impact on society [33] (Fig. 3).

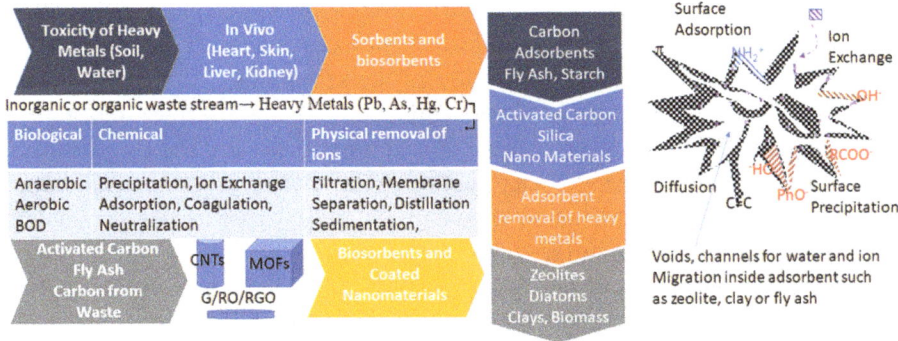

Fig. 3: The relationship between in vivo, in vitro *interactivity*, and in situ coating of nanomaterials in terms of effect (top panel), waste flow and remediation (middle panel), and application of natural and structured carbon supports (bottom panel and middle vertical column). The type of water migration within the macroparticle, surface modification, and conversation of different valent metal ions is shown using the particle schematic (right-hand side figure).

Green synthesis of nanomaterials

Nanotechnology has been widely regarded as one of the most important advances in science since the early 1990s. Its varied applications and rapidly increasing demand have led to novel approaches to producing higher quality nanomaterials. Traditional synthesis methods were used in the early phases, and they depended on both carcinogenic chemicals and high energy input to synthesize nano-sized materials. The pollution caused by standard synthesis processes requires the implementation of ecologically safer synthesis technologies. As the consequences of climate change become increasingly apparent, the scientific community is working hard to find answers to the damage created by harmful manufacturing practices. Green nanomaterial synthesis approaches use natural biological processes to produce nanomaterials.

The term "nano-sized technology" refers to materials with diameters ranging from 1 to 100 nm [34]. Though the size of the compound is what defines it as a nanomaterial, its shape and geometry also play an important influence on its properties. Nanomaterials have been used in practically every industry, including, but not limited to, electronics, agriculture, and medicine. Nanotechnology enables NPs to change the materials created for usage, resulting in significant improvements in thermal, mechanical, and barrier qualities. The fine-tuned creation of diverse NP morphologies, including spheres, rods, quantum dots, and particles, provides applications [35].

There are two types of nanomaterial synthesis methods: standard methods and greenways. However, the major detrimental consequences of using these old procedures are obvious. Organic solvents are widely used to synthesize nanomaterials, providing a significant neurobehavioral and reproductive risk throughout the process [36]; moreover, high pressure and heat conditions may contribute to hazardous working conditions. Concern about volatile vapor and excessive carbon dioxide generation, which contributes significantly to the greenhouse impact, is a high-priority unfavorable outcome of these syntheses [37]. Overall, these procedures offer irreparable hazards to scientists doing the synthesis and the environment. These possible negative consequences outweigh the advantages of standard nanomaterial manufacturing processes, shown in Fig. 4, for experimental parameters and fabrication of different surfaces for catalysis.

Different morphologies of NiO-based systems can be fabricated using hydrothermal synthesis with different alkali metal ions and experimental conditions. The morphology and dimensions can be controlled with experimental parameters such as solvent system, annealing temperature, salt concentration, the ratio of reactants, and mixing time. NiO nanorods are preferred (Fig. 4a) when nickel is precipitated from the hydroxide carbonate on nickel foam and hydrolysis using urea and high-temperature annealing in argon. When the urea is replaced with hexamethylenetetramine, Ni hydroxide nanosheets are fabricated (Fig. 4b). This hydrothermal urea hydrolysis approach can be used with Co_3O_4 with ammonium fluoride and high-temperature annealing to generate nanoarrays of cobalt oxides (Fig. 4c, d). If the reaction temperature and reaction time is increased from 100 to 120 °C and from 6 to 12 h, respectively, then a cobalt oxide nanowire structure (Fig. 4d) is obtained instead of a nanocage (Fig. 4c). The hydrothermal method is suitable for generating one-dimensional (1D) or two-dimensional (2D) metal oxide arrays with strong interactions with the substrate and has applications in catalysis and environmental support.

Due to these reasons, traditional synthesis methods have gone out of favor, paving the way for green synthesis. With the present climate problem, it is critical to create innovative and forward-thinking approaches that adhere to the key principles of green chemistry. Green synthesis is a method of creating clean, safe, cost-effective, and environmentally friendly nanomaterials. It offers nanomaterial advantages ranging from antibacterial characteristics to natural reducing and stabilizing qualities.

Regardless of their eventual shape, many nanosystems begin as zero-dimensional nanomaterials. These systems may be created utilizing various techniques and substrates, including bacteria, yeast, fungus, plant material, live plants, viruses, and pure enzymes [38].

Bacteria used in the green synthesis of nanomaterials are part of a vast category of unicellular organisms with cell walls but no organelles or an organized nucleus. Although certain bacterial strains are extremely harmful, many others occur naturally in the body and provide little to no risk to those who deal with them. Additionally,

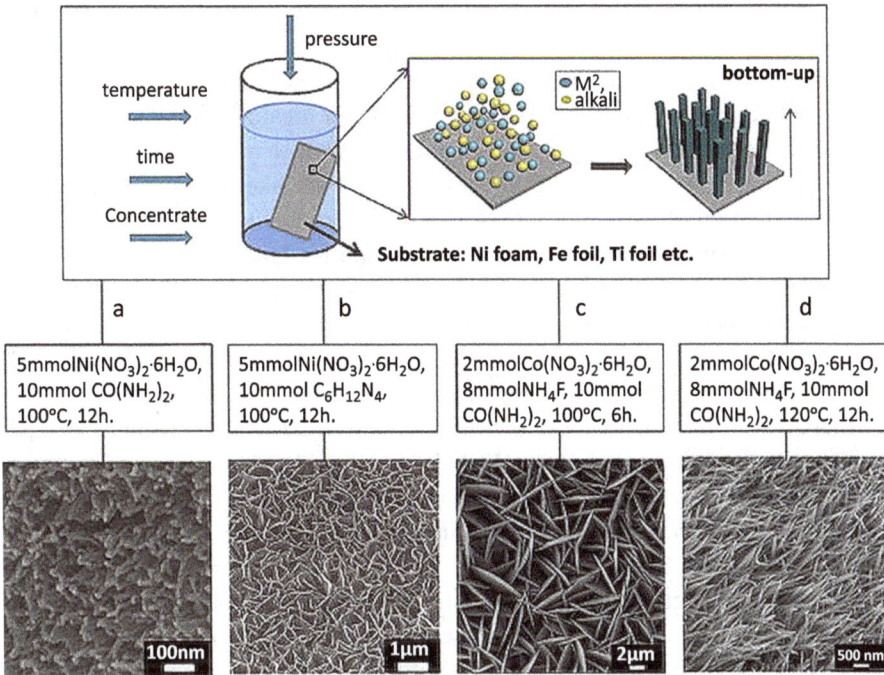

a	b	c	d
5mmolNi(NO$_3$)$_2$·6H$_2$O, 10mmol CO(NH$_2$)$_2$, 100°C, 12h.	5mmolNi(NO$_3$)$_2$·6H$_2$O, 10mmol C$_6$H$_{12}$N$_4$, 100°C, 12h.	2mmolCo(NO$_3$)$_2$·6H$_2$O, 8mmolNH$_4$F, 10mmol CO(NH$_2$)$_2$, 100°C, 6h.	2mmolCo(NO$_3$)$_2$·6H$_2$O, 8mmolNH$_4$F, 10mmol CO(NH$_2$)$_2$, 120°C, 12h.

Fig. 4: Hydrothermal synthesis in fabricating nickel and cobalt oxide nanorods or sheets. For alternations to the synthesis, parameters result in four unique morphologies for nickel oxide (NiO) nanorod design (a) or nickel hydroxide rough sponge (b), cobalt oxide finger sheets (c), or toothbrush-type wires [63].

many of these strains, such as *E. coli* and *Bacillus subtilis*, are exceedingly easy to cultivate and have a genetic code that may be easily manipulated.

Yeasts, similar to bacteria, are unicellular creatures that belong to the fungal family. The species *Saccharomyces cerevisiae*, which converts carbohydrates to alcohols and carbon dioxide, is the most traditional and widely used yeast. This species is employed in baking and producing alcoholic beverages via fermentation. The utilization of yeast cells allows for the production of nanosystems that are impossible with bacteria. Yeast species have been used to create silver, gold, cadmium sulfide, lead sulfide, ferrous oxide, selenium, and antimony NPs. Nanosystems can be produced by utilizing living cells or cell extracts with more typical nanomaterial composites, such as silver and gold.

Fungi, which technically include yeast, are eukaryotic creatures that obtain nourishment by secreting digestive enzymes into their immediate surroundings and then absorbing the dissolved molecules. However, their distinguishing feature is chitin, a long-chain polymer and glucose derivative that strengthens their cell walls. In addition to possessing chitin, fungi cell walls can aid in the creation of

NPs of various forms, sizes, and compositions. Enzymes and protein residues can produce NPs both intracellularly and extracellularly.

Algae are photosynthetic, eukaryotic creatures that are not generally perceived as plants. Chlorophyll-containing single or multicellular creatures thrive on water but lack the actual stems, leaves, and vascular structures that characterize plants. Furthermore, their impact on humans can range from therapies like *Spirulina*, which has a high concentration of natural nutrients, to species like *Anabaena*, which are fatal if their cells or toxins are consumed. Many algal species have been found in recent years for their ability to catalyze the production of nanomaterials. When it comes to the bioactive molecules used in the manufacturing of NPs, algae use the same but somewhat different compounds as other groups. Polysaccharides and protein residues in algae are capable of reducing and stabilizing NPs. One significant advantage of using algae is the availability of a wide range of phytochemicals. Certain algal species, such as *Sargassum tenerrimum*, include amino acids, alkaloids, carbohydrates, flavonoids, saponins, sterols, tannins, and phenolic compounds. Once purified, each of the chemicals may be used to modify further the size, shape, and active qualities of the nanomaterial [38].

Given the right circumstances, NPs and other zero-dimensional structures can generate higher level structures outside the 100 nm range after being produced from their respective chemicals. Nanowires, nanotubes, nanorods (1D), and nanosheets are examples of higher order structures (2D). These nanosystems vary from their less ordered counterparts in terms of functional features, such as the capacity to conduct electricity, making them good electrodes in batteries [39]. Higher ordered nanosystems like NPs and other 1D nanomaterials may be manufactured using ecologically benign agents such as plant material, the most commonly employed agent in system synthesis. Metals such as silver, lead, gold, and other elements such as carbon can be used to create highly ordered structures.

Noble metal nanomaterials

Noble metal nanomaterials (NMNs) are perfect foundations for designing and modifying nanoscale structures for specific technological applications due to their fascinating physical and chemical characteristics. Controlling the size, shape, design, composition, hybrid, and microstructure of NMNs, in particular, is critical for unveiling new or improved capabilities, and application potentials such as fuel cells and analytical sensors due to different reactive surfaces, such as graphene, graphene oxide, or nitride-modified surfaces, are shown in Fig. 5.

In theory, controlling these parameters can properly adjust nanomaterials' physical and chemical properties (NMNs) using noble metals. However, the flexibility and extent of change are sensitive to specific parameters. For example, Au NPs have a size-dependent surface plasmon resonance (SPR) feature and display visible

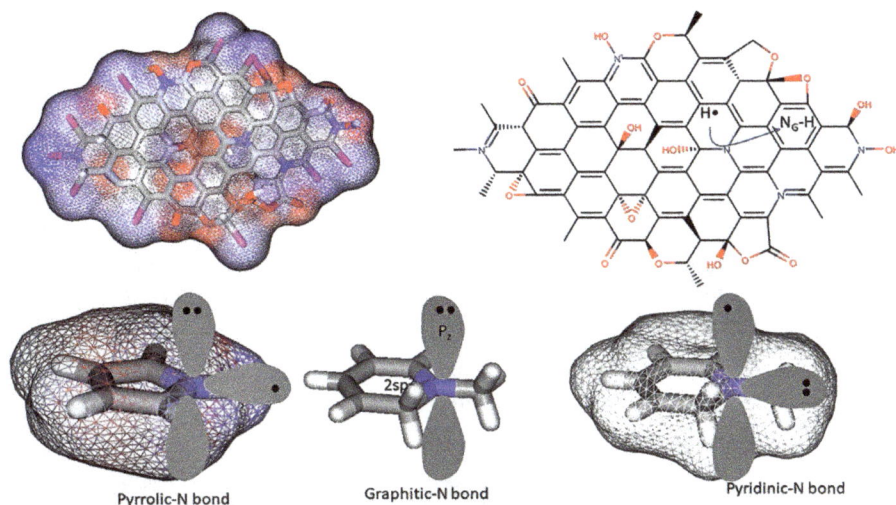

Pyrrolic-N bond   Graphitic-N bond   Pyridinic-N bond

Fig. 5: A polar space volume of composite graphene, graphene oxide, and graphene modified from different hybridized nitrogen (left) with the surface (top right). The space volume for hybridization of pyrrolic nitrogen–carbon (left bottom), graphitic nitrogen (bottom middle), or pyridinic nitrogen (bottom right) gives rise to different hybridization along the P_z-axis and also a different sp^2–sp^2 type of interactions along with the sigma bond. These surface polarities, charges, and energetics enable different surface chemistries, such as radical hydrogen attachment and potential role in photocatalysis, heavy metal remediation, and green process chemistries in wastewater management.

SPR absorption. However, gold nanorods, gold nanocages, and hollow gold nanospheres have high near-infrared (NIR) absorption [39]. Because blood and soft tissue in the NIR range are largely transparent in this region, collateral damage to the surrounding healthy tissue is avoided. These unique gold nanostructures with NIR absorption are particularly significant for photothermal treatment and bioimaging in this region. Pt nanomaterials with high-index facets of complex morphologies or multicompositions have been proven to exhibit higher electrocatalytic activities toward small-molecule oxidation and oxygen reduction reactions (two key reactions in the field of the fuel cell) than the commercial catalysts [40]; Ag nanostructures with the proper size, complex sharp structure, or more edges and corners have higher surface-enhanced Raman scattering activity than spherical Ag NPs [41]; certain noble nanoclusters (Au, Ag, and Pt in particular), consisting of several to roughly a hundred atoms and possessing sizes comparable to the Fermi wavelength of electrons, can exhibit molecule-like properties and strong size-dependent fluorescent emission [42]. Therefore, controlling these critical parameters offers a strong chance to expand their application potential in catalysis, electronics, photonics, sensing, imaging, medicine, and other domains. An example of this approach is shown in Fig. 6, where horseradish peroxide is attached on graphene oxide under neutral pH conditions by heteroatom interactions with oxygen in carbonyl, epoxy, and hydroxyl functionalities. The enzyme that is directly observed is coated into the graphene oxide surface (Fig. 6a) using atomic force microscopy, high-angle annular dark-field scanning transmission electron microscopy (Fig. 6b), electron energy loss spectroscopy (EELS) for cobalt (Fig. 6c), and extended X-ray absorption fine structure (EXAFS) for cobalt molybdenum disulfide (Co-MoS$_2$), and the energetics of catalyzed dissociation of hydrogen peroxide on the surface of cobalt-modified cobalt molybdenum disulfide (Fig. 6e). The superior catalysis of nitrogen-doped molybdenum disulfide sheets relative to the underivatized surface is shown in Fig. 6f. A related example of B-doped Fe–N–C single-atom catalysis (Fig. 6g) is an example of the versatility of the proposed approach not only for transmission-based metal/metal oxides but also for MOFs using their 2D arrays for biomineralization.

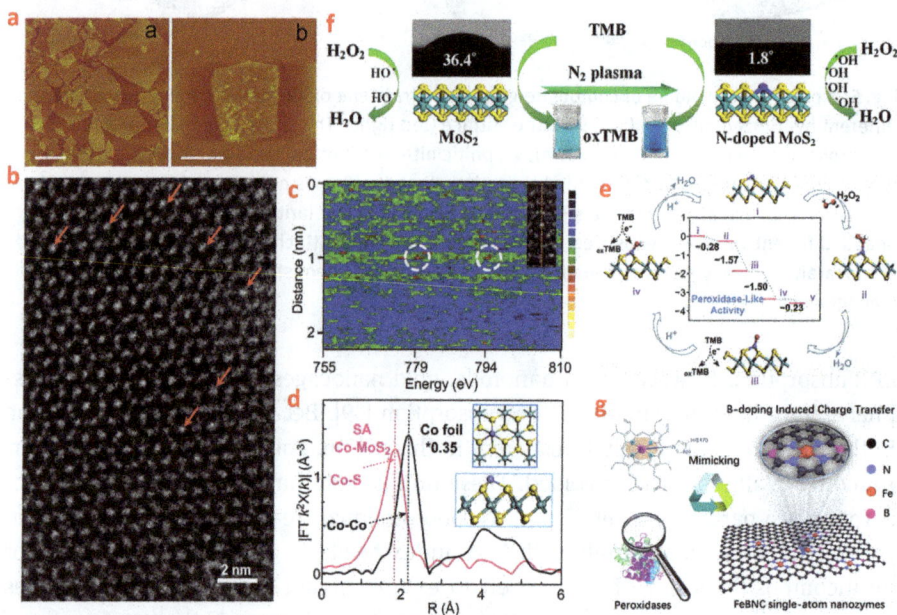

Fig. 6: Demonstration of the utility of graphene oxide (GO) immobilization [64] of horseradish peroxidase (HRP) on (a) surface of GO with HRP from atomic force spectroscopy as white regions [65]; (b) high-angle annular dark-field scanning transmission electron microscopy of cobalt (red arrows) on a cobalt–molybdenum sulfide exhibiting peroxidase-like heterocatalysis from single-atom Co-MoS$_2$; (c) EELS for Co L2 and L3 edges of 794 and 779 eV, respectively, confirming single-atom (SA) catalysis with the MoS$_2$ scaffold; (d) EXAFS of SA Co-MoS$_2$ with Co–S coordination number of 3.4 and oxidation state of Co^{3+}; (e) density functional theory calculation diagram for the heterocatalysis of the Co-MoS$_2$ SA site for the energetics and dissociation of H$_2$O$_2$ on the surface of the catalyst [66]; (f) the scaffolding of the MoS$_2$ supports the peroxidase-like activity without N-doped MoS$_2$ exhibiting higher catalyticity than the undoped structure; (g) schematic of B-doped graphene with nitrogen and iron as a single-atom catalyst for the peroxidase-type activity [67].

With growing environmental concerns and the rapid depletion of fossil fuels [43], the development of new technology for producing new alternative energy conversion and storage devices, such as solar cells, supercapacitors, and lithium-ion batteries (LIBs), is critical for resolving the current energy issues [44]. Fuel cells have been extensively researched as an ecologically friendly energy devices because of their multiple advantages: high energy density, simplicity of handling liquid, minimal environmental impacts, and potential microfuel cell applications. At the moment, enormous research efforts are being directed toward developing efficient, high-performance fuel cells [45]. Pt and Pt-based NPs are still necessary and are the most effective catalysts for fuel cells in all NMNs. However, one of the greatest barriers to the commercialization of fuel cells is the high cost and dependability of the Pt nanocatalysts utilized [45]. Thus, enhanced catalyst design and synthesis to satisfy the needs of lowering Pt loading levels while boosting the activity and stability of the Pt-based catalyst are highly sought.

The nanoanalytical sensing system is a burgeoning interdisciplinary field that combines the inherent characteristics of analytical techniques (e.g., high sensitivity, rapid detection, and low cost) with the unique electronic, optical, magnetic, mechanical, and catalytic properties of nanomaterials to become one of the most exciting sectors [46]. With the progressive development of new or improved features of NMNs, many analytical methodologies or tactics for building high-sensitivity sensors for detecting varied targets have been created. Nanomaterial-based analytical sensing devices offer significant benefits and potential for detecting various targets, which are crucial in environmental contamination, serious illnesses, human health, and food safety, among many others [47].

Advanced techniques for the synthesis of NMNs with controllable size, shape, composition, hybrid, architecture, and microstructure, for example, as well as the design of high-efficiency NMNs-based nanoelectrocatalysts for fuel cell and electrochemical sensors, and the development of analytical strategies for the construction of NMN-based colorimetric and fluorescent sensors are among the new research contributions [48]. These findings show that NMNs can provide many attractive possibilities in an exceedingly diverse environment for fostering the rapid development of several research topics. At the moment, one of the greatest barriers to further NMN research is the lack of a scalable manufacturing process for synthesizing high-quality NMNs with adjustable size and form [49]. It is challenging to create monodisperse small-size NMN rich in high-index facets.

Aside from single-component metal nanostructures, controlled synthesis of bimetallic nanomaterials in various topologies has lately been a hot issue. However, managing their microstructure and catalytically for long periods remains a challenge. The ultimate objective of controlled NMN synthesis is to create low-cost, high-throughput, environmentally friendly methodologies (one-pot) for preparing NMNs with appropriate fine structures, such as size, shape, composition, architecture, and microstructure. In light of this lofty ambition, a greater knowledge of NMN development processes may give some avenues and possibilities. Some in situ experimental approaches may open up new paths for investigating the creation process of NMNs generated using various synthetic methods shown in Fig. 7 for cobalt hydroxide and oxide systems. When a two-step hydrothermal process with thermal annealing is utilized, a Co_3O_4@Ni–Co–O is generated using cobalt hydroxide nanosheets grown on nickel foam by a primary hydrothermal synthesis step. A second hydrothermal step is introduced with additional Ni and alkali metal ions and annealing Ni–Co–O nanorods on Co_3O_4 nanosheets (Fig. 7a), with the thickness related to the concentration of Ni^{2+}. If *mesoporous cobalt carbonate hydroxide* (MPCHH) is introduced instead and coprecipitation occurs under hydrothermal conditions on nickel foam with $Al(OH)_3$ and Co^{3+}, then a layered double hydroxide (LDH) nanosheet is generated, which can be made porous by immersion in NaOH (Fig. 7b).

The need for highly active metal catalysts in fuel cell and electrochemical sensing applications has sparked intense interest in the development of monodisperse, single-component, small-size NPs of Pt, Pd, and Au, as well as multimetallic NPs and metal NP-based hybrids. However, the bulk of NMN-based catalysts only increases the fuel cell performance in two ways: activity or stability, not both. In nature, microbes over these hurdles generate stable fuel production through oxidation of sugars and carbon using metal and metal oxide-based catalysts. The design of novel metal catalysts must utilize nature-derived solutions for synthesizing new monometallic or multimetallic nanostructures with low cost and high activity and durability.

Given this, three types of NMN-based nanocatalysts should be preferred: The first is to create monodisperse, small-size metal NPs with rich high-index facets; the second is to create multimetallic NPs with diverse morphologies and compositions (which should also have rich high-index facets); and the third is to control the uniform distribution and effectively connect of monometallic or multimetallic NPs with controllable size, shape, and microstructure on the surface of advanced supports [47]. The development of matching hybrids with well-defined morphologies is critical for improving their performance in fuel cell applications. Another intriguing aspect to mention is that for electroanalytical chemists, it is critical to search for new NMN-based electrode materials to construct high-performance electrochemical sensors with high sensitivity and selectivity, which is preferable to be achieved through new analytical techniques. The promising paths for analytical chemists specializing in NMN-based colorimetric and fluorescent sensors will center on the molecular engineering of NMNs or new metal fluorescent materials for building a range of analytical sensors via novel signal conversion mechanisms. An example of a tailored approach is used in Fig. 8. In the synthesis of $Co_{3-x}Fe_xO_4$, a single hydrothermal step with Co^{2+} and F^- ions, urea, at 120 °C for 12 h is followed by an annealing step. A hexagonal LDH platform structure is formed (CoFe-LDH). These are grown on an iron substrate by hydrothermal reaction (Fig. 8b and e) and can result in orthorhombic cobalt hydroxide carbonate (CoFe-HC) nanowires that are epitaxially oriented from the hexagonal edge of the CoFe-LDH platelet (Fig. 8c and f). These wires are grown in all directions at 60° between adjacent branches of the

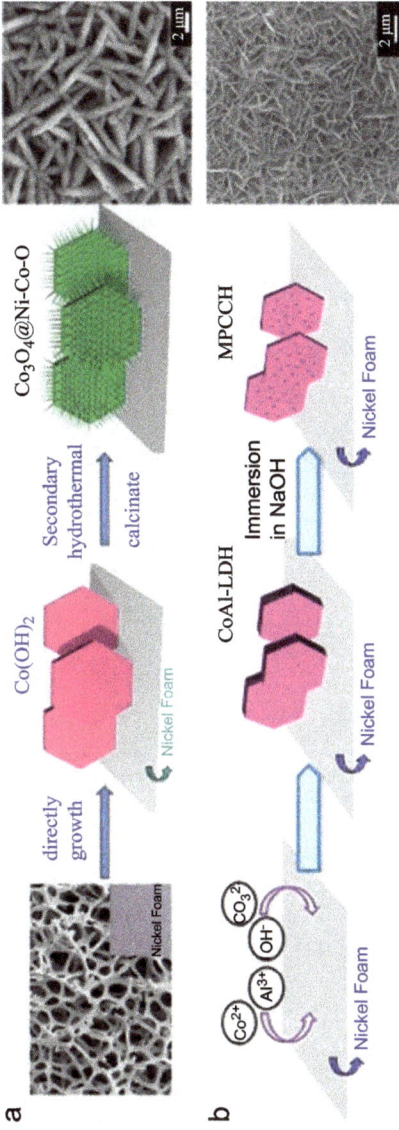

Fig. 7: Schematic of hydrothermal fabrication of hierarchical nanoarrays through alteration of metal oxide salt and annealing or addition of ions in second hydrothermal steps [63]. (a) The synthesis of Co_3O_4@Ni–Co–O nanosheets using a two-stage hydrothermal and annealing process [68], and (b) the fabrication of mesoporous cobalt carbonate hydroxide nanosheets through hydrothermal synthesis using Co ions and layered double hydroxide from $Al(OH)_3$ followed by NaOH immersion to facilitate the etching to thinner sheets and formation of porous sheets [69].

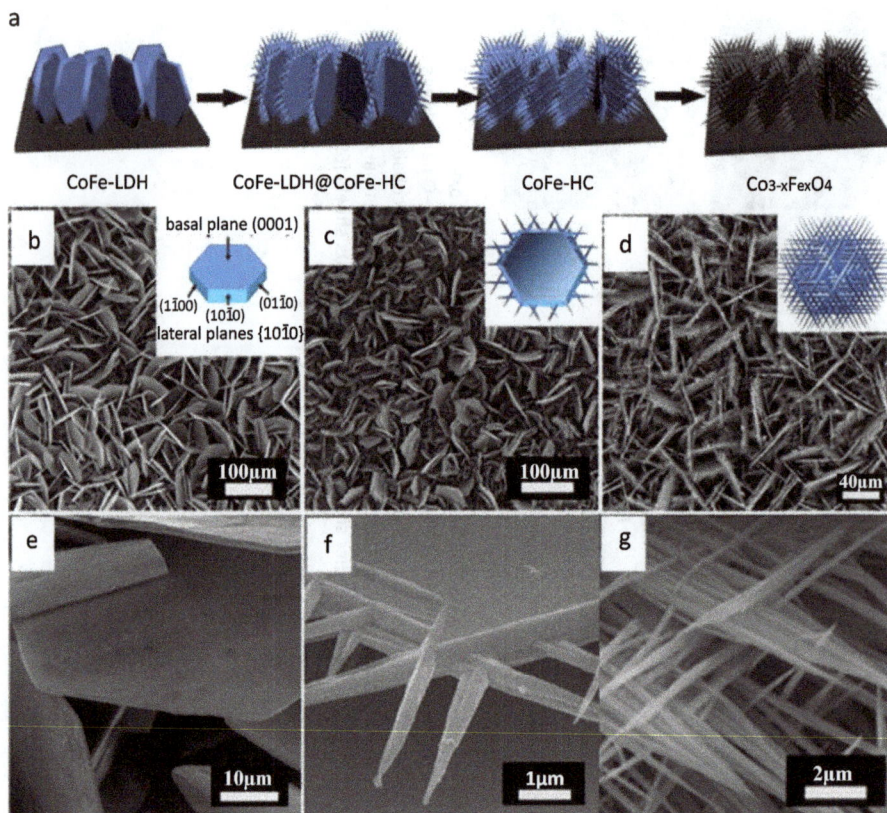

Fig. 8: An example of fabrication of hierarchical metal oxides using hydrothermal synthesis [63]. (a) Scheme for the growth of single-step hydrothermal synthesis of $Co_{3-x}Fe_xO_4$ arrays on an iron substrate and corresponding SEM micrographs at 3h (b and e), 4 h (c and f), and 12 h (d and g) with an illustration of the crystal morphologies overlaid in inserts b, c, and d crystallites [70].

platelet. When the material is calculated, the final hierarchical $Co_{3-x}Fe_xO_4$ array is obtained.

Rare-earth-containing perovskite nanomaterials

Perovskites have been widely employed as star materials in optics, photovoltaics, electronics, magnetics, catalysis, sensing, and other applications. However, several intrinsic flaws, such as low efficiency (power conversion efficiency and external quantum efficiency (EQE)) and poor stability (against water, oxygen, and ultraviolet (UV) light, among others), restrict their practical uses [50]. Nanostructured materials and the incorporation of rare earth (RE) ions are efficient ways to improve their characteristics and widen their applications.

CaTiO$_3$ was the first recognized perovskite, and this formula was later derived into a plethora of new forms, including ABX$_3$, A$_3$B$_2$X$_9$, A$_2$BB$_0$X$_6$, A$_2$BX$_6$, A$_4$BX$_6$, in which A and B are cations (A has a bigger radius than B), and X is a halogen or oxygen anion. Two different anions and metal cations form halide perovskites and oxide perovskites. These two forms of perovskites exhibit numerous similarities in their crystal structures, including the presence of BX$_6$ octahedra and A atoms in the interstitial voids of surrounding octahedra [51]. Perovskite materials have a wide range of characteristics and use due to their different compositions and structures: Luminescent perovskites have been employed in lighting, displays, sensing, biological imaging, and other applications [52].

With a growing emphasis on the advancement of nanotechnology and multidisciplinary research, scientists have been attempting to downscale perovskite structures into the nanoregime to improve their performance and use [53]. When compared to bulk perovskite materials, perovskite nanomaterials (PNMs, which include NPs, nanowires, and nanofilms) have several advantages: for the fabrication of thin films and flexible devices, PNMs have high processability for catalysis, PNMs have rich and controllable facets and active sites, and PNMs have outstanding photoelectromagnetic properties due to the small-size effect and quantum effect [54]. These new properties improve the performance of perovskites and expand their applications.

Some halide PNMs have poor stability (against water, oxygen, and heat) and restricted adjustability of optical and electrical characteristics. In contrast, some oxide PNMs have inferior catalytic, optical, and electromagnetic performances [55]. To address these problems, scientists have begun adding foreign metal ions into PNMs. RE elements, sometimes known as "modern industry's vitamins," are widely employed as dopants or components to control various materials' unique physical and chemical characteristics. Some halide PNMs exhibit low stability (against water and oxalic acid). RE ions have flexible redox properties and distinctive luminous and electromagnetic capabilities due to their changing valence states and electronic configurations. As a result, REs have been frequently introduced into perovskite nanostructures to increase their performance and expand their applications. Furthermore, researchers created a range of RE-based perovskite functional materials. Both RE-doped and RE-based PNMs perform satisfactorily [50].

It is feasible to enhance the absorption efficiency of solar light in the NIR region by adding RE into perovskite frameworks, hence increasing the photo-electrochemical efficiency (PECE). The addition of RE ions can also modify the bandgap of perovskites and improve device performance. Another major technique is to substitute lead in perovskites. The low EQE and poor stability of halide perovskites further limit their use as light sources. Improving the EQE and stability of perovskite-based lighting sources is thus both basic and practical. Exploration of white-light lighting sources is likewise needed, and currently available white-light perovskite-based systems are quite limited.

Some halide perovskites have strong photocatalytic activity in catalytic applications but low stability. Exploration of RE-containing halide perovskites is thus a potential field. Although oxide perovskites have been widely employed in various catalytic processes such as oxygen reduction reaction, oxygen evolution reaction, nitric oxide (ORR, OER, NO_x) removal, the catalytic performances still need to be improved. The insertion of RE ions with varying radii and valences into perovskite structures results in defects, strains, and oxygen vacancies, which improves catalytic efficiency in ORR, carbon, or nitrogen transformation processes. Although oxide perovskites offer great stability when employed as magnetic materials and detectors, they are rarely used in biomedical magnetic resonance imaging.

Furthermore, the bulk of magnetic and spin devices are still in the fundamental research stage, needing significant effort to get to practical applications, as shown in Fig. 9 of lead halide perovskites, where material properties depend on carrier recombination. The nonradiative component undergoes Shockley–Read–Hall and Auger recombination (Fig. 9a) and is weaker in perovskite solar cells under air mass 1.5 G illumination than crystalline silica cells. The doping with RE ions would decrease the Cl vacancies and lessen nonradiative recombination, increasing photoluminescence quantum yield (Fig. 9b and c), such as incorporating Yb^{3+} ions at Pb^{2+} sites. The excitonic emissions may occur due to in-band hot-exciton relaxation and exciton trapping to the bandgap defect states generated through doping. The RE ion insert into the lattice leads to a near band-edge state (Fig. 9d and e). These RE ion induced near band-edge states provide a channel for emission instead of nonradiative recombination and increase photoluminescence. The optimal ratio of RE redox ions leads to a decrease in defect density and increased emission lifetimes (Fig. 9f and g).

Selective RE doping has improved the dielectric, piezoelectric, thermoelectric, and ferroelectric characteristics of perovskites. However, in terms of material compositional design, RE-doped perovskites have significant hurdles in meeting the criteria of high-temperature piezoelectrics and sensors in commercial applications. It is critical to research and create perovskite materials with high transition temperatures and high dielectric and piezoelectric constants. RE-based oxide perovskites are promising solid oxide fuel cell electrode materials with high activity and energy conversion efficiency (β, 60 %). However, the efficiency remains significantly lower than the theoretical value.

Furthermore, the high working temperature and acidic (or basic) environment provide significant problems to electrode stability. All of these critical concerns must be addressed. Although several very active catalysts have been developed for metal-air batteries, the construction of high-capacity, reliable, and efficient batteries remain problematic. As a result, it is critical to investigate dependable battery and solar cell assembly procedures and appropriate electrode materials. An example is shown in Fig. 10. The thin perovskite film doped with Nd^{3+} into Pb^{2+} vacancies with correct crystalline orientation in perovskite films can result in superior device performance, as shown in Fig. 10 for a film with increased optimization and morphology without holes (Fig. 10a–d). The reduction in defect density results in enhanced charge carrier

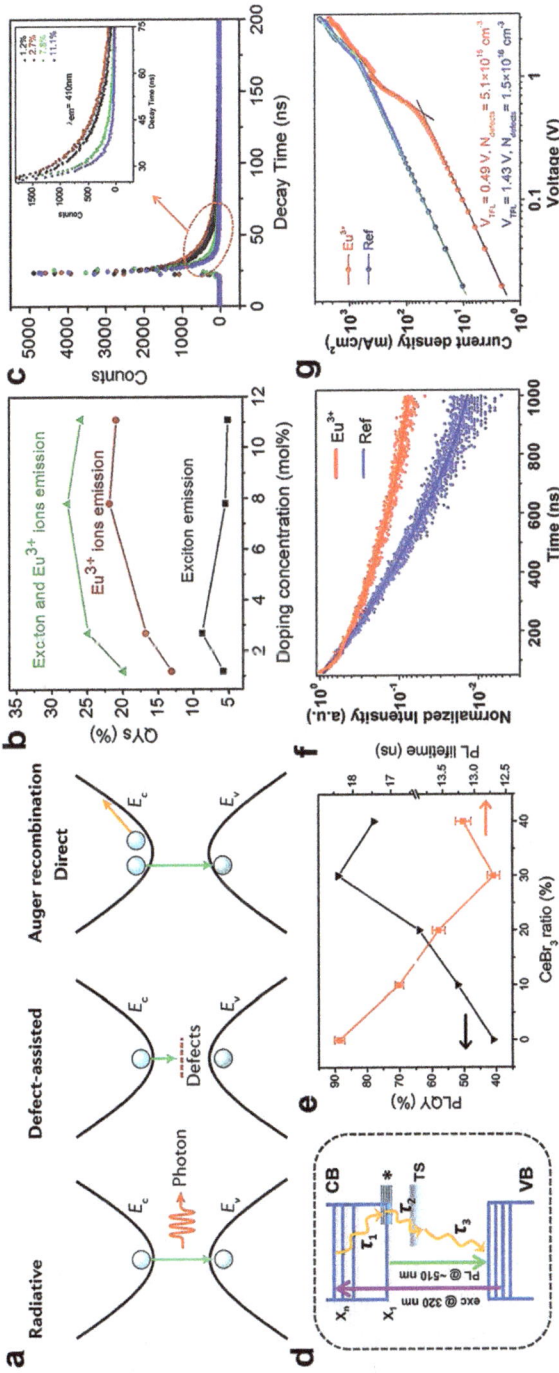

Fig. 9: The RE-coupled lead halide perovskites [71] with (a) known modes of recombination [72], (b) photoluminescence quantum yield, and (c) emission profile of the exciton for Eu^{3+} (410 nm) ion-doped $CsPbCl_3$ nanocrystal at different RE doping concentrations [73]. (d) The plausible transitions for the photoluminescence kinetics for $Cs^{3+}:CsPbBr_3$ nanocrystal, (e) the average photoluminescence (PL, red) and the photoluminescence quantum yield (PLQY, black) as a function of RE dopant concentration [74]. (f) The time-resolved PL spectra for the RE-doped perovskites and (g) space charge—limited current for the perovskites (ref) and Ee^{3+}/Eu^{2+} ion pair thin film [76].

and device performance. In our example, the efficiency was 21.15 % with minimal J–V photocurrent hysteresis. Incorporating n-type atoms such as Sb^{3+} and In^{3+} into a thin film with Pb^{2+} enables ions to redistribute into the perovskite absorber (Fig. 10e) due to electrostatic interactions between the mismatched ions. This approach will lead to forming a gradient heterostructure with a well-matched band edge of the TiO_2 compact layer enabling charge separation and enhancing the fill factor in optimized cells of up to 84 %. As stated earlier (Fig. 9), incorporating RE will improve film quality with a prolonged carrier lifetime and reduced defect density. These led to a reduction of nonradiative recombination (Fig. 10f and g) and improved $J_{current\ density}$–$V_{voltage}$ photocurrent, and efficiency as high as 10.14 % at a $V_{open\ circuit}$ of 1.594 V under one sun simulated radiation (Fig. 10h) with the RE ions located at grain boundaries of the perovskite absorber later and limited doping into the perovskite lattice for improved film quality and prolonged carrier lifetimes.

The manufacture of flexible electronics, enabled by the growth of wearable technology, offers new potential for perovskite materials. The large-scale production of flexible solar cells, flexible sensors, flexible capacitors, and batteries, in particular, has accelerated the development of RE-containing PNMs. Several issues, including cost, environmental protection, and market demand, must be balanced during the batch manufacturing process. Although lead-based halide and oxide perovskites have great features in luminous materials, solar cells, and dielectric materials, Pb is detrimental to the environment and human health.

Lead-free perovskite materials are now promising for achieving comparable or even superior photovoltaic performance. The integration of RE has been shown to control the valence band and the light absorption capabilities of lead-free perovskite materials. However, the catalytic activity of certain RE-containing perovskite materials has surpassed that of noble metals, although the stability of these oxides remains weak. Green chemistry standards of synthesis of catalysts without toxic byproducts, using perovskite material, must use ecological and, when necessary biodegradable raw ingredients, with high-efficiency synthesis and carbon-neutral energy balance.

1D nanomaterials: applications in sodium-ion batteries

With the global energy crisis worsening and the environment deteriorating, it is critical to creating energy storage technologies (ESSs). Among the different ESSs, rechargeable batteries are the most successful technology for producing green energy from stored materials and converting chemical energy into electrical energy in a sustainable manner [56]. LIBs, the most promising ESSs, have seized the present global rechargeable battery markets because of their tremendous energy and power potential, particularly in portable electronic devices [57]. However, lithium deposits in the Earth's crust are relatively limited. The distribution of lithium resources is

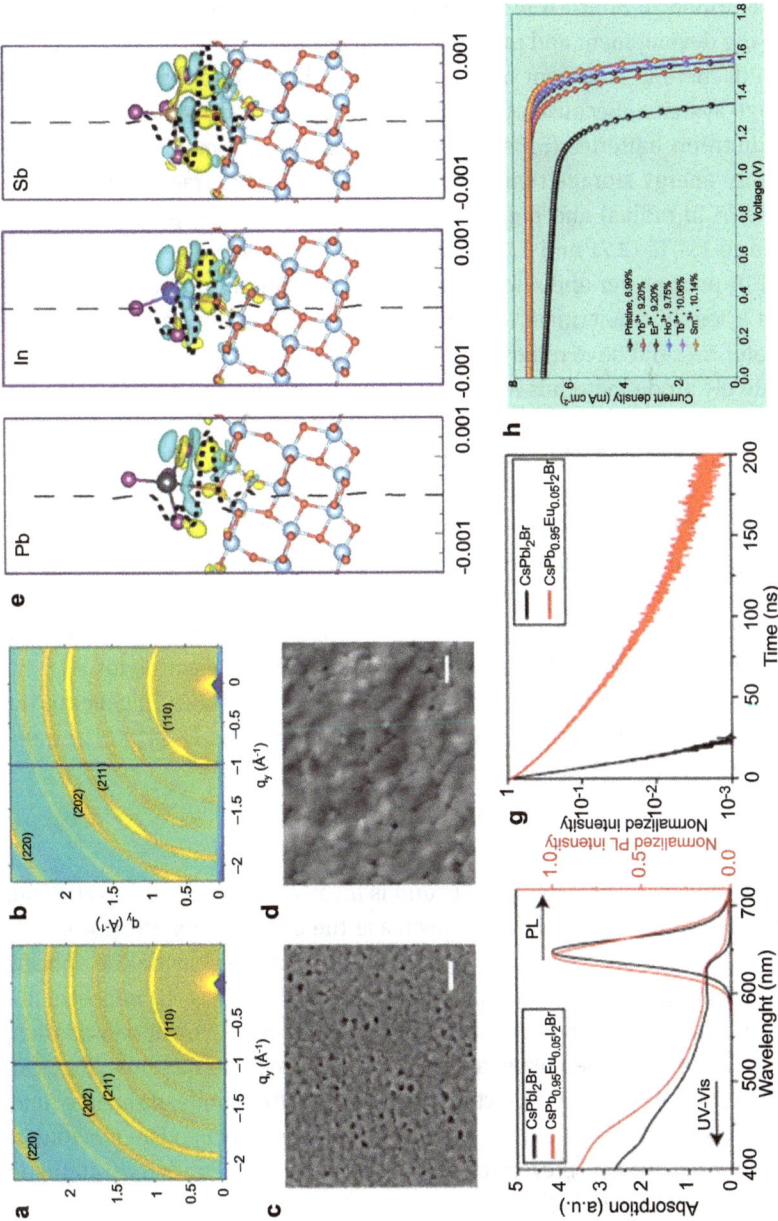

Fig. 10: The effect of RE dopants on the film quality of perovskite thin-film absorbers [71]: (a) the grazing-incidence wide-angle X-ray scattering spectra and scanning electron micrographs for MAPbI$_3$:xNd^{3+} thin reference (a, c) and doped (b, d). (e) The simulated charge density profile between Pb^{2+}, In^{3+}, and Sb^{3+} dopants and the (101) surface of TiO$_2$, respectively [77]. (f) The normalized UV–visible absorption, photoluminescence emission spectra, and (g) time-resolved photoluminescence decays for CsPbI$_2$Br and CsPb$_{0.95}$Eu$_{0.05}$I$_2$Br films [78], (h) the J–V curve device profile of CsPbBr$_3$ solar cells with different Ln^{3+} dopants [79].

mostly concentrated in South America, resulting in a high lithium price that severely limits the development and implementation of LIBs.

As a result, there is an urgent need for low-cost, high-performance alternative energy storage systems. Because of the copious supplies and inexpensive cost of sodium, sodium-ion batteries (SIBs) are more promising for medium- and large-scale stationary energy storage than LIBs. However, the wider radius of Na$^+$ than Li$^+$ (1.02 vs. 0.76 in radius) and higher standard electrochemical potential of Na$^+$/Na compared to Li$^+$/Li (2.71 and 3.04 V vs. standard hydrogen electrode, respectively) result in poor power and energy densities, impeding future development of SIBs [58]. As a result, it is critical to developing suitable electrode materials that can host greater Na$^+$ and have rapid ion diffusion kinetics.

Nanoscale electrode materials are gaining popularity due to their tiny size, increased specific surface area, and ease of stress relaxation procedures. The tiny NPs can lessen the time it takes for Na$^+$ to diffuse. The huge specific surface area enhances the electrode/electrolyte contact area and charges storage via electrical double-layer and surface redox processes [59].

One-dimensional nanostructured electrode materials have several benefits for achieving high capacity, long-term cycling, and higher rate performance in SIBs. As mentioned in the literature, the power density of SIBs is determined by the kinetic diffusion of electrode materials, which consists of diffusion in electrolytes and diffusion in electrode materials. The latter is the most important stage. In general, the diffusion of Na$^+$ in electrode materials is related to the diffusion length (L) and diffusion coefficient (D), as follows:

$$\tau = L^2/D \tag{1}$$

where diffusion time τ is proportional to L^2 and is inversely proportional to D. Small NPs in 1D nanostructured materials can decrease the diffusion length (L) of Na$^+$, reducing the Na$^+$ diffusion time in electrode materials and increasing the specific capacity and power density of SIBs. Compared to other electrodes, 1D nanostructured materials provide direct current channels, advantageous for electrical transmission. Also, because 1D nanostructured materials have a large specific surface area, they can increase the electrode/electrolyte contact area and shorten the charge–discharge time. Plus, in charge/discharge operations, 1D nanomaterials can accommodate the volume change of electrode materials, preventing pulverization and aggregation, resulting in long-term cycling performance [60].

The advantage of stress relaxation is that it reduces the volume fluctuation of electrode materials during cycling operations. Among the many nanoscale materials, 1D nanomaterials such as nanowires, nanofibers, nanobelts, nanorods, and nanotubes have been identified as a potential class of materials in ESSs. The distinct structure of 1D nanomaterials allows for easy electrical and ionic transport and great tolerance to stress shift, which contributes to the high performance of

ESSs [61]. Even though various publications on 1D nanostructured materials have introduced their applications in ESSs, a study comprehensively summarizing the manufacturing and application of 1D nanomaterials in SIBs is still required.

A large variety of 1D nanostructured electrodes for SIBs have been developed and fabricated, and they display excellent electrochemical performance. The exceptional performance can be attributed to the characteristics of 1D nanomaterials, which include short ion diffusion paths, high mechanical strength, and large surface areas. Such properties effectively improve Na^+ diffusion kinetics, reduce volume expansion during charge and discharge (especially in anode materials), improve electrode structure stability, and increase the utilization rate of active materials, thereby effectively improving the cycling and rate performance of SIBs. Despite significant progress, there are still various barriers and significant development potential for the manufacturing and application of 1D nanomaterials.

All approaches for producing 1D nanomaterials tend toward being obstacles. The electrospinning technique is currently incapable of producing homogenous nanofibers with diameters less than 50 nm. Organic solvents that are poisonous and corrosive are typically used to manufacture precursor solutions. The hydrothermal technique is poorly regulated in the synthesis process, and repeatability is poor. The CVD and ALD pathways are quite restricted. CVD, for example, can only create a few types of 1D nanomaterials (e.g., semiconductor carbide and nitride). Because of its high cost, the ALD process can only generate 1D nanomaterials in the presence of a template or cover numerous thin layers on the surface of original 1D nanomaterials [62]. These constraints and issues must be addressed soon. With the increasing need for portable and wearable gadgets, electrospun flexible energy storage devices or 1D nanomaterials built on flexible conductive substrates might be a lucrative and exciting field. So far, large-scale manufacturing of 1D nanomaterials has not been fully realized. The key explanation might be the poor controllability and reproducibility of the synthesis process.

Furthermore, high production costs and stringent criteria, particularly synthesis pathways, are major roadblocks to large-scale manufacturing [60]. The nanomaterial for a device for fuel and solar cell catalysts is illustrated in Fig. 11. Here upconverting NPs (UCNPs) are integrated into the mesoporous electron transport layer of perovskite solar cells to enhance device performance.

An example is the introduction of monodisperse β-$NaYF_4$:Yb^{3+}/Er^{3+} UCNPs into $MAPbI_3$ as a form of modified mesoporous TiO_2 instead of mesoporous titania, resulting in performance under NIR radiation. The modified one enables the device to produce a photovoltage ($V_{\text{open circuit}}$ of 0.89 V, a short-circuit current density (J_{SC}) of 0.74 mA/cm^2, a filling factor of 53.9 %, and a power conversion efficiency of 0.35 %) relative to the unmodified mesoporous TiO_2 layer (reference) which did not produce a response (Fig. 11a). By introducing UCNPs into the interfaces between the perovskite absorbed and carrier transport layers, photocurrent, and absorbance, a response can be broadened into the NIR and UV regions (Fig. 11b and c). The NIR-to-UV receptivity was accomplished using β-NaYF$_4$: Yb^{3+}/Er^{3+}, β-NaYF$_4$: Yb^{3+}/Tm^{3+}, and silver undercoat to convert NIR to visible light and Eu(2-thenoyltrifluoro acetone)$_2$(1,10-phenanthroline) methacrylate acid down conversion layer to employ UV light. A short-circuit current density of 27.1 mA/cm^2 was achieved under 1.5 W/cm^2 illumination. The fluorescence spectra indicated energy transfer from the β-NaYF$_4$:Yb^{3+}/Tm^{3+} to the conduction band of TiO$_2$ through the nano-heterojunctions under the excitation at 980 nm (Fig. 11d), demonstrating the efficient electron transfer of TiO$_2$ and perovskite layers occurred due to the high light scattering of the core–shell UCNPs (Fig. 11e). The incorporation of these elements enhanced the upconversion luminescence efficiency of β-NaYF$_4$: Yb^{3+}/Er^{3+}@NaYF$_4$ core–shell UCNPs, with Sc^{3+} to interact with the various energy levels, as shown by the emissions lines at 541 and 654 nm, resulting in increased short-circuit current density and improved the device efficiency of 20.19 % (Fig. 11f). This was confirmed by energy dispersion spectrum mapping data of the β-NaYF$_4$:20 %Yb^{3+}/2 %Er^{3+}/8 %Sc^{3+}@NaYF$_4$ (NYES-30) film, which confirms the presence of RE elements in the mesoporous TiO$_2$ layer (Fig. 11g). The light-harvesting obtained from the incident-photon-to-current conversion efficiency spectra in the 300–800 nm region through better charge carriers collection and a lower charge recombination for the NYES-30 device relative to the pristine device (Fig. 11g). Electrochemical impedance spectroscopy measurement measured the charge carrier transportation in the solar cell. The Nyquist plots for different RE-doped devices (NYES-15, NYES-30, NYES-60, and NYES-100) and undoped pristine cell were investigated. The charge-transfer resistance (R_{ct}) at high frequency between the hole transport layer/Au electrode. The low-frequency charge recombination resistance (R_{rec}) was measured between the mesoporous/perovskite layer interface. The R_s are the series resistance of the electrodes and the external circuit. The NYES-30 device shows the highest R_{ec} as compared to the pristine cell, indicating a suppressed charge recombination and an enhanced charge migration for the perovskite film (short-circuit current density of 22.91 mA/cm^2, the open-circuit voltage of 1.14 V, fill factor of 77.04 %, the power conversion efficiency of 20.19 %, R_s of 17.72 Ω, R_{ct} of 67.11 Ω, and R_{rec} of 76.43 Ω) for the tested devices. A similar approach could fabricate catalysts of fuel cells or battery materials by taking advantage of intrinsic properties, co-doping, and structural supports for optimal charge carrier transport or redox reactions for high current and power performance at the device level.

Additional specialized fabrication and application must be researched and optimized for an ideal SIB application. The needed synthesis techniques, which are quite sophisticated, must be reduced and adapted for large-scale and low-cost manufacturing. We are confident that the future of 1D nanomaterials will be bright in terms of energy storage and conversion. A significant SIB research and deployment problem is developing SIBs with high energy density, safety, and long-term cycle capacity. Electrolyte tuning may provide an opportunity to improve the electrochemical performance of SIBs.

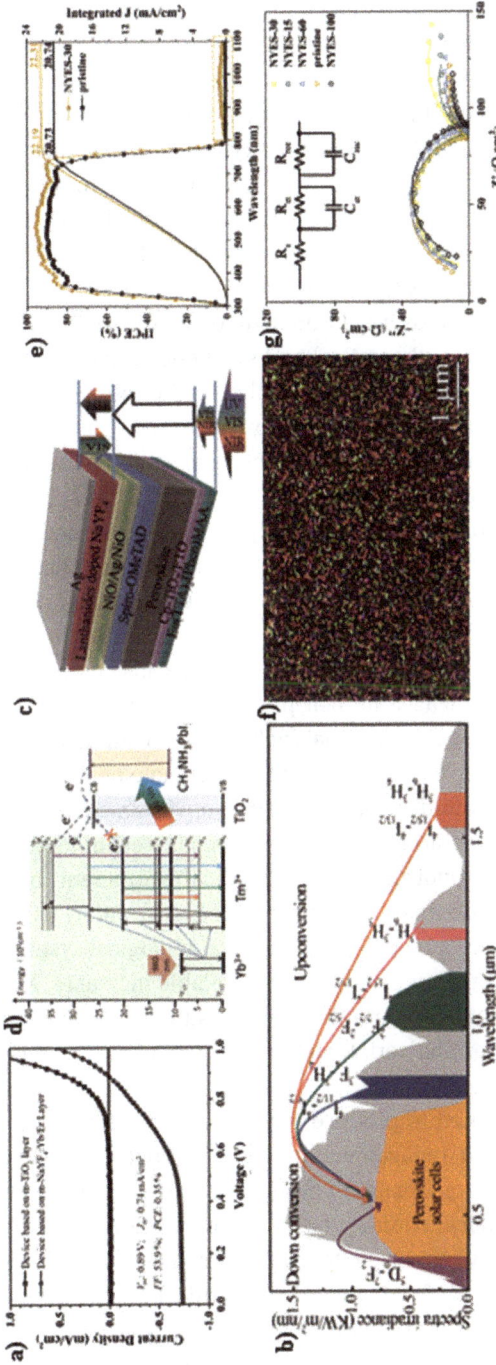

Fig. 11: The optical properties of lead halide perovskite films [71], (a) The J–V curve of the MAPbI$_3$-based solar cell using β-NaYF4:Yb^{3+}/Er^{3+} upconverting nanoparticles (UCNPs) under 980 nm NIR light [80]. (b) The air-mass 1.5G spectrum fraction absorbed by the solar cell and associated spectral regions through upconversion (UC) and downconversion (DC) pathways. (c) The 3D cutaway structure of the layered perovskite solar cell with spectral responsiveness [81]. (d) The energy transfer scheme in TiO$_2$ layer with β-NaYF4:Yb^{3+}/Tm^{3+} @TiO$_2$ core–shell and perovskite absorbed [82]. (e) The external quantum efficiency spectra for the MAPbI$_3$-based solar cells with new and NaYF4:Yb^{3+}/Er^{3+}/Sc $^{3+}$@NaYF4 core–shell UCNPs, (f) energy-dispersive spectroscopy micrograph of the β-NaYF4:Yb^{3+}/Er^{3+}/Sc^{3+}@NaYF4 UCNPs (30 % dopant to TiO$_2$), and (g) the curves from the Nyquist plots for the perovskite solar cell UC mesoporous layers with different core–shell UCNP amounts, and the corresponding equivalent circuit in the insert [83].

The above examples are what we believe as emerging areas, but a few areas I think will become more important in the future as the reach of nanoscience becomes ubiquitous. The potential acute and chronic toxicity of nano- and energy materials in the environment, their disposal, and recycling will become more pronounced as widespread usage. The fabrication of more stable MOFs for heavy metal, carbon dioxide capture and conversion, or catalysis of microplastics in water and soil across wide pH scales, and lastly, scale-up from milligram to kilogram are future challenges that need to be addressed before nanomaterials can be distributed globally to meet the societal impact of more energy, more water resources, and food medicine nexus.

Allied interdisciplinary research is necessary to understand and utilize the field of novel perovskite materials truly. Greater theoretical and synthetic understanding will promote perovskites' application and subsequently enhance the research field to benefit the global materials' engineering and development market.

The area and applications of nanomaterials are growing, and more research is being undertaken. Traditional nanomaterial production processes (such as sol–gel, chemical vapor deposition, laser ablation, flame spray pyrolysis, ultrasound, and hydrothermal) are environmentally hazardous due to toxic chemicals and high energy needs. The use of these technologies is both stupid and irresponsible, given their detrimental consequences on the environment during the current period of climate change. Green synthesis methods are equally effective as traditional ones and cause little or no harm to the environment or the people engaged in their production. As evidenced by the use of NPs in SARS-CoV-2 vaccinations, nanomaterials will transform our daily life.

Humanity is on the verge of disaster due to uncontrollable climate change. We believe it is critical to conduct this groundbreaking research environmentally responsible manner with these two considerations in mind. Green nanotechnology research will lead to the development of more significant medical equipment, new supercomputer conductors, and the examination of space, the last frontier, using revolutionized sensors. The use of active molecules from natural biological systems is becoming more widespread as the number of organisms available increases. Manipulation of active molecules in nanomaterial production leads to refinement and accuracy of nanomaterial shape and antibacterial, stabilizing, reducing, and capping characteristics.

Nanosystem synthesis may be tailored to produce tiny NPs, micron-long nanowires, or even nanosheets for specialized applications. Overall, identifying the active molecules in the green synthesis of nanomaterials enables continuous improvements and modification of physical and chemical characteristics of nanomaterials that are useful to the larger scientific community. As the globe adapts to climate change, further advancement in the green synthesis of nanomaterials is critical for preserving environmental health, energy, and scientific research integrity.

All the hard work is collaborative, and appreciation is given to Christene Smith at De Gruyter. Her passion and attention to detail allowed this process to occur seamlessly. The invaluable assistance of Stella Mueller and her colleagues of the editorial staff is also not forgotten or underestimated. With one exception, any errors, omissions, or failings are mine and not of the editorial staff or subject matter experts or, for that matter, my fellow coeditors.

Lastly, we acknowledge Texas A&M University-Kingsville, Texas A&M Energy Institute, Texas A&M University Materials Characterization Facility, and Microscopy & Imaging Center. I also acknowledge the contributions of the corresponding authors, their coauthors, contributors, collaborators, students, postdoctoral research associates, colleagues, and their institutions in this project:

Sajid Bashir as coeditor (Texas A&M University-Kingsville, TAMUK, The Department of Chemistry, 700 University Blvd., Kingsville, TX, 78363-8202),

Jingbo Louise Liu as coeditor (TAMUK, The Department of Chemistry and TAMU Energy Institute, Frederick E. Giesecke Engineering Research Bldg. 3372 TAMU College Station, TX 77843-3372),

Xuan Wang as coeditor (Texas A&M University Higher Education Center at McAllen, 6200 Tres Lagos Blvd. McAllen, TX 78504,

Ashraf Abedin as coauthor (Cain Department of Chemical Engineering, Louisiana State University, 3307 Patrick F. Taylor Hall, Baton Rouge, LA 7080), and

Sai Raghuveer Chava as coauthor (TAMUK, The Department of Chemistry).

References

[1] Peralta-Videa, J. R., Zhao, L., Lopez-Moreno, M. L., de la Rosa, G., Hong, J. & Gardea-Torresdey, J. L. (2011). Nanomaterials and the environment: a review for the biennium 2008–2010. Journal of Hazardous Materials, 186(1), 1–15.

[2] Edelstein, A. S. & Cammaratra, R. C. (1998). Nanomaterials: synthesis, properties, and applications, CRC press, Baton Rouge, FL.

[3] Burda, C., Chen, X., Narayanan, R. & El-Sayed, M. A. (2005). Chemistry and properties of nanocrystals of different shapes. Chemical Reviews, 105(4), 1025–1102.

[4] Zhang, Q., Uchaker, E., Candelaria, S. L. & Cao, G. (2013). Nanomaterials for energy conversion and storage. Chemical Society Reviews, 42(7), 3127–3171.

[5] Khin, M. M., Nair, A. S., Babu, V. J., Murugan, R. & Ramakrishna, S. (2012). A review on nanomaterials for environmental remediation. Energy & Environmental Science, 5(8), 8075–8109.

[6] Kumar, S., Ahlawat, W., Kumar, R. & Dilbaghi, N. (2015). Graphene, carbon nanotubes, zinc oxide, and gold as elite nanomaterials for fabrication of biosensors for healthcare. Biosensors & Bioelectronics, 70, 498–503.

[7] Schodek, D. L., Ferreira, P. & Ashby, M. F. (2009). Nanomaterials, nanotechnologies, and design: an introduction for engineers and architects, Butterworth-Heinemann, Woburn, MA.

[8] Oki, T. & Kanae, S. (2006). Global hydrological cycles and world water resources. Science, 313(5790), 1068–1072.

[9] Anjum, M., Miandad, R., Waqas, M., Gehany, F. & Barakat, M. A. (2019). Remediation of wastewater using various nanomaterials. Arabian Journal of Chemistry, 12(8), 4897–4919.

[10] Vayssieres, L. (2004). On the design of advanced metal oxide nanomaterials. International Journal of Nanotechnology, 1(1-2), 1–41.
[11] Ruiz-Hitzky, E., Ariga, K. & Lvov, Y. M. Eds., (2008). Bio-inorganic hybrid nanomaterials: strategies, synthesis, characterization, and applications, John Wiley & Sons, Hoboken, NJ.
[12] Lim, J. Y., Mubarak, N. M., Abdullah, E. C., Nizamuddin, S. & Khalid, M. (2018). Recent trends in the synthesis of graphene and graphene oxide-based nanomaterials for removal of heavy metals – A review. Journal of Industrial and Engineering Chemistry, 66, 29–44.
[13] Zhao, G., Wen, T., Chen, C. & Wang, X. (2012). Synthesis of graphene-based nanomaterials and their application in energy-related and environmental-related areas. RSC Advances, 2 (25), 9286–9303.
[14] Pang, H., Wu, Y., Wang, X., Hu, B. & Wang, X. (2019). Recent advances in composites of graphene and layered double hydroxides for water remediation: a review. Chemistry–An Asian Journal, 14(15), 2542–2552.
[15] Baby, R., Saifullah, B. & Hussein, M. Z. (2019). Carbon nanomaterials for the treatment of heavy metal-contaminated water and environmental remediation. Nanoscale Research Letters, 14(1), 1–17.
[16] Zhang, W., Shi, X., Zhang, Y., Gu, W., Li, B. & Xian, Y. (2013). Synthesis of water-soluble magnetic graphene nanocomposites for recyclable removal of heavy metal ions. Journal of Materials Chemistry A, 1(5), 1745–1753.
[17] Du, X., Yao, Y. & Liu, J. (2013). Structural architecture and magnetism control of metal oxides using surface grafting techniques. Journal of Nanoparticle Research, 15(7), 1–8.
[18] Li, C., Li, Z., Du, X., Du, C. & Liu, J. (2012). Structural characterization and magnetic property of iron-platinum nanoparticles fabricated by pulse electrodeposition. Rare Metals, 31(1), 31–34.
[19] Schwaminger, S. P., Surya, R., Filser, S., Wimmer, A., Weigl, F., Fraga-García, P. & Berensmeier, S. (2017). Formation of iron oxide nanoparticles for the photooxidation of water: alteration of finite size effects from ferrihydrite to hematite. Scientific Reports, 7(1), 1–9.
[20] Lewandowska-Łańcucka, J., Staszewska, M., Szuwarzyński, M., Kępczyński, M., Romek, M., Tokarz, W. . . . Nowakowska, M. (2014). Synthesis and characterization of the superparamagnetic iron oxide nanoparticles modified with cationic chitosan and coated with silica shell. Journal of Alloys and Compounds, 586, 45–51.
[21] Medina-Ramírez, I., Liu, J. L., Hernández-Ramírez, A., Romo-Bernal, C., Pedroza-Herrera, G., Jáuregui-Rincón, J. & Gracia-Pinilla, M. A. (2014). Synthesis, characterization, photocatalytic evaluation, and toxicity studies of TiO_2–Fe^{3+} nanocatalyst. Journal of Materials Science, 49 (15), 5309–5323.
[22] Bashir, S. & Liu, J. (2015). Nanomaterials and their application. Advanced nanomaterials and their applications in renewable energy (pp. 1–50), Amsterdam, The Netherlands: Elsevier Inc.
[23] Lachheb, H., Puzenat, E., Houas, A., Ksibi, M., Elaloui, E., Guillard, C. & Herrmann, J. M. (2002). Photocatalytic degradation of various types of dyes (Alizarin S, Crocein Orange G, Methyl Red, Congo Red, Methylene Blue) in water by UV-irradiated titania. Applied Catalysis. B, Environmental, 39(1), 75–90.
[24] Chamakura, K., Perez-Ballestero, R., Luo, Z., Bashir, S. & Liu, J. (2011). Comparison of bactericidal activities of silver nanoparticles with common chemical disinfectants. Colloids and Surfaces. B, Biointerfaces, 84(1), 88–96.
[25] Wang, C., Shi, H., Zhang, P. & Li, Y. (2011). Synthesis and characterization of kaolinite/TiO_2 nano-photocatalysts. Applied Clay Science, 53(4), 646–649.
[26] Zou, R., Zhang, Z., Yu, L., Tian, Q., Chen, Z. & Hu, J. (2011). A general approach for the growth of metal oxide nanorod arrays on graphene sheets and their applications. Chemistry-A European Journal, 17(49), 13912–13917.

[27] Abdel-Messih, M. F., Ahmed, M. A. & El-Sayed, A. S. (2013). Photocatalytic decolorization of Rhodamine B dye using novel mesoporous SnO_2–TiO_2 nano mixed oxides prepared by sol-gel method. Journal of Photochemistry and Photobiology. A, Chemistry, 260, 1–8.

[28] Basavarajappa, P. S., Patil, S. B., Ganganagappa, N., Reddy, K. R., Raghu, A. V. & Reddy, C. V. (2020). Recent progress in metal-doped TiO_2, non-metal-doped/codoped TiO_2, and TiO_2 nanostructured hybrids for enhanced photocatalysis. International Journal of Hydrogen Energy, 45(13), 7764–7778.

[29] Kumaravel, V., Imam, M. D., Badreldin, A., Chava, R. K., Do, J. Y., Kang, M. & Abdel-Wahab, A. (2019). Photocatalytic hydrogen production: role of sacrificial reagents on the activity of oxide, carbon, and sulfide catalysts. Catalysts, 9(3), 276.

[30] Chen, P., Xiao, T. Y., Li, H. H., Yang, J. J., Wang, Z., Yao, H. B. & Yu, S. H. (2012). Nitrogen-doped graphene/ZnSe nanocomposites: hydrothermal synthesis and their enhanced electrochemical and photocatalytic activities. ACS Nano, 6(1), 712–719.

[31] Rani, L., Kaushal, J., Srivastav, A. L. & Mahajan, P. (2020). A critical review on recent developments in MOF adsorbents for the elimination of toxic heavy metals from aqueous solutions. Environmental Science and Pollution Research, 27(36), 44771–44796.

[32] Liu, J., Wang, Z., Luo, Z. & Bashir, S. (2013). Effective bactericidal performance of silver-decorated titania nanocomposites. Dalton Transactions, 42(6), 2158–2166.

[33] Bashir, S., Chava, S. R., Yuan, D., Palakurthi, S. & Liu, J. (2020). Metal-organic frameworks and exemplified cytotoxicity evaluation. In: Metal-organic frameworks for biomedical applications (pp. 347–381, editor Madoud Mozafari). Woodhead Publishing, Sawston, Cambridge.

[34] Sajanlal, P. R., Sreeprasad, T. S., Samal, A. K. & Pradeep, T. (2011). Anisotropic nanomaterials: structure, growth, assembly, and functions. Nano Reviews, 2(1), 5883.

[35] Devatha, C. P. & Thalla, A. K. (2018). Green synthesis of nanomaterials. In: synthesis of inorganic nanomaterials (pp. 169–184, Editors Sneha Mohan Bhagyaraj, Oluwatobi Samuel Oluwafemi, Nandakumar Kalarikkal, and Sabu Thomas). Woodhead Publishing, Sawston, Cambridge.

[36] Joshi, D. R. & Adhikari, N. (2019). An overview on common organic solvents and their toxicity. Journal of Pharmaceutical Research International, 28(3), 1–18.

[37] Caramazana, P., Dunne, P., Gimeno-Fabra, M., McKechnie, J. & Lester, E. (2018). A review of the environmental impact of nanomaterial synthesis using continuous flow hydrothermal synthesis. Current Opinion in Green and Sustainable Chemistry, 12, 57–62.

[38] Huston, M., DeBella, M., DiBella, M. & Gupta, A. (2021). Green Synthesis of Nanomaterials. Nanomaterials, 11(8), 2130.

[39] Zhao, H., Yuan, W. & Liu, G. (2015). Hierarchical electrode design of high-capacity alloy nanomaterials for lithium-ion batteries. Nano Today, 10(2), 193–212.

[40] Tian, N., Zhou, Z. Y., Sun, S. G., Ding, Y. & Wang, Z. L. (2007). Synthesis of tetrahexahedral platinum nanocrystals with high-index facets and high electro-oxidation activity. Science, 316(5825), 732–735.

[41] Mulvihill, M. J., Ling, X. Y., Henzie, J. & Yang, P. (2010). Anisotropic etching of silver nanoparticles for plasmonic structures capable of single-particle SERS. Journal of the American Chemical Society, 132(1), 268–274.

[42] Xu, H. & Suslick, K. S. (2010). Water-Soluble fluorescent silver nanoclusters. Advanced Materials, 22(10), 1078–1082.

[43] Abedin, A. & Spivey, J. J. (2021). Direct catalytic low-temperature conversion of co 2 and methane to oxygenates. In: Advances in sustainable energy (pp. 227–250, Editors Yong-jun Gao, Weixin Song, Jingbo Louise liu, and Sajid Bashir). Springer, Cham., Berlin, Germany.

[44] Li, Y. & Somorjai, G. A. (2010). Nanoscale advances in catalysis and energy applications. Nano Letters, 10(7), 2289–2295.

[45] Mazumder, V., Lee, Y. & Sun, S. (2010). Recent development of active nanoparticle catalysts for fuel cell reactions. Advanced Functional Materials, 20(8), 1224–1231.

[46] Guo, S. & Dong, S. (2009). Biomolecule-nanoparticle hybrids for electrochemical biosensors. TrAC Trends in Analytical Chemistry, 28(1), 96–109.

[47] Guo, S. & Wang, E. (2011). Noble metal nanomaterials: controllable synthesis and application in fuel cells and analytical sensors. Nano Today, 6(3), 240–264.

[48] Wang, L. & Yamauchi, Y. (2009). Block copolymer mediated synthesis of dendritic platinum nanoparticles. Journal of the American Chemical Society, 131(26), 9152–9153.

[49] Guo, S. & Wang, E. (2007). Synthesis and electrochemical applications of gold nanoparticles. Analytica Chimica Acta, 598(2), 181–192.

[50] Zeng, Z., Xu, Y., Zhang, Z., Gao, Z., Luo, M., Yin, Z. . . . Yan, C. (2020). Rare-earth-containing perovskite nanomaterials: design, synthesis, properties, and applications. Chemical Society Reviews, 49(4), 1109–1143.

[51] Yin, W. J., Weng, B., Ge, J., Sun, Q., Li, Z. & Yan, Y. (2019). Oxide perovskites, double perovskites, and derivatives for electrocatalysis, photocatalysis, and photovoltaics. Energy & Environmental Science, 12(2), 442–462.

[52] Wang, L., Zhou, H., Hu, J., Huang, B., Sun, M., Dong, B. . . . Yan, C. H. (2019). A Eu3+-Eu2+ ion redox shuttle imparts operational durability to Pb-I perovskite solar cells. Science, 363 (6424), 265–270.

[53] Khalfin, S. & Bekenstein, Y. (2019). Advances in lead-free double perovskite nanocrystals, engineering bandgaps, and enhancing stability through composition tunability. Nanoscale, 11(18), 8665–8679.

[54] Milstein, T. J., Kluherz, K. T., Kroupa, D. M., Erickson, C. S., De Yoreo, J. J. & Gamelin, D. R. (2019). Anion Exchange and the Quantum-Cutting Energy Threshold in Ytterbium-Doped CsPb (Cl1–x Br x) 3 Perovskite Nanocrystals. Nano Letters, 19(3), 1931–1937.

[55] Saha, R., Sundaresan, A. & Rao, C. N. R. (2014). Novel features of multiferroic and magnetoelectric ferrites and chromites exhibiting magnetically driven ferroelectricity. Materials Horizons, 1(1), 20–31.

[56] Winter, M. & Brodd, R. J. (2004). What are batteries, fuel cells, and supercapacitors?. Chemical Reviews, 104(10), 4245–4270.

[57] Goodenough, J. B. & Park, K. S. (2013). The Li-ion rechargeable battery: a perspective. Journal of the American Chemical Society, 135(4), 1167–1176.

[58] Kundu, D., Talaie, E., Duffort, V. & Nazar, L. F. (2015). The emerging chemistry of sodium-ion batteries for electrochemical energy storage. Angewandte Chemie International Edition, 54 (11), 3431–3448.

[59] Brezesinski, T., Wang, J., Tolbert, S. H. & Dunn, B. (2010). Ordered mesoporous α-MoO3 with iso-oriented nanocrystalline walls for thin-film pseudocapacitors. Nature Materials, 9(2), 146–151.

[60] Jin, T., Han, Q., Wang, Y. & Jiao, L. (2018). 1D nanomaterials: design, synthesis, and applications in sodium-ion batteries. Small, 14(2), 1703086–1 to 1703086–26.

[61] Taylor, G. I. (1964). Disintegration of water drops in an electric field. Proceedings of the Royal Society of London. Series A, Mathematical and Physical Sciences, 280(1382), 383–397.

[62] Guan, C., Wang, X., Zhang, Q., Fan, Z., Zhang, H. & Fan, H. J. (2014). Highly stable and reversible lithium storage in SnO2 nanowires surface coated with a uniform hollow shell by atomic layer deposition. Nano Letters, 14(8), 4852–4858.

[63] Yang, Q., Lu, Z., Liu, J., Lei, X., Chang, Z., Luo, L. & Sun, X. (2013). Metal oxide and hydroxide nanoarrays: hydrothermal synthesis and applications as supercapacitors and nanocatalysts. Progress in Natural Science: Materials International, 23(4), 351–366.

[64] Lyu, Z., Ding, S., Du, D., Qiu, K., Liu, J., Hayashi, K. . . . Lin, Y. (2022). Recent advances in biomedical applications of 2D nanomaterials with peroxidase-like properties. Advanced Drug Delivery Reviews, 185 (June 2022), 114269–1 to 114269–12.

[65] Zhang, F., Zheng, B., Zhang, J., Huang, X., Liu, H., Guo, S. & Zhang, J. (2010). Horseradish peroxidase immobilized on graphene oxide: physical properties and applications in phenolic compound removal. The Journal of Physical Chemistry C, 114(18), 8469–8473.

[66] Wang, Y., Qi, K., Yu, S., Jia, G., Cheng, Z., Zheng, L. . . . Zheng, W. (2019). Revealing the intrinsic peroxidase-like catalytic mechanism of heterogeneous single-atom Co–MoS2. Nano-micro Letters, 11(1), 1–13.

[67] Jiao, L., Xu, W., Zhang, Y., Wu, Y., Gu, W., Ge, X. . . . Guo, S. (2020). Boron-doped Fe-NC single-atom nanozymes specifically boost peroxidase-like activity. Nano Today, 35, 100971–1 to 100971–10.

[68] Lu, Z., Yang, Q., Zhu, W., Chang, Z., Liu, J., Sun, X. . . . Duan, X. (2012). Hierarchical Co3O4@ Ni-Co-O supercapacitor electrodes with ultrahigh specific capacitance per area. Nano Research, 5(5), 369–378.

[69] Lu, Z., Zhu, W., Lei, X., Williams, G. R., O'Hare, D., Chang, Z. . . . Duan, X. (2012). High pseudocapacitive cobalt carbonate hydroxide films derived from CoAl layered double hydroxides. Nanoscale, 4(12), 3640–3643.

[70] Sun, J., Li, Y., Liu, X., Yang, Q., Liu, J., Sun, X. . . . Duan, X. (2012). Hierarchical cobalt iron oxide nanoarrays as structured catalysts. Chemical Communications, 48(28), 3379–3381.

[71] Chen, Y., Liu, S., Zhou, N., Li, N., Zhou, H., Sun, L. D. & Yan, C. H. (2021). An overview of rare Earth coupled lead halide perovskite and its application in photovoltaics and light-emitting devices. Progress in Materials Science, 120, 100737–1 to 100737–29.

[72] Luo, D., Su, R., Zhang, W., Gong, Q. & Zhu, R. (2020). Minimizing non-radiative recombination losses in perovskite solar cells. Nature Reviews Materials, 5(1), 44–60.

[73] Pan, G., Bai, X., Yang, D., Chen, X., Jing, P., Qu, S. . . . Song, H. (2017). Doping lanthanide into perovskite nanocrystals: highly improved and expanded optical properties. Nano Letters, 17(12), 8005–8011.

[74] Yao, J. S., Ge, J., Han, B. N., Wang, K. H., Yao, H. B., Yu, H. L. . . . Yu, S. H. (2018). Ce3+-doping to modulate photoluminescence kinetics for efficient CsPbBr3 nanocrystals based light-emitting diodes. Journal of the American Chemical Society, 140(10), 3626–3634.

[75] Wang, L., Zhou, H., Hu, J., Huang, B., Sun, M., Dong, B. . . . Yan, C. H. (2019). A Eu3+-Eu2+ ion redox shuttle imparts operational durability to Pb-I perovskite solar cells. Science, 363 (6424), 265–270.

[76] Wang, K., Zheng, L., Zhu, T., Yao, X., Yi, C., Zhang, X. . . . Gong, X. (2019). Efficient perovskite solar cells by hybrid perovskites incorporated with heterovalent neodymium cations. Nano Energy, 61, 352–360.

[77] Qiao, H. W., Yang, S., Wang, Y., Chen, X., Wen, T. Y., Tang, L. J. . . . Yang, H. G. (2019). A gradient heterostructure based on tolerance factor in high-performance perovskite solar cells with 0.84 fill factor. Advanced Materials, 31(5), 1804217–1 to 1804217–6.

[78] Xiang, W., Wang, Z., Kubicki, D. J., Tress, W., Luo, J., Prochowicz, D. . . . Hagfeldt, A. (2019). Europium-doped CsPbI2Br for stable and highly efficient inorganic perovskite solar cells. Joule, 3(1), 205–214.

[79] Duan, J., Zhao, Y., Yang, X., Wang, Y., He, B. & Tang, Q. (2018). Lanthanide Ions Doped CsPbBr3 Halides for HTM-Free 10.14 %-Efficiency Inorganic Perovskite Solar Cell with an Ultrahigh Open-Circuit Voltage of 1.594 V. Advanced Energy Materials, 8(31), 1802346–1 to 1802346–9.

[80] He, M., Pang, X., Liu, X., Jiang, B., He, Y., Snaith, H. & Lin, Z. (2016). Monodisperse dual-functional upconversion nanoparticles enabled near-infrared organolead halide perovskite solar cells. Angewandte Chemie, 128(13), 4352–4356.

[81] Li, H., Chen, C., Jin, J., Bi, W., Zhang, B., Chen, X. . . . Song, H. (2018). Near-infrared and ultraviolet to visible photon conversion for full-spectrum response perovskite solar cells. Nano Energy, 50, 699–709.

[82] Liang, J., Gao, H., Yi, M., Shi, W., Liu, Y., Zhang, Z. & Mao, Y. (2018). β-NaYF4: yb3+, Tm3+@ TiO2 core-shell nanoparticles incorporated into the mesoporous layer for high-efficiency perovskite solar cells. Electrochimica Acta, 261, 14–22.

[83] Guo, Q., Wu, J., Yang, Y., Liu, X., Jia, J., Dong, J. . . . Huang, Y. (2019). High-performance perovskite solar cells based on β-NaYF4: yb3+/Er3+/Sc3+@ NaYF4 core-shell upconversion nanoparticles. Journal of Power Sources, 426, 178–187.

Section 1: **Overview of nanoscience and nanochemistry**

Sajid Bashir, William Houf, Luis Villanueva, Jianhong Ren,
Jingbo Louise Liu

Chapter 1
Nanochemistry: development of nanomaterials

1.1 Introduction

Nanotechnology, a multidisciplinary field, has been systematically investigated to understand, manipulate, and tune materials from either atomic or molecular levels [1]. The design, construction, and utilization of functional structures are the critical perspectives to tackle with at least one characteristic dimension measured in nanometer (nm) [2]. Nanomaterials and systems exhibit physical, chemical, electronic, magnetic, and biological properties [3]. These properties and phenomena of nanomaterials allow their diversified applications, such as water, energy, medicine, and food resource utilization, with the highest impact on the life quality of human beings [4–6]. Scientists and professionals from different disciplines have revolutionized today's research breakthroughs in lowering energy consumption, developing molecular medicine, and deploying artificial intelligence (AI) systems [7–9]. Nanoscience, engineering, and nanotechnology involve materials imaging, property measurements, structural characterization, modeling, and manipulating matter at the nanoscale.

Acknowledgments: This work was supported by the Petroleum Research Fund of the American Chemical Society (53827-UR10), Texas A&M Energy Institute, and the Robert Welch Foundation (Departmental Grant, AC-0006) to analyze the data written in this chapter. The technical support from Texas A&M University-Kingsville and Texas A&M Energy Institute is also duly acknowledged. The authors also gratefully acknowledge the helpful discussions with colleagues' comments and reviewers' suggestions, which have improved the presentation.

Author contribution: S. Bashir and J. Liu collectively conceived the projects and oversaw their progress. S. Bashir was in charge of the submission processing of all chapters. W. Houf, an undergraduate student, and L. Villanueva, a graduate student, participated as protégés to obtain material synthesis and characterization knowledge to improve their core knowledge and hands-on experience. J. Liu wrote the first draft of all authors for editing.

Sajid Bashir, William Houf, Luis Villanueva, The Department of Chemistry, Texas A&M University-Kingsville, MSC161, 700 University Blvd., Kingsville, TX 78363, USA
Jianhong Ren, The Department of Environmental Engineering, Texas A&M University-Kingsville, MSC213, 917 W. Avenue B, Kingsville, TX 78363, USA
Jingbo Louise Liu, The Department of Chemistry, Texas A&M University-Kingsville, MSC161, 700 University Blvd., Kingsville, TX 78363, USA; The Department of Environmental Engineering, Texas A&M University-Kingsville, MSC213, 917 W. Avenue B, Kingsville, TX 78363, USA; Texas A&M Energy Institute, 1617 Research Parkway, Suite 308, College Station, TX 77843-3372, USA, e-mail: jingbo.liu@tamuk.edu and e-mail: jingbo.liu@tamu.edu

https://doi.org/10.1515/9783110739879-002

From the size perspective, nanotechnology focuses on understanding and controlling the more interactive and resilient materials at dimensions roughly 1–100 nm [10]. Nanotechnological research and development have been directed toward in-depth understanding and innovation in efficient and large-scale nanomanufacturing. According to the end demand, the advances in nanotechnology will enable its industrial and military applications in diversified fields, such as medicine, catalysis, sustainable energy, consumer products, and electronics [11, 12].

The chapter discusses several key features of nanotechnology, including bottom-up or top-down approaches to create tunable structures (Scheme 1.1) [13, 14]. The fabrication variables can be optimized using AI (such as ANN) and laboratory automation techniques based on experimental data. These structures in the range of 1–100 nm exhibit different behavior than those in isolated molecules or bulk materials [15, 16]. The shapes, architectures, and components of nanomaterials and systems are critical factors in enhancing biological, chemical, electronic, and physical properties, and further displaying unique phenomena and processes due to their nanoscale size [17]. Under certain circumstances, nanomaterials' new properties and behaviors may be unpredictable from behaviors observed at bulk scales. The most important changes in behavior and activities of nanomaterials mainly result from the observed intrinsic nature of nanocomponents instead of a sole reduction in the order of magnitude size [18]. These observations include size confinement, the predominance of interfacial phenomena, and quantum mechanics correlated to the electron configurations and orbital diagrams [19]. After more than five decades of intensive and systematic research on nanoscience, scientists and engineers revolutionized the materials design and production of more resilient materials and system architectures by functionalization, intercalation, doping, and grafting [20, 21]. These different approaches allow new features to be added to an engineered material by changing the surface chemistry and other properties [22]. Further decrease in magnitudes of nanostructures enabled the development of unique properties of a series of newly developed materials, such as carbon nanotubes, quantum wires and dots, thin films, DNA-based structures (to control stimuli responsiveness), and laser emitters (polyhedral oligomeric silsesquioxanes) [23–25]. These compositions and forms of nanomaterials and devices herald a revolutionary age for science, engineering, and technology [26].

With an emphasis on the dimension and importance of quantum mechanical effects of nanomaterials, the National Nanotechnology Initiative defines nanotech as "the manipulation of matter with at least one dimension sized from 1 to 100 nanometers" [27]. Nanotechnology has become more and more essential in modern scientific fields and has constantly been evolving to meet commercial and academic interests [28, 29]. The societal need for nanotechnology fosters scientists to continue new research and to provide new insights to the scientific community. Nanotechnology can be extended from conventional device physics to completely new approaches. Research discoveries on molecular self-assembly depict that nanotechnological research opens a new paradigm to advance SARS-CoV-2 diagnosis, treatment, and vaccine formulation [30, 31]. However, as a double-edged sword, nanotechnology raises problematic issues, such as safe operation, toxicity, and ecological and environmental concerns (see Chapter 12) [32–34]. Buerki-Thurnherr and her team reported that the engineered nanomaterials pose a risk for "pregnancy, fetal development, and offspring health later in life" [35]. The origin and mechanisms induced by nanomaterials will be a cornerstone, and further understanding of its function will guide the design and produce safe and sustainable nanomaterials. The risk of nanomaterials posed to surface water and soil near the point sources has drawn significant attention due to the nanomaterial-induced toxicity, which can be intrigued by sunlight irradiation, natural organic matter, and mineral particles [34]. The other potential impacts of nanomaterials on global economics and speculation about "various doomsday scenarios" have drawn scientists, policy-makers, and the public's attention [36]. These concerns have led to a debate among advocacy groups and governments on whether special regulation of nanotechnology is warranted [37, 38], and Scheme 1.1.

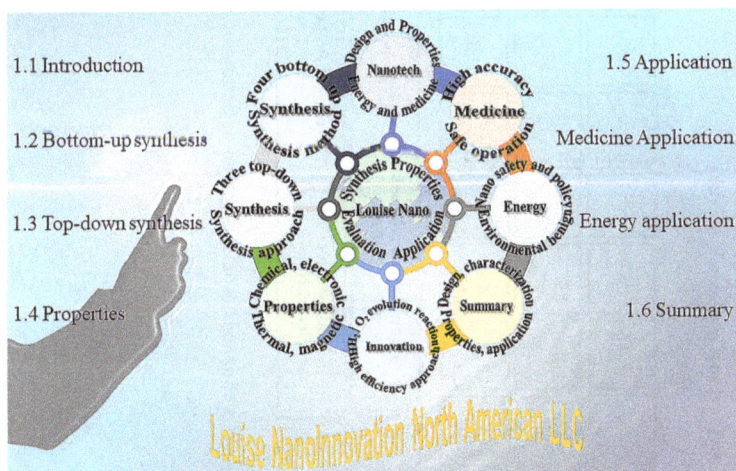

Scheme 1.1: In the flowchart of this chapter, six areas will be described in detail. This chapter summarizes nanotechnology, four bottom-up and three top-down syntheses, four major properties, and their application in energy and medicine fields.

1.2 Bottom-up synthesis of nanomaterials

More than 72 nanomaterial formulations have been developed in Dr. Liu's and her collaborators' laboratories (Fig. 1.1). They are composed of metals, metal oxides, metal borides, perovskite-based ceramics, carbon nanotube and graphene, metal-organic frameworks (MOF), and polymers, which can be used as fuel cells and battery catalysts; water remediation, cancer theranostics, and disinfection; carbon capture

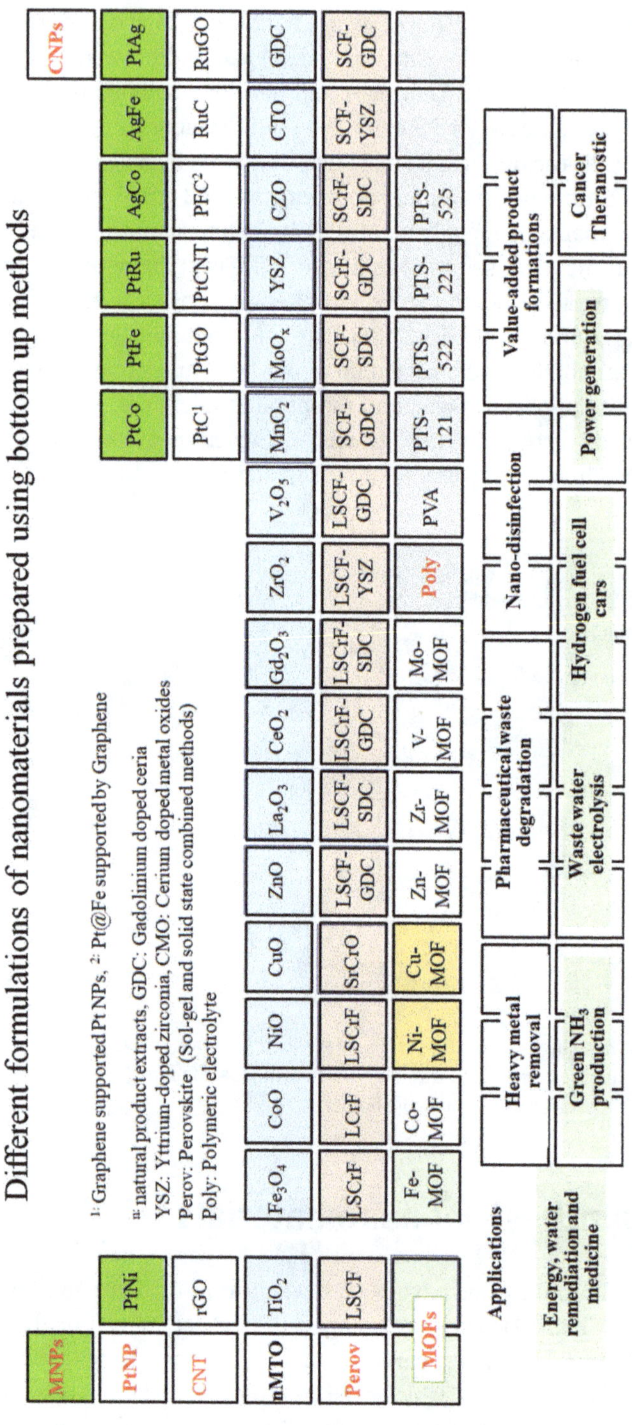

Different formulations of nanomaterials prepared using bottom up methods

1: Graphene supported Pt NPs, 2: Pt@Fe supported by Graphene
n: natural product extracts, GDC: Gadolinium doped ceria
YSZ: Yttrium-doped zirconia, CMO: Cerium doped metal oxides
Perov: Perovskite (Sol-gel and solid state combined methods)
Poly: Polymeric electrolyte

CNPs

MNPs / PtNi										PtCo	PtFe	PtRu	AgCo	AgFe	PtAg
PtNP / rGO										PtC^1	PtGO	PtCNT	PFC^2	RuC	RuGO
CNT															
nMTO / TiO_2	Fe_3O_4	CoO	NiO	CuO	ZnO	La_2O_3	CeO_2	Gd_2O_3	ZrO_2	V_2O_5	MnO_2	MoO_x	YSZ	CZO	CTO	GDC
Perov / LSCF	LSCrF	LCrF	LSCF	SrCrO	LSCF-GDC	LSCrF-SDC	LSCrF-GDC	LSCrF-SDC	LSCF-YSZ	LSCF-GDC	SCF-GDC	SCF-SDC	SCrF-GDC	SCrF-SDC	SCF-YSZ	SCF-GDC
MOFs	Fe-MOF	Co-MOF	Ni-MOF	Cu-MOF	Zn-MOF	Zr-MOF	V-MOF	Mo-MOF	PVA	**Poly**	PTS-121	PTS-522	PTS-221	PTS-525		

Applications

- Energy, water remediation and medicine
- Heavy metal removal
- Green NH_3 production
- Pharmaceutical waste degradation
- Waste water electrolysis
- Nano-disinfection
- Hydrogen fuel cell cars
- Power generation
- Value-added product formations
- Cancer Theranostic

Fig. 1.1: The periodic table of different formulations of nanomaterials, prepared by bottom-up methods.

and sequestration; microwave (MW) absorption; as well as electrolytes. These materials were proved to show high resistance and responses to external signal changes and can self-regulate their structures and properties. The four different bottom-up synthesis methods (sol–gel, microemulsion, hydrothermal, and chemical vapor deposition (CVD)) that are used to produce nanostructural materials with high chemical reactivities are discussed in this section. These methods were used to produce these materials with high product yields and showed potential to be scaled up and commercialized. The team implemented AI during the materials design to optimize synthesis variables based on the accumulative experimental research data.

1.2.1 Sol–gel method

The sol–gel synthesis is a chemical process to disperse solid substances into a liquid medium, forming a continuous three-dimensional (3D) network extending throughout the liquid. The sol–gel method is one of the bottom-up wet-chemical techniques used to produce nanostructured materials [39]. During this process, metal alkoxide or ionic compound solutions have been commonly used as the starting materials to form a heterogeneous mixture. The above-dispersed substance (known as a sol) normally has a size varying from 10^{-7} to 10^{-9} m (100–1 nm). Due to its high molecular polarity, the water or alcohol–water mixture is commonly used as the dispersing agent. The liquid-phased sol can become a gel (xerogel under ambient conditions) via constant Brownian motion upon hydrolysis and condensation during the sol–gel transition [40]. The high surface tension of these dispersed nanoparticles allows them to "glue" together when collisions occur. The nanoparticles can either form bonding or agglomerate by electrostatic forces [41]. During these two steps, a continuous inorganic lattice containing the bonding of M-O-M or M-H-M will be formed as a skeleton for the final products [42]. Upon further polycondensation and solvent removal, 3D with steady and well-aligned networks can be formed and encase the liquid stage due to colloidal particles' aggregation and/or flocculation [43]. Upon completion of the sol–gel transition, the viscosity of the 3D gelation approaches infinity. Eventually, immobility will be achieved due to the networking extension following this gelation at the gel point [44]. The synthesis variables affecting sol-gel chemistry are activity (pH), solvent, reaction temperature, gelation duration, and catalysts [45]. Their influences are briefly discussed in the following.

1. Activity effect: The sol–gel colloidal chemistry involves water as a dispersing agent, sensitive to pH. The hydrolysis step of most of the precursors is highly affected by acidity. As mentioned above, the metallo-organic compounds (normally alkoxides) can be used as precursors. In these alkoxide $(M(OR)_n)$ structures, M stands for a metal or a metalloid, and R is a hydrophobic alkyl group. The alkoxide compounds are dispersed in a solvent, resulting in different possible reactions under a controlled pH value. In the low pH range from 2.5 to 4.5, a generalized trend showed that gelation time decreased with increasing pH, temperature, and water. On the other hand, when pH increases, such as in ammonia conditions, mesoporous xerogels can be produced through cross-linking. In this basic condition, the hydrolysis of alkoxides to form ul-trafine nuclei as the seed for gelation can be the rate-limiting step [46]. A study on polysilsesquioxanes gelation affected by the pH (1.3–12; Fig. 1.2A) was chosen [47]. It was found that slow gelation was observed in the pH range of 4.5–5.2 (Fig. 1.2B). The reaction rate of this system showed an "inverse volcano effect," with the decrease in kinetics with pH increase, and then the rate raised after a trough was achieved at about neutral conditions (Fig. 1.2C). Although the difference between the alkyl and aryl groups showed a certain degree of effect, the presence of silanols is the major factor that causes the pH values to change and then the reaction rate. The NMR data on the degree of condensation of the hexylene-bridged polysilsesquioxane (Fig. 1.2D) indicate a slow condensation rate in different pH ranges [48]. Another observation is that the porous xerogels can be formed when the degree of condensation is >75 %.

2A

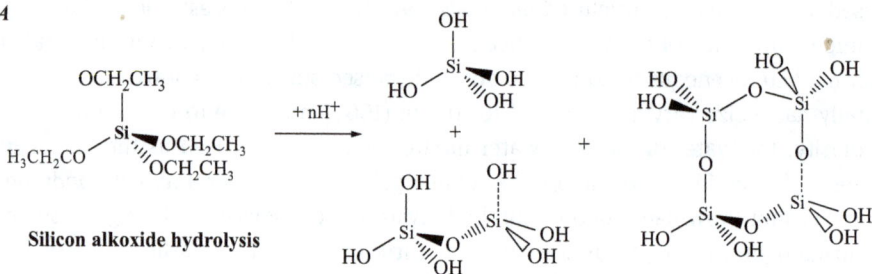

2B

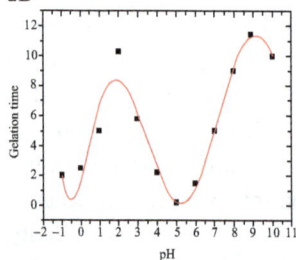

2C

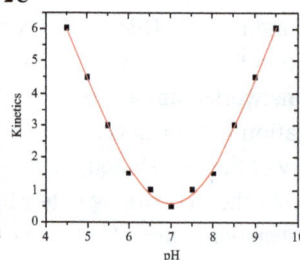

2D

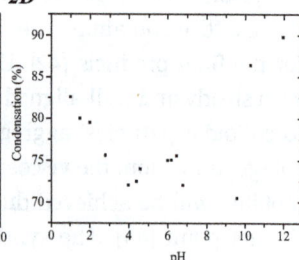

Fig. 1.2: The pH effect on the sol–gel transition of polysilsesquioxanes: (A) the silicon alkoxide hydrolysis; (B) gelation–pH relationship; (C) minimum in reaction rate–pH profile; and (D) condensation degree dependent on SOH acidity.

2. Solvent effect: There are two classifications of solvents used in sol–gel chemistry: protic and aprotic. The solvents affect the reaction rate due to their intrinsic natures, such as the polarity and complexity of molecules. In particular, the rate-determining step of the gelation responds to the properties of the solvent. The protic solvents (water and ethanol) are molecules that normally contain -OH and -NH groups, capable of forming hydrogen bonding. On the other hand, the aprotic solvents (such as chloroform) are not capable of forming a hydrogen bond or acting as hydrogen bond donors. The solvent polarity is a critical intrinsic factor related to its dielectric constant. The polar solvents normally display high dielectric constants. The nonpolar solvent with low dielectric constants is due to the low electronegativity differences between two bonded atoms or the cancelation of bond dipole moments.

In the polymerization process, the sol precursors agglomerate into nanoparticles via solvent removal. The sol–gel procedures provided advantages in obtaining metastable materials, achieving superior purity and compositional homogeneity of the products. Moderate temperatures with simple laboratory equipment can achieve the above goal, influencing the particle morphology during the chemical transformation of the molecular precursor to the final products. A highly polar molecule (H_2O) has been widely used as a solvent in the sol–gel method, where three reaction types (hydrolysis, condensation, and aggregation) occur almost simultaneously. A nonaqueous approach using organic solvents was found to overcome some of the aqueous limitations and provide controllable morphologies and ensure the reproducibility of the final products, as reported by Bilecka and Niederberger [49]. The organic solvent showed multiple roles, acting as the oxygen-supplying agent for the metal oxide, stabilizing effect, and displaying slow reaction rates [50]. As the oxygen-supplying agent, the organic solvent strongly influences the particle size, shape, surface, and assembly properties (Fig. 1.3) [51]. The slow reaction rates using organic solvents resulted from the moderate reactivity of the C–O bond. In combination with the stabilizing effect of the organic species, the organic solvent provides oxygen sources for oxide nanomaterials and leads to the formation of highly crystalline products. The characterization results showed that these products exhibited uniform particle morphologies and crystallite sizes. The nonaqueous solvent can donate electrons to the metal d-orbitals, increasing the bonding between metal and oxygen. In short, the fundamental role of organic reaction and the intrinsic nature of nonaqueous solvents provide pathways for tailoring particle size, shape, composition, and surface properties.

3. Temperature effect: The chemical kinetics of the different reactions involved in the formation of nanoparticles and the assembly of nanoparticles into a gel network is accelerated by temperature. In general, the sol-to-gel transition (gelation) represented a specific extent of reaction, which can be described using the Arrhenius equation (eq. (1.1A)) and/or its logarithm form (eq. (1.1B)). The gelation time (τ_g) is affected by temperature. The temperature increase will result in an increase in the kinetic energy of the reactive precursors. The increased kinetic energy leads to more frequent collisions and causes an increase in the reaction rate:

$$k = Ae^{-(E_a/RT)} \qquad (1.1A)$$

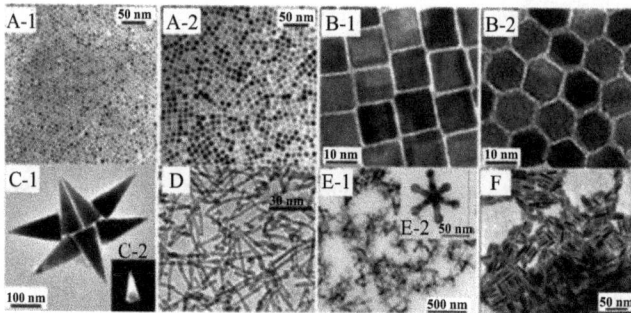

Fig. 1.3: TEM images of different nanomaterials: (A-1) spinel cobalt ferrite ($CoFe_2O_4$) spherical nanoparticles were used as seeds (8-nm) and (A-2) cubelike nanoparticles. Reproduced from [51]. Copyright 2004 American Chemical Society. (B-1) Cubelike $MnFe_2O_4$ nanoparticles and (B-2) polyhedron-shaped nanoparticles. Reproduced from [52]. Copyright 2004 American Chemical Society. (C-1) ZnO nanocrystals with conelike and (C-2) zoom-in image of one crystal. Reproduced from [53]. Copyright 2005 Wiley-VCH. (D) Sol–gel-derived TiO_2 nanorods with near-monodispersion single-crystal manganese oxide. Reproduced from [54]. Copyright 2006 Wiley-VCH. (E-1) Single-crystal manganese oxide (MnO) multipods and (E-2) zoom-in image of a hexapod. Reproduced from [55]. Copyright 2005 American Chemical Society. (F) In situ one-pot synthesis of one-dimensional tungsten oxide nanorods. Reproduced from [56]. Copyright 2005 American Chemical Society.

$$\ln k = \ln A - \frac{E_a}{R}\left(\frac{1}{T}\right) \tag{1.1B}$$

where k represents the rate constant, A is the frequency factor for the reaction ($A = pz$, p is the orientation probability factor, and Z is the collision frequency); E_a is the activation energy (kJ/mol), which exceeds a certain energy threshold for a reaction to happen; T is the absolute temperature (K).

As shown in Tab. 1.1, the low activation energy and high reaction temperature correspond to the high fraction (f) of collisions at a higher temperature. If a reaction shows sufficient energy equal to or greater than E_a at a certain temperature, the f increases, resulting in an effective collision. As a result, the reaction rate increases. Particles have more energy at higher temperatures and collide more often and more effectively [57].

Tab. 1.1: The relationship between activation energy, temperature, and fraction of collision for reaction.

Activation energy (kJ/mol) at 298 K	Fraction of collisions	Temperature (K) at 50 kJ/mol	Fraction of collisions
50	1.70×10^{-9}	298	1.70×10^{-9}
75	7.03×10^{-14}	308	3.29×10^{-9}
100	2.90×10^{-18}	318	$6.12.70 \times 10^{-9}$

Research data showed that gelation might take weeks or months if the temperature is too low (Fig. 1.4). On the contrary, high temperature results in a fast reaction rate, where the gel network occurs quickly. The rapid gelation normally resulted in solid precipitate out of the liquid due to the formation of large clusters. During gelation, the exothermic reaction favors the growth of nanoparticles and gel networks, further leading to a faster reaction, and more precipitate will be produced. Twej reported that the gelling time of silica alkoxide obtained by the sol–gel process was shorter at a higher temperature under different pH values and hydrolysis ratios (Fig. 1.4A) [58]. The large fraction of collisions has enough energy to exceed activation energy, leading to reaction (Fig. 1.4B). At high temperatures, the high kinetic energy of reactant molecules is correlated to the high frequency of effective collision, meaning a larger amount of fraction of collision will exceed the activation energy, as shown in the green-colored region in Fig. 1.4B.

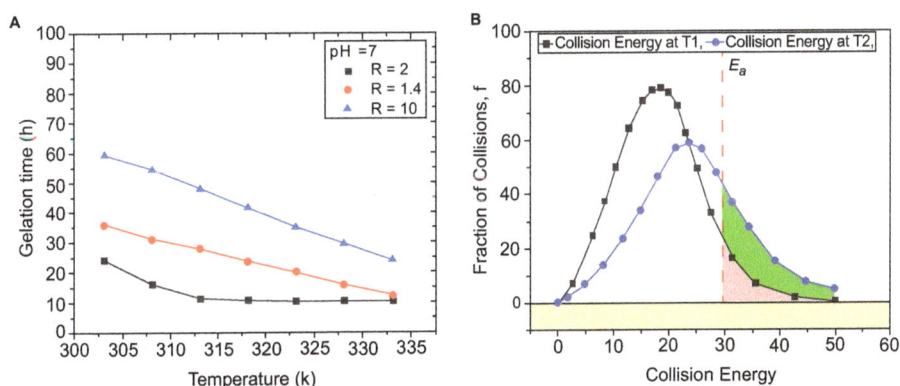

Fig. 1.4: The temperature effect on sol–gel chemistry. (A) The gelation time is dependent on the temperature, and (B) the relationship between collision fraction and collision energy highly depends on the temperatures.

4. Time effect: In general, the slow gelation time will allow for better alignment and arrangement of the gel network, leading to a uniform structure (Fig. 1.5A). However, the gel type might display different steps in its formation over different timescales. As mentioned above, the fast reaction normally results in precipitation due to the rapid formation of gel networks.

5. Use of catalysts: A catalyst usually accelerates a chemical reaction, including the gelation process. Homogeneous acids (H^+) and bases (OH^-) can be used as catalysts in sol–gel chemistry, while the gelation may occur by different mechanisms (Fig. 1.5B). In both aqueous and nonaqueous, the sol–gel chemistry is usually pH-sensitive. The use of a catalyst will lower the activation energy, leading to a different pathway for a reaction due to the large fraction of effective collision.

6. Mixing/agitation: Even a mixture of sol reactants is important to ensure the homogenization of the chemical reactions in the solution as sol precursors transit into an integrated gel. The magnetic or mechanical agitation allows for event heat and/or mass transfer, further improving contact between different components. A reduction in the agglomeration can be prevented due to the continuous movement of the molecules; meanwhile, the intrinsic natures of reactants can be altered. However, sometimes the gel network may be disrupted by mixing, causing fragmentation of domains.

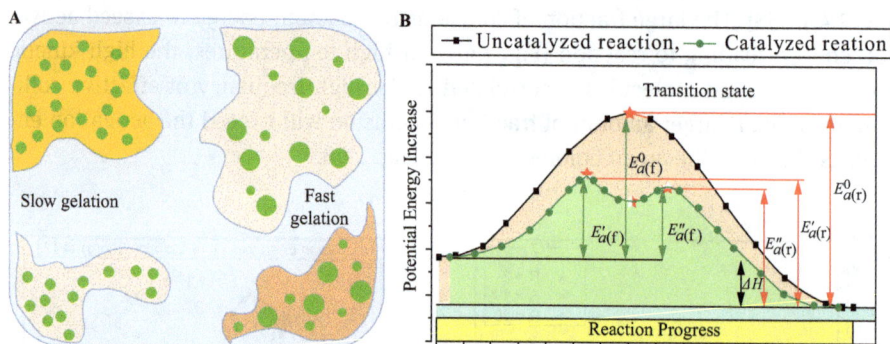

Fig. 1.5: The time and catalyst effects on sol–gel chemistry: (A) the gelation time affects the particle morphology and size and (B) the schematic of the different pathways of reaction upon the use of catalysts.

The sol–gel-derived nanomaterials with multiple constituents can be tuned and functionalized by varying chemical stoichiometry, controlling the number of sols of various components, and optimizing the synthesis parameters [59]. The homogeneity of these engineered nanomaterials using the sol–gel method can be controlled from the atomic and molecular levels [60]. The factors that influence the structure and properties of the final products have been discussed briefly in this chapter. The mechanism and kinetics of the sol–gel method and three key steps will be discussed in Chapter 2: (1) hydrolysis of alkoxides, (2) polycondensation to form networks, and (3) gelation to form a solid phase. Liu's lab has widely used this method to prepare different formulations of nanomaterials used as photocatalysts for water remediation, electrocatalysts for energy storage and conversion, and fingerprint development agents.

1.2.2 Microemulsion synthesis

Based on the definition by the International Union of Pure and Applied Chemistry (IUPAC), the microemulsion refers to a "dispersion composed of water, oil, and surfactant(s)" [61]. This system is "isotropic and thermodynamically stable," where the diameter of the dispersed domain ranges from 1 to 100 nm. Usually, these dispersed particles are sized 10–50 nm, and the surfactants combine with cosurfactants [62]. In the ternary microemulsion, water and "oil" (any nonwater-soluble liquids) are present with a surfactant. There are three basic microemulsion families: oil dispersed in water (o/w), water dispersed in oil (w/o), and bicontinuous. Different from the formation of traditional colloids, these microemulsions can be produced upon simple mixing of the components without applying high shear [63]. Microemulsion synthesis has become popular due to its versatility in controlling particle sizes and properties precisely. A series of monodispersed polymer spherical substances (e.g., polystyrene, poly(methyl methacrylate), and poly(hydroxyethyl methacrylate)) can be prepared using this versatile and scalable microemulsion approach. This feasible wet chemistry enables the production of a great variety of materials combined with other synthesis methods as invented in Dr. Liu's and her collaborator's laboratories. One group of materials is electrocatalysts which can be used for green hydrogen production via water electrolysis, nanocomposites used in oxygen reduction reaction, and the perovskite-based catalyst for green NH_3 production using solid oxide electrolysis cells. Other materials include photocatalysts for water remediation, flexible ternary metal oxide for emerging energy storage using capacitors, porous materials (e.g., MOF) for carbon capture and conversion, metal oxide catalysts to convert plastic wastes into hydrogen and nanotubes, biocatalysts used in microbial fuel cells, CNT-based sensing elements for in situ COVID-19 detection, nanogold-based materials for cancer imaging and therapy and materials used in lithium battery. This section discusses the reaction during three key steps in microemulsion, polymerization, cross-link, and carbonization. The different types of surfactants and their functions will also be discussed to guide their selection.

A simplified flowchart using the microemulsion to produce carbon-based materials is chosen and shown in Fig. 1.6A. Three key subsequent steps such as polymerization, cross-linking, and carbonization will occur. A monolayer at the interface between two immiscible phases of water and oil can be formed due to the presence of surfactant molecules. The hydrophilic "tails" of the surfactant allow for interaction between the aqueous phase, while the hydrophobic ends can be dissolved in the oil phase. The mechanism of microemulsion formation can ensure the tunable size and geometry, controllable morphology, homogeneity, and surface area [64]. The materials produced by microemulsion chemistry are summarized in Fig. 1.6B, and the phase diagram is shown in Fig. 1.6C. The critical factors that affect the microemulsion chemistry, pH, surfactants, catalyst used, and temperature effects have a similar theory as the sol–gel method and will be discussed in Chapter 2.

As shown in the ternary phase diagram (Fig. 1.6C), the microemulsion domains are composed of three components, two immiscible liquids (such as water and oil), and a surfactant or cosurfactant. Totally, five regions will be observed: (a) molecular solution (red-colored), (b) water-in-oil (w/o) emulsion (blue), (c) bicontinuous (b/c) structure (yellow), (d) oil-in-water (o/w) emulsion (purple), and (e) conventional two-phase emulsion (green). In the oil–water interfacial layer, the b/c microstructure has a zero-mean curvature, meaning the flatness on average. The disordered sponge-like structure microemulsions enable the incorporation of the vinyl monomer into different phases (dispersed, continuous, and bicontinuous) during the polymerization. As a result, this incorporation makes the microemulsion practical to design and prepare versatile microemulsion polymers.

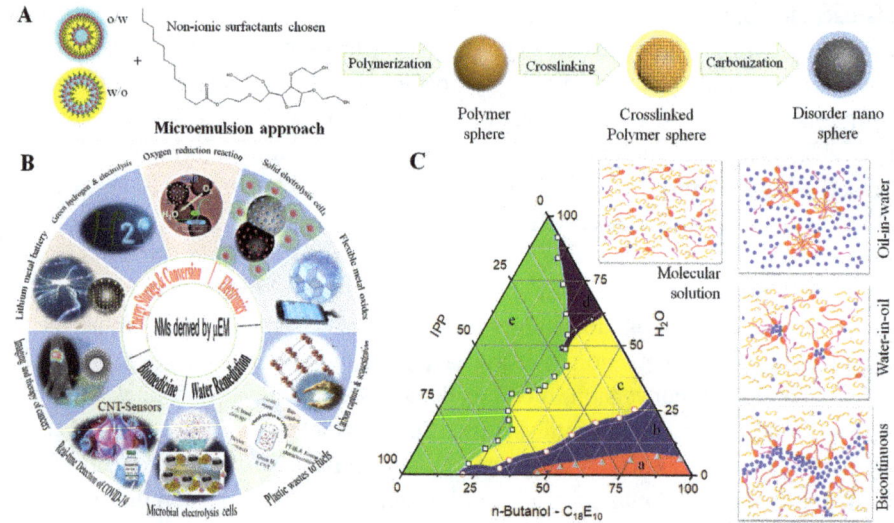

Fig. 1.6: The microemulsion chemistry: (A) the simplified flowchart shows three key steps in the microemulsion synthesis of carbon-based materials; (B) the summary of nanomaterials produced using microemulsion (bioimaging graphs were reproduced from [65], Copyright 2010 Elsevier Press; the real-time detection of COVID-19 image was reproduced from [66]; other data are collected from Liu's group); and (C) a typical ternary phase diagram demonstrates the three families of emulsion systems, oil-in-water (o/w), water-in-oil (w/o), and bicontinuous (b/c) phase (data were reproduced from [67]; J. Liu did the plot).

The surfactants or cosurfactants are ionic and/or covalent compounds, enabling a synergistic effect for producing stable emulsion systems. When a cosurfactant is used, optimizing the molar ratio between surfactant and cosurfactant is required to form a single component and treated as a single "pseudo-component." These three components (water, oil, and surfactant) are found at the triangle's apex, corresponding to the volume fraction of 100 %. The component's volume fraction will be decreased when it moves away from the corner. The different points within the triangle correspond to the compositions of the three components in the mixture. The region composed of these points represents the "phase behavior" of the microemulsion system under constant temperature and pressure [68].

Polymerization: The emulsion polymerization is known as radical polymerization, where an emulsion incorporates water, monomer, and surfactant [69]. The process usually involves monomers, initiators, a dispersion medium, and a colloid stabilizer, allowing a formation of a homogeneous system. Three commonly used emulsions are o/w, w/o, and bicontinuous emulsion. In the o/w emulsion, the droplets of hydrophobic oil monomer will be emulsified in a continuous water phase in the presence of surfactants. An emulsifier (also called a stabilizer, such as polyvinyl alcohols) will prevent particles from agglomeration or coagulation. This polymerization is heterogeneous in nature, whose reaction mechanisms and kinetics dramatically differ depending on the systems. This free radical polymerization mechanism involves three subsequent steps such as initiation, propagation, and termination, as follows:

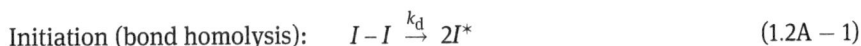

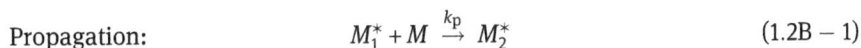

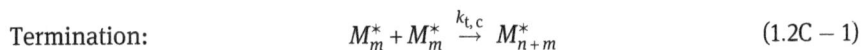

Initiation (bond homolysis): $\quad I-I \xrightarrow{k_d} 2I^*$ $\qquad$ (1.2A – 1)

$$I^* + M \xrightarrow{k_i} M_1^* + I \qquad (1.2A - 2)$$

Propagation: $\qquad\qquad M_1^* + M \xrightarrow{k_p} M_2^* \qquad (1.2B - 1)$

$$M_2^* + M \xrightarrow{k_p} M_3^* \qquad (1.2B - 2)$$

$$M_2^* + M \xrightarrow{k_p} M_3^* \qquad (1.2B - 3)$$

$$M_n^* + M \xrightarrow{k_\mu} M_{n+1}^* \qquad (1.2B - 4)$$

Termination: $\qquad\qquad M_m^* + M_m^* \xrightarrow{k_{t,c}} M_{n+m}^* \qquad (1.2C - 1)$

$$M_m^* + M_n^* \xrightarrow{k_{t,d}} M_m + M_n \qquad (1.2C - 2)$$

where $I–I$ represents the initiator to start the homolytic bond cleavage under light or thermal decomposition; I^* is the initiator radical; M is the reactant monomer; M_1^* is a free radical with different monomer units; and M_m is the terminated polymeric chain. The constant values are: k_d for homolysis rate constant of initiator; k_i for the rate constant of initiation of monomer; k_p for the rate constant of propagation; $k_{t,c}$ for the rate constant of combined termination; $k_{t,d}$ for the rate constant of disproportionation termination of two different radicals. The green region shows the differences between ideal kinetics and limiting conditions.

Cross-link: In the formation of microemulsion, the cross-link refers to the intermolecular or intramolecular attraction of two or more molecules through forming a covalent bond. Based on the IUPAC definition, the cross-link involves "a small region in a macromolecule from which at least four chains emanate, formed by reactions involving sites or groups on existing macromolecules or by interactions between existing macromolecules (Fig. 1.7)." These small regions may be "an atom, a group of atoms, or a number of branch points connected by bonds, groups of atoms, or oligomeric chains." The cross-link mainly refers to a covalent structure, but weaker chemical interactions occur. The portions of crystallites, and even physical interactions and

A

B

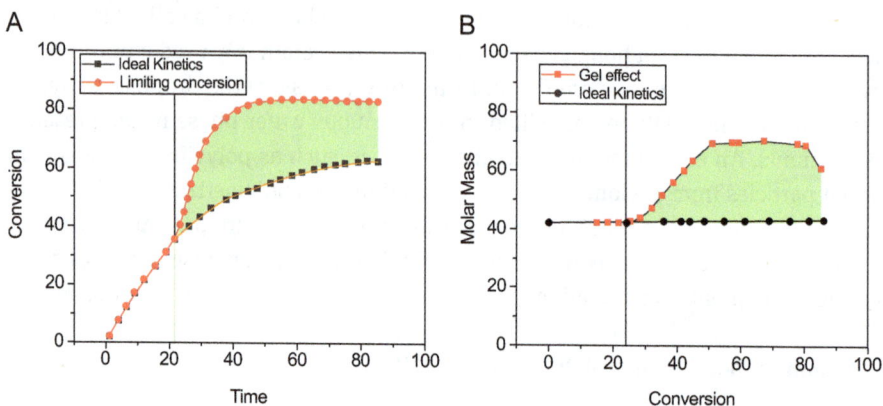

Fig. 1.7: The microemulsion chemistry: (A) the conversion dependence on time during the polymerization and (B) The molar mass depends on the conversion process (imaging graphs were reproduced from [68]).

entanglements may also be observed [70]. There are two major types of cross-linking: physical cross-links are formed by weak interactions (intermolecular forces) and chemical cross-links are formed through covalent bonding. The oxidative cross-link typically occurs when the reactants are exposed to atmospheric oxygen. The factors of reaction temperature, pressure, acidity, and irradiation can initiate cross-links during chemical reactions. A commonly known example is that mixing an unpolymerized resin results in a chemical reaction when applied to cross-linking reagents. Thermoplastic materials (e.g., C type of cross-linked polyethylene (PE)) can be formed through cross-linking when the electron beam is used [71]. Type A cross-linked PE can be produced when oxidizing agents (such as hydrogen peroxide (H_2O_2)) are used during extrusion. The cross-linked membrane (PTS) used as an electrolyte in microbial fuel cells was chosen to demonstrate the chemical and hydrogen bonding formation (Fig. 1.8).

Carbonization: Carbonization is an exothermic process to convert organic matter (such as plants and/or animal remains) into carbon. This process is completed through a destructive distillation, an application of pyrolysis. Carbonization is a pyrolytic reaction that decomposes unprocessed materials by treating organic matter at high temperatures, normally without air or other reagents, catalysts, or solvents. Different reactions concurrently take place during this complex process. The reaction, "dehydrogenation, condensation, hydrogen transfer, and isomerization," normally happens during this carbonization due to its fast reaction rate. The degree of carbonization and the amount of the foreign element residual highly depend on the final pyrolysis temperature. The amount of heat applied controls the degree of carbonization and the residual content of foreign elements. Carbonization can be self-sustainable as a source of energy without carbon dioxide production. For example, glucose carbonization releases about 237 cal/g. The intensive heat can result in the

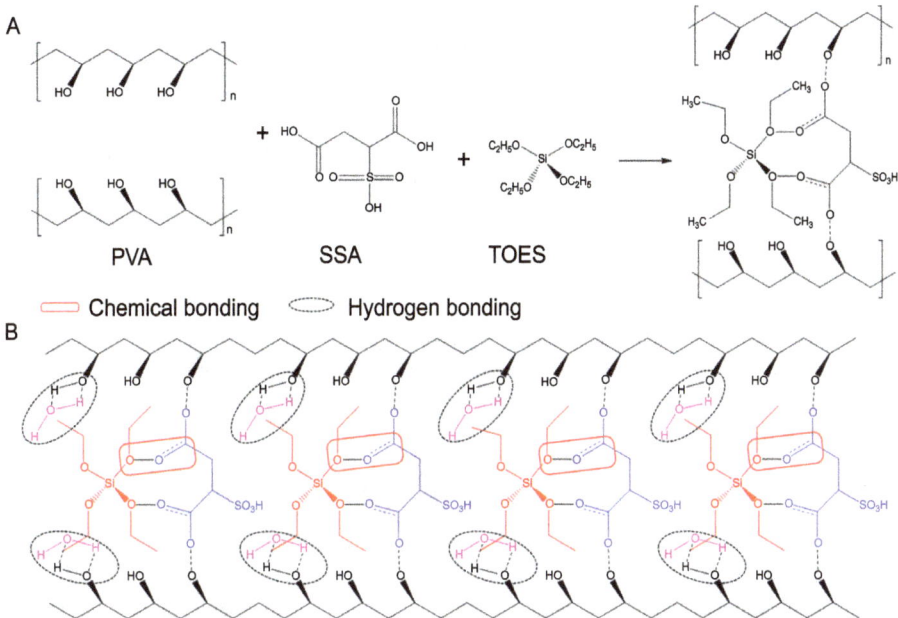

Fig. 1.8: The cross-linking of PTS formation using microemulsion chemistry: (A) the monomer formation through covalent bonding and (B) the tetramer formation shows chemical bonding (the red-squared region) and intermolecular force (the black oval area; imaging graphs were reproduced from [72]).

carbonization of many organic objects, turning them into solid carbon quickly when exposed to sudden searing heat [73]. A series of carbonization processes and a simplified apparatus are listed in Fig. 1.9 [74].

Carbonization occurs under pressure and can be classified into three groups: (a) under pressure resulting from gases decomposed from the reactants, (b) under hydrothermal conditions, and (3) reduction of CO_2 under pressure. The yield and morphology of resultant carbon-based materials are the focus under carbonization conditions. The reaction variables to produce different carbon materials include temperature, pressure, reactor volume, and precursors' chemical composition. Carbonization is a slow exothermic reaction and is one of the pyrolysis processes. Typically, carbonization occurs during heating the biomass in an oxygen-free or oxygen-limited environment. The final product yield can be maximized, and morphology can be tuned by varying the reaction conditions. In the case of biomass conversion into highly carbonaceous materials, hydrosolvothermal may be a wise approach. Although carbonization has been used to convert dead vegetation into coal, this slow but practical process produces nanocarbon-based materials. It is critical to note that carbonization may not occur during microemulsion synthesis unless carbon coating is required.

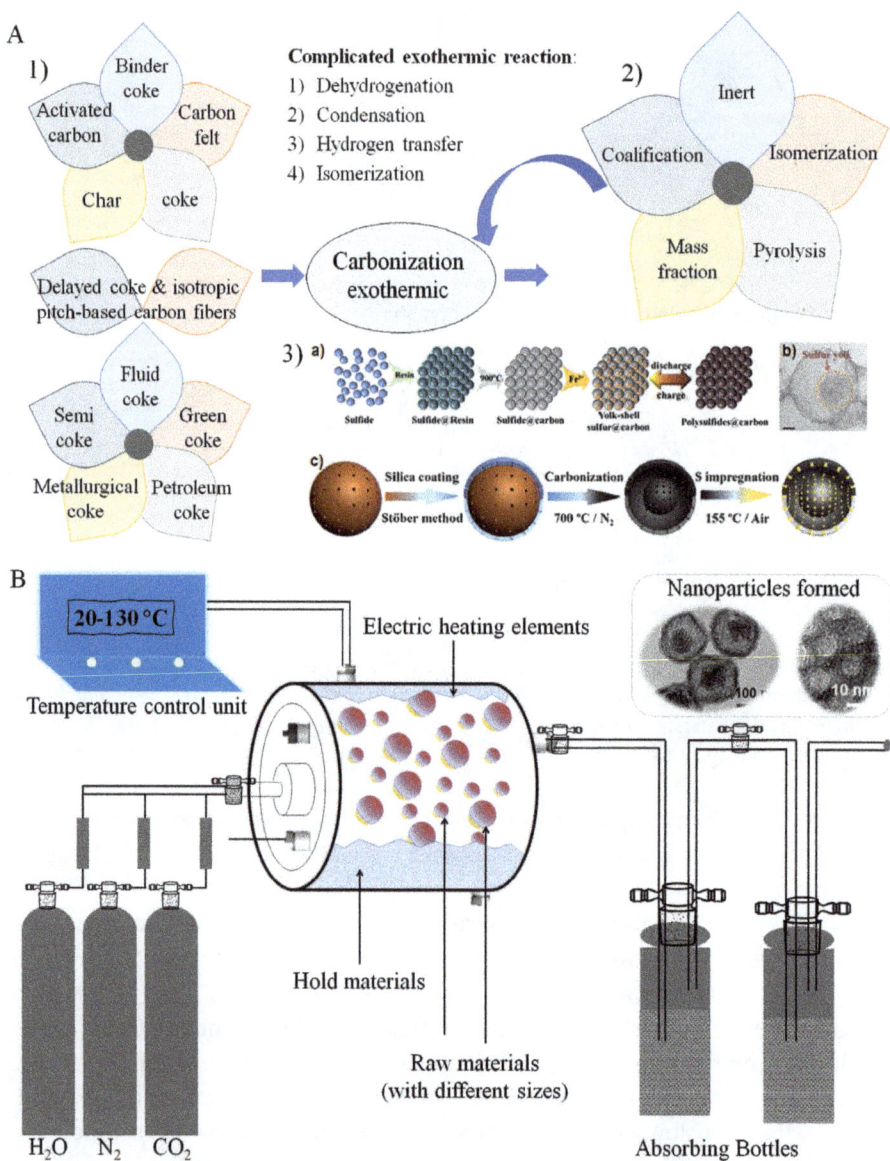

A

1)

Binder coke

Activated carbon

Carbon felt

Char coke

Complicated exothermic reaction:
1) Dehydrogenation
2) Condensation
3) Hydrogen transfer
4) Isomerization

2)

Inert

Coalification

Isomerization

Mass fraction

Pyrolysis

Delayed coke & isotropic pitch-based carbon fibers

Carbonization exothermic

Fluid coke

Semi coke

Green coke

Metallurgical coke

Petroleum coke

3) a)

Sulfide   Sulfide@Resin   Sulfide@carbon   Yolk-shell sulfur@carbon   Polysulfides@carbon

b) Sulfur voil

c) Silica coating
Stöber method

Carbonization
700 °C / N₂

S impregnation
155 °C / Air

B

20-130 °C

Temperature control unit

Electric heating elements

Nanoparticles formed

100 nm   10 nm

Hold materials

Raw materials (with different sizes)

Absorbing Bottles

H₂O   N₂   CO₂

Fig. 1.9: The carbonization progress in producing carbon-based nanomaterials: (A) different procedures of carbonization (images in A-1 and A-2 are adapted from [73], and the image in A-3 and sulfur nanoparticles encapsulated inside a hollow carbon nanosphere were reproduced from [75]); (B) an apparatus used during carbonization (adapted from [76], and nanoparticle images were reproduced from [73]).

Several factors are affecting the emulsion morphology, porosity, and properties. The effects of initiators, addition methods of monomer, and surfactants' roles will be discussed in the following sections. Other factors (volume ratio of reactants, molar mass, and reaction temperature) will be discussed in Chapter 2.

Effect of initiators: The water-soluble initiator can be used in emulsion polymerization to generate oligomeric radicals in the continuous aqueous phase. These initiators are hydrophilic and often anionic (such as SO_4^{2-} derived from $Na_2S_2O_8$). With the increase in hydrophobicity of these oligomeric radicals, the initiator can enter the monomer to swollen polymer particles. During the latex particle formation, the negatively charged sulfate end remained on the particle surface. It constrained the movement of free radicals into the interior, resulting in the encapsulation of seed particles. The nonuniform distribution of free radicals in the reaction loci significantly affects the morphology and size of the final products [77]. Unfortunately, the formation of a perfect core/shell structure or uniformly distributed particles cannot be ensured without optimizing other factors; for example, the reactant polarity and reaction parameters. If the oil-soluble initiator is applied, the free radicals distribute more evenly in the particles (such as latex). The emulsion polymerization will result in a morphological structure, where the post-formed polymer can be present near the latex particle surface layer and inside the composite particle [78]. The different initiators have been investigated to evaluate their effects on the emulsion polymerization of styrene. Examples are potassium persulfate ($K_2S_2O_8$), 2,2'-azobisisobutyronitrile (AIBN, $[(CH_3)_2C]_2N_2$, alcohol-soluble), and 4,4'-azobis(4-cyanovaleric acid) (ACVA, water-soluble organic compound, less hydrophilic than $K_2S_2O_8$, $HOCOCH_2CH_2C(CH_3)(CN)N = NC(CH_3)(CN)CH_2CH_2COOH$).

Potassium persulfate decomposes in water to form negatively charged sulfate radicals (eq. (1.3A–1)). These free radicals will further react with the monomers (M), which are dissolved in water and form soap-type free radicals, as shown in (eq. (1.3A–2)). The various morphological structures of latex particles were formed when the potassium persulfate was used to initiate free radicals [79]. These radicals can be associated with dissolved emulsifier molecules to form micelles. They can migrate into existing micelle droplets to form large particles by adding to the growing polymer chains. In this case, polymerization occurs in the water phase due to the solubility of monomers and initiators. However, the monomers do not provide polymerization loci due to their encapsulation by the surfactant molecules. The encapsulated monomer droplets with negative charges are virtually impossible to penetrate by the negatively charged initiator molecules:

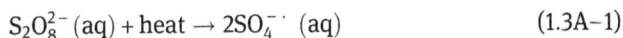

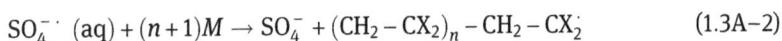

$$S_2O_8^{2-}\,(aq) + heat \rightarrow 2SO_4^{-}{}^{\cdot}\,(aq) \tag{1.3A–1}$$

$$SO_4^{-}{}^{\cdot}\,(aq) + (n+1)M \rightarrow SO_4^{-} + (CH_2 - CX_2)_n - CH_2 - CX_2^{\cdot} \tag{1.3A–2}$$

On the other hand, it was found that inverted core/shell latex particles were obtained when AIBN or ACVA was used as the initiator due to their non-water-soluble or less water-soluble nature. The thermal or light initiation reactions for both AIBN and ACVA were shown in eqs. (1.3B) and (1.3C). The radical formation of AIBN normally occurs at 85 °C within 30 min, while the ACVA at 69 °C within 10 h. The previous study indicated that the initiator concentration and polymerization temperature largely determined the morphology of the subsequent products [79].

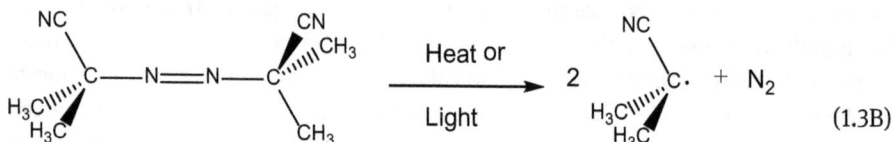

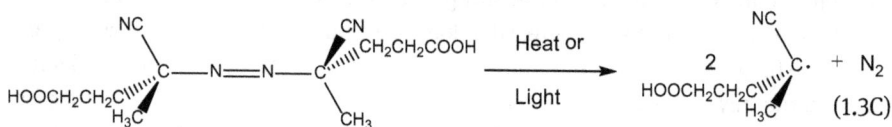

$$\underset{\substack{H_3C\\H_3C}}{\overset{NC}{C}}-N=N-\underset{CH_3}{\overset{CN}{C}}\underset{}{\overset{CH_3}{}} \xrightarrow[\text{Light}]{\text{Heat or}} 2 \; \underset{\substack{H_3C\\H_3C}}{\overset{NC}{C}}\cdot \; + \; N_2 \qquad (1.3B)$$

$$\underset{\substack{HOOCH_2CH_2C\\H_3C}}{\overset{NC}{C}}-N=N-\underset{CH_3}{\overset{CN}{C}}\overset{CH_2CH_2COOH}{} \xrightarrow[\text{Light}]{\text{Heat or}} 2 \; \underset{\substack{HOOCH_2CH_2C\\H_3C}}{\overset{NC}{C}}\cdot \; + \; N_2 \qquad (1.3C)$$

Effect of addition methods of monomer: In the process of emulsion polymerization, the sequence of monomer addition to the system is critical to determine the molecular weight, degree of grafting, and morphology of the final products. It is difficult to control the uniformity of polymerization due to those unavoidable side reactions of irreversible radical termination. As reported by Luo and coworkers, "over 500 kg mol^{-1} polystyrene with high livingness and low dispersity could be synthesized by a facile two-stage reversible addition-fragmentation transfer emulsion polymerization" [80]. This method enabled the high monomer conversion (90 %) within 10 h. Both initiator concentration and the monomer chain transfer constant were significantly lower than the well-accepted values in the conventional method. All the oligomeric reversible addition-fragmentation transfer (RAFT) agents and the initiator were charged in the first stage (initiation to target at 30 kg/mol). A fraction of the styrene monomer was added to the above system. After 100 min initiation, the remaining styrene monomer, water, and strong base (such as NaOH) were fed for polymerization. The polymerization kinetics and evolutions of molecular weights and dispersities with monomer conversions are shown in Fig. 1.10.

It can be seen that the two-stage synthetic strategy results in a great reduction of dispersities while the polymerization rates remained comparable to those in the batch cases. This report introduced the styrene monomer separately at two stages, maintaining the same monomer concentrations. The two-stage method avoids dealing with the low RAFT transfer rate or slow feeding rate. As a result, monomer concentration and propagation rate have been improved compared with the traditional RAFT reaction. Data (Fig. 1.10) showed that the molecular weight of the high living polymer exceeded 500 kg/mol, and low dispersity (ca. 1.5) can be obtained within an adequate polymerization time. Continuous feed stream, reactor configuration, mean residence time, and precursor concentration are also critical to determine the

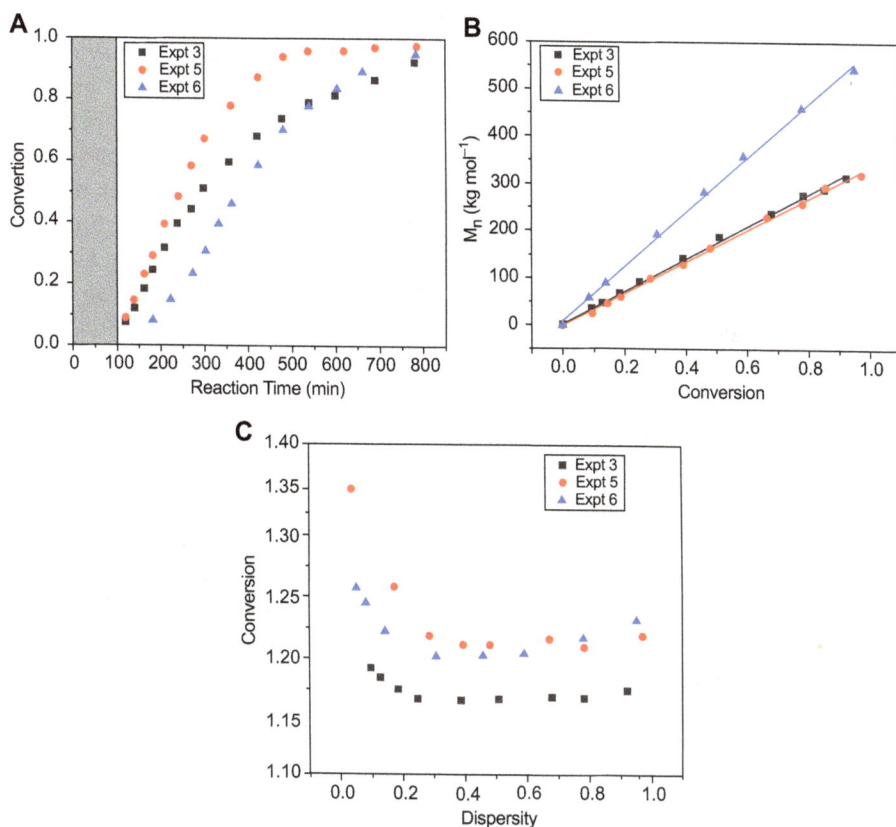

Fig. 1.10: Experimental results of two-stage polymerization of styrene. Experiments 3 and 5 targeted about 350 kg/mol and experiment 6 at 640 kg/mol. (A) Styrene monomer conversion as a function of polymerization time; (B) the molecular weight of the polymer as a function of monomer conversion; and (C) the dispersity versus monomer conversion. Data were adapted with permission from [80].

particle size distribution and to obtain various morphological structures of the effluent product.

Surfactant/cosurfactant: In the emulsion system, the incompatibility between oil-soluble and water-soluble phases results in thermodynamic instability. The use of surfactants during emulsion formation can minimize the above issues and prevent nanoparticles from agglomerating. The introduction of surfactants can lower the Gibbs free energy ($\Delta G = \Delta H - T\Delta S$, $\Delta G = \gamma\Delta A - T\Delta S$) since molecules adsorbed on the emulsion droplets result in a decrease in the interfacial tension ($\gamma\Delta A$). Surfactants are amphiphilic organic molecules that simultaneously comprise hydrophilic (head) and hydrophobic (tail) groups to lower surface or interfacial tension [81]. Since surfactants contain water-soluble and oil-soluble components, they can be dissolved in water and oil. Surfactants will diffuse in water and be adsorbed at interfaces between air and water or at the interface between oil and water. These hydrophobic tails may extend out of the water phase into the air, or the oil phase, while the hydrophilic head remains in the water phase (Fig. 1.11A). This schematic diagram shows the formation of a micelle, where the oil-soluble tails remain within the inner layer. The lipophilicity of the surfactant allows for stronger attraction of these tails to the oil phase than interaction with water. The polar heads within the surfactant enable a strong attraction with polar water molecules to form a hydrophilic outer layer. Therefore, a barrier between micelles inhibits oil droplets from agglomerating into larger clusters. In other cases, the reverse micelles (Fig. 1.11B) can be formed when the polar ends face the inner side of the water droplets, where the nonpolar tails are located on the outer side. These reverse micelles are also water droplets (sized from 1 to 10 nm), but the surfactant molecules are organized differently from the micelles.

The polar ends of surfactants facing the inner side can solubilize water, while the nonpolar ends face outside toward the organic phase. Although the surfactant molecules have single or double tails, Fig. 1.11 only showed the single-tailed schematics. The formation processes of both micelles and reverse micelles can be reversible due to the weak intermolecular attractive forces, such as London dispersion (Fig. 1.11C), dipole–dipole moments (Fig. 1.11D), and hydrogen bond (Fig. 1.11E). The London dispersion can be normally found between nonpolar molecules due to instantaneous polarization of electron clouds. It is an instantaneous dipole–induced dipole attraction between adjacent atoms. The energy magnitude is relatively low, ranging from 0.05 to 40 kJ/mol [82]. The dipole–dipole moments are attractive forces between polar molecules, from the positive end of one molecule to the negative end of another

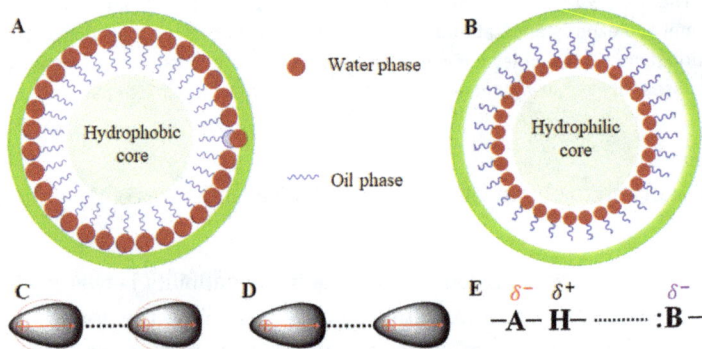

Fig. 1.11: The formation of micelles and reverse micelles through intermolecular forces. (A) The schematic diagram of micelle with hydrophobic core (with a single tail); (B) the diagram of reverse micelles with hydrophilic core; (C) the London dispersion attraction due to polarizable electron clouds between nonpolar molecules; (D) the dipole–dipole attraction due to dipole charges between polar molecules; and (E). hydrogen bond formation via a polar bond to H-dipole charges (note: A and B must be one of the N, O, or F atoms). These relatively weak van der Waals attractive forces allow for the reversible formation of micelles and reverse micelles.

polar molecule. This attraction is also relatively low, ranging from 5 to 25 kJ/mol [83]. The hydrogen bond (in short, H-bond) is a primarily electrostatic attraction between an H atom and another atom with more electronegativity. The atom can be one of the oxygen (O), nitrogen (N), or fluorine (F), which bears a lone pair of electrons as an H bond donor [84].

Most surfactants comprise a polyether chain with a high polar anionic group at the terminal. The polyether groups usually consist of ethoxylated sequences inserted to increase the surfactant hydrophilicity due to the polarity of carbon–oxygen bonds. Conversely, the nonpolar sequences in polypropylene oxides can increase the lipophilic character. The surfactants are classified into two groups (ionic and nonionic) based on their polar head groups. There are three types of ionic surfactants: anionic with a negative charge at the polar head, cationic with a positive charge, and zwitterionic (also known as amphoteric) with two oppositely charged heads. The nonionic surfactants contain ether $[-(CH_2CH_2O)_nOH]$ and/or hydroxyl [–OH] hydrophilic groups, which are not ionized at any acidity. They are also nonelectrolytes, which are less sensitive to the pH variation of the medium. They can be used to stabilize o/w and w/o emulsions due to their less sensitive properties. They are composed of positively charged heads and can be used in beauty products. Cationic detergents are extremely harsh and can disrupt the cell membranes of bacteria and viruses. The anionic surfactants comprise negatively charged heads such as sulfonate ($R\text{-}SO_3^-$), phosphate (PO_4^{3-}), sulfate (SO_4^{2-}), and carboxylates (RCO_2^-). The zwitterionic (also known as amphoteric, shown in Fig. 1.12) surfactants consist of negative and positive charged heads attached to the same molecule. The anionic groups can be variable and are composed of sulfonates and betaines, while the cationic functional groups are based on primary, secondary, or tertiary amines or quaternary ammonium cations. These zwitterionic surfactants are sensitive to pH changes, and they behave as anionic or cationic under different pH conditions. The above amphoteric molecules tend to diffuse toward the interface between water and oil phases and reside therein.

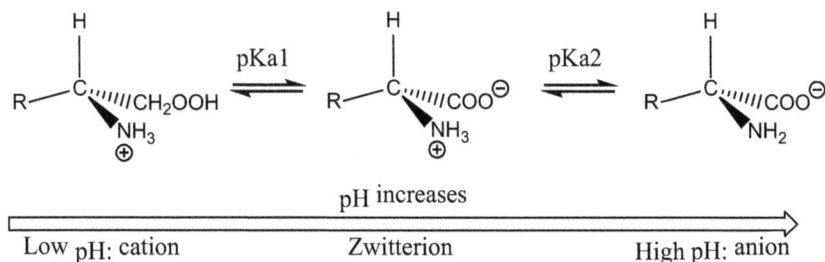

Fig. 1.12: The zwitterionic surfactants and their ionization under different pH values. The amphoteric molecules achieve an isoionic point at a certain pH value when the positive and negative charges in the hydrophilic portion are internally neutralized.

In general, the surfactant molecules increase the wettability and stable region of the microemulsion, although different surfactants have unique functions. The major function of surfactants is to stabilize the particle nucleation during the early stage of emulsion formation. The further growth and polymerization of particles acquire surfactant molecules from those species adsorbed on the monomer droplets or dissolved in the continuous aqueous phase. As a result, the microemulsion structure can be stabilized using the surfactant/cosurfactant molecules. At the water/oil interface, the stoichiometry of the water/surfactant in the microemulsion system is the critical factor in determining the micelles' size and shape, as discussed above. The examples and definitions of four families of surfactants are tabulated in Tab. 1.2.

Tab. 1.2: The summary of compounds and their functions in microemulsion.

Surfactants	Examples	Definition and functions
Nonionic	Polysorbates (Tweens®) Sorbitans (Spans®) PEGs Laureth Poloxamer (Pluronics®)	They are of no charge to the molecules and are more commonly found in emulsifiers. Nonionic detergents are super harsh and are rarely used in skincare. Their functions are: To act as nonionic amphiphiles, induce packing at the interface layer of the microemulsion and increase the adsorption capacity at the interface
Cationic	Cetrimonium bromide Benzalkonium chloride (BAC) Cetylpyridinium chloride (CPC) Benzethonium chloride (BZT) Quaternary ammonium cations Alkyltrimethylammonium salts Cetyl trimethylammonium bromide (CTAB) Cetyl trimethylammonium chloride (CTAC)	They are composed of positively charged heads and can be used in beauty products. Cationic detergents are extremely harsh. Their functions are: To disrupt cell membranes of bacteria and viruses
Anionic	Sodium lauryl sulfate (SLS, $CH_3(CH_2)_{10}CH_2(OCH_2CH_2)_nOSO_3Na$) Alkyl carboxylates (soaps) Sodium lauryl ether sulfate (SLES) Sodium myreth sulfate Dioctyl sodium sulfosuccinate (DOSS) Perfluorooctanesulfonate (PFOS) Linear alkylbenzene sulfonates (LABs) Perfluorobutanesulfonate Alkyl-aryl ether phosphates	Negatively charged, extremely effective, and harsh, higher incidence of irritation and lathers well and makes foams. Their functions are mainly to induce intermolecular attraction.

Tab. 1.2 (continued)

Surfactants	Examples	Definition and functions
Zwitterionic	Cocamidopropyl betaine (CAPB, a mixture of organic compounds derived from coconut oil and dimethylaminopropylamine, $C_{19}H_{38}N_2O_3$)	Both negatively and positively charged, the final charge depending on the pH values, mild and less irritating but foaming less
	Sodium cocoamphoacetate (RC(O)NH $(CH_2)_2N(CH_2CH_2OH)CH_2COONa$)	To attract oil and water
		To help the different ingredients of a formula to blend
Trade name	Nomenclature/chemical formula	Application [85]
	Nonionic	
Triton^{TM} -X-100	Polyoxyethylene glycol octylphenol ethers/C_8H_{17}-(C_6H_4)-$(O$-$C_2H_4)_{1-25}$-OH	Wetting agents (for coating)
Nonoxynol-9	Polyoxyethylene glycol alkylphenol ethers/C_9H_{19}-(C_6H_4)-$(O$-$C_2H_4)_{1-25}$-OH	Spermicide
Polysorbate	Polyoxyethylene glycol sorbitan alkyl esters	Food ingredient
Span$^®$	Sorbitan alkyl esters	Polishes, fragrance carriers
Poloxamers, Tergitol^{TM}, Antarox$^®$	Block copolymers of polyethylene glycol and polypropylene glycol	Diversified applications
	Anionic	
Pentex 99	Dioctyl sodium sulfosuccinate (DOSS)/ $C_{20}H_{37}NaO_7S$	Wetting agent (coatings, toothpaste)
PFOS	Perfluorooctanesulfonate (PFOS)/ $C_8HF_{17}O_3S$	Scotchguard^{TM}, Skydrol^{TM}
Calsoft$^®$	Linear alkylbenzene sulfonates	Laundry detergents, dishwasher detergents
Texapon$^®$	Sodium lauryl ether sulfate	Shampoos, bath products
Darvan$^®$	Lignosulfonate	Concrete plasticizer, plasterboard, DMSO
	Sodium stearate	Handsoap, HI&I products

The appropriate selection of ionic surfactant allows for the conformation changes and tunability of the final products by packing these nonionic amphiphiles at the interface layer of microemulsion droplets. The anionic surfactant introduction to nonionic systems results in an attraction between surfactants and oil phases due to the van der Waals forces. This process will lead to the formation of mixed micelles. The commonly used hydrophilic groups can be either ionic or nonionic, for example, SO_4^{2-},

SO_3^{2-}, $-COOH$, and $^+N(CH_3)_3$, $-O-(CH_2-CH_2-O)_nH$. The widely used hydrophobic groups include the alkyl chains ($-C_nH_{2n+1}$) and alkyl chains ($-C_nH_{2n+1}-C_6H_4-$). The commercially available surfactants and their applications are listed in Tab. 1.2. These surface-active agents used in typical emulsion polymerization are normally lower than 5 %. However, these surfactants play a crucial role in the particle nucleation and growth processes. These surfactants and cosurfactants are critical factors determining the final product's morphology, properties, and applications. The authors summarized the commonly used surfactants and their functionalities in Tab. 1.2. Four different major types of surfactants include nonionic, anionic, cationic, and amphoteric, with diversified composition and polarity.

Giant surfactant: One special case was discussed using amphiphilic bottlebrush block copolymers (BBCPs) giant surfactants to produce photonic pigments and their ability to control micellization self-assembly mechanisms [86]. The amphiphilic BBCPs have been reported to form much larger micelles (a few tens of nanometers) in water than those obtained with traditional surfactants or BCPs. The team led by Parker invented a method using toluene-in-water (t/w) to "induce a controlled swelling of reverse BBCP micelles" to overcome the above limitation. In this method, the (polynorbornene-*g*-polystyrene)-*b*-(polynorbornene-*g*-polyethylene oxide) [P(PS-NB)-*b*-P(PEO-NB)] was dissolved in anhydrous $C_6H_5CH_3$ and emulsified in water. The addition of water during emulsification has been absorbed evenly by toluene droplets, resulting in a significant swelling of BBCP micelles. The so-called soft confinement within these t/w emulsions resulted in internal aqueous droplets whose size ranges in visible light. Subsequently, highly porous microparticles with tightly packed nanodroplets were formed due to toluene evaporation for about 30 min via self-assembly templation.

The structural coloration of the photonic pigments (Fig. 1.13A) was then created, corresponding to the "short-range order of the pores within the BBCP scaffold." The pore size and colors of the resultant photonic pigments can be tuned by adjusting the amount of water available to the micelles. The different toluene-to-water ratios led to temporary disruption of the microdroplet's interface due to high shear. The reflected color in the visible electromagnetic radiation can be tuned by increasing the water amount (Fig. 1.13B and C). It was reported that the wide range of colors changed from blue to red using this BBCP ($M_W = 290$ kDa, $f_{PEO} = 50$ vol %, DP = 72). In stark contrast to "lamellar BBCP photonic systems" and the "strongly directional character of the reflection from photonic pigments," the short-order coherent scattering led to the coloration of the resulting pigments, as shown by the optical microscopic analysis upon epi-illumination in reflection mode (Fig. 1.13D). These photonic pigments displayed several key advantages, "suppressed iridescence and improved color purity, removing the need for refractive index matching or the inclusion of a broadband absorber."

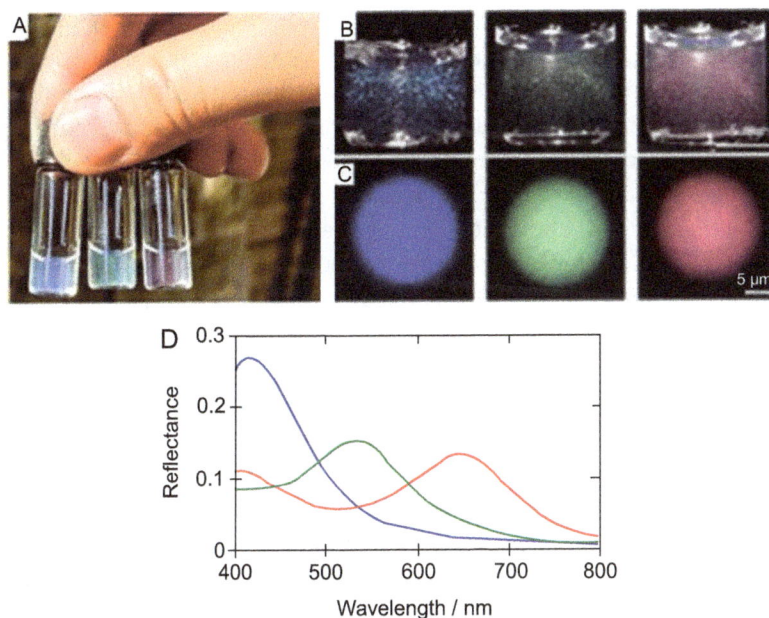

Fig. 1.13: The analyses of the BBCP photonic pigments. (A) The aqueous dispersions of three formulations of pigments (blue, green, and red) illuminated under natural sunlight; (B) photographs showing the above three dispersions (0.03 wt %) under direct illumination; (C) microscopic images of microspheres of these three pigments; and (D) reflectance spectra for the microspheres using a Lambertian diffuser and measured from a light intensity from a white light source (images were reproduced from [86]).

1.2.3 Hydrosolvothermal synthesis

The hydrothermal and solvothermal wet-chemistry methods are widely used to provide nanomaterials (such as metal oxides used for supercapacitors [87] and MOFs) [88]. A hydrothermal synthesis is a wet-chemical approach to preparing crystalline materials, which are water-soluble. The reactions normally occur at higher temperatures and pressures in an autoclave composed of steel vessels. The polar molecule water acts as the solvent and provides a medium for nucleation and crystal growth. A temperature gradient can be maintained in the growth chamber, where the hotter regions allow for the dissolution of solute and the opposite region for the deposition of crystals [89]. Like hydrothermal synthesis, the solvothermal method uses organic compounds as a solvent to produce chemical compounds from a nonaqueous precursor solution. Typical solvents include acetone (CH_3OCH_3), acetonitrile (CH_3CN), dimethylformamide ((CH_3)$_2$NCH), diethyl formamide ((CH_3CH_2)$_2$NCH), ethanol (CH_3CH_2OH), methanol (CH_3OH), and various alcohols [90]. It was reported that almost 70 % of MOFs have been prepared by a solvothermal method [91]. In this solvothermal synthesis, a closed system will be used, where escalating pressure and temperature will be employed, either above or under the boiling point of the preferred solvents. This solvothermal is a subcritical synthesis controlled under the boiling point when the condition is controlled [92]. Otherwise, the chemical approach is called the supercritical solvothermal method [93].

When a mixture of water and organic is used as a solvent, the method is known as hydrosolvothermal. The laboratory-scale process can be carried out in the sealed autoclave to sustain various temperatures (up to 300 °C) and pressures (up to 15 MPa). The autoclave normally consists of stainless steel with Teflon-lined reaction chamber (Fig. 1.14A). The mixed solvent can dissolve the different reactant precursors due to their solubilities in different solvents. The hydrosolvothermal chemistry draws attention due to the advantages of potential for flexibility, green and scalable nature to produce various formulations of products used in a supercapacitor, and carbon capture [94].

However, the reactants under hydrosolvothermal conditions can experience quick and unexpected changes, either physically or chemically. Therefore, some unexpected reactions may hinder the formation of designed nanoparticles, indicating that inert gases may be needed to prevent this side reaction. Another drawback of this method is that it normally requires high temperature and prolonged duration (300 °C, >72 h or longer). Regardless, well-reticulated single crystals and other tunable structures can be formed using the hydro/solvothermal reactions [95]. Three key steps to produce final products include dissolution of reagents in solvents, nucleation of crystalline matters, and crystal growth to form the resulting product (Fig. 1.14B–D). The products' structures, properties, functionalities, and compositions can be achieved by varying and controlling the synthesis parameters, such as temperature, reaction duration, and pressure. The physical parameters of reactants (e.g., intermolecular forces and dipole moment), concentration, and acidity are other factors that play critical roles in defining the structure and properties of the end products. The hydrothermal approaches showed high solvation power, high compressibility, and mass transport of these solvents. Therefore, different reactions can result in a series of final products, as follows [89]:

1) Synthesis of new phases or stabilization of complexes (such as MOFs)
2) Crystal growth of several inorganic compounds (oxide and borides)
3) Preparation of finely distributed materials (nanoparticles)
4) Leaching of ores in metal extraction
5) Decomposition, alteration, corrosion, and etching

Nucleation: The nucleation is defined as progress when the crystal reaches a critical size at a threshold to create a new structure or phase in a vapor, solution, or liquid. Beyond this critical point, the energy benefit of growth exceeds the cost. The nucleation is the first spontaneous step to form a new thermodynamic phase. This new structure starts from metastability with a higher intermediate energetic state. The

A
Screw
Spring
Pressure plate
Teflon lid
Solvents
Autoclave body
Reagents
Bottom disc

**B Dissolution of reatants
in solvents**
– Polarity
– volume
– concentration
Commonly used solvents:
– H_2O (highly polar molecule)
– Diethyl formamide
– Acetone
– Methanol
– Dimethyl formamide
– Acetonitrile
– Ethanol

C Nucleation
Variable Control:
– Pressure
– Temperature
– Reaction duration

D Crystallization

Tunable Structure:
– Dimension
– Functionalization
– Reaction duration

Dehydration
-H_2O

H^+
Protonation

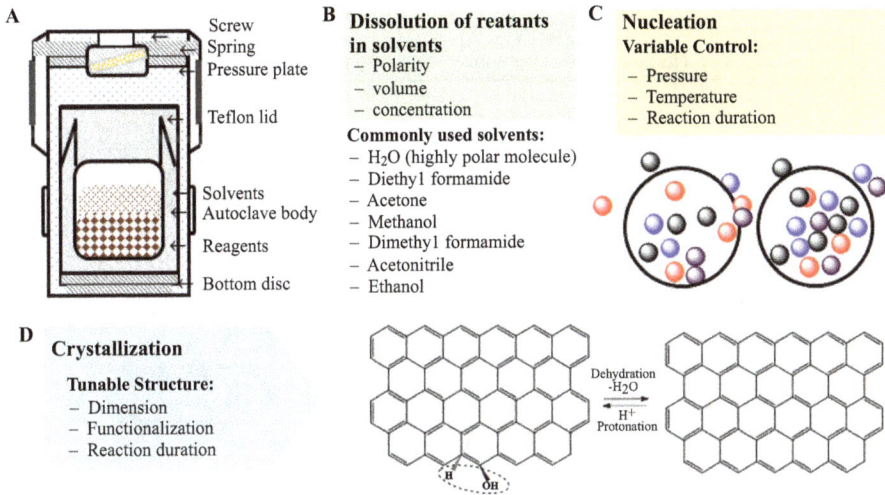

Fig. 1.14: The hydrothermal synthesis. (A) The schematic diagram of autoclave; (B) dissolution of solute into the water and other preferred solvents; (C) nucleation procedure; and (D) crystallization to form a final product (adopted from [102] with permission).

time to nucleate, which varies significantly from instantaneous to prolonged, determines the kinetics of nucleation for the new phase to form. Traditionally, there have been two theories to describe this nucleation: classical and two-step models (Fig. 1.15A). The classical nucleation theory has been used to explain and quantify nucleation kinetics. It can be broken down into two steps: (1) formation of crystalline seed due to collision of ions or molecules and (2) expanding the crystal lattice to form particles when the other solvated ions fall into place [95].

The traditional nucleation mechanism posits that nucleation starts forming perfectly spherical nuclei at threshold size. Based on classical theory, the rate of nucleation can be expressed by eq. (1.4A). This rate expression can be considered as a product of the number of nucleation sites multiplied by the probability that the threshold mass or size ratio is reached, around which the propagation can proceed. This threshold is the number of nuclei species at the nucleation barrier, governed by the system's free energy [96]. The probability of a nucleus formation at the loci is inversely exponentially proportional to the Gibbs energy. If the Gibbs energy is large and positive, the probability of forming a nucleus is very low, and nucleation will be slow. At a given time, the nucleation rate will be slow or insignificant:

$$R_N = N_s Z j \exp\left(-\frac{\Delta G^*}{k_B T}\right) \qquad (1.4A)$$

where R_N is the nucleation rate; ΔG^* stands for Gibbs free energy to overcome the nucleation barrier; N_s is the number of nucleation sites; Z, is the *Zeldovich* factor giving the probability of forming a new phase instead of re-dissolved into solution;

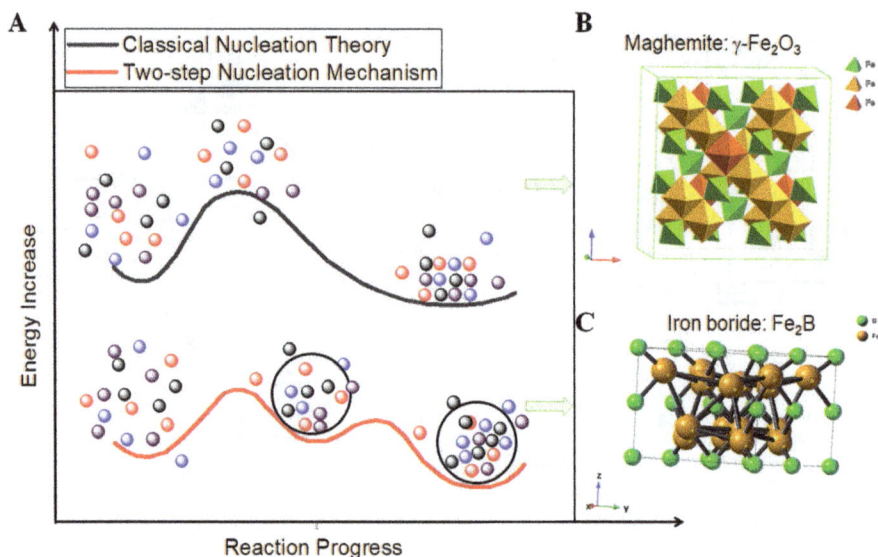

Fig. 1.15: The nucleation mechanisms and select formations of two inorganic compounds: (A) the classical and two-step nucleation theories depict crystal formation with energy changes; (B) maghemite compound; and (C) iron boride.

j is the rate where molecules are attached to the nucleus; k_B is the Boltzmann constant; and T is the absolute temperature.

Recently, a team led by Miao developed four-dimensional (4D) atomic electron tomography (AET) to capture the nucleation of FePt nanoparticles with a chemically disordered face-centered cubic (fcc) structure. This 4D AET technology was used to study nucleation at its early stage, including three spatial dimensions and one dimension (1D) of time [97]. The research data showed that nuclei are irregularly shaped according to the traditional approximation at the early stage. Each nucleus has a core of one to a few atoms with the maximum order parameter. The parameter gradient was defined as the distance between the points from the core to the boundary of the nucleus. Their study indicated that "growth, fluctuation, dissolution, merging, and/or division" occurred during the nucleation. The order parameter distribution and its gradient collectively regulate the structure and dynamics of the same nuclei. The FePt nanoparticles were used as a model system, which underwent irregularity of shape at their early stage nuclei (Fig. 1.16). The total number of atoms in the FePt nanoparticles was slightly changed (Fig. 1.16A, a–c) under different annealing durations due to the atomic diffusion between nanoparticles. The data also showed that a portion of atoms at the surface and subsurface were rearranged to form $L1_0$ phases with no change in the Pt-rich core. Authors proposed that vacancy-mediated atomic diffusion plays a critical role during annealing to create vacancies on or near the surface due to its lowered energy requirement.

The results from Miao offer three broad conclusions. (1) That early nucleation is governed by Oh symmetry, that the particles are anisotropic and spherical. When the particles are not spherical, local inhomogeneity in nucleation is due to non-uniform particle distribution and shape, anisotropy of physical constraints between the interfacial region and water tension; (2) the nuclei sites are composed of a few atoms, which enable maximum order parameter and order parameter gradient to be optimized from the center to the boundary layer. These boundary conditions result in the generation of a diffuse interface between the nucleation event and the mother liquor or initiatory phase; and (3) the nucleation processes which propagate are highly dynamic and consist of fluctuations in atom migrations, dissolution of nucleation bodies, division or aggregation that are governed by the order parameter distribution and the resulting gradient parameter. Each order parameter has a distribution, and the associated gradient represents various metastable states between which the nucleus can be distributed and eventually settle to continue the nucleation process. Peter Ercius from the Molecular Foundry of the Lawrence Berkeley

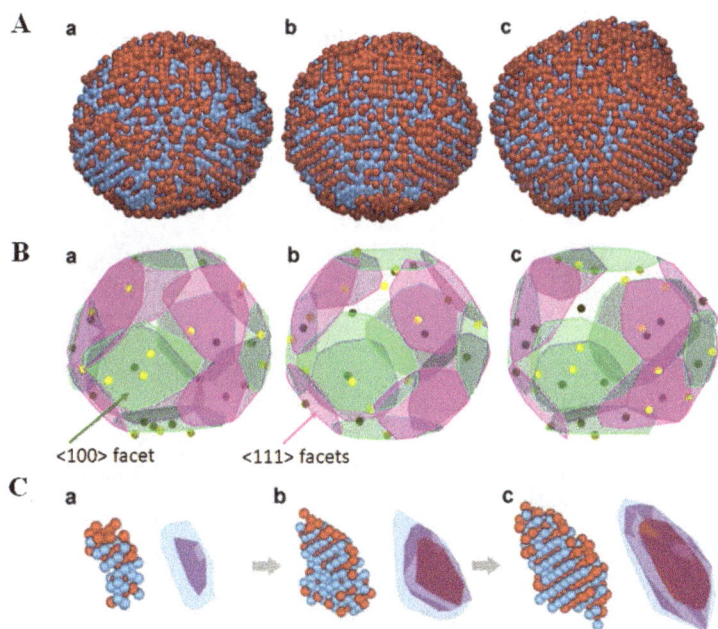

Fig. 1.16: Four-dimensional (4D) atomic motion captured by AET. (A) Three-dimensional atomic models (Fe in red and Pt in blue) of FePt nanoparticles with accumulated annealing time; (B) heterogeneous nucleation sites and distribution of the nucleation sites with an accumulated annealing time; (C) experimental observation of the same nuclei undergoing growth, fluctuation, dissolution, merging, or division at 4D atomic resolution (image was reproduced from [97]) (note: heat treatment time: a. 9 min, b. 16 min, and c. 26 min; the lighter colored dots are closer to the front side and the darker dots are closer to the backside of the nanoparticle, and red represents Fe and blue represents Pt).

National Laboratory reported that Miao and his team "have gained a never-before-seen view of nucleation, capturing how the atoms rearrange at 4D atomic resolution." The atomic positions in the nanoparticles were captured using the new AET approach to demonstrate the nucleation process after different annealing times.

Crystal growth: Crystal growth converts gas or liquid into a solid, resulting in a phase transition. The growth mechanism depends on the initial phase, the state of the surface, and the strength of the driving force. During this major stage of the crystallization process, the crystalline lattice arrangement will occur due to new atoms, ions, or polymer strings [98]. This action will yield single or multiple crystalline phase(s) with highly ordered close packing of atoms and molecules. Subsequently, the nuclei size increase will lead to further crystal growth to achieve critical cluster size. Surface-catalyzed nucleation as an initial stage will occur, followed by the seeded crystallization as shown in Fig. 1.14C and D. During this dynamic process, an equilibrium of solute will be achieved through precipitation out of and redissolution back into solution. Supersaturation is the critical driving force of crystallization, attributed to the solubility of a reactant at an equilibrium point between precipitation and recrystallization. Two different crystallization types are consistent normal growth (Fig. 1.17A) and nonuniform lateral growth (Fig. 1.17B).

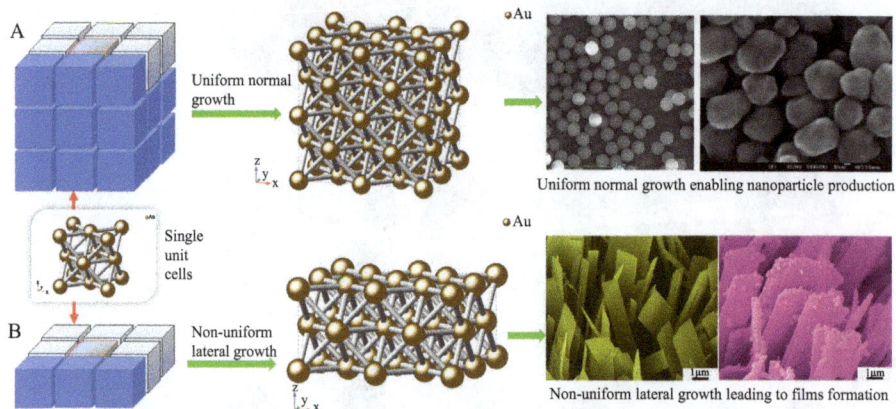

Fig. 1.17: The crystal growth mechanism: (A) normal uniform growth to produce nanoparticles (Au nanoparticles were chosen as examples); (B) nonuniform lateral growth to produce layered films (MnO_2-grafted V_2O_5 supercapacitor electrodes as examples; copyright is obtained from [99] with permission).

Uniform normal growth: The surface advances itself isotropically without a stepwise growth mechanism during uniform growth, which is considered an "entire surface motion." This uniform "normal growth" is based on the time sequence of an element or an atom on the surface. As long as the thermodynamic driving force is sufficient, all different elements or atoms on the surface (red-colored square) can continuously grow and change to advance the interface. During this continuous growth mechanism, the surface diffusion may simultaneously require change through unit cell transformation to generate large-scale superlattice over several successive layers. When the crystals grow uniformly over large areas to form a successive new layer, nanoparticles will be formed to minimize the surface tension (Fig. 1.17A).

Nonuniform lateral growth: During nonuniform growth, surface advancement occurs by the lateral motion of each step. These steps can be one interplanar or integral multiple spacing in height. Without undergoing any changes, some surface atoms will not advance normal growth to themselves, except during the passage of a step. During this lateral growth, two adjacent regions are parallel and show identical configurations, corresponding to the displacement from different regions by an integral number of lattice planes. This nonuniform lateral growth is considered a "geometrical motion of steps" (Fig. 1.17B) instead of an "entire surface motion" [100] (Fig. 1.17A). There is no change in this lateral growth mode except a step that passes via a continual change. Surface diffusion and driving force are two criteria used to predict the types of the mechanism during crystal growth.

A diffuse surface undergoes continuous change from one phase to another, occurring over several atomic planes and being "confined to a depth of one interplanar distance (Fig. 1.17B)" [100]. This step contrasts the "sharp surface" for which the discontinuous changes in properties will happen. The driving force for this diffuse surface is considered a requirement for the appearance of lateral growth. This lateral growth mechanism will be found at the metastable equilibrium of the surface composition in the presence of driving forces. Until the passage of the next step, this metastable condition tends to remain. This configuration in the crystalline medium will be kept identical in the following steps until the height step is further advanced. On the other hand, if the surface cannot reach equilibrium conditions, it will continue to expand its regions, leading to nonuniform lateral growth. Therefore, it can be said that the large driving forces facilitate uniform movement. On the contrary, the heterogeneous nucleation mechanism (or screw dislocation) will favor the presence of driving forces. The diffusiveness of the interface is the critical driving force for crystallization. The crystal properties, including mechanical, chemical, and electrical characteristics, are pertinent to the end products. Thermodynamics and kinetics are the fundamentals for further investigations. In the late nineteenth century, the "necessary thermodynamic apparatus" was invented by Josiah Willard Gibbs to study heterogeneous equilibrium. Gibbs indicated that surface energy (or surface tension) applies to solids and liquids as a driving force to form large clusters. He also proposed that an "anisotropic surface free energy implied a nonspherical equilibrium shape." Thermodynamically, this nonspherical equilibrium can define the shape of particles, which can minimize the total surface free energy [101].

Water as solvent: When water is chosen as the solvent, the method is defined as a hydrothermal process. This hydrothermal synthesis is the most widely used wet chemistry due to its similar nature to the geological phenomena, ability to synthesize

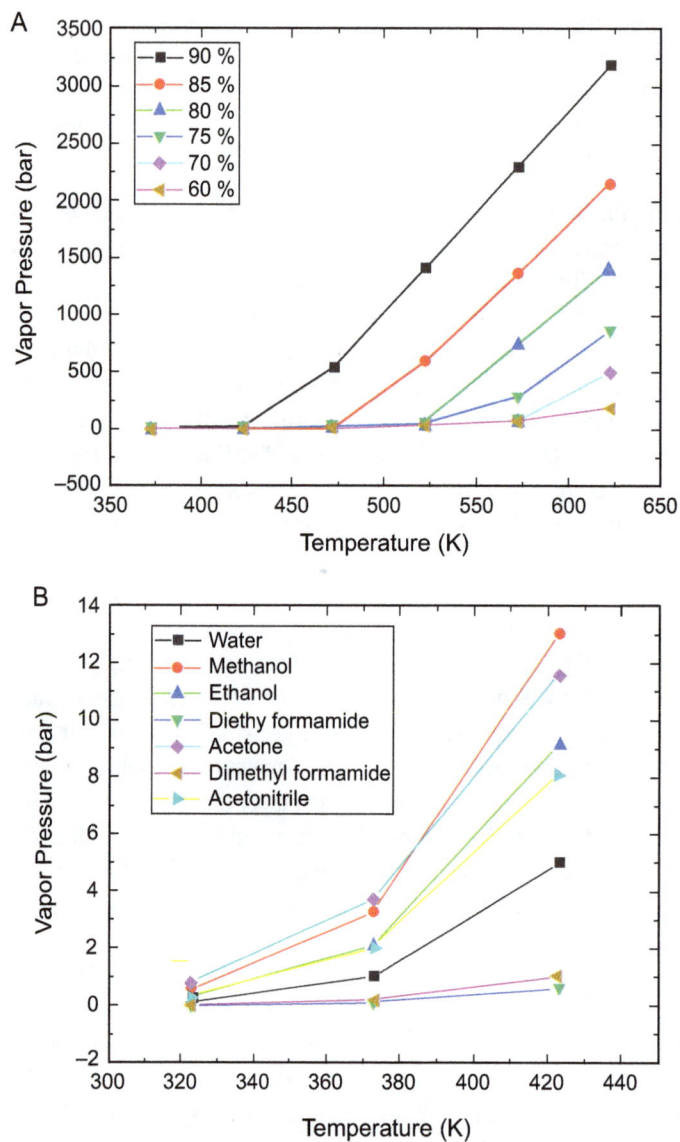

Fig. 1.18: The relationship between vapor pressure and temperature of solvents: (A) the phase change of water as a solvent under different humidity (data were derived from the NIST Reference Database Version 9.0); and (B) the phase change data for water and six organic solvents as a comparison (pressures were calculated using the Antonine equation ($\log_{10} P = A - B/(C + T)$), the constant values of A, B, and C are cited from the NIST WebBook and figures from [102] with permission).

hydroxylated materials, low-temperature conditions, and environmental friendliness. The properties of water, such as thermal conductivity, viscosity, surface tension, and dielectric constant, highly depend on different compositions and types of materials. Other organic solvents can be selected depending on the solubility of starting materials (reactants) and their mixture with water. The products' structures and reactivities can be tuned by controlling the synthesis variables, such as solvent volume, solution concentration, reaction temperature, and acidity. The pressure–temperature plots (Fig. 1.18A) show the relationship between vapor pressure inside an autoclave and temperature concerning the percentage fill of water. The graphs indicated that the hydrosolvothermal reaction pressure could be controlled by programming the temperature and altering the solvent volume. Low pressure can be obtained when a small amount of solvent is introduced into the autoclave reactor (Fig. 1.14A).

On the other hand, a high solvent volume can create high pressure even at low temperatures in the same reactor. Figure 1.18B indicated pressure variation as temperature changes regarding different select organic solvents. In generalization, the high temperature corresponds to the high vapor pressure of the solvents and reactants, which will result in a more effective collision, further improving the reaction rate. The organic solvents (e.g., methanol, CH_3OH) showed the highest vapor pressure, while dimethyl formamide ((CH_3)$_2$NCH) produced the lowest pressure under the same reaction temperature. We did not find a clear trend between the vapor pressure and the molecular dielectric constant or the thermal conductivity due to the molecular structure's complexity and thermal motion behavior. These data provided a guideline for selecting solvent, which depends on the individual reactions. It is safe to conclude that the less polar molecule with a larger molar mass provides lower vapor pressure under the same conditions.

1.2.4 Chemical vapor deposition

CVD is also known as vacuum deposition, which produces high-quality and high-performance solid materials in the activated environment. Industrial products using this CVD include cutting tools, filters, optoelectronics, refractory fibers, semiconductors, optoelectronics, optics, cutting tools, refractory fibers, or filters. As stated by Dr. Pierson, "CVD is no longer a laboratory curiosity but a major technology on par with other major technological disciplines" [103]. The CVD facility comprises three key parts (as shown in Fig. 1.19A): (1) a chemical vapor reactant supply system; (2) a CVD reactor, and (3) an effluent gas treatment system. The precursor supply system provides gaseous reactants delivered to the reactor. In the CVD reactor, the substrate will be heated to suitable conditions, allowing for the deposition of final products [104]. The effluent gas treatment system consists of exhaust gases and a vacuum system, where the pressure, temperature, and vacuum variables can be controlled during deposition. In this CVD process, the vapor precursors undergo dissociation or chemical reactions under heat, light, and plasma exposure. The CVD method can produce different formulations of high-purity, well-controlled structure, and composition at molecular levels. These materials include ultrafine powder (zero dimension (0D)), nanowires/nanotubes (1D), nanosheets and wells (two-dimensional (2D)), and thin film (3D) structures. It was found that ultrafine particles in the nanoscale can be formed when the CVD reaction occurs in the gaseous state due to its homogeneity [105]. Otherwise, a dense film will be obtained if the reaction occurs near or on the heated substrate due to its heterogeneous nature [106]. The CVD experimental setup requires complicated variables and procedures, including a selection of activated methods, precursor choice, and pressure/volume control. The activated methods include heat, plasma, light, and ions. Based on different external energy, the CVD method can be classified into thermal CVD, plasma-assisted CVD, MW-assisted CVD, photo/laser-assisted CVD, and ion-assisted CVD. Atomic layer deposition could operate via thermal or plasma mode. Those methods normally acquire high vacuum systems.

CVD deposition is a complex approach, producing materials with different shapes and compositions. The seven key steps [106] are stated as follows (as shown in Fig. 1.19B):

1. *Vaporization and generation of gaseous reactants*: Evaporation or sublimation occurs when the precursor is exposed to heat, light, or plasma.
2. *Transport of gaseous reactants*: The reactant was then delivered into the reactor, where the chemical reaction occurs. The gaseous phases allow for an even distribution of reactants, leading to a high effective collision.
3. *Gas-phase reactions*: The precursors in the reaction zone produce reactive intermediates at a sufficiently high temperature in the gas phase; intermediates undergo chemical reactions. Stable solid fine particles can be produced in the homogeneous gaseous phases, while by-products can be formed. At lower temperatures than dissociation, the intermediate species will be formed. They will consequently undergo deposition, diffusion, and removal steps.
4. *Adsorption of reactants*: The precursors will be absorbed on the heated substrate surface. If heterogeneous reactions occur at interfaces between gas and solid substrate, deposition and by-product will be formed.
5. *Surface diffusion*: The deposited substances diffuse or migrate along the heated substrate, allowing for nucleation and crystallization. The growth of film or particles will occur due to surface chemical reactions.
6. *By-product removal*: Those gaseous by-products will be removed from the boundary layer via diffusion and convection to improve the quality of products.
7. *Desorption and mass transport*: The by-product and remaining fragments of the unreacted species will be removed from the reaction zone.

Choice of precursors: Metal halides and hydrides are commonly used reactants. Metal-organic precursors are commonly used in metal-organic CVD.

Deposition pressure: Ultra-low pressure is normally needed to depose products using molecular-beam epitaxy and chemical-beam epitaxy. CVD is a chemical process used to produce high-purity and high-performance materials. This CVD involves thermal decomposition or other chemical reactions of gas-phase species, occurring at elevated temperatures ranging from 500 to 1,000 °C. This procedure includes a quartz substrate exposed to a volatile precursor to deposit the target product.

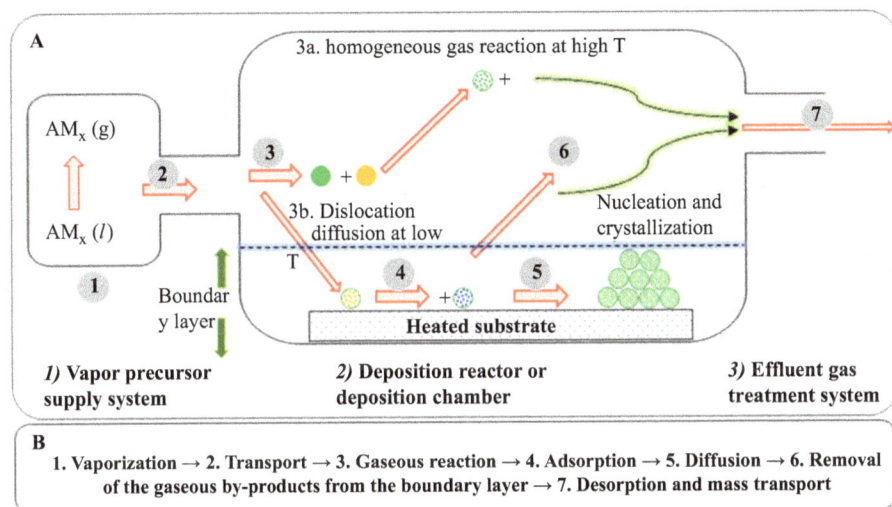

Fig. 1.19: A schematic representation of three key parts in the CVD apparatus: vapor precursor supply system, deposition reactor (chamber), and effluent gas treatment system. The image was adapted from [107] with permission.

The CVD-deposited materials vary from monocrystalline, polycrystalline, amorphous, and epitaxial. Materials with different elemental compositions include silicon, carbon fiber, filament, silica (SiO_2), silicon–germanium (Si–Ge), tungsten (W), silicon carbide (SiC), titanium nitride (Ti_3N_4), various high-K dielectrics, and synthetic diamond [108]. In the conventional CVD technique, the whole substrate is heated, and the reactant gas or vapor is flown over the substrate surface, where chemical reactions occur, and the film is deposited. This process is very slow (the film growth rate is 100–1,000 Å/min) and is not an energy-efficient process since the whole substrate is heated, but only the hot surface is used to carry out the film deposition. In this technique, the substrate temperature can be as high as 1,500 K, depending on the chemical reaction, temperature of the precursors, and the properties of the film. Since the diffusion coefficients of dopants become significant when the temperature exceeds 1,100 K, such high temperatures affect the quality of semiconductor films because of the change in the impurity concentration profiles due to the diffusion of dopants into the film [104]. Also, contaminants inside the CVD chamber or the wafer material can diffuse into the film grown at high temperatures. The volatilities of various species used for deposition compound semiconductor films are different. Therefore, at high temperatures, the species will evaporate in different amounts during the film deposition process, which can affect the properties of the film. Physical damage to the wafers at high temperatures can also cause defects in the devices produced by the conventional CVD method.

The CVD procedure includes a substrate (also known as a wafer) exposed to a volatile precursor to deposit the target product. The form of CVD deposit materials can be varied from monocrystalline, polycrystalline, amorphous, and epitaxial. Materials produced by CVD are highly diversified, including silicon and carbon nanotubes [105].

Liu and her collaborators employed the two-chambered CVD (Fig. 1.20) method to synthesize the aligned carbon nanotubes (ACNT). In this process, a two-compartment tube furnace was utilized. The reactants were vaporized in the first compartment and maintained in the second one to deposit the ACNT on the surface of a glass substrate. This process or two-furnace step is used to control the length and diameter, which are critical to gas transport and water management in the fuel cell [106]. The precipitation of carbon leads to the formation of tubular carbon solids in an sp^2 structure, which provides the molecules with their unique strength. Our results show that the impurity of carbon nanotubes can be controlled in a bifurcated furnace. Moreover, in our study, ferrocene was used as ACNT's seeding material.

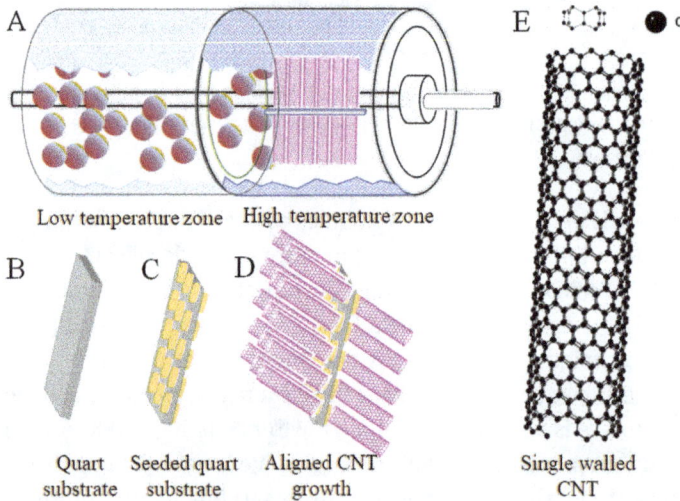

Fig. 1.20: Chemical vapor deposition produces aligned carbon nanotubes used as cathode catalysts for proton exchange membrane fuel cells. (A) The schematic diagram of the CVD furnace with two temperature zone; (B) substrate used for nanotube growth; (C) the metallic Fe as a seed increases nanotube growth in a highly aligned manner; (D) the nanotube growth; and (E) carbon unit cells' (P1 (triclinic), allowing for chirality; polarity) buckytube.

Here, the oxygen product partially reacted with ferrocene ($Fe(C_5H_5)_2$), which decreased the efficiency of the fuel cell due to oxygen loss but maintained ACNT integrity. The xylene–ferrocene solution was used as the precursor, where xylene (or dimethyl benzene ($(CH_3)_2C_6H_4$) is the carbon source. The ferrocene provides the iron metal nanoparticles as the seed catalysts for the growth of nanotubes. Three 5 cm^2 quartz substrates are placed inside the quartz reaction tube. The tube is placed in a two-stage furnace and tightly sealed from air. The first stage of the furnace is at a temperature of 225 °C, which suffices to vaporize the solution. The second stage is held at 725 °C and is used to carbonize the vaporized solution, depositing the iron nanoparticles on the quartz substrates and allowing the carbon nanotubes to grow along with the iron seeds [106]. The solution with the chemicals is injected into the reaction tube in the low-temperature stage, using argon and hydrogen as the carrier gases, at flow rates of 100 and 50 mL/min, respectively. The injection rate of chemicals is maintained at 0.225 and 0.250 mL/min, respectively [109].

1.3 Top-down synthesis of nanomaterials

The top-down process is normally considered a physical or chemical process to break down particles into smaller particles by applying the external force. This synthesis implies that the nanostructures can be obtained by removing crystal planes present on the substrate. In other words, the building blocks are removed or etched out from the substrate to form the nanostructure. Normally, grinding systems (wet and dry) convert regular-sized materials into nanoparticles. The dry systems are obtained by breaking down the bulk particles using a ball mill, jet mill, and roller mill. The decreased condensation of the particles prevents them from growing and controls the size to less than 3 μm. On the other hand, the wet grinding can be carried out using the centrifugal fluid mill, vibratory ball mill, flow conduit bead mill, and wet jet mill. When differentiated by both wet and dry methods, the grinding shows advantages in preventing agglomeration and forming desirable size particles to produce nanostructured materials further.

In top-down nanofabrication, the nanoscopic structures can be obtained using physical- or chemical-based processes (Fig. 1.21) [110]. The physical top-down approach relies on optical, electron beam, and ion beam lithography. This method has been intensively utilized in the microelectronic industry, such as dynamic random-access memories [111]. The chemical top-down approaches involve chemical etching and heat application reactions. Compared to the most physical top-down methods, the top-down chemical techniques do not require highly controlled systems (such as a cleanroom). The nonexpensive facility will be sufficient to build and maintain nanostructures, attracting more attention for in-depth investigation. This approach also offers synthetic versatility to produce a wide variety of 3D nanostructures with tunable architectures. These novel nanoscale architectures are promising in diversified applications from biomedicine, catalysis, electronics, energy storage and conversion, and sensors and actuators.

In the last few decades, top-down approaches have been widely used to produce nanostructured materials, optimizing fabrication variables, facility modification, affordability for manipulation, and tunability of product features. The spotlight in this section will be the newly emerging fabrication route with an in-depth understanding of the relationship between nanostructure and preparation parameters. Different top-down methods' affordability and ease of use will also be discussed for future improvement. These final products with novel nanoscale architectures exhibit unique chemical, electronic, and biological properties that render the diversified applications of these materials as mentioned above. The top-down approaches include physical methods (lithographic tools and chemical-based processes, such as templated etching, selective dealloying, anisotropic dissolution, and thermal decomposition). This section briefly discusses the bottom-up methods: physical top-down method (Section 1.3.1; ball milling, lithography, and machining) and chemical top-down methods (Section 1.3.2). Four different approaches to the chemical process include templated etching, selective dealloying, anisotropic dissolution, and thermal decomposition.

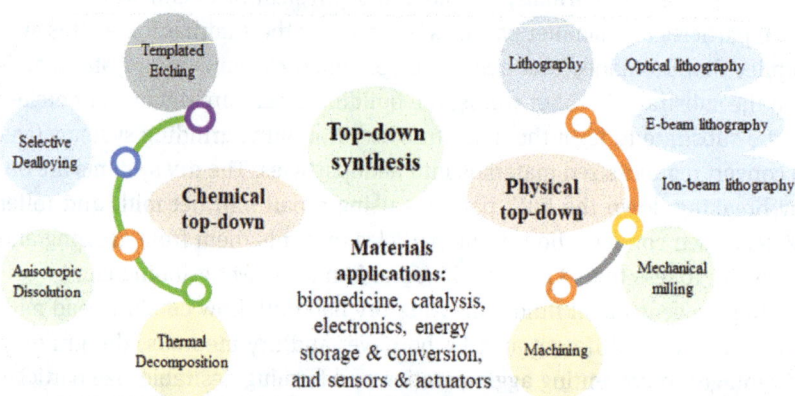

Fig. 1.21: A summary of top-down synthesis and chemical- and physical-based processes to produce various materials used in biomedicine, catalysis, electronics, energy storage and conversion, and sensors and actuators.

1.3.1 Physical top-down approach

The top-down approach has been widely used to fabricate nanostructures used in the microelectronics industry. The major advantage of the top-down synthesis to form joint circuits lies in creating orderly arranged structures to eliminate further assemblage steps. This section discusses about ball milling, lithography, and machining methods.

Ball milling: This is an operation to grind materials into extremely fine powders, such as nanoscale. One nanofabrication process is known as high-energy ball milling with industrial importance. This method is also called mechanical attrition or mechanical alloying. Through ball milling, coarse-grained powder materials (such as metals, ceramics, and polymers) are mechanically crushed into fine powders in rotating drums composed of hard steel or tungsten carbide (WC) balls. The controlled atmospheric conditions (e.g., the introduction of inert gas Ar) can be applied to prevent unexpected reactions. Grain size can be reduced via the formation and organization of grain boundaries within particles. Different components can be mechanically alloyed together [112]. This ball milling can also achieve nanosized dispersion of one phase in another. However, the so-produced microstructures and phases can be thermodynamically metastable and transient. The milling technique can be operated on a large scale, showing industrial interests. The milling of materials has been a major component of the mineral, ceramic processing, and powder metallurgy industries. The objectives of milling include particle size reduction, mixing or blending, and particle shape changes.

The typical milling method used for these purposes has been the tumbler ball mill. This method consists of a cylindrical container rotating along its axis, in which balls impact the powder charge. The balls may roll down the surface of the chamber in a series of parallel layers, or they may fall freely and impact the powder and balls beneath them [113]. An agitator's stirring occurs with a vertical rotating shaft with horizontal arms. A differential movement between the balls and the powder can result from this motion. Therefore, a high degree of surface contact can be achieved in tumbler ball mills. The rotation speed of the tumbler mill cylinder and the attritor shaft are the key factors in determining the kinetic energy imparted to the milling media [114]. Tumbler mills can provide high energy if a large shaft diameter is used and milling operates at a critical speed for a short period to "pin" balls to the chamber wall. Another milling method, attrition mills, has been replaced by the large bumbler ball milling to achieve large-scale production of commercial alloys. Another type of mill, the vibratory tube mill, has been used for pilot-size production [115]. The motion of the balls and particles in a vibratory mill was complicated. The cylindrical container is vibrated, and the impact forces on the powders in the milling chamber highly depend on the rate of milling, the amplitude of vibration, and the mass of the milling medium [116]. Our previous study using ball milling to produce electrocatalysts used for green NH_3 synthesis indicated improved milling efficiency using two different grinding balls due to increased contact surface area (Fig. 1.22A). The simple and polar molecule methanol (CH_3OH) was found to facilitate its ion–dipole intermolecular forces between reactants; as a result, the complexity of the electrocatalysts can be tuned from intra- and intermolecular perspectives. During the milling procedure, the elastic and plastic deformation were assumed to increase the strong frictional forces concurrently. This impact is the driving force to grind the material down to smaller sizes, leading to a crystalline formation with high homogeneity. The three milling methods such as tumble ball mill, vibratory tube, and mill attrition mill will be discussed in Chapter 2.

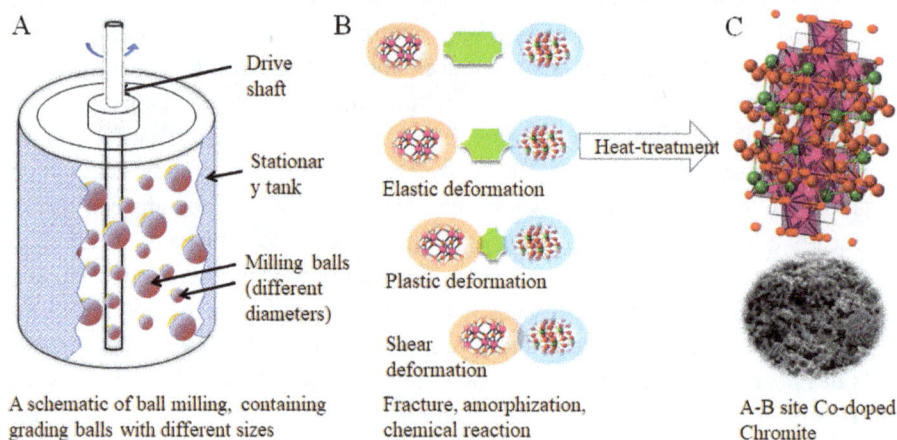

Fig. 1.22: Schematic diagram of (A) a ball milling process; (B) shear and impact forces generated in the milling chamber; and (C) the catalysts were formed in solid oxide electrolysis cells (adapted from [117] with permission).

High-energy milling forces can be obtained using high vibrational frequencies and small vibration amplitudes. The energy of the milling media depends on the internal mechanics of the specific mill. The power is supplied to drive the milling chamber and the balls' composition, size, and distribution. Since the balls' kinetic energy is a function of their mass and velocity, dense materials (steel or tungsten carbide) are preferable to ceramic balls. The size and size distribution should be optimized for the given mill [116]. Too dense packing of balls reduces the mean free path of the ball motion, while a dilute distribution minimizes the collision frequency. Empirically, a mass ratio between ball and powder of 5–10 is typically used and is effective. One important process variable is the temperature of the mill that is induced in the powders by the kinetic energy of the milling media. This temperature may be critical for the reactions or transformations in the powders during milling. However, the precision of milling temperatures induced by the powder surfaces remains improved. The milling of materials has been a major component of the ceramic processing and powder metallurgy industries. Through milling, the particle size can be reduced, mixing or blending can be improved, and particle shape can be changed. High-energy milling is often used to fabricate solid-state alloy mechanically. Benjamin and his colleagues developed this process at the international nickel company, and they provided a detailed description of this high-energy milling method [118]. This dry, high-energy ball milling process has been used to produce composite metal alloy powders, and the structure and size of the powders can be controlled well. Mechanical alloying is usually carried out in milling equipment capable of high-energy compressive impact forces such as attrition mills, vibrating ball mills, and shaker mills [119]. The milling machine will stress the maximum number of individual powders during comminution. The fracture of starting materials will be initiated with minimum energy. The motion of the milling medium and charge can vary depending on the types of mills. The movement and trajectories of individual balls, the mass of balls, and the degree of energy applied will highly influence shear, attrition, and compression forces acting on powder particles [120].

Lithographic processes: Lithography is a "planographic printmaking process" to fabricate nanosized materials or patterns. A design or structure can be drawn onto a flat surface, such as on the surface of a stone or a pre-prepared metal plate. Nano-imprint lithography is a novel method. Lithography uses visible or ultraviolet (UV) light, X-rays, electrons, or ions to project an image containing the desired pattern onto a surface [121]. These different techniques are generally termed photolithography, X-ray lithography, electron beam lithography, or ion beam lithography, depending on the radiation employed. The pattern transferring process includes four key steps: (1) solution-based wet chemical etching procedures, (2) dry etching in a reactive plasma, (3) doping using ion implantation techniques, and (4) thin-film deposition [122].

Optical lithography is the most traditionally used top-down synthesis (Fig. 1.23) to create fine-scale patterning of integrated circuits. The UV light will transfer the desired pattern from a mask to a photoresist semiconductor substrate (such as Si wafer) [111]. Subsequently, the predesigned pattern can be transferred to the photoresist substrate, coated, semiconductor, and often etched through a chemical process. Projection lenses can enhance the resolution, producing clear images (Fig. 1.23B).

Optical lithography is a top-down method that projects an image into a silicon wafer using a photon-based technique. This method has been widely used in electronic nanostructure production by the semiconductor industry due to its ease of use and feasibility. The wavelength used in nanoelectronic fabrication determines the minimum attainable feature size. The mercury (Hg)-based lamps that emit near UV are used as the illuminating source during the early years. Moore's law states that "the number of transistors on a microchip doubles about every two years, though the cost of computers is halved." To keep pace with Moore's prediction, researchers use shorter wavelength light sources, including excimer lasers, krypton fluoride (KrF) laser (248 nm), argon fluoride (ArF) laser (193 nm), fluorine (F_2, 157 nm), and xenon monochloride (XeCl) laser (308 nm). These lasers produce light with shorter wavelengths in the far-UV region, which are extensively utilized for two decades [111].

Photolithography can use UV light sources in microelectronic fabrication. The electron beam (e^- beam) and X-ray lithography are two other alternatives that have

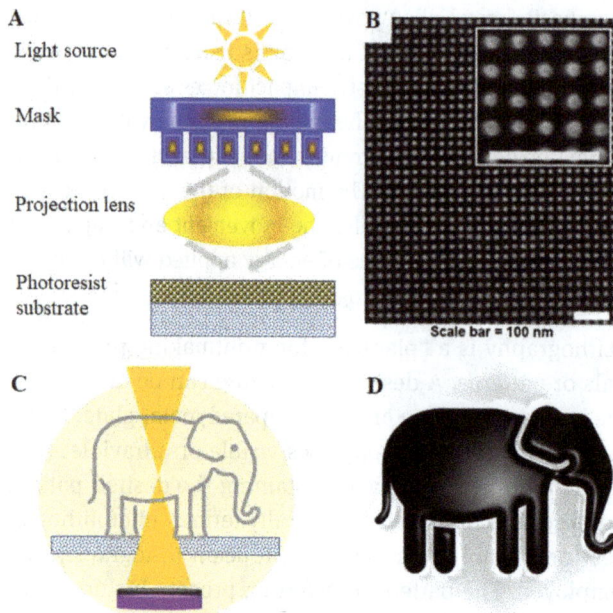

Fig. 1.23: Physical top-down method to fabricate nanostructures. (A) Simplified schematic diagram showing projection optics for the lithographic process; (B) scanning electron microscopic image of silicon (Si) nanodot array produced by electron beam lithography (reprinted with permission from [123]; Copyright 2011, American AIP). (C) simulated scheme of an elephant sculpture fabricated by direct laser; (D) projected 3D image of the elephant (adopted from [143] with permission.

attracted considerable attention in the MEMS and nanofabrication areas [124]. These methods will be discussed in Chapter 2 (top-down synthesis).

Machining: This is one of the top-down procedures to produce patterned and topographically constrained surfaces. The focused ion beams (FIBs) and high-intensity lasers are commonly used to produce and tune materials at micron to nanoscales. In its early stage, starting from 1970, the FIB was used to manufacture electronic components and microdevices by 3D pattern layered deposition, ablation, and repatterning [125]. The patterning is done by ejecting precise liquid droplets using nozzle thrusters of a specific geometry and angle to disperse charged metal droplets. Krohn first documented ion emission from a liquid metal source in 1961 [126]. His research on liquid metal ion sources (LMISs) for applications in outer space [127] was quickly proved to be useful for many other areas, especially semiconductors and materials science. Commercialization of the FIB was not far behind in the 1980s, mainly oriented toward the rapidly growing semiconductor industry. The modern FIB system utilizes an LMIS at the top of its column to produce ions (Fig. 1.24). Gallium (Ga) is usually used for its low melting point (30 °C), high mass, low volatility, and low vapor pressure because it is easily distinguished from other elements. The Ga source is heated to near evaporation, allowing to liquefy and flow down the tungsten needle where it can remain in a liquid state for weeks without further heating, owing to its supercooling properties [128]. The extractor, an annular electrode centered just below the tip of the needle, is held at a voltage of −6 kV relative to the source, drawing liquid Ga into a Taylor cone. An electric field on 10^{10} V/m between the tip and aperture induces field evaporation, causing ion emission and acceleration down the column. The source is typically operated at low emission currents of 1–3 μA to produce a stable beam [129]. Another suppressor electrode works alongside the extractor at around +2 kV to maintain a constant beam current. The suppressor and extractor act as "fine" and "coarse" controls regulating the ion extraction current.

As a result, the creation of ions and ion–ion in the beam and ion–surface interactions is an important consideration to ensure consistency and intricate patterning. The beam flux can be varied by changing the voltage between the electrodes and the rate of liquefying; the second parameter of ion–surface interaction is more variable since the final pattern depends on ion energy, ion flux, and ion–surface interaction angle. Upon collisions, the ions are refracted (slow down within the solid), while some are reflected (backscattered). Since the ions are charged atoms, they have a larger radius than electrons (as in an electron beam) and have a lower probability of transmission than electrons but an increased probability of interaction with atoms at the surface. Upon two-body elastic collision, energy changes are due to the large mass difference between the ion and electron (~1850:1) along the collision pathway. The collision disperses energy but not momentum due to the ions' higher ($\sim \times 400$) mass. The elastic collision process is distinct from surfacing collisions, which are many-body collisions resulting in inelastic collisions and radiative decay [130].

In an elastic collision, kinetic energy and momentum are conserved (during a nuclear collision). When the incident ions strike the target atoms, the energy and momentum transferred to the atoms on the material's surface cause a sufficient disturbance to remove them from their aligned positions. This sputtering effect, which does not occur when an electron beam is used, allows the FIB to precisely remove atoms from the material's surface in a controlled manner. The incident ions are also scattered, some becoming implanted in the material and others getting backscattered. Furthermore, collisions resulting from ion beam bombardment induce many secondary processes such as recoil and sputtering constituent atoms, electron excitation, emission, defect formation, and photon emission (Fig. 1.24) [131]. These collisions lead to thermal- and radiation-induced diffusion, which contributes to various phenomena with the constituent elements, including phase transformations, crystallization, amorphization, and track formation, thereby propagating the process toward the formation of materials. Also, ion implantation and sputtering alter the sample's surface morphology, resulting in craters, facets, pyramids, grooves, ridges, and "blisters," or the formation of a porous surface. Electronic collisions produce secondary electrons and X-rays, apart from sputtering effects from elastic collisions. In inelastic ion–electron collisions, the electrons are excited and ionized. Additionally, the incident ions transfer energy to the surface atoms and electrons, producing secondary electrons, ions, and electron ejection. If the incident ion has insufficient energy to penetrate deep inside the atom, the inner-shell electrons screen the nuclear charge, and this screening effect by electrons must be considered for these nuclear collisions [132].

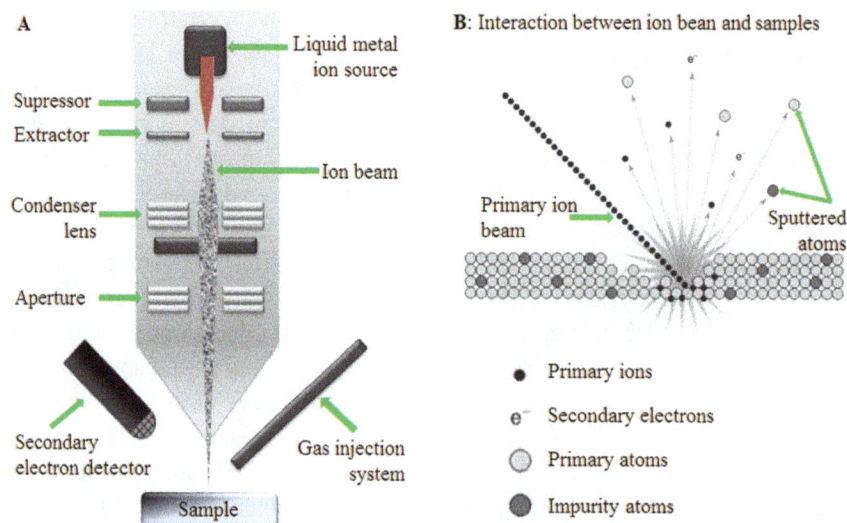

Fig. 1.24: A schematic diagram of focused ion beams (FIBs): (A) the generic steps of FIB to produce nanomaterials and (B) interaction between ion beams and samples.

FIBs can deliver tens of nanoamps of current to a sample and image the sample with a spot size on a nanometer. FIB is an inherently destructive technique since the specimen, upon rasterization, is transformed. Due to the spattering capability of FIB, it is used as a micromachining tool [133]. Machining is an operation to achieve intricate 3D patterning of a material. FIB and high-intensity lasers have been used to directly pattern or shape materials at micron and submicron levels. FIB is a semiconductor and materials science technique for site-specific analysis, deposition, and ablation. The focused beam of gallium (Ga) is easy to build an LMIS. Ga metal is placed in contact with a tungsten needle and heated. Gallium wets tungsten, and a high electric field ($>10^8$ V/cm) causes ionization and field emission of Ga atoms. FIB can be incorporated into a system with electron and ion beam columns [134].

1.3.2 Chemical top-down synthesis

The photolithographic top-down approach to produce 65 nm commercial transistors has evidenced significant improvement over the last decades. Although this method leads to precisely controlled shape and size, its manufacturing expenses have reached unprecedented. Therefore, follow alternative strategies to produce patterned silicon architectures for academia and industry uses. One of the new processes compatible with existing Si-based fabrication techniques was the chemical process. The chemical top-down approach has become more attractive in producing nanoscale patterns. This method shows advantages in crafting more sophisticated nanostructures in solids and solutions and ordered arrays of nanostructures with 3D features. Four different approaches (Figs. 1.21 and 1.25) discussed in this section are: (1) templated etching, (2) selective dealloying, (3) anisotropic dissolution, and (4) thermal decomposition.

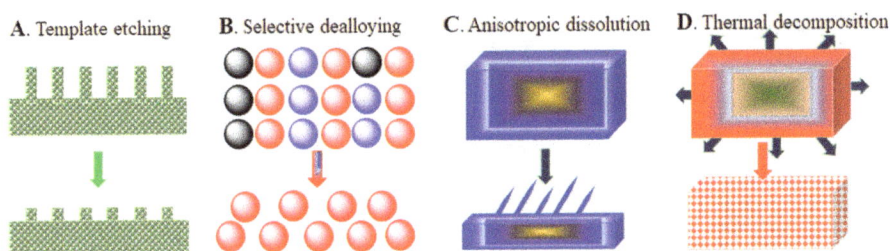

Fig. 1.25: A summary of top-down chemical synthesis: (A) the templated etching, (B) selective dealloying, (C) anisotropic dissolution, and (D) thermal decomposition.

Templated etching: Templated etching top-down is a technique to produce a nano-structured pattern, which can be crafted on the substrate surface (such as silicon), followed by subsequent template removal. Polymer self-assembly enabled the semi-conductor formation of a uniform pattern with large areas and interfaces. The "spontaneous phase separation of block copolymers into nanoscale domains" provides an effective and feasible approach to producing periodic patterns. The template is normally used to direct the substrate etching through a chemical reaction during the templated etching. An amphiphilic block copolymer, poly(styrene-block-4-vinyl pyridine) (PS-b-P4VP, [Mn = 128 400-b-33 500 g/mol]) template (Fig. 1.26A), was used to tune the nanoscaled structure into a 3D etch pit array as reported by Buriak et al. [135]. This molecular amphiphilicity is critical in directing surface chemistry in a spatially defined manner. The chemical specificities of different building blocks lead to localized reactions on the substrate surface. Their results indicated that the silicon (Si) surface chemistry could be tuned by controlling the reagents' spatial location and concentration. The "self-assembled block copolymer-based quasihexagonal template arrays" can direct aqueous fluoride anion-based etching of Si surfaces. This processing occurred at ambient conditions within minutes, leading to patterned nanoscale Si–H-terminated etch pits on Si surface. The surface without etching was "terminated by the native oxide," suggesting that these etch pits and top Si interface can be further chemically functionalized (Fig. 1.26B). The plane and sectional morphology of the above amphiphilic copolymer-coated Si surface were evaluated using scanning electron microscopy (SEM; Fig. 1.26C). The Si

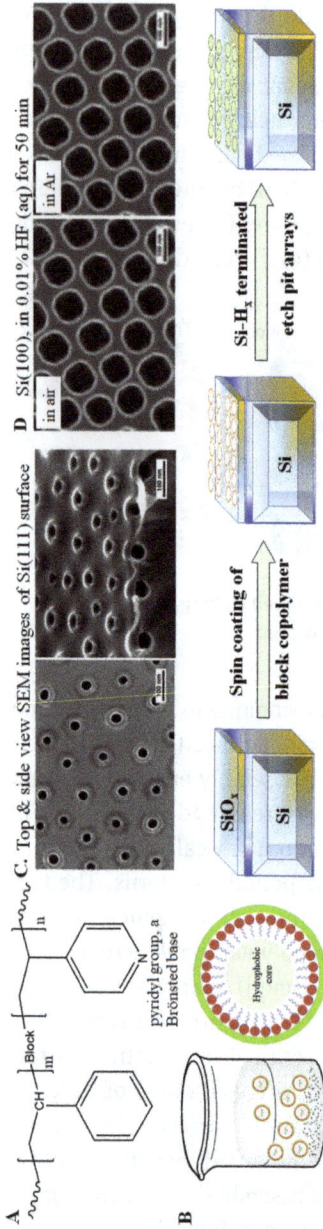

Fig. 1.26: A summary of templated etching top-down synthesis: (A) the formula of an amphiphilic block copolymer, poly(styrene-block-4-vinyl pyridine) (PS-b-P4VP); (B) schematic diagram to prepare "etch pits on a Si(100) surface" using an amphiphilic template composed of PS-b-P4VP; (C) SEM top and side views of the copolymer template-coated Si(111) surface after immersion into 0.01 % HF solution for 40 min; and (D) The impact of Si crystal orientation (100 was chosen) on geometric shape evolution of the patterned etch pits. Figures (B–D) were adapted from [135] with permission.

(111) surface was etched in 0.01 % HF (aq) for different duration, demonstrating surface changes.

The PS-b-P4VP-coated Si wafer topology varied significantly depending on exposure duration in HF solution and single-crystalline orientation. The image indicated that "truncated square and pseudohexagonal-shaped etch pits" were formed, where the center-to-center spacing was averaged at 125 nm. The shape and size were identical to the "parent block copolymer monolayer." The top and cross-sectional images of the etched Si surface were collected to evaluate the impact of the Si orientation on the morphology. The Si(100) orientation was chosen to demonstrate in this chapter. The observation revealed the 3D characteristics of Si(111), Si (100), and possibly transitional Si(311) faces. The corner rounding was more clearly observed from the images than in samples with a top surface mask. It was also found that oxygen's role is minor due to its anisotropic oxidant nature based on the subtle difference in etch pit shapes on Si(100) (Fig. 1.26D).

Selective dealloying: Selective dealloying (also known as selective leaching) is a procedure for demetalification. This demetalification as a parting and selective corrosion normally can be observed in some "solid solution alloys." A chemical component in the alloys is preferentially leached out from the material at suitable conditions. Due to the redox potential, more active or less noble metal (LNM) will be removed from the alloy through a galvanic redox corrosion mechanism. Recently, selective dealloying was used to generate a nanoporous metal from the alloys. The metallic alloy can be selectively "unalloyed" through chemical reactions, removing the most reactive metal constituent. The reactive sequence of different metals is shown in Tab. 1.3.

Typically, the elements that undergo selective dealloying or removal are reasonably active metals, such as commonly used aluminum (Al), cobalt (Co), chromium (Cr), iron (Fe), and zinc (Zn). Recently, noble metals have been produced using selective dealloying to create nanoporous structures. Chen and his team reported that nanoporous gold (NPG) was formed by selective removal, named "desilvering" of silver–gold (Ag–Au, $Au_{35}Ag_{65}$) alloys. This desilvering technique was the most feasible approach in nanofabrication to tune the pore structures [137]. The SEM images of desilvering NPG from $Au_{35}Ag_{65}$ using nitric acid (HNO_3) showed that the nanopore size increased with dealloying time and temperature (Fig. 1.27A). This team also reported that a series of transition metals with nanopores had been fabricated from their Al-based alloys in alkaline basic or acidic solutions under free corrosion conditions. These elements include three noble metals such as palladium (Pd), platinum (Pt), and silver, and one transition metal, copper (Cu). Upon dissolution of Al in either basic or acidic solution, the electron tomography image of the metal showed well-tuned pore structures (Fig. 1.27B). The nanoporous metals showed one to two orders of magnitude higher electrical resistivity than their bulk counterparts. Initially, the LNM component is dissolved in a "quasi-layer-by-layer mode" to form step edges during the dealloying process. Due to the surface diffusion, the more noble metal

Tab. 1.3: The reactive sequences of metals and their standard reduction potentials [136].

Reactivity	Metal elements	Ions	Standard reduction potential, E^o (V)
1	Lithium (Li)	Li^+	$Li^+(aq) + e^- \rightarrow Li(s)$, $E^o = -3.04$
2	Rubidium (Rb)	Rb^+	$Rb^+(aq) + e^- \rightarrow Rb(s)$, $E^o = -2.98$
3	Potassium (K)	K^+	$K^+(aq) + e^- \rightarrow K(s)$, $E^o = -2.93$
4	Barium (Ba)	Ba^{2+}	$Ba^{2+} + 2e^- \rightarrow Ba(s)$, $E^o = -2.92$
5	Cesium (Cs)	Cs^+	$Cs^+(aq) + e^- \rightarrow Cs(s)$, $E^o = -2.92$
6	Barium (Ba)	Ba^{2+}	$Ba^{2+}(aq) + 2e^- \rightarrow Ba(s)$, $E^o = -2.91$
7	Strontium (Sr)	Sr^{2+}	$Sr^{2+}(aq) + 2e^- \rightarrow Sr(s)$, $E^o = -2.89$
7	Calcium (Ca)	Ca^{2+}	$Ca^{2+}(aq) + 2e^- \rightarrow Ca(s)$, $E^o = -2.84$
8	Sodium (Na)	Na^+	$Na^+(aq) + e^- \rightarrow Na(s)$, $E^o = -2.71$
9	$Mg(OH)_2$	Mg^{2+}	$Mg(OH)_2(s) + 2e^- \rightarrow Na(s) + 2OH^-(aq)$, $E^o = -2.69$
10	Lanthanum (La)	La^{3+}	$La^{3+} + 3e^- \rightarrow La(s)$, $E^o = -2.38$
11	Magnesium (Mg)	Mg^{2+}	$Mg^{2+}(aq) + 2e^- \rightarrow Mg(s)$, $E^o = -2.36$
12	Cerium (Ce)	Ce^{3+}	$Ce^{3+} + 3e^- \rightarrow Ce(s)$, $E^o = -2.34$
13	Beryllium (Be)	Be^{2+}	$Be^{2+} + 2e^- \rightarrow Be(s)$, $E^o = -1.99$
14	Aluminum (Al)	Al^{3+}	$Al^{3+}(aq) + 3e^- \rightarrow Al(s)$, $E^o = -1.68$
15	Uranium (U)	U^{3+}	$U^{3+} + 3e^- \rightarrow U(s)$, $E^o = -1.66$
16	Manganese (Mn)	Mn^{2+}	$Mn^{2+} + 2e^- \rightarrow Mn(s)$, $E^o = -1.17$
17	Chromium (Cr)	Cr^{2+}	$Cr^{2+}(aq) + 2e^- \rightarrow Cr(s)$, $E^o = -0.90$
18	Zinc (Zn)	Zn^{2+}	$Zn^{2+}(aq) + 2e^- \rightarrow Zn(s)$, $E^o = -0.76$
19	Chromium (Cr)	Cr^{3+}	$Cr^{3+}(aq) + 3e^- \rightarrow Cr(s)$, $E^o = -0.74$
20	Selenium (Se)	Se^{2+}	$Se^{2+}(aq) + 2e^- \rightarrow Se(s)$, $E^o = -0.67$ (in 1.0 M NaOH)
21	Iron (Fe)	Fe^{2+}	$Fe^{2+}(aq) + 2e^- \rightarrow Fe(s)$, $E^o = -0.44$
22	Cadmium (Cd)	Cd^{2+}	$Cd^{2+}(aq) + 2e^- \rightarrow Cd(s)$, $E^o = -0.40$
23	Cobalt (Co)	Co^{2+}	$Co^{2+}(aq) + 2e^- \rightarrow Co(s)$, $E^o = -0.28$
24	Nickel (Ni)	Ni^{2+}	$Ni^{2+}(aq) + 2e^- \rightarrow Ni(s)$, $E^o = -0.26$
25	Molybdenum (Mo)	Mo^{3+}	$Mo^{3+} + 3e^- \rightarrow Mo(s)$, $E^o = -0.20$
26	Titanium (Ti)	Ti^{2+}	$Ti^{2+} + 2e^- \rightarrow Ti(s)$, $E^o = -0.16$
27	Tin (Sn)	Sn^{2+}	$Sn^{2+} + 2e^- \rightarrow Sn(s)$, $E^o = -0.136$
28	Lead (Pb)	Pb^{2+}	$Pb^{2+} + 2e^- \rightarrow Pb(s)$, $E^o = -0.126$

Tab. 1.3 (continued)

Reactivity	Metal elements	Ions	Standard reduction potential, E^o (V)
29	Iron (Fe)	Fe^{3+}	$Fe^{3+}(aq) + 2e^- \rightarrow Fe(s)$, $E^o = -0.037$
30	Antimony (Sb)	Sb^{3+}	$Sb + 3\,H^+ + 3e^-$ $SbH_3(g)$, $E^o = -0.510$
31	Bismuth (Bi)	Bi^{3+}	$Bi^{3+} + 2e^- \rightarrow Bi(s)$, $E^o = +0.317$
32	Copper (Cu)	Cu^{2+}	$Ni^{2+}(aq) + 2e^- \rightarrow Ni(s)$, $E^o = +0.34$
33	Thallium (Tl)	Tl^{3+}	$Tl^{3+} + 3e^- \rightarrow Tl(s)$, $E^o = +0.742$
34	Platinum (Pt)	Pt^{2+}	$Pt^{2+} + 2e^- \rightarrow Pt(s)$, $E^o = +1.2$
35	Mercury (Hg)	Hg^{2+}	$Hg^+ + 2e^- \rightarrow Hg(l)$, $E^o = +0.85$
36	Silver (Ag)	Ag^+	$Ag^+ + e^- \rightarrow Ag(s)$, $E^o = +0.78$
37	Gold (Au)	Au^{3+}	$Au^{3+} + 3e^- \rightarrow Au(s)$, $E^o = +1.52$
38	Gold (Au)	Au^+	$Au^+ + e^- \rightarrow Au(s)$, $E^o = +1.83$

(MNM) elements are expected to be etched out of the original lattice sites and aggregate at the step edges. With the continuous dissolution of more LNM atoms, MNM atoms are needed to passivate the steps of the resulting MN-capped hills. The nucleation of MNM clusters will further widen the base perimeters of the above MN-capped hill, allowing for a decrease in the hill spaces [138]. Therefore, the pores are extended into the bulk materials, forming a porous network within the core-shelled structures. The MNMs were leached out to act as shells with tunable pore structures, as shown in Fig. 1.27B.

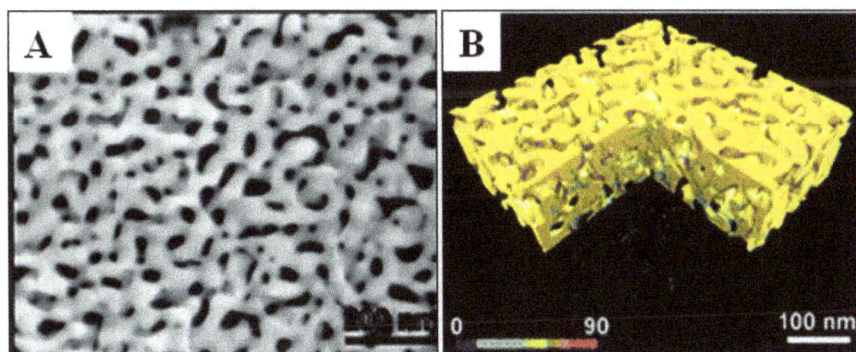

Fig. 1.27: A desilvering method to produce porous nanogold (NPG) from $Au_{35}Ag_{65}$ alloy using HNO_3 under free corrosion conditions: (A) SEM image showing the pore size increase with dealloying and temperature; (B) electron tomographic image, confirming the SEM observation. Reprinted with permission from [137]. Copyright 2008 by APS.

Erlebacher and coworkers have established a kinetic Monte Carlo model to evaluate physical mechanism and porosity evolution during dealloying a binary metal alloy [139]. The authors proposed a continuum model to demonstrate the intrinsic dynamical pattern formation of nanoporous structures in metals. The difference in reduction potential between metals is the driving force for noble metals to aggregate into 2D clusters. The phase separation process (also known as spinodal decomposition) was found to occur at the interface of the solid–electrolyte interface. During the chemical etching, the surface area grows, enabling the evolution of the porosity, which shows characteristics in length and pore size. Through chemical tailoring, the porosity of NPG is anticipated to be suitable for biological sensing applications.

Anisotropic dissolution: This is a process in which the materials can be dissolved in a solvent, allowing for different properties in different directions. Contrary to the nucleation and growth from supersaturated solutions, the separation of larger crystals into their constituents in a controlled manner will create materials with a well-defined nanostructure. This procedure normally occurs in "undersaturated solutions" via anisotropic dissolution of preferred crystalline planes and facets (Fig. 1.25C). Yu and coworkers reported that well-aligned nanosheet arrays composed of calcium carbonate ($CaCO_3$, calcite) were formed using this newly developed anisotropic dissolution [140]. Yu and his team carried out this anisotropic dissolution using a 1 % aqueous formamide solution of the calcite microcrystals, followed by heating at 65 °C for 10 min and 1 day, respectively. Complex calcite, the most stable of $CaCO_3$ polymorphs, was naturally produced through biomineralization, such as protein-directed nucleation and growth due to the binding between peptide and crystallographic facets of calcite. Yu and his team invented a dissolution method to produce densely packed nanoarrays. Their research discoveries (Fig. 1.28A) indicated that stable {104} planes consist of calcium cation (Ca^{2+}) and carbonate anions in alternating order. This plane was found to be more preferentially dissolute than {012} planes. As a result, the higher energy {012} planes are fully composed of calcium ions, largely produced on the nanoarray surfaces. The calcite shows a highly anisotropic crystal structure, including CO_3 "rotor" groups arranged in layers, while the Ca cation resides between different layers (Fig. 1.28B). In addition, the high electron density within the CO_3 layers (purple-colored sub-building units) gives calcite a high refractive index. The low density perpendicular to the layers is mainly composed of Ca (indigo color, #8de6f5). The intrinsic anisotropic nature of $CaCO_3$ from the perspective of unit cells lays the foundation for forming nanosheet arrays.

Thermal decomposition: Thermal decomposition or thermolysis is a chemical decomposition when heat is applied to the system. When the materials are decomposed at a certain temperature (called decomposition temperature), the chemical bonds are cleaved, and new bonds will be formed through rearrangements of electron transfer or sharing. Usually, thermolysis is endothermic due to the employment of heat

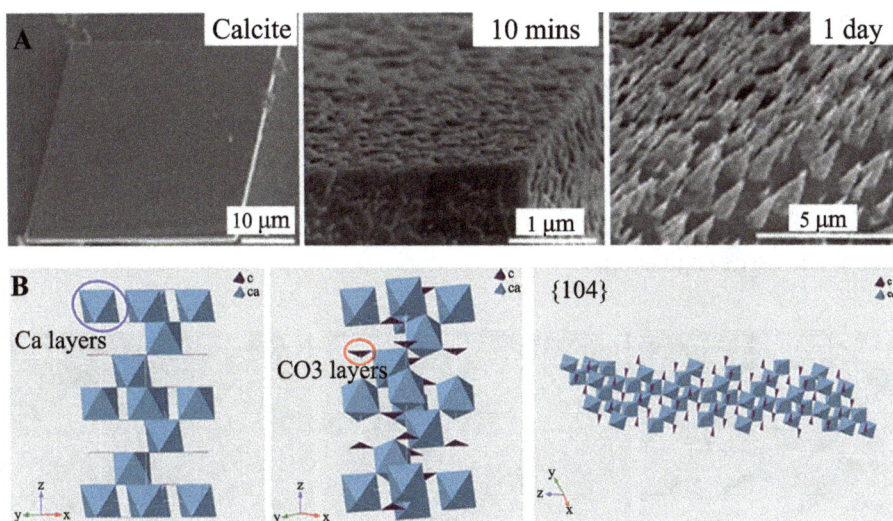

Fig. 1.28: Demonstration of anisotropic dissolution approach: (A) formation of calcite nanosheet arrays (reproduced with permission from [140]; Copyright 2010 Wiley-VCH), and (B) crystalline structure of calcite (space group: $\bar{R}3c$, trigonal, $a = 4.9900$ Å, $b = 7.0595$ Å, and $c = 367.873$ Å^3).

treatment. Upon thermal decomposition, the large starting inorganic solids (bulk and microstructure) can be effectively transformed into nanostructured substances with tunable pores and particle sizes (Fig. 1.25D). These chemical routes become attractive due to their cost-effectiveness in forming nanomaterials with a potential for large-scale fabrication. The researchers demonstrated this by converting calcium carbonate ($CaCO_3$) microcrystals into highly porous materials at 900 °C for 1 h (Fig. 1.29A), as reported by Han and coworkers [141]. It can be seen that $CaCO_3$ crystals sized at ~5 μm were decomposed into nanoporous calcium oxide (CaO) using thermolysis. These nanoporous CaO particles were subsequently cooled to be retransformed into nanoporous $CaCO_3$ in the carbon dioxide atmosphere. It was found that the large size (~50 μm) of starting $CaCO_3$ crystals enabled two different pore types, namely nano-pores and micrometer-sized channels. The rapid CO_2 release rate is critical for generating these different sized and shaped pores. Han and his team from the National University of Singapore reported the "noncrystalline pellets of different calcium compounds." Their research data showed that calcium contents in the starting materials affect the porosity of final products (Fig. 1.29B and C) [142]. This team employed different resources and compositions of Ca using different compounds, calcium carbonate ($CaCO_3$, 40 % Ca), calcium citrate ($Ca_3(C_6H_5O_7)_2$, 21 % Ca), and calcium lactobionate ($Ca(C_{12}H_{21}O_{12})_2$, 5 % Ca). These compounds were pressed into pellets and heated at 1,000 °C for 1 h to produce porous CaO and release CO_2. The system was cooled down to ambient conditions to re-form porous $CaCO_3$. The pore size of the final product "$CaCO_3$" increases based on different reactants. It was found that the smaller weight percentage of Ca in the reactants ($CaCO_3$) correlated with the small pore size due to the small amount of CO_2 release (Fig. 1.29B). The two organic-based compounds, calcium citrate and lactobionate (Fig. 1.29C), possess many ligands composed of carboxylate- and hydroxyl-functionalized groups, resulting in CO_2 and water vapor released in large amounts. The huge empty spaces will remain till the gas release, leading to large pore size. The use of organic compounds enables the control of the porosity or pore size of the final products, which can be used in catalysis, separation, and tissue engineering.

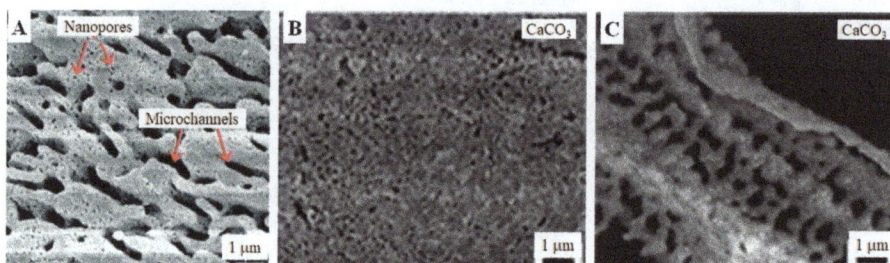

Fig. 1.29: Microscopic image of porous $CaCO_3$ was formed using thermolysis. (A) SEM image of porous $CaCO_3$ was produced from large microcrystals by thermolysis at 900 °C for 1 h and subsequently cooled in air. Nanopores and microchannels were formed based on the CO_2 release rate and amount (adapted from [142]). Moreover, (B) SEM image of porous $CaCO_3$ was produced using calcium carbonate ($CaCO_3$, Ca % = 40) as a reactant with high Ca content, and (C) SEM image of porous $CaCO_3$ was produced using calcium lactobionate ($Ca(C_{12}H_{21}O_{12})_2$, Ca % = 5) as a reactant with low Ca content ((B) and (C) were adapted from [144]).

1.4 Properties of nanomaterials

Nanotechnology is working with nanoscaled materials, which enable scientists, engineers, and other professionals to understand and utilize their unique properties [143]. These properties include chemical, electro-optical, magnetic, physical, and thermal characteristics of materials that can be observed at nanodimension. When particle size decreases to less than 100 nm, materials' properties will change significantly compared with the large-scale bulk materials. At the nanoscale, the quantum effects dominate, ruling the materials' properties and performances. Researchers can fine-tune the materials' properties based on the end application. It has been observed that the size of the nanomaterials can affect their fluorescence, porosity, conductivity, and reactivity. This chapter discusses different nanostructured materials' chemical, magnetic, electro-optical, and thermal properties.

Because of the ultra-high surface area, nanomaterials will be exposed more significantly to the surrounding environment. Therefore, these materials' chemical reactions (or reactivity) can be greatly enhanced, especially in the catalysis field. A commonly used catalytic converter in a car contains nanoscaled platinum alloy coated on the surface of porous ceramics. As a result, the gasoline combustion efficiency can be improved. Further, automobile exhaust and air pollution can be reduced or even eliminated. In addition, engineers have made full use of the increased reactivity of nanomaterials to design batteries with lighter weight and prolonged life span, fuel cells with high electrochemical behavior, and catalysts used in emerging energy generation and storage systems [144].

A cubic solid is used to show that particles have phenomenally high surface areas with a size decrease (Fig. 1.30A). We can assume that there is a solid cube with a unit cell lattice of 1 cm on each side. The total surface area of this "*centi-cube*" can be calculated to be 6 cm^2 (eqs. (1. 5A) and (1.5B)). If this cube is disassembled into smaller cubes with 1 mm on a side, 1,000 cubes ($10 \times 10 \times 10$) are obtained. Each small "*minicube*" cube has a surface area of 6 mm^2, for a total surface area of 6,000 mm^2 (or 60 cm^2, eq. (1.5C)). With further decrease of the size to 1 nm, the total cubes needed will be 10^{21} (1 cm = 10^7 nm, $10^7 \times 10^7 \times 10^7 = 10^{21}$). The surface area of each "*nanocube*" can be calculated as 6 nm^2. The total surface area of these 10^{21} cubes will be 6×10^{21} nm^2 or 6×10^7 cm^2. It can be seen that the surface area of the cubic nanoparticles can be large than the area of a football field:

$$A_{1\ cm} = 6(1\ cm \times 1\ cm) = 6\ cm^2 \tag{1.5A}$$

$$A_{1\ mm} = 6(1\ mm \times 1\ mm) \times 1,000 = 6,000\ mm^2 \tag{1.5B}$$

$$A_{1\ nm} = 6(1\ nm \times 1\ nm) \times 10^{21} = 6 \times 10^{21}\ nm^2 \tag{1.5C}$$

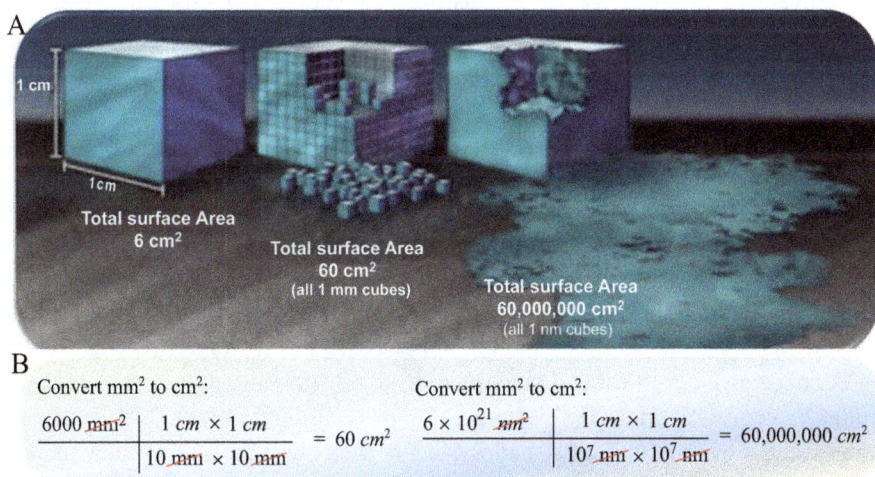

Fig. 1.30: Illustration demonstrating the effect of the increased surface area of nanomaterials: (A) Nanoscale materials have phenomenally larger surface-area-to-volume ratio than bulk materials and (B) the conversion of mm^2 and nm^2 to cm^2.

1.4.1 Chemical properties

There is a correlation between chemical structure and electronic properties for nanoscale and microscale materials. Any changes in the structure due to the changes in the particle size will alter the electronic property because of the intrinsic connectivity. The minimum energy required to remove an outermost electron is the first ionization energy (IE1), commonly high for smaller atomic groups than bulk particles due to the electron affinity. Ultrafine powders used in catalysis can enhance the reaction rate, selectivity, and efficacy of a chemical reaction in combustion or in the synthesis, where the reduction of wastage and pollution is observed. For example, gold (Au) nanoparticles generally possess fcc crystalline structure ($Fm\overline{3}m$, cubic, $a = 4.0786$ Å, cell volume = 67.847 Å^3, 4 sites/unit cell, Fig. 1.31A). However, when the size of Au is decreased to 5 nm, Au exists as an icosahedron (Fig. 1.31B) and displays an improved catalytic reactivity (Fig. 1.31C) [145]. Scrimin and coworkers reported the self-assembly of triazacyclonane ($C_6H_{15}N_3$)-functionalized thiols (R–S–H) on the surface of nanogold to mimic the nuclease systems (Fig. 1.31C). Their data showed that complexation with Zn^{2+} to form a chelating "Au MPC1" was tuned into a powerful catalyst [146]. The nanogold enabled transphosphorylation (Fig. 1.31C1) of an RNA model substrate, known as 2-hydroxypropyl p-nitrophenyl phosphate ($C_9H_{11}NO_7P^-$, HPNPP). The "Au MPC1" catalyst showed an improved cleavage activity compared to its monomeric reference complex [147].

Further studies on the catalytic activity of Au MPCs with different formulations indicated that a "low molar fraction of the catalytic unit triazacyclonane-Zn^{2+}" resulted in a decrease in catalytic activity. This observation suggested that this enhanced catalytical reactivity "originates from the cooperative effect between two neighboring catalytic units." The rate constant (k) for HPNPP cleavage by Au MPC1 highly depends on the amount of Zn^{2+} ions (Fig. 1.31C2). A different gold-containing catalyst using an elaborate ligand (a bis-(2-amino-pyridinyl-6-methyl)amine (BAPA) functionalized thiol), named "Au MPC2," was used for the transphosphorylation of BPNPP. The "Au MPC2" exhibited 100-fold higher efficiency for the cleavage of bis-p-nitrophenyl phosphate (BNPP), shown in Fig. 1.31C3. This improvement is attributed to the "cooperation between metal Lewis acid activation and hydrogen bonding" by the BAPA–Zn^{2+} complex, allowing for an improved "hydrolytic activity toward phosphate diesters" [148]. After the first plasmid insertion, deoxyribonucleic acid (DNA) is cleaved to form II (less torque) to form a linear strand (III).

A successive cut to form II generates linear III, the most released least twist strain, by concatenating 100 single-strand cuts. Coordination with Zn^{2+} results in high catalysis by the "Au MPC2" nanosystem at the bond cleavage site at standard conditions (pH 7.0 and 37 °C for 24 h) with the DNA plasmid chelating with Zn^{2+} and Au MPC2 catalyst to affect cleavage (Fig. 1.31C4), while no reaction was observed at all with monomeric-Zn^{2+} complex. At a higher concentration of Zn^{2+} (15 μM), the form III DNA (linear) was 50 % greater than nicked DNA (unwinds to form II), taking advantage of the multivalent catalyst MPC2 and differential chelating and binding of the ligand 6-benzoylacetyl-2-pyridine carboxylic acid.

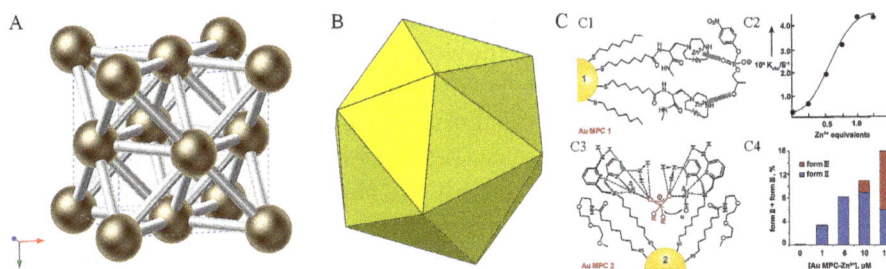

Fig. 1.31: Crystalline structure and reactivity of gold (Au): (A) face-centered cubic (fcc) crystalline structure (Fm̄3 m, cubic, a = 4.0786 Å, cell volume = 67.847 Å^3, 4 sites/unit cell; (B) icosahedron of nanogold with the size of 5 nm or smaller; and (C) an improved catalytic reactivity of nanogold [147, 148].

The electronic properties influence topography and dimensionality, with the energy to remove an electron (ionization potential) higher for small atomic clusters than the corresponding bulk material. Furthermore, the ionization potential exhibits observed fluctuations when cluster size varies. This effect is correlated to chemical reactivity, such as the reaction of Fe_n clusters with hydrogen gas. Nanoscale structures such as nanoparticles and nanolayers have very high special surface-area-to-volume ratios and different crystallographic structures. Therefore, the radical alteration in chemical reactivity may result from these changes in either dimension or crystalline structure [149]. Catalysis using ultrafine nanoscale systems can increase reaction kinetics, selectivity, and efficiency of chemical reactions. This phenomenon is normally observed when catalyst-assisted combustion occurs. Consequently, environmental waste and pollution were reduced or mitigated simultaneously. When the size of gold (Au) nanoparticles was reduced to smaller than about 5 nm in diameter, they were known to adopt icosahedral structures rather than the normal fcc arrangement [150]. An extraordinary increase in the catalytic activity accompanies this structural change. Furthermore, nanoscale catalytic supports with controlled pore sizes can select the products and reactants based on their physical size and thus ease of transport to and from internal reaction sites within the nanoporous structure. Additionally, nanoparticles often exhibit new chemistries as distinct from their larger particulate counterparts. For example, many new medicines are insoluble in water when in micron-sized particles but dissolve easily when in a nanostructured form [151].

When atoms come into proximity with other atoms in a solid, most electrons remain localized and associated with a particular atom. However, some outer electrons will become involved in bonding with adjacent atoms. The atomic energy-level diagram is modified upon the formation of bonds due to the hybridization of atomic orbitals (AOs). Briefly, the well-defined outer electron states of the atom overlap with those of neighboring atoms and become broadened into energy bands. For a simple diatomic molecule, the two outermost AOs overlap to produce two molecular orbitals (MOs). The formation of MOs can be viewed as a linear combination (addition or subtraction) of the two constituent AOs. Solid can be considered as a cluster composed of a large number of molecules [152]. The electronic energy levels of a 1D solid (a linear chain of atoms) were affected when the number of atoms within the solid was increased (Fig. 1.32). The electronic band structure of solids can combine the electron waves in a periodic crystalline potential. The Drude–Lorentz free electron model for metals is used for demonstration. In this model, a metallic solid is assumed to consist of a close-packed lattice in which cations are surrounded by an electron sea or cloud formed from the ionization of the outer shell (valence) electrons [153]. Therefore, a classical kinetic gas theory can treat the valence electrons as gas molecules moving inside a container. This model applies to these so-called free electron metals, the electropositive metals of groups I and II, and aluminum. The model was useful for explaining many of the fundamental properties of metals, such as high electrical and thermal conductivities, optical opacity, reflectivity, ductility, and alloying properties. However, a realistic approach treats the free electrons in metal quantum mechanics and considers their wave-like properties [154].

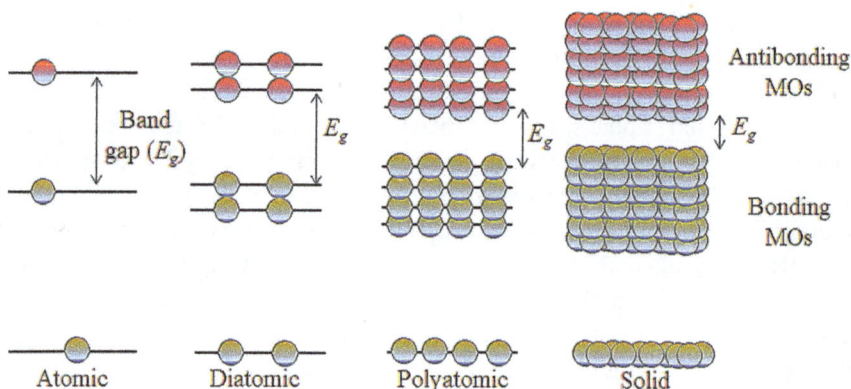

Fig. 1.32: The electron energy-level diagram for a progressively larger linear chain of atoms shows molecular orbitals' broadening into energy bands for a one-dimensional solid [154].

It is assumed that the free valence electrons are constrained within a potential well to prevent electrons from leaving the metal (the "particle-in-a-box" model) [155]. The box boundary conditions require the wave functions to vanish at the edges of the crystal (or "box"). The Schrödinger equation gives the allowed wave functions then correspond to certain wavelengths [156]. For a 1D box of length L (nm), the permitted wavelengths are $\lambda_n = 2L/n$, where $n = 1, 2, 3$ is the quantum number of the state; the permitted wave vectors $k_n = 2\pi/\lambda$ are given by $k_n = n\pi/L$. This simple particle-in-a-box model results in a set of wave functions, and the corresponding energy is given by eq. (1.6) and Fig. 1.33:

$$\Psi_n = (2/L)^{1/2} \sin\left(\frac{n\pi x}{L}\right) \tag{1.6A}$$

$$E_n = n^2 h^2/8mL^2 \tag{1.6B}$$

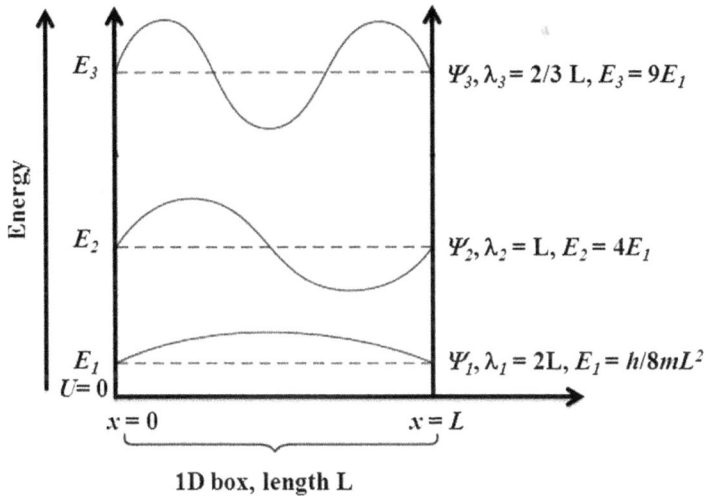

Fig. 1.33: The energy-level diagram also shows some of the allowed wave functions for an electron confined to a one-dimensional potential well [157].

It is commonly known that nanoscaled materials have a very high surface-area-to-volume ratios. The crystalline structure was also found to be different from the bulk structures. These two findings led to an extraordinary increase in chemical reactivity, selectivity, and efficiency of chemical reactions. Recently, nanocatalysis has drawn significant attention by using finely divided nanoscale systems since this processing can solve environmental pollution, as discussed above [156]. Furthermore, nanoscale catalytic supports with tunable internal and external porosity can selectively speed up the chemical reactions based on their dimension and functionality. Importantly, nanoparticles usually display unique intrinsic chemistries, differentiating them from their particulate counterparts sized at the micro- or macrolevel. Recently, many newly developed medicines have become more water-soluble and have high bicompatibility compared with micron-sized counterparts. Applying bifunctional synthesis methodologies will open a new paradigm for nanomaterial application [158].

1.4.2 Magnetic properties

Magnetism is a class of physical phenomena resulting from the external magnetic forces applied to other materials, particularly magnetic materials. Magnetism originated from electric currents by the atom rotation. When the external magnet is applied, the magnetic moments of the substance will be reorientated to align along the direction of an external field [159]. Due to the state of the electron pairing, we can distinguish the material into paramagnetism and diamagnetism. A substance with one or more unpaired electrons exhibits paramagnetism. An external magnetic field will attract this material with paramagnetism [160]. On the other hand, the species with all electrons paired will repel the applied magnetic field. Therefore, diamagnetism can be observed. Figure 1.34A–C shows the change of apparent mass upon application of the magnetic fields. The nanoparticles that exhibit magnetic properties have extensive applications in catalysis, biotechnology, biomedicine, reverberation imaging, ecological remediation, and storing information [161]. Magnetic nanoparticles can be created by unionizing suitable systems of attractive species with different structures [162].

This chapter investigated iron oxide nanoparticles (Fe_3O_4 nanoparticles) with ferromagnetic properties [163]. Those materials are intrinsically magnetically ordered and develop spontaneous magnetization even though an external magnetic field is not applied. The ordering mechanism is the quantum mechanical exchange interaction [164]. This study focuses on the synthesis of these Fe_3O_4 magnetic nanoparticles. One of the main concerns in producing magnetic particles is size. Magnetic particles with ultrafine size tend to agglomerate, which will lower their vitality because of their high surface tension. However, this vitality is directly linked with these nanoparticles' high accessible surface region and volume degree. On the other side, these metallic or magnetic nanoparticles are synthetically very dynamic and can be oxidized in the air resulting in loss of magnetic property and dispersibility [165]. To widen the applications of these magnetic nanoparticles, the author treated these particles with natural product extracts, which behave like surfactants, polymers, and other inorganic layers (such as silica or carbon). This treatment also prevents the nanoparticles from debasement and protects them from losing attractive properties. It is also an important outer layer, or the shell of these nanoparticles must be chosen based on the functions, and the applications as the shell can be made from different ligands [166]. The functionalized nanoparticles can be used in diversified fields, as mentioned above. This functionalization enhances high scattering, high reactivity, and simple detachments. Now, these magnetic nanomaterials should be blended using different techniques which do not involve oxidation and corrosive disintegration, as the security of the particles is essential [167]. The Fe_3O_4 magnetic nanoparticles functionalized by natural plants were found to have the potential to imitate wound healing.

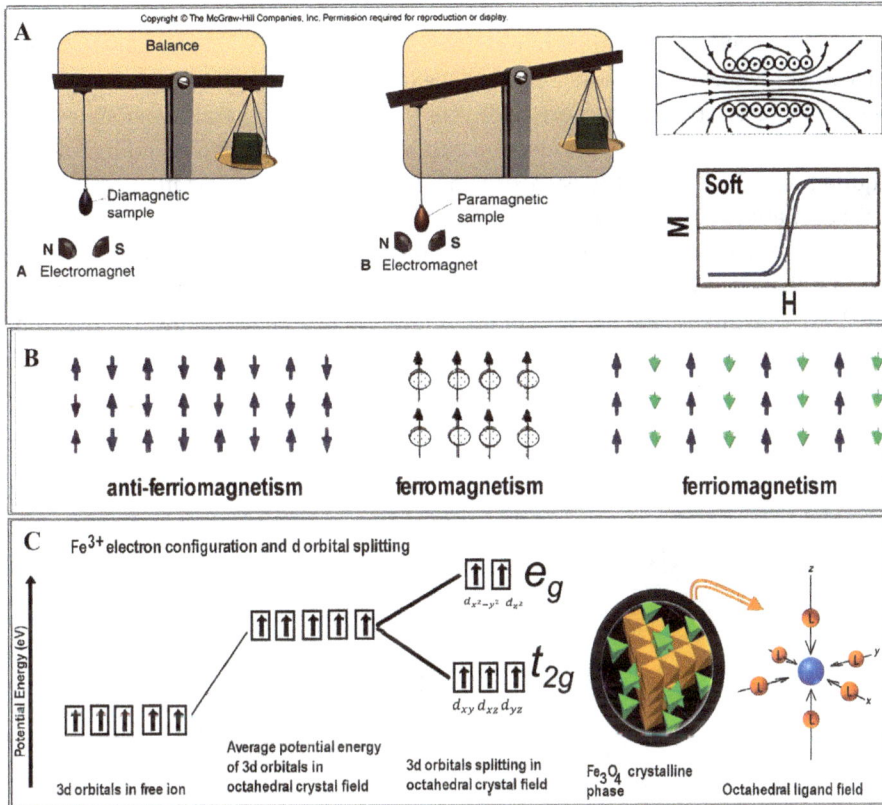

Fig. 1.34: A demonstration of magnetic property: (A) apparent mass differences to determine diamagnetic and paramagnetic properties; (B) the types of ferromagnetism; and (C) the d-orbital splitting, resulting in high spin contributing to the magnetism (Fe^{3+} is used as an example) [154].

1.4.3 Electro-optical properties

The Bohr model of the atom depicts the atom as a small, positively charged nucleus surrounded by electrons that travel in circular orbits around the nucleus, similar in structure to the solar system, using the hydrogen atom as a model. Nuclei provide electrostatic attraction to the electrons within the atoms rather than gravity. The Bohr model suggests that only certain electron orbits or specific energy or quanta shells are allowed. The shells in a Bohr-type atom occupy 3D space according to the quantum number, n (not visible), to the corresponding spectroscopic labels K, L, M, and N (visible), where the smallest shell closest to the nuclei has a quantum number of 1 ($n = 1$), corresponding to the K transition between the S and the next orbital. Similarly, the successive orbitals, labeled by n as 2, 3, 4 [168], are used to describe chemical stability across the periodic table in terms of half-full or complete shells. Each Bohr's shell can contain $2\,n^2$ electrons, starting with the K shell ($n=1$), which can contain two electrons, whereas other shells occupying large volumes can accommodate more electrons. The next shell, the L shell ($n=2$), can accommodate eight electrons. Once more than one electron can be occupied, additional information is required to uniquely identify each system, such as angular momentum, spin state, and occupancy, which are quantized of differing energies relative to the vacuum level (zero of the energy scale) that represents the potential energy of a free electron far from the atom [169]. In order to correspond with atomic emission spectra measured experimentally, the energies of these levels E_n are negative (i.e., the electrons are bound to the atom) and are proportional to $1/n^2$, summarized in Fig. 1.35.

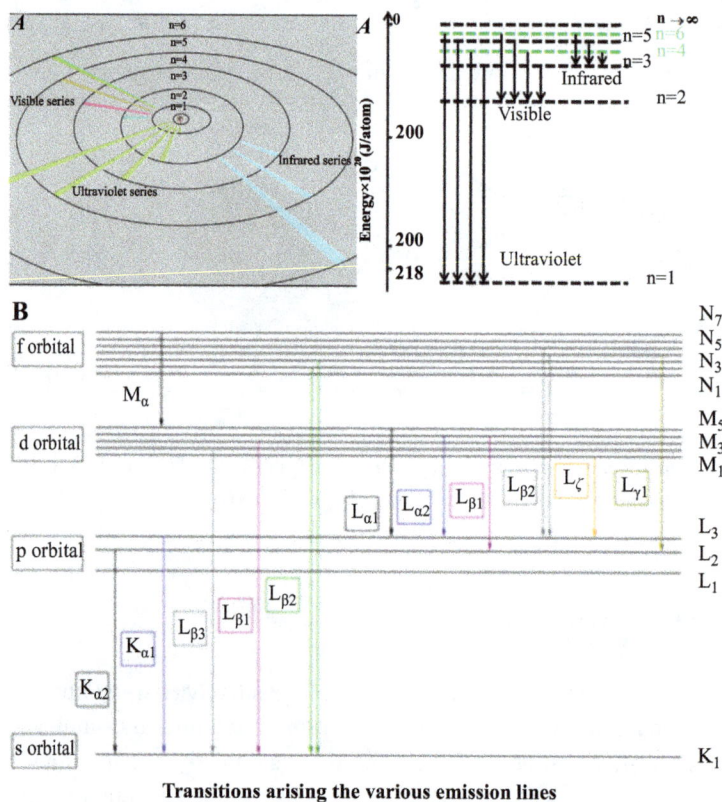

Fig. 1.35: (A) Bohr model depicting the electron distribution around the nuclei and energy distribution of different orbitals; and (B) demonstration of transitions arising from the various emission lines [154].

Electron occupancy is at times described in discrete quanta with the hydrogen atom, although experimental data also suggest that electrons have many wave-like characteristics due to their subatomic size. The wave nature of the electrons and the dearth of the scattering electron are correlated with the electronic properties of nanoparticles. These electronic properties change when particle size and shape increase or decrease. Three-dimensional particles possess complete discrete energies (electrons are confined in all dimensions). Normally, this phenomenon will be observed if the particle sizes are comparable to de Broglie's wavelength of electrons [170]. Conductors become insulators below a critical size length due to the separation of conduction and valence bands. When the system length scale is reduced, changes in electronic properties are related mainly to the increasing influence of the wave-like property of the electrons and the scarcity of scattering centers based on quantum mechanical effects [170]. When the particle size is comparable with the de Broglie wavelength ($\lambda = h/mv$, where h represents Planck's constant, m is the mass of the particle, and v is the velocity) of the electrons, the apparent differences between the energy states can be observed, where the electrons are confined in all 3D [171]. The electrons can move freely between adjacent nanostructures, defined as the quantum tunneling effect, since the barriers are lower than expected under classical mechanics of larger bodies, enabling movement. If an external voltage is applied to the nanosystem, the discrete energy levels can thus be aligned. Therefore, the resonant tunneling effect occurs, causing a rapid increase in the tunneling current [172] utilized in the scanning tunneling microscope.

Electrons are highly confined in all dimensions within 0D materials (e.g., quantum dots). In this case, the conduction is highly sensitive to the presence of other charge carriers. As a result, the charge state is also sensitive to those carriers. These phenomena are known as "Coulomb blockade effects." In this case, conduction mainly involves single electrons. It requires only a small amount of energy to overcome the energy barrier from valence and conduction bands, allowing for the operation of a switch, transistor, or memory element. Based on the effects, we can develop different devices and systems used in electronics and optoelectronics. Recent research indicated that these new systems could open a new approach to process information using resonant tunneling and single-electron transistors [173].

For a large-scale material, the density of states is proportional to $E^{1/2}$. The dimensionality of the system decreases will highly affect the density of states (Fig. 1.36). This observation results from the associated "reduction of degrees of freedom in wave vector space." In terms of purely dimensional space, 3D, 2D, and 1D systems, the density of states is proportional to k^{n-2}, where E is the state's energy, k is the constant, and n is the number of dimensions in the presence of the nanosystem. For parabolic bands, the density of states is dependent on $E^{(n-2)/2}$ [174]. For a quasi-low-dimensional system, the density of states takes the $E^{(n-2)/2}$ dependence in each subband. These quasi-low-dimensional materials include semiconductor quantum wells or quantum wire, in which a series of confined subbands is formed. There is no continuous distribution of states for a quantum dot due to the confinement in all dimensions. The density of states takes the form of a spectrum of discrete energy values, similar to that found for individual atoms [175].

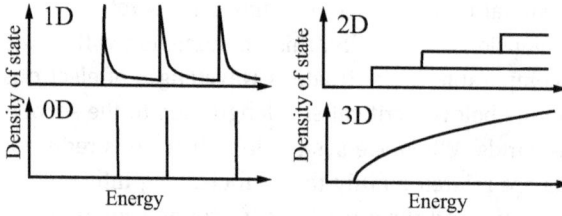

Fig. 1.36: The electronic density of states for a bulk semiconductor (3D), a quantum well (2D), a quantum wire (1D), and a quantum dot (0D) [176].

It was found that the density of states highly affects the optical and electronic properties of the majority of nanostructured materials. Particularly, nanoscaled semiconductors exhibit a strong dependence on their dimensionality. For example, in terms of electrical transport, the available number of states is an important factor in determining the density of states [177]. It was also found that the strength of optical transition is proportional to the density of states. This density is correlated with the initial point in the valence band and the final points in the conduction band, creating a joint density of states. The adsorption is found to be corresponding to the joint density, indicating that the nanostructured materials display different optical and electronic properties [178].

1.4.4 Thermal properties

Thermal conductivity: The theory of heat conduction by lattice waves in solid materials provides insight into understanding how conductivity is influenced by lattice defects (such as point defects and stacking fault), grain boundaries, and extended imperfections. At high temperatures, conduction by electromagnetic radiation within the solid can be controlled. The thermal conductivity (κ, W/K) by mobile carriers, which can be waves or particles, can be expressed as follows:

$$\kappa = \tfrac{1}{3}\, C v \iota \tag{1.4}$$

where C is their specific heat per unit volume (J/m^3 K), v represents the speed of mobile carriers (m/s), and ι is their mean free path (nm). Assume that the carriers result from waves (lattice waves or electromagnetic waves), ranging over a spectrum of frequencies f. This frequency can be generalized as follows:

$$\kappa = \tfrac{1}{3} \int C(f) v l\,(f) d(f) v \tag{1.5}$$

where $C(f)df$ is the contribution to the specific heat per unit volume from waves in the frequency interval df, v is the group velocity of the waves, and $l(f)$ is their attenuation length, usually a function of frequency. In eq. (1.5), f_m denotes an effective upper limit to the spectrum because either v or l becomes very small for f/f_m. The f_m is the upper frequency limit of the acoustic branch. The energy content of waves consists of quanta, phonons, or photons, respectively. In addition, these quanta can be considered as particles to carry and transfer heat. The present treatment emphasizes the wave nature of the carriers of heat [179].

Lattice waves: Lattice waves are considered either elastic or ultrasonic waves. However, their spectrum can be extended to very high frequencies, f. In this case, their wavelength λ corresponds to atomic dimensions. The relationship between frequencies (f, Hz) and the wavelength (λ, m) must be modified.

At low frequencies, the frequencies are inversely proportional to the wavelength: $\lambda = v/f$. However, at high frequencies, the above relation is no longer applicable due to the discrete atomic structure of the lattice. The displacement field of the wave has the form of a progressive wave; therefore, the displacement $u_{(r,t)}$ varies with position r and time t, as follows:

$$\mu_{(r,t)} \propto \exp\left(i\,q\,r - i\,\omega\,t\right) \qquad (1.6)$$

where the wave vector (q) has magnitude $2\pi/\lambda$ Moreover, points in the propagation direction, the angular frequency (ω), can be expressed by $\omega = 2\pi f$.

The angular frequency (ω) is no longer proportional to the wave vector (q) at high frequencies. The schematic relation between frequency and wave vector for a discrete lattice is shown in Fig. 1.5 [180].

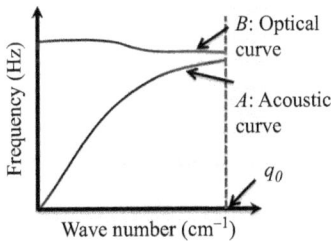

Fig. 1.37: Dependence of angular frequency (ω) on the wave vector (q) for a discrete lattice in the principal direction. (A) The acoustic branch is the only type in simple lattices, and (B) an optical branch is present in complex crystals, describing the vibration of atoms within molecular units [181].

The slope of the curve (shown in Fig. 1.37) defines the group velocity, which is related to the angular frequencies and wave vector: $v = d\omega/dq$. This relationship indicated that the wave speed transports energy. The Debye theory of lattice vibrations states that v can be assumed to be a constant but that the lattice wave spectrum is terminated at the Debye frequency f_D. This frequency is chosen so that the number of normal modes agrees with the number of atoms. On the other hand, the spectrum can be divided into acoustic modes of the advanced waveform (shown in curve A in Fig. 1.37) if the basic structural unit contains N atoms of varying mass. The spectrum can be divided into optical modes (shown in curve B in Fig. 1.37) if the vibrations of N atoms in each structural unit are relative to each other. With upper frequency $f_m = f_D/N^{1/3}$, only the former has appreciable group velocity and transports heat. This approximation is justified and can explain the relationship between frequencies and the number of atoms [182].

In the limit, $T > hf_m/k$ is another interesting case. Under this circumstance, each wave or normal mode reaches thermal equilibrium. The energy equals kT and C $(T) \propto f^2$. Here, k and h are the Boltzmann (m^2 kg/s^{2}K) and Planck (m^2 kg/s) constants. Figure 1.38 demonstrates a difference between acoustic and optic vibrations schematically. The optic vibrations result from the motion of the atoms in the lattice. The term "optic" refers to their strong interaction with infrared electromagnetic waves in the case of ionic solids [183].

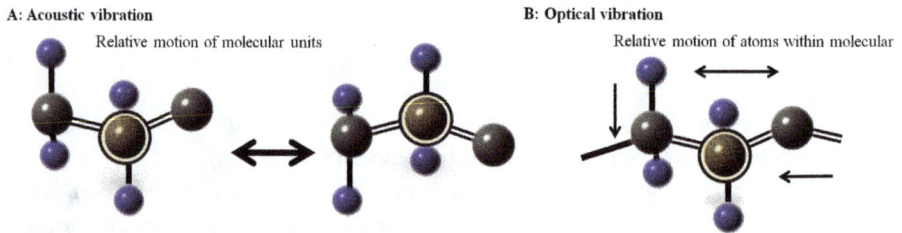

A: Acoustic vibration

Relative motion of molecular units

B: Optical vibration

Relative motion of atoms within molecular

Fig. 1.38: In crystals composed of molecular units, such as ceramic oxides, the lattice waves are composed of (A) acoustic vibration and (B) optical vibration [181].

Interaction processes: In an ideally perfect crystal, it can be assumed harmonic, structurally perfect, and without external boundaries, the lattice waves would be normal modes. This approximation indicates that the energy content of each wave would remain constant. Thermal equilibrium between the waves could not be established [184]. However, real solids were found to have deviated from this ideal behavior. The lattice forces have anharmonic components, indicating the linear elasticity departed. It is also known as strain energy, which is cubic in strain. The lattice defects always exist in real solids, and so are boundaries. These defects cause some energy exchange between the waves. This exchange results in the deviation of true normal from ideal modes. The increased number of normal modes results in insufficient randomness in these interactions, leading to thermal equilibrium. In order to understand their thermal equilibrium properties, we anticipate to know the existence of interactions between atoms and molecules. The in-depth understanding of these interactions will provide insight into transport properties and estimation of the attenuation lengths.

Thermal conductivity is limited by various interaction processes, which transfer energy between the waves and establish thermal equilibrium. Figure 1.39 illustrates a schematic of the interaction process, which corresponds to the different types of departures from ideal behavior. The intrinsic processes cause energy exchange between triplets of waves (also known as three-phonon interactions). The energy exchange maintains frequency conservation, as in all nonlinear processes, and wave vector selection rules. The resulting intrinsic attenuation length is expressed as follows:

$$l_i(f, T) = Bf^{-2}T^{-1} \tag{1.7}$$

where $B \propto \mu \alpha^3 v f m$, with μ being the shear modulus (Pa) and α^3 the volume per atom (nm^3). It is thus inversely proportional to the mean square thermal strain $\kappa T/\mu \alpha^3$. The intrinsic conductivity (W/m K) is then expressed as follows:

$$\kappa_i = \frac{{}^3/_4 N^{-\frac{2}{3}} \mu v^2}{f_D T} \tag{1.8}$$

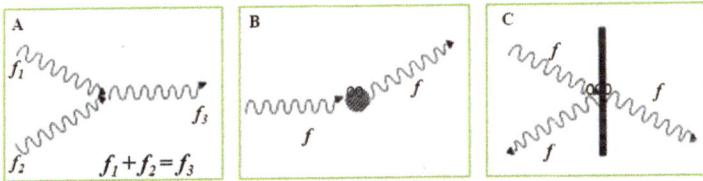

Fig. 1.39: Interaction processes between lattice waves: (A) energy exchange between wave triplets due to anharmonic interactions; (B) scattering of waves by lattice defects; and (C) reflection of waves by boundaries and interfaces [181].

Equation (1.8) well expresses the thermal conductivity of structurally perfect dielectric crystals near and above their Debye temperature $h f_D/k$. The factors, such as intermolecular forces, large atomic masses, and structural complexity (large N), tend to reduce the intrinsic thermal conductivity.

Note that κ can be expressed as $\kappa = (f) \, df$, indicating the integrand $\kappa_i \propto C(f) l_i (f)$ is independent of f in the intrinsic case. Therefore, equal frequency intervals will equally contribute to the intrinsic conductivity [182]. The thermal conductivity kappa contrasts to the specific heat, which for acoustic modes is mainly due to their highest frequencies, since for $f < f_m$, $C(f) \propto f^2$. This observation is shown schematically in Fig. 1.40.

Furthermore, the lattice imperfections will reduce the effective attenuation length. Different imperfections will result in different frequency dependence, accordingly. It was found that the point defects scatter as the fourth power of frequency, with an inverse attenuation length of $1/l_p(f) = A \, f^4$, where A depends on their nature and concentration. The grain boundaries were found to scatter independently of the frequency with attenuation length L comparable to the grain size.

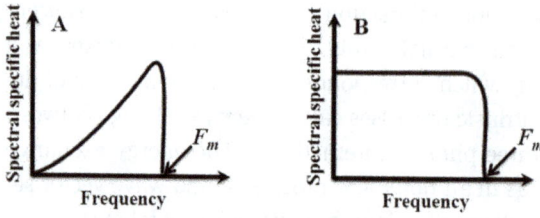

Fig. 1.40: Spectral contributions from lattice waves as a function of frequency: (A) the vibrational specific heat as a function of frequency, and (B) the intrinsic thermal conductivity as a function of frequency [181].

The inverse attenuation length $1/l(f)$ is composed of the sum of these processes. Since point defects and grain boundaries scatter mainly in different frequency ranges, their conductivity reductions are approximately accumulative, as expressed in eq. (1.9) and is shown in Fig. 1.41:

$$\kappa = \delta_{\kappa l} - \delta_{\kappa B} - \delta_{\kappa P} \tag{1.9}$$

where $\delta \kappa_B$ and $\delta \kappa_P$ are the reductions due to boundaries and point defects. Point defects are small volume regions α^3 in which the value of the wave velocity v is locally changed by δ_v. Since the intrinsic attenuation length of (f, T) has a different dependence on wave frequency f than that due to point defects, it is convenient to define a frequency f_0 at which the two attenuation lengths are equal, that is, $l_i (f_0, T) = l_p(f_0)$. Similarly, one can define a frequency f_B such that $l_i(f_B, T) = L$, where L is the grain diameter:

$$\delta_{\kappa_B} = \kappa_i \left(\frac{f_B}{f_m}\right) \arctan\left(\frac{f_m}{f_B}\right) \tag{1.9A}$$

$$\delta_{\kappa P} = \kappa_i \left[1 - \left(\frac{f_0}{f_m}\right) \arctan\left(\frac{f_m}{f_0}\right)\right] \tag{1.9B}$$

Note that the equation for $\delta \kappa_P$ will be applicable if $f_0 > f_m$.

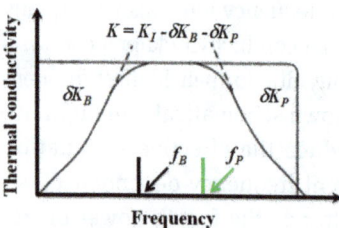

Fig. 1.41: Reduction in thermal conductivity, which is the area under the curve, at high frequencies due to point defects ($\delta \kappa_P$) and at low frequencies due to grain boundaries and other extended imperfections ($\delta \kappa_B$) [181].

1.5 Applications of nanomaterials

When grain sizes of materials decrease in a billionth (10^{-9}) of a meter, they are called nanomaterials or nanocrystalline materials. These nanomaterials exhibit very attractive and useful properties, as discussed in Section 1.4. Due to their unique properties, nanomaterials are applied in biomedicine, catalysis, cosmetics, electronics, energy storage, aerospace components, and textiles. In this section, five categories will be discussed to emphasize nanomaterials used for plastic waste remediation (Section 1.5.1), synergistic combination cancer immunotherapy (Section 1.5.2), revolutionization of eye care (Section 1.5.3), treatment of traumatic injuries (Section 1.5.4), and shale gas storage (Section 1.5.5).

1.5.1 Nanomaterials used for plastic waste conversion

Nowadays, the world faces two major problems: demand for energy and planet pollution due to increased anthropogenic activities. Massive production of plastics which is used to improve the quality of human life over the past century has resulted in global plastic pollution. Microplastics (MPs) (<5 mm) and nanoplastics (1–1,000 nm) are of special concern due to their harm to ecosystems. Several studies have confirmed MP contamination in agroecosystems. A recent study shows that MPs with sizes up to 2 μm are found in edible plants, including vegetables. The Lower Rio Grande Valley in South Texas is heavily dependent upon irrigation water delivered from the Rio Grande River to support the production of staple crops. No studies have been conducted to investigate MPs present in the irrigation water and the crops.

Their high persistent characteristics of Micro nanoplastics (MNPs) and release of chemicals/additives have been posing cascading impacts on living organisms across the globe. Natural connectivity of all the environmental compartments leads to migration/dispersion of MNPs from one sector to another, such as irrigation water to farmland soils. The quantification of the dispersion of MNPs across the environmental compartments will provide a guideline for the remediation of plastic pollutants and their conversion to useful materials and fuel supplies [185]. This activity will evaluate the properties, distribution, and remediation of MNPs collected from soil compartments along the Lower Rio Grande Valley. The morphology can be determined using transmission electron microscopy (TEM) or SEM. The high-resolution images will be obtained to determine the surface morphology, fine internal structures, particle shape, and size. The chemical constituents, function groups, and their bonding forces will be analyzed using X-ray photoelectron spectroscopy (XPS), Fourier-transform infrared spectroscopy, and nuclear magnetic resonance (NMR). The proposed strategy to convert these MNPs in irrigation water into "gray" H_2 and graphene will be monitored using a hydrogen fuel cell model car through its electrochemical performance. The evaluation of PE films and Nylon fibers will be discussed further. The chemical congeners/additives (dioxins and polycyclic aromatic hydrocarbons) and endogenetic chemicals in MNPs will be evaluated using similar approaches. These chemicals consist of hetero- and homo-monomeric units. Representative chemicals include bis(2-ethylhexyl) phthalate, polybrominated diphenyl ethers, bisphenol A, and triclosan.

Area selection and sample collection: Ten farmlands along the Lower Rio Grande Valley in South Texas will be investigated and distributed among Brownsville, Harlingen, Weslaco, Pharr, McAllen, Edinburg, Mission, San Juan, and Rio Grande City metropolitan areas (Fig. 1.42A). The farmland soil texture type and crops (corn, fruits, and cotton) will be used for survey purposes. The samples with a point size of 0.3×0.3 m^2 were collected from June to August each year to determine the total mean abundance of microplastics, mesoplastics, and nanoplastics. Firstly, crops and large residues (>5 cm) were removed from the sampling area's soil surface. The depth of the soil surface was controlled by 0–50 cm. A sample of 500 g was collected from different layers from nine different locations. The sample size varied from 200 to 300 and was used for characterization. Saturated sodium bromide aqueous solution will be applied to extract MNPs. The MNP filtrates were digested using a 30 % H$_2$O$_2$ solution for purification. The abundance of MNPs was determined among selected sampling sites in Lower Rio Grande Valley and decreased from surface to deep soil layers. The procedure of MNP purification in different depths is shown in Fig. 1.42B. The team will establish a protocol to collect and measure MNPs along Lower Rio Grande Valley through this activity.

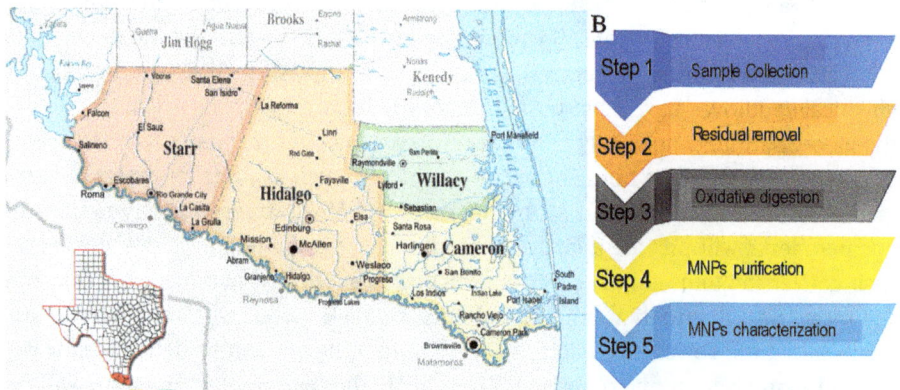

Fig. 1.42: The sample collection and purification: (A) geographic sites of MNPs along the Lower Rio Grande Valley (image is adapted from [186]); and (B) a procedure of MNP sampling and characterization.

Constituents' evaluation: The purified MNPs were measured by micro-FTIR to determine the absorption or transmittance. The spectrum wavenumber was controlled from 675 to 4,000 cm^{-1}, with a collection time of 3 s for each measurement. The spectra database was used to determine the type of MNP pollutants further. The NMR was collected to determine chemical shift and spin–spin coupling, further identifying the chemical structures of MNPs. The ChemDraw Professional was used to make structure-spectrum assignments. The reported FTIR and ^{1}H estimated NMR spectra of PE (low Mw) and polyamide (Nylon 6/6 Pellets) are shown as an illustration (Fig. 1.43A and B). The abundance of MNPs was calculated by mass $= \rho \times V$. The reported abundance of MNPs by Luo et al. is shown in Fig. 1.43C. This activity will establish the MNP's mass density and abundance dependence on different sample locations and soil layers.

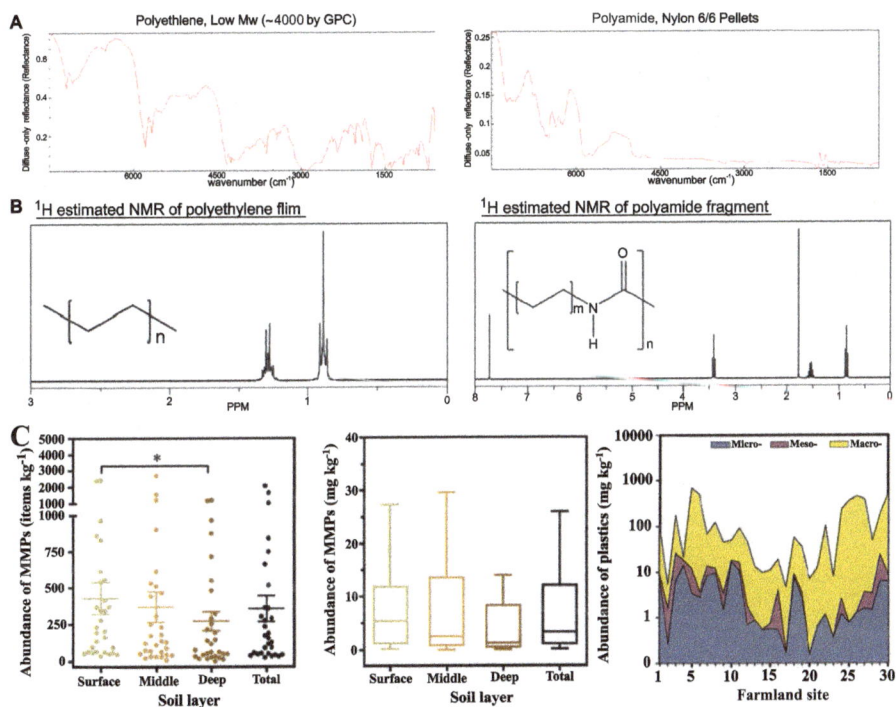

Fig. 1.43: The constituent's measurement of MNPs, polyethylene, and polyamide: (A) the Fourier-transform infrared spectra (data were adapted from NIST, https://webbook.nist.gov/); (B) the ChemNMR ^{1}H estimation; and (C) The distribution of MNPs in the soil layer by count (left), by mass (middle) and by distribution on farmland (right hand side) (adapted from [187]).

Morphology characterization: TEM and SEM will be used to obtain images of MNPs, for example, Nylon 6 fabrics. Using a carbon-coated copper grid, an electrospinning technique was applied to a substrate for the Nylon fiber deposition. The surface topography can be viewed through surface images and fine internal structures. The morphology and diameter of Nylon 6 nanofibers examined by Rajendran et al. are shown in Fig. 1.44A and B [188], demonstrating the homogeneity of nanofibers. The image and particle distribution will provide a guide for MNP remediation. Using X-ray photo-electron spectroscopy (XPS) and Energy dispersive X-Ray (EDX), the elemental composition can be further quantified or semiquantified, as shown in Fig. 1.44C and D. In this activity, the C_{1s} emission can reflect the elemental content of these collected MNPs, while the O- and N-elemental contents can also be quantified. In addition, the functional groups deposited onto the surface MNPs can also be estimated to evaluate the strength of relevant chemical bonds and the new functional groups.

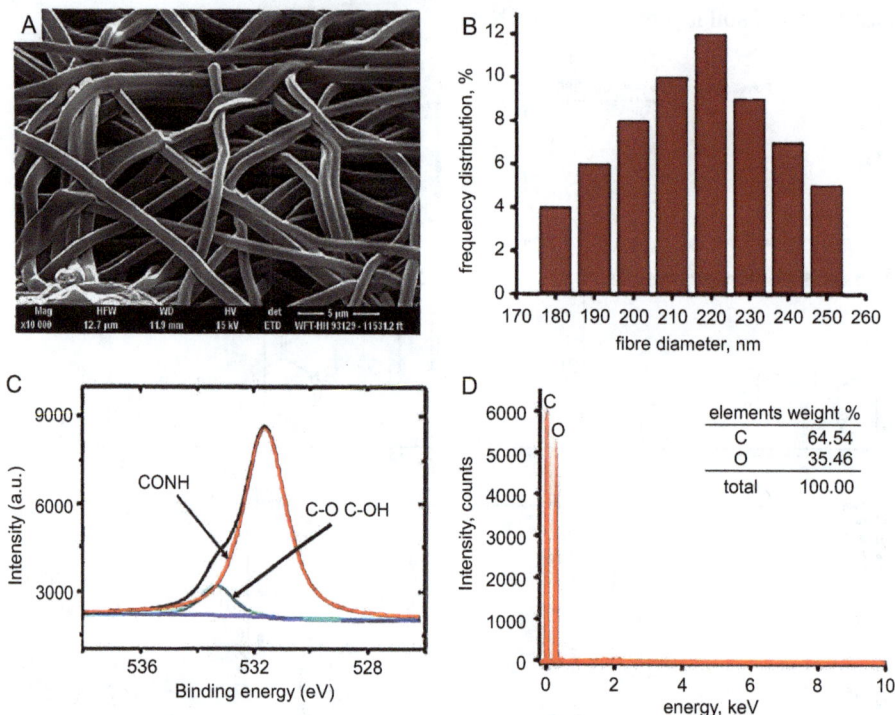

Fig. 1.44: The nanostructural characterization and elemental composition of MNPs (Nylon 6 fiber as an example): (A) the TEM image; (B) the distribution of Nylon fibers; (C) the XPS elemental analysis; (D) the EDX analysis of the above fiber (data in A, B, and D were adapted from [188] and C from [189]).

Remediation strategy: It has been reported that PE and polypropene into hydrogen and porous carbon materials (conductive carbon nanosheets) [190]. This research will apply metal oxide (V_2O_5–MnO_2) composites as catalysts to convert PE into graphene through three steps: dissolution of PE into toluene, MW-assisted catalytical pyrolysis, followed by carbonization (Fig. 1.45A). C–H bond cleavage is proposed to release hydrogen atoms during carbonization. This atomic hydrogen will be combined to form hydrogen gas (H_2), defined as gray H_2 (GH_2). The so-called G-H_2 can be used as fuel supplies and graphene as the electrode catalyst to improve the electrochemical performance of microbial fuel cells (Fig. 1.45B).

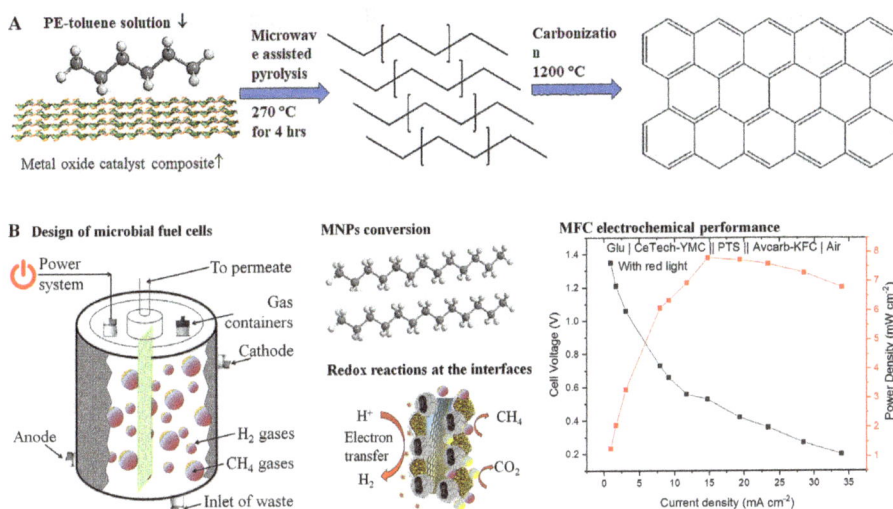

Fig. 1.45: The nonstructural characterization and elemental composition of MNPs (Nylon 6 fiber as an example): (A) the TEM image; (B) the distribution of Nylon fibers; (C) the XPS elemental analysis; (D) the EDX analysis of the above fiber (data in A, B, and D were adapted from [188] and C from [189]).

1.5.2 Nanomaterial-based synergistic combination cancer immunotherapy

Although conventional treatments (e.g., chemotherapy, radiotherapy, and surgery) play critical roles in tumor therapy, cancer immunotherapy becomes attritive due to its advantages to "achieving precision medicine" and "preventing recurrence and metastasis" [191]. The new therapeutic immunotherapy has been used to fight cancer by harnessing the immune system in vivo antibodies generated by the patient [192]. Nanomaterials provide effective multiple treatments for immunotherapy to enhance patients' outcomes [193]. The bioengineered nanomaterials improved the treatment efficacy using combined cancer treatments. This synergistic "cancer combination immunotherapy" showed the potential to stimulate the design strategies in versatile use for clinical application transformation.

Principle for immune therapy: Nobel Laureate in Medicine, Dr. James P. Allison, from the University of California, Berkeley, studied the cytotoxic T-lymphocyte-associated protein 4 (CTLA-4). He found that "CTLA-4 functions as a brake on T cells" and developed an antibody binding with CTLA-4 to block its function. Allison and coworkers researched mice with cancer and treated them with the "antibodies that inhibit the brake and unlock antitumor T-cell activity" [194]. The team continued their activities to "develop the strategy into a therapy for humans." In 2010, a clinical study indicated that patients with skin cancer (melanoma) made significant improvements. Their mechanism (Fig. 1.46) indicated that the "activation of T cells requires that the T-cell receptor binds to structures on other immune cells recognized as *non-self*." In addition, a protein functioning as a T-cell accelerator is also required to activate T cells. Their results suggested that CTLA-4 functions as a brake on T cells, inhibiting the accelerator.

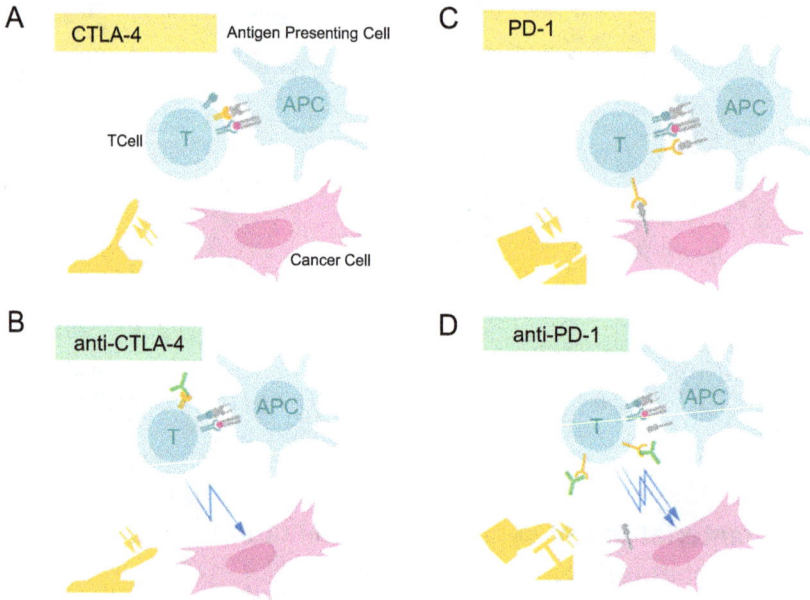

Fig. 1.46: Mechanism of antibody binding to CTLA-4 and blocking its function: (A) the activation of T cells; (B) antibodies (green) against CTLA-4 block the function of the brake leading to activation of T cells and attack on cancer cells; (C) PD-1 is another T-cell brake that inhibits T-cell activation; and (D) antibodies against PD-1 inhibit the function of the brake, leading to activation of T cells and highly efficient attack on cancer cells [195].

Classification of cancer immunotherapy: In immunotherapy, the patient's immune system has been used to trigger the antitumor response, further providing personalized vaccines [196]. Cancer immunotherapy includes various treatment modalities (Fig. 1.47A), lowering the side effect by targeting cancer cells and regulating the immune system. With its development, immunotherapy strategies were used to harness the antitumor immune response available for the routine management of cancer (Fig. 1.47B). Nowadays, this immunotherapy enables the rapid mapping of the mutations within a genome, rational selection of vaccine targets, and on-demand production of a therapy customized to an individual patient's tumor. Since its inception, enhancement immunotherapies have become more and more attractive to activate and increase the immune response [196]. However, cancers have immune evasion mechanisms to delay, weaken, and even prevent antitumor immunity. The principle of blocking inhibitory surface receptors emphasizes that the immune system is normal and stable. In addition, newly developed immunomodulation strategies that aim to reverse the tumor microenvironment are known as vaccine/adjuvant strategies [197]. Figure 1.47A summarizes the common anticancer therapeutic modalities, including chemotherapy, high-intensity focused ultrasound, photodynamic therapy, photochemotherapy, therapy radiotherapy, and gene therapy. Nanomaterials can potentially combine these therapeutic strategies with this new immunotherapy in a tumor microenvironment (TEM). A combinatorial approach is used to design, develop, and deploy effective treatments to regulate the immune response and destroy active tumors without comprising the body's innate defense. Five typical immunotherapies are used to treat cancer (Fig. 1.47B). These immunotherapy modalities include cytokines, blockade at checkpoint sites, adoptive T-cell and cell transfer, and antibodies and antigen interactions, both monoclonal and polyclonal. However, they often induce greater side effects. Several small signaling molecules, such as cytokines, are used by cells. These include chemokines, interferons, tumor necrosis factor-alpha, and other related molecules such as lymphokines [198] which initiate a cellular response with interferons and interleukins stimulating the immune response (Fig. 1.47B-a). These innate and programmed responses can be strengthened by applying therapeutic vaccines, which assist with the programmed response by bolstering the immune response to certain tumors (Fig. 1.47B-b) [199].

In an antigen-initiated infection, the immune system generates antibodies, which take time. A method to accelerate the immune response is adoptive cell transfer (ACT), which exhibits specific antigenic epitopes and can enable the biosynthesis of antibodies to be accelerated, with acute coronary syndromes (ACS) being from the patient (e.g., liver cells) or from a different patient with the same immune disease (Fig. 1.47B-c).

These ACTs exhibit "T cells with specific chimeric antigen receptors" and can be stored outside the body in media and transferred when required, assisting the overall immune response.

For traditional checkpoint inhibitors, the low selectivity may result in immune-related toxicities. The combination synergistic immunomodulation may be more efficient at blocking the interactions between negative regulators and T cells. Checkpoint blockade is used for self-tolerance to protect cells from attack by the immune system but can be tailored to recognize specific tumor cells which are attacked and destroyed (Fig. 1.47B-d). Monoclonal antibodies' function is to bind to specific antigens on the surface of cells and enable these cells to be better identified to the immune system for lysis and inactivation. Monoclonals enable a subset of surfaces to be tagged (Fig. 1.47B-e).

Nanomaterials for cancer immunotherapy: Nanomaterials with different formulations have been widely used in cancer diagnosis and therapy [200]. The side effect of nanomaterials plays a critical role in avoiding kidney clearance due to their capability to penetrate tumor tissues selectively. As a result, the drug loaded inside nanomaterials showed longer blood retention time, decreased toxicity, and improved

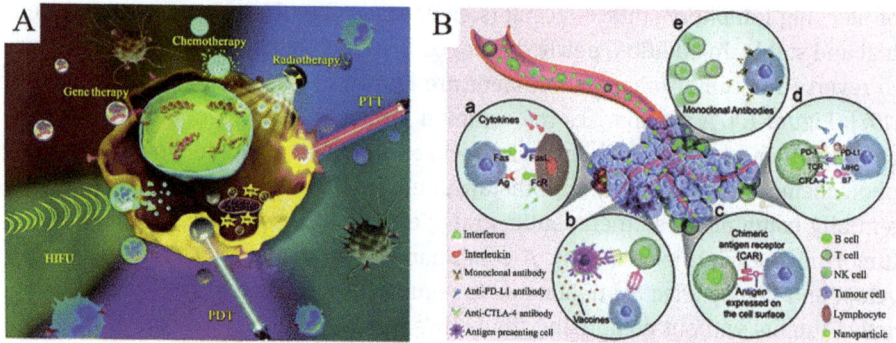

Fig. 1.47: Cancer treatment strategy and typical immunotherapies: (A) schematic diagram using common anticancer therapeutic strategies; and (B) five typical types of immunotherapies to treat cancer. They are cytokines (a), therapeutic vaccines (b), adoptive cell transfer (c), checkpoint blockade (d), and binding between monoclonal antibodies and tumor antigens (e) [199].

tumor distribution. Functionalization and modification of nanomaterials enabled their versatile application through varying surface chemistry to advance their uptake by targeted cells [201]. In addition, the nanomaterials can be used for delivering cytotoxic drugs or imaging agents to provide high-precision diagnosis and treatment efficacy [202]. The innovative pioneer research indicates that the application of nanomaterials in cancer immunotherapy provides a new platform for both cancer vaccination and immunosuppressive TME modulation (Fig. 1.48). When applied in cancer vaccine design, nanomaterials possess several advantages: coencapsulation of antigen and adjuvant, dendritic cell uptake and targeting, inherent adjuvant effect, lymph node drainage, and antigen presentation. Antigens are components of foreign cells, are composed of proteins and nucleic acids, and can be attached to the surface of nanoparticles and then injected into a site to initiate and strengthen the immune response. This enhancement can be achieved via innate immune potentiation or delivery of the antigen due to the specific physicochemical characteristics of nanomaterials.

Various nanomaterials have been designed to carry active substances, alter the formulations, and modulate the immunological tumor microenvironment (TME). Although nanomaterial-based TME modulation made significant progress, several limitations must be overcome before being applied in clinical trials. These problematic issues relate to limited tumor tissue penetration, immune toxicity, and tumor heterogeneity. The TME is highly tissue and cell-specific. It is often associated with specific tumor types of cells which target specific tumors, such as myeloid-derived suppressor cells (MDSCs), tumor-associated macrophages (TAMs), tumor-associated fibroblasts, and soluble small molecular mediators, which are used as target checkpoints. The nanomaterial can be modified with the appropriate antigen, drug, or ligand to promote cellular disruption of the extracellular matrix, plasma membrane tumor vascular systems to enable these tissues to become more porous for easier

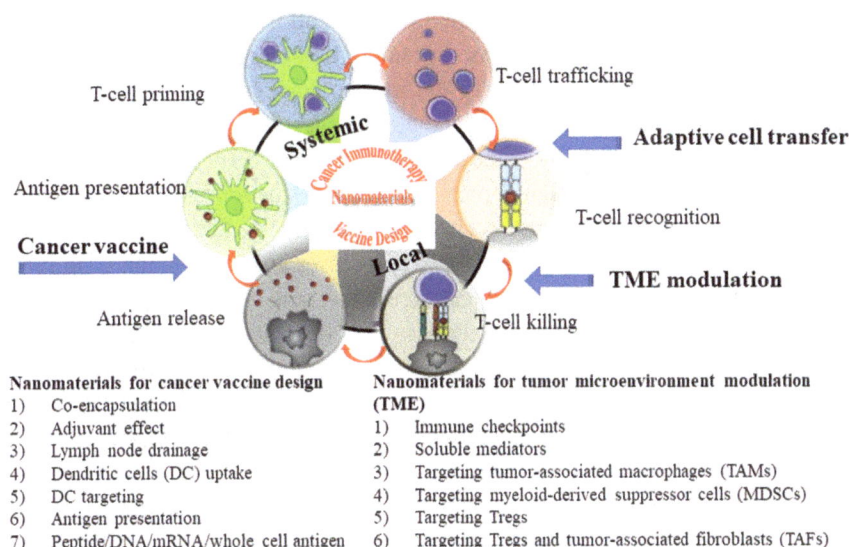

Nanomaterials for cancer vaccine design
1) Co-encapsulation
2) Adjuvant effect
3) Lymph node drainage
4) Dendritic cells (DC) uptake
5) DC targeting
6) Antigen presentation
7) Peptide/DNA/mRNA/whole cell antigen

Nanomaterials for tumor microenvironment modulation (TME)
1) Immune checkpoints
2) Soluble mediators
3) Targeting tumor-associated macrophages (TAMs)
4) Targeting myeloid-derived suppressor cells (MDSCs)
5) Targeting Tregs
6) Targeting Tregs and tumor-associated fibroblasts (TAFs)

Fig. 1.48: Nanomaterials for cancer vaccine design and tumor microenvironment modulation to balance the cancer–immunity cycle (modified from [203], with permission from Elsevier).

perfusion of drugs or other nanoparticles, and lower oxygen content to promote hypoxia, slow cell growth, and affect cell death. Modified nanoparticles have been shown to change the cell populations of cancer-associated fibroblasts by altering MDSCs and associated TAMs and regulatory T cells to inhibit cellular proliferation and promote cell death of tumor cells through these targeted approaches. Such approaches have resulted in the development of more effective chemotherapeutics and lesser adverse immune responses and relapses.

1.5.3 Nanomaterial used to revolutionize eye care

Due to the unique anatomy and physiology of the eye, efficient ocular drug delivery is a great challenge to researchers and pharmacologists [204]. Although conventional noninvasive and invasive treatments, such as eye drops, injections, and implants, are commonly used, the current treatments suffer from low bioavailability or severe adverse ocular effects. Alternatively, emerging nanoscience and nanotechnology play an important role in developing novel strategies for ocular disease therapy [205]. Various active molecules have been designed to associate with nanocarriers to overcome ocular barriers and intimately interact with specific ocular tissues. The human eye is a visualization spheroid organ about 24 mm in circumference (Fig. 1.49A). The eye consists of the anterior and posterior segments, including vitreous humor, which has various biological barriers to protect it from being harmed by foreign substances.

Nanotechnology provides a feasible, rapid, and straightforward eye-care treatment method due to its easy access to its exposed position. Nanodrugs must be bioadherent to prevent being rapidly cleared through the nasolacrimal system [206]. The small amounts of medication delivered to the ocular surface may reduce the risk of blurring, and there is the possibility that the nanoparticles may enter corneal cells prolonging contact time [207]. The Swiss-designed prototype contact lens sensor (Fig. 1.49B) contains strain gauges that monitor corneal curvature changes resulting from intraocular pressure variations. It can be seen that a microprocessor and an antenna were integrated into the soft contact. This contact lens sensor can collect wireless powering and communication. The composition of the nanodrug carriers must be biodegradable and compatible with securing the safety of nanoparticle therapeutics before their potential use. Various biodegradable artificial polymers and natural polymers (e.g., chitosan, gelatin, albumin, and sodium alginate) promise drug delivery. Piedad Calvo and coworkers studied the bioavailability of the indomethacin ($C_{19}H_{16}ClNO_4$)-based nanodrug. This nonsteroidal, anti-inflammatory drug was encapsulated in nanoparticles, which showed triple in vitro corneal penetration compared to a conventional solution [208]. Recently, the so-called nanoknives were fabricated from silicone to make "very small and precise cuts." This procedure avoided molecular and cellular damage to the surrounding tissue of the treated eye relative to a surgeon's blade [209]. The nanoblades with varying shapes are sized from 100 to several hundred nanometers. The shape of the "nanoknife" can be determined by applying individual instruments. The commercial version of the contact lens sensor called the Triggerfish (Sensimed AG, Lausanne, Switzerland) received the CE designation in 2009 and differed slightly from the earlier prototype (Fig. 1.49B) [210].

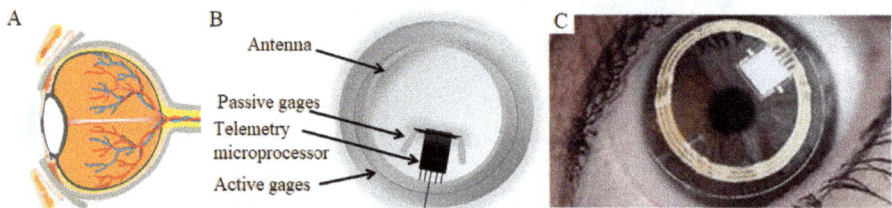

Fig. 1.49: Nanomaterials for eye care: (A) globular structure anatomy of a human eye; (B) the Swiss-designed prototype contact lens sensor contains strain gauges for the corneal curvature sensitive to intraocular pressure variations. A microprocessor and an antenna facilitate wireless communication to computer readout. (C) The commercial version of the contact lens sensor, the Triggerfish (Sensimed AG, Lausanne, Switzerland), received the CE designation in 2009 (modified from [210] with permission from Jobson Medical Information LLC).

Nanomaterials have a high surface-area-to-volume ratio. This attribution can be beneficial in reducing or eliminating reactive species, which are established as a cause of cataracts and other ocular diseases [211]. Nanoparticles are colloidal systems with several advantages over conventional delivery. They improve delivery efficiency and are particularly useful in chronic diseases like glaucoma, uveitis, and ocular neovascularization [212]. Molecules such as growth factors can be delivered via nanoparticles. It was found that numerous photoreceptor cells are subject to rapid death when exposed to bright light over a longer period. To visualize these damaged cells, Chen, McGinnis, and coworkers carried out experiments on animals killed 5 days after light exposure. Their eyes were processed for histological analysis using the TUNEL (terminal deoxynucleotidyl transferase dUTP nick end labeling) assay. This fluorescent assay reveals those cells that have been damaged by the light and have progressed along an apoptotic cell death pathway (Fig. 1.50). The results demonstrate

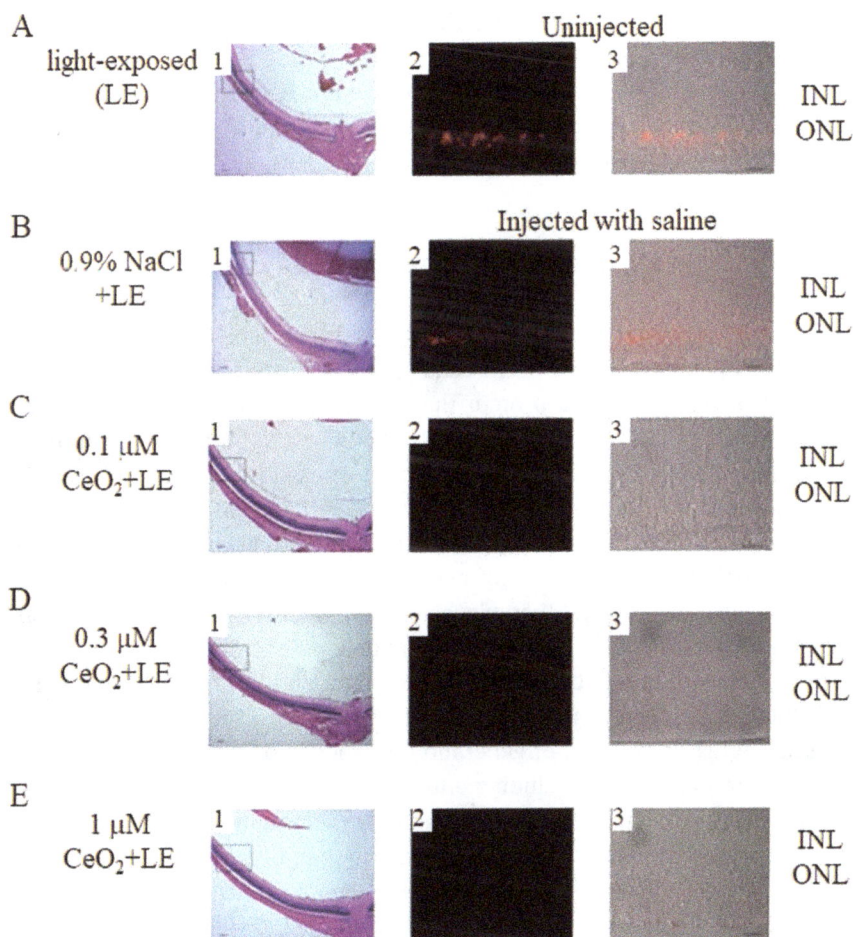

Fig. 1.50: Nanoceria particles prevent TUNEL-positive photoreceptor cells' appearance, which occurs days after exposure to damaging light, resulting in TUNEL-positive cells in those retinas that were either uninjected. Images were adapted with permission from [211].

that exposure to bright light results in TUNEL-positive cells in those retinas from eyes that were either uninfected (Fig. 1.50A-2 and A-3) or injected with saline (Fig. 1.50B-2 and B-3). Nanoparticles composed of CeO_2 with different concentrations (0.1, 0.3, 1.0 µM) were used to protect the retinas and prevent TUNEL-positive cells' appearance (Fig. 1.50C-2, D-2, and E-2) [211]. These data suggested that CeO_2 nanoparticles, even at lower concentrations, effectively inhibit light-induced injury to the photoreceptor cells due to prolonged exposure. The results of eyes without CeO_2 nanoparticle injection were shown in Fig. 1.50A-1 and B-1, while the injected eyes were evaluated at day 0, exposed to light on day 3, and the experiment ended on day 8 (Fig. 1.50 C-1, D-1, and E-1). The CeO_2 nanoparticles with a diameter of 5 nm showed the effective capability to scavenge reactive oxygen intermediates. The high surface-to-volume ratio of nanoparticles allows rapid radical neutralization regeneration without re-addition of the agent. Chen and colleagues have demonstrated that intravitreal injection with CeO_2 nanoparticles effectively prevents light damage in rodents. The nanotechnology invented by Chen demonstrated the potential to treat other ocular conditions related to oxidative damage, such as diabetic retinopathy.

1.5.4 Nanoparticles used to treat traumatic injuries

Our mother nature has created perfect nanoscaled arts of biology over the millennia. Naturally, the inner working of the cells occurs at the nanodimension, such as the diameter of hemoglobin, a protein that carries oxygen through the human body, is 5.5 nm (Fig. 1.51). The strand of the building block of life, DNA, is approximately 2.0 nm. The iron heme used for oxygen capture is the protein (Fig. 1.51B–D) for a quaternary alpha–beta motif. The oxygen binds the metal axially under high pressure. It is released under low oxygen tension in the tissues. The natural nanoscale of biology is the driving force for many medical researchers to design tools, invent treatments, and conduct therapies with high precision and accuracy to personalize individual treatment compared to conventional methods [212]. Nanopharmacy and nanomedicine is the application of metal within organic or inorganic scaffolds tailored for specific interactions with host tissues, such as the release of agents, scavenging of radicals, or generation of reactive oxygen species to inactivate microbes or tumor cells, including antimicrobial properties to reduce the risk of infection. Due to the great number of possible combinations of metal and ligand or coating, nanomaterials can be fine-tuned, unlike their bulk counterparts for specific biological applications, making the usage of nanomaterials more ubiquitous.

Blunt trauma is a major cause of death, especially in individuals under 45 years of age, and is often due to ischemia, loss of blood, and calcium flux. To aid recovery, the potential to inject nanoparticles at the site of trauma is an attractive option, where blood flow is restricted, and oxygen is released to enable the patient to be

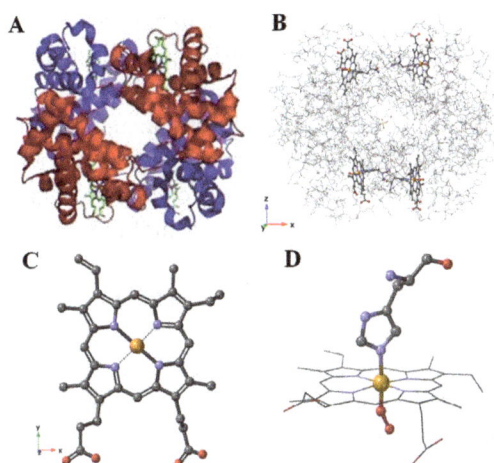

Fig. 1.51: The hemoglobin, a naturally occurring nanoscale protein, found in the blood:
(A) computer simulation of the structure [213]; (B) the hemoglobin molecule with the four
Fe–porphyrin subunits ($C_{5908}Fe_8N_{1560}O_{2073}P_3S_{24}$); (C) the Fe–porphyrin subunit ($C_{34}FeN_4O_4$); and
(D) Fe–porphyrin with axial ligand was taken from the complete hemoglobin structure ($C_{40}FeN_7O_7$)
(source: Protein Data Bank; ID Code: 1HHO; Revision Level: 1HHOA; Author: B. Shaanan [214]).

moved to a more secure location, such as a hospital. However, the role of nanoparticle circulation and toxicity is unknown [215].

A recent study in rats examined the effect of different sized polymer nanoparticles, their circulation, and their interaction with platelets, the cells promoting blood clotting. The study demonstrated that particles of 150 nm in diameter were the most effective to stop bleeding. These particles were less likely to travel to other organs or away from the injection site. Particles larger than 150 did travel to other organs, such as the lungs, and would not be effective in promoting blood clotting [216]. "With nanosystems, there is always some accumulation in the liver and the spleen, but we would like more of the active system to accumulate at the wound than at these filtration sites in the body," says Paula Hammond, an MIT Institute Professor, head of the Department of Chemical Engineering, and a member of MIT's Koch Institute for Integrative Cancer Research [217].

Dr. Hammond concluded that intravenous nanoparticle hemostats offer a potentially attractive approach to promoting hemostasis, particularly for inaccessible wounds such as noncompressible torso hemorrhage [218]. In their work, particle size was tuned over a range of <100–500 nm, and its effect on nanoparticle–platelet interactions were systematically assessed using in vitro and in vivo experiments. Smaller particles bound a larger percentage of platelets per mass of particle delivered, while larger particles resulted in higher particle accumulation on a surface of platelets and collagen. Intermediate particles led to the greatest platelet content in platelet–nanoparticle aggregates, indicating that they may recruit more platelets to the wound. In biodistribution studies, smaller and intermediate nanoparticles exhibited longer circulation lifetimes, while larger nanoparticles resulted in higher pulmonary accumulation. The particles were then challenged in a 2-h lethal inferior vena cava puncture model. Intermediate nanoparticles significantly increased survival and injury-specific targeting relative to saline and unfunctionalized particle controls [219]. The in vitro and in vivo results suggest that platelet content in aggregates and extended nanoparticle circulation lifetimes are instrumental in enhancing hemostasis [220].

They used a murine lethal injury model to investigate the effects of nanoparticles. The particles were injected into uninjured mice to assess their circulation lifetimes and organ distributions. The mice were imaged live using in vivo imaging system for 3 h and observed for several days to ensure no deleterious side effects following the injection. Figure 1.52 shows the timeline for blood sampling and imaging postinjection. When injected to achieve an estimated final therapeutic concentration of approximately 1 mg/mL in mouse blood, particles of different sizes were found to have different retention times and organ accumulation (Fig. 1.52b and c): specifically increased splenic accumulation in the larger particle sizes versus predominantly liver accumulation in the <100 nm size group.

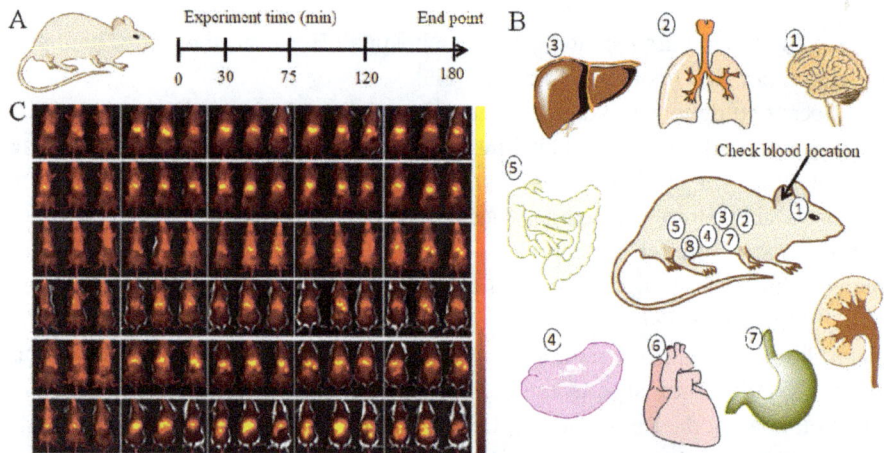

Fig. 1.52: In vivo biodistribution of Cy7-labeled nanoparticles and a free dye control in BALB/c mice. (A) Dosing timeline and (B) organic evaluated (adapted from [214, 220]). (C) Organ imaging using Cy7-labeled hemostatic nanoparticles and free dye control over 3 h. Dark red/black denote lower fluorescence levels, while yellow indicates accumulation in a particular area as a function of nanoparticle size.

Size effects: Hemostatic nanoparticles can slow blood flow and assist blood clotting through interaction with platelet cofactors. The nanoparticles are coated with soft biopolymers such as polyethylene glycol and poly-lactic-co-glycolic acid as a conjugated surface to reduce nanoparticle toxicity and attract platelets [221]. The coating for the nanoparticles was the initial focus and not the actual size. The group investigated the size and how variation in size affects distribution. Dr. Hong reported that "We were trying to look at how the size of the nanoparticle affects its interactions with the wound, which is an area that has not been explored with the polymer nanoparticles used as hemostats before."

While nanoparticles can initiate blood clotting (Fig. 1.53), they also migrate to other organs and potentially cause stress. The group examines low diameter (<100 nm), intermediate (140–220 nm), and large (500–650 nm) nanoparticles. The stop-flow study to examine nanoparticle-to-platelet interaction shows that the smallest nanoparticles are bound to platelets with the greatest percentage. In a follow-up experiment, they examined the role of platelet surface coverage and nanoparticle interaction and demonstrated that the largest nanoparticles interacted with the platelet-bound surface the best. The data were converted to the mass ratio of platelets to nanoparticles, and they found that the highest ratio was nanoparticles of intermediate diameter. Dr. Hong said, "If you attract a bunch of nanoparticles and block platelet binding because they clump onto each other, that is not very useful. We want platelets to come in".

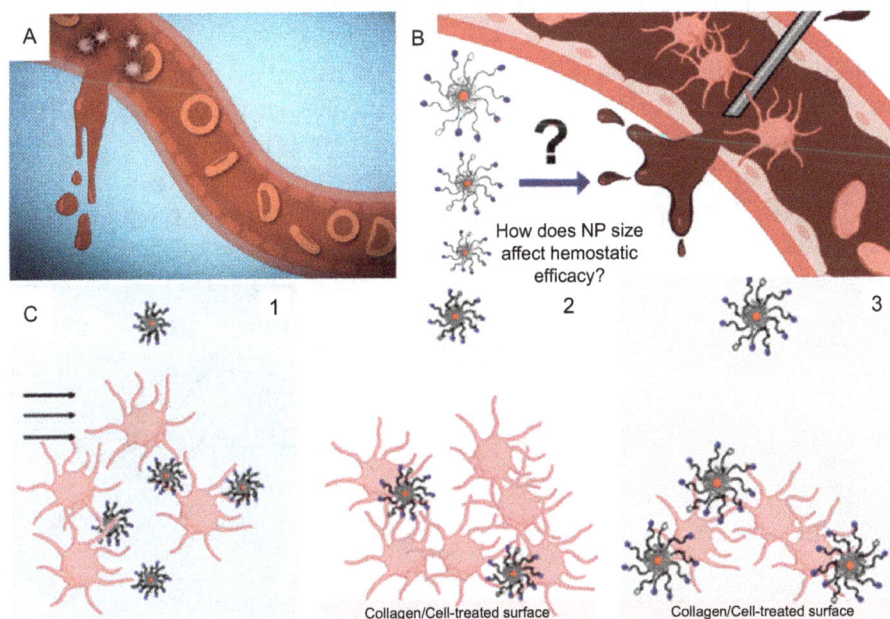

Fig. 1.53: Use of nanoparticles (NPs) to treat trauma: (A) a schematic of blooding; (B) NP size effect on hemostatic efficacy; and (C) summary of in vitro data to evaluate NP size on lethal injury model. (C1) NPs sized <100 nm resulted in the greatest number of specifically bound platelets underflow; (C2) intermediate-sized NPs (140–220 nm) resulted in the greatest number of platelets in a platelet–particle surface aggregate; and (C2) large NPs (500–650 nm) resulted in the greatest polymer mass accumulation onto platelet–particle surface aggregates. The image was adapted from Biorender [215].

Also, "When we did that experiment, we found that the intermediate particle size was the one that ended up with the greatest platelet content."

Stopping the bleeding: In the in vivo model, mice were injected with these three-sized nanoparticles. It was determined that the largest diameter nanoparticles would accumulate in other organs, limiting their on-site circulation time and half-life. Using an internal injury rat model, the team found that the intermediate diameter nanoparticles were optimal in promoting blood clotting and the highest accumulation at the trauma site. Dr. Hammond said. "This study suggests that the bigger nanoparticles are not necessarily the system we want to focus on, which was unclear from the previous work." Hammond says that "turning our attention to this medium-size range can open up some new doors."

Dr. Hong commented that "These particles are meant to address preventable deaths. They are not a cure-all for internal bleeding, but they are meant to give a person extra hours until they get to a hospital to receive adequate treatment", summarised in Fig. 1.53.

1.5.5 Nanomaterials used for shale gas storage

Shale gas introduction: As an important source of natural gas, Shale gas has been studied in many countries. The USA (Fig. 1.54A) and Canada have significant shale gas production, although other countries have many recoverable resources [222]. The Eagle Ford in Southern Texas enhanced the confidence in producing natural gas from shale (Fig. 1.54B) [223]. It was found that the Eagle Ford is predominantly composed of organic matter-rich fossiliferous marine shales and marls. It was projected that the natural gas from the shale resources will grow from 42 BCF/D in 2015 to 168 BCF/D in 2040,

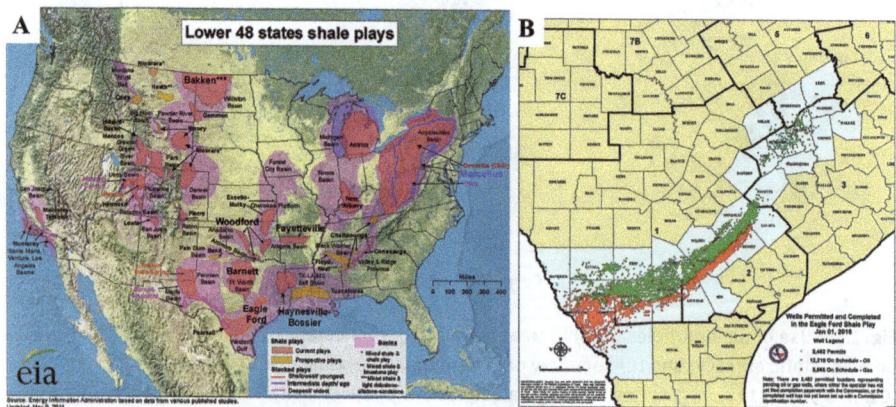

Fig. 1.54: Shale gas reserves, estimated by the U.S. Energy Information Administration (EIA), estimate that in 2020 (A) the national shale oil and shale gas resources, and (B) the Texas Eagle Ford shale gas reserve.

fourfold. By the end of the forecast period, shale will account for 30 % of the world's natural gas production. Therefore, it is critical to develop materials to improve shale gas storage [224].

Porous materials used for shale gas storage: The shale is a fine-grained sedimentary rock formed from the compaction of silt and clay-size mineral particles. One of the major components of shale gas is methane (CH_4). Suitable material designed for efficient methane storage is the key perspective in this review [225]. To improve methane (CH_4) storage, its physical and electronic properties need to be fully understood, such as at room temperature, CH_4 is a gas. It can be subjected to a liquid-like state using supercritical storage conditions. Another main characteristic of methane ($T_c = 190.6$ K, $P_c = 45.8$ atm) indicated that CH_4 could not be liquefied by compression above T_c. It is challenging to store a large CH_4 in a given volume for portable applications, such as gas-driven automobiles [226]. Therefore, preparing porous materials becomes important to achieve the above end application of CH_4. This report presented a series of materials (MOFs and caged porous composites) prepared by a feasible wet-chemistry approach. The gas uptake was also measured to evaluate their porosity and surface area. The ideal materials for methane storage need to display the following properties: (1) high surface area, (2) large pore volume, (3) low framework density, and (4) a strong interaction with the adsorbed methane. Recently, the US DOE announced that the volumetric and gravimetric targets for CH_4 onboard storage are 263 cc (STP)/cc and 699 cc (STP)/g, respectively. In reality, a 25 % loss in volumetric capacity will be lost after the porous materials are packed into the fuel tank [227]. We also implement a strategy by introducing amphiphilic groups to produce caged porous materials to reach this goal. Unfortunately, the pore size was decreased due to this introduction.

Experimental procedure: A feasible wet chemistry was used to produce a series of shielded porous metal oxides and MOFs for CH_4 storage. The synthesis method variables were optimized to control nanocomposites' porosity, size, and shape [228]. Those so-prepared materials were characterized using state-of-the-art techniques (TEM, XRD, XPS, and BET) to evaluate their nanostructure, porosity, and gas uptake. The materials' safety was also evaluated by measurement of toxicology, focusing on the cellular assays to determine cellular homeostasis developed using ultrafine particles [229].

Materials' preparation: The colloidal method produced Ag-MOFs, Fe_3O_4 composites, and ZnO composites. In this method, the monodispersed system is found to have a high energy state, which is equivalent to the free energy required to fabricate the increased surface area due to the attractive van der Waals interaction resulting in particle growth [231]. Therefore, amphiphilic dispersing agents were introduced to prevent particle growth by balancing attractive and repulsive forces between the particles. The second method to prevent aggregation resulted from steric stabilization [232]. Herein, the long-chain molecules (e.g., starch and gum arabic) are chemically bonded or physically attracted to colloidal particles, creating a repulsive force due to the steric hindrance. Polymeric surfactant chains attach to more than two particles at low concentrations, resulting in bridging flocculation [233].

Materials characterization: The morphology and elemental composition of the nano-composite were characterized using TEM (FEI Company, Tecnai F20-G^2, Hillsboro, OR). The TEM (Fig. 1.55A–C) images of three series of nanocomposites indicate that the shapes of these fabricated materials (a total of 89 formulations were prepared) were controlled as rod-shaped crystals with an average diameter of 10 nm and length of 78 nm (gum arabic-shielded Fe_3O_4, Fig. 1.55A), cubic (aloe vera extract-modified ZnO, Fig. 1.55B), and fiber (starch-modified ZnO, Fig. 1.55C) depending on the shielding agents and heat-treatment conditions. The silver-based MOFs showed a highly crystalline structure [234]. However, Cu, Co, and Zn metal-centered MOFs did now show the desired structure nor the gas up-takes. The amount of MOFs also needs to be scaled up for the gas storage study.

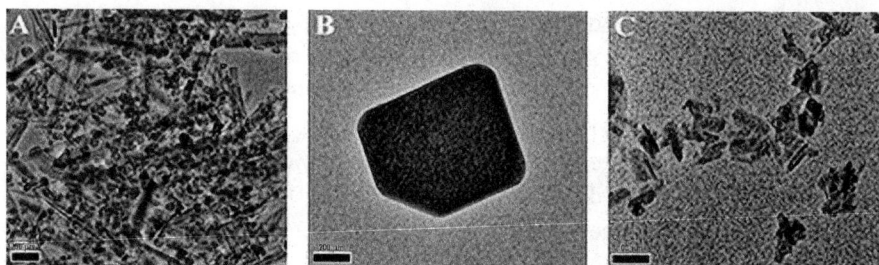

Fig. 1.55: The morphology and elemental composition of nanocomposite were characterized using transmission electron microscopy: (A) gum arabic-shielded Fe_3O_4 nanoparticles; (B) aloe vera extract-modified ZnO nanoparticles; and (C) starch-modified ZnO nanoparticles.

The study of powder X-ray diffraction (Bruker-AXS D8 Vario X-ray Powder Diffractometer, Madison, WI) indicated that nanocomposites and Ag-MOFs were highly crystalline. The gum arabic-shielded Fe_3O_4 (Fig. 1.56A) displayed an inverse spinel structure (PDF 65–3,107, $a = 8.3905$ Å, $\alpha = 90°$). The natural-product-modified ZnO nanocomposites (Fig. 1.56B) were tetragonal and well-indexed with the standard (PDF 89–0511 and PDF 05–0664 ($a = 3.249$, $c = 5.205$, $\alpha = 90°$). Briefly, the nanocomposites were generated by a colloidal chemistry approach to produce well-defined and highly crystallized compounds. The nucleation of the crystal units was promoted when the solvent molecules were evaporated from the solution. This controllable nucleation favored the formation of composites with tunable size and monodispersity. Natural product extracts play a critical role in the shape of nanocomposites [230].

XPS (Omicron ESCA +, Scienta Omicron, Inc., Denver, CO) determined the elemental composition depending on the photoelectron emission [235]. The chemical state of Ag, Fe, Zn, C, and O atoms can be determined from the binding energy obtained from photoelectron spectroscopy. XPS can be used as a "fingerprint" identification of elements in the nanocomposite according to the associated bonding information derived

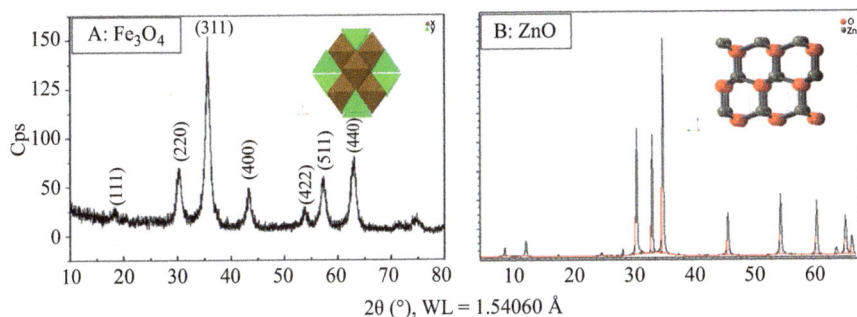

Fig. 1.56: The crystallographic analyses of the nanocomposite obtained using X-ray powder diffraction: (A) gum arabic-shielded Fe_3O_4 nanoparticles and (B) aloe vera extract-modified ZnO nanoparticles (note: different natural product extracts show no effect on the crystalline phase or crystallinity of ZnO nanoparticles).

from these chemical shifts. The XPS spectra of the nanocomposite (Fig. 1.57A–C) indicated the presence of Fe, Zn, and Ag; and O and C elements stem from the composite, dispersing, or reducing agents [236].

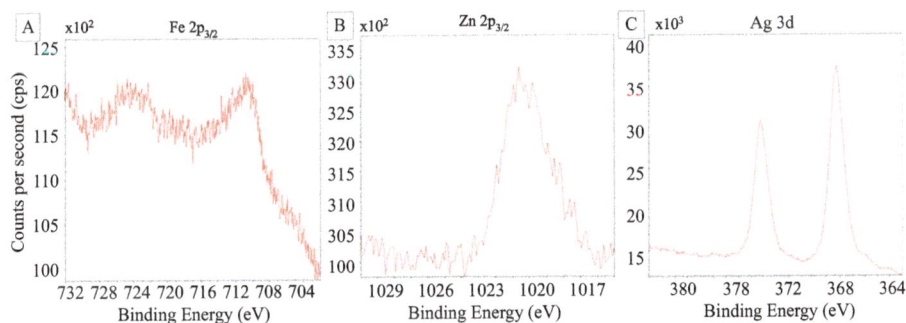

Fig. 1.57: The elemental analyses of the nanocomposite were obtained using X-ray electron–photon spectroscopy (XPS); (A) element Fe from Fe_3O_4 nanoparticles, (B) element Zn from ZnO nanoparticles, and (C) element Ag from Ag^0 nanoparticles.

Gas uptake (CH_4, N_2, and H_2): The BET absorption model (Micromeritics ASAP 2020, Norcross, GA) and related measurement techniques will be applied. CH_4, CO_2, H_2, and N_2 evaluated the catalyst capacity for selective gas adsorption. The measurement temperatures are controlled at 87, 175, and 231 K using liquid argon, methanol–liquid nitrogen, and acetonitrile–dry ice bath. Before adsorption measurement, the samples will be activated using the "outgas" function of the adsorption analyzer for 12 h at 80 °C. Then these activated samples will be used in gas adsorption measurement in the presence of purity grade. Table 1.4 shows the CH_4 storage data. The pore size distributions in the nanometer scale allow for gases to penetrate effectively. However, the porosity decreased due to the introduction of functional groups (such as alpha-acetal from starch's coil structure). The Fe_3O_4 nanoparticles were chosen to showcase the CH_4 uptakes (Fig. 1.58A). It was found that the reproducibility of the adsorption exceeds our expectations, suggesting a stable structure (Fig. 1.58B and C) and pores.

Tab. 1.4: A summary of CH_4 storage using natural product-coated metal oxide nanoparticles.

Samples	CS (m²/g)	PV (cm³/g)	P	D (g/cm³)	EU (g/L)	Q_{st} (kJ/mol)
GA–Fe_3O_4	536.8	0.73	67.6	0.167	4.32	7.54
ST–ZnO	536.8	0.73	67.6	0.167	3.15	7.54
AV–ZnO	536.8	0.73	67.6	0.167	3.44	7.54

CS, calculated surface area; PV, pore volume; P, porosity; D, density; EU, excess CH_4 uptake.

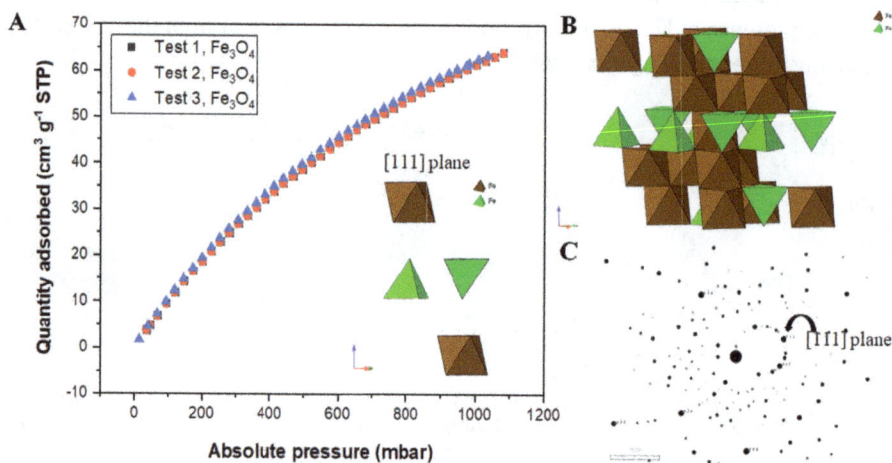

Fig. 1.58: The CH_4 uptake using Fe_3O_4 nanoparticles as a demonstration: (A) gas update results; (B) Fe_3O_4 111 plane showing a stable crystalline structure; (C) Laue plane front scattering, showing high crystallinity of the nanoparticles with well-maintained integrity.

1.6 Summary

In this chapter, the authors discussed three main areas: synthesis and characterization of nanomaterials using various chemistries, their characterization using advanced spectroscopies and microscopy, and selection of diverse examples of their applications in energy storage and medicine. The chapter outlines the design criteria and how these materials could be fabricated, characterized, and applied to precise applications, including correlation with animal studies. The aim was to give an overview of diverse applications and approaches rather than an exhaustive list and to document the physical and chemical properties of nanomaterials required to achieve the end goals.

References

[1]   Yao, K. & Liu, Y. (2014). Plasmonic metamaterials. Nanotechnology Reviews, 3(2), 177–210.
[2]   Poole, C. P. & Owens, F. J. Eds. (2003). Introduction to nanotechnology (p. 371). New York: John Wiley & Sons.
[3]   Roco, M. C. (2003). Nanotechnology: convergence with modern biology and medicine. Current Opinion in Biotechnology, 14(3), 337–346.
[4]   Street, A., Sustich, R., Duncan, J. & Savage, N. Eds. (2014). Nanotechnology applications for clean water: solutions for improving water quality (p. 651). New York: William Andrew.
[5]   Zang, L. Ed. (2011). Energy efficiency and renewable energy through nanotechnology (p. 451). Berlin: Springer.
[6]   Gutiérrez, F. J., Mussons, M. L., Gatón, P. & Rojo, R. (2011). Nanotechnology and food industry. Scientific Health and Social Aspects of the Food Industry, 95–128.
[7]   Chen, Y. X., Lavacchi, A., Miller, H. A., Bevilacqua, M., Filippi, J., Innocenti, M. & Vizza, F. (2014). Nanotechnology makes biomass electrolysis more energy efficient than water electrolysis. Nature Communications, 5(1), 1–6.
[8]   Chakraborty, A. K., Roy, T. & Mondal, S. (2016). Development of DNA nanotechnology and uses in molecular medicine and biology. Insights Biomed, 1(2), 13.
[9]   Adir, O., Poley, M., Chen, G., Froim, S., Krinsky, N., Shklover, J. . . . Schroeder, A. (2020). Integrating artificial intelligence and nanotechnology for precision cancer medicine. Advanced Materials, 32(13), 1901989.
[10] Roduner, E. (2006). Size matters: why nanomaterials are different. Chemical Society Reviews, 35(7), 583–592.
[11] Charitidis, C. A., Georgiou, P., Koklioti, M. A., Trompeta, A. F. & Markakis, V. (2014). Manufacturing nanomaterials: from research to industry. Manufacturing Review, 1, 11.
[12] Kumar, N. & Dixit, A. (2019). Nanomaterials-enabled lightweight military platforms. In: Nanotechnology for defence applications (pp. 205–254). Berlin: Springer, Cham.
[13] Khan, F. A. (2020). Synthesis of nanomaterials: methods & technology. In: Applications of nanomaterials in human health (pp. 15–21). Singapore: Springer.
[14] Arole, V. M. & Munde, S. V. (2014). Fabrication of nanomaterials by top-down and bottom-up approaches-an overview. Journal of Material Science, 1, 89–93.
[15] Grimsdale, A. C. & Müllen, K. (2005). The chemistry of organic nanomaterials. Angewandte Chemie International Edition, 44(35), 5592–5629.
[16] Chhabra, H. & Kumar, M. (2020). Size and shape-dependent equation of state for nanomaterials with application to bulk materials. Journal of Physics and Chemistry of Solids, 139, 109308.
[17] Gogotsi, Y. (2006). Nanomaterials handbook (pp. 1–800). New York: CRC Press.
[18] Ramesh, K. T. (2009). Nanomaterials. In: Nanomaterials (pp. 1–20). Boston: Springer.
[19] Stein, C. J. & Reiher, M. (2017). Automated identification of relevant frontier orbitals for chemical compounds and processes. CHIMIA International Journal for Chemistry, 71(4), 170–176.
[20] Wang, M., Wang, T., Cai, P. & Chen, X. (2019). Nanomaterials discovery and design through machine learning. Small Methods, 3(5), 1900025.

[21] Baglioni, P., Carretti, E. & Chelazzi, D. (2015). Nanomaterials in art conservation. Nature Nanotechnology, 10(4), 287–290.
[22] Martin, A. L., Li, B. & Gillies, E. R. (2009). Surface functionalization of nanomaterials with dendritic groups: enhanced binding to biological targets. Journal of the American Chemical Society, 131(2), 734–741.
[23] Willander, M., Nur, O., Zhao, Q. X., Yang, L. L., Lorenz, M., Cao, B. Q. . . . Dang, D. L. S. (2009). Zinc oxide nanorod-based photonic devices: recent progress in growth, light-emitting diodes, and lasers. Nanotechnology, 20(33), 332001.
[24] Wang, Z. G. & Ding, B. (2013). DNA-based self-assembly for functional nano-materials. Advanced Materials, 25(28), 3905–3914.
[25] Costela, A., Garcia-Moreno, I., Cerdan, L., Martin, V., Garcia, O. & Sastre, R. (2009). Dye-doped POSS solutions: random nanomaterials for laser emission. Advanced Materials, 21(41), 4163–4166.
[26] Bonnell, D. (2010). The next decade of nanoscience and nanotechnology. ACS Nano, 4(11), 6293–6294.
[27] Tarafdar, J. C., Sharma, S. & Raliya, R. (2013). Nanotechnology: interdisciplinary science of applications. African Journal of Biotechnology, 12, 3.
[28] Bell, C. & Marrapese, M. (2011). Nanotechnology standards and international legal considerations. In: Nanotechnology standards (pp. 239–255). New York: Springer.
[29] Hunt, G. & Mehta, M. Eds. (2013). Nanotechnology:" Risk, Ethics and Law" (p. 289). Sterling: Routledge.
[30] Li, S., Duan, S., Zha, Z., Pan, J., Sun, L., Liu, M. & Qiu, X. (2020). Structural phase transitions of molecular self-assembly driven by nonbonded metal adatoms. ACS Nano, 14(5), 6331–6338.
[31] Upadhyay, S. K., Dan, S., Girdhar, M. & Rastogi, K. (2021). Recent advancement in SARS-CoV-2 diagnosis, treatment, and vaccine formulation: a new paradigm of nanotechnology in strategic combating of COVID-19 pandemic. Current Pharmacology Reports, 1–14.
[32] Gottardo, S., Mech, A., Drbohlavova, J., Malyska, A., Bøwadt, S., Sintes, J. R. & Rauscher, H. (2021). Towards safe and sustainable innovation in nanotechnology: state-of-play for smart nanomaterials. NanoImpact, 100297.
[33] Sharifi, S., Behzadi, S., Laurent, S., Forrest, M. L., Stroeve, P. & Mahmoudi, M. (2012). Toxicity of nanomaterials. Chemical Society Reviews, 41(6), 2323–2343.
[34] Zhao, J., Lin, M., Wang, Z., Cao, X. & Xing, B. (2021). Engineered nanomaterials in the environment: are they safe?. Critical Reviews in Environmental Science and Technology, 51(14), 1443–1478.
[35] Dugershaw, B. B., Aengenheister, L., Hansen, S. S. K., Hougaard, K. S. & Buerki-Thurnherr, T. (2020). Recent insights on indirect mechanisms in developmental toxicity of nanomaterials. Particle and Fibre Toxicology, 17(1), 1–22.
[36] Aulic, S., Laurini, E., Marson, D., Skoko, N., Fermeglia, M. & Pricl, S. (2021). Regulatory, safety, and toxicological concerns of nanomaterials with their manufacturing issues. In: Nano-pharmacokinetics and theranostics (pp. 93–115). New York: Academic Press.
[37] Chen, K., Han, H., Tuguntaev, R. G., Wang, P., Guo, W., Huang, J. & Liang, X. J. (2021). Applications and regulations of nanotechnology-based innovative in-vitro diagnostics. View, 2(2), 20200091.
[38] Engelmann, W., von Hohendorff, R. & Leal, D. W. S. (2021). Regulations for using nanotechnology in food and medical products. In Biopolymer-based nano films (pp. 387–411). New York: Elsevier.
[39] Serrano, E., Linares, N., Garcia-Martinez, J. & Berenguer, J. R. (2013). Sol-Gel Coordination Chemistry: building Catalysts from the Bottom-Up. ChemCatChem, 5(4), 844–860.

[40] Arole, V. M. & Munde, S. V. (2014). Fabrication of nanomaterials by top-down and bottom-up approaches – an overview. Journal of Material Science, 1, 89–93.

[41] Gonçalves, M. (2018). Sol-gel silica nanoparticles in medicine: a natural choice. Design, synthesis, and products. Molecules, 23(8), 2021.

[42] Parashar, M., Shukla, V. K. & Singh, R. (2020). Metal oxide nanoparticles via sol-gel method: a review on synthesis, characterization, and applications. Journal of Materials Science: Materials in Electronics, 31(5), 3729–3749.

[43] Rajput, N. (2015). Methods of preparation of the nanoparticles-a review. International Journal of Advances in Engineering & Technology, 7(6), 1806.

[44] Rochman, N. T. & Akwalia, P. R. (2017, May). Fabrication and characterization of Zinc Oxide (ZnO) nanoparticle by sol-gel method. In Journal of physics: conference series (Vol. 853, No. 1, p. 012041). IOP Publishing.

[45] Dimitriev, Y., Ivanova, Y. & Iordanova, R. (2008). History of sol-gel science and technology. Journal of the University of Chemical Technology and Metallurgy, 43(2), 181–192.

[46] Shea, K. J. & Loy, D. A. (2001). A mechanistic investigation of gelation. The sol-gel polymerization of precursors to bridged polysilsesquioxanes. Accounts of Chemical Research, 34(9), 707–716.

[47] Loy, D. A., Mather, B., Straumanis, A. R., Baugher, C., Schneider, D. A., Sanchez, A. & Shea, K. J. (2004). Effect of pH on the gelation time of hexylene-bridged polysilsesquioxanes. Chemistry of Materials, 16(11), 2041–2043.

[48] Shea, K. J. & Loy, D. A. (2001). Bridged polysilsesquioxanes. Molecular-engineered hybrid organic-inorganic materials. Chemistry of Materials, 13(10), 3306–3319.

[49] Bilecka, I. & Niederberger, M. (2010). New developments in the nonaqueous and/or non-hydrolytic sol-gel synthesis of inorganic nanoparticles. Electrochimica Acta, 55(26), 7717–7725.

[50] Niederberger, M. (2007). Nonaqueous sol-gel routes to metal oxide nanoparticles. Accounts of Chemical Research, 40(9), 793–800.

[51] Song, Q. & Zhang, Z. J. (2004). Shape control and associated magnetic properties of spinel cobalt ferrite nanocrystals. Journal of the American Chemical Society, 126(19), 6164–6168.

[52] Zeng, H., Rice, P. M., Wang, S. X. & Sun, S. (2004). Shape-controlled synthesis and shape-induced texture of MnFe2O4 nanoparticles. Journal of the American Chemical Society, 126(37), 11458–11459.

[53] Joo, J., Kwon, S. G., Yu, J. H. & Hyeon, T. (2005). Synthesis of ZnO nanocrystals with cone, hexagonal cone, and rod shapes via non-hydrolytic ester elimination sol-gel reactions. Advanced Materials, 17(15), 1873–1877.

[54] Li, X. L., Peng, Q., Yi, J. X., Wang, X. & Li, Y. (2006). Near monodisperse TiO2 nanoparticles and nanorods. Chemistry – A European Journal, 12(8), 2383–2391.

[55] Zitoun, D., Pinna, N., Frolet, N. & Belin, C. (2005). Single crystal manganese oxide multipods by oriented attachment. Journal of the American Chemical Society, 127(43), 15034–15035.

[56] Seo, J. W., Jun, Y. W., Ko, S. J. & Cheon, J. (2005). In situ one-pot synthesis of 1-dimensional transition metal oxide nanocrystals. The Journal of Physical Chemistry. B, 109(12), 5389–5391.

[57] Silberberg, M. S., Amateis, P., Venkateswaran, R. & Chen, L. (1996). Chemistry: the molecular nature of matter and change (pp. 1062). St. Louis, MO: Mosby.

[58] Twej, W. A. (2009). Temperature influence on the gelation process of tetraethylorthosilicate using the sol-gel technique. Iraqi Journal of Science, 50(1), 43–49.

[59] Styskalik, A., Skoda, D., Barnes, C. E. & Pinkas, J. (2017). The power of non-hydrolytic sol-gel chemistry: a review. Catalysts, 7(6), 168.

[60] Esposito, S. (2019). Traditional" sol-gel chemistry as a powerful tool for the preparation of supported metal and metal oxide catalysts. Materials, 12(4), 668.

[61] Shinoda, K. & Saito, H. (1968). The effect of temperature on the phase equilibria and the types of dispersions of the ternary system composed of water, cyclohexane, and nonionic surfactant. Journal of Colloid and Interface Science, 26(1), 70–74.

[62] Shinoda, K. (1967). The correlation between the dissolution state of nonionic surfactant and the type of dispersion stabilized with the surfactant. Journal of Colloid and Interface Science, 24(1), 4–9.

[63] Sajjadi, S., Zerfa, M. & Brooks, B. W. (2002). Dynamic behavior of drops in oil/water/oil dispersions. Chemical Engineering Science, 57(4), 663–675.

[64] Esquena, J. (2016). Water-in-water (W/W) emulsions. Current Opinion in Colloid & Interface Science, 25, 109–119.

[65] Chang, Y. J., Chang, C. H., Yu, C. Y., Chang, T. J., Chen, L. C., Chen, M. H. . . . Ting, G. (2010). Therapeutic efficacy and microSPECT/CT imaging of 188Re-DXR-liposome in a C26 murine colon carcinoma solid tumor model. Nuclear Medicine and Biology, 37(1), 95–104.

[66] 2020's Top Nanotechnologies for Life, https://statnano.com/news/68539/2020-Top-Nanotechnologies-for-Life#ixzz7F80cHrC5;andhttps://statnano.com/news/67755/Real-time-Detection-of-COVID-19-Aided-by-CNT-based-Electrochemical-Sensors; and Miripour, Z. S., Sarrami-Forooshani, R., Sanati, H., Makarem, J., Taheri, M. S., Shojaeian, F. . . . Abdolahad, M. (2020). Real-time diagnosis of reactive oxygen species (ROS) in fresh sputum by electrochemical tracing; a correlation between COVID-19 and viral-induced ROS in lung/respiratory epithelium during this pandemic. Biosensors and Bioelectronics, 165, 112435; and the United States Patent Application Publication No.: US 2018 / 0299401 A1 Abdolahad et al. Pub. Date: Oct. 18, 2018.

[67] Chiappisi, L., Noirez, L. & Gradzielski, M. (2016). A journey through the phase diagram of a pharmaceutically relevant microemulsion system. Journal of Colloid and Interface Science, 473, 52–59.

[68] Chern, C. S. (2008). Principles and applications of emulsion polymerization (pp. 1–50) 154–174. New York: John Wiley & Sons.

[69] Slomkowski, S., Alemán, J. V., Gilbert, R. G., Hess, M., Horie, K., Jones, R. G. . . . Stepto, R. F. (2011). Terminology of polymers and polymerization processes in dispersed systems (IUPAC Recommendations 2011). Pure and Applied Chemistry, 83(12), 2229–2259.

[70] Jenkins, A. D., Kratochvíl, P., Stepto, R. F. T. & Suter, U. W. (1996). Glossary of basic terms in polymer science (IUPAC Recommendations 1996). Pure and Applied Chemistry, 68(12), 2287–2311.

[71] El-Sayed, S. M. & Madani, M. (2008). Effect of dosage on the conduction of electron-beam cross-linked thermoplastic elastomeric films from a blend of LDPE and EVA copolymer. Materials and Manufacturing Processes, 23(2), 162–167.

[72] Bashir, S., Houf, W., Liu, J. L. & Mulvaney, S. P. (2021). 3D Conducting Polymeric Membrane and Scaffold Saccharomyces cerevisiae Biofilms to Enhance Energy Conversion in Microbial Fuel Cells. ACS Applied Materials & Interfaces.

[73] McNaught, A. D. & Wilkinson, A. (2019). Compendium of chemical terminology. IUPAC recommendations (P. 1622); https://goldbook.iupac.org/

[74] Lewis, I. C. (1982). Chemistry of carbonization. Carbon, 20(6), 519–529.

[75] Tian, H., Liang, J. & Liu, J. (2019). Nanoengineering carbon spheres as nanoreactors for sustainable energy applications. Advanced Materials, 31(50), 1903886.

[76] Li, W., Yang, K., Peng, J., Zhang, L., Guo, S. & Xia, H. (2008). Effects of carbonization temperatures on characteristics of porosity in coconut shell chars and activated carbons derived from carbonized coconut shell chars. Industrial Crops and Products, 28(2), 190–198.

[77] Chern, C. S. & Poehlein, G. W. (1987). Polymerization in nonuniform latex particles: distribution of free radicals. Journal of Polymer Science. Part A, Polymer Chemistry, 25(2), 617–635.

[78] Merkel, M. P., Dimonie, V. L., El-Aasser, M. S. & Vanderhoff, J. W. (1987). Morphology and grafting reactions in core/shell latexes. Journal of Polymer Science. Part A, Polymer Chemistry, 25(5), 1219–1233.

[79] Chern, C. S. & Liou, Y. C. (1999). Styrene miniemulsion polymerization is initiated by 2, 2′-azobisisobutyronitrile. Journal of Polymer Science. Part A, Polymer Chemistry, 37(14), 2537–2550.

[80] Yan, K., Gao, X. & Luo, Y. (2015). Well-defined high molecular weight polystyrene with high rates and high livingness synthesized via two-stage RAFT emulsion polymerization. Macromolecular Rapid Communications, 36(13), 1277–1282.

[81] Shinoda, K. & Lindman, B. (1987). Organized surfactant systems: microemulsions. Langmuir, 3(2), 135–149.

[82] Strauss, M. A. & Wegner, H. A. (2019). Molecular systems for the quantification of London dispersion interactions. European Journal of Organic Chemistry, 2019(2–3), 295–302.

[83] Talapin, D. V., Shevchenko, E. V., Murray, C. B., Titov, A. V. & Kral, P. (2007). Dipole-dipole interactions in nanoparticle superlattices. Nano Letters, 7(5), 1213–1219.

[84] Novoa, J. J. & Sosa, C. (1995). Evaluation of the density functional approximation on the computation of hydrogen bond interactions. The Journal of Physical Chemistry, 99(43), 15837–15845.

[85] Hirsch, M. (2021).Classifying Surfactants for use in coatings formulation https://knowledge.ulprospector.com/11661/pc-classifying-surfactants-for-use-in-coatings-formulation/

[86] Zhao, T. H., Jacucci, G., Chen, X., Song, D. P., Vignolini, S. & Parker, R. M. (2020). Angular-independent photonic pigments via the controlled micellization of amphiphilic bottlebrush block copolymers. Advanced Materials, 32(47), 2002681.

[87] Wang, J., Zheng, F., Yu, Y., Hu, P., Li, M., Wang, J. . . . Liu, J. L. (2021). Symmetric supercapacitors composed of ternary metal oxides (NiO/V2O5/MnO2) nanoribbon electrodes with high energy storage performance. Chemical Engineering Journal, 426, 131804.

[88] Farha, O. K. & Hupp, J. T. (2010). Rational design, synthesis, purification, and activation of metal-organic framework materials. Accounts of Chemical Research, 43(8), 1166–1175.

[89] Byrappa, K. & Yoshimura, M. (2012). Handbook of hydrothermal technology (pp. 404). New York: William Andrew.

[90] Demazeau, G. (2008). Solvothermal reactions: an original route for the synthesis of novel materials. Journal of Materials Science, 43(7), 2104–2114.

[91] Sun, Y. & Zhou, H. C. (2015). Recent progress in the synthesis of metal-organic frameworks. Science and Technology of Advanced Materials.

[92] Walton, R. I. (2002). Subcritical solvothermal synthesis of condensed inorganic materials. Chemical Society Reviews, 31(4), 230–238.

[93] Julian, I., Roedern, M. B., Hueso, J. L., Irusta, S., Baden, A. K., Mallada, R. . . . Santamaria, J. (2020). Supercritical solvothermal synthesis under reducing conditions to increase stability and durability of Mo/ZSM-5 catalysts in methane dehydroaromatization. Applied Catalysis. B, Environmental, 263, 118360.

[94] Call, T. P., Carey, T., Bombelli, P., Lea-Smith, D. J., Hooper, P., Howe, C. J. & Torrisi, F. (2017). Platinum-free, graphene-based anodes and air cathodes for single chamber microbial fuel cells. Journal of Materials Chemistry A, 5(45), 23872–23886.

[95] Kemsley, J. (2015). Illuminating crystal nucleation. Chemical and Engineering News, 93(2), 28–29.

[96] Duft, D. & Leisner, T. (2004). Laboratory evidence for volume-dominated nucleation of ice in supercooled water microdroplets. Atmospheric Chemistry and Physics, 4(7), 1997–2000.

[97] Zhou, J., Yang, Y., Yang, Y., Kim, D. S., Yuan, A., Tian, X. . . . Miao, J. (2019). Observing crystal nucleation in four dimensions using atomic electron tomography. Nature, 570(7762), 500–503.

[98] Cölfen, H. & Antonietti, M. (2005). Mesocrystals: inorganic superstructures made by highly parallel crystallization and controlled alignment. Angewandte Chemie International Edition, 44(35), 5576–5591.

[99] Wang, J., Zheng, F., Yu, Y., Hu, P., Li, M., Wang, J. . . . Liu, J. L. (2021). Symmetric supercapacitors composed of ternary metal oxides (NiO/V2O5/MnO2) nanoribbon electrodes with high energy storage performance. Chemical Engineering Journal, 426, 131804.

[100] Vance, J. E. (1959). Growth and perfection of crystals. Journal of the American Chemical Society, 81(13), 3489–3490.

[101] Gibbs, J. W. (1879). On the equilibrium of heterogeneous substances. 300–320.

[102] Bashir, S., Mulvaney, S. P., Houf, W., Villanueva, L., Wang, Z., Buck, G. & Liu, J. L. (2021). Microbial fuel cells: design and evaluation of catalysts and device. Advances in sustainable energy (pp. 681–764). Cham: Springer.

[103] Pierson, H. O. (1999). Handbook of chemical vapor deposition: principles, technology, and applications (pp. 1–505). New York: William Andrew.

[104] Kumar, M. & Ando, Y. (2010). Chemical vapor deposition of carbon nanotubes: a review on growth mechanism and mass production. Journal of Nanoscience and Nanotechnology, 10(6), 3739–3758.

[105] Hampden-Smith, M. J. & Kodas, T. T. (1995). Chemical vapor deposition of metals: part 1. An overview of CVD processes. Chemical Vapor Deposition, 1(1), 8–23.

[106] Jones, A. C. & Hitchman, M. L. Eds. (2009). Chemical vapour deposition: precursors, processes and applications (p. 1–36). London: Royal society of chemistry.

[107] Choy, K. L. (2003). Chemical vapour deposition of coatings. Progress in Materials Science, 48(2), 57–170.

[108] Zhou, Y., Bao, Q., Tang, L. A. L., Zhong, Y. & Loh, K. P. (2009). Hydrothermal dehydration for the "green" reduction of exfoliated graphene oxide to graphene and demonstration of tunable optical limiting properties. Chemistry of Materials, 21(13), 2950–2956.

[109] Choy, K. L. Ed. (2019). Chemical vapour deposition (CVD): advances, technology, and applications (p. 1–391). London: CRC Press.

[110] Yu, H. D., Regulacio, M. D., Ye, E. & Han, M. Y. (2013). Chemical routes to top-down nanofabrication. Chemical Society Reviews, 42(14), 6006–6018.

[111] Ito, T. & Okazaki, S. (2000). Pushing the limits of lithography. Nature, 406(6799), 1027–1031.

[112] Acharya, A. (2000). A distinct element approach to ball mill mechanics. Communications in Numerical Methods in Engineering, 16(11), 743–753.

[113] Belenguer, A. M., Lampronti, G. I., Wales, D. J. & Sanders, J. K. (2014). Direct observation of intermediates in a thermodynamically controlled solid-state dynamic covalent reaction. Journal of the American Chemical Society, 136(46), 16156–16166.

[114] Hasa, D., Miniussi, E. & Jones, W. (2016). Mechanochemical synthesis of multicomponent crystals: one liquid for one polymorph? A myth to dispel. Crystal Growth & Design, 16(8), 4582–4588.

[115] Boey, F. Y. C., Yuan, Z. & Khor, K. A. (1998). Mechanical alloying for the effective dispersion of sub-micron SiCp reinforcements in Al-Li alloy composite. Materials Science and Engineering: A, 252(2), 276–287.

[116] Koch, C. C. (1993). The synthesis and structure of nanocrystalline materials produced by mechanical attrition: a review. Nanostructured Materials, 2(2), 109–129.

[117] Li, R., Liu, X., He, G., Hu, P., Zhen, Q., Liu, J. L. & Bashir, S. (2021). Green catalytic synthesis of ammonia using solid oxide electrolysis cells composed of multicomponent materials. Catalysis Today, 374, 102–116.

[118] Baláž, P. & Dutková, E. (2009). Fine milling in applied mechanochemistry. Minerals Engineering, 22(7–8), 681–694.

[119] Thorwirth, R., Bernhardt, F., Stolle, A., Ondruschka, B. & Asghari, J. (2010). Switchable selectivity during oxidation of anilines in a ball mill. Chemistry – A European Journal, 16(44), 13236–13242.

[120] Moraes, J., Alves, F. S. & Franco, C. M. (2013). Effect of ball milling on structural and physicochemical characteristics of cassava and Peruvian carrot starches. Starch-Stärke, 65 (3-4), 200–209.

[121] Iankov, D., Zuerbig, V., Pletschen, W., Giese, C., Iannucci, R., Ambacher, O. & Lebedev, V. (2014). Processing of nanoscale gaps for a boron-doped nanocrystalline diamond-based MEMS. Procedia Engineering, 87, 903–906.

[122] Abramova, V., Slesarev, A. S. & Tour, J. M. (2015). Meniscus-Mask Lithography for fabrication of narrow nanowires. Nano Letters, 15(5), 2933–2937.

[123] Lee, M. H., Kim, H. M., Cho, S. Y., Lim, K., Park, S. Y., Jong Lee, J. & Kim, K. B. (2011). Fabrication of ultra-high-density nanodot array patterns (~ 3 Tbits/in. 2) using electron-beam lithography. Journal of Vacuum Science & Technology B, Nanotechnology and Microelectronics: Materials, Processing, Measurement, and Phenomena, 29(6), 061602.

[124] Karitans, V., Kundzins, K., Laizane, E., Ozolinsh, M. & Ekimane, L. (2012). Applicability of a binary amplitude mask for creating correctors of higher-order ocular aberrations in a photo resistive layer. Optical Engineering, 51(7), 078001.

[125] Davim, J. P. Ed. (2008). Machining: fundamentals and recent advances (p. 1–359). Girona: Springer Science & Business Media.

[126] Phaneuf, M. W. (1999). Applications of focused ion beam microscopy to materials science specimens. Micron, 30(3), 277–288.

[127] Madou, M. J. (2018). Fundamentals of microfabrication: the science of miniaturization (pp. 615–666). London: CRC Press.

[128] Plank, H. (2015). Focused particle beam nano-machining: the next evolution step towards simulation aided process prediction. Nanotechnology, 26(5), 050501.

[129] Repetto, L., Firpo, G. & Valbusa, U. (2008). Applications of the focused ion beam in material science. Materiali in Tehnologije, 42(4), 143–149.

[130] Simmons, H. L. (2011). Olin's construction: principles, materials, and methods (pp. 1083–1160). New Jersey: John Wiley & Sons.

[131] El-Hofy, H. A. G. (2005). Advanced machining processes: nontraditional and hybrid machining processes (pp. 77–180). New York: McGraw Hill Professional.

[132] Kubis, A. J., Vandervelde, T. E., Bean, J. C., Dunn, D. N. & Hull, R. (2006). Analysis of the three-dimensional ordering of epitaxial Ge quantum dots using focused ion beam tomography. Applied Physics Letters, 88(26), 263103.

[133] Li, J., Malis, T. & Dionne, S. (2006). Recent advances in FIB–TEM specimen preparation techniques. Materials Characterization, 57(1), 64–70.

[134] Giannuzzi, L. A., Kempshall, B. W., Schwarz, S. M., Lomness, J. K., Prenitzer, B. I. & Stevie, F. A. (2005). FIB lift-out specimen preparation techniques. Introduction to focused ion beams (pp. 201–228). Boston, MA: Springer.

[135] Qiao, Y., Wang, D. & Buriak, J. M. (2007). Block copolymer templated etching on silicon. Nano Letters, 7(2), 464–469.

[136] Potentials from https://www.nist.gov/system/files/documents/2019/04/02/jpcrd355.pdf; andhttps://chem.libretexts.org/Ancillary_Materials/Reference/Reference_Tables/Electro chemistry_Tables/P2%3A_Standard_Reduction_Potentials_by_Value

[137] Fujita, T., Okada, H., Koyama, K., Watanabe, K., Maekawa, S. & Chen, M. W. (2008). The unusually small electrical resistance of three-dimensional nanoporous gold in external magnetic fields. Physical Review Letters, 101(16), 166601.

[138] Yu, H. D., Regulacio, M. D., Ye, E. & Han, M. Y. (2013). Chemical routes to top-down nanofabrication. Chemical Society Reviews, 42(14), 6006–6018.

[139] Erlebacher, J., Aziz, M. J., Karma, A., Dimitrov, N. & Sieradzki, K. (2001). Evolution of nanoporosity in dealloying. Nature, 410(6827), 450–453.

[140] Yu, H. D., Yang, D., Wang, D. & Han, M. Y. (2010). Top-Down Fabrication of Calcite Nanoshoot Arrays by Crystal Dissolution. Advanced Materials, 22(29), 3181–3184.

[141] Yu, H., Wang, D. & Han, M. Y. (2007). Top-down solid-phase fabrication of nanoporous cadmium oxide architectures. Journal of the American Chemical Society, 129(8), 2333–2337.

[142] Yu, H. D., Zhang, Z. Y., Win, K. Y., Chan, J., Teoh, S. H. & Han, M. Y. (2010). Bioinspired fabrication of 3D hierarchical porous nanomicrostructures of calcium carbonate for bone regeneration. Chemical Communications, 46(35), 6578–6580.

[143] Zhang, Y. L., Chen, Q. D., Xia, H. & Sun, H. B. (2010). Designable 3D nanofabrication by femtosecond laser direct writing. Nano Today, 5(5), 435–448.

[144] Yu, H. D., Tee, S. Y. & Han, M. Y. (2013). Preparation of porosity-controlled calcium carbonate by thermal decomposition of volume content-variable calcium carboxylate derivatives. Chemical Communications, 49(39), 4229–4231.

[145] Lin, Y., Ren, J. & Qu, X. (2014). Nano-gold as artificial enzymes: hidden talents. Advanced Materials, 26(25), 4200–4217.

[146] Bonomi, R., Cazzolaro, A., Sansone, A., Scrimin, P. & Prins, L. J. (2011). Detection of enzyme activity through catalytic signal amplification with functionalized gold nanoparticles. Angewandte Chemie International Edition, 50(10), 2307–2312.

[147] Manea, F., Houillon, F. B., Pasquato, L. & Scrimin, P. (2004). Nanozymes: gold-nanoparticle-based transphosphorylation catalysts. Angewandte Chemie, 116(45), 6291–6295.

[148] Bonomi, R., Selvestrel, F., Lombardo, V., Sissi, C., Polizzi, S., Mancin, F. . . . Scrimin, P. (2008). Phosphate diester and DNA hydrolysis by a multivalent, nanoparticle-based catalyst. Journal of the American Chemical Society, 130(47), 15744–15745.

[149] Jana, N. R., Gearheart, L., Obare, S. O. & Murphy, C. J. (2002). The anisotropic chemical reactivity of gold spheroids and nanorods. Langmuir, 18(3), 922–927.

[150] Uzun, O., Hu, Y., Verma, A., Chen, S., Centrone, A. & Stellacci, F. (2008). Water-soluble amphiphilic gold nanoparticles with structured ligand shells. Chemical Communications, (2), 196–198.

[151] Guardia, P., Pérez, N., Labarta, A. & Batlle, X. (2010). Controlled synthesis of iron oxide nanoparticles over a wide size range. Langmuir, 26(8), 5843–5847.

[152] Sondheimer, E. H. (2001). The mean free path of electrons in metals. Advances in Physics, 50(6), 499–537.

[153] Brack, M. (1993). The physics of simple metal clusters: self-consistent jellium model and semiclassical approaches. Reviews of Modern Physics, 65(3), 677.

[154] Neth, E. J., Flowers, P., Theopold, K., Langley, R. & Robinson, W. R. (2018). Chemistry: Atoms First (OpenStax).

[155] Cummings, F. E. (1977). The particle in a box is not simple. American Journal of Physics, 45(2), 158–160.

[156] Makri, N. & Miller, W. H. (1989). A semiclassical tunneling model for use in classical trajectory simulations. The Journal of Chemical Physics, 91(7), 4026–4036.

[157] LibreTexts (Textmap of Petrucci's Book) Hill, J. W., Petrucci, R. H. & Mosher, M. D. (1996). General chemistry. Vol. 2, New Jersey: Prentice-Hall.

[158] Mitragotri, S., Anderson, D. G., Chen, X., Chow, E. K., Ho, D., Kabanov, A. V. . . . Xu, C. (2015). Accelerating the translation of nanomaterials in biomedicine. ACS Nano, 9(7), 6644–6654.

[159] Zhang, X., Brynda, M., Britt, R. D., Carroll, E. C., Larsen, D. S., Louie, A. Y. & Kauzlarich, S. M. (2007). Synthesis and characterization of manganese-doped silicon nanoparticles: bifunctional paramagnetic-optical nanomaterial. Journal of the American Chemical Society, 129(35), 10668–10669.

[160] Charpentier, S., Kassiba, A., Emery, J. & Cauchetier, M. (1999). Investigation of the paramagnetic centres and electronic properties of silicon carbide nanomaterials. Journal of Physics: Condensed Matter, 11(25), 4887.

[161] Pankhurst, Q. A., Connolly, J., Jones, S. K. & Dobson, J. (2003). Applications of magnetic nanoparticles in biomedicine. Journal of Physics D: Applied Physics, 36(13), R167.

[162] Du, X., Yao, Y. & Liu, J. (2013). Structural architecture and magnetism control of metal oxides using surface grafting techniques. Journal of Nanoparticle Research, 15(7), 1–8.

[163] Hong, R., Li, J., Wang, J. & Li, H. (2007). Comparison of schemes for preparing magnetic Fe3O4 nanoparticles. China Particuology, 5(1–2), 186–191.

[164] Li, Y., Jiang, R., Liu, T., Lv, H., Zhou, L. & Zhang, X. (2014). One-pot synthesis of grass-like Fe3O4 nanostructures by a novel microemulsion-assisted solvothermal method. Ceramics International, 40(1), 1059–1063.

[165] Kulkarni, S. A., Sawadh, P. S., Palei, P. K. & Kokate, K. K. (2014). Effect of synthesis route on the structural, optical, and magnetic properties of Fe3O4 nanoparticles. Ceramics International, 40(1), 1945–1949.

[166] Meng, H., Zhang, Z., Zhao, F., Qiu, T. & Yang, J. (2013). Orthogonal optimization design for preparation of Fe3O4 nanoparticles via chemical coprecipitation. Applied Surface Science, 280, 679–685.

[167] Choi, S. U. & Eastman, J. A. (1995). Enhancing thermal conductivity of fluids with nanoparticles (No. ANL/MSD/CP-84938; CONF-951135-29). Argonne National Lab. (ANL), Argonne, IL (United States).

[168] Haendler, B. L. (1982). Presenting the Bohr atom. Journal of Chemical Education, 59(5), 372.

[169] Jacobson, J., Björk, G., Chuang, I. & Yamamoto, Y. (1995). Photonic de Broglie waves. Physical Review Letters, 74(24), 4835.

[170] Arndt, M., Nairz, O., Vos-Andreae, J., Keller, C., Van der Zouw, G. & Zeilinger, A. (1999). Wave-particle duality of C60 molecules. nature, 401(6754), 680–682.

[171] Steuernagel, O. (2002). de Broglie wavelength reduction for a multiphoton wave packet. Physical Review. A, 65(3), 033820.

[172] Xiao, M., Martin, I., Yablonovitch, E. & Jiang, H. W. (2004). Electrical detection of the spin resonance of a single electron in a silicon field-effect transistor. Nature, 430(6998), 435–439.

[173] Lehmann, G. & Taut, M. (1972). On the numerical calculation of the density of states and related properties. Physica Status Solidi (B), 54(2), 469–477.

[174] Lang, J. K., Baer, Y. & Cox, P. A. (1981). Study of the 4f and valence band density of states in rare-earth metals. II. Experiment and results. Journal of Physics F: Metal Physics, 11(1), 121.

[175] McMillan, W. L. & Rowell, J. M. (1965). Lead phonon spectrum calculated from the superconducting density of states. Physical Review Letters, 14(4), 108.

[176] Britney spears Physics.

[177] Wegner, F. (1981). Bounds on the density of states in disordered systems. Zeitschrift Für Physik B Condensed Matter, 44(1), 9–15.

[178] Teja, A. S. & Koh, P. Y. (2009). Synthesis, properties, and applications of magnetic iron oxide nanoparticles. Progress in Crystal Growth and Characterization of Materials, 55(1–2), 22–45.

[179] Klemens, P. G. (1951). The thermal conductivity of dielectric solids at low temperatures (theoretical). Proceedings of the Royal Society of London. Series A. Mathematical and Physical Sciences, 208(1092),108–133.

[180] Cuevas, J. C. & García-Vidal, F. J. (2018). Radiative heat transfer. Acs Photonics, 5(10), 3896–3915.

[181] Liu, J. L. & Bashir, S. (2015). Advanced nanomaterials and their applications in renewable energy (pp. 1–44). New York: Elsevier.

[182] Howell, J. R., Mengüç, M. P., Daun, K. & Siegel, R. (2020). Thermal radiation heat transfer (pp. 1–52). Baton Rouge: CRC press.

[183] Alivisatos, A. P., Harris, T. D., Carroll, P. J., Steigerwald, M. L. & Brus, L. E. (1989). Electron–vibration coupling in semiconductor clusters studied by resonance Raman spectroscopy. The Journal of Chemical Physics, 90(7), 3463–3468.

[184] Cuenot, S., Frétigny, C., Demoustier-Champagne, S. & Nysten, B. (2004). Surface tension affects the mechanical properties of nanomaterials measured by atomic force microscopy. Physical Review B, 69(16), 165410.

[185] Kumar, M., Chen, H., Sarsaiya, S., Qin, S., Liu, H., Awasthi, M. K. . . . Taherzadeh, M. J. (2021). A review of current research trends on micro-and nano-plastics as an emerging threat to the global environment. Journal of Hazardous Materials, 409, 124967.

[186] https://www.lawyersgunsmoneyblog.com/2021/05/racial-polarization-and-american-politics

[187] Hu, J., He, D., Zhang, X., Li, X., Chen, Y., Wei, G. . . . Luo, Y. (2022). National-scale distribution of micro (meso) plastics in farmland soils across China: implications for environmental impacts. Journal of Hazardous Materials, 424, 127283.

[188] Dhineshbabu, N. R., Manivasakan, P. & Rajendran, V. (2014). Hydrophobic and thermal behavior of nylon 6 nanofibre web deposited on cotton fabric through electrospinning. Micro & Nano Letters, 9(8), 519–522.

[189] Peng, M., Li, L., Xiong, J., Hua, K., Wang, S. & Shao, T. (2017). Study on surface properties of polyamide 66 using atmospheric glow-like discharge plasma treatment. Coatings, 7(8), 123.

[190] Choi, D., Yeo, J. S., Joh, H. I. & Lee, S. (2018). Carbon nanosheet from polyethylene thin film as a transparent conducting film: "Upcycling" of waste to organic photovoltaics application. ACS Sustainable Chemistry & Engineering, 6(9), 12463–12470.

[191] Gabrilovich, D. I. & Nagaraj, S. (2009). Myeloid-derived suppressor cells as regulators of the immune system. Nature Reviews. Immunology, 9(3), 162–174.

[192] Sang, W., Zhang, Z., Dai, Y. & Chen, X. (2019). Recent advances in nanomaterial-based synergistic combination cancer immunotherapy. Chemical Society Reviews, 48(14), 3771–3810.

[193] Song, W., Musetti, S. N. & Huang, L. (2017). Nanomaterials for cancer immunotherapy. Biomaterials, 148, 16–30.

[194] Leach, D. R., Krummel, M. F. & Allison, J. P. (1996). Enhancement of antitumor immunity by CTLA-4 blockade. Science, 271(5256), 1734–1736.

[195] https://www.nobelprize.org/prizes/medicine/2018/press-release/

[196] Sahin, U. & Türeci, Ö. (2018). Personalized vaccines for cancer immunotherapy. Science, 359(6382), 1355–1360.

[197] Khalil, D. N., Smith, E. L., Brentjens, R. J. & Wolchok, J. D. (2016). The future of cancer treatment: immunomodulation, CARs, and combination immunotherapy. Nature Reviews. Clinical Oncology, 13(5), 273–290.

[198] Xu, X., Lu, H. & Lee, R. (2020). Near-infrared light-triggered photo/immuno-therapy toward cancers. Frontiers in Bioengineering and Biotechnology, 8, 488.

[199] Sang, W., Zhang, Z., Dai, Y. & Chen, X. (2019). Recent advances in nanomaterial-based synergistic combination cancer immunotherapy. Chemical Society Reviews, 48(14), 3771–3810.

[200] Peer, D., Karp, J. M., Hong, S., Farokhzad, O. C., Margalit, R. & Langer, R. (2020). Nanocarriers as an emerging platform for cancer therapy. Nano-Enabled Medical Applications, 61–91.

[201] Chen, Y., Wang, L. & Shi, J. (2016). Two-dimensional non-carbonaceous materials-enabled efficient photothermal cancer therapy. Nano Today, 11(3), 292–308.

[202] Irvine, D. J., Swartz, M. A. & Szeto, G. L. (2013). Engineering synthetic vaccines using cues from natural immunity. Nature Materials, 12(11), 978–990.

[203] Chen, D. S. & Mellman, I. (2013). Oncology meets immunology: the cancer-immunity cycle. Immunity, 39(1), 1–10.

[204] Weng, Y., Liu, J., Jin, S., Guo, W., Liang, X. & Hu, Z. (2017). Nanotechnology-based strategies for the treatment of ocular disease. Acta Pharmaceutica Sinica B, 7(3), 281–291.

[205] Sang, W., Zhang, Z., Dai, Y. & Chen, X. (2019). Recent advances in nanomaterial-based synergistic combination cancer immunotherapy. Chemical Society Reviews, 48(14), 3771–3810.

[206] Li, Y. J., Luo, L. J., Harroun, S. G., Wei, S. C., Unnikrishnan, B., Chang, H. T. . . . Huang, C. C. (2019). Synergistically dual-functional nano eye drops for simultaneous anti-inflammatory and anti-oxidative treatment of dry eye disease. Nanoscale, 11(12), 5580–5594.

[207] Mobaraki, M., Soltani, M., Zare Harofte, S., Zoudani, L., Daliri, R., E., Aghamirsalim, M. & Raahemifar, K. (2020). Biodegradable nanoparticle for cornea drug delivery: focus review. Pharmaceutics, 12(12), 1232.

[208] Calvo, P., Vila-Jato, J. L. & Alonso, M. J. (1996). Comparative in vitro evaluation of several colloidal systems, nanoparticles, nanocapsules, and nanoemulsions, as ocular drug carriers. Journal of Pharmaceutical Sciences, 85(5), 530–536.

[209] Chang, W. C., Hawkes, E. A., Kliot, M. & Sretavan, D. W. (2007). In vivo use of a nano knife for axon microsurgery. Neurosurgery, 61(4), 683–692.

[210] William, D. & Townsend, O. D. (2011) How Nanotechnology Will Revolutionize Eye Care, Nov 18, https://www.reviewofcontactlenses.com/issue/november-2011-1691

[211] Chen, J., Patil, S., Seal, S. & McGinnis, J. F. (2006). Rare earth nanoparticles prevent retinal degeneration induced by intracellular peroxides. Nature Nanotechnology, 1(2), 142–150.

[212] Pajic, B., Pajic-Eggspuchler, B. & Haefliger, I. (2011). Continuous IOP fluctuation recording in normal-tension glaucoma patients. Current Eye Research, 36(12), 1129–1138.

[213] Nanotechnology Initiative, https://www.nano.gov

[214] Perutz, M. F., Fermi, G., Luisi, B., Shaanan, B. & Liddington, R. C. (1987). Stereochemistry of cooperative mechanisms in hemoglobin. Accounts of Chemical Research, 20(9), 309–321.

[215] Hong, C., Alser, O., Gebran, A., He, Y., Joo, W., Kokoroskos, N. . . . Hammond, P. T. (2022). Modulating nanoparticle size to understand factors affecting hemostatic efficacy and maximizing survival in a lethal inferior vena cava injury model. ACS Nano, https://doi.org/10.1021/acsnano.1c09108.

[216] Bakhaidar, R., O'Neill, S. & Ramtoola, Z. (2020). PLGA-PEG nanoparticles show minimal risks of interference with the platelet function of human platelet-rich plasma. International Journal of Molecular Sciences, 21(24), 9716.

[217] Hammond, S. M., Aartsma-Rus, A., Alves, S., Borgos, S. E., Buijsen, R. A., Collin, R. W. . . . Arechavala-Gomeza, V. (2021). Delivery of oligonucleotide-based therapeutics: challenges and opportunities. EMBO Molecular Medicine, 13(4), e13243.

[218] Hong, C., Olsen, B. D. & Hammond, P. T. (2022). A review of treatments for non-compressible torso hemorrhage (NCTH) and internal bleeding. Biomaterials, 121432.

[219] Hammond, P. T. (2004). Form and function in multilayer assembly: new applications at the nanoscale. Advanced Materials, 16(15), 1271–1293.

[220] Venkataraman, S., Hedrick, J. L., Ong, Z. Y., Yang, C., Ee, P. L. R., Hammond, P. T. & Yang, Y. Y. (2011). The effects of polymeric nanostructure shape on drug delivery. Advanced Drug Delivery Reviews, 63(14–15), 1228–1246.

[221] Huerta, T. S., Devarajan, A., Tsaava, T., Rishi, A., Cotero, V., Puleo, C. . . . Chavan, S. S. (2021). Targeted peripheral focused ultrasound stimulation attenuates obesity-induced metabolic and inflammatory dysfunctions. Scientific Reports, 11(1), 1–12.

[222] Nandlal, K. & Weijermars, R. (2022). Shale well factory model reviewed: eagle Ford case study. Journal of Petroleum Science and Engineering, 212, 110158.

[223] Xu, B., Haghighi, M. & Cooke, D. (2012, March). Production data analysis in Eagle Ford shale gas reservoir. In SPE/EAGE European Unconventional Resources Conference & Exhibition-From Potential to Production (pp. cp–285). European Association of Geoscientists & Engineers.

[224] Raterman, K., Liu, Y., Roy, B., Friehauf, K., Thompson, B. & Janssen, A. (2020, December). Analysis of a Multi-Well Eagle Ford Pilot. In Unconventional Resources Technology Conference, 20–22 July 2020 (pp. 19–38). Unconventional Resources Technology Conference (URTEC).

[225] Colosimo, F., Thomas, R., Lloyd, J. R., Taylor, K. G., Boothman, C., Smith, A. D. . . . Kalin, R. M. (2016). Biogenic methane in shale gas and coal bed methane: a review of current knowledge and gaps. International Journal of Coal Geology, 165, 106–120.

[226] Howarth, R. W. (2019). Ideas and perspectives: is shale gas a major driver of a recent increase in global atmospheric methane?. Biogeosciences, 16(15), 3033–3046.

[227] Makal, T. A., Li, J. R., Lu, W. & Zhou, H. C. (2012). Methane storage in advanced porous materials. Chemical Society Reviews, 41(23), 7761–7779.

[228] Eddaoudi, M., Kim, J., Rosi, N., Vodak, D., Wachter, J., O'Keeffe, M. & Yaghi, O. M. (2002). Systematic design of pore size and functionality in isoreticular MOFs and their application in methane storage. Science, 295(5554), 469–472.

[229] Bashir, S., Hanumandla, P., Huang, H. Y. & Liu, J. L. (2018). Nanostructured materials for advanced energy conversion and storage devices: safety implications at end-of-life disposal. Nanostructured materials for next-generation energy storage and conversion (pp. 517–542). Berlin, Heidelberg: Springer.

[230] Liu, J. L. & Bashir, S. (2015). Advanced nanomaterials and their applications in renewable energy.

[231] Niederberger, M. (2007). Nonaqueous sol-gel routes to metal oxide nanoparticles. Accounts of Chemical Research, 40(9), 793–800.

[232] Niederberger, M. & Garnweitner, G. (2006). Organic reaction pathways in the nonaqueous synthesis of metal oxide nanoparticles. Chemistry – A European Journal, 12(28), 7282–7302.

[233] Loy, D. A., Mather, B., Straumanis, A. R., Baugher, C., Schneider, D. A., Sanchez, A. & Shea, K. J. (2004). Effects of pH on the gelation time in the sol-gel polymerization of 1, 6-bis (trimethoxysilyl) hexane (No. LA-UR-04-2104). Los Alamos National Lab. (LANL), Los Alamos, NM (United States).

[234] Abdi-Khanghah, M., Adelizadeh, M., Naserzadeh, Z. & Barati, H. (2018). Methane hydrate formation in the presence of ZnO nanoparticle and SDS: application to transportation and storage. Journal of Natural Gas Science and Engineering, 54, 120–130.

[235] Song, Q. & Zhang, Z. J. (2004). Shape control and associated magnetic properties of spinel cobalt ferrite nanocrystals. Journal of the American Chemical Society, 126(19), 6164–6168.

[236] Zeng, H., Rice, P. M., Wang, S. X. & Sun, S. (2004). Shape-controlled synthesis and shape-induced texture of MnFe2O4 nanoparticles. Journal of the American Chemical Society, 126(37), 11458–11459.

Section 2: **Focus on synthesis methods**

S. Bashir, C. Li, Y. Huang, S. Palakurthi, Z. Wang, Y. Olaseni,
P. Villarreal, W. Houf, M. Alexander, J. Ren, J. Liu

Chapter 2A
Wet-chemistry-derived nanomaterials and their multidisciplinary applications

2A.1 Bottom-up synthesis

In general, the bottom-up method is a synthesis of nanoparticles (NPs) through chemical reactions with starting materials in the form of atoms, ions, or molecules. The molecular precursor was further decomposed to generate atoms or ions subject

Acknowledgments: This work was supported by the Petroleum Research Fund of the American Chemical Society (53827-UR10), Texas A&M Energy Institute, and the Robert Welch Foundation (Departmental Grant, AC-0006) to analyze the data and write this chapter. The technical support from Texas A&M University-Kingsville and Texas A&M Energy Institute is also duly acknowledged. The authors also gratefully acknowledge the helpful discussions with colleagues, comments, and reviewers' suggestions, which have improved the presentation.

Authors' contribution: S. Bashir and J. Liu collectively conceived this project and oversaw its progress. S. Bashir characterized the RPE and cancer toxicity. C. Li conceived and supervised the projects of nanostructured photocatalysts used for pharmaceutical waste treatment. Z. Wang carried out the ANN deep learning on fingerprint development. Y. Huang, S. Palakurthi, Y. Olaseni, P. Villarreal, and W. Houf have been involved as protégés to obtain knowledge on nanotechnology research. M. Alexander and J. Ren participated in water bioremediation projects. J. Liu performed the synthesis of polymeric membranes and constructed the MFC devices. She also wrote and revised this manuscript based on the inputs from all coauthors.

S. Bashir, Y. Huang, S. Palakurthi, Y. Olaseni, P. Villarreal, W. Houf, The Department of Chemistry, Texas A&M University-Kingsville, MSC161, 700 University Blvd., Kingsville, TX, USA
C. Li, School of Materials Science and Engineering, Lanzhou University of Technology, 287 Langongping Rd, Qilihe District, Lanzhou, Gansu, PR China
M. Alexander, The Department of Chemical Engineering, Texas A&M University-Kingsville, MSC 161, 700 University Blvd., Kingsville, TX, USA
Z. Wang, The Department of Electrical Engineering and Computer Science, Texas A&M University-Kingsville, MSC 192, 700 University Blvd., Kingsville, TX, USA
J. Ren, The Department of Environmental Engineering, Texas A&M University-Kingsville, MSC 213, 700 University Blvd., Kingsville, TX, USA
J. Liu, The Department of Chemistry, Texas A&M University-Kingsville, MSC161, 700 University Blvd., Kingsville, TX, USA; The Department of Electrical Engineering and Computer Science, Texas A&M University-Kingsville, MSC 192, 700 University Blvd., Kingsville, TX, USA; Texas A&M Energy Institute, 1617 Research Parkway, Suite 308, College Station, TX 77843-3372, USA,
e-mail: jingbo.liu@tamuk.edu and jingbo.liu@tamu.edu

https://doi.org/10.1515/9783110739879-003

to nucleation and growth into monodispersed colloids or clusters [1]. This synthesis approach involves the assembly of atoms or molecules into nanostructured arrays due to attractive forces. In nanotechnology, various bottom-up methods have been widely used due to their advantages of producing nanostructures with fewer defects, controlling chemical composition homogeneity, and tuning size and morphology [2]. The bottom-up approaches evaluate "specific characteristics and micro attributes of an individual stock" [3]. The final products were in the nanoscale due to assembling basic units into a cluster with enlarged structures.

Bottom-up (sometimes also called self-assembly) approaches were used for the nanofabrication using chemical or physical forces [4]. This method is widely used to construct desired structures from atom to atom. In the "bottom-up" method, the resources of starting materials were in the state of gases, liquids, or solids. The bottom-up processes essentially encompass chemical synthesis, producing a controlled deposition and materials growth [5]. The associated chemical synthesis was achieved through reactions between solid, liquid, or gaseous states. Solid precursors (or reactants) are mixed by intimate contact with different reactants in the solid-state synthesis. However, this bottom-up approach showed disadvantages, such as complexity, time-consuming, and less effectiveness. The bottom-up method embraces different approaches and a simplified mechanism [6], as summarized in Scheme 2A.1. In this section, bottom-up syntheses used in our laboratory are briefly discussed.

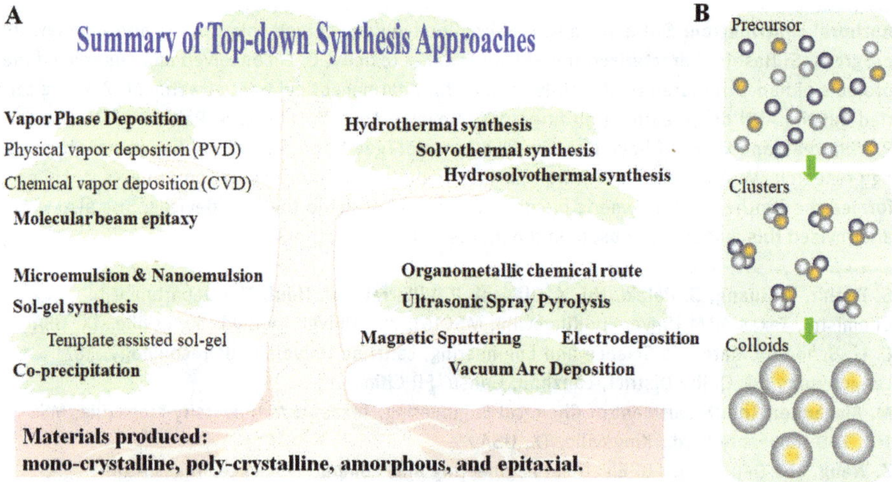

Scheme 2A.1: Bottom-up synthesis used to produce monocrystalline, polycrystalline, amorphous, and epitaxial materials, (A) Summary of different methods, and (B) a simplified mechanism shows the molecular precursors subject to decomposition, nucleation (re-nucleation), and growth to form clusters and colloids.

2A.2 Colloidal synthesis

A colloid is a heterogeneous mixture of microscopically dispersed insoluble particles suspended throughout dispersing media. This mixture has one or more visible boundaries between the different components. Colloids show two properties, the Tyndall effect and Brownian motion. Both dispersed substances (suspended particles) and dispersing medium (continuous phase) were solid, liquid, or gas. Colloids were categorized into six families (Tab. 2A.1). The suspended particles have small sizes, ranging from 1 nm to 1 μm [7]. Colloidal synthesis relies on the precipitation of nanosized particles within a continuous fluid solvent matrix to form a colloidal sol (a sol is a misnomer but is commonly used). A finely and monodispersed system is in a high energy state equivalent to the free energy required to produce increased surface area. The colloidal materials tend to aggregate due to the attractive van der Waals forces (Fig. 2A.1A). The presence and magnitude of an energy barrier to agglomeration depend on the balance of attractive and repulsive forces between the particles. The energy of a colloidal solution arises from the Brownian motion (Fig. 2A.1B) and in the order of $K_b T$.

Tab. 2A.1: A summary of different colloids based on the dispersed phases [8].

Type of colloid	Dispersed phase	Dispersing medium	Examples
Aerosol	Liquid	Gas	Fog
Aerosol	Solid	Gas	Smoke
Foam	Gas	Liquid	Whipped egg
Solid form	Gas	Solid	Marshmallow
Emulsion	Liquid	Liquid	Soy milk
Solid emulsion	Liquid	Solid	Butter, cheese
Sol	Solid	Liquid	Cell fluid
Solid sol	Solid	Solid	Opal, gemstones

2A.2.1 Colloidal stabilization

Stabilizing colloidal dispersions against aggregation is important for application in diversified fields, such as sustainable energy, the food industry, or personal care products. The aggregation includes coagulation when irreversible particle growth and flocculation when reversible. Often the system is an oil-in-water dispersion that was stabilized by adding interfacial and active components such as amphiphiles or proteins [9]. These components segregate to the oil-water interface and stabilize emulsions by reducing interfacial tension. The enhanced rigidity and elasticity of the membrane that forms also help prevent coalescence. Colloidal sols found in paints and pastes also need to be stabilized for long shelf life; this was achieved several ways [10]. First, the

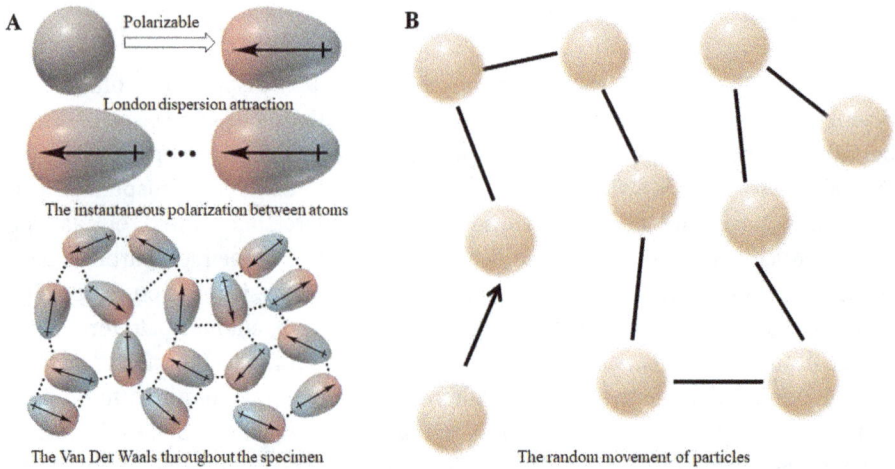

Fig. 2A.1: The characteristics of colloids, (A) van der Waals attraction through instantaneous polarization of nonpolar molecules, (B) the Brownian motion is due to the thermal movement of particles.

balance between the repulsive electrostatic and attractive van der Waals contribution to the total potential energy was adjusted for charged colloidal particles in an electrolyte medium. A barrier to aggregation is created. The phenomenon of charge stabilization was analyzed using the Derjaguin–Landau–Verwey–Overbeek (DLVO) theory [11]. The DLVO theory suggests that a particle's stability in solution depends on its total potential energy function V_T. This theory recognizes that V_T is the balance of several competing contributions (eq. 2A.1A):

$$V_T = V_A + V_R + V_S \tag{2A.1A}$$

where V_A and V_R are the attractive and repulsive contributions; V_S is the potential energy due to the solvent; it usually only makes a marginal contribution to the total potential energy over the last few nanometers of separation. Much more important is the balance between attraction and repulsion. Potentially, these two forces are much larger and operate over a much larger distance. The attractive force is associated with particle separation (D) and was determined using eq. (2A.1B):

$$V_A = -A/\left(12\,\pi D^2\right) \tag{2A.1B}$$

where A is the Hamaker constant and D is the particle separation.

The repulsive potential V_R is a far more complex function, which were estimated by eq. (2A.1C):

$$V_R = 2\alpha\varepsilon r\xi^2 \exp(-\kappa D) \tag{2A.1C}$$

where r is the particle radius, α is the solvent permeability, κ is a function of the ionic composition, and ξ is the zeta potential [12].

The DLVO theory predicts the stability of a colloidal system, which was determined by the sum of these van der Waals attractive (V_A) and electrical double layer (DL) repulsive (V_R) forces, as mentioned above. Both forces exist between particles as they approach each other due to their Brownian motion. This DLVO theory proposes that an energy barrier resulting from the repulsive force prevents two particles from approaching and adhering together (Fig. 2A.1A). However, if the particles collide with sufficient energy to overcome that barrier, the attractive force will pull them into contact where they adhere strongly and irreversibly together, resulting in coagulation. Therefore, the colloidal system was stable if the particles had a sufficiently high repulsion to resist flocculation. However, if a repulsion mechanism no longer exists when flocculation or coagulation will eventually take place [13].

There are five postulations in the DLVO theory: 1) infinite flat solid surface, 2) uniform surface charge density, 3) no re-distribution of surface charge (surface electric potential remains constant), 4) no change of concentration profiles of both counterions and surface charge determining ions (the electric potential remains constant), and 5) solvent exerts influences via dielectric constant (no chemical reactions between the particles and solvent). These assumptions will result in deviations when the DLVO theory is applied to explain observations. This theory is widely used to provide guideline control stability of the colloidal system. The colloids were stable by adjusting the repulsive and attractive forces when the repulsion was more influential than attraction. Two fundamental mechanisms, steric and electrostatic repulsion, are the major factors to affect dispersion stability [14].

Steric repulsion: This repulsive force involves polymeric dispersing agents added to the system to stabilize the particles. The polymer with long chairs was adsorbed onto the particle surface, preventing the particles from close contact. If enough polymeric molecules were adsorbed, the thickness of the coating would be sufficient to overcome the van der Waals attraction. Therefore, the particles were separated by steric repulsions between the polymer layers (Fig. 2A.2A).

Electrostatic repulsion: This mechanism is also called charge stabilization due to the distribution of charge species. These repulsive Coulomb forces on particle interaction can counterbalance van der Waals forces (Fig. 2A.2B). When two charges are

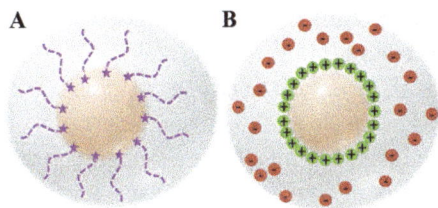

Fig. 2A.2: The fundamental mechanisms that affect colloidal dispersion stability, (A) the simplified scheme of steric repulsion and (B) electrostatic repulsion is used to stabilize colloids.

moving on parallel paths, electrostatic repulsion occurs. These moving entities will form two currents in parallel and are attracted to each other due to the magnetic forces. The magnetic forces reduce the electrostatic repulsion; on the other hand, the space-charge limit was increased [15].

In our research lab, steric stabilization has been widely used to control the size and morphology of the NPs. The fatty acid, a family of long-chain molecules, was chemically attached or physically attracted to colloidal particles, creating a repulsive force due to the steric hindrance (Fig. 2A.3A). The chains act as a dispersing agent and interpenetrate when the particles approach. This steric stabilization is highly applicable in nonaqueous media, unlike charge stabilization. A wide range of particles was produced with controllable size distribution and uniformity. The concentration and characteristics of surfactants need to be selected carefully to achieve steric stabilization (see Chapter 1, Section 1.2.2). At low concentrations, it was found that polymeric surfactant chains can attach themselves to two (or more) particles, resulting in flocculation. At higher polymer concentrations, the nonadsorbed polymeric molecules can lead to depletion flocculation, as recognized by Asakura and Oosawa [16].

In general, the colloids with high zeta potential (ζ) are stable for the long term due to the high degree of electrostatic repulsion between adjacent. Consequently, the charged particles in dispersion are separated to avoid close contact or agglomeration (Fig. 2A.3B). As the indicator of colloidal stability, the small potential indicated that attractive forces might exceed this repulsion, causing the collapse of dispersion, further leading to flocculation. Either high negative or positive zeta potential confers the stabilities of colloids [17]. The schematic diagram (Fig. 2A.3C) showed the relationship between free energy and particle agglomeration, according to DLVO theory. The net energy was determined using the sum of the DL repulsion and the van der Waals attractions. The particles experience both forces, while the repulsive forces dominate as particles approach each other. The net energy showed an exponential increase with particle agglomeration.

Most colloidal dispersions in aqueous media carry an electric charge on the surface; the origin of the charge depends on the surrounding medium or the nature of these particles. There are three types of origins of surface charge, ionization of surface group, ions loss from the crystal lattice, and adsorption of charged species. The ionization of surface groups mainly depends on their acidity or basic strengths. When the acidic groups on the particle surface are dissociated, the negatively charged surface is obtained. The particle surface can absorb some surfactants; in this case, if it is a cationic surfactant, then the positively charged surface was obtained. On the other hand, the anionic surfactants will cause the negatively charged surface of particles [19].

Theoretically speaking, the zeta potential is the potential difference between the dispersion medium and the stationary layer of fluid attached to the dispersed particle. This electrokinetic potential occurred in the interfacial DL L [20], particularly at the location of the slipping plane relative to a point in the bulk fluid away from the interface (Fig. 2A.4A). The liquid layer around the particles was observed, a Stern

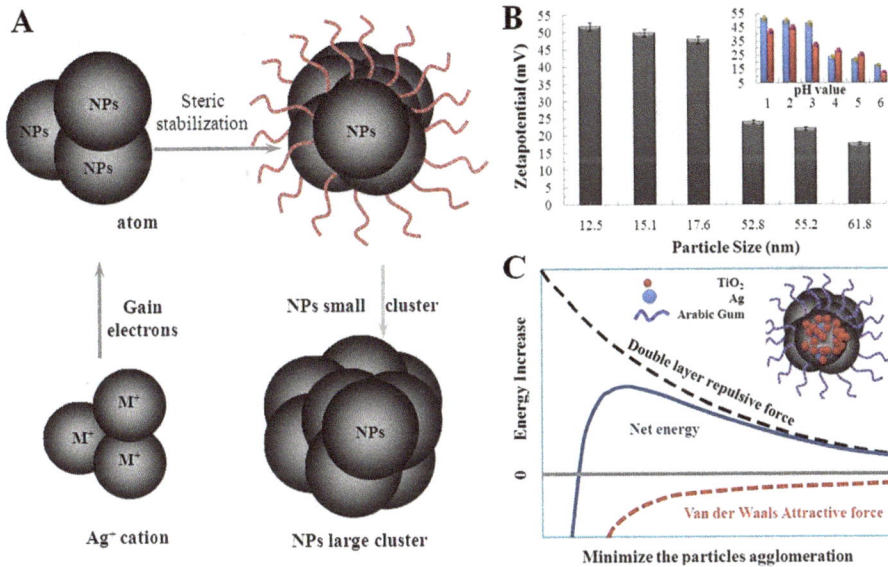

Fig. 2A.3: Diagram of stabilizing colloidal dispersions against aggregation by adding interfacially active components. (A) The steric stabilization, resulting in uniform nanoparticle formation, (B) zeta potential as a measure of ultrafine particle stability, and (C) the net energy depends on the balance of attraction and repulsion forces. Image adapted from [18].

layer (inner region) and a Gouy layer (the diffuse layer in the outer region), as shown in Fig. 2A.4B. The ions are bonded with the particles solidly; conversely, ions are loosely distributed in the diffusion layer. Only the ions within the boundary move when the particles move, caused by gravity. However, ions beyond the boundary stay with the liquid dispersant.

As mentioned above, the zeta (ζ) potential is the electrokinetic potential at the slipping plane. The magnitude of zeta potential can indicate the stability of particles. Because particles with great positive or negative zeta potential tend to repel each other and the aggregate behavior will not be present in this situation. The aggregation can easily happen when the zeta potential is small, owing to repel force absent. Generally speaking, the absolute value of zeta potential over 30 mV contributes toward a stable colloid [21]. Two factors, net electrical charge and location of the slipping plane will determine the magnitude of zeta potential. As shown in Fig. 2A.4B, zeta potential is different from either the Stern potential or electric surface potential due to their different locations.

Electrostatic stabilization of charged colloidal particles is another important method to control particle size. In this mechanism, four perspectives are 1) surface charge density, 2) electric potential at the proximity of the solid surface, 3) van der Waals attraction potential, and 4) Interactions between two particles: DLVO theory (see the section of Derjaguin, Landau, Verwey and Overbeek theory (DLVO theory).

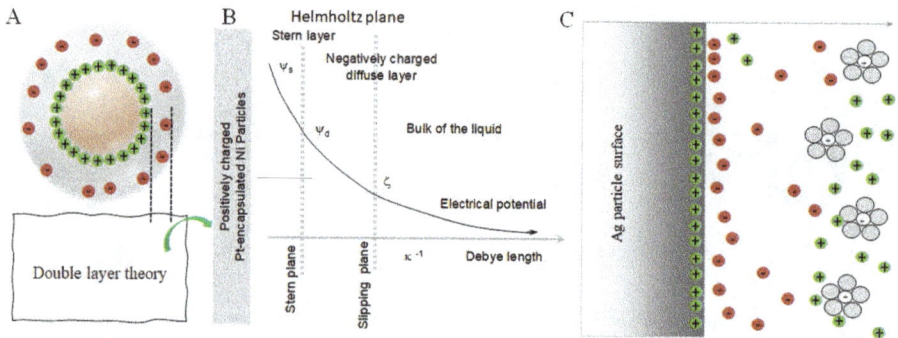

Fig. 2A.4: Schematic demonstration showing the ionic concentration and potential difference, (A) double-layer theory is used to explain the zeta potential and ion distribution in a colloidal suspension. (B) The function of distance from the charged surface of a particle suspended in a dispersion medium, and (C) a case study of Ag nanoparticle colloidal suspension with a positive surface charge. Images were adapted from [22].

Surface charge density: When a solid is dispersed in a polar solvent (also known as an electrolyte solution), a charge is developed on the surface through one or more mechanisms. The mechanisms include preferential adsorption of ions, dissociation of the surface charged species, the isomorphic substitution of ions, accumulation or depletion of electrons at the surface, and physical adsorption of charged species onto the surface. The Nernst equation is commonly used to determine the electric potential (E) at the proximity of a solid surface for a given system:

$$E = E_0 + (R_g T / n_i F) Ln(a_i) \tag{2A.2A}$$

where E_0 is the standard electrode potential (in V); R_g, the ideal gas constant (0.02816 atm L/mol K); T, the absolution temperature in Kelvin; F, the Faraday constant (96,485 C/mol or kJ/mol); and a_i represents the valence state of ions (dimensionless).

In particular, the surface charge in oxides depends on the types of ions adsorbed on the solid surface. This surface charge was derived from ions' preferential dissolution or deposition, which is the charge determining ions (also called co-ions). Two ions, proton (H^+ or hydronium H_3O^+) and hydroxide (OH^-), are the commonly known determining co-ions in the typical metal oxide system. The surface change and concentrations of the co-ions are inter-correlated to each other. The point of zero charges (PZC) and acidity (pH values) are two key factors determining the surface potential of the typical metal oxide system (eq. 2A.2B). The PZC values of oxide systems in an aqueous solution are tabulated in Tab. 2A.2. The system pH was calculated using eq. (2A.2C):

$$E = 2.303 R_g T (PZT - pH)/F \tag{2A.2B}$$

$$pH = \log([H_3O]^+) \tag{2A.2C}$$

Tab. 2A.2: A summary of zero charge (PZC) points of metal oxides in aqueous [23].

Metal oxides	PZC	Metal oxides	PZC
Fe_2O_3	8.6	TiO_2 (calcined)	3.2
ZnO	8.0	TiO_2	6.0
Cr_2O_3	8.4	SnO_2	4.5
ZrO_2	6.7	WO_3	0.5
V_2O_5	1.0–2.0	β-MnO_2	7.3
SiO_2	2.5	δ-MnO_2	1.5
SiO_2 (quart)	3.7	Al-O-Si	6.0

The electric potential at the interface of the solid surface: When a surface charge density of a solid surface is established, there is an electrostatic force between the solid surface and the charged species in the proximity to segregate positive and negatively charged species. However, Brownian motion (Fig. 2A.1B) and entropic force also exist, which homogenize the distribution of various species in the solution. Two opposite charged ions, surface charge determining ions and counterions, always exist in the solution. Although the system of interest maintains its charge neutrality, distributions of charge determining ions and counterions are inhomogeneous and vary in the solid surface's proximity. The distributions of both ions are mainly controlled by combining the following forces: Coulombic force (also known as electrostatic force), entropic force or dispersion, and Brownian motion [23].

The net ion concentration depends on the magnitude of determining ion and counter ions. If the surface is positively charged, the determining ions of the surface charge show the lowest concentration near the solid surface and increase with the distance from the surface. On the contrary, the counterions show the highest concentration near the solid surface and decrease with the distance from the surface. The so-called double-layer structure was formed due to these inhomogeneous ion distributions in the proximity of the solid surface (Fig. 2A.4B) [24]. The DL consists of the Stern and Gouy layers (also called the diffuse DLs). The Stern layer is formed between the solid surface and the Helmholtz plane. In this Stern layer, the electric potential decreased linearly through the tightly bound layer of solvent and counterions. Beyond the Helmholtz plane, the Gouy layer was observed until the counter ions reached the average concentration in the solvent. In this Gouy diffuse layer, the counter ions diffuse freely, and the electric potential does not reduce linearly. The electric potential (E) change is inversely exponential to the Debye-Hückel screening strength ($1/\kappa$) and was approximated in eq. (2A.3A). The Debye-Hückel screening strength, corresponding to the solvent dielectric constant valence of counterion and counterion concentrations, was determined using eq. (2A.3B):

$$E \propto e^{(-\kappa(h-H))} \tag{2A.3A}$$

$$\kappa = \left\{ (F^2 \sum_i C_i Z_i^2) / \varepsilon_r \varepsilon_0 R_g T \right\}^2 \tag{2A.3B}$$

where the κ^{-1} is known as Debye-Hückel screening strength; ε_r, the dielectric constant of the solvent; ε_0, the permittivity of vacuum; and C_i, the concentration of counterion of type i; Z_i, the valence of counterion of type i.

Based on the above exponential equations (eqs. (2A.3A) and (2A.3B)), the increase in counterion concentration results in a deduction of electric potential at the proximity of the solid surface. When the valence state (Z_i) increases, and the electric potential (E) also exponentially decreases. On the other hand, the increase in the dielectric constant of the solvent (ε_r) and absolute temperature collectively increases electric potential (E). The thickness of the DLs also highly depends on the valence and concentration of the counterion and the dielectric constant of solvents [25]. The Gouy diffuse layer ended at the zero point of electrical potential, where its distance is infinite from the solid surface. However, the thickness of the DL is approximately 10 nm in reality; otherwise, the system's stability will be subject to susceptibility. The above discussion is assumed that the solid surface is flat and the electrolyte solutions are applicable. If the surface is smooth enough, the above discussion is applicable to explain the observation of curved surfaces due to their constant surface charge densities. If the particles are spherical, these assumptions are valid because the particles are evenly dispersed in an electrolyte solution. The charge distribution of the particles is not inter-influenced due to their large distance. In reality, the interactions between particles are much more complex than in the above discussion. The electric potential and surface charge are two factors that influence the particle interaction, such as electrostatic repulsion (ϕ_R), as approximated by eq. (2A.3C). This electrostatic repulsion arises from the electric surface change when the particles come into contact. The double-layer thickness can result in attenuation to a different extent. The DL has no overlap when the particles are located far away, indicating that the electrostatic repulsion was negligible. When the particles approach each other, the repulsion dominates and keeps the particles separated, preventing agglomeration:

$$\phi_R = 2\pi\varepsilon_r\varepsilon_0 RE^2 \exp(-S\kappa) \tag{2A.3C}$$

Van der Waals attraction potential: If the particles are sized in micrometers or less (known as NPs), they are readily dispersed in a solvent. Under this condition, the van der Waals attraction force and Brownian motion play important roles [26]. The gravity influence of these NPs was ignored due to their ultra-small size. The van der Waals attraction is an intermolecular nonbonding force that is relatively weak compared to the bonding forces. This attractive force only exists in a short distance, normally among a couple of adjacent molecules. The combined van der

Waals attraction and Brownian motion result in particle agglomeration. The Brownian motion is associated with the thermal motion of molecules in the particles, and the van der Waals attraction is the sum of the molecule interaction. The total interaction energy was calculated using eq. (2A.4A). The negative sign (–) in the equation indicates the intrinsic nature of the attractive forces between two adjacent particles [27]:

$$\phi_A = -A/6 \left\{ 2r^2/(S^2 + 4rS) + 2r^2/(S^2 + 4rS + 4r^2) + Ln\left[(S^2 + 4rS)/(S^2 + 4rS + 4r^2) \right] \right\}$$

(2A.4A)

where ϕ_A stands for the total interaction energy or attraction potential; A, a positive constant termed the Hamaker constant (in the order of 10^{-19} to 10^{-20} J, the values of representative materials listed in Tab. 2A.3); r, stands for the radius of the spherical particles; and S, a distance of two separated particles [28].

Tab. 2A.3: Hamaker constants (α_i (10^{-20} J)) in a vacuum and across the water (vacuum/water) of inorganic materials interacting against four materials [30].

Materials	Hamaker constants (α_i (10^{-20} J))			
	Silica	Silicon nitride	Alumina	Mica
BaTiO$_3$	10.1/0.62	16.5/4.84	15.2/3.55	12.4/1.98
BeO	9.67/0.95	15.4/3.87	14.8/3.50	11.9/2.06
C (diamond)	13.7/1.71	22.0/7.94	21.1/7.05	17.0/4.03
CaCO$_3$	8.07/0.69	12.9/2.53	12.3/2.17	9.94/1.35
CaF$_2$	6.70/0.45	10.6/1.17	10.3/1.10	8.26/0.73
CdS	8.03/0.72	13.1/3.12	12.0/2.15	9.86/1.43
KCl	5.94/0.37	9.53/0.73	9.00/0.51	7.31/0.46
MgAl$_2$O$_4$	9.05/0.85	14.5/3.39	13.8/2.97	11.2/1.79
MgF$_2$	6.15/0.36	9.74/0.66	9.42/0.69	7.57/0.50
MgO	8.84/0.81	14.2/3.26	13.5/2.79	10.9/1.69
Mica	8.01/0.69	12.8/2.45	12.2/2.15	9.86/1.34
NaC1	6.45/0.44	10.3/1.17	9.77/0.88	7.93/0.66
PbS	5.37/−0.08	8.88/0.64	7.90/−0.20	6.57/−0.03
6 H-SiC	12.6/1.52	20.3/7.22	19.2/6.05	15.5/3.54
β-Si$_3$N$_4$	10.8/1.17	17.3/5.13	16.5/4.43	13.3/2.61
SiO$_2$ (quartz)	7.59/0.63	12.1/2.07	11.6/1.83	9.35/1.16
SrTiO$_3$	9.44/0.57	15.4/4.02	14.2/2.98	11.6/1.69
TiO$_2$	9.46/0.69	15.4/4.26	14.2/3.11	11.6/1.83
Y$_2$O$_3$	9.24/0.89	14.9/3.80	14.0/3.11	11.4/1.89
ZnO	7.38/0.58	12.0/2.30	11.1/1.58	9.06/1.08
ZnS (cubic)	9.69/1.02	15.7/4.56	14.6/3.55	11.9/2.19
3Y-ZrO$_2$	11.4/1.25	18.4/5.89	17.4/4.95	14.1/2.89

Note: Data collected at room temperature, calculated using Lifshitz theory [31].

A simplified equation (eq. (2A.4B)) was used if the distance of particles is significantly smaller than the radius of the particle with equal sizes ($S \ll r$):

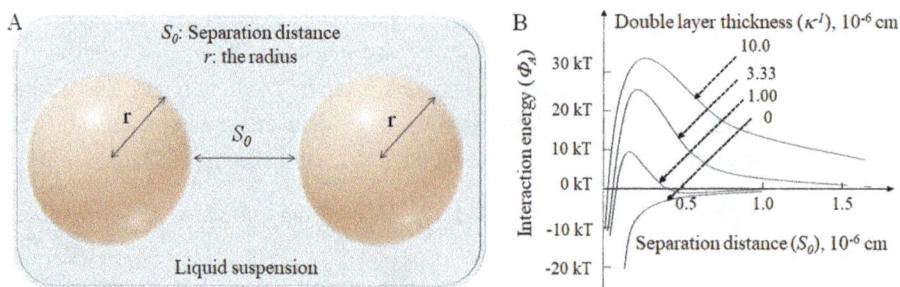

Fig. 2A.5: Electrostatic stabilization, (A) van der Waals interactions between two molecules over two spherical particles with equal size, (B) relationship between electric potential and separation distance between two particles.

$$\phi_A = -Ar/12S \tag{2A.4B}$$

If the two spherical particles have different radii, r_1 and r_2, the equation to calculate total interaction energy was simplified as follows:

$$\phi_A = -Ar_1r_2/6S(r_1+r_2) \tag{2A.4C}$$

If two parallel particles are flat in shape with a plate thickness of δ, the total interaction energy per unit area were determined using the following equation:.

$$\phi_A = -A/12\pi\left[(S)^{-2}+(S+2\delta)^{-2}+(S+\delta)^{-2}\right] \tag{2A.4D}$$

If two identical spheres are separated with a distance of S interact, the total interaction energy per unit area was determined using (and shown in Figs 2A.5A and 2A.5B)

$$\phi_A = -A/12\pi(S)^{-2} \tag{2A.4F}$$

2A.2.2 Kinetic modeling

Collodial compositional chemistries of NPs have drawn researchers' attention to fundamental and applied research due to their unique optical, electrical, and catalytic properties, as reported by Ren and Duan et al. [31]. These nanomaterials were composed of metals [32], metal oxides [33], ceramics [34], semiconductors [35], and metal-organic frameworks (MOFs) [36]. The colloidal-derived NPs were precisely tuned to control their chemical composition and architectures, such as particle size, shape, crystal facet, and surface chemistry [37]. In most cases, the ligands with small molecular mass play critical roles in modulating the nanomaterial growth due to their interaction with precursors and NPs. As a result, the structure and morphology of the final NPs with multicomponent were well-controlled [38]. The composition and morphology of colloidal inorganic compounds highly depend on kinetics and thermodynamics [39]. For decades, the mechanisms of the nanoparticle nucleation and growth have been described by the LaMer burst nucleation (Fig. 2A.6A) [40], followed by Ostwald ripening (Fig. 2A.6B) [41] to describe the change in the particles size. Originally, this model was proposed by Reiss [42], with well-accepted postulation being developed by Lifshitz–Slyozov–Wagner, LSW theory [43]. This theory was dominantly used to explain nucleation and growth until Watzky and Finke formulated the two-step approaches. In 1997, Watzky and Finke proposed a two-step mechanism (Fig. 2A.6C), slow continuous nucleation followed by a rapid autocatalytic surface growth, to understand the formation of iridium (Ir) NPs [44]. The "pseudo-temporal separation of nucleation and growth" was observed owing to rapid growth compared to slow nucleation. However, this Finke–Watzky two-step model did not consider the role of ligand when it is used to provide a mechanism to explain the overlap of nucleation and growth [45].

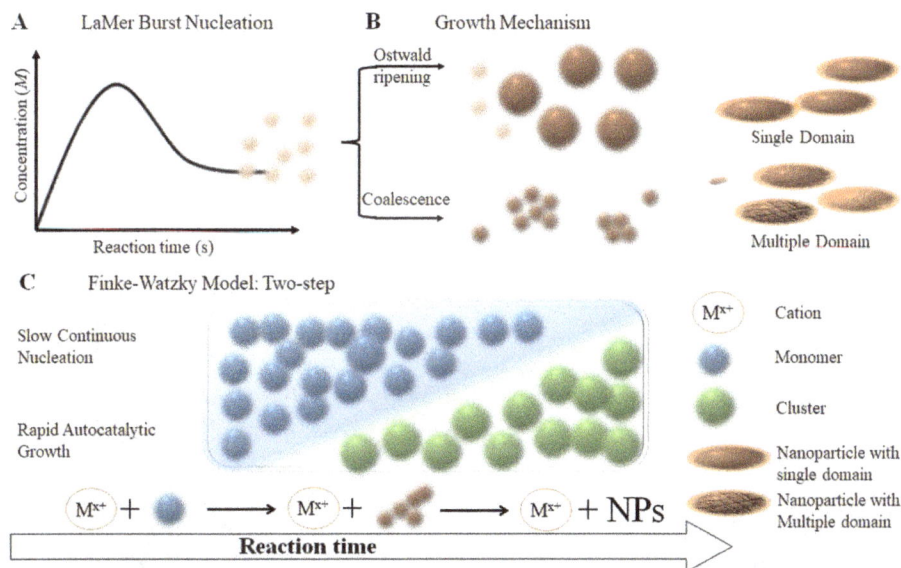

Fig. 2A.6: The mechanism of nanoparticle formation, (A) traditional LaMer Burst nucleation model, (B) Ostwald ripening, and (C) Finke–Watzky two-step model.

Mozaffari, Li, and coworkers reported a combined in situ small-angle X-ray scattering and kinetic modeling of colloidal metal nanoparticle size control [46]. The palladium (II) acetate (Pd(CH$_3$CO$_2$)$_2$, tetragonal structure) and trioctylphosphine (P(C$_8$H$_{17}$)$_3$, abbreviated TOP) precursor were used for kinetic evaluation. This method quantitatively captured the roles of ligand-metal binding in controlling the synthesis nucleation and growth on the surface of NPs. Their research data showed that the kinetic rate constant depends on the particle size and number [47]. The solvents used in the colloidal metal synthesis also affect the metal-ligand binding and the coverage of Pd NPs. It was found that toluene as a solvent produces small particles (averaged at 1.4 nm), while the pyridine resulted in a larger particle size of 4.3 nm [48]. Metal concentration, ligand-to-metal ratio, and solvents affect the nucleation and growth kinetics of the metal NPs. The author proposed a "ligand-based model" (Fig. 2A.7 and eqs. (2A.5A–D)), taking "the metal-ligand binding as reversible reactions" into consideration. This model is an advancement of the two-step mechanism proposed by Finke and Watzky, which has been considered an improved mechanism to explain metal nanoparticle synthesis compared with the traditional LaMer model [49].

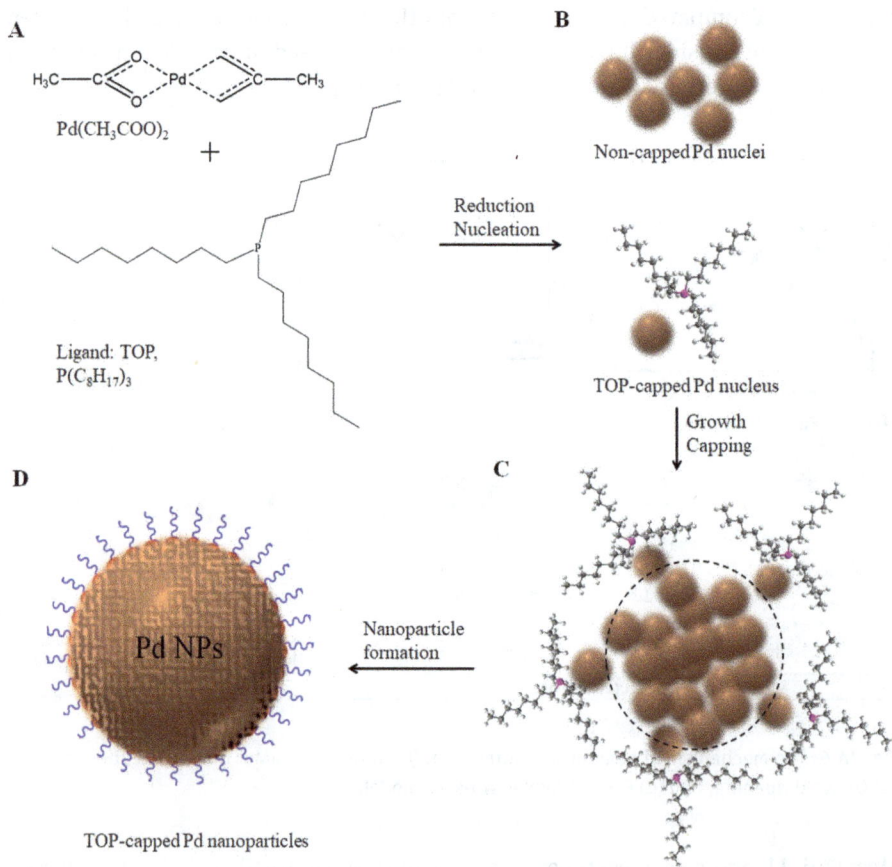

Fig. 2A.7: The "ligand-based model" schematic diagram proposed by Mozaffari and coworkers, (A) the Pd-containing reactant and TOP as capping agent; (B) the noncapped Pd nucleation and TOP-capped Pd nucleation; (C) the continuous growth and capping; and (D) the Pd nanoparticle formation.

In this "ligand-base model," the authors proposed two more reversible steps [39] and slow nucleation and rapid catalytical growth, as proposed by Finke–Watsky. There have three postulations in this ligand-based model, 1) the fraction of atoms that become core atoms are not considered, 2) Autocatalytic growth and binding of ligands are independent of nanoparticle size and ligand coverage, and 3) the stoichiometry of the ligand–metal precursor binding is approximated to be unit 1. This four-step model (eqs. (2A.5A) and (2A.5D)) provides a more accurate and thorough explanation for metal nanoparticle formation [50]. The first two equations (2A.5A) and (2A.5D) are aligned with the slow nucleation and rapid autocatalytic surface growth in the Finke–Watzky model. The reaction rate constant ($k_{1-nul.}$) in step 1 (eq. (2A.5A)) is used to evaluate the combined pseudo-elementary reduction–nucleation. The reaction rate constant ($k_{2-G.}$) in step 2 (eq. (2A.5B)) is for the autocatalytic surface growth. Similarly, in step 3 (eq. (2A.5C)), the $k_{3-F.}$ is the forward reaction rate constant for the precursor–ligand binding, while the $k_{4-F.}$ is associated with the ligand-particle binding in the forward reaction (eq. (2A.5D)). Two equilibrium constants (K_{3-eq} and correspond to ligand-precursor (eq. (2A.5C)) and ligand-nanoparticle (eq. (2A.5D)) binding processes, respectively [51]:

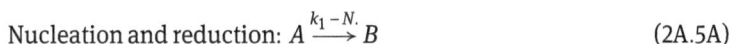

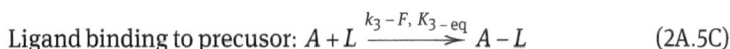

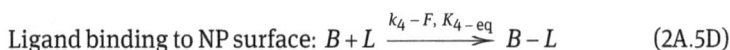

$$\text{Nucleation and reduction: } A \xrightarrow{k_1 - N.} B \tag{2A.5A}$$

$$\text{Autocatalytical surface growth: } A + B \xrightarrow{k_2 - G.} C \tag{2A.5B}$$

$$\text{Ligand binding to precusor: } A + L \xrightarrow{k_3 - F, K_3 - eq} A - L \tag{2A.5C}$$

$$\text{Ligand binding to NP surface: } B + L \xrightarrow{k_4 - F, K_4 - eq} B - L \tag{2A.5D}$$

The authors carried out experiments to examine the relations between the reaction kinetics and TOP-Pd interaction during the Pd nucleation and growth. It was found that the rate of Pd nucleation was fast at its early formation stage (Fig. 2A.8A); within 2,000 s, a rapid decrease was noticed, followed by a subsequent slow decrease. During the particle growth, it was found that the growth rate of the Pd nucleus drastically increased until reaching a maximum in 1,800–2,000 s. Afterward, an exponential decrease and the slight decrease were found, depending on prolonged reaction time (Fig. 2A.8B). The ratio between growth and nucleation showed an essentially similar trend as the growth rate (Fig. 2A.8C). With the precursor ($Pd(CH_3COO)_2$) consumption, its concentration decreases, decreasing the autocatalytic surface growth rate. The percentage of TOP-capped Pd NPs on the surface is attributed to the ligand's dissociation rate (TOP), which is the reversible reaction of eq. (2A.5C). In addition, with the Pd formation (denoted as B), its percentage on the surface increases, causing the growth rate to decrease. The TOP capping showed an influence on the nucleation and growth of NPs. The "ligand-based model" includes metal-ligand interactions corresponding to the number of free surface sites B. The ligand-nanoparticle binding consequently affects the growth/nucleation ratio. The model enabled the accurate

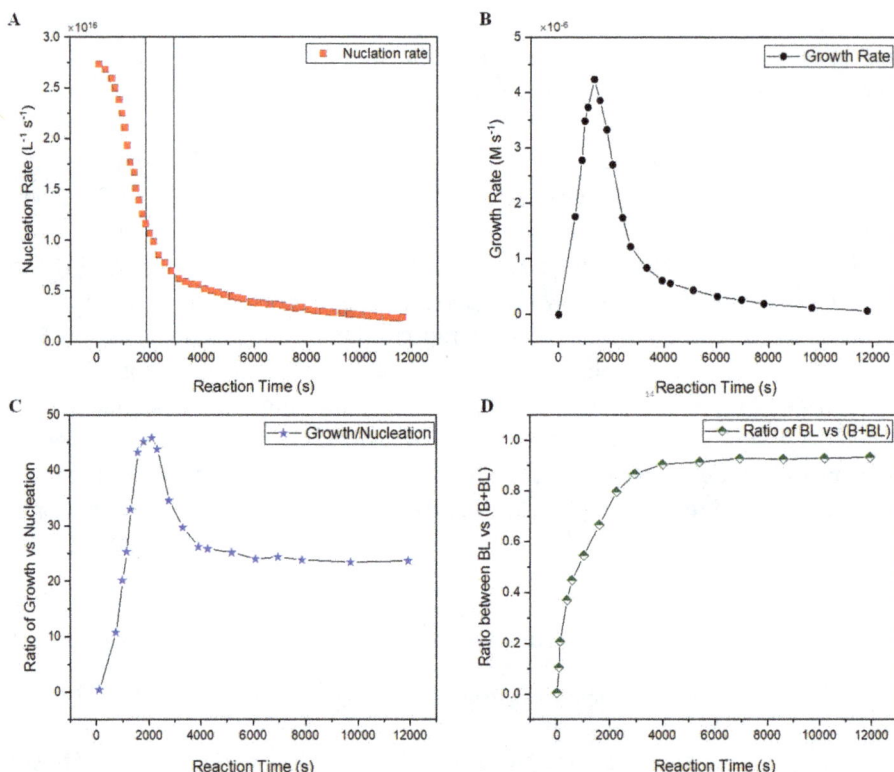

Fig. 2A.8: The formation mechanism of Pd nanoparticles, (A) nucleation rate; (B) growth rate; (C) ratio between growth versus nucleation; and (D) (BL)/(B + BL) ratio (BL: precursor-ligand binding). Reaction conditions: 10 mM Pd(OAc)$_2$ in 50:50 toluene:hexanol, trioctylphosphine: Pd = 2:1, T = 100 °C. The images were adapted with permission from [39].

predictions of an increase in both the "fraction of precursor coordinated with TOP (AL) and in the coverage of TOP on the particle surface (BL)," as reported by Mozaffari, Li, and coworkers [39].

The shape and architecture of NPs were tuned using different templates and ligands via colloidal synthesis. However, the small-molecule-assisted colloidal synthesis strategy showed intrinsic drawbacks and technical hindrances, such as (1) lattice mismatch in the multicomponent structures, (2) unable to produce unique structures, and (3) lack of stability. Due to the lattice mismatch in the multicomponent materials, the lattice strain was created between neighboring lattice units, hindering the integration of diverse components into large structures [52]. The core–gap–shell nanostructures with highly ordered arrays are not obtainable due to the fixed interparticle distance and the gap between the core and shell [53]. Research indicated that interparticle distance between two metallic entities affects the optical properties of the designated core–shell metallic structures. Researchers employed functional macromolecules to interact with precursors, chemically and physically, to resolve these problematic issues. These attractions between inorganic precursors and NPs allowed for producing nanocrystals and the construction of multicomponent nanostructures [54].

2A.2.2 Macropolymer templation

This section summarizes the structures and morphologies using colloidal wet chemistry to produce inorganic NPs with functional and reactive macromolecules as templating agents (Fig. 2A.9). The macromolecules and polymeric colloidal systems exhibit intrinsic reactivity for modulation of particle growth. Three families of polymeric agents were chosen to highlight the achievement using "macromolecule-enabled colloidal synthesis of nanomaterials," as reported by Duan and co-workers [31]. Based on the functions of macromolecules involved in the colloidal synthesis, the polymers are categorized into three types: (1) synthetic block copolymer (Fig. 2A.9A) [55, 56], (2) bioinspired multifunctional adhesive polymers (Fig. 2A.9B) [57], and (3) biomacromolecules (Fig. 2.9BC) [58]. The properties of these functional polymers highly depend on the "pendant functional groups on their monomeric units." The chemical interactions between the monomer and inorganic precursors showed similarity to the binding between small molecules. The fundamental mechanism applied in "small molecule-assisted synthesis" was applied to the system using a functional macropolymer as a template. These macromolecules displayed several intrinsic advantages when used in colloidal synthesis. First, the monomer's covalent coordination between functional groups gives a "multivalency effect," providing multiple reactive sites. These binding sites allow the immobilization of inorganic precursors into the polymer backbones [59]. The large size of the functional macropolymer and coordination between precursors in the backbone of polymeric groups restrict the precursors' diffusion. The chemical transformation of starting materials (or inorganic precursors) was confined within a nanoscale region. Second, the multiple anchoring groups provided a pathway to stabilize the capping ligands, which showed affinity to the surface NPs, as shown in Fig. 2A.7C. The diversity of surface chemistry using functional polymer provides diversified morphology of the final products, such as nanosphere, nanorod, nanosheet, core–gap–shell, core satellite, and nano frameworks (Fig. 2A.9D). In particular, the core–shell structure composed of dissimilar materials was produced using macromolecule-assisted colloidal synthesis. The large lattice mismatch was overcome using the multifunctional groups to stabilize the surface reactivities of the NPs. Third, multicomponent nanomaterials were produced using functional groups by integrating the different configurations. In this case, the functional macropolymer act as directing ligand to adjust the interfacial energy between adjacent monomers [60]. The growth rate of different components was integrated, producing nanomaterials with multi-components. Additionally, the polymeric modulations provide a different pathway to "immobilize inorganic precursors within their backbones." As a result, the templated growth of NPs resulted in the formation of single-crystalline NPs. The macro-polymer components are integrated into the lattice structure or become parts of the lattice, as observed in monodispersed colloids [61].

In the colloidal suspension, the macro-sized functional polymers (Fig. 2A.9B) occupy large volumes in collapsed and stretching conditions and act as "a spacing layer" [62]. As a result, the multicomponents in the nanomaterials were separated to avoid agglomeration due to thermal motion. The structural formula of the polymer is the determining factor of the thickness of these "spacing layers," which were adjusted from angstroms and nanoscales, depending on the molar mass and solubility in the medium of these polymers. Smart macro-sized polymers were responsive to external signal changes, such as pressure, temperature, the intensity of light, and electrical or magnetic to adjust their structure and properties [63]. These functional polymers were used to tune the inorganic nanostructures by manipulating their molecular geometry, surface chemistry, and elemental compositions. The precise and accurate nano-synthesis control was achieved by applying macro-polymers as spacing layers. In addition, the functional building blocks provide another pathway to synthesize nanomaterials by varying their structure through a "living polymerization" strategy (Fig. 2A.9C) [64]. These designed polymeric structures and architectures were manipulated into the form of nanosphere, nanosheets, nanofibers, and core shells (Fig. 2A.9D). Since 1999, "click" chemistry has been used to engineer nanostructures with different formulations using bioconjugation through joining specific biomolecules via chemical bonding and intermolecular interactions. The intrinsic reactivities of the naturally occurring polymers have been utilized to generate colloidal products by mincing natural evolutions. The "click" chemistry was useful in chemoproteomic, pharmacological, and biomimetic fields, allowing for the detection, localization, and qualification of biomolecules [65, 66].

The macro-sized polymeric templation enabled precise and accurate control of nanomaterials with tailorable catalytical, chemical, electrical, magnetic, and optical properties. The complex polymeric geometries were tailored using naturally existing biomacromolecules' intrinsic properties and functionality and natural product extracts. The "click" chemistry integrates a series of macromolecules via either chemical bonding or intermolecular interaction to produce structure-controlled nanomaterials with different formulations and architectures [67] (Fig. 2A.9D).

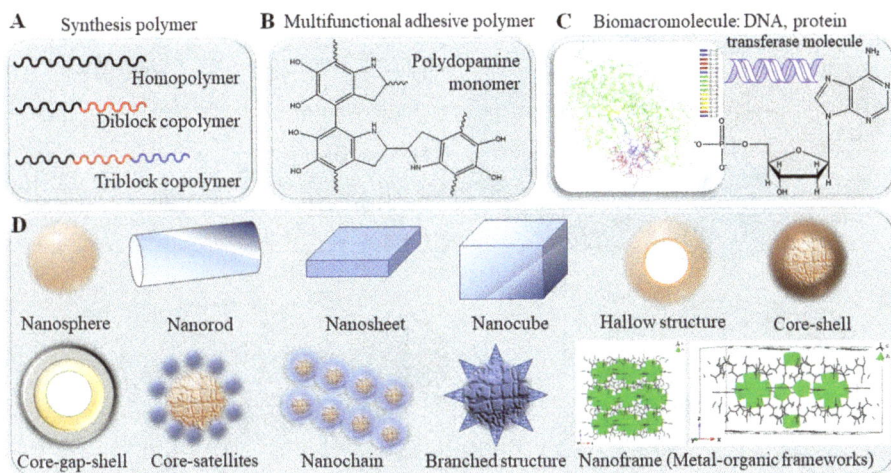

Fig. 2A.9: Schematic illustration using functional macromolecule colloidal synthesis to tune the architecture of nanoparticles, (A) synthetic polymers (homopolymer, biblock copolymer, and triblock copolymer), (B) multifunctional adhesive polymer providing anchoring effect and multiple binding sites, (C) naturally existing biomacromolecules, for example, deoxyribonucleic acid (DNA) and proteins, (D) 1D, 2D, and 3D structures including biomimetic structures.

2A.3 Water remediation

The United Nations has designated 2005–2015 as the "Water for Life" decade and estimates that one-fifth of the world population faces a water shortage [68]. This complex problem requires multifaceted approaches to reducing water stress on photosynthesis, water use efficiency, and irrigation water productivity [69]. The traditional debate on water scarcity focuses on surface water and groundwater (called blue) while assessing the scarcity of evapotranspiration (green) water [70]. Both natural and anthropogenic factors collectively result in water stress, including population explosion, industrialization, and globalization [71]. Green water is the main source of food, feed, fiber, timber, and bioenergy, although blue water is used for irrigation [72, 73]. Therefore, we must understand factors causing water stress, focusing on freshwater scarcity. Three variables, green water footprint (WF_g), maximum sustainable level to green water footprint ($WF_{g,m}$), and green water scarcity (WS_g), are used to express green water availability and appropriation.

The green water footprint (WF_g) reflects the human appropriation of its flow to crop production, livestock grazing, wood production, and urban area uses (Fig. 2A.10A). To support current water consumption and maintain healthy biodiversity, the maximum sustainable level of green water footprint ($WF_{g,m}$) was about 18 × 10^3 km^3/year [74]. It was found that 56 % of the green water flow available has already

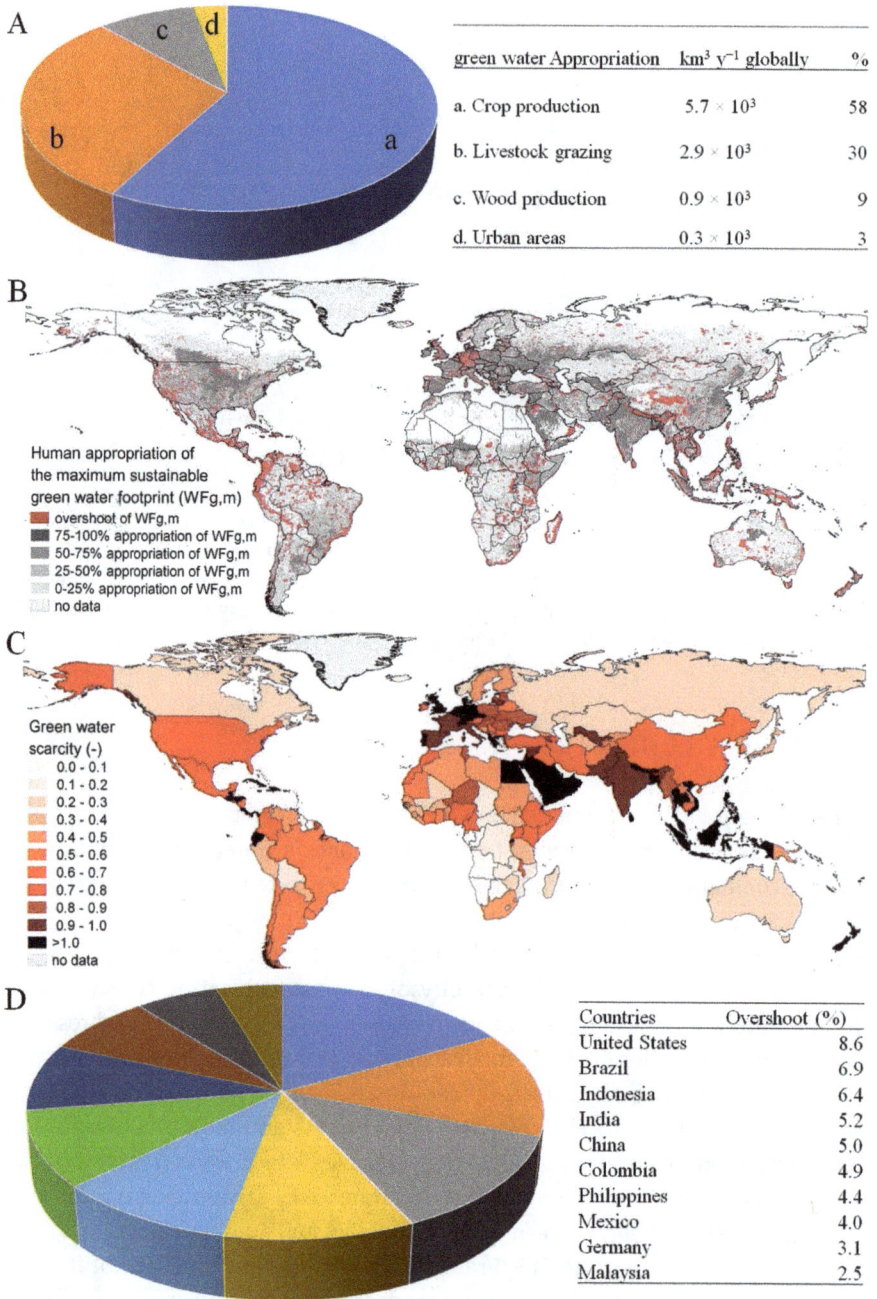

A

green water Appropriation	km³ y⁻¹ globally	%
a. Crop production	5.7×10^3	58
b. Livestock grazing	2.9×10^3	30
c. Wood production	0.9×10^3	9
d. Urban areas	0.3×10^3	3

B

Human appropriation of
the maximum sustainable
green water footprint (WFg,m)

- overshoot of WFg,m
- 75-100% appropriation of WFg,m
- 50-75% appropriation of WFg,m
- 25-50% appropriation of WFg,m
- 0-25% appropriation of WFg,m
- no data

C

Green water
scarcity (-)
- 0.0 - 0.1
- 0.1 - 0.2
- 0.2 - 0.3
- 0.3 - 0.4
- 0.4 - 0.5
- 0.5 - 0.6
- 0.6 - 0.7
- 0.7 - 0.8
- 0.8 - 0.9
- 0.9 - 1.0
- >1.0
- no data

D

Countries	Overshoot (%)
United States	8.6
Brazil	6.9
Indonesia	6.4
India	5.2
China	5.0
Colombia	4.9
Philippines	4.4
Mexico	4.0
Germany	3.1
Malaysia	2.5

Fig. 2A.10: Green water appropriation and maximum sustainable green water footprint, (A) human appropriation of the green water flow, (B) the degree of human appropriation of the WF$_{g,m}$, at 5 × 5 arc-minute grid cell resolution, (C) green water scarcity (WS$_g$), and (D) half the overshoot occurs in just 10 countries (Images were adapted from [78] with permission).

been allocated to human activities world's sustainably available, as shown in the spatially explicit map (Fig. 2A.10B) [75]. The darker shades of gray (WF_g) were close to ($WF_{g,m}$), meaning that the sustainable available green water flow has been almost allocated to human use. Furthermore, the lighter shades of gray suggested a local increase in WF_g. The green water scarcity (WS_g) expressed the availability of the sustainably available green water flow and was estimated using the national aggregate ratio of $WF_g/WF_{g,m}$ (Fig. 2A.10C). In Europe, Central America, the Middle East, and South Asia, the green water scarcity is close to greater than 1, indicating that their sustainably available green water flow has been fully allocated to human activities. The overshoot in the country of interest is canceled out by remaining potential in another part of this country. Over half the overshoot occurred in just 10 countries (Fig. 2A.10D). The countries, including the United Kingdom, Germany, Indonesia, and New Zealand, are known for ample rainfall [76]. Their $WF_{g,m}$ is limited and mostly or fully allocated to human activities. In addition, the green water flows are reserved for nature, being appropriated already mainly for agriculture. In short, humans and nature share the world's limited green water flow, in which its explicit allocation is imperative to maintain sustainable green water supplies [77].

Contamination of water bodies has led to a serious problem for human beings, living organisms, and any other life forms, jeopardizing the sustainable future of humanity and the conservation of ecosystems [79, 80]. Xeniobiotics, microplastics, dyes, industrial waste, and discarded pharmaceuticals pose a great threat to clean water, human consumption, and aquatic life in green and blue water bodies [81]. Exposure to toxic metals (mercury, cadmium, chromium, lead) for a longer time will cause various disorders in the human body, including decreased immunological defenses, disabilities associated with malnutrition, and a high rate of upper gastrointestinal cancer [82]. Therefore, water remediation requires appropriate actions to eliminate heavy metals and metalloids, degrade organic compounds, and disinfect biological contaminants from water to acceptable levels to achieve a pollution-free environment. Attention has also been focused on developing green sustainable technologies that must be esthetic, eco-friendly, socially acceptable, and economical [83]. Water remediation is any process to improve water quality for an appropriation of specific end-uses, inducing drinking, industrial supply, irrigation, river flow maintenance, and water recreation [84]. Water remediation technology generally involves the chemical and physical treatment, biological treatment, and other technologies (such as desalination and portable water purification based on different constituents) [85], as shown in Fig. 2A.11.

Recently, green technologies (bioremediation) have drawn significant attention to using microorganisms to transform hazardous contaminants into a less toxic or nontoxic state and further purify water [86]. Bioremediation is decontamination processing using plants, fungi, algae, bacteria, and other microorganisms

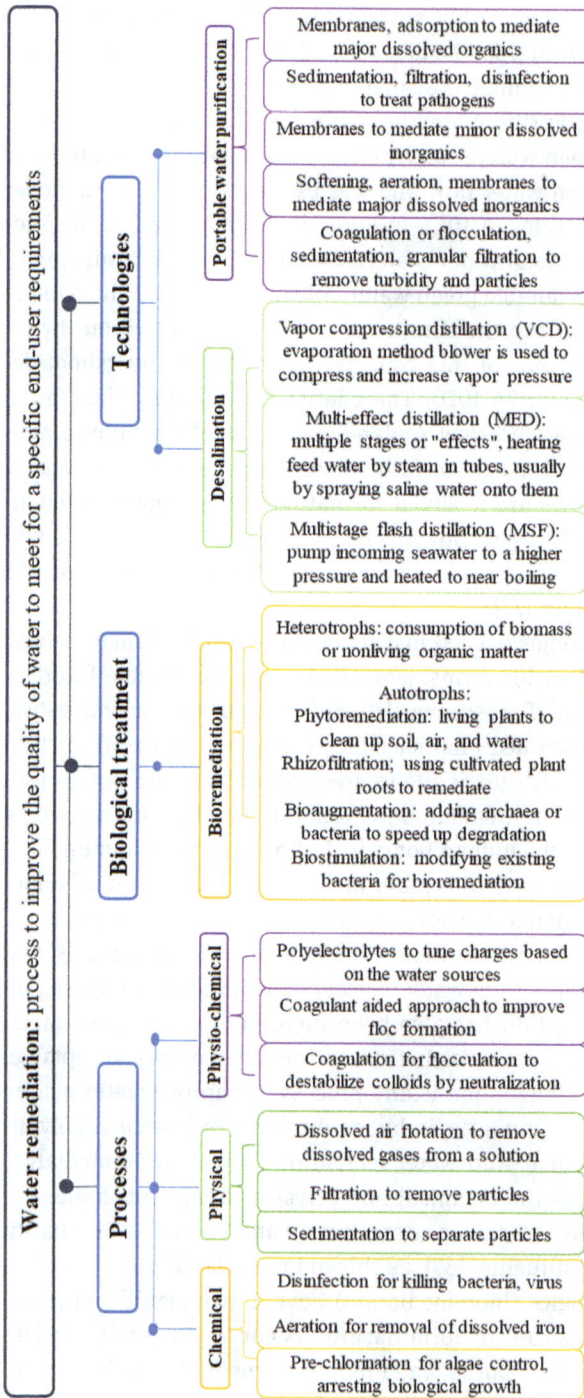

Fig. 2A.11: A summary of water remediation technologies using chemical, physical, physiochemical, biological, and desalination methods.

Water remediation: process to improve the quality of water to meet for a specific end-user requirements

Technologies

Portable water purification
- Membranes, adsorption to mediate major dissolved organics
- Sedimentation, filtration, disinfection to treat pathogens
- Membranes to mediate minor dissolved inorganics
- Softening, aeration, membranes to mediate major dissolved inorganics
- Coagulation or flocculation, sedimentation, granular filtration to remove turbidity and particles

Desalination
- Vapor compression distillation (VCD): evaporation method blower is used to compress and increase vapor pressure
- Multi-effect distillation (MED): multiple stages or "effects", heating feed water by steam in tubes, usually by spraying saline water onto them
- Multistage flash distillation (MSF): pump incoming seawater to a higher pressure and heated to near boiling

Biological treatment

Bioremediation
- Heterotrophs: consumption of biomass or nonliving organic matter
- Autotrophs:
  Phytoremediation: living plants to clean up soil, air, and water
  Rhizofiltration; using cultivated plant roots to remediate
  Bioaugmentation: adding archaea or bacteria to speed up degradation
  Biostimulation: modifying existing bacteria for bioremediation

Processes

Physio-chemical
- Polyelectrolytes to tune charges based on the water sources
- Coagulant aided approach to improve floc formation
- Coagulation for flocculation to destabilize colloids by neutralization

Physical
- Dissolved air flotation to remove dissolved gases from a solution
- Filtration to remove particles
- Sedimentation to separate particles

Chemical
- Disinfection for killing bacteria, virus
- Aeration for removal of dissolved iron
- Pre-chlorination for algae control, arresting biological growth

to improve water quality. Natural or modified products can also modify the treatment procedure to secure clean water. Bioremediation uses plants and microorganisms to cleavage contaminants into nontoxic or less harmful substances [86]. Bioremediation can purify groundwater, soil, sludge-carrying pesticides, hydrocarbons, organic compounds, and pharmaceutical chemicals. Bioremediation consists of autotrophs and heterotrophic methods. The former involves four categories, (1) phytoremediation using living plants to clean up soil, air, and water; (2) rhizofiltration using cultivated plant roots to remediate water; (3) bioaugmentation by adding archaea or bacteria to speed up the degradation of contaminants; and (4) biostimulation through modifying existing bacteria for bioremediation [87].

Our investigations focused on investigating phytoremediation, where the detoxification is carried out through green plants and their biomass as nanoparticle surface coatings. The bioadsorbent-tailored NPs were effective in removing toxic metals by complexation [88]. Periphyton biofilms composed of prokaryotic *cocci* have also been used to treat the aquatic ecosystem by removing chemical oxygen demand (COD), limiting aerobic growth, and cell inactivation, similar to radiation at UV 254 nm on organic matter. Unlike UV irradiation, periphyton adsorbents can also remove toxic metals [89]. Natural product extracts also emerge as potential compounds to remediate toxic metals from industrial effluents [90]. We selected natural extracts from polysaccharides, crab shells, and papaya wood. Using these product extracts, the modification of porous materials such as zeolites showed selectivity for ammonia, transition metals, bacteria, and radioactive elements [88]. Nanomaterials as potential antimicrobial agents have been widely used to remove metals and organic contaminants. Various heavy metals were extracted using ionic liquids as excellent solvents to achieve metal-free water environments. The surfactant modification for removing oxo ions, such as chromate and oxalate, was feasible to remove multiple components from waste water [91]. The inorganic-organic hybrid polymeric materials prepared using the click-chemistry approach were found application for water purification [92]. Such materials will lead to a better and more sustainable treatment of polluted water. Most of these disinfection strategies have drawbacks, such as incomplete disinfection [93]. This chapter aims to find a feasible and effective solution to eliminating impurities from contaminated water using three different approaches, photocatalysis (Section 2A.3.1), heavy metal absorption (Section 2A.3.2), and bioinspired remediation (Section 2A.3.3).

2A.3.1 Photocatalysis in water remediation

Nanomaterials with high surface area, surface-to-volume ratio, and porosity will enhance catalyst reactivity and self-regulation of structure and chemical properties at the macroscale level of engineering [94]. The potential advantage of responsive nanosystems is facilitating treatment at the source of waste generation rather than degradation at large centralized treatment systems [95]. The leading-edge design of photocatalysts with rapid response to external stimuli beyond current approaches for water remediation and pharmaceutical-waste treatment requires sound knowledge and a data-driven approach to ensure photocatalytic activity with high persistence [96]. A wet-chemistry method can fabricate mesoporous nanomaterials to uniformly "disperse" one component into the supporting substrate under moderate conditions to create a tunable interface and morphology on the nanoscale [97]. Photocatalytic functionality is tuned with changes in band structure that governs the reactivity of photocatalysts at a wavelength window, where light energy is used to drive catalysis to yield reactive species (such as •OH radicals) by affecting the reactions between photoelectrons and the primary oxidant (O_2 or O_3) as well as holes at the interfacial region (e.g., H_2O) [98]. Although these complex relationships between synthesis, structures, and functions of engineered nanomaterials have been investigated systematically, the specific parameters for titania-based catalysis that self-regulation remains were improved [99]. Therefore, optimizing fabrication variables and tailoring the band structure of photocatalytic systems with maintaining the structure and properties is critical to establishing their robust relationship between structure and function. It is crucial to develop catalysts with high persistence and responsiveness [100]. Due to the relationship between the number of charge carriers and their reactivity in photocatalytic reactions, it is important to choose catalytical components and optimize their design and synthesis variables, using an extended network as a supporting scaffold, such as graphene [101].

In collaboration with Dr. C. Li, we developed sol–gel chemistry that produced TiO_2 NPs with polyvinylpyrrolidone (PVP) templation and graphene oxides (GOs) to produce nanocomposites [103]. The crystalline phases of TiO_2 were controlled to be anatase, using porogen to tune the porosity and pore size, further achieving high degradation efficacy of pharmaceutical waste through photocatalysis. These heterogenous composites were sintered at 400 °C to obtain crystalline TiO_2 while GO was converted to reduced graphene oxide (rGO) under ultraviolet radiation. Through four steps, hydrolysis, templation, crystallization, and reduction, the uniformly distributed nanocomposites were used to advance the degradation rate of tetracycline (TCH). Antibiotics (TCH) were used in food and feeding to promote animal husbandry to treat infections caused by bacteria. It is important to point out the natural product extract; the citric acid acts as both a tri-topic ligand and dispersing agent in the sol–gel process. The use of citric acid facilitated the formation of a coordinative complex between the Ti metal center and carboxylic groups. This coordination successfully prevented the rapid hydrolysis of $Ti(O^nBu)_4$ and thus inhibited particle growth during heat treatment. The water-soluble polymer PVP K30 (PVP-K30) was used to tune the polymerization phase of the Ti-precursor due to its intrinsic hygroscopic nature, adhesive properties, and stable acidity [104]. The five-numbered lactam, K30 strand with Guinier length (5.43 ± 0.01 nm) and contour length (34.31 ± 0.51 nm), was chosen due to its intrinsic natures, such as linearity and easiness to bond with polar bond, O-Ti. At the same time, its molar mass averaged 40,000 g/mol to secure the formation of mesopore sizes. The uniform pore diameters and particle sizes were obtained, attributed to the above intrinsic properties of PVP-K30, particularly the pyrrolidone ring, high polarity, cohesivity, and resultant propensity. The ion-dipole moment between the Ti atom in the precursor and the O atom in PVP resulted in polymer-anchored TiO_2 hybrid materials. The continuous inorganic lattice, such as Ti-O-Ti or Ti-H-Ti [105], was formed to create an oxide matrix through poly-condensation responses of a sub-atomic precedent [106]. Different amounts of PVP-K30 were introduced through controlled stoichiometry to tune the pore structures. The above "wet-gel" was converted into anatase TiO_2 polycrystals with further drying and heat treatment [107]. Based on the crystal field theory, the five d-orbitals of the Ti atom were subject to the energy increase when Ti was coordinatively bonded with bonding atoms, such as oxygen elements. It was anticipated that a rapid conversion from tetrahedral to octahedral units due to the decrease in splitting energy resulted in thermally stable anatase crystalline phases upon heat treatment. The last step was the reduction of GO under ultraviolet, as demonstrated in our previous publication [103]. The Hummer method derived from graphene was ideal for incorporating the PVP-templated TiO_2 matrix; however, the functional groups decreased photocatalytic reactivities. The graphene substrate anchors PVP-templated TiO_2 through chemical bonding, mainly through dipole-dipole intermolecular forces [108]. This relatively weak attraction was readily overcome under heat treatment to initiate anatase TiO_2 formation along the graphene surfaces. Followed by the ultraviolet radiation, the GO was reduced to remove a controlled amount of functional groups and impurities [109]. During this reduction, the hybridization of carbon atoms was converted to sp^2 from sp^3, resulting in unoccupied p_z orbitals, which enhanced the electron transfer. The steric hindrance of citric acid and cohesivity of PVP-K30 prevented the TiO_2 crystal growth at high heating temperatures. The homogeneity and architecture of RGO-supported TiO_2 nanocatalysts were tuned, enabling fine-tuning precision and accuracy of the light photoreceptor to be active at UV-VL regimes. Multiscale texturation and up-scaling production were achieved by combining sol–gel chemistry and heat treatment, followed by a post-reduction thermal-dynamically.

The crystalline structure obtained using X-ray diffraction showed that the reduced GO-TiO_2 composite and PVP-templated RGO-TiO_2 (Fig. 2A.12A) were well-indexed with standard XRD patterns. At the diffraction angle of 13.10°, the diffraction pattern of GO was found, corresponding to the (001) plane. Upon reduction under UV radiation, diffraction at 26.60° was detected due to the formation of reduced GO. Although the reduced graphene was obtained in this approach, it is XRD radiation (~26°) overlapped

A

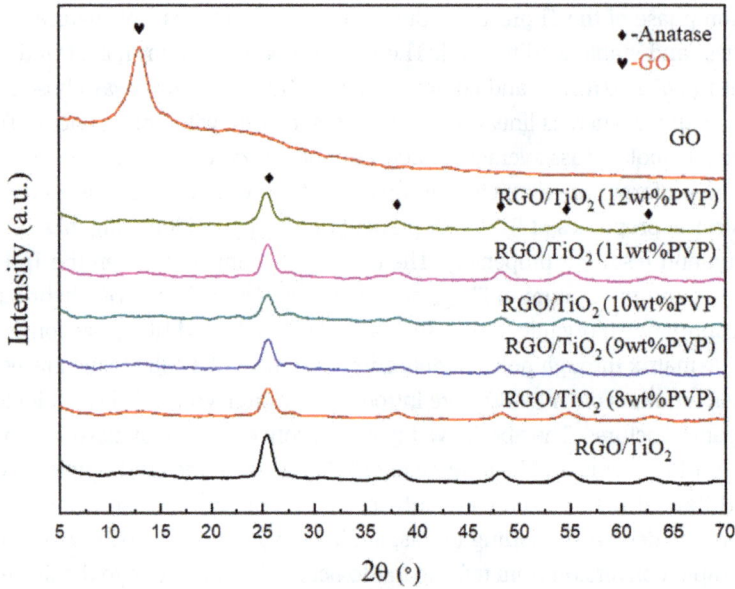

B

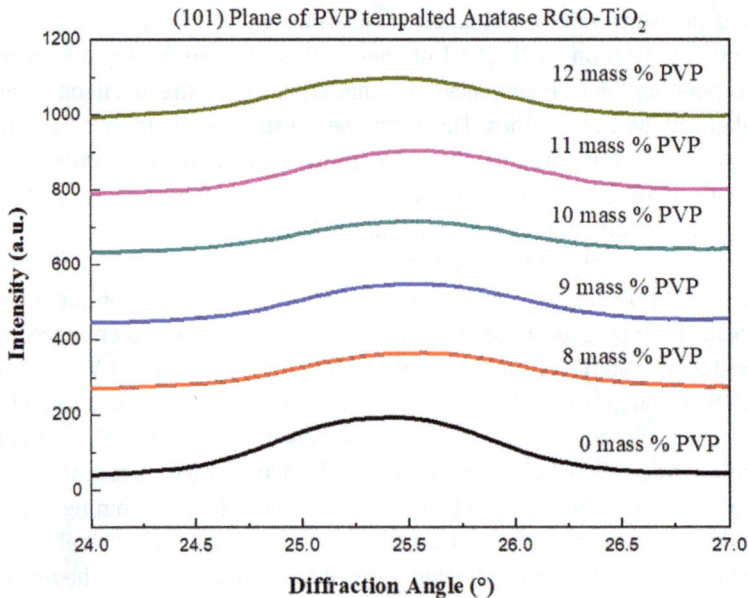

Fig. 2A.12: XRD patterns of different TiO$_2$-based hybrid photocatalysts are mediated with PVP:
(A) formation of GO, RGO-TiO$_2$ composite, PVP-K30-templated RGO-TiO$_2$ composite
and (B) the William–Hall plot estimates the crystallite size and effective strain
of the RGO-TiO$_2$ composite.

with the major peak (101) of anatase TiO_2. The XRD results were well-indexed with standard JCPDS, 89-4921 (tetragonal, SG I41/amd (141), lattice parameters: $a = 3.777$ Å, $c = 9.501$ Å; $\alpha = 90°$, cell volume = 136.300 Å^3) in agreement with published data [110]. The distance decrease of the graphene layer (0.9–0.4) was found with an increase in C/O ratio and decreased C sp^3 hybridization, in which the sp^3 hybridization of carbon was partially recovered into the sp^2 hybridization [111]. The diffraction peaks of different samples appeared at 25.40° (101), 28.10° (004), 48.20° (200), 54.20° (105), and 62.70° (204), indicating that the tetragonal anatase TiO_2 was obtained with high purity and crystallinity. The in-depth study indicated that the (101) diffraction angle and PVP-K30 amount form a volcano relationship. An increase of PVP concentrations from 8 to 11 mass % results in the peak red-shift of the main peak at (101) plane (Fig. 2A.12B) until the amount reaches 12 mass %. This shift of the thermo-dynamically stable peak of TiO_2 corresponded to the tensile stress (<1.1 %), which was found to occur between the boundary of the PVP template and RGO-TiO_2 composite.

The effective strain, ε, was calculated to be 0.964 by William–Hall plot (Fig. 2A.13A), indicating that the lattice expansion occurs along the (101) plane. These defects serve as the reactive chemical sites to allow for TCH decomposition [103]. The positive slope of Fig. 2A.13A suggested that the lattice expansion occurred with the increased PVP-K30 amount linearly. Another observation from the XRD data is the broadening of the major peaks (101) and others, suggesting particle size deduction. The Debby–Scherrer (D–S) equation (eq. 2A.6) was used to calculate the crystallite size of the PVP-templated nanocomposites by neglecting the strain effect [112]. The William–Hall (W–H) equation (eq. 2A.7) calculated the crystallite sizes and lattice strains [113]. It was found that the effective strain is close to 1, indicating that the TiO_2 anatase structure was maintained well, while the positive slope is associated with cell expansion. The size correlation to the amount of PVP was shown in Fig. 2A.13B, indicating that the particle size and amount of PVP-K30 polymer demonstrated a volcano relationship. The XRD data on particle sizes also confirmed that a volcano trend of the ratiometry curve, indicative of the optimized amount of PVP K30 (11 mass %, as shown by the red star in Fig. 2A.13B), resulting in the smallest size (12.5 nm for RGO-TiO_2 and 11.0 nm for PVP-templated RGO-TiO_2) and lowest effective strain. With the PVP-K30 amount increase, an exponential increase in particle size (yellow region) was found, followed by a cliff drop. As a result, the particle growth of nanosized TiO_2 was prevented due to this polarity effect and steric hindrance [114]. With the increase of PVP above the 11 mass %, the modification was found otherwise in terms of the trend. Accordingly, this size decrease results in the photocatalytic reactivity increase, as shown by the TCH degradation. It is critical that both Debby–Scherrer and William–Hall's methods only estimated the crystallite sizes and effective strain due to the complex function of elastic constants and dislocation contrast factors for the polymer templated TiO_2-based catalysts [115].

A

Equation	y = a + b*x
Plot	InterpolatedY1
Weight	No Weighting
Intercept	8.31833 ± 2.87015
Slope	1.03955 ± 2.4248E
Residual Sum of Squa	6.42729E-9
Pearson's r	1
R-Square (COD)	1
Adj. R-Square	1

PVP Amount (mass %)	2θ (°)
0	25.44
8	25.48
9	25.50
10	25.54
11	25.56
12	25.44

B

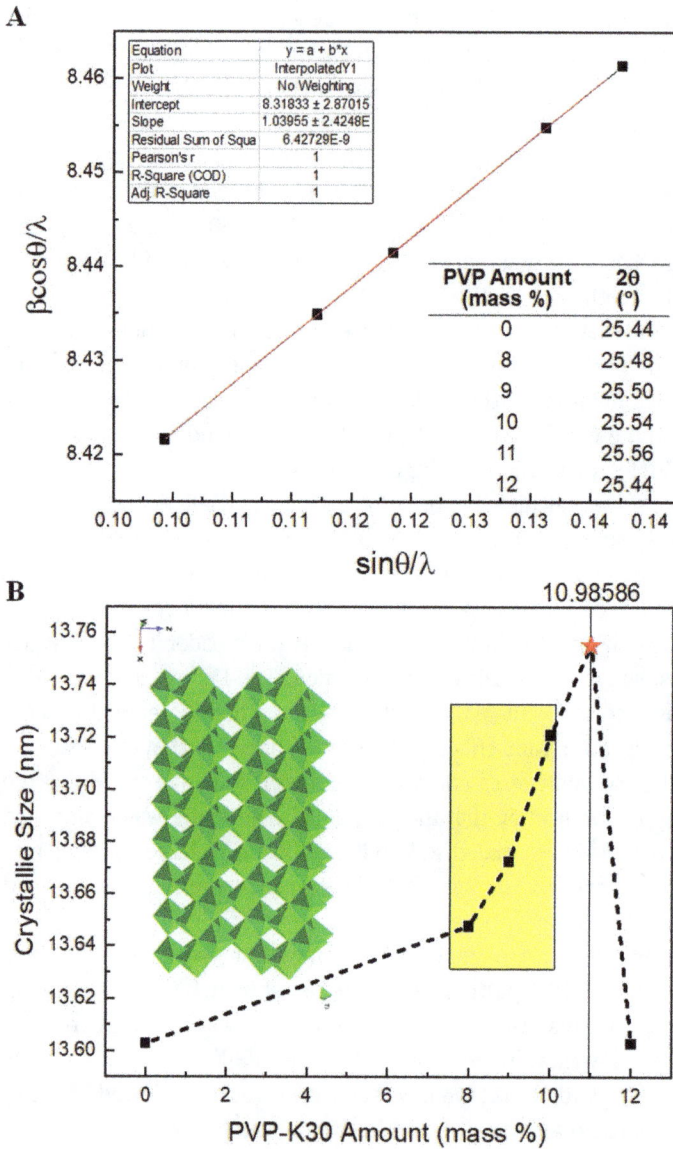

Fig. 2A.13: XRD patterns of different TiO_2-based hybrid photocatalysts mediated with PVP: (A) the William–Hall plot estimates the crystallite size and effective strain of RGO-TiO_2 composite (the inserted table showed the diffraction angle when peak splits) and (B) the relationship between crystallite size and PVP-K30 amount (the inserted diagram showed the anatase TiO_2 of 8 × 2 × 2 supercells).

$$D = K\lambda/\beta\cos\theta \qquad (2A.6)$$

$$\beta\cos\theta/\lambda = 1/D + \varepsilon \times \sin\theta/\lambda \qquad (2A.7)$$

where β is the full width at half maximum, θ is the diffraction angle, λ is the X-ray wavelength, D is the average crystallite size, and ε is the effective strain. In all the cases, the Scherrer formula yields a lower size value than the size extracted from the W–H plot, as expected [116].

The high-resolution transmission electron microscopic (TEM) images (Fig. 2A.13) showed a well-crystallized particle and pore size uniformity of the RGO-TiO$_2$ nanocomposite treated at different concentrations of PVP (6–14 mass %). It was found that the 11 mass % of PVP was the optimal amount of porogen, along with 5 mass % of RGO as support, and all catalysts were under this condition. The RGO-TiO$_2$ without templating has an average size of 10.5 nm ($\pm$0.15 nm) with high integrity and is distributed on the surface of the RGO. The well-distinctive motifs of nanocomposites (Fig. 2A.14A) were obtained after heat treatment, while some of their edges and corners were observed. These unique regions were found acting as the reactive sites and subject to polarization due to the polarity of the PVP polymeric template. The rGOs (Fig. 2A.14B) were found as ultrathin layers with corrugated configurations to host the TiO$_2$ NPs (Fig. 2A.14C). The ring pattern (Fig. 2A.14D) was well-aligned with anatase TiO$_2$ polycrystals, while the selected area electron diffraction patterns (Fig. 2A.14E) confirmed that highly crystalline catalysts were obtained. The simulated SAED patterns with different sample thicknesses, including Laue front-plate, back-plate, and cylindrical scattering (Figs. 2-14G-H), were applied to understand the topotactic relationships of different components in the nanocomposites. The TEM characterization of fine structures confirmed that the configurations of anatase RGO-TiO$_2$ composites with different formulations have been subject to minor changes with detectable local stress and crystallite size variations via polymeric templating. The in-situ templation using the nonionic polar polymer PVP-K30 successfully tuned the morphology and fine structures of the final products. Optimizing synthesis variables can regulate the relationship between nanostructures and catalytical reactivities.

The Barrett, Joyner, and Halenda (BJH) analyses were carried out to determine the pore structure of the PVP templated RGO-TiO$_2$ photocatalysts. It was evidenced that the in-situ templation generated a contiguous network of mesopores [117]. The instrumentation analyses provided an in-depth understanding of porous nanocomposites' physical and chemical properties as photocatalysts. The CrystalMaker was also used to demonstrate the proposed structure of RGO-TiO$_2$ composites. The pore sizes (Fig. 2A.15A) of the RGO$_5$-TiO$_2$ and PVP-K30 templated RGO$_5$-TiO$_2$ were about 3.40(65) nm to 4.81(56) nm based on the BJH model. The diagram in Fig. 2A.15B showed the supercell ($4 \times 2 \times 2$) of TiO$_2$ with unit cell volume: 2,180.80 Å^3, filled space: 996.93 Å^3 (45.71 %) per unit cell, void space: 1,183.87 Å^3 (54.29 %) per unit cell. The graphene layers were used as substrates to direct the TiO$_2$ nanoparticle

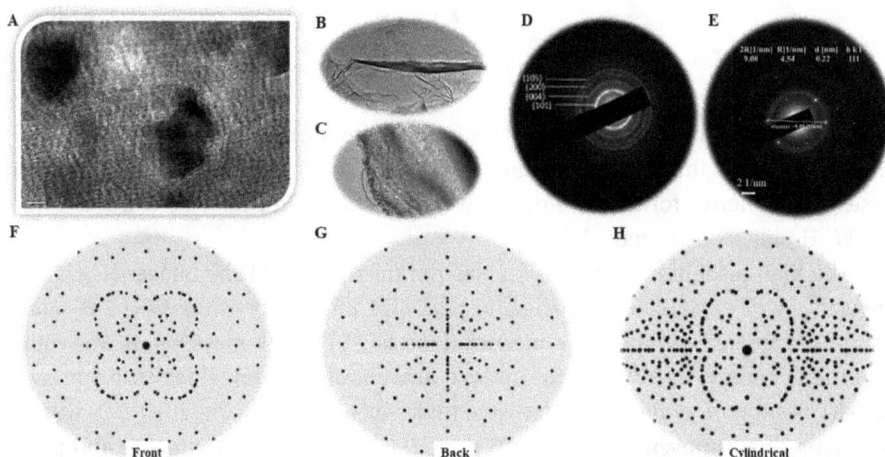

Fig. 2A.14: High-resolution TEM images of PVP templated RGO-TiO$_2$ photocatalysts: (A) select RGO-coordinated (RGO$_5$, 5 mass %) TiO$_2$, (B) morphology of RGO thin layer, (C) TEM image of the templated RGO$_5$-TiO$_2$ catalyst, showing TiO$_2$ distributed along the curvature of RGO, (D) The ring pattern resulted from TiO$_2$ in polycrystal cluster, (E) Selected area of electron diffraction patterns, (F, G) Simulated SAED results, the Laue plate scattering from the front, back, and cylindrical, indicating highly crystallized TiO$_2$ formation upon RGO-coordination and PVP-templating (images were adapted with permission from [103]).

orientation. Based on the BJH analysis, the PVP-K30 porogen successfully enlarged pores, allowing for interaction between photocatalysts and N$_2$ adsorbates, further multiplying the layered adsorption of N$_2$ molecules. These mesopores provide suitable adsorption space and channels to promote the rapid mass diffusion of the aqueous TCH toward the interfaces, further enhancing the degradation reaction with the highest efficiency of 94.22 % when the 11-mass % PVP was used under simulated sunlight (SS). The degradation efficiencies of an antibiotic TCH using PVP-K30-templated RGO-TiO$_2$ photocatalysts were evaluated to determine the optimal variable of the pharmaceutical waste treatments. This advancement in the current state-of-the-art TiO$_2$-based photocatalysts in pharmaceutical waste management under visible lights without extra energy consumption [103].

The PVP-K30-templated RGO$_5$-TiO$_2$ composites were evaluated using Raman, photoluminescence (PL), and UV-visible (UV-vis) spectroscopic methods (Fig. 2A.16) to validate the crystalline phase, crystallinity, and defects. Raman data in complementary with XRD analysis indicated anatase TiO$_2$ and RGO were formed (Fig. 2A.16A and B). The anatase TiO$_2$ showed four bands, rising from the (101) O–Ti–O symmetrical stretching and aligned with principal peak E_{1g} (142.20 cm^{-1}) and E_g (636.40 cm^{-1}) of anatase TiO$_2$ [118]. The B_{1g} (395.40 cm^{-1}) and A_{1g} (512.60 cm^{-1}) bands are well-indexed to the (001) O–Ti–O bending variation. As shown in the green region (Fig. 2A.16B), the ratio of the intensity of D and G bands (I_D/I_G) has been

A

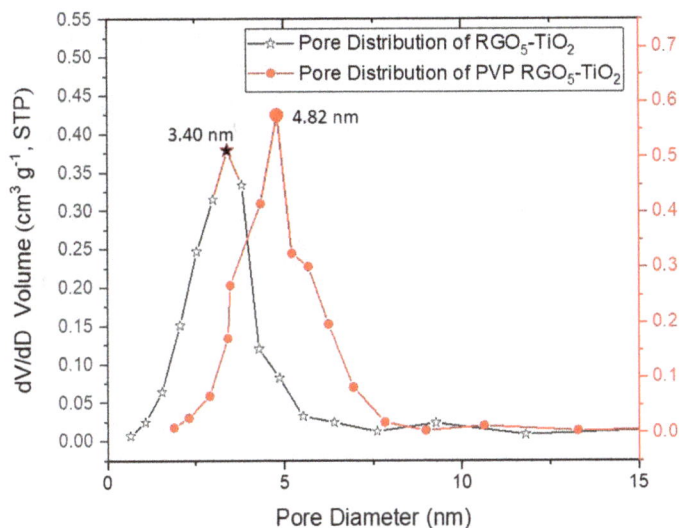

B

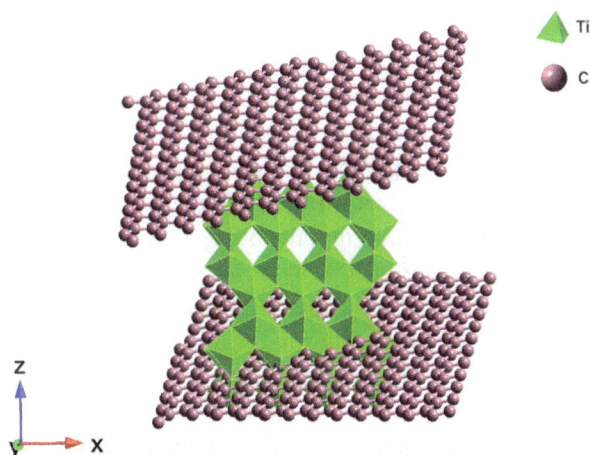

Fig. 2A.15: Barrett, Joyner, and Halenda (BJH) analysis of PVP templated RGO-TiO$_2$ photocatalysts, (A) mesopore size distribution of RGO$_5$-TiO$_2$ composite with an average size of 3.40 nm, and (B) mesopore size distribution of PVP-K30 templated RGO$_5$-TiO$_2$ composite with an average size of 4.82 nm.

used as a metric to evaluate the graphene disorder, such as arising from ripples, edges, charged impurities, and the presence of domain boundaries [119]. Our study indicated that (I_D/I_G) was decreased from 1.186 (GO) to 1.074 (RGO), indicating the decrease in graphene disorder after GO was partially reduced. The most important observation about this structural deformation of graphene affects the electrical conductance of the RGO-TiO$_2$ and their chemical interaction [103].

The PL() spectra were obtained to evaluate recombination mechanisms and bandgap of RGO-TiO$_2$ photocatalysts [120]. The date for GO and rGO supported TiO$_2$ (11.00 mass % PVP-K30), and pure TiO$_2$ (Fig. 2A.16C) indicated that the broad emission bands were detected in the range of 3.54–2.48 eV (corresponding to 350–500 nm). The anatase TiO$_2$ displayed three emission lines, where the blue PL intensity centered at 2.658 eV, while the violet PL peak split into two emissions with intensity centered at 3.139 and 3.003 eV. The aggregation-caused quenching of TiO$_2$ emission was observed upon introducing aromatic graphene and polymeric PVP-K30 [121]. Our data indicated that the recombination of electrons (e$^-$) and holes (h$^+$) resulted from the synergistic effect of the graphene support and PVP-K30 templating. Hypothetically, self-trapped holes are located at oxygen atoms in the lattice of the TiO$_6$ subunit, while trapped electrons occupy mid-gap states positioned below the Fermi level. This approach showed an increase in the GO amount, and the intensity of PL decreased accordingly, as demonstrated in our previous work [103].

As a complementary method to PL spectroscopy mentioned above, UV-vis diffuse reflectance spectroscopy (DRS) was used to evaluate the optical bandgap of the PVP-K30 templated RGO-TiO$_2$. The results (Fig. 2A.16D) indicated that both RGO coordination and PVP templating resulted in a peak intensity decrease. These adsorption decreases also correlated with the observation from blue PL and violet PL, implying that the lifetimes of charge carriers were expanded [122]. The DRS peak boarding was also detected, proving defects and particle size decrease of the RGO-TiO$_2$. The sharp absorption edge was also obtained, which was used to calculate the band gap of three chosen photocatalysts. It was found that the bandgap was lowered with the introduction of both graphene and PVP-K30. The absorption edge (428.2 nm) of RGO$_5$-TiO$_2$ nanocomposite showed a red-shift from 417.1 nm (TiO$_2$ alone), while the PVP-K30-templated RGO$_5$-TiO$_2$ showed a further red-shift to 446.8 nm. The corresponding band gap energies were calculated to be 2.97(3), 2.89(6), and 2.77(5) eV for the three photocatalysts, as evidence of achieving high degradation efficiency under wide light sources. These energy data were attributed to an enhancement of Ti–O–C band upward vibration (bending and stretching), resulting in an effective separation of electrons and holes on the interfaces or along the edges of three components [123].

The antibiotic TCH ($C_{22}H_{24}N_2O_8$) is a very adaptive molecule capable of easily modifying itself through tautomerism in response to various chemical environments [124]. It was chosen as a model molecule to test the RGO-TiO$_2$ photocatalysis under ultra-violet (UV), SS, varying pH values, Ca^{2+} concentrations, and degradation duration. Considering four deprotonations, we can identify 64 diversified tautomers of TCH. Among these tautomeric species, two tautomers of TCH with highlighted differences (Fig. 2A.17A and B) were demonstrated to understand their bonding and geometry during the degradation processing when equilibrium was achieved. When these TCH molecules undergo different deprotonations or bond cleavage pathways, the pK_a

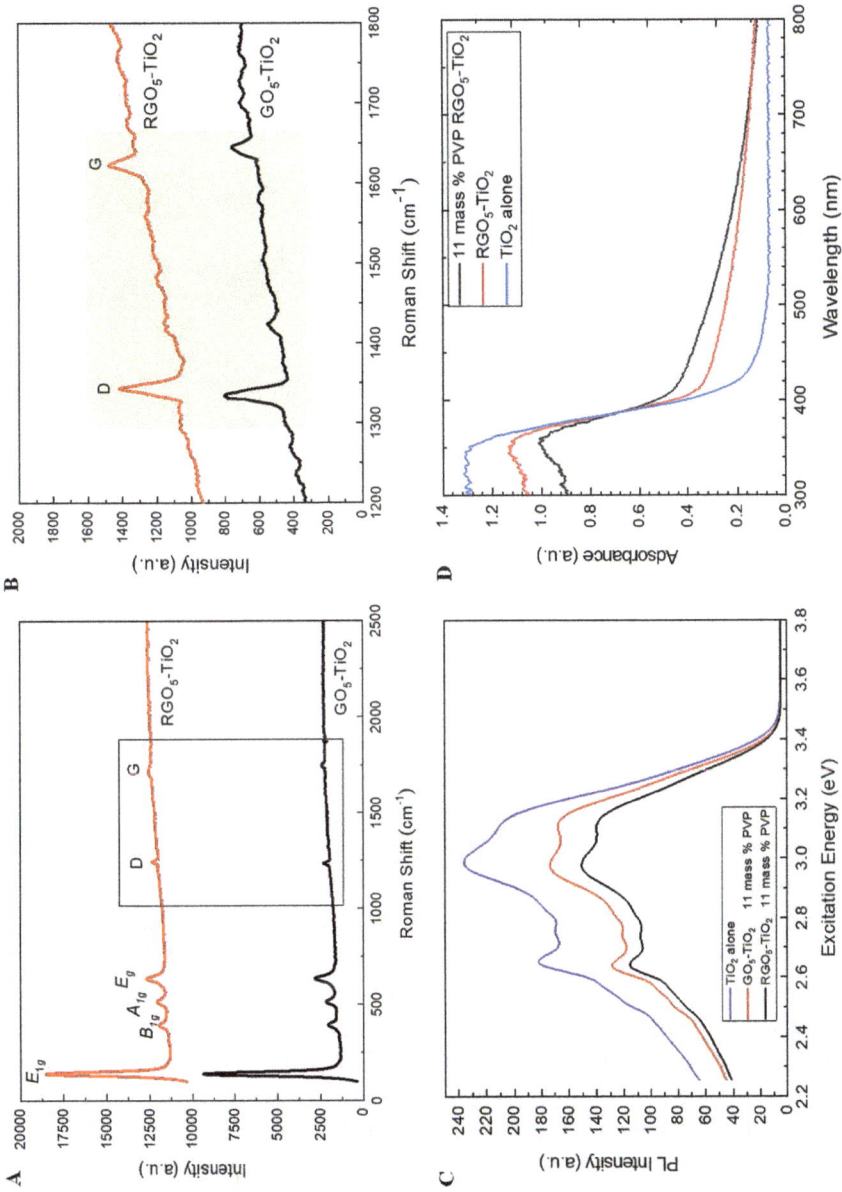

Fig. 2A.16: Spectroscopic analysis of TiO$_2$-based nanocomposites: (A) Raman spectra of 11 mass % PVP-K30 templated GO$_5$-TiO$_2$ catalysts, demonstration of vibration mode of TiO$_2$, (B) the D and G bands resulted from GO and rGO, (C) The PL of three select TiO$_2$-based catalysts, and (D) the ultraviolet visible spectra of three select TiO$_2$-based catalysts.

can vary, influencing the RGO-TiO$_2$ photocatalytical reactivities. TCH tautomerism is essential for its chemical behavior under different environmental conditions. This comprehensive molecule's spatial orientation, bond cleavage, and degrees of protonation affect the deterioration efficacy, as discussed in the following sessions.

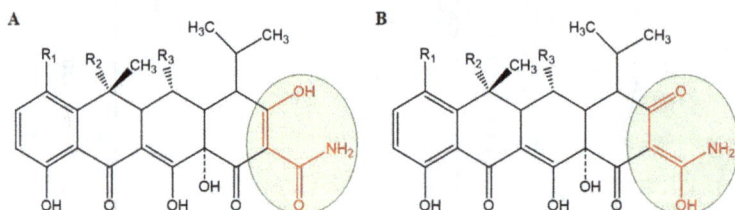

Fig. 2A.17: Two tautomers of tetracycline with tautomerization among 64 possible configurations.

The degradation rate (η) using the templated RGO-TiO$_2$ photocatalysts (Figs. 2A.18A and 2A.16B) were investigated under UV and SS conditions. The generalization lies in the fact that PVP addition resulted in an increase η by 9.8 % under both UV and SS conditions due to the increase in pore size and surface area. The UV light exposure showed a higher degradation rate (about 3.9 %) than SS radiation to break down the TCH. The graphene substrate successfully provided "in-situ" riveting sites to guide the TiO$_2$ nucleation and facilitated the hybridization of p-orbital from carbon and titanium d-orbital [125]. As a result, the electrons were delocalized among the neighboring atoms, resulting in an extended life span of excitons. The band gap energy decrease by 034 V from 3.21 to 2.87 eV led to the visible light adsorption by TiO$_2$-based catalysts, allowing for degradation under visible light conditions.

The mesoporous template PVP has been used to tune the pore structure and surface areas of the RGO-TiO$_2$ photocatalyst by varying its hydrophilicity, dispersibility, and cohesivity. It was found that the degradation rate depends on the PVP-K30 amount, showing "volcano" relationships under both UV and SS conditions for different durations of radiation. Suppose the PVP-K30 is less than 11 mass %, its increase increases in degradation rate. Otherwise, the higher amount of PVP corresponds to a decrease (the SS data shown in Fig. 2A.19A under different treatment durations). The experimental data showed that RGO-TiO$_2$ catalysts with 11 mass % of PVP-K30 maximized the degradation, averaging 93.36 % and 91.34 % under UV and SS conditions. The further increase of PVP to 12 mass% resulted in a deduction of approximately 5.29 % under both UV and SS radiations. It was proposed that the Ti from the reactants and O from PVP formed ion-dipole intermolecular forces, anchoring the TiO$_2$ nucleation and particle growth (Fig. 2A.19B). This water-soluble PVP templating was critical to tuning the pore size and surface areas and improving the catalytical reactivity of these RGO-TiO$_2$ photocatalysts, further enhancing the degradation of degradation antibiotics.

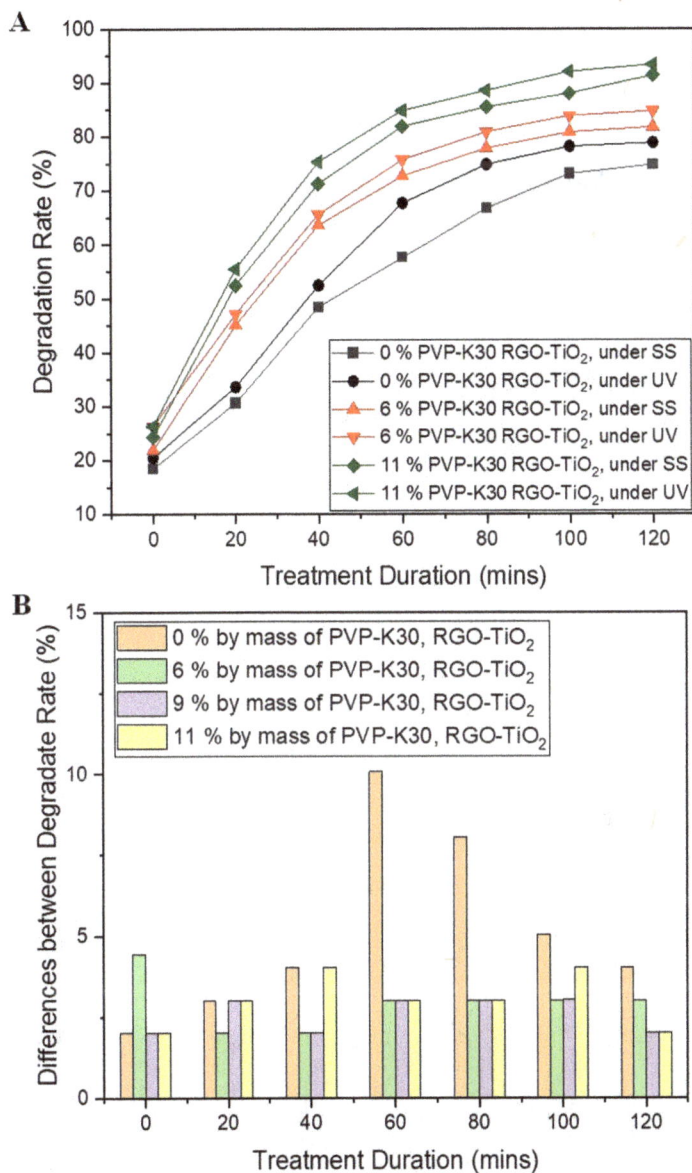

Fig. 2A.18: The degradation rate of tetracycline using PVP-K30 templated RGO_5-TiO_2 photocatalyst with different adoption amounts of PVP, where three formulations were chosen for comparison: (A) degradation rate of catalysts under ultraviolet light (UV) and simulated sunlight (SS), (B) differences between degradation rate indicated the UV exposure shaving a higher efficacy; however, the SS provided an applicable photocatalysis for pharmaceutical waste remediation.

A

B

Polyvinylpyrrolidone templation to tune the mesopores and structure

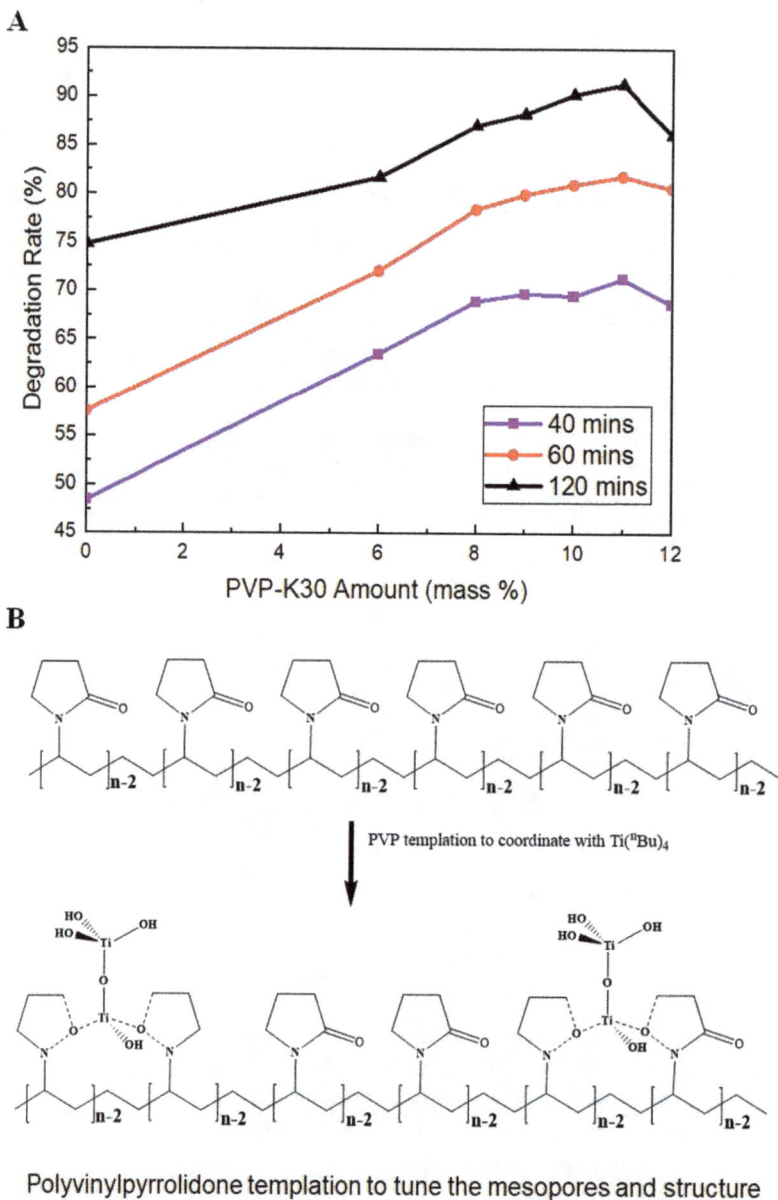

Fig. 2A.19: The degradation rate of PVP-K30 templated RGO-TiO$_2$ photocatalyst with different adoption amounts of PVP: (A) degradation rate of catalysts under simulated sunlight and (B) the PVK-K30 templation through intermolecular forces.

The pH value of the TCH solutions varied from 3, 5, 7, and 9, where a volcano relationship between the degradation rate and pH was also found (Fig. 2A.20A). Under neutral conditions pH = 7, the photocatalytic efficiency achieved its maximum, 93.36 %,

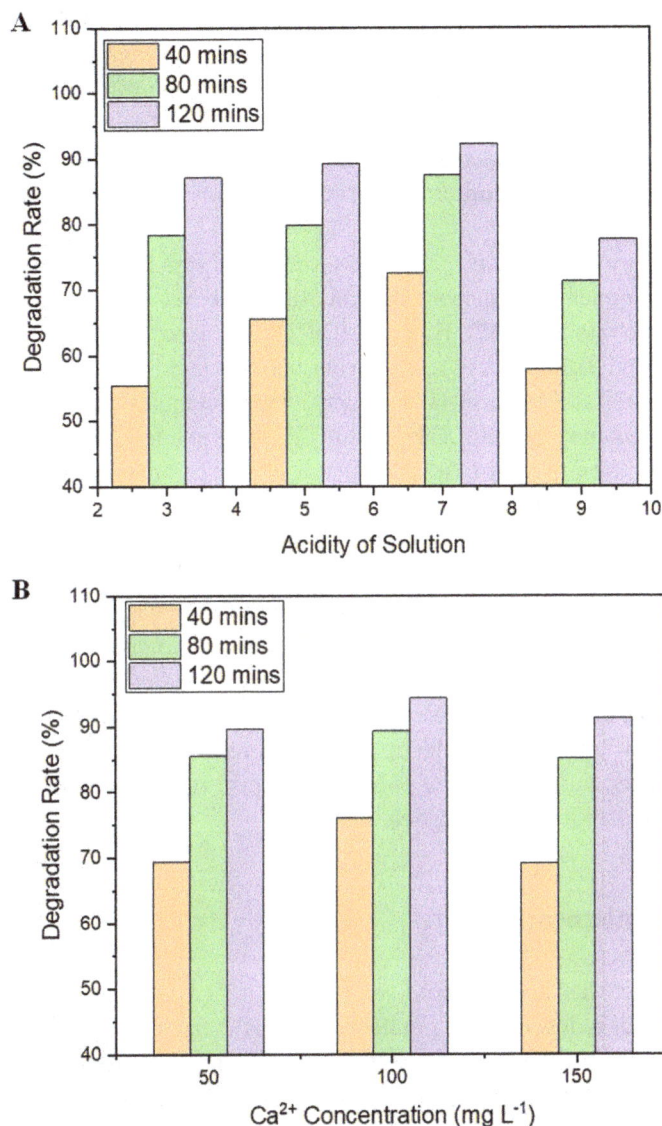

Fig. 2A.20: The degradation rate of RGO-TiO$_2$ photocatalyst templated by PVP-K30 (11 % by mass) under simulated sunlight: (A) effect of pH on the degradation rate and (B) effect of Ca^{2+} concentration on the degradation rate.

and 91.34 % under both UV and SS conditions, respectively. This relationship was attributed to the amphoteric nature of TiO$_2$ with a theoretical isoelectric point of about 6.4 and the high degree of freedom attributed to the TCH tautomers (64 possibilities, Fig. 2A.19) based on protonation. At neutral conditions (pH = 7), there are five tautomers of TCH, among which two were neutral isomers, accounting for about 61 [103].

Under neutral conditions, the weakened repulsion between RGO-TiO$_2$ and TCH corresponded to fastened TCH adsorption, leading to degradation improvement. TiO$_2$ attracted negatively charged hydroxide ions (OH–) at basic conditions. The repulsion between negatively charged species prevented the TCH from being adsorbed, and further degradation efficiency decreased. The amphoteric propensity of the catalyst and the complexity of TCH conformations must be considered when designing photocatalysts.

The concentration of calcium cation ([Ca^{2+}]) as a water hardness indicator was found to influence TCH degradation efficiency (Fig. 2A.21B). There has no distinctive trend between degradation rate and [Ca^{2+}]. However, [Ca^{2+}] = 100.00 mg/L, the photocatalytic efficiency achieved 93.36 % and 91.34 % under both UV and SS conditions, respectively. These Ca^{2+} cations act as a metallic center to form a complex with the N end of the 2-dimethylamino and phenolic OH group in TCH, causing internal electron transfer from TCH to electron-deficient cations. At the lower Ca^{2+} concentration range, hydrated Ca(H$_2$O)$_6$$^{2+}$ cations were formed, causing spatial hindrance. The hydrated Ca^{2+} cations were prevented from contacting the TCH oxygen negative pole or occupying the RGO-TiO$_2$ reactive sites. As a result, the degradation of TCH using photocatalyst was hindered due to the particular site occupation by Ca^{2+}.

On the other hand, concentrations greater than 100.00 mg/L increased TCH degradation. This observation suggested that hard water can facilitate TCH degradation using the RGO-TiO$_2$ photocatalyst. The TCH molecules (Fig. 2A.19) were activated to the radial formation by losing electrons. These unstable radicals were ready to be further oxidized to small molecules by the hydroxyl radicals and other reactive oxygen species existing in the water solution.

2A.3.2 Heavy metal absorption

The heavy metal toxic waste has become more abundant with increased energy demand for more than two decades, causing problematic global health and environmental issues. We developed a series of catalysts shielded by natural products to remediate wastewater. The hazardous ions composed of AsO$_4$$^{3-}$ and CrO$_4$$^{2-}$ have been absorbed using bio-encapsulated Fe$_3$O$_4$ magnetic NPs (MNPs) [88]. These MNPs were encapsulated with natural product extracts (such as polyphenols from pomegranate and chitosan from shrimp shells) to improve their biocompatibility and mitigate their toxicity. The results demonstrated the minimization of oxidative stress and cellular damage to the human retinal pigmented epithelium cells when the encapsulated MNPs were used. These functionalized NMPs also showed high removal efficiency of sodium hydrogen arsenate (Na$_2$HAsO$_4$). We further optimized the molecular structure of a series of MNPs by cost-effective colloidal chemistry (Section 2A.2), studied their thermodynamics and interactions with the heavy metal ions, and affirmed the mechanism under either electrochemistry reaction or physiosorption approaches. Three key perspectives are discussed in this section: (1) material design and characterization to improve the robustness and coverage of MNPs; (2) removal of heavy metal ions from solid or aqueous waste to provide an in-depth understanding of the thermodynamics and interactions between AsO_4^{3-} or CrO_4^{2-} with Fe_3O_4 MNPs; and (3) mechanistic analyses using spectroscopic method(s) to determine the oxidation number change or bond vibration.

These two types of nanocomposites (Fe_3O_4 and $Ca_{10}(PO_4)_6(OH)_2$) were chosen for water remediation. The Fe_3O_4-based nanomaterials were prepared using the wet-chemistry method, using natural product extracts as dispersing agents to tune their particle size, shape, and functionalization. The colloidal chemistry, sol–gel, and inverse emulsion approaches have been employed to control the composition and architecture of the nanoproducts from the atomic or molecular levels. The structure-diversified core–shell, yolk–shell, Janus, and needle-shaped nanocomposite were obtained [126]. We also aim to produce recyclable nanocomposites with mesoporous, hollow, or multilayered structures to improve the permeability and absorptivity of the nanocomposites. The structural characterization will provide feedback and guideline for us to optimize the synthesis variable further to enhance the stability, configuration, and reactivities of these nanocomposites when applied in the remediation. Their surface grain, particle size, pore size, wettability, and interaction with contaminants were evaluated using nano-characterization [127]. The $Ca_{10}(PO_4)_6(OH)_2$ was studied using CrystalMaker to determine its electron density, crystallographic structure, and functionalization.

An inverse emulsion synthesis was carried out to produce core-shelled mixed Fe_3O_4-based magnetic nanocomposites (MIMNs). The starting materials are $Fe(NO_3)_3$ and $Fe(NO_3)_2$ to synthesize mixed oxide magnetic NPs under N_2 protection. Natural product extracts include Chitin and chitosan obtained from the fishery waste shrimp shell extract. Other compounds, namely citronellal, citrate, and limonene extract from orange peels, are also used to prepare MIMNs. These compounds are chosen due to their nontoxic nature and ability to recycle natural products, further minimizing energy and resource consumption. Chitin (2-(acetylamino)-2-deoxy-D-glucose, also named β-(1–4)-poly-N-acetyl-D-glucosamine), is the second most abundant long-chain polymer, an amide derivative of glucose. This molecule is widely distributed in nature as an amino-polysaccharide after cellulose [128]. Chitin occurs as ordered macro fibrils and is the major structural component in the exoskeletons of the crustaceans, crabs, shrimps, and the cell walls of fungi (Fig. 2A.21). For biomedical applications, chitin is usually converted to its deacetylated derivative, chitosan [129].

As a nontoxic and linear polysaccharide copolymer, chitosan is the main component of a crustacean's shell. It is composed of β-(1→4)-linked N-acetyl-D-glucosamine (acetylated unit, $C_6H_{13}NO_5$) and N-acetyl-glucosamine ($C_8H_{15}NO_6$). Many benefits of chitosan were found, including antibacterial activity and antioxidation of metal ions [130]. Chitosan was soluble in acidic solutions as a deacetylated chitin form with varying deacetylation degrees [131]. Citronellal ($C_{10}H_{18}O$, also called radical) is the monoterpenoid

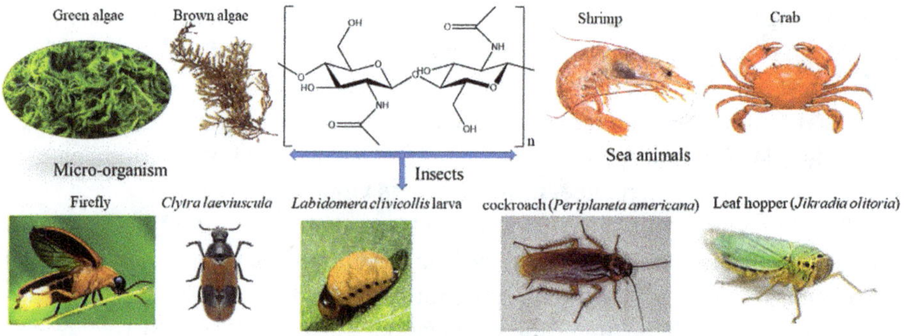

Fig. 2A.21: The chitin is an integral structural component present in different organisms (images were from shutterstock.com with permission).

aldehyde and primary reductive compound. It exists in plants Cymbopogon as the reason for oranges' special fragrance, giving citronella oil its distinctive lemon scent [132]. In this study, citronellal has been extracted from waste orange peel and used as a dispersing agent to improve MIMN's dispersion and biocompatibility. Citric acid is another reducing and capping agent used in MIMN's synthesis, inhibiting nanoparticle growth and agglomeration. It exists in a variety of fruits and vegetables. These naturally existing compounds are ideal candidates for fabricating biomimetic materials with high networking polymeric scaffolds. Using these natural products includes high porosity, good biodegradability and biocompatibility, predictable degradation rate, and maintainable structure integrity, (with select examples summarized in table 2A.4).

Tab. 2A.4: A summary of natural product extracts and their properties.

Name	The formula of the natural product extract	Properties
Chitin	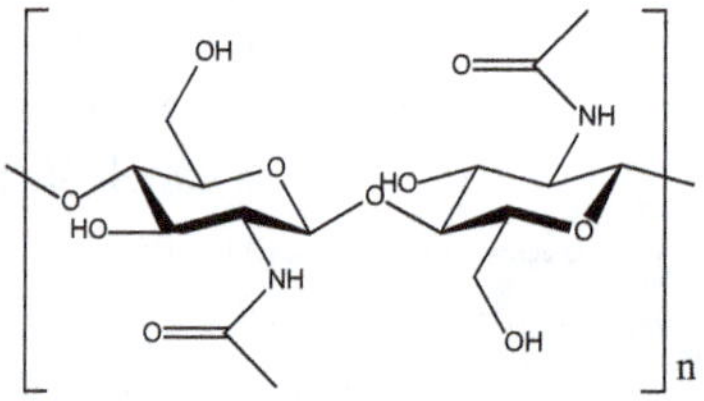	Abundant amino polysaccharide polymer Biodegradability, Lack of toxicity, Anti-fungal effects, Wound healing acceleration immune system stimulation
Chitosan	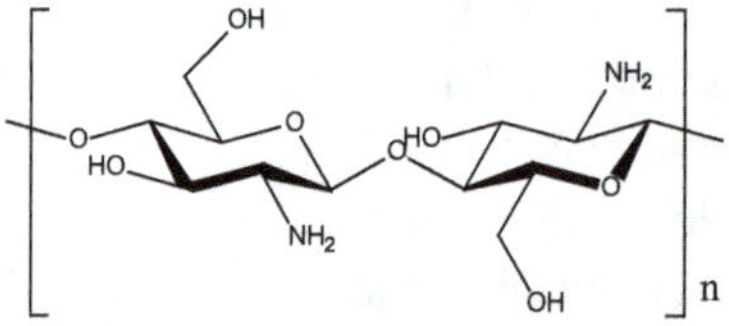	Protonation in neutral solution Water-soluble and a bioadhesive Binding to negatively charged surfaces Making crosslinked polymeric networks Enhancing the transport of polar drugs

Tab. 2A.4 (continued)

Name	The formula of the natural product extract	Properties
Citronellal		Insect repellent properties High repellent effectiveness against mosquitoes Strong antifungal qualities Low or very low-volatile Combustible substance Poorly flammable (flash point > 60 up to 93 °C)
Citric acid		Colorless weak organic acid Intermediate in the citric acid cycle Acidifier Flavoring and chelating agent

More than 20 known iron oxides and oxyhydroxides (Tab. 2A.5) play critical roles in geological and biological processes. The iron oxides' structures and cell volumes change significantly based on temperature and pressure. For example, iron peroxide showed 24 different phases under different synthesis conditions. Table 2A.5 summarizes iron metal oxides' selected crystallographic structure and lattice parameters.

Tab. 2A.5: The summary of iron oxides and their crystallographic structures Fe_2O_3 [133].

#	Nomenclature	Crystalline strucutre	Lattice parameters
		Oxide of Fe^{2+}	
1	Iron(II) oxide (FeO)		Wüstite [134] Space group: Fm$\bar{3}$m Crystal System: Cubic a: 4.3320 Å α: 90° Cell volume: 81.295 Å^3 Rock salt (Halite) structure Fe and O in octahedral coordination

Tab. 2A.5 (continued)

#	Nomenclature	Crystalline strucutre	Lattice parameters
2	Iron peroxide (FeO_2)		Space Group: P 1 Allows Chirality; Polar Crystal System: Triclinic a: 10.0830 Å b: 2.9695 Å c: 10.2054 Å $\alpha = \gamma$: 90.000° β: 90.195° Cell Volume: 305.563 Å^3
		Mixed oxides of Fe^{2+} and Fe^{3+}	
3	Iron(II, III) oxide, magnetite (Fe_3O_4)		Space Group: F d3m [135] Crystal System: Cubic a: 8.3941 Å Cell Volume: 591.456 Å^3 Spinel-type structure. Fe_1 is octahedrally coordinated by O Fe_2 is tetrahedrally coordinated by O.
3'	Iron(II, III) oxide (Fe_3O_4) Formed at 44 GPa		Space Group: Bbmm Crystal System: Orthorhombic a: 9.3090(30) Å b: 9.2820(20) Å c: 2.6944(9) Å Cell Volume: 232.813 Å^3
4	Fe_4O_5 Formed under high pressure, P = 10 GPa		Space Group: Cmcm [136] Crystal System: Orthorhombic a: 2.8430 Å b: 9.7000 Å c: 12.2900 Å Cell Volume: 338.923 Å^3

Tab. 2A.5 (continued)

#	Nomenclature	Crystalline strucutre	Lattice parameters
5	Fe_5O_7 (Homologous series nFeO mFe$_2$O$_3$, formed at 41 GPa)		Space Group: C2/m Crystal System: Monoclinic a: 9.2080(70) Å b: 2.7327(10) Å c: 8.2700(50) Å β: 105.500(80)° Cell Volume: 200.527 Å^3 Prisms are connected through common triangular faces Octahedra connect only via shared edges Triangular face-shared prisms and edge-shared octahedra
6	$Fe_{25}O_{32}$ (formed at 80 GPa)		Space Group: P $\bar{6}$2 $\bar{m}$ Crystal System: Hexagonal a: 13.4275(16) Å c: 2.6289(4) Å Cell Volume: 410.483 Å^3 Edge-shared one-capped prisms; belongs neither to the homologous series nor adopts any other known structural motif
7	$Fe_{13}O_{19}$		Space Group: P 1 Allows Chirality; Polar Crystal System: Triclinic a: 22.1917 Å b: 2.9108 Å c: 11.0205 Å $\alpha = \gamma$: 90.000° β: 118.470°
		Oxide of Fe^{3+}	
8	α-Fe$_2$O$_3$: alpha phase, hematite		Space Group: R$\bar{3}$c Crystal System: Trigonal a: 5.0380 Å c: 13.7720 Å Cell Volume: 302.722 Å^3

Tab. 2A.5 (continued)

#	Nomenclature	Crystalline strucutre		Lattice parameters
9	Fe$_{1.76}$ H$_{.06}$ O$_3$, hematite proto (formed at T = 313 °C)		Fe, H, O	Space Group: $R\,\bar{3}c$ Crystal System: Trigonal a: 5.0145 Å c: 13.6920 Å Cell Volume: 298.162 Å^3
10	Fe$_2$O$_3$		Fe, O	Space Group: $P\,1$ Crystal System: Triclinic a: 5.1447 Å b: 5.2884 Å c: 7.5114 Å $\alpha = \beta = \gamma$: 90.000° Cell Volume: 204.363 Å^3
11	β-Fe$_2$O$_3$: beta phase		Fe, O	Space Group: Ia$\bar{3}$ (P 1) Allows Chirality; Polar Crystal System: Triclinic $a = b = c$: 9.5561 Å $\alpha = \beta = \gamma$: 90.000° Cell Volume: 872.658 Å^3
12	γ-Fe$_2$O$_3$: gamma phase, maghemite		Fe3, Fe2, Fe, O	Space Group: P 43 3 2 [137] Allows Chirality Crystal System: Cubic a: 8.3474 Å Cell Volume: 581.639 Å^3
13	η-Fe$_2$O$_3$ at 64 GPa		Fe, O	Space Group: Cmcm [138] Crystal System: Orthorhombic a: 2.6400(60) Å b: 8.6390(90) Å c: 6.4140(140) Å Cell Volume: 146.284 Å^3

Tab. 2A.5 (continued)

#	Nomenclature	Crystalline strucutre	Lattice parameters
14	θ-Fe$_2$O$_3$ (formed at 74 GPa)	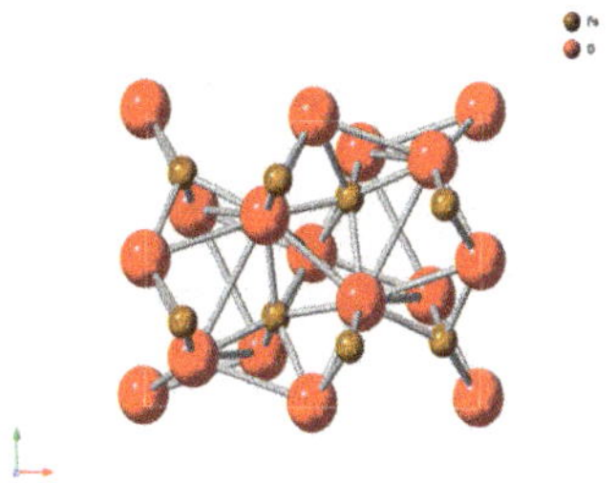	Space Group: Aba2 [139] Polar Crystal System: Orthorhombic a: 6.5240(90) Å b: 4.7020(30) Å c: 4.6030(70) Å Cell Volume: 141.201 Å^3
15	τ-Fe$_2$O$_3$ (formed at 41 GPa)	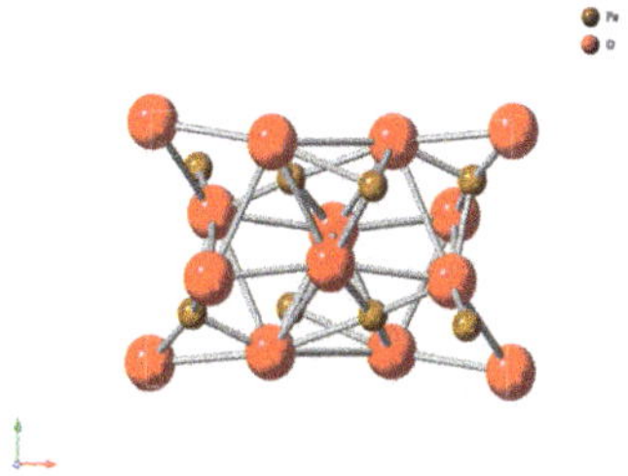	Space Group: Pbcn Crystal System: Orthorhombic a: 7.0620(100) Å b: 4.8108(13) Å c: 5.0019(8) Å Cell Volume: 169.934 Å^3

Oxide-hydroxides

#	Nomenclature	Crystalline strucutre	Lattice parameters
	Goethite (α-FeOOH)	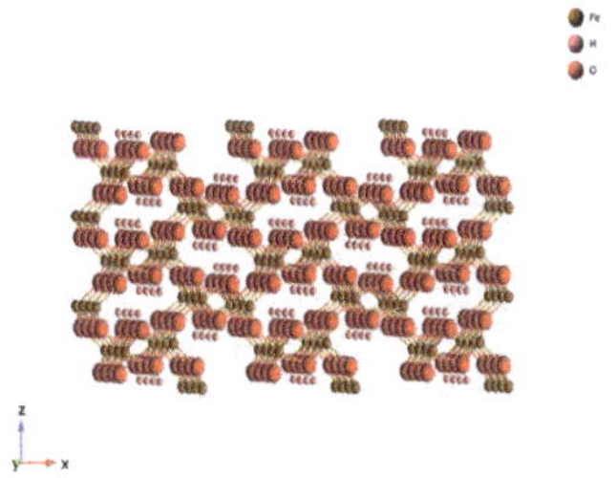	Space Group: Pnma Crystal System: Orthorhombic a: 9.9500 Å b: 3.0100 Å c: 4.6200 Å Cell Volume: 138.367 Å^3 all atoms in 4c positions
14	Akaganéite (β-FeOOH), an iron oxyhydroxide mineral (Fe$_{7.6}$H$_{9.28}$Ni$_{0.4}$O$_{16}$Cl$_{1.16}$)		Space Group: I 2/m [140] Crystal System: Monoclinic a: 10.5870 Å b: 3.0311 Å c: 10.5150 Å $β$: 90.030° Cell Volume: 337.429 Å^3 hollandite structure octahedral chains of FeO$_6$ square channels along the structure's y-axis

Tab. 2A.5 (continued)

#	Nomenclature	Crystalline strucutre	Lattice parameters
15	Lepidocrocite (γ-FeOOH), an iron oxo-hydroxide mineral		Space Group: Bbmm [141] Crystal System: Orthorhombic a: 12.4000 Å b: 3.8700 Å c: 3.0600 Å Cell Volume: 146.843 Å^3 FeO6 octahedral groups Share edges along the structure's y- and z-axes Form crenulated octahedral sheets
16	Feroxyhyte (δ-FeOOH), an iron oxyhydroxide mineral		Space Group: P $\bar{3}$m1 [142] Crystal System: Trigonal a: 2.9500 Å c: 4.5600 Å Cell Volume: 34.367 Å^3 Edge-sharing FeO$_6$ groups Connected to form hexagonal sheets parallel to (001) Adjacent sheets share octahedral faces along the z-axis

The simulated X-ray powder diffraction patterns of Fe_3O_4 (Hermann Mauguin, Fd3m [227]; Hall F4d23$\bar{1}$d, Point Group, m$\bar{3}$m; Crystal System, cubic). Moreover, Fe_2O_3 (maghemite standard) was shown in Fig. 2A.22A and B. These crystallographic data have been used to guide the materials synthesis to produce highly crystalline structures used for heavy metal removal. It was found that these well-crystallized nanomaterials were able to be dispersed in the aqueous solution well and maintained their sub-building units for a prolonged period [143].

Band structure and density of states (DOS) of Fe_3O_4 and Fe_2O_3 were shown in Fig. 2A.23A and B, respectively. The electronic band structure of these metal oxides can describe the range of energy levels. The electrons occupied the orbitals with lower energy, and the higher energy was occupied afterward. There is a gap between the highest occupied molecular orbital and lowest unoccupied molecular orbitals, defined as the band gaps or forbidden bands. Except for the first element, hydrogen (H), all atoms have more than one electron. The electrons occupy the individual atomic orbitals within a single isolated atom with discrete energy levels. When different atoms bonded with each to form a molecule, the atomic orbitals will overlap through hybridization to split these discrete energy levels. The adjacent energy levels are close together for the nanomaterials and are considered an energy band continuum [145].

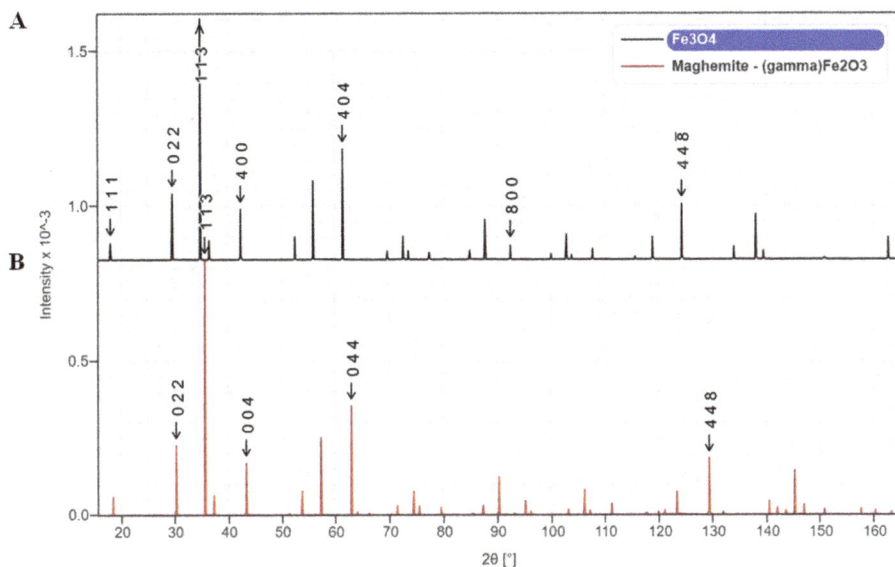

Fig. 2A.22: The X-ray powder diffraction patterns were simulated using Crystal Maker to demonstrate lattice plane: (A) Fe_3O_4 structure [144] and (B) Fe_2O_3 from [137].

The density-functional theory (DFT) has been widely used to calculate and determine the electronic structure under the ground state. The physicists, chemists, and materials scientists use the DFT to investigate many-body systems, especially atoms, molecules, and condensed phases [146]. The DFT has been built using force constants and basis set extensions based on the Bohr atom concept but taking account of the extended forces. These Generalized Gradient approximations (GGAs) for exchange and correlation-energy expressions into mathematical functionals accurately predict the electron density used to evaluate the properties of a "many-electron system." [147] Specific basis and extended sets use these functionals for in-depth understanding and design of catalytic processes, drug delivery, electron transport, energy conversion, and other problems in science and technology [148]. One of the functions is spatially dependent on electron density. This computational quantum mechanical modeling method is one of the most popular and versatile techniques available for computational studies [149]. Systems' DOS can explain the proportion of states occupied at each energy. The DOS is a mathematical representation showing the electronic distribution by a probability density function [150]. The various states occupied by the system provided this function over the space and time domains. Two figures (Fig. 2A.23A and B) showed electronic band structures and DOS for Fe_3O_4 and Fe_2O_3 systems [151]. The DOS is directly related to the dispersion relations at a specific energy level. The DOS corresponded to many available states for electrons to be allocated.

X-ray absorption spectroscopy (XAS) is an analytical technique to evaluate solid-state materials' local geometric or electronic structure [155]. The experiment is

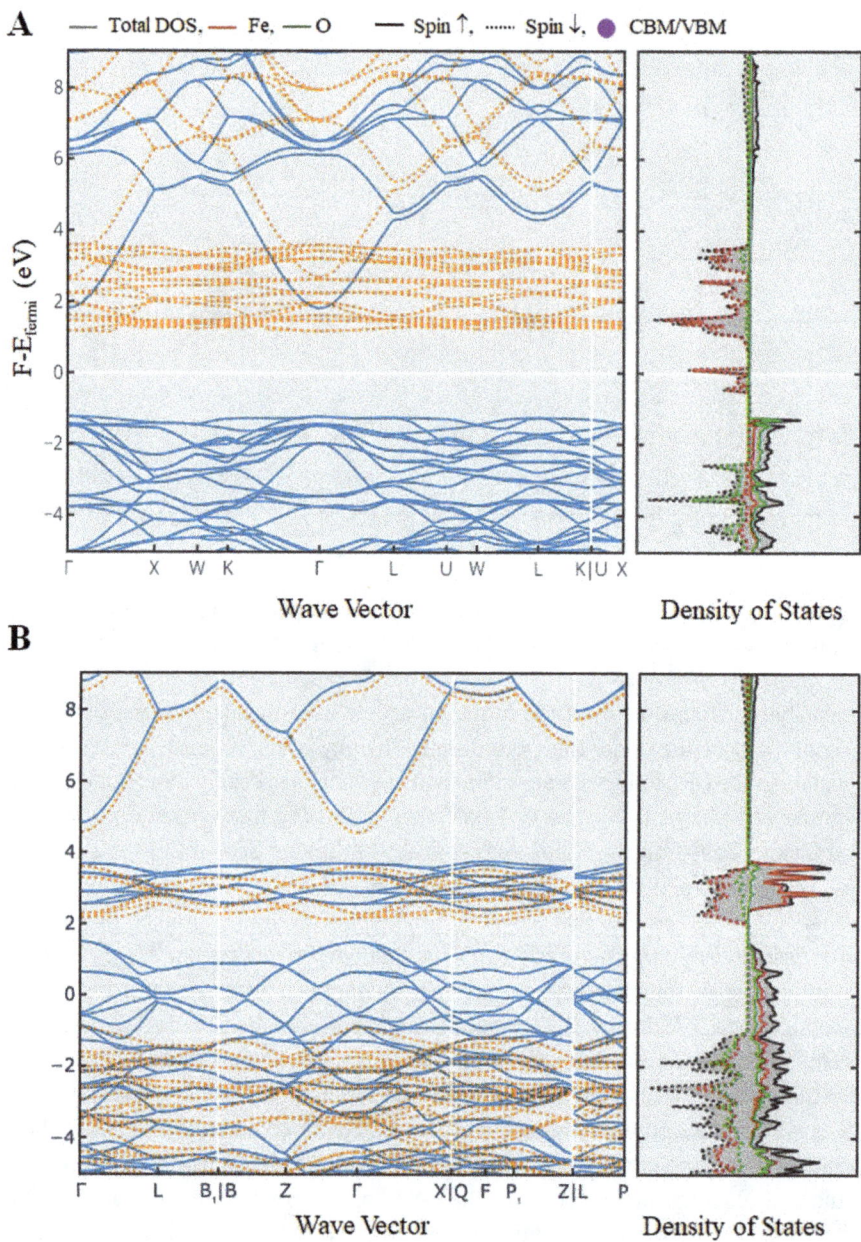

Fig. 2A.23: Band structure and density of states using semi-local density functional theory (DFT): (A) Fe_3O_4 results [152] and (B) Fe_2O_3 results [153] (note: semi-local DFT method tends to underestimate bandgaps severely) [154].

usually performed at synchrotron radiation facilities, which provide intense and tunable X-ray beams. Samples were in the gas phase, solutions, or solids [156]. XAS data are obtained by tuning the photon energy [157] using a crystalline monochromator to a range where core electrons were excited (0.1–100 keV) [158].

In our investigation [88], Ms. Huang used gum arabic (N-(8-(2-hydroxybenzoyl) amino)caprylate) as a dispersing and capping agent due to its weakly acidity. This hetero-polymer is composed of a neutral hetero-polysaccharide with branched chains. The gum arabic was found in salts composed of polysaccharides, acids, and countercations such as magnesium (Mg^{2+}), sodium (Na^+), and potassium (K^+). The units of l-arabinose (aldopentose, $C_5H_{10}O_5$), l-rhamnose (deoxy sugar, $C_6H_{12}O_5$), d-glucuronic acid (uronic acid, $C_6H_{10}O_7$), and 1,3-linked β-D-galactopyranosyl are the backbones in gum arabic (Figs. 2A.24A, B, and 2A.25A). It has been used to control the hydrolysis rate and tune the architecture of the final catalysts. A key component, sodium 2-[(7-carboxyheptyl)-C-hydroxycarbonimidoyl]benzen-1-olate ($C_{15}H_{20}NNaO_4$), was used for the demonstration of the templation of the Fe_3O_4 MIMNs (Figs. 2A.24B, C and 2A.25B, C). Based on the simulated structural results, it was hypothesized that the O from gum arabic could occupy the lattice site of the Fe_3O_4 sub-building units through the common element oxygen to prevent particle growth (Fig. 2A.25D). The NPs are evaluated using Tecnai F20-G2 TEM equipped with X-ray energy dispersive spectroscopy to determine the morphology. The TEM images showed that uniform distributed NPs were obtained (Fig. 2A.25E). It was also found that the capping agent can also result in different shapes of NPs observed based on the different chemical compounds (Fig. 2A.25F) and synthesis conditions.

X-ray photoelectron spectroscopy (XPS) was used to evaluate the surface and in-depth chemical composition of the core/shell MIMNs (Fig. 2A.26). The Fe element's mixed oxidation states (Fe^{2+} and Fe^{3+}) were found based on the principal emission. The characterization data of self-synthesis core/shell Fe_3O_4 MIMNs with a coating by gum arabic were well-indexed with the standard Fe_3O_4. The Si peak found in this XPS data was caused by glassware or the impurity of the starting materials. Besides the small binding energy difference between Fe_2O_3 and Fe_3O_4, the existence of a satellite between Fe $2p_{1/2}$ and Fe $2p_{3/2}$ of Fe_2O_3 can also be the other distinguishing index. Based on the previous research (Fig. 2A.26), there was a satellite peak in the XPS data of Fe_2O_3 (Fig. 2A.27A), but there was no satellite peak in Fe_3O_4 (Fig. 2A.27B) [160]. In the previous study, Ertl et al. reported that Fe $2p_{3/2}$ in the standard Fe_3O_4 specimen does not have a satellite peak [161]. The presence of the satellite peak at 718 eV is used as an indicator of the formation of Fe_2O_3 rather than Fe_3O_4. In the Fe_3O_4, the peak positions of Fe $2p_{3/2}$ and Fe $2p_{1/2}$ are located at 710.6 (S.D. = 0.05) and 724.1 eV (S.D. = 0.07), respectively. The Fe^{2+}:Fe^{3+} ratio should be 1:2 (0.33:0.67) in Fe_3O_4 (also expressed as FeO Fe_2O_3), calculated by the mean relative areas of each constituent peak. Our data align with the published results for Fe_2O_3 (Fig. 2A.28A) and Fe_3O_4 (Fig. 2A.28B), indicating that the mixed iron oxides were produced using gum arabic as a capping agent under nitrogen gas protection.

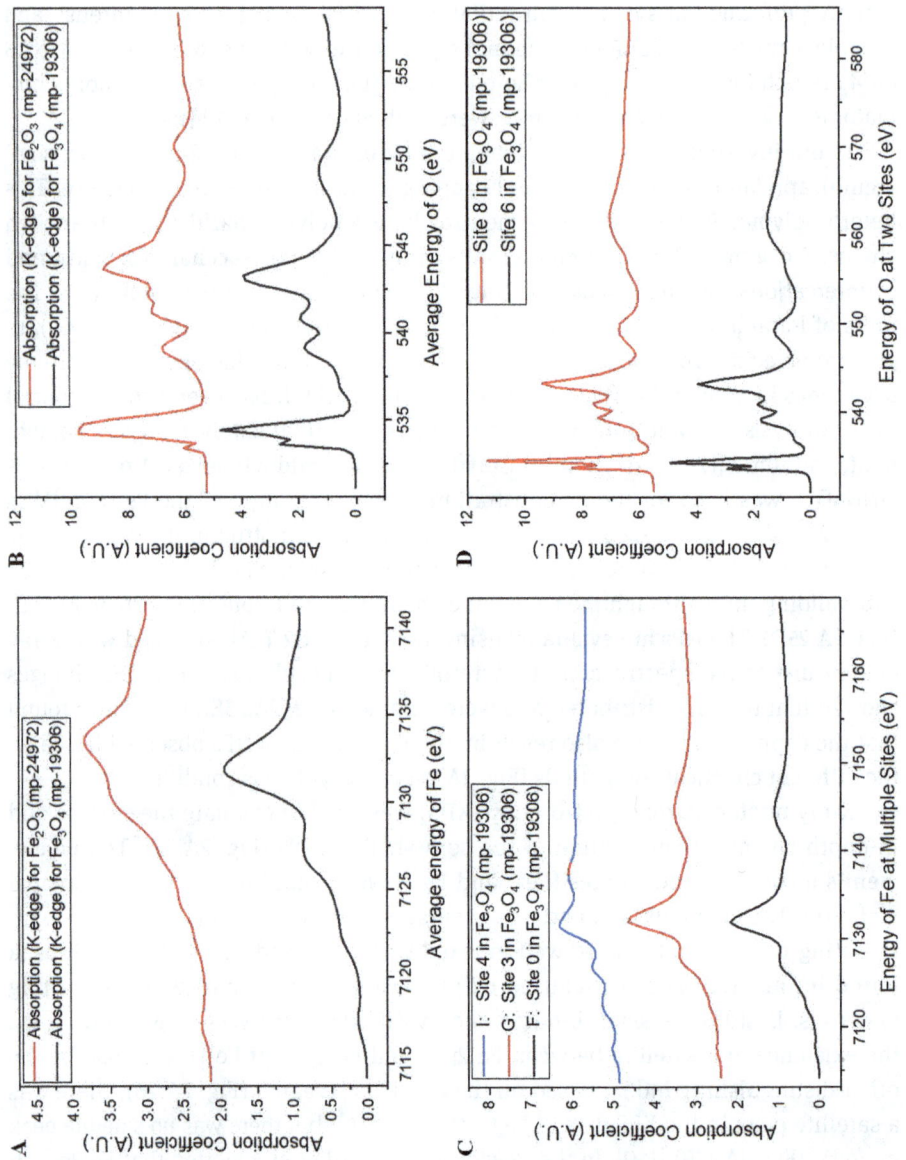

Fig. 2A.24: X-ray absorption spectra of Fe_2O_3 and Fe_3O_4 were obtained from the Materials Project, an open-access database: (A) the average K-edge absorption of Fe element in Fe_2O_3 and Fe_3O_4, (B) the average K-edge absorption of O element in Fe_2O_3 and Fe_3O_4, (C) K-edge absorption of Fe element in Fe_3O_4 at three different sites, and (D) K-edge absorption of O element in Fe_3O_4 at two different sites (averaged spectra are third-order spline interpolations ranged over the intersection of E-axis ranges of site spectra) [159].

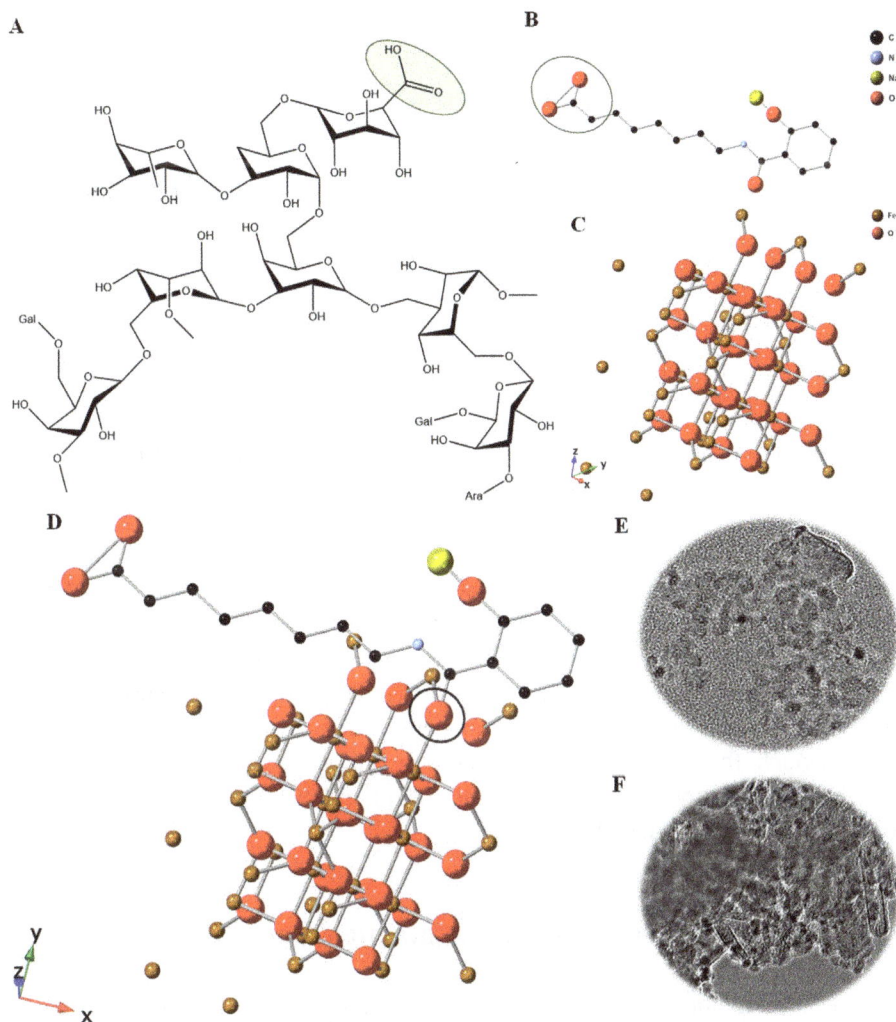

Fig. 2A.25: The formation of Fe_3O_4 nanoparticles using gum arabic as a capping agent: (A) the skeleton of gum arabic, (B) the key unit in gum arabic chelated to Fe_3O_4 surface oxygen corner, (C) Fe_3O_4 unit cell, (D) the conjugation between gum arabic and Fe_3O_4 to form chelated nanoparticles, resulting in spatial hindrance, and (E) the spherical structure of gum-arabic-modified Fe_3O_4 NPs. The rod-shaped morphology was obtained for the combination of Fe_2O_3 and FeOOH.

The atomic absorption spectroscopy evaluated the arsenic removal using the MINMs as adsorbents (Fig. 2A.28). The "plant-extract and sea-animal extract encapsulated Fe_3O_4" nanocomposites showed that the "biota-on-surface" strategy can improve the sequestration of heavy metals and bioremediation pharmaceutical waste. The AsO_4^{3-} removal using Fe_3O_4 NPs coated by natural product extract was increased by 25 % compared with the nonshielded ones. The water remediation efficiency

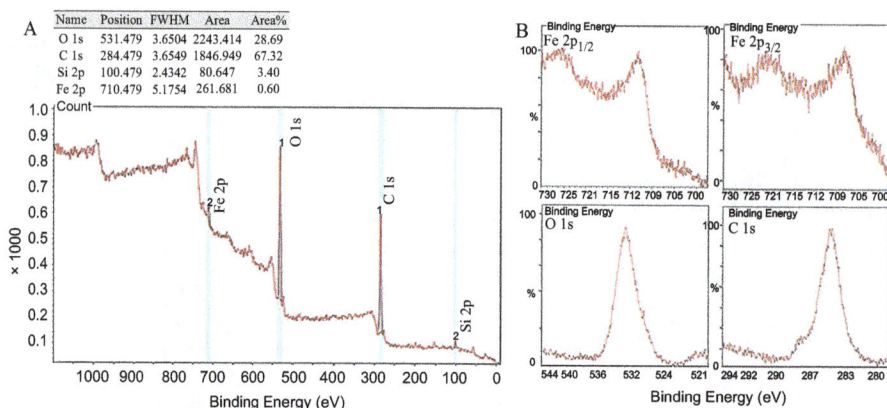

Fig. 2A.26: The XPS analysis of gum-arabic-modified magnetite nanoparticles: (A) survey of the MIMNs and (B) binding energy of Fe ($2p_{1/2}$, $2p_{3/2}$), O 1s, and C 1s.

achieved 100 % when the As(V) concentration was lower than 10 %. The result showed that the higher arsenic concentration attained the lower removal capacity in Fig. 2A.28A. However, the reason the lowest water remediation capacity occurred at 10 ppm still needs further studies because it did not correspond to the general efficiency trend. Shrimp shell extract core/shell MNPs obtained slightly more arsenic removal than gum arabic MNPs because chitosan contains an amino group. Based on the introduction of this paragraph, the amino group is the reason for chitosan absorbing various substances. The removal amount of arsenic oxoanions did not show a distinctive trend relative to concentrations (Fig. 2A.28B).

2A.3.3 Bioinspired remediation: a practical approach

Our objective in bioinspired water remediation was to produce three formulations of interactive nanocomposites (INCs) to respond to the variations of environmental stimuli and self-regulate their nanostructures and biochemical properties. These interactive nanomaterials aim to improve the bioremediation efficacy of organic pollutants and a heterogeneous mixture of polycyclic aromatic hydrocarbons (PAHs) and heavy metals. These INCs are biota-anchored porous nanocomposites with hollow core–shell structures and metal-organic frameworks (MOFs). The first formulation is composed of β-cyclodextrin-oriented $CeO_{2-\delta}$ thin film (β-CDC) used for biodegradation of benzene, toluene, ethylbenzene, and xylenes (BTEX), trichloroethylene (TCE) under aerobic conditions. The second is iron-centered MOFs with positive surface functionalization for bioremediation of a heterogeneous PAHs and Cr(VI) mixture. The third is Zn bio-MOFs with negative surface modification to enhance bio-sequestration and degradation of PAHs and Cd(II) mixture. The naturally existing biota, such as *Pseudomonas*

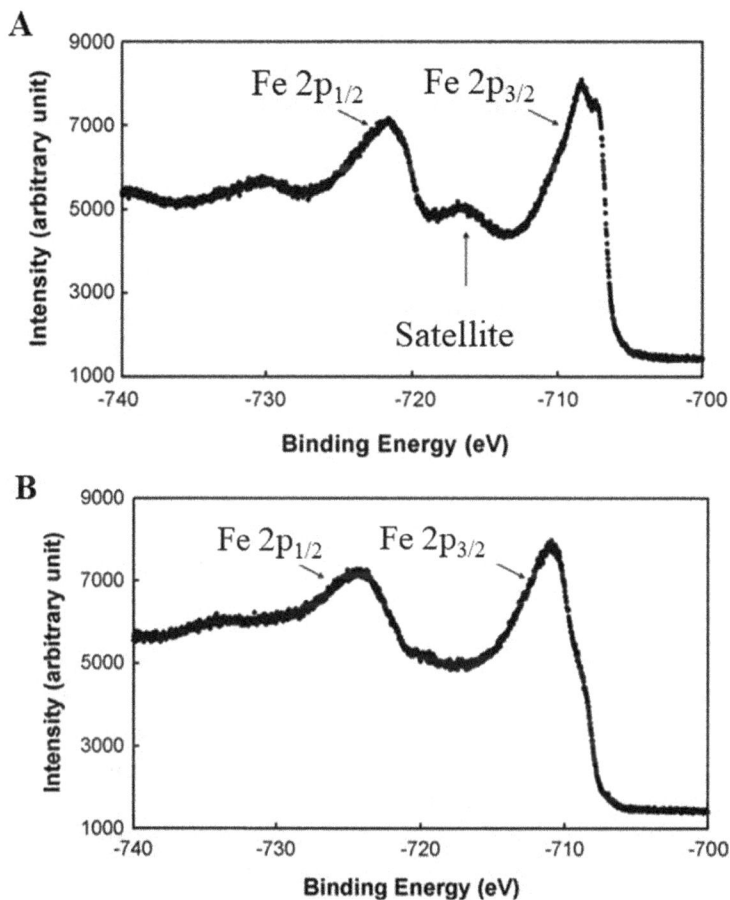

Fig. 2A.27: The XPS analysis of iron oxides: (A) The XPS spectrum of Fe 2p from the fractured surface of the Fe_2O_3 standard sample and (B) The XPS spectrum of Fe 2p from the fractured surface of the Fe_3O_4 standard sample (images were adapted from [160] with permission).

aeruginosa, were aligned along the surface of membrane or frameworks to provide metabolic pathways for the remediation of organic compounds. These "biota-on-surface" structures were tuned based on the environmental conditions when used as interfacial biocatalysts with the optimized amphiphilic balance between shell and core and host and guest. As discussed previously, our earlier study on "plant-encapsulated Fe_3O_4" and "citrate-coordinated TiO_2" nanocomposites indicated that this biota-on-surface strategy enhanced the adsorption capacity of heavy metal ions (Hg^{2+} and AsO_4^{3-}) [126] and bioremediation of pharmaceutical waste (TCH) with a dynamic and responsive reconfiguration to adapt the condition changes [103]. It is anticipated that the bacteria-anchored synthetic INCs act as "bioreactors" to provide biota protection, facilitate O_2 diffusion, and collectively enhance remediation efficacy. The working

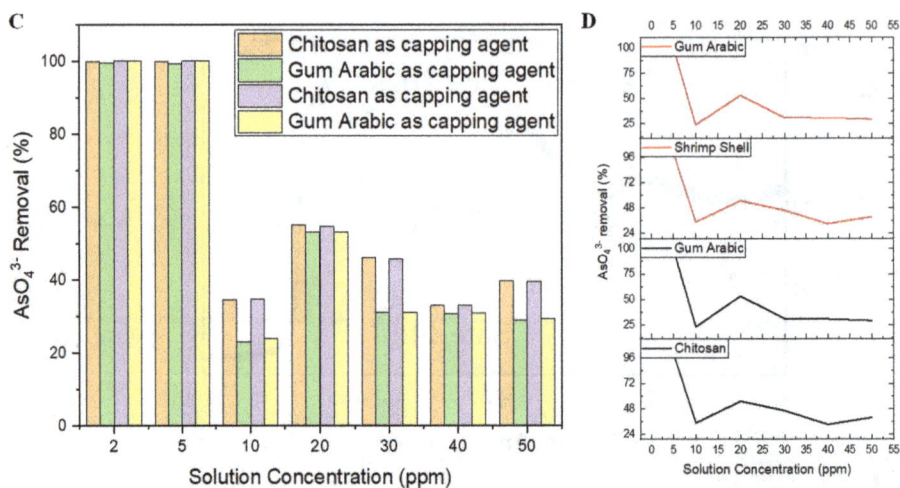

Fig. 2A.28: The formation of Fe_3O_4 nanoparticles using gum arabic as a capping agent: (A) the skeleton of gum arabic; (B) the key unit in gum arabic chelated to Fe_3O_4 surface oxygen corner; (C) Fe_3O_4 unit cell; and (D) The removal efficiency of heavy metals using natural binding and scaffolding agents was also compared.

hypotheses of these studies lie in that (1) the reversibility between the two crystalline phases of cerium oxides is desirable for O_2 release via the formation of oxygen vacancies, enabling isometric cubic crystal unit; carried out by an encapsulated cargo (β-CDC) into biological environments and (2) the functionalization of MOF motifs allows for fine-tuning of the surface charges and internal and external porosity, maximizing bioremediation efficacy of pollutant mixtures via immobilization of charge carriers at the heterojunctions of host-guest systems.

Bio-remediation: The bioremediation rate of the CDCs and MOFs were measured in aqueous solutions composed of organic compounds and mixtures of heavy metals and PAHs. The aqueous model solutions of contaminants (20 ppm) were prepared, into which INCs colloids were introduced. Then the mixture was agitated for 50 min in the darkness to establish contaminant adsorption-desorption equilibrium. The different variables include degradation duration, pH values, and pollutant concentrations. At 20 min intervals, an aliquot (0.3–1.0 mL) was filtered to remove INCs. The supernatant was collected to measure the absorbance or fluorescence using a Tecan M200 Pro to determine the degradation efficiency.

Metal ions sequestration: Both Langmuir and Freundlich isotherms were carried out to describe the adsorption behavior of heavy metals using functionalized MOFs. A self-generated host-guest system was tested to evaluate its bioremediation efficacy. The in situ analyses used the pilot data from these reactivities of different formulations of INCs. These different formulations of INCs are responsive to environmental changes and are self-regulative in their structures and performances. The metal oxide-based INCs improve the bioremediation of organic compounds via spontaneous catabolism. The MOFs-based INCs address metal sequestration, forming a self-generated host-guest system to immobilize the metallic pollutants. The naturally existing bacteria are anticipated to align along the material surface, acting as the bioreactor favorable to electron transfer. The nontoxic nature of these materials on neuron (RPE) cells mitigates or eliminates potential secondary pollution caused by these INCs. Thus, biota-on-surface nanomaterials were accomplished to better the environmental control of hazardous wastes and bioremediation of commingled contaminants.

Three activities were carried out to improve water quality using different formulations of interactive nanomaterial composites (INCs). (1) Fluorescence changes were measured to evaluate the bioremediation efficacy of the BTEX, PAHs, and TCE in a biphysic system composed of water-organic solvents using a Tecan M200 PRO; (2) The adsorption capacity of heavy metal ions by MOFs (host-guest system) and the bioremediation efficacy of the PAHs were measured in the same approach as mentioned above; and (3) The oxidation state, crystalline phase, and morphology were tested using XPS, powder, and single X-ray diffraction and TEM. Based on our literature survey, INCs with self-regulating behavior have not been reported. Their efficacy in bioremediation will open a new paradigm for the environmental control of organic and heavy metal pollutants. A reversible and rapid lattice switch and space occupancy upon oxidation state changes between Ce^{4+} and Ce^{3+} is anticipated to allow for electron transfer freely, avoiding high lattice distortion. Due to electrostatic or van der Waals attraction, the heavy metal ions existing in wastewater and soil are absorbed by MOFs. Therefore, functionality and porosity must be actuated to improve the adsorption capability of metal cations and polyatomic anions. The biodegradation efficacy of organic waste and the biological organism is expected to be enhanced using this self-generated host-guest system due to the synergistic effect.

As we all know, nature has inspired researchers to design and develop materials with functionalization and fine-structural tuning in different reacting atmospheres [162]. An appropriate selection of solvent molecules, sub-building units, and switchable functional groups allow the preparation of smart and responsive materials [163]. Water is a widely used solvent to create versatile aqueous solutions due to its high polarity nature and uneven electron distribution, which enable this simple molecule to dissolve a majority of inorganic and some organic compounds [164]. Its powerful solvation ability will facilitate the self-assembly, biomimicry, and complexity of produced materials from intra- and intermolecular perspectives [165]. The major advances in designing and constructing interactive and structure-tunable porous nanomaterials using wet chemistry are cost-effective and user-friendly [166]. The homogeneity and architecture of end-products were tuned from the molecular level, enabling fine-tuning precision and accuracy [167]. Multiscale texturation and upscaling production were achieved using sol–gel chemistry [168] and self-assembly to obtain thin films using cyclodextrin as cargo to disperse the CeO_2 uniformly.$_\delta$ NPs, as found in our previous study [169]. The combined solvothermal and microwave methods also facilitated rapid nucleation and miniaturization of MOFs [170], followed by a post-functionalization to modify their surface properties [171]. The final products were manifested as particles, tubes, hollow spheres, and thin films [172]. These synthesis methods maintain the starting materials' intrinsic biological, physical, and chemical properties, inorganic compounds, polymeric surfactants, or hybrid matrices [173]. Therefore, the porosity, pore size, surface area, particle size, and distribution were tuned to obtain well-defined porous polymeric networks, connectivity between interfaces, and accessibility to reactive sites [174].

The inner transition metal oxide, nonstoichiometric $CeO_{2-\delta}$ has been chosen as a core for the hollow core-shelled INCs, used for in situ bioremediation of organic compounds due to its advantages, low-cost, environmental protection friendliness, and high chemical/structure stability. It was reported that CeO_2 hollow spheres enabled the removal of dyes such as Congo red and acid orange [175]. Adsorption and separation of heavy metal ions and dye using CeO_2 NPs with high adsorption capacity were also reported due to the large pore/cavity volumes and facile mass transport [176]. Moreover, these CeO_2 hollow spheres can easily be regenerated through a centrifugation-calcination procedure with a limited loss in adsorption performance [173, 177]. Introducing a pendant beta-cyclodextrin (β-CDC) unit into the $CeO_{2-\delta}$ sol–gel synthesis allowed self-assembly into a Ce-containing hydrogel through complexation between both components through ion-dipole attraction [173, 178]. The redox state of $CeO_{2-\delta}$ can alter its binding affinity to β-CDC, seducing polymer chains with a changeable modulus [179], surface charges, and accessibility to pores [169]. As a result, the O_2 release (1.5×10^{-4} mmoles O_2/L hr) was modulated based on the needs of the existing biota. Iron-centered MOFs were selected for their proven toxic heavy metal sequestration properties [180], further generating a host-guest system, which is anticipated to synergistically improve the biodegradation of aromatic compounds (BTEX and PAHs) [181, 182]. The MOFs are crystalline frameworks of metal ions and organic ligands [183]. The pores inside an open MOF were stable and tunable to high surface areas, various topologies, and network types via increasing linker length and size, targeting specific topologies, and further addressing metal-organic vertices to produce bio-MOFs [184]. The carboxylate linkers were widely used to prepare MOFs with microporous and mesoporous cavities [185]. Uniform and tunable pore sizes and functionalization readiness offer great MOFs application [186]. Cavities up to 4 nm have been introduced to hierarchically assembled MOF motifs by adopting an angular, semi-flexible tetra-topic ligand [187]. The resulting MOFs show permanent porosity and exhibit stepwise sorption isotherms for select metal ions by utilizing ion exchange, changing the functional group of ligands, and introducing unsaturated metal centers [188]. The functionalized MOFs can prevent dissociation of the host-guest complex due to the possible ion exchange or covalent cross-linking [189]. A recent study indicated bacterial attachment to PAHs that serve as carbon and energy sources, depending on the solubility of PAHs [190].

Since bacteria initiate PAH degradation by intracellular dioxygenases, PAHs were oxidized into cis-dihydrodiols and further into catechols or derivatives by incorporating both atoms of O_2 [191]. This approach demonstrates that harmful metal ions were immobilized by surface adsorption, and their toxicity was moderated. Consequently, a self-generated host-guest system was formed to mitigate damaging biota and provide reactive sites for synergistic PAH degradation.

The "plant-extract encapsulated Fe_3O_4" and "citrate-coordinated TiO_2" nanocomposites indicated that the "biota-on-surface" strategy could improve the sequestration of heavy metal and bioremediation of pharmaceutical waste. The AsO_4^{3-} removal using plant-Fe_3O_4 NPs was increased by 25 % compared with the nonshielded iron nanomaterials. Using citrate-TiO_2 NPs (20 ppm), the TCH degradation rate achieved a 92.5 % reduction within 2 h under SS [103]. Their toxicity was evaluated using retinal pigment epithelium (RPE) cells in the US Air Force research laboratory, Fort Sam Houston, indicating they are user-friendly [126]. Sol–gel wet chemistry is one of the most feasible and effective methods to develop interactive $CeO_{2-\delta}$ nanofilms involving bacteria aligned along the film surface. The Ce oxidation number was changed using electrochemically active biota to release O_2 controllably, with the β-CDC matrix as the nutrient supplier. The synthesis variables to produce different INCs have been optimized in our group, and 12 formulations were found biocompatible based on the cytotoxicity evaluation on RPE cells. This strategy will modify the current sol–gel synthesis to produce nonstoichiometric $CeO_{2-\delta}$ into which 1–5 mass % of carbon graphene quantum dots is introduced to facilitate the Ce^{4+} to Ce^{3+} redox reaction for controllable release of O_2. There are four key variables to control β-CDC structure using Ce(IV) alkoxides as starting materials: hydrolysis inhibition, hollow sphere templating, crystallization and rapid shift of redox states, and formation of "biota-on-surface." This method offered ultra-small particle size, long triple phase boundary, homogeneity at the molecular scale, and provision of O_2 and C sources for bacteria to activate the BTEX, PAHs, and TCE spontaneous degradation. This proposed family of innovative and biocompatible INCs is expected to enhance the bioremediation efficiency of in situ hazardous substances in groundwater and contaminated soil in the junction of using those naturally existing bacteria to construct "biota-on-surface" structures (Fig. 2A.29).

The controllable O_2 release is proposed through vacancy formation within the CeO_2 lattice while maintaining its crystalline structure. The rapid phase shift will not produce enough O_2 for bacteria to react to unsaturated covalent bonds. Two alternative approaches were also implemented: (1) carbon quantum dots (1–5 % by mass) were introduced to initiate the redox reaction of $Ce^{4+/3+}$; (2) Aloin-encapsulated MO_2 composite were tested to facilitate the spontaneous oxidation of C = C double bond and C–Cl single bond. Lectin is a family of carbohydrate-binding proteins. Recently, seven lectins have been identified: C-type, L-type, P-type, M-type, fibrinogen-like domain lectins, galectins, and calnexin/calreticulin. These lectins with different structures showed diverse expression patterns and multiple functions in the immune response. The lectin C was chosen to show the biota-on-surface concepts, allowing nanomaterials to be sensible and responsive to external signal changes and self-regulate their structures. Although the mechanism of how lectins bind to and communicate with the cells remains to be investigated, the binding triggered a response to environmental changes. This example focused on forming lectin-anchored nanoparticles to remove biological pollutants through selective chemical binding.

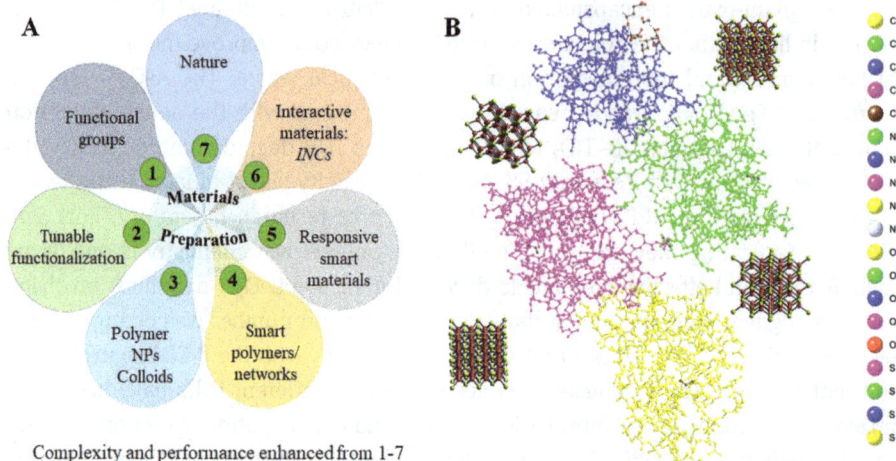

Fig. 2A.29: Creation of "biota-on-surface" structure with improved remediation of organic wastes with a sensible response to external stimuli and tunable geometry: (A) wet chemistry synthesis to balance complexity and performance of INCs and (B) "biota-on-surface" lectin-CeO$_2$ structure offering spontaneous metabolic pathways.

The functionalized MOFs are proposed to be used for heavy metal sequestration and PAH biodegradation. These highly porous MOFs are attractive for outer coordination and intermolecular interactions to immobilize harmful contaminants. The proposed MOFs are iron (Fe$^{2+/3+}$) and Zn^{2+} as central elements, dicarboxylic benzoic acid (BDC), BDC derivatives, and adenine as ligands. The Zn-MOFs have been widely studied and produced using a solvothermal method, followed by a post-modification to tune the surface charge and porosity. A mixture of H$_2$O and DMSO was used as a solvent, and the temperature was controlled at 120 ± 5 °C for 24–72 hrs. The biocompatible and robust bio-MOF-1 (C$_{104}$H$_{64}$N$_{20}$O$_{25}$Zn$_8$) produced by An et al. [192] demonstrated large open cavities. In this MOF structure, polyhedral model, benzene rings are represented by black hexagons; green pentagons represent imidazolate (C3N2) rings, and pyrimidine (C$_4$N$_2$) rings are represented by blue hexagons (Fig. 2A.30A). A series of organic pollutants (Fig. 2A.30B) was oxidized through adsorption, and heavy metals were removed through ion exchange.

This geometry enables a post-modification by amphiphilic groups into the ligands or MOF surfaces. Both protic and aprotic channels for metal adsorption were achieved in the synthesis under mild conditions. Our previous results demonstrate that bioremediation of xenobiotics was accomplished using microorganisms via engineered biota-on-surface nanomaterials (Fig. 2A.31C). If cytochrome % activity > whole-cell (WC), indicating favorable e⁻ transfer, bioremediation occurs spontaneously (Fig.2A.31D). Thermodynamically, the Gibbs free energy is negative; however, the reaction is low due to steric hindrance of the functional groups that limit the kinetics of the bio-remediation of BTEX compounds. We aimed to increase the reaction rate under ambient conditions using a biota-on-surface strategy. The immobility of harmful metal ions using MOF-based materials and their subsequent host-guest formation will be problematic, causing damage to biota and low biodegradation efficacy. Natural products were found to be more resistant to these poisonous cations.

A. BioMOF with large cavities: Metal ion immobilization

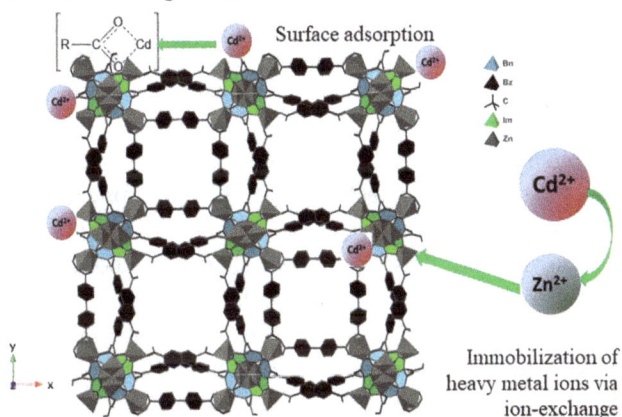

B. BTEX

Immobilization of heavy metal ions via ion-exchange

C. Oxidation of C=C double bonds

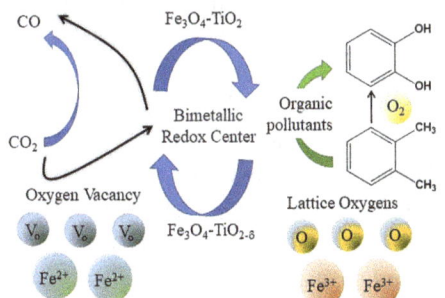

Aromatic compound oxidation ↔ Reduction of metals oxides

$\Delta G_0^r = -2 \times 96485 \text{ J/mol·V} \times 0.224\text{V}$
$\Delta G_0^r = -43.19 \text{ kJ (per mole of each cytochrome)}$

D. Cytochrome activity: remediation spontaneity

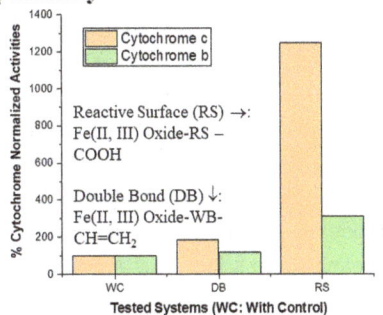

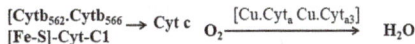

Reactive Surface (RS) →:
Fe(II, III) Oxide-RS – COOH

Double Bond (DB) ↓:
Fe(II, III) Oxide-WB- CH=CH₂

$[Cytb_{562}.Cytb_{566}] \rightarrow$ Cyt c $\xrightarrow[O_2]{[Cu.Cyt_a Cu.Cyt_{a3}]}$ H₂O
$[Fe-S]$-Cyt-C1

Fig. 2A.30: Evaluation of pharmaceutical waste degradation using INCs. (A) Metallic ion immobilization using bio-MOFs, as bioreactors to break down aromatic compounds and remove heavy metals through ion exchange; (B) a series of organic compounds used as the model molecules, (C) proposed redox reaction of organic compounds; (D) % cytochrome normalized activity indicating spontaneous bioremediation.

Mimic natural performance; we will realize spontaneous bioremediation of commingled contaminants.

Liu and her collaborator carried out a preliminary study on nanobioremediation (NBR) of PAHs in the soil in various levels of heavy metals to improve in situ remediation effectiveness. PAHs, widespread environmental pollutants including benzene rings arranged in linear, angular, or cluster ways, are considered priority pollutants by the U.S. Environmental Protection Agency due to their toxicity, mutagenicity, and carcinogenicity [193–195]. In many superfund sites across the US, heavy metals often occur as co-contaminants with PAHs and are reported to exert adverse effects on the biodegradation of the target pollutant [196]. To lower the abundance of PAHs in the soil at specific locations, more research and investment are required to establish field-deployable approaches. These methodologies yield real-time tox and species data and can assist in designing NPs to remediate the effects of PAHs in soil [197, 198]. In particular, nanoparticle-enabled *in situ* remediation techniques represent a key area of contemporary scientific advancement to remove pollutants from various media such as soils, surfaces, or groundwater. As a result, there has been a growing interest in the synthesis and assessment of biomolecules functionalized NPs via, for example, surface functionalization to increase the sensitivity, selectivity, reliability, and practicality of environmental nanoremediation [199–201]. However, several challenges have been encountered during in situ employment of these technologies, such as poor mobility, transport, and distribution of these NPs in field applications [200]. In addition, the application of bioremediation in the natural environment is often significantly influenced by the field conditions; thus, in situ environmental remediation demands significant efforts on a complete understanding of the biochemical mechanisms, environmental constraints, and the byproducts of these mechanisms to avoid the creation of the unpredictable negative remediation outcomes. From the above technical hurdles, our approach focused on the NBR of PAHs in the soil in the presence of heavy metals as co-contaminants. PAHs often occur together with heavy metals in Superfund sites. Thus, successful implementation of biodegradation of PAHs often depends on the understanding of their biodegradation in the presence of heavy metals [202].

Many researchers have highlighted the benefit of bioremediation of the combined pollution of heavy metals and PAHs. For example, Lu et al. [203] found that adding a moderate dosage of pyrene altered the microbial population and promoted microbial prosperity in soils, thus relieving metal-induced stress and favoring the phytomass yield. However, some other researchers, such as Gauthier et al. [204], described significant deleterious effects of PAHs-metal mixtures upon microbes. Also, most of the current studies have focused on the remediation outcomes of the combined pollution. Mechanisms of interaction effects among heavy metals, PAHs, and microbes are still not deeply understood. Bacteria from the genus Pseudomonas are one of the microorganisms that can effectively decompose organic pollutants through co-metabolism in the natural water and soil environment. Patel et al. [202] concluded that a bacterial consortium consisting of four strains, including *Achromobacter* sp. BAB239,

Pseudomonas sp. DV-AL2, *Enterobacter* sp. BAB240, and *Pseudomonas* sp. BAB241 was able to degrade naphthalene (1,000 ppm) at a wide range of pHs with an excellent degradation rate of 80 mg/h and other pollutants. Brito et al. [205] indicated that a bacterial consortium *Pseudomonas* sp. is very efficient in degrading pyrene. Zhang et al. [206] stated that the inoculation with a bacterial mixture containing *Pseudomonas stutzeri* (91.7 %) and *Candidatus kuenenia* (2.3 %) removed 54 % of the added phenanthrene (45 mg/L) under anaerobic conditions. This bacteria mixture has been identified as present within the functional microbes during the bioremediation of landfill leachate with high-strength nutrients and organic pollutants. The engineering of biotechnological systems employing these diverse microbial comminutes holds a key role in the efficient and sustainable treatment of PAH pollution. However, the microbial synergistic mechanisms and interactions for the efficient removal of PAHs in heavy metal co-contaminated environments are still not well understood [207].

NBR, the integration of traditional bioremediation with NPs, as an innovative method for effective, efficient, and eventually sustainable remediation, has received greater attention recently. Recent findings have suggested that inoculation with multiple microbial communities and biofunctionalized nanomaterials can enhance microbial activity, synergistically enhance PAH removal, and relieve heavy metal stress while producing high-end materials [208, 209]. Various laboratory and field application studies on NBR in saturated soils or groundwater have been conducted [210]. Most of these studies focused on the use of elemental or zero-valent metals in nanoscale form, such as iron, nickel, and palladium; however, problems involving the reactivity, the useful life, in situ transport processes, clustering tendency, and accelerating deposition of NPs, and their negative effects on microbes have been reported [210]. In addition, studies on the effects of NP behaviors in the soil pores, adsorption on mineral particles, interaction with soil microorganisms, and biogeochemistry of the contaminated site such as particle size, moisture, pH, presence of nutrients, organic matter, native microorganism, and type of clay mineral on the NBR have been very limited [197, 210]. Thus, the successful in-situ application of this biotechnology can only be possible after all of these limitations and problems are sufficiently addressed.

Fe_3O_4 NPs coated with chitosan in removing chromium (VI) from wastewater were evaluated by us using batch adsorption experiments. Results have shown that the chitosan-coated Fe_3O_4 NPs have a better adsorption capacity than uncoated Fe_3O_4 NPs reported in the literature, and chromium removal by the surface-coated NPs is highly dependent on solution pH and the NP dosage. The removal of organic refractory chemicals, such as bisphenol-A (BPA) and estrone (E1), in a sequencing batch reactor inoculated with anammox bacteria has also been evaluated by us. Results have shown that BPA and E1 have a negligible effect on microbial activity while removing 78 % and 99 % of BPA and E1, respectively.

2A.4 Biological and forensic application

2A.4.1 Cancer theranostics

The current theranostic barriers are low specificity, rapid instability in biological fluids of the target drug, and removal from the host organs through several biological modalities [211]. It was recently found that MOFs are ideally suited to meet these challenges. MOFs are suitable as drug carriers due to their high surface-area-to-volume ratio, allowing for low drug loading and fewer side effects due to their host, crystallinity, and stability in biological fluids [212]. The small size of the crystal can minimize an immune response and large surface area for fast kinetics (controlled drug release or selective capture of messengers like NO) [213]. Nitric oxide (NO) is important, but it is central to cell survival. High NO levels lead to mitochondrial membrane polarization, calcium overload, energy (adenosine triphosphate) depletion, lactate dehydrogenase, and cytochrome c release [214]. Nitrix oxide cross-talk also leads to the activation of caspase 3 and induction of apoptosis [215]. Low NO levels promote vasodilation, cell proliferation, enhanced wound healing, decreased phosphorylation of Protein kinase B (PKB, also known as Akt), and control of pro-apoptotic factors [216]. Recent research also suggests that NO levels be attenuated (decreased) or potentiated (increased) by irradiation with near-infrared/red light (NIR/RL), a phenomenon known as phototherapy (PT) [217]. The beneficial effects of NIR/RL are thought to occur via cytochrome c oxidase [218]. The Ni-based MOFs will also enable us to examine this biochemical pathway through changes in singlet oxygen species, reactive oxygen species (ROS) that alter mitochondrial membrane potential and affect respiration through changes in cytochrome c NO reductase activity (as well as its more common oxidase activity) [219].

A potential difficulty is poor drug storage or lack of interaction at the release site. A silica shell (for improved stability) and amino acid RDG tag binding were included for site-specificity, and targeted release was attempted. Our preliminary use of MOFs suggests singlet oxygen generation comparable to methylene blue (MB), indicating that the MOF architecture will be incorporated to offset poor drug or NO release by promoting SOS generation as supplemental photodynamic therapy (PT) [220]. Other cell lines, such as RPE (ATCC® CRL-4000), will be conducted using standard protocols. Cell morphology was examined using confocal microscopy. Cell proliferation/cytotoxicity assays such as lactate dehydrogenase were used for the 3-(4,5-dimethyl-2-thiazolyl)-2,5-diphenyl-2H-tetrazolium bromide (MTT) assay and caspase-3/7 activity assay as an alternative to SOS/ROS [221]. These tests yield information on the mechanism of cell death and target organelle (cytoplasm, mitochondria, or nucleus) by comparing biomarkers for cell death and mapping (EELS, confocal microscopy) data coupled to MOF design [127]. The PT assay was based on the recovery of RPE cells to H_2O_2 wounding by comparing RPE (in the dark, d),

RPE + H_2O_2 (d), and RPE + PT RPE + H_2O_2 + PT where PT is continuous-wave red light (2.88 J/cm^2).

The multi-metal entities were potentially adaptable as nodes for constructing diverse molecular architectures and the stability and magnetism in the supramolecular systems resulting in new properties and functions [222]. Potentially, the stability, magnetism, and adaptability of multi-metal entities as nodes for constructing diverse molecular architectures could result in supramolecular systems with new properties and functions. An antineoplastic drug (SN-38, a topoisomerase I inhibitor) can lead to low metabolism, resulting in diarrhea and toxicity in some patients [223]. These adverse effects were reduced through greater drug loading within the MOF superstructure and slower release. The drug metabolism is lower than current loads, resulting in low glucuronidation and toxicity. Nanostructured MOFs effectively encapsulate and release the drug through pH-triggered bio-degradation and thermal hysteresis by applying an external magnetic field or slow-release from the target organ [224]. If suitably tuned, the MOFs will be activated by NIR radiation facilitation photodynamic therapy and charge transfer (CT), potentially in the same MOF structures [225].

Nanoscaled MOFs (NMOFs) offer improved pharmacokinetic properties similar to those of polymer encapsulated drugs [226]. A size range between 35–70 nm was generated with encapsulation of sugars or silica to improve stability and further post-synthetic modification by functionalization with alpha-beta integrin (RDG) as a delivery vector [227]. The IC_{50} of such functionalized NMOFs is expected to be lower than uncoated MOFs. In addition, by functionalization with a specific peptide vector such as RDG, the NMOF can facilitate receptor-mediated endocytosis and activation (by reducing the environment of the cancer cells) from prodrug to anticancer drug, potentially lowering the admitted doses and side effects [228]. An alternative approach is to engineer the MOF to release NO and promote cell death upon activation using a visible light laser [229]. The development of functionalized MOF-based probes can provide a new paradigm for understanding cellular mechanisms related to oxidative stress and the treatment of cancerous cells. Our approach focused on these three activities, *1)* preparation of a series of new MOFs using a feasible wet-chemistry method; *2)* characterization of the structures of these MOFs using state-of-the-art instrumentation; and *3)* embedment of drugs into the MOFs and use them for cancer theranostics (NO release/adsorption, ROS, and SOS test). The team directed new MOF applications in cancer theranostics and disinfection science [230]. Some of her MOFs were evaluated as NO, SOS, and ROS probes using the capabilities of the US Air Force Research Laboratory (AFRL) based at Fort Sam Houston.

Traditional approaches include the selective attachment of NPs to cancerous and healthy cells. Drug dose and release gradient, side effects, drug clearance, potential immunogenicity, nonspecific drug release, or low bioavailability are potential issues in any disease treatment [231]. However, MOFs with tunable structures and functionality could minimize many of these issues [232]. The metal within the

MOFs will be used for diagnosis or therapy using intrinsic magnetic, optical, or pH-related properties [233]. We aimed to use prepared MOFs' integrated therapeutic and diagnostic capabilities in a theranostic approach targeting cancer cells. Metastatic and drug-resistant cancers and cancer stem cells pose the greatest challenge for targeted therapy [192]. This research investigates the theranostic application of MOFs to overcome the biological barriers to reaching cancerous cells due to their crystalline nature, ultrahigh surface area, tunable porosity, drug loading, controlled control release, and high stability biological fluids [234]. In addition, the prepared MOFs allows for the integration of therapeutic and diagnostic platforms to be integrated into this theranostic approach, target cancer cells specifically, and examine the role of nitric oxide (NO) in any phototherapy (PT) [235]. In particular, metastatic cancers, drug-resistant cancers, and cancer stem cells impose the greatest therapeutic challenge for targeted therapy [236]. In immunotherapy, targeted therapy was achieved with appropriately designed drug delivery vehicles such as NPs, adult stem cells, or [thymus-matured] T cells [237]. We implemented a strategy to treat drug-resistant tumors using combination therapeutic agents with different mechanisms for a synergistic effect to overcome the drawbacks mentioned [238]. In this example, functionalized MOFs were used in cancer theranostics.

To understand the mechanism of interaction between the cancerous cells and nanomaterials. Different cancers affect different biochemical pathways, and many cancers "uncouple" cell proliferation and metabolism, allowing cells to divide uncontrollably and use valuable metabolic resources [239]. MOFs can track biochemical changes and delivery of drugs of known efficacy to aid in understanding the biogenesis, cell-to-cell communication, and control of apoptosis [240]. The application of NO (a neuro-messenger) affects many biochemical pathways that have been identified as essential in low-intensity visible light therapy [241]. Low light illumination therapy takes advantage of NO's deregulation that mirrors the mitochondrial deregulation characterizing cells in a cancer-like state instead of the normal state where mitochondria appear tolerant to apoptotic impulses [242]. Using MOFs as probes to capture and release NO and then examine other "reporter" ions allows us to analyze the mechanisms [243]. In most studies, since inhibitors and site-specific dyes are used in tandem, MOFs allow the integration of these two different molecules into a single MOF with dual or multiple purposes. We designed a quintuple-faceted MOF for *imaging* (virtue of metal center fluorescence), *quantification* of NO (via dichlorodihydro-fluorescein diacetate based loading), *diagnosis* (electron energy loss spectroscopy mapping of MOF within the cell), and *therapy* (singlet oxygen species (SOS) or reactive oxygen species (ROS)) (or PT (using near-infrared to red light (NIR to RL)). Thus, versatile MOFs can effectively treat disease and serve as a NO probe to better understand the mechanism behind observed PT [244].

A feasible solvothermal synthesis has been used to produce series MOFs using four metallic cations (Fe^{3+}, Ni^{2+}, Cu^{2+}, and Zr^{4+}) as the central elements due to their magnetic and fluorescent properties. Five groups of ligands (chosen from Zhou's ligand library) with different sizes and bridging angles were selected to tune the MOF's architectures and properties. Amphiprotic groups (such as carboxylic and amine) were added to functionalize the structure and increase their flexibility to connect with the cancerous cell surfaces. To construct MOFs for cancer theranostics, we used pyrene for its stability under pH and moisture conditions and strong fluorescence photolysis as the organic linker selected MOFs for PT and CT). Based on the "hard and soft (Lewis) acids and bases" theory, Fe^{3+} and Zr^{4+} cation with a high valence and closed-shell electron configuration could form a strong coordinative covalent bond with carboxylate and produce fluorescence, along with the their intrinsic magnetism [245].

Similarly, the Ni and Cu MOFs were prepared using the solvothermal method to control pore size for NO capture and release. Morphological and crystalline structure analyses were conducted using a TEM and atomic force microscopy (Fig. 2A.31). Images showed that well-distinctive domains were received, where the edges and corners were the reactive sites for cancer diagnosis and treatment.

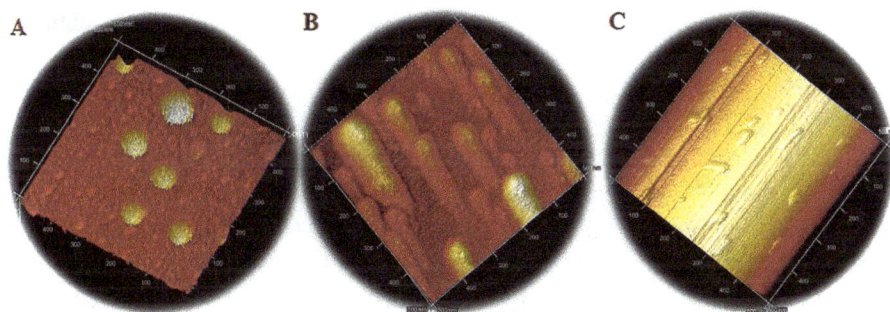

Fig. 2A.31: The atomic force microscopic analysis of MOFs motifs.

Commercially available SN-38 complexes were embedded into the magnetic MOFs through double injection. Under strong agitation, an aliquot of SN38 powder has been introduced into the above MOF's DMSO solution. The resulting suspension was then filtered through a 0.45 μm syringe filter to remove the unbound drug. The drug loading was estimated using high-performance liquid chromatography from phosphate-buffered saline (PBS) extracts over a two-week duration, the maximum time that SN-38 will exist when encapsulated. DAF (50 μg in DMSO) was impregnated into MOF using pressure and temperature injection with an anticipated release of 50 μM/h as a fluorescence integrator (signal increases with time). In vitro cytotoxicity of Zr-MOFs was tested on ovarian cancer cell line A2780 (Fig. 2A.32) and retinal pigmented epithelium (RPE) cells. The MTT assay measured cell viability. SOS green [246] and ROS [247] measured oxidative stress [248]. Data were collected using a microplate reader with excitation and emission at the appropriate wavelengths. The viability of cells exposed to the MOFs was expressed as a percentage of the viability of cells grown in a normal growth medium (Fig. 2A.32A). The MOF-based materials have applications in health care as catalysts for water purification and NO probes to treat cancerous cells. The MOF motifs exhibit crystalline structures with high surface area and large pore volumes. These factors enable them to adsorb functional molecules or anti-cancer drugs. These substances were adsorbed on the external surface or open channels of the MOFs pr trapped inside the pores of the MOF framework. The high efficacy for medical diagnosis, treatment, and imagining was obtained using MOFs alone, even without loading the SN-38 drugs. The loading between 5 and 15 mass % of drug within the MOFs was achieved by adjusting the open porous structures. The MOFs were stable in PBS and the biological matrices of similar pH will biodegrade at lower pH, typical of rapidly dividing cells ($IC_{50} < 10$ µM, Fig. 2A.33A) [249]. The MOFs were excreted in the urine, with the Fe recycled by the body, similar to iron supplements. The aim was not to upload ultrahigh doses of the drug but with moderate loads and slow release. The slow release facilitates a rapid drawdown of the cancerous cell population without common side effects such as diarrhea. The TEM images (Fig. 2A.32C) and a schematic diagram (Fig. 2A.2D) showed a well-rectified MOF structure with high porosity, allowing for improvement of cancer theragnostic efficiency. Zr-PCN MOFs display an isoreticular structure of (P6/mmm group with $a = 39.38$ Å and $c = 16.48$ Å, $\alpha = 90°$) [250]. The secondary building unit comprises a Zr 6-octahedron involving eight μ_3-OH, eight normal OH, and eight carboxylates. The 3D framework with two types of 1D channels oriented along the c axis was controlled based on the end application. TEM morphology showed that Zr-MOFs displayed well-defined and highly crystallized coordinative compounds. These NMOFs with versatile vectors provided theragnostic applications with improved efficacy toward ovarian cancer cell lines. Percentage cell viability with normal A 2780 cell line (Fig. 2A.33E) revealed that cells are ~100 % viable below a concentration of 1×10^{-8} g/mL. The MOF concentration was controlled from 100 µM to 0.1 nM. After being treated using MOF motifs, the A 2780 cells swelled significantly, indicating cell death. Developed nanotechnology platforms with high target specificity and minimum collateral damage and immune reaction. It is hypothesized that the magnetic properties of Fe-MOFs play a critical role in cancer theranostic study due to the hybridization of d orbitals of metal with 2p orbital from oxygen and the promotion of hyperthermia to the magnetically induced spinning of the metal center [251].

Theranostics combines a diagnostic tool with targeted therapy and expands our understanding of cell function [252]. Engineered nanomaterials are well suited for the theranostics approach [253]. In particular, MOFs allow for controlled drug loading. The flexibility of ligands and metal availability in fabricating MOFs and the possibility for differential drug loading are suitable for imaging, quantification, diagnosis, and therapy [254]. We anticipate that MOFs are neither the cause of oxidative stress (via SOS/ROS) nor cytotoxic (via MTT assays); however, once the drug is released or

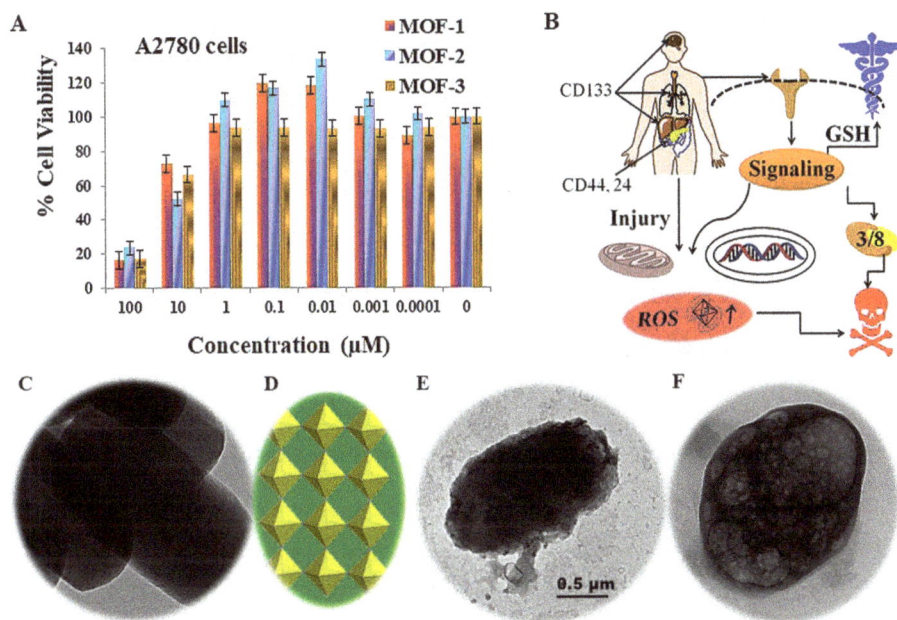

Fig. 2A.32: Three metal-organic frameworks used as nanodrugs targeting cancer cells:
(A) percentage cell viability at different concentrations of MOFs with normal A 2780 cell line revealed that cells are ~100 % viable below a concentration of 1×10^{-8} g/mL; (A) viability of Fe-MOFs;
(B) mechanism of inactivation of cancer cells; (C) TEM image of Zr-MOF; (D) a diagram showing highly ordered and porous MOF motifs; (E) intact A2780 cells as control; and (F) damaged A2780 cells. With permission from [259].

the MOF bound, cell proliferation is expected to decrease in a dose-dependent manner, resulting in cell death via ROS or inhibition of cell proliferation pathways [255] Our data (Fig. 2A.33) indicated that MOFs cytotoxicity could be tuned (low Fe-ABBT alone to high with H_2O_2) [256]. We expect to design MOFs with steady NO release throughout exposure (0–4 h). Nitric oxide release was determined using a NO-specific dye, 4-amino-5-methylamino-2′,7′-difluoro-fluorescein diacetate (DAF-FM dye), measured using differences in nonfluorescent and (unbound) and fluorescent benzotriazole (NO bound) species upon light irradiation using a plate reader (Tecan 200Pro). Both MOFs using ligands of 2′-hydroxy-[1,1′:3′,1″-ter phenyl]-4,4″,5′-tricarboxylic acid (HTTA) and 5,5′-(propane-1,3-diylbis(oxy))diiso-phthalic acid (PDDA) show lower toxicity and low inhibition of cell division ($m = 0.01$, $b = 17.55$; $m = 0.01$, $b = 33.57$, respectively, where m = rate of inhibition and b = intercept).

The MOF projects summarized in the last segments demonstrate a new paradigm in cancer theranostics using organometallic chemistry as a therapeutic strategy. There have been three broad, innovative research discoveries: (1) the functionalized MOFs display a unique surface structure and internal pores, allowing for drug entrapment and release. Highly specific epitopic tagging of flexible, tailored MOFs to be

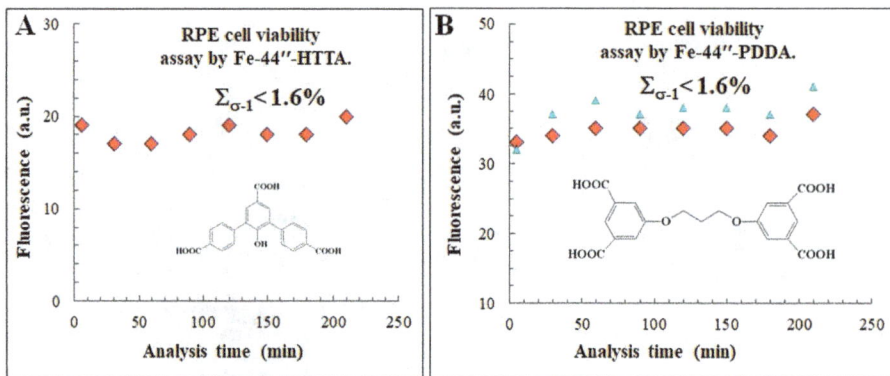

Fig. 2A.33: Evaluation of MOF toxicity against RPE cells. MOFs incorporating different drugs yield different toxicities: (A) RPE cell viability assay by Fe-44″-HTTA and (B) RPE cell viability assay by Fe-44″-PDDA.

selectively conjugated with the cell surface. Through surface-to-surface interactions, the embedded drugs entrapped within the MOF structure were released to the cell membrane of the cancerous cell membranes to promote cell death; (2) NO capture

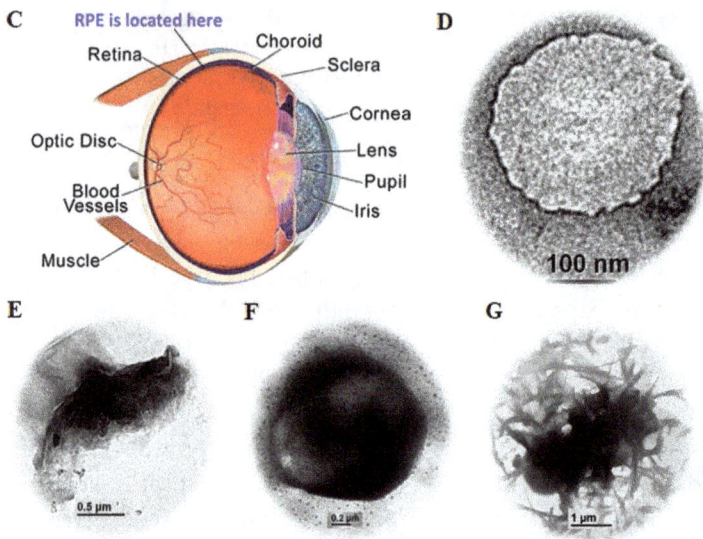

Fig. 2A.34: Efficacy of MOFs against retinal pigment epithelium (RPE) whole cells (WC): low (1–20 ppm) doses: (A) diagram of the RPE whole cell as a control, (B) TEM image of the RPE whole cell, (C) TEM image of the damaged RPE WCs by Fe-HTTA, (D) TEM image of the damaged RPE WCs by Fe-PDDA, (E) TEM Image of RPE WC+Zr-MOF after 1 h, showing plasma membrane peeling, (F) TEM Image of RPE WC+Zr-MOF at 0 h, showing healthy cell, and (G) TEM Image of RPE WC+Zr-MOF after 6 h, showing microfibrils and cellular shrinkage.

and controlled release provide a better understanding of the mechanism of interaction between MOFs and the cancerous (or oxidized) cell membranes. The progression was inferred due to the release of secondary messengers upon cellular stress and was utilized to measure the development of the treatment with the minimal side effects; (3) the potential to construct NO probes with NIR to RL capabilities which have yet to be documented to assist PT-based therapies by enabling "deep scans" via NIR and therapies via RL. The example shown in Fig. 2A.34 is the efficacy of MOFs against RPE whole cells (WC): low (1–20 ppm) doses.

The "comprehensive review of public health effects of energy fuel cycles in Europe" indicated that natural gas causes 1–11 deaths per TWh with three deaths per TWh [257]. In the USA, the casualty was between 10 and 20 from oil and gas extraction. Although the physicist Richard A. Muller indicated "that the public health benefits from shale gas far outweigh its environmental costs," it is important to evaluate the safety of materials used in shale gas storage [258]. In our evaluation, RPE wild-type cells were preliminarily analyzed against known standards. These standards were cells re-suspended in PBS supplemented with 5 mM $CaCl_2$ and $MgCl_2$ (set to a relative percent change of 0 %), hydrogen peroxide (HP, 3 %), and sodium nitroprusside (NitroP, 50 µM in H_2O). The nanocomposites (ZnO-22 (10 mg/mL), ZnO-28, Fe_3O_4-T_2), and MOFs (Ag-A19) were used to test their toxicity (Fig. 2A.35) using dimethyl

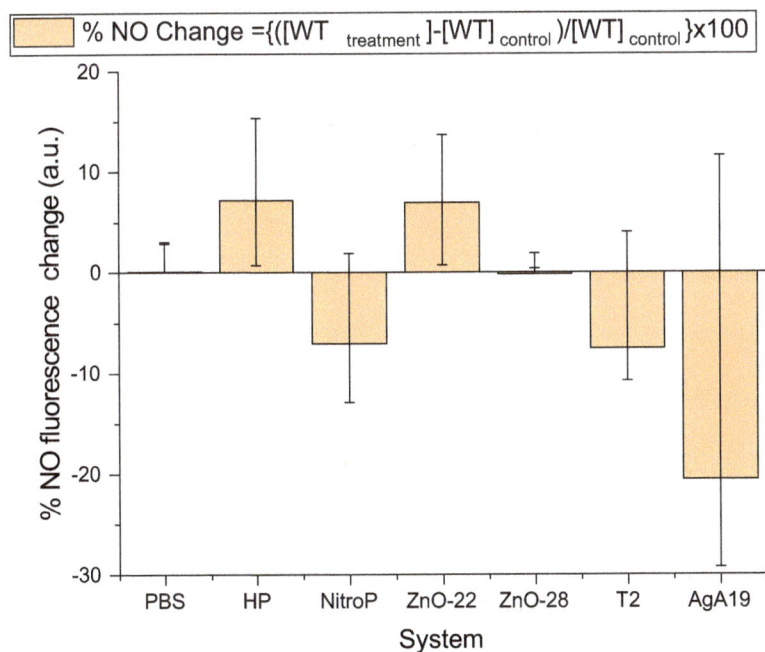

Fig. 2A.35: Nitric oxide fluorescence as a difference between WC cells in PBS (control) and treated cells with nanoparticles with independent error bars ($n = 8$).

sulfoxide (DMSO) as solvent. The standards were grouped into three chemical groups: a control group (PBS), an oxidizing (HP), and (NO) releasing agent or agent that facilitates the generation of NO (NitroP) were evaluated, respectively.

Compared with cells with no treatment (PBS group), the addition of HP resulted in a relative increase in NO. The addition of NO-releasing agents is complicated by the NO feedback loop, where high NO levels result in a biological switching off of NO synthase, the primary enzyme responsible for biological NO. The addition of sodium nitroprusside releases NO and, under biological conditions of high NO, results in the activation of the negative feedback loop, lowering NO levels. The nanocomposite Zn-28 has almost no effect in terms of NO generation and appears to be nontoxic. ZnO-22 appears to be slightly oxidizing, not sufficient to cause cellular damage, but sufficient to cause an increase in NO, which protects against further oxidative stress. The NO changes after the addition of Fe_3O_4-T2 appear to indicate that the agent can translocate to the mitochondria and directly affect the redox states of the respiratory heme and copper centers, similar to sodium nitroprusside. The Ag-A19 motif was destabilizing and caused cross-linking in the mitochondria membrane due to the silver center, resulting in a NO drop.

2A.4.2 Fingerprint development

The initial goal of this research direction is to enhance the capacity, capability, and readiness in the field for the detection and development of latent fingerprints from the porous and nonporous surfaces that are currently problematic. The technical barriers are partial, overlapping, or aged prints, while administrative hurdles lack sophisticated crime detection equipment, personnel, and training in using the latest detection techniques. The team has been focused on designing and implementing nanotechnology-oriented research and artificial intelligence to facilitate latent print and forensic crime scene analysis and provide reliable evidence for criminal justice. Three technical activities have been carried out: (1) Hyperspectral Scanner Development using green light to enhance detection of latent fingerprints using chemical fluorescence of chemical components in latent print residue; (2) Latent Print Development Agent Synthesis using a chemical approach to obtain two novel classes of agents composed of covalent organometallic polyhedra (COP); (3) Automated Fingerprint Identification using artificial neural networks and fingerprint aging analysis to enhance decision-making in matching low-quality print images. These applied research methodologies are used to design an expert fingerprint matching and identification system. The matching prints were interfaced with the Automated Fingerprint Identification System (AFIS). The practice aspects were strengthened through specific Nanoscience and Forensic Science courses.

Latent print and forensic crime scene analysis are complex areas of law enforcement that rely on applied and developmental goals to push the practice forward.

Current barriers unique to latent print identification are technical factors such as detecting faint residues through technology or chemical development agents. The evaluation of latent fingerprint age, resolution of partial and overlapping prints, and eventual matching of prints to a suspect are the additional biggest obstacles that must be overcome to facilitate criminal investigations and delivery of justice. These barriers are of particular concern to KCSO, which has identified four areas. These areas are the abilities to (1) resolve poor quality prints, (2) detect latent prints, (3) determine the age, and (4) identify likely suspects. The first two areas were addressed using technology, whereby the fluorescence arising from the fingerprint residual was measured. Knowledge of latent fingerprint age and synchronicity of a print is based on the chemical decay in the latent print [260]. The inability to resolve partial or overlapping prints affects the ability to confirm or refute a suspect's alibi. Multiple factors [261] influence the outcome of latent fingerprint detection. These are the amount of print residue, the surface texture, and the type of DAs used [262]. The sensitivity factor is a metric of the DA, where the signal intensity limits detection through fluorescence [263]. This low detection sensitivity will be resolved by developing a more sensitive hyperspectral (HS) scanner to provide high resolution [264] or the synthesis of nanostructured composites with different functional groups that generate a higher quantum yield [265].

A HS scanner would compensate for small print residue by scanning for lines simultaneously and producing a three-dimensional information cube. Each fluorescence wavelength would be linked to a particular chemical component in the latent print residue. An alternative approach is to design agents with amphiphilic topology, which generate a high quantum yield, using latent print residues from the porous and nonporous substrates [266]. Using the DAs with polyhedra will provide increased binding sites leading to improved detection by fluorescence and enabling increased sensitivity to the latent print residues [267]. The new latent fingerprint DAs have the potential to replace iodine or ninhydrin usable on a variety of surfaces [268].

An ANN-based expert system resolved the last two concerns, even though the latent prints are difficult to develop. This difficulty is due to limited print residue resulting in low print resolution, incomplete ridges, partial bifurcations, or even a lack of a core, hindering expert print matching. Automated searching and matching of prints with ridges but without minutiae are barriers to correct print identification. An ANN-based expert system will improve suspect identification probability within a shortened time from the massive database before the detailed matching is implemented. The contrast enhancement can reveal ridges and bifurcations, while spatial filtering can remove noise between distinct points. This approach simulates what human experts will undertake by utilizing additional information not observable in the prints. The last concern was addressed by implementing high-quality training and professional development in Nanoscience and Forensic Science [269]. The tools will enable small police departments to gain expertise in fingerprint suspect matching, similar to the more experienced police departments. Crime detection requires experience and know-how, which are problematic for latent fingerprint detection and evaluation due to numerous variables in print quality. The ability to train under standardized conditions and develop a toolkit library to meet various scenarios will benefit detectives' in-field operations. This benefit can only be accomplished using standard prints with artificial sweat for print duplication. Current problems with using artificial sweat are two folds: (a) the composition of commonly used sweat does not resemble sweat in healthy adults due to lack of metal cations and (b) the sweat "concentration" is adjusted through dilution of the cartridges with water. This dilution does not reflect latent prints in the touch pressure and finger sweat, determining print homogeneity and spread uniformity. The latter was addressed by creating more realistic sweat and specific courses tailored to law enforcement needs.

The team focused on three activities to improve fingerprint extractions' development accuracy and precision.

HS Scanner Development Workflow: The HS scanner was constructed and evaluated for print detection on porous and nonporous surfaces. The detection without DAs can utilize the print and latent print residue(s). The detection threshold can also be compared and contrasted against eccrine secretion DAs. The data was analyzed for probability match using the five-point grade scale (Fig. 2A.36A) examined using ANOVA statistical significance.

Construction: The scanner was constructed based on an existing design with an imaging spectrograph and charge-coupled camera orthogonal to a second monitored image, capturing portions that undergo HS imaging (Fig. 2A.36B). A green-emitting diode can illuminate the surfaces. An HS image is generated where every exposed surface is digitized across the wavelength of the interesting visible light. The images are stored as a pixelated HS dataset. An instrument designed by Mitaka Kohki Co., Japan, has been used as our model [270]. The image is obtained by moving a spectrograph and charge-coupled equipped camera (Pika XC2, Resonon, USA). The spectrograph has a spatial resolution of 1.3 nm × 13 nm slit over 440–800 nm. A surface with latent prints was illuminated and focused onto the entrance slit of the spectrograph. The output spectrum is focused onto the camera, enabling all points within the frame (1600 × 1600 × 447 line, scan, and spectral pixels) to be recorded at a frame rate of 30 frames-per-second. Depending on the distance between surface and lens, the image area was as small as 30 × 30 mm, a spatial resolution of about 46 μm, a spectral resolution of 2.0 nm, and an acquisition time of 50 s. The acquisition time can increase with a larger area, while spatial resolution can decrease.

A second orthogonal camera (Pika L, Resonon, USA) was used to assist in capturing part of the image undergoing HS imaging. The underlying (latent) fingerprint surface was illuminated and filtered using a polarizer (VWR, USA) with a 532 nm power at 2.0–4.0 W. A continuous-wave laser was focused onto the surfaces using a cylindrical lens. The incident light was polarized linearly and diffused and reflected with the polarized perpendicular to the incident light (Fig. 2A.1B). The light is selected

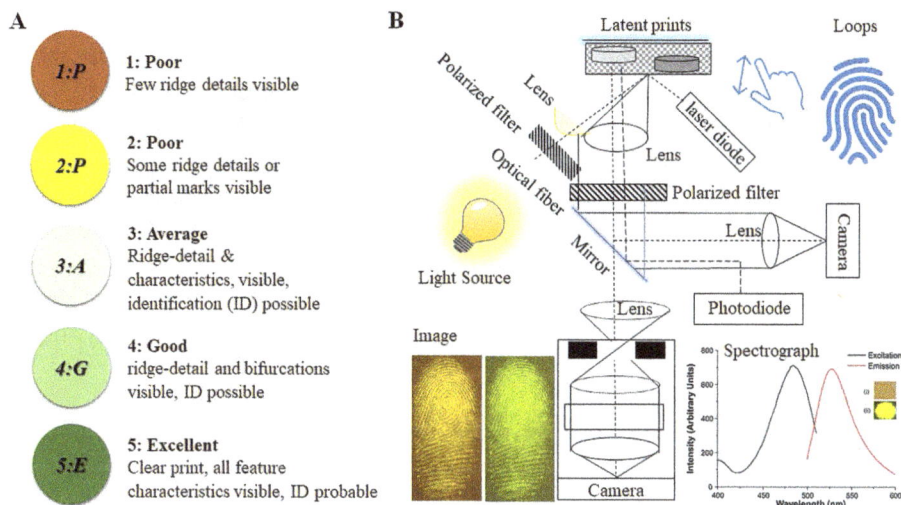

Fig. 2A.36: The hyperspectral scanner (HS) development workflow: (A) five-point-grade scale for latent fingerprints, and (B) plan of HS scanner to detect partial or overlapping prints.

by the polarizer linear to the detector focusing, automatically controlled using a laser diode with an oscillation wavelength of 850 nm. Accordingly, a photodiode and long-pass cutoff filter (Sigma Koki, Japan) were selected for the wavelength of interest.

Spectral processing: The fluorescence data regarding latent prints were processed as a ratio (β) of intensity difference (ΔI) or I between acquired and reference spectra, where the reference is fluorescence reflected from the standard surface. Under high background, fluorescence illuminated between 560 and 660 nm with a background at $_{600}\beta$ is evaluated using the calculation as eq. (2A.8):

$$\beta = (I_{560} - I_{600})/(I_{620} - I_{600}) \tag{2A.8}$$

The surfaces with low background fluorescence were treated using eq. (2A.9) as a first approximation:

$$\beta = I_{560}/I_{620} \tag{2A.9}$$

These proposed expressions can differentiate surfaces with small print residues and overlapping prints. Scanning of prints at a window of wavelengths ($\Delta\lambda$), processed as subtracted intensity ratio, can form a generated HS dataset.

Quality control procedure: Fingerprint images were ranked using the quality assessment scale (Fig. 2A.36A) described by Verdue [271]. The grade "3: A" or higher was examined, where ridgelines and junctions are recognized without laceration or incomplete regions. An expected print and related spectra from ridges and furrows are shown in Fig. 2A.37A. While HS scanning is useful for individual prints, the technique will extract valuable information about partial or different prints. This fluorescence is due to the composition of the print residue being composed of hydrophobic domains, which undergo dipole transitions. Each residue would emit energy, determined by its atomic arrangements and internal electronic states at a unique wavelength. If the residue is excited using a particular wavelength, excitation reflects the components of each fingerprint as a fluorescence spectrum. The pressure exerted on the surface results in different residue deposition. This uniqueness, known as the aging of latent fingerprints, is reflected by its fluorescence spectra [272].

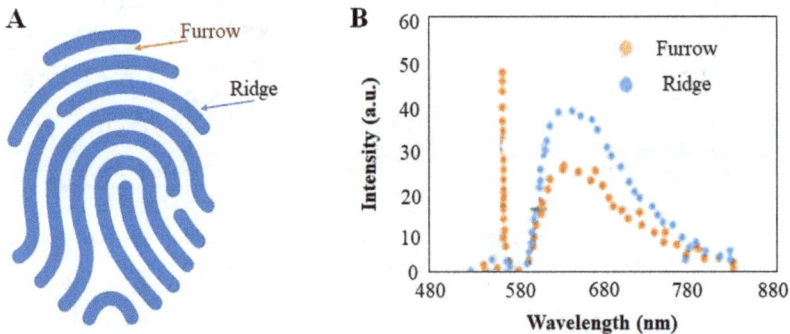

Fig. 2A.37: Quality control to improve resolutions of fingerprint images, (A) a schematic image of fingerprint and (B) simulated excitation wavelength dependence of fluorescence on latent residue shown in an image of the ridge (print) and furrow (background) with a green light (bandwidth: 10 nm) with a red filter.

These temporal changes in the fluorescence spectra between "initial" and "final" prints were identified even if their ridge or bifurcations are incomplete in cases of incomplete ridges; the dataset stores the spectral-temporal coordinates. An individual with the same fingerprint and secretions will result in an incomplete or overlapping print, which cannot be identified by standard means. Each print generates its fluorescence due to electron transition from the ground to excited states. Through the deconvolution of superposition coordinates of both initial and final prints, the images will be "reconstructed." Therefore, these prints can extract useful information due to different dipole transitions, representing an "HS fingerprint" (Fig. 2A.38). The HS scanning was successful for unique prints and practical to reconstruct partial or overlapping prints. Background subtraction through spectral difference (eqs. (2A.8) and (2A.9)) is expected to resolve some of the latent prints, depending on the ratio (β) of intensity difference (ΔI) between acquired and reference spectra, as stated above. In cases where no print was extracted, the fluorescence spectra can serve as a basis for identity verification.

Fingerprint acquisition procedure: The sample collection protocol was based on the guidelines recommended by the International Fingerprint Research Group [273]. The

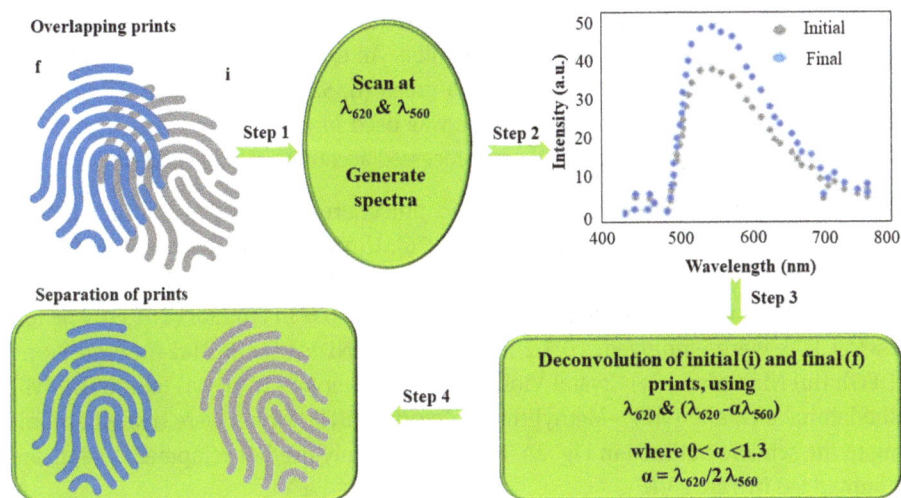

Fig. 2A.38: The HS fingerprint schematic corresponds to the different dipole transitions of residues through four key steps to improve image resolution and separation.

purpose of obtaining three types of deposits is to evaluate the sensitivity of the HS scanner by studying the fluorescence of different deposits. Eccrine and sebum-rich latent fingermark deposits from three donors were obtained on microscope glass slides, pre-cleaned with ethanol and ultrapure water, then dried under nitrogen. Immediately before sample collection, both hands were thoroughly washed with soap and warm water for 2–3 min. Moreover, eccrine deposits were obtained after 30 min without touching any objects between the hand washing stage and deposition [274]. For sebum-rich deposits, donors can press their fingers together and run the fingertips across their nose and forehead. Afterward, the fingerprints were placed lightly on the glass slides downward for 5–10 s. The same procedure was repeated to collect fingermarks on the black electrical tape, a polyethylene bag, and a soda can as examples of nonporous surfaces, and copy, gloss papers, and card stock as porous surfaces [275].

Spectroscopic characterization and imaging procedure: Fluorescence spectra of the developed surface were obtained using a spectrophotometer with an average of 10 scans and with excitation and emission slit widths of 10 nm. The surface was photographed in absorbance and luminescence modes. The calibration for white balance was set up to obtain the best contrast using either white light (WL), ultraviolet (UV) light, or filtered light similar to actual observations (compensated with safety goggles). The images were saved as fluorescent marks through a suitable light (WL or UV), and the use of colored (blue (B), green (G), yellow (Y)) filters, and saved as an eight-bit grayscale TIFF file format. For both modes, the focal length was kept at 60 mm, exposure and white balance set to automatic, and aperture corresponding to f11, mimicking a film sensitivity of ISO 200. In absorbance mode, the shutter speed was controlled at 1/20 s and in luminescence mode at 1 s to obtain the highest quality image on either dry or wet exhibit. An excitation wavelength of 490–510 nm was used in luminescence mode with a green (G, 529 nm) or orange (O, 620 nm) filter to improve image contrast, while WL was used in absorbance mode. The prints were scored using grade scoring [271] and processed according to the workflow.

Latent fingerprint visualization: Standard procedures to visualize the fingerprints has been described by Lee and Gaensslen [276]. Ultrapure (18 MΩ) water (H_2O) and another solvent (analytical grade) were utilized as solvents for dissolution. Fluorescence quantum yield was measured using rhodamine 6 G in ethanol (C_2H_6O). The following DAs were employed: (i) 1,2-Indanedione (IND); (ii) 1,8-Diazafluoren-9-one (DFO); (iii) Ninhydrin, (iv) Crystal Violet (CV); (v) Silver NPs (Ag); (vi) Iodine (I_2); (vii) Rhodamine 6 G; and (viii) 5-Methyl-thioninhydrin (MNT), which were treated according to the scheme outlined in Fig. 2A.39. The commonly used development agents are summarized in Tab. 2A.6.

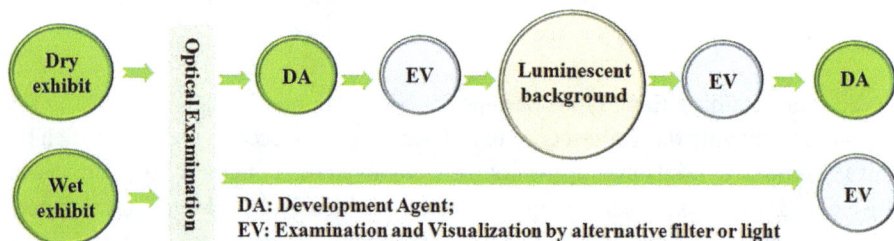

Fig. 2A.39: The sequences for enhancing latent fingermarks on porous and nonporous surfaces for the wet or dry exhibit.

Latent print development agent synthesis workflow: The new DAs were synthesized using solvothermal chemistry with carboxylic acid or succinic acid as organic linkers and multiaromatic ring structures as fluorescent dyes. In another approach, microwave-assisted synthesis was used to generate metal-organic frameworks as DAs composed of pyrene derivatives. The ligands were coordinated with preferred metals for charge stabilization and enhanced detection based on metal d-orbital vacancy in both methods [278]. This activity consists of three specific deliverables to tailor DAs to detect prints for nonporous and porous surfaces. Three perspectives are discussed in detail. **Nanostructured covalent organometallic polyhedron DAs to detect fingerprints from the nonporous surfaces:** A solvothermal chemistry has been widely employed to prepare COPs (Fig. 2A.40) [279]. A complex was formed between metal cation (Mn +) and multicoordinate ligands (M-COPs). This coupling can introduce hydrophilic active sizes into the substructure. These substructures are highly stable, enabling the COPs to be stable in air and water, enhancing print residue detection and selectivity due to various carboxylates that can interact with amino acids in fingerprint residues [280]. The interaction

Tab. 2A.6: A summary of development agents used to extract fingerprints.

#	Nomenclature/properties	The molecular or crystalline structure
1	Rhodamine 6 G chloride IUPAC name: 9-[2-(ethoxycarbonyl)phenyl]-*N*-ethyl -6-(ethylamino)-2,7-dimethyl-3 H-xanthen-3- iminium chloride highly fluorescent	
2	1,2-Indanedione (IND) Other names: Indan-1,2-dione 1 H-Indene-1,2(3 H)-dione 3 H-Indene-1,2-dione Vicinal diketone	
3	1,8-Diazafluoren-9-one (DFO) IUPAC name: 9 H-Cyclopenta[1,2-b:4,3-b′]dipyridin-9-one Aromatic ketone	
4	Ninhydrin IUPAC name: 2,2-Dihydroxy-1 H-indene-1,3(2H)-dione	
5	Crystal violet IUPAC name: 4-{Bis[4-(dimethylamino)phenyl] methylidene}-*N*,*N*-dimethylcyclohexa-2,5- dien-1-iminium chloride	

Tab. 2A.6 (continued)

#	Nomenclature/properties	The molecular or crystalline structure
6	5-Methyl-thioninhydrin (MNT)	
7	Silver (Ag) Space group: Fm3m Crystal system: cubic a: 4.0855 Å Cell volume: 68.192 Å^3 Asymmetric unit: 1 site Unit cell: 4 sites/unit cell Density: 10.5099 g/cm^3 [277].	
8	Iodine Space group: *Bmab* Crystal system: orthorhombic a: 7.2701 Å b: 9.7934 Å c: 4.7900 Å Cell volume: 341.046 Å^3 Asymmetric unit: 1 site Unit cell: 8 sites/unit cell Density: 4.9430 g/cm^3 [141]	

between prints and M-COPs is expected to occur within 5–30 s. The sensitivity and visibility of latent prints are aimed to increase by impregnating fluorescent dye into COP's pore post-modification [281]. This technique allows for encapsulation of M-COPs by dyes to emit different colors when 'hemiketalized' with the amino acid in fingerprint residue. The ligands, dyes, and metal ions, which can produce COP DAs to improve sensitivity, selectivity, and detection speed, are tabulated in Tab. 2A.7. Dr. H.-C produced these ligands with different bridging angles and molar mass. Zhou's group.

The solvothermal wet chemistry (Fig. 2A.40A) produced metal-centered COPs. The ligand bridging angles (shown in Tab. 2A.7) were varied to tune the pore structure and

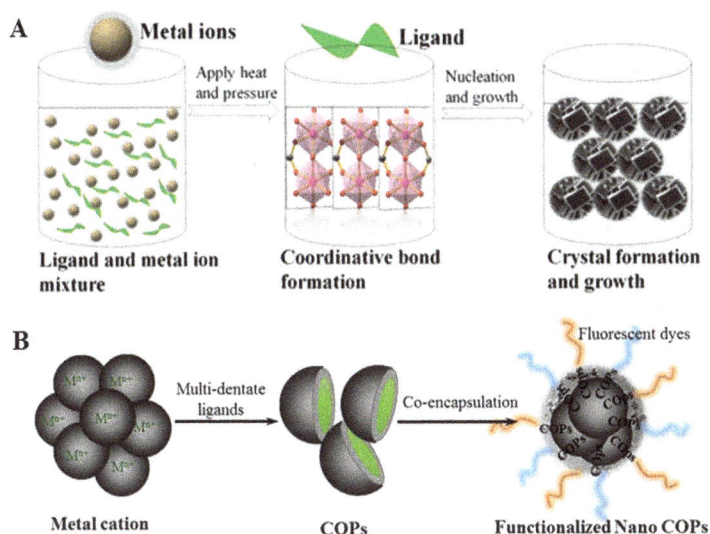

Fig. 2A.40: The COP synthesis and post-modification: (A) A schematic diagram of solvothermal wet-chemistry to produce COPs; and (B) A double-jet approach to preparing dye encapsulated COPs (increase adhesion between inorganic and organic interfaces).

Tab. 2A.7: The series of multidentate ligands were covalently bonded with three fluorescent dyes using different metal ions.

Bridging angle: ~0°	Bridging angle: ~60°
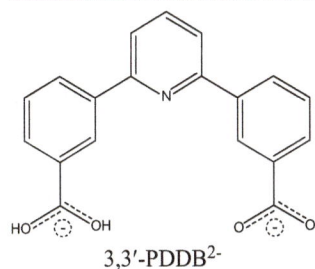 3,3'-PDDB^{2-}	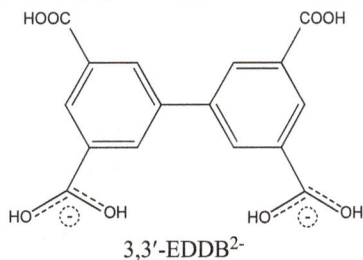 3,3'-EDDB^{2-}
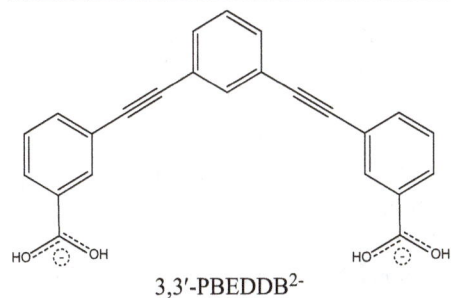 3,3'-PBEDDB^{2-}	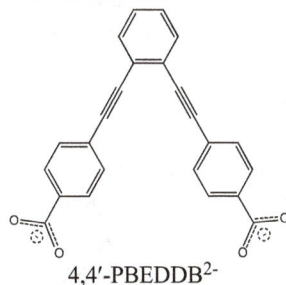 4,4'-PBEDDB^{2-}

Tab. 2A.7 (continued)

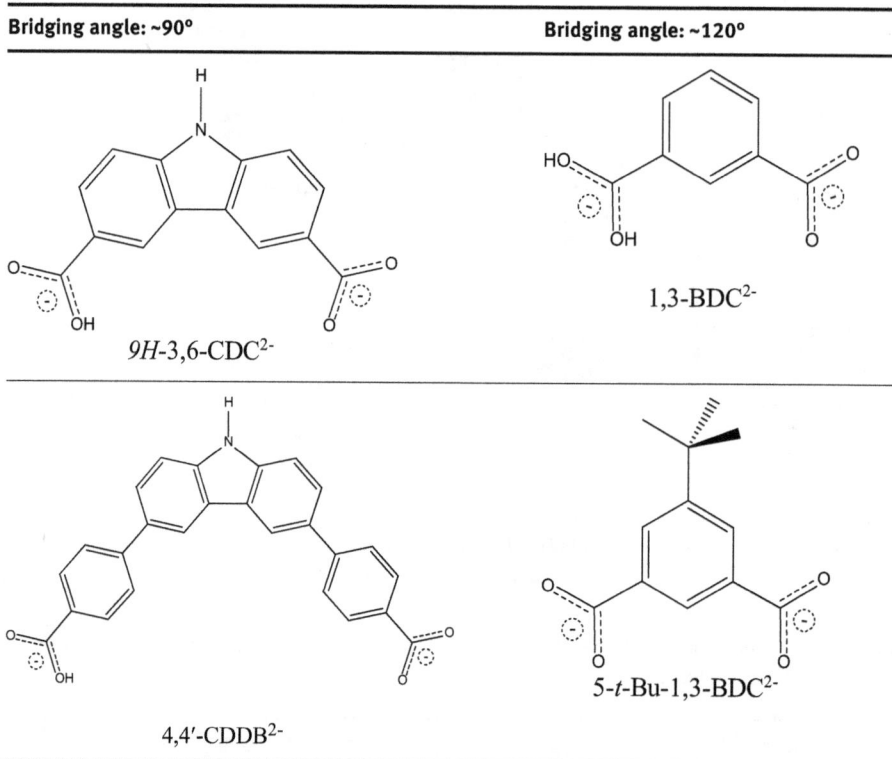

Bridging angle: ~90°	Bridging angle: ~120°

9H-3,6-CDC²⁻   1,3-BDC²⁻

4,4′-CDDB²⁻   5-t-Bu-1,3-BDC²⁻

tomography of different MOFs with transition elements as the center. These binding ligands (also called linkers) were dissolved in organic solvents (such as alcohols or DMSO) and the metal ions in aqueous using glass vials. The mixtures were ultra-sonicated to ensure that the different raw materials with various solubilities could form similar solutions. The reaction was controlled between 25 and 120 °C for 72 h to form well-reticulated crystals or colloidal suspension [282].

COP encapsulation: Commonly used fluorescent dyes with different functional groups (Tab. 2A.7) were dissolved in ethanol or other chosen solvents. Then these dyes were embedded into the COPs through the double injection under intense agi-tation (Fig. 2A.40B) [283]. An aliquot of a dye solution with different concentrations (0.05–0.5 M) was used. The resulting suspension was filtered through a 0.45 μm sy-ringe filter to remove the un-embedded dye. The filtrate was further diluted with H_2O/CH_3CH_2OH and analyzed for complexation efficiency using UV-vis spectros-copy at wavelength of 200–1100 nm. Quantitative estimation of loading efficiency of dye was determined using the high-performance liquid chromatography method using a C8 column in reverse phase. Estimating the dye loading efficiency can enable us to optimize the preparation parameters to produce a suitable dye encapsulated COPs formulation [284].

Fluorescent metal-organic framework DAs to detect fingerprints from porous surfaces: DAs composed of polycyclic aromatic hydrocarbons (PAHs) were produced [285]. Pyrene derivatives (Sigma-Aldrich) are chosen as molecular probes to design novel DAs for fingerprint detection. PAH fluorescence is sensitive to solvent polarity [286]. Emission bands are unaffected due to the pyrene excited state having a different transition dipole than its ground state. Our previous study demonstrated that highly stable zirconium (Zr)-MOFs were fluorescent with negligible toxicity (at 0.25 ppm) to humans as examined by in-vitro bioassays using retinal pigment epithelium cells. M-MOFs were constructed using chemical synthesis pyrene linkers with high fluorescence and were selected as the functional group. For example, a 4,4′,4″4‴-(pyrene-1,3,6,8-tetryl) tetra benzoic acid (H_4TBAPy) containing pyrene as the backbone were chosen as the binding ligands [287]. The "hard and soft (Lewis) acids and bases" theory suggested that Zr^{4+} cation with high valence can form a strong coordinative covalent bond with carboxylate.

As a result, fluorescent frameworks were produced. Microwave synthesis can generate MOFs between H_4TBAPy and Zr^{4+} (Fig. 2A.41A). Microwaves with a broad range of frequencies (300 MHz to 300 GHz) were used to generate thermal excitations. The microwave radiations shorten the reaction times, provide a rapid response rate of crystal nucleation and growth, and generate high yields of crystals with the least byproducts. Our previous study indicated that microwave irradiation for less than one minute was optimal in producing well-crystallized. The charge attraction developed during microwave radiation is the driving force for nucleation, leading to crystallization, and growth. The fluorescent MOFs (with different formulations) with highly porous structures (Fig. 2A.41B) were generated. Functional groups of $-NH_2$ and $-C = O$ were introduced into the MOF pores. These groups can react with 1° and 2° amines in amino acids present in the papillary exudate, forming hemiketals. The fluorescent emission yield was used to evaluate the fingerprint detection thresholds quantitatively [288]. Two MOF motifs were chosen as examples, which are anticipated to improve the accuracy and precision of fingerprint development. The MOF-10 (crystal system: triclinic, a: 17.1470 Å, b: 23.3220 Å, c: 25.2550 Å, $\alpha = \beta = \gamma$: 90.000°; cell volume: 10,099.533 $Å^3$) is a MOF-5 framework, composed of a "strut" that connects the zinc oxide "joints" in this framework structure. The MOF-5 (space group: $Fm\bar{3}m$, crystal system: cubic, a: 25.8247(4) Å, cell volume: 17,222.883 $Å^3$, asymmetric unit: 8 sites, unit cell: 428 sites/unit cell (0.0246 atoms/$Å^3$), density: 0.5938 g/cm^3). Lock et al. [289] reported groups of ZrO_4 tetrahedra interlinked via benzo-ethanoic acid units. A MOF-114 (space group: C 2/c, crystal system: monoclinic, a: 22.2410(20) Å, b: 12.7034(12) Å, c: 17.6543(16) Å, β: 126.361(2)°, cell volume: 4016.809 $Å^3$) is composed of the disordered C4S site, which has been shown as an average site, reported by Omar M. Yaghi et al. [290].

A

Random distribution of Molecules or ions	Alignment due RF effects	Crystallization
Reactants	**Microwave radiation**	**TEM image of MOFs**

Ions under microwave radiation

Attraction between positive and negative poles

B MOF-10 MOF-114

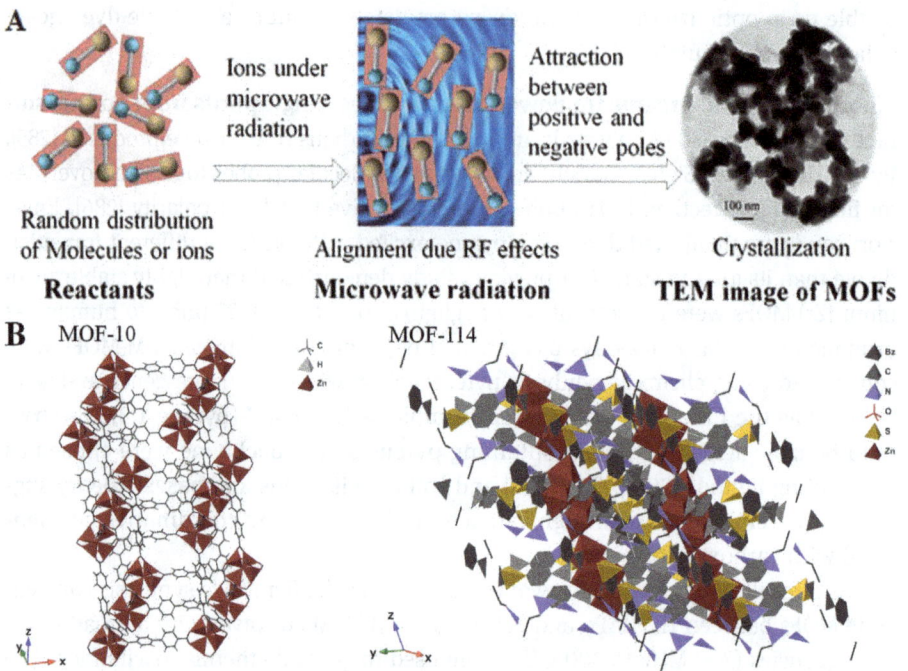

Fig. 2A.41: The synthesis and characterization of COPs (MOFs as an example): (A) the synthesis of MOFs using microwave-assisted solvothermal chemistry and (B) two existing MOF motifs with different pores and geometries were chosen as examples.

Fabrication of aerosol system to easily smooth depositions of DAs on surfaces: To convert COPs and MOFs into an aerosol, we can use an adopted procedure described by Fox [291]. An aerosol composed of MOF with high porosity and surface areas can hold 2–50 mg of the newly prepared DAs suspended in a 100–250 mL water/ethanol mixture (1:1 v/v). Organic solvents, such as ethanol, can facilitate solvent vaporization and increase fingerprint development speed. The Argon inert gas with a pressure controlled at the 2–6 bar was introduced for aerosolization under ambient conditions (Fig. 2A.42) [292]. The supercell dimensions with $2 \times 2 \times 2$ (Fig. 2A.42A) were generated as the multiples of the existing unit cells along x, y, and z. By controlling the solvent–COP ratio, aerosol was produced to collect figure prints (Fig. 2A.42B). The X-ray powder diffraction, X-ray energy-dispersive spectroscopy, and porosity map of MOF-114 were simulated and shown in Fig. 2A.43 as a guideline for future evaluation of COPs.

Automated fingerprint identification workflow: Enhanced fingerprint matching by ANNs was implemented using NIST Spectral Database 4 as a training module. The Microsoft Visual Studio Professional and MatLab packages were used to design, implement, and deliver a Windows-compatible graphical user interface (GUI) [293]. This activity has two specific deliverables: Latent fingerprint aging analysis by HS

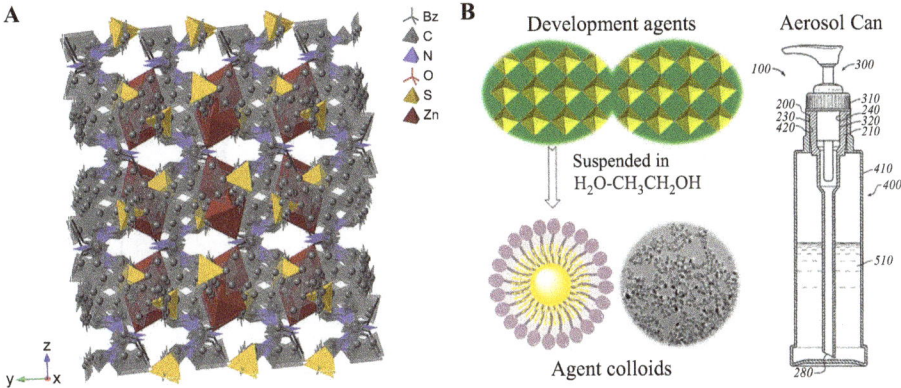

Fig. 2A.42: A production flow of an aerosol system: (A) MOF-114 with enlarged unit cells as development agents (DA) of fingerprints; (B) aerosol used to enable smooth depositions of DAs on surfaces.

overlay; and Enhanced matching using artificial neural networks. These two features are discussed in the following.

Latent fingerprint aging analysis by HS overlay: The fingerprint aging is determined by a series of spectral images. These prints acquired at variable time lags were used to improve recognition accuracy. Aging is mostly influenced by sweat composition, environmental factors, surface type, fingerprint contamination, resolution, and measured area [294]. The overlapped print spectral separation depends on the temporal degeneration difference of each spectrum [295]. The intensity differences between the "initial" and "final" fingerprints become significant between 560 and 640 nm. Therefore, single fingerprint images (Fig. 2A.38) were separated using eqs. (2A.8) or (2A.9) from the overlay [296]. The quality of images derived from HS data should be high enough for an ANN to compare with reference images. The first stage in overlay processing is the normalization of the picture and removing outliers at the beginning ($t = 0$ ms, eqs. (2A.10A), (2A.10B), and (2A.10C)). The second stage was the re-measurement of fluorescence at different time points until the t_{max} was achieved. From the t_{max} value, the fingerprinting age was derived. Dr. Wang can apply two additional transforms to the acquired images to make an optimal spatial and temporal resolution. The tendency of the ridgeline is improved by dilation of the original pixel:

$$F_1 = \frac{1}{x_{max}y_{max}} \sum_{x=1,y=1}^{x=x_{max},y=y_{max}} \text{Binarize}(I_{C\_DN}(x,y), \text{thresh} = 0.8) \quad (2A.10A)$$

$$F_2 = \frac{1}{x_{max}y_{max}} \sum_{x=1,y=1}^{x=x_{max},y=y_{max}} \text{Binarize}(I_{C\_SN}(x,y), \text{thresh} = 0.8) \quad (2A.10B)$$

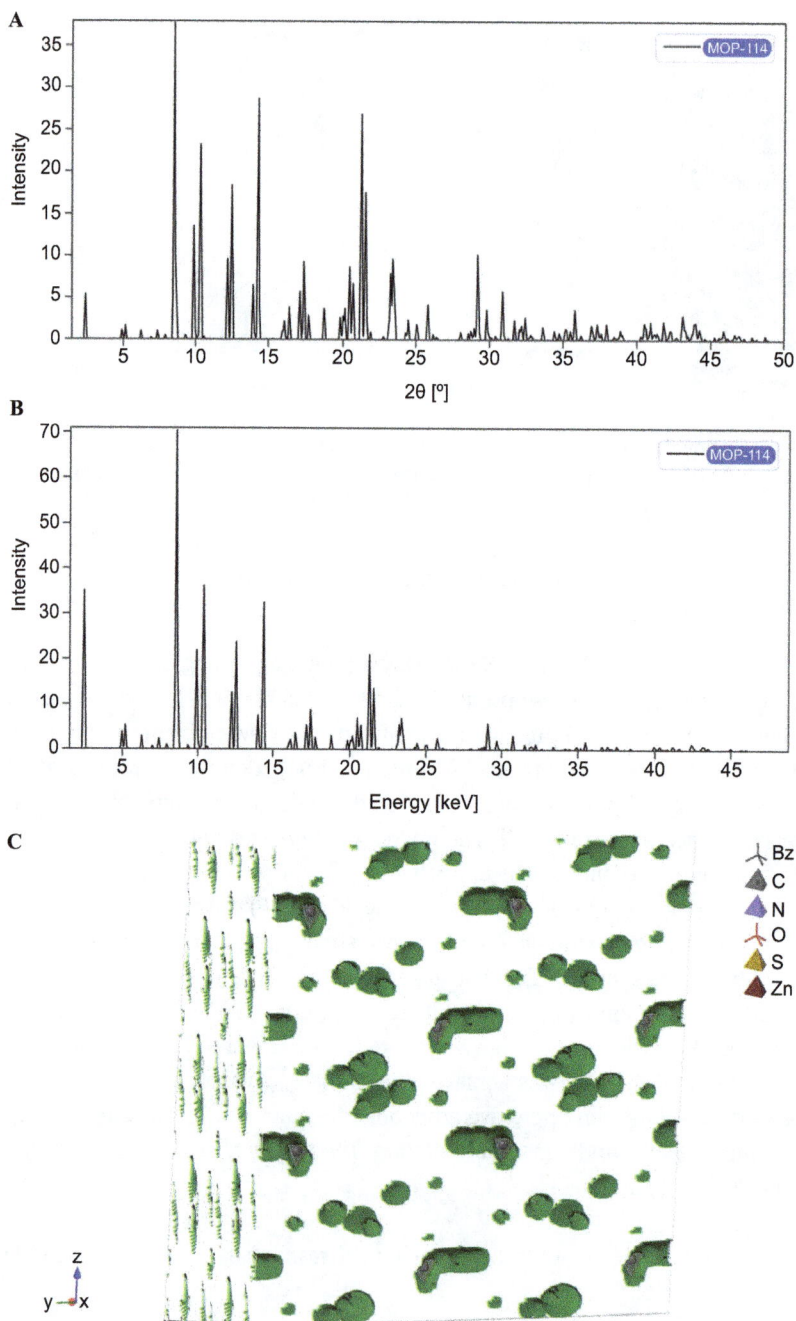

Fig. 2A.43: The simulated diffraction pattern and porosity mapping of MOF-114 with supercell dimension of 2 × 2 × 2: (A) X-ray powder diffraction pattern, (B) energy-dispersive spectroscopic diffraction pattern, and (C) porosity mapping.

$$F_3 = \frac{1}{x_{max}y_{max}} \sum_{x=1, y=1}^{x=x_{max}, y=y_{max}} \text{Binarize}\left(I_{M(x,y)} \cdot I_{C\_DN}(x,y), \text{thresh} = 0.8\right) \qquad (2A.10C)$$

where I_C is cropping; I_{C_DN} is the dynamic normalization; I_{C_SN} is the static normalization; and $I_M = \text{Dilate}(i,j) \cdot \text{Binarize}(I_{C\_DN}, \text{thresh} = 0.8)$.

Age determination is currently considered a significant barrier to effective law enforcement. The aging behavior of latent fingerprints was determined with greater certainty through the above processing. By using the bin and sort algorithms, the decay in the spectra pattern was correlated to the approximate age of the fingerprint.

Enhanced matching using artificial neural networks: One bottleneck in crime scene detection of a latent fingerprint is the poor quality of generated print images. We can characterize images with grade "2:P" (Fig. 2A.36A) using the ANN expert system with varying degrees of intelligence to guarantee picture quality. These criteria are applied to each extracted fingerprint image in the spectral separation trial [297, 298]. The automated processing of the latent images (Fig. 2A.44A and B) and the reference images (Fig. 2A.44C) determines the regions of the pictures with well-defined ridge flow [299]. Processing steps include the binarization and skeletonization of the images and marking quality areas in the pictures. Quality areas of print are defined, and high-contrast print images are created. The print image is rendered as a high-contrast image that captures ridge flow (with black ridges and white furrows) based on phase measurements of bands of light and dark patterns and masks out areas where constant quality ridge flow is not present. An examiner can manually refine the clarity of the high-contrast image portion of the automated process when the result of that process is not adequate. ANN is used to determine the fingerprint patterns, expediting candidate searching in a massive database. The reference candidates with a high match ranking of the pattern can continue the warp analysis with the latent image to achieve convergence. An accurate warp of a latent image to the mated or nonmated reference image is automatically created [300]. All ridges and furrows within the high-contrast latent and the reference images are thinned to one-pixel skeletons. The cubic and higher-order Bezier curves can approximate small segments of ridge and furrow skeletons. A cubic Bezier curve, defined by two endpoints and two internal control points, was altered by manipulating these four control points. Higher-order Bezier curves have more control points and are used as required to fit more complex curvature on the skeleton [301]. A new Bezier curve segment was added to provide finer coverage in the reference collection, starting at a shorter interval and a skeleton for a reference image than the range for a latent image [34].

The accuracy of warp and selection of the overlay (i.e., "best" warp) between the latent print image and each reference image is quantified at the level of single pixels within the skeletonized latent image. The basis for defining the pixel level quantification is the unique pairing of a skeletonized reference image (Fig. 2A.44C) using a

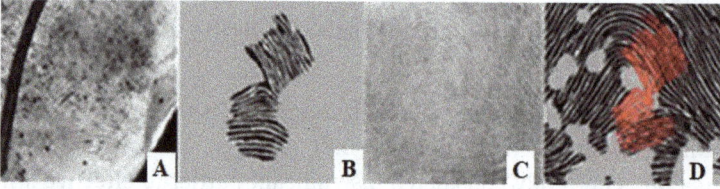

Fig. 2A.44: Spectroscopic and image of prints: (A) select latent images; (B) the high-contrast masked image; (C) the true mate reference print with poor clarity; and (D) the best warp of the latent onto the quality masked true mate reference print.

small Bezier curve overlaid to the latent image (Fig. 2A.44A). The matching criterion for Bezier curves is a distance measure for the similarity of Bezier curves. The reference images, according to the latent print overlays, are prioritized. Once an optimal warp provides the overlay between the latent and each reference image is obtained, the overlays are compared for accuracy, and the reference images are scored for final ranking (Fig. 2A.44D). A scoring algorithm is applied for indexing and ranking the overlays to the set of candidate reference images [302]. A competitive score between warps w i and w j at the latent skeleton pixel ω is defined as $S_{i,j} = -\log[d_i(\omega)/d_j(\omega)]$. Here, $d_w(\omega)$ is the minimum distance between all slight Bezier curves, containing the skeletonized latent image pixel ω and the warp w inverted paired short Bezier curves from the skeleton of the reference image. Microsoft Visual Studio Professional was used to code a GUI package incorporating the ANN algorithms trained in NIST DB 4 (freeware). Its advantage lies in that the smaller police departments can automate expertise before AFIS interfacing and identification are installed. Recognition results (Fig. 2A.45A) of the reference images (Fig. 2A.44B) were recorded into the database for accurate verification. A comparison of different chelation and binding agents is shown in Fig. 2A.46 illustrating how effective agent of contrast agents can greatly assist print resolution and accurate identification using database match artifical intelligence engines.

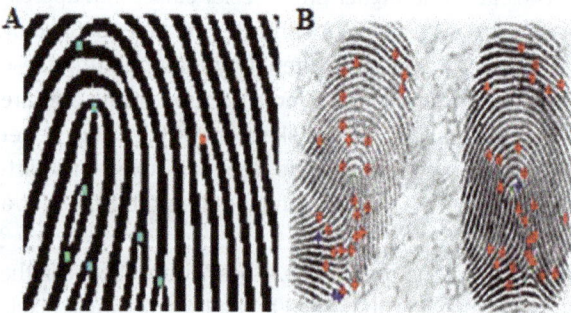

Fig. 2A.45: Pattern matching of latent fingerprints: (A) pattern recognition and (B) alignment and matching of the latent and reference fingerprints using ANN-based algorithms.

Professional development and training workflow: Multiple fingerprint generation is difficult due to variability in touch pressure, sweat production, and the tendency to lose sweat by successive impressions. Artificial sweat and 2D printers were used to create standardized prints to overcome the high barriers to obtaining uniform latent fingerprints. This procedure was utilized in the technician's training to enhance their expertise and experience in crime investigation. This activity has two deliverables as described below: Using a 2D printer with artificial sweat to generate standardized prints; Creating a specialty certificate in Nanoscience and Forensic Science.

Use of a 2D printer with artificial sweat to generate standardized prints: We can formulate a sweat composition based on a procedure described by Schwarz *et al.* [303]. This method was found to have two drawbacks, lacking metal cation and inhomogeneity of sweat concentration. Our approach can generate a realistic sweat composition and control print quality through copy numbers instead of sweat dilution. Standard fingerprints from Liu's team were obtained for analysis. The visible fingerprints were saved as print image pairs in the 2,000 eight-bit grayscale TIFF format. The digital TIFF image was made "transparent" to remove artifacts between the ridges. Then gamma correction (default 2.5) and the image contrast output parameters (default 0) in the picture processing software were adjusted to the digital image quality (grade 5-excellent) is obtained. These digital fingerprints are then reproduced on white cotton print paper (A4 size) and evaluated using standard DAs described in the Latent fingerprint visualization section. Dr. Bashir can cut out the prints previously sprayed with Ag NPs to examine reproduction efficiency. The print cuts were immersed in methanol to release Ag-bound print residues. The Ag excitation was measured by UV-vis spectroscopy from the resulting suspension. Herein, the reproducibility of the latent prints was examined through variations of the Ag resonance peak at 410–425 nm.

Artificial sweat composition: All L-amino acids were purchased from VWR International. The structure can consist of an equimolar concentration of serine, glycine, alanine, leucine, threonine, histidine, valine, asparagine, lysine, sodium chloride, calcium chloride, and magnesium chloride. The average concentration of the amino acids was controlled at 30 mM and salts at 115 mM. The prints were evaluated using DFO, Ninhydrin, MTN, Ag NPs, and IND. The cartridges were filled with sweat and prints using an Epson Artisan 50 Color Inkjet Printer with T0781 refillable cartridges.

Creation of a specialty certificate in Nanoscience and Forensic Science: The proposed course can cover five topic areas in the Nanoscience certificate and six themes for the Forensic Science certificate at undergraduate and graduate levels for law enforcement personnel. The enrolled students shall select 3 for the graduate or 4 for the undergraduate certificate, transcribed into their academic records. Artificial sweat and 2D printing can generate standardized prints evaluated using different DAs (Fig. 2A.45). Enrollees and volunteers to contribute their fingerprints have been

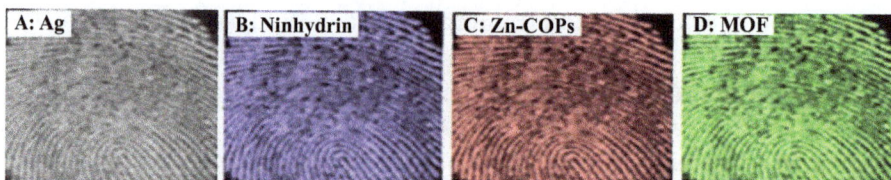

Fig. 2A.46: Incorporation of new development agents (Das) into a class for fingerprint technicians: (A) standard print generated using artificial sweat and 2D printing; (B–D) eccrine development agents were tested on different surfaces and as part of classes for a certificate in material or forensic science, (B) ninhydrin as DA, (C) Zn-COP (MOF-10) as DA, and (D) MOG-114 as DA.

chosen for undergraduate or graduate under Dr. S. Bashir and Dr. J. Louise Liu. This professional training can introduce a new paradigm for forensic evidence collection.

2A.5 Summary

In this chapter, the authors discussed photocatalysis, water remediation, heavy metal remediation, and the application of metal-organ frameworks. The centrality of water was discussed, and how critical it is to convert gray wastewater to potable water. The effect of heavy metals and other organic pollutants was discussed, and how to use photocatalysts that generate holes and electrons to promote carbon-bond fission and small molecule generation for decarbonization of organics, as well biosorption of heavy metals using biota-anchored systems. The MOF as drug delivery, single-atom catalyst, and energy transfer reagent was briefly described as well as more detailed examples of the biotoxicity within retinal epithelium cells and application in forensic science such as detection and imaging of fingerprints, alongside machine pattern recognition and HS imaging, indicate that nanoscience and engineering coupled with computer (bio)informatics will accompany the next revolution in health care, the environment, and energy applications.

References

[1] Sabatier, P. A. (1986). Top-down and bottom-up approaches to implementation research: a critical analysis and suggested synthesis. Journal of Public Policy, 6(1), 21–48.
[2] Lu, J., Elam, J. W. & Stair, P. C. (2016). Atomic layer deposition – Sequential self-limiting surface reactions for advanced catalyst "bottom-up" synthesis. Surface Science Reports, 71(2), 410–472.
[3] Langille, M. R., Personick, M. L., Zhang, J. & Mirkin, C. A. (2011). Bottom-up synthesis of gold octahedra with tailorable hollow features. Journal of the American Chemical Society, 133(27), 10414–10417.

[4]   Perry IV, J. J., Kravtsov, V. C., McManus, G. J. & Zaworotko, M. J. (2007). A bottom-up
        synthesis that does not start at the bottom: quadruple covalent cross-linking nanoscale
        faceted polyhedra. Journal of the American Chemical Society, 129(33), 10076–10077.
[5]   Wang, Y. & Xia, Y. (2004). Bottom-up and top-down approaches to synthesizing
        monodispersed spherical colloids of low melting-point metals. Nano Letters, 4(10),
        2047–2050.
[6]   de Oliveira, P. F., Torresi, R. M., Emmerling, F. & Camargo, P. H. (2020). Challenges and
        opportunities in the bottom-up mechanochemical synthesis of noble metal nanoparticles.
        Journal of Materials Chemistry A, 8(32), 16114–16141.
[7]   Wang, L., Sun, Y., Zhuang, L., Wu, A. & Wei, G. (2016). Bottom-up synthesis and sensor
        applications of biomimetic nanostructures. Materials, 9(1), 53.
[8]   Silberberg, M. (2017). Chemistry: the molecular nature of matter and change with advanced
        topics (pp. 2–514), New York: McGraw-Hill Education.
[9]   Bijlard, A. C., Wald, S., Crespy, D., Taden, A., Wurm, F. R. & Landfester, K. (2017). Functional
        colloidal stabilization. Advanced Materials Interfaces, 4(1), 1600443.
[10]   Piacenza, E., Presentato, A. & Turner, R. J. (2018). Stability of biogenic metal (loid)
        nanomaterials related to the colloidal stabilization theory of chemical nanostructures.
        Critical Reviews in Biotechnology, 38(8), 1137–1156.
[11]   Lyklema, J., Van Leeuwen, H. P. & Minor, M. (1999). DLVO-theory, a dynamic re-
        interpretation. Advances in Colloid and Interface Science, 83(1–3), 33–69.
[12]   Ninham, B. W. (1999). On progress in forces since the DLVO theory. Advances in Colloid and
        Interface Science, 83(1–3), 1–17.
[13]   Dahirel, V. & Jardat, M. (2010). Effective interactions between charged nanoparticles in
        water: what is left from the DLVO theory?. Current Opinion in Colloid & Interface Science,
        15(1–2), 2–7.
[14]   Missana, T. & Adell, A. (2000). On the applicability of DLVO theory to the prediction of clay
        colloids stability. Journal of Colloid and Interface Science, 230(1), 150–156.
[15]   Mishchuk, N. A. (2011). The model of hydrophobic attraction in the framework of classical
        DLVO forces. Advances in Colloid and Interface Science, 168(1–2), 149–166.
[16]   Asakura, S. & Oosawa, F. (1954). On the interaction between two bodies immersed in a
        solution of macromolecules. The Journal of Chemical Physics, 22(7), 1255–1256.
[17]   Hanaor, D., Michelazzi, M., Leonelli, C. & Sorrell, C. C. (2012). The effects of carboxylic acids
        on the aqueous dispersion and electrophoretic deposition of ZrO2. Journal of the European
        Ceramic Society, 32(1), 235–244.
[18]   Liu, J. L. & Bashir, S. (2015). Advanced nanomaterials and their applications in renewable
        energy (pp. 57), New York: Elsevier.
[19]   Butler, J. A. V. (1948). Theory of the stability of lyophobic colloids. Nature, 162(4113),
        315–316.
[20]   Hiemenz, P. C. & Rajagopalan, R. (2016). Principles of Colloid and Surface Chemistry, revised
        and expanded (pp. 461–498), Boca Raton: CRC Press.
[21]   Hunter, R. J. (2013). Zeta potential in colloid science: principles and applications (Vol. 2)
        (pp. 59–141), New York: Academic Press.
[22]   Nič, M., Jirát, J., Košata, B., Jenkins, A. & McNaught, A. (2009). IUPAC Compendium of
        chemical terminology, IUPAC, Research Triangle Park, NC.
[23]   https://depts.washington.edu/solgel/pages/courses/MSE_502/Electrostatic_Stabilization.
        html
[24]   Delgado, Á. V., González-Caballero, F., Hunter, R. J., Koopal, L. K. & Lyklema, J. (2007).
        Measurement and interpretation of electrokinetic phenomena. Journal of Colloid and
        Interface Science, 309(2), 194–224.

[25] Brinker, C. J. & Scherer, G. W. (2013). Sol-gel science: the physics and chemistry of sol-gel processing (pp. 235–302), New York: Academic Press.

[26] Cao, G. (2004). Nanostructures & nanomaterials: synthesis, properties & applications (pp. 7–10), London: Imperial College Press.

[27] Probstein, R. F. (2005). Physicochemical hydrodynamics: an introduction (pp. 237–273), New York: John Wiley & Sons.

[28] Myers, D. (2020). Surfactant science and technology (pp. 115–159), New York: John Wiley & Sons.

[29] Pierre, A. C. (2020). Introduction to sol-gel processing (pp. 15–64), Switzerland: Springer Nature.

[30] Bergström, L. (1997). Hamaker constants of inorganic materials. Advances in Colloid and Interface Science, 70, 125–169.

[31] Lu, D., Zhou, J., Hou, S., Xiong, Q., Chen, Y., Pu, K. . . . Duan, H. (2019). Functional Macromolecule-Enabled Colloidal Synthesis: from Nanoparticle Engineering to Multifunctionality. Advanced Materials, 31(44), 1902733.

[32] Polte, J. (2015). Fundamental growth principles of colloidal metal nanoparticles–a new perspective. CrystEngComm, 17(36), 6809–6830.

[33] Korotcenkov, G., Thomas, S., Sunny, A. T. & Prajitha, V. (2019). Colloidal metal oxide nanoparticles: synthesis, characterization and applications (pp. 15–22), Oxford, UK: Elsevier.

[34] Moreno, R. (2012). Colloidal processing of ceramics and composites. Advances in Applied Ceramics, 111(5–6), 246–253.

[35] El-Sayed, M. A. (2004). Small is different: shape-, size-, and composition-dependent properties of some colloidal semiconductor nanocrystals. Accounts of Chemical Research, 37(5), 326–333.

[36] Li, X., Iocozzia, J., Chen, Y., Zhao, S., Cui, X., Wang, W. . . . Lin, Z. (2018). From a precision synthesis of block copolymers to properties and applications of nanoparticles. Angewandte Chemie International Edition, 57(8), 2046–2070.

[37] Yang, D. & Fonseca, L. F. (2013). Wet-chemical approaches to porous nanowires with linear, spiral, and meshy topologies. Nano Letters, 13(11), 5642–5646.

[38] Xia, Y., Xiong, Y., Lim, B. & Skrabalak, S. E. (2009). Shape-controlled synthesis of metal nanocrystals: simple chemistry meets complex physics?. Angewandte Chemie International Edition, 48(1), 60–103.

[39] Mozaffari, S., Li, W., Thompson, C., Ivanov, S., Seifert, S., Lee, B. . . . Karim, A. M. (2017). Colloidal nanoparticle size control: experimental and kinetic modeling investigation of the ligand–metal binding role in controlling the nucleation and growth kinetics. Nanoscale, 9(36), 13772–13785.

[40] LaMer, V. K. (1952). Kinetics in phase transitions. Industrial & Engineering Chemistry Research, 44(6), 1270–1277.

[41] Ostwald, W. (1900). Über die vermeintliche Isomerie des roten und gelben Quecksilberoxyds und die Oberflächenspannung fester Körper. Zeitschrift Für Physikalische Chemie, 34(1), 495–503.

[42] Reiss, H. (1951). The growth of uniform colloidal dispersions. The Journal of Chemical Physics, 19(4), 482–487.

[43] Lifshitz, I. M. & Slyozov, V. V. (1961). The kinetics of precipitation from supersaturated solid solutions. Journal of Physics and Chemistry of Solids, 19(1–2), 35–50.

[44] Watzky, M. A. & Finke, R. G. (1997). Transition metal nanocluster formation kinetic and mechanistic studies. A new mechanism when hydrogen is the reductant: slow, continuous nucleation and fast autocatalytic surface growth. Journal of the American Chemical Society, 119(43), 10382–10400.

[45] Xia, Y., Xiong, Y., Lim, B. & Skrabalak, S. E. (2009). Shape-controlled synthesis of metal nanocrystals: simple chemistry meets complex physics?. Angewandte Chemie International Edition, 48(1), 60–103.

[46] Mozaffari, S., Li, W., Thompson, C., Ivanov, S., Seifert, S., Lee, B. . . . Karim, A. M. (2018). Ligand-mediated nucleation and growth of palladium metal nanoparticles. JoVE (Journal of Visualized Experiments), (136), e57667.

[47] Li, W., Taylor, M. G., Bayerl, D., Mozaffari, S., Dixit, M., Ivanov, S. . . . Karim, A. M. (2021). Solvent manipulation of the pre-reduction metal-ligand complex and particle-ligand binding for controlled synthesis of Pd nanoparticles. Nanoscale, 13(1), 206–217.

[48] Mozaffari, S., Li, W., Dixit, M., Seifert, S., Lee, B., Kovarik, L. . . . Karim, A. M. (2019). The role of nanoparticle size and ligand coverage in size focuses on colloidal metal nanoparticles. Nanoscale Advances, 1(10), 4052–4066.

[49] Watzky, M. A. & Finke, R. G. (1997). Nanocluster size-control and "magic number" investigations. Experimental tests of the "living-metal polymer" concept and mechanism-based size-control predictions led to the syntheses of Iridium (0) nanoclusters centering about four sequential magic numbers. Chemistry of Materials, 9(12), 3083–3095.

[50] Morris, A. M., Watzky, M. A., Agar, J. N. & Finke, R. G. (2008). Fitting neurological protein aggregation kinetic data via a 2-step, Minimal/"Ockham's Razor" Model: the Finke– Watzky mechanism of nucleation followed by autocatalytic surface growth. Biochemistry, 47(8), 2413–2427.

[51] Watzky, M. A., Finney, E. E. & Finke, R. G. (2008). Transition-metal nanocluster size vs. formation time and the catalytically effective nucleus number: a mechanism-based treatment. Journal of the American Chemical Society, 130(36), 11959–11969.

[52] Feng, Y., He, J., Wang, H., Tay, Y. Y., Sun, H., Zhu, L. & Chen, H. (2012). An unconventional role of ligand in continuously tuning of metal-metal interfacial strain. Journal of the American Chemical Society, 134(4), 2004–2007.

[53] Yoo, S., Kim, J., Kim, J. M., Son, J., Lee, S., Hilal, H. . . . Park, S. (2020). Three-dimensional gold nanosphere hexamers linked with metal bridges: near-field focusing for single-particle surface-enhanced Raman scattering. Journal of the American Chemical Society, 142(36), 15412–15419.

[54] Bi, D., Yi, C., Luo, J., Décoppet, J. D., Zhang, F., Zakeeruddin, S. M. . . . Grätzel, M. (2016). Polymer-templated nucleation and crystal growth of perovskite films for solar cells with efficiency greater than 21%. Nature Energy, 1(10), 1–5.

[55] Jones, M. R., Osberg, K. D., Macfarlane, R. J., Langille, M. R. & Mirkin, C. A. (2011). Templated techniques for the synthesis and assembly of plasmonic nanostructures. Chemical Reviews, 111(6), 3736–3827.

[56] Jones, M. R., Macfarlane, R. J., Prigodich, A. E., Patel, P. C. & Mirkin, C. A. (2011). Nanoparticle shape anisotropy dictates the collective behavior of surface-bound ligands. Journal of the American Chemical Society, 133(46), 18865–18869.

[57] Lee, H., Dellatore, S. M., Miller, W. M. & Messersmith, P. B. (2007). Mussel-inspired surface chemistry for multifunctional coatings. Science, 318(5849), 426–430.

[58] Liu, N. & Liedl, T. (2018). DNA-assembled advanced plasmonic architectures. Chemical Reviews, 118(6), 3032–3053.

[59] Zhulina, E. B., Borisov, O. V. & Birshtein, T. M. (1999). Polyelectrolyte brush interaction with multivalent ions. Macromolecules, 32(24), 8189–8196.

[60] Lin, M., Kim, G. H., Kim, J. H., Oh, J. W. & Nam, J. M. (2017). Transformative heterointerface evolution and plasmonic tuning of anisotropic trimetallic nanoparticles. Journal of the American Chemical Society, 139(30), 10180–10183.

[61] Pang, X., Zhao, L., Han, W., Xin, X. & Lin, Z. (2013). A general and robust strategy for the synthesis of nearly monodisperse colloidal nanocrystals. Nature Nanotechnology, 8(6), 426–431.

[62] Chen, T., Yang, M., Wang, X., Tan, L. H. & Chen, H. (2008). Controlled assembly of eccentrically encapsulated gold nanoparticles. Journal of the American Chemical Society, 130 (36), 11858–11859.

[63] Gil, E. S. & Hudson, S. M. (2004). Stimuli-responsive polymers and their bioconjugates. Progress in Polymer Science, 29(12), 1173–1222.

[64] Zoppe, J. O., Ataman, N. C., Mocny, P., Wang, J., Moraes, J. & Klok, H. A. (2017). Surface-initiated controlled radical polymerization: state-of-the-art opportunities and challenges in surface and interface engineering with polymer brushes. Chemical Reviews, 117(3), 1105–1318.

[65] Hein, C. D., Liu, X. M. & Wang, D. (2008). Click chemistry is a powerful tool for pharmaceutical sciences. Pharmaceutical Research, 25(10), 2216–2230.

[66] Moses, J. E. & Moorhouse, A. D. (2007). The growing applications of click chemistry. Chemical Society Reviews, 36(8), 1249–1262.

[67] Sapsford, K. E., Algar, W. R., Berti, L., Gemmill, K. B., Casey, B. J., Oh, E. . . . Medintz, I. L. (2013). Functionalizing nanoparticles with biological molecules: developing chemistries that facilitate nanotechnology. Chemical Reviews, 113(3), 1904–2074.

[68] https://www.un.org/waterforlifedecade/

[69] Caturegli, L., Matteoli, S., Gaetani, M., Grossi, N., Magni, S., Minelli, A. . . . Volterrani, M. (2020). Effects of water stress on the spectral reflectance of bermudagrass. Scientific Reports, 10(1), 1–12.

[70] Schyns, J. F., Hoekstra, A. Y., Booij, M. J., Hogeboom, R. J. & Mekonnen, M. M. (2019). Limits to the world's green water resources for food, feed, fiber, timber, and bioenergy. Proceedings of the National Academy of Sciences, 116(11), 4893–4898.

[71] Vanham, D., Leip, A., Galli, A., Kastner, T., Bruckner, M., Uwizeye, A. . . . Hoekstra, A. Y. (2019). Environmental footprint family to address local to planetary sustainability and deliver on the SDGs. Science of the Total Environment, 693, 133642.

[72] Mekonnen, M. M. & Hoekstra, A. Y. (2020). Sustainability of the blue water footprint of crops. Advances in Water Resources, 143, 103679.

[73] Mekonnen, M. M. & Hoekstra, A. Y. (2020). Bluewater footprint linked to national consumption and international trade is unsustainable. Nature Food, 1(12), 792–800.

[74] Mekonnen, M. M. & Hoekstra, A. Y. (2011). The green, blue, and grey water footprint of crops and derived crop products. Hydrology and Earth System Sciences, 15(5), 1577–1600.

[75] Mekonnen, M. M. & Hoekstra, A. Y. (2010). A global and high-resolution assessment of the green, blue, and grey water footprint of wheat. Hydrology and Earth System Sciences, 14(7), 1259–1276.

[76] Niccolucci, V., Tiezzi, E., Pulselli, F. M. & Capineri, C. (2012). Biocapacity vs. Ecological Footprint of world regions: a geopolitical interpretation. Ecological Indicators, 16, 23–30.

[77] Fu, W., Turner, J. C., Zhao, J. & Du, G. (2015). Ecological footprint (EF): an expanded role in calculating resource productivity (RP) using China and the G20 member countries as examples. Ecological Indicators, 48, 464–471.

[78] Schyns, J. F., Hoekstra, A. Y., Booij, M. J., Hogeboom, R. J. & Mekonnen, M. M. (2019). Limits to the world's green water resources for food, feed, fiber, timber, and bioenergy. Proceedings of the National Academy of Sciences, 116(11), 4893–4898.

[79] IPCC (2022). IPCC Sixth Assessment Report (AR6): Climate Change 2022-Impacts, Adaptation and Vulnerability, https://www.ipcc.ch/report/ar6/wg2/

[80] Birkmann, J., Jamshed, A., McMillan, J. M., Feldmeyer, D., Totin, E., Solecki, W. . . . Alegría, A. (2022). Understanding human vulnerability to climate change: a global perspective on index validation for adaptation planning. Science of the Total Environment, 803, 150065.

[81] Ismail, M., Akhtar, K., Khan, M. I., Kamal, T., Khan, M. A., M Asiri, A. . . . Khan, S. B. (2019). Pollution, toxicity, and carcinogenicity of organic dyes and their catalytic bio-remediation. Current Pharmaceutical Design, 25(34), 3645–3663.

[82] Richardson, J. B., Dancy, B. C., Horton, C. L., Lee, Y. S., Madejczyk, M. S., Xu, Z. Z. . . . Lewis, J. A. (2018). Exposure to toxic metals triggers unique responses from the rat gut microbiota. Scientific Reports, 8(1), 1–12.

[83] Singh, N. B., Nagpal, G. & Agrawal, S. (2018). Water purification by using adsorbents: a review. Environmental Technology & Innovation, 11, 187–240.

[84] Lu, F. & Astruc, D. (2020). Nanocatalysts and other nanomaterials for water remediation from organic pollutants. Coordination Chemistry Reviews, 408, 213180.

[85] Ghadimi, M., Zangenehtabar, S. & Homaeigohar, S. (2020). An overview of the water remediation potential of nanomaterials and their ecotoxicological impacts. Water, 12(4), 1150.

[86] Ziegler, P., Sree, K. S. & Appenroth, K. J. (2016). Duckweeds for water remediation and toxicity testing. Toxicological and Environmental Chemistry, 98(10), 1127–1154.

[87] Guerra, F. D., Attia, M. F., Whitehead, D. C. & Alexis, F. (2018). Nanotechnology for environmental remediation: materials and applications. Molecules, 23(7), 1760.

[88] Huang, H. Y. (2021). Water Remediation by Natural Products Core/Shell Magnetite Nanoparticles, Doctoral dissertation, Texas A&M University-Kingsville.

[89] Shabbir, S., Faheem, M., Ali, N., Kerr, P. G. & Wu, Y. (2017). Periphyton biofilms: a novel and natural biological system for the effective removal of sulfonated azo dye methyl orange by synergistic mechanism. Chemosphere, 167, 236–246.

[90] Liu, J., Wang, F., Wu, W., Wan, J., Yang, J., Xiang, S. & Wu, Y. (2018). Biosorption of high-concentration Cu (II) by periphytic biofilms and the development of a fiber periphyton bioreactor (FPBR). Bioresource Technology, 248, 127–134.

[91] Flanigan, D., Sandoval, R., Bridgers, K. & Hristovski, K. (2013). Bolaform amphiphile modified granular activated carbon media for removal of strong acid oxo-anions from water: nitrate. Journal of Environmental Chemical Engineering, 1(4), 1188–1193.

[92] Bozbas, S. K., Ay, U. & Kayan, A. (2013). Novel inorganic-organic hybrid polymers to remove heavy metals from aqueous solution. Desalination and Water Treatment, 51(37–39), 7208–7215.

[93] Gyürék, L. L. & Finch, G. R. (1998). Modeling water treatment chemical disinfection kinetics. Journal of Environmental Engineering, 124(9), 783–793.

[94] Li, Q. & Lu, G. (2008). Controlled synthesis and photocatalytic investigation of different-shaped one-dimensional titanic acid nanomaterials. Journal of Power Sources, 185(1), 577–583.

[95] Myadav, S. & Jaiswar, G. (2017). Review on undoped/doped TiO2 nanomaterial; synthesis and photocatalytic and antimicrobial activity. Journal of the Chinese Chemical Society, 64(1), 103–116.

[96] Liu, J., Wang, Z., Luo, Z. & Bashir, S. (2013). Effective bactericidal performance of silver-decorated titania nano-composites. Dalton Transactions, 42(6), 2158–2166.

[97] Savage, N. & Diallo, M. S. (2005). Nanomaterials and water purification: opportunities and challenges. Journal of Nanoparticle Research, 7(4), 331–342.

[98] Chen, P., Wang, F., Chen, Z. F., Zhang, Q., Su, Y., Shen, L. . . . Liu, G. (2017). Study on the photocatalytic mechanism and detoxicity of gemfibrozil by a sunlight-driven TiO2/carbon

dots photocatalyst: the significant roles of reactive oxygen species. Applied Catalysis. B, Environmental, 204, 250–259.

[99] Miklos, D. B., Remy, C., Jekel, M., Linden, K. G., Drewes, J. E. & Hübner, U. (2018). Evaluation of advanced oxidation processes for water and wastewater treatment–A critical review. Water Research, 139, 118–131.

[100] Dewil, R., Mantzavinos, D., Poulios, I. & Rodrigo, M. A. (2017). New perspectives for advanced oxidation processes. Journal of Environmental Management, 195, 93–99.

[101] Barzegar, M. H., Ghaedi, M., Avargani, V. M., Sabzehmeidani, M. M., Sadeghfar, F. & Jannesar, R. (2019). Electrochemical synthesis of Zn: znO/Ni2P and efficient photocatalytic degradation of Auramine O in aqueous solution under multi-variable experimental design optimization. Polyhedron, 165, 1–8.

[102] Wang, H., Mi, X., Li, Y. & Zhan, S. (2020). 3D graphene-based macrostructures for water treatment. Advanced Materials, 32(3), 1806843.

[103] Li, C., Hu, R., Lu, X., Bashir, S. & Liu, J. L. (2020). Efficiency enhancement of photocatalytic degradation of tetracycline using reduced graphene oxide coordinated titania nanoplatelet. Catalysis Today, 350, 171–183.

[104] Sun, P., Lee, W. N., Zhang, R. & Huang, C. H. (2016). Degradation of DEET and caffeine under UV/chlorine and simulated sunlight/chlorine conditions. Environmental Science & Technology, 50(24), 13265–13273.

[105] Liu, J. L. & Bashir, S. (2015). Advanced nanomaterials and their applications in renewable energy, Vols 1–50, New York: Elsevier.

[106] Barrera, M. C., Escobar, J., José, A., Cortés, M. A., Viniegra, M. & Hernández, A. (2006). Effect of solvothermal treatment temperature on the properties of sol-gel ZrO2–TiO2 mixed oxides as HDS catalyst supports. Catalysis Today, 116(4), 498–504.

[107] Antonelli, D. M. (1999). Synthesis of phosphorus-free mesoporous titania via templating with amine surfactants. Microporous and Mesoporous Materials, 30(2–3), 315–319.

[108] Lee, G. & Cho, K. (2009). Electronic structures of zigzag graphene nanoribbons with edge hydrogenation and oxidation. Physical Review B, 79(16), 165440.

[109] Spyrou, K. & Rudolf, P. (2014). An introduction to graphene. Functionalization of Graphene, 1–20.

[110] Wang, Y., Liu, M., Liu, Y., Luo, J., Lu, X. & Sun, J. (2017). A novel mica-titania@ graphene core-shell structured antistatic composite pearlescent pigment. Dyes and Pigments, 136, 197–204.

[111] Lv, T., Wu, M., Guo, M., Liu, Q. & Jia, L. (2019). Self-assembly photocatalytic reduction synthesis of graphene-encapsulated LaNiO3 nanoreactor with high efficiency and stability for photocatalytic water splitting to hydrogen. Chemical Engineering Journal, 356, 580–591.

[112] Mustapha, S., Tijani, J. O., Ndamitso, M. M., Abdulkareem, A. S., Shuaib, D. T., Amigun, A. T. & Abubakar, H. L. (2021). Facile synthesis and characterization of TiO2 nanoparticles: x-ray peak profile analysis using Williamson–Hall and Debye–Scherrer methods. International Nano Letters, 11(3), 241–261.

[113] Santara, B., Giri, P. K., Imakita, K. & Fujii, M. (2014). Microscopic origin of lattice contraction and expansion in undoped rutile TiO2 nanostructures. Journal of Physics D: Applied Physics, 47(21), 215302.

[114] Kamble, S. P., Sawant, S. B., Schouten, J. C. & Pangarkar, V. G. (2003). Photocatalytic and photochemical degradation of aniline using concentrated solar radiation. Journal of Chemical Technology & Biotechnology: international Research in Process. Environmental & Clean Technology, 78(8), 865–872.

[115] Hathway, T. & Jenks, W. S. (2008). Effects of sintering of TiO2 particles on the mechanisms of photocatalytic degradation of organic molecules in water. Journal of Photochemistry and Photobiology. A, Chemistry, 200(2–3), 216–224.

[116] Patterson, A. L. (1939). The Scherrer formula for X-ray particle size determination. Physical Review, 56(10), 978.

[117] Villarroel-Rocha, J., Barrera, D. & Sapag, K. (2014). Introducing a self-consistent test and the corresponding modification in the Barrett, Joyner, and Halenda method for pore-size determination. Microporous and Mesoporous Materials, 200, 68–78.

[118] Ohsaka, T., Izumi, F. & Fujiki, Y. (1978). Raman spectrum of anatase, TiO2. Journal of Raman Spectroscopy, 7(6), 321–324.

[119] Minitha, C. R. & Rajendrakumar, R. T. (2013). Synthesis and characterization of reduced graphene oxide. In: Advanced materials research (pp. 56–60), Vol. 678, Trans Tech Publications Ltd.

[120] Luo, Z., Vora, P. M., Mele, E. J., Johnson, A. C. & Kikkawa, J. M. (2009). Photoluminescence and bandgap modulation in graphene oxide. Applied Physics Letters, 94(11), 111909.

[121] Ramanan, V., Siddaiah, B., Raji, K. & Ramamurthy, P. (2018). Green synthesis of multi-functionalized, nitrogen-doped, highly fluorescent carbon dots from waste expanded polystyrene and its application in the fluorimetric detection of Au3+ ions in aqueous media. ACS Sustainable Chemistry & Engineering, 6(2), 1627–1638.

[122] Kozak, O., Sudolska, M., Pramanik, G., Cigler, P., Otyepka, M. & Zboril, R. (2016). Photoluminescent carbon nanostructures. Chemistry of Materials, 28(12), 4085–4128.

[123] Liang, Y. T., Vijayan, B. K., Lyandres, O., Gray, K. A. & Hersam, M. C. (2012). Effect of dimensionality on the photocatalytic behavior of carbon–titania nanosheet composites: charge transfer at nanomaterial interfaces. The Journal of Physical Chemistry Letters, 3(13), 1760–1765.

[124] Duarte, H. A., Carvalho, S., Paniago, E. B. & Simas, A. M. (1999). Importance of tautomers in the chemical behavior of tetracyclines. Journal of Pharmaceutical Sciences, 88(1), 111–120.

[125] Kiarii, E. M., Govender, K. K., Ndung'u, P. G. & Govender, P. P. (2017). A DFT study on the effect of supporting titania on silica graphene, epoxy graphene, and carbon nanotubes-Interfacial properties and optical response. Computational Condensed Matter, 13, 6–15.

[126] Bashir, S., Hanumandla, P., Huang, H. Y. & Liu, J. L. (2018). Nanostructured materials for advanced energy conversion and storage devices: safety implications at end-of-life disposal. In Nanostructured Materials for Next-Generation Energy Storage and Conversion (pp. 517–542), Berlin, Heidelberg: Springer.

[127] Liu, J. L. & Bashir, S. (2015). Advanced nanomaterials and their applications in renewable energy (pp. 367–405), New York: Elsevier.

[128] Elieh-Ali-Komi, D. & Hamblin, M. R. (2016). Chitin and chitosan: production and application of versatile biomedical nanomaterials. International Journal of Advanced Research, 4(3), 411.

[129] Jayakumar, R., Prabaharan, M., Kumar, P. S., Nair, S. V. & Tamura, H. (2011). Biomaterials based on chitin and chitosan in wound dressing applications. Biotechnology Advances, 29(3), 322–337.

[130] Fernandes, J. C., Tavaria, F. K., Soares, J. C., Ramos, Ó. S., Monteiro, M. J., Pintado, M. E. & Malcata, F. X. (2008). Antimicrobial effects of chitosan and chitooligosaccharides, upon Staphylococcus aureus and Escherichia coli, in food model systems. Food Microbiology, 25(7), 922–928.

[131] Park, B. K. & Kim, M. M. (2010). Applications of chitin and its derivatives in biological medicine. International Journal of Molecular Sciences, 11(12), 5152–5164.

[132] https://gestis.dguv.de/data?name=491257&lang=en

[133] Bykova, E., Dubrovinsky, L., Dubrovinskaia, N., Bykov, M., McCammon, C., Ovsyannikov, S. V. . . . Prakapenka, V. (2016). The structural complexity of simple Fe2O3 at high pressures and temperatures. Nature Communications, 7(1), 1–6.

[134] Jette, E. R. & Foote, F. (1933). An X-Ray Study of the Wüstite (FeO) Solid Solutions. The Journal of Chemical Physics, 1(1), 29–36.

[135] Fleet, M. E. (1986). The structure of magnetite: symmetry of cubic spinels. Journal of Solid State Chemistry, 62(1), 75–82.

[136] Lavina, B., Dera, P., Kim, E., Meng, Y., Downs, R. T., Weck, P. F. . . . Zhao, Y. (2011). Discovery of the recoverable high-pressure iron oxide Fe4O5. Proceedings of the National Academy of Sciences, 108(42), 17281–17285.

[137] Shmakov, A. N., Kryukova, G. N., Tsybulya, S. V., Chuvilin, A. L. & Solovyeva, L. P. (1995). Vacancy ordering in γ-Fe2O3: synchrotron X-ray powder diffraction and high-resolution electron microscopy studies. Journal of Applied Crystallography, 28(2), 141–145.

[138] Szytuła, A., Burewicz, A., Dimitrijević, Ž., Kraśnicki, S., Rżany, H., Todorović, J. . . . Wolski, W. (1968). Neutron diffraction studies of α-FeOOH. Physica Status Solidi (B), 26(2), 429–434.

[139] Bykova, E., Dubrovinsky, L., Dubrovinskaia, N., Bykov, M., McCammon, C., Ovsyannikov, S. V. . . . Prakapenka, V. (2016). The structural complexity of simple Fe2O3 at high pressures and temperatures. Nature Communications, 7(1), 1–6.

[140] Post, J. E., Heaney, P. J., Von Dreele, R. B. & Hanson, J. C. (2003). Neutron and temperature-resolved synchrotron X-ray powder diffraction study of akaganéite. American Mineralogist, 88(5–6), 782–788.

[141] Wyckoff, R. W. G. & Wyckoff, R. W. (1963). Crystal structures (pp. 312), Vol. 1, New York: Interscience Publishers.

[142] Patrat, G., De Bergevin, F., Pernet, M. & Joubert, J. C. (1983). Structure locale de δ-FeOOH. Acta Crystallographica. Section B, Structural Science, 39(2), 165–170.

[143] Olsen, J. S., Cousins, C. S. G., Gerward, L., Jhans, H. T. & Sheldon, B. J. (1991). A study of the crystal structure of Fe2O3 in the pressure range up to 65 GPa using synchrotron radiation. Physica Scripta, 43(3), 327.

[144] https://materialsproject.org/materials/mp-19306/)

[145] Van Zeghbroeck, B. J. (2011). Principles of semiconductor devices (pp. 35–115), Wilmington, DE: Prentice-Hall.

[146] Gill, P. M. & von Rague Schleyer, P. (1994). Density functional theory (DFT), Hartree-Fock (HF), and the self-consistent field. Journal of Chemical Physics, 100, 5066–5075.

[147] Kohn, W., Becke, A. D. & Parr, R. G. (1996). Density functional theory of electronic structure. The Journal of Physical Chemistry, 100(31), 12974–12980.

[148] Cohen, A. J., Mori-Sánchez, P. & Yang, W. (2012). Challenges for density functional theory. Chemical Reviews, 112(1), 289–320.

[149] Perdew, J. P. & Zunger, A. (1981). Self-interaction correction to density-functional approximations for many-electron systems. Physical Review B, 23(10), 5048.

[150] Monkhorst, H. J. (1979). Hartree-Fock density of states for extended systems. Physical Review B, 20(4), 1504.

[151] Mathew, K., Zheng, C., Winston, D., Chen, C., Dozier, A., Rehr, J. J. . . . Persson, K. A. (2018). High-throughput computational X-ray absorption spectroscopy. Scientific Data, 5(1), 1–8.

[152] https://materialsproject.org/materials/mp-19306/

[153] https://materialsproject.org/materials/mp-19770/

[154] Marsman, M., Paier, J., Stroppa, A. & Kresse, G. (2008). Hybrid functionals are applied to extended systems. Journal of Physics: Condensed Matter, 20(6), 064201.

[155] Evans, J. (2018). X-ray absorption spectroscopy for the chemical and materials sciences (pp. 1–32), New York: John Wiley & Sons.

[156] Lamberti, C. (2004). The use of synchrotron radiation techniques in the characterization of strained semiconductor heterostructures and thin films. Surface Science Reports, 53(1–5), 1–197.

[157] Joseph, D., Basu, S., Jha, S. N. & Bhattacharyya, D. (2012). Chemical shifts of KX-ray absorption edges on copper in different compounds by X-ray absorption spectroscopy (XAS) with Synchrotron radiation. Nuclear Instruments & Methods in Physics Research. Section B, Beam Interactions with Materials and Atoms, 274, 126–128.

[158] Brown, G. E. & Sturchio, N. C. (2002). An overview of synchrotron radiation applications to low-temperature geochemistry and environmental science.

[159] Mathew, K., Zheng, C., Winston, D., Chen, C., Dozier, A., Rehr, J. J. . . . Persson, K. A. (2018). High-throughput computational X-ray absorption spectroscopy. Scientific Data, 5(1), 1–8.

[160] Yamashita, T. & Hayes, P. (2008). Analysis of XPS spectra of Fe2+ and Fe3+ ions in oxide materials. Applied Surface Science, 254(8), 2441–2449.

[161] Muhler, M., Schlögl, R. & Ertl, G. (1992). The nature of the iron oxide-based catalyst for dehydrogenation of ethylbenzene to styrene 2. Surface chemistry of the active phase. Journal of Catalysis, 138(2), 413–444.

[162] McCune, J. A., Mommer, S., Parkins, C. C. & Scherman, O. A. (2020). Design Principles for Aqueous Interactive Materials: lessons from Small Molecules and Stimuli-Responsive Systems. Advanced Materials, 32(20), 1906890.

[163] Calandra, P., Caschera, D., Liveri, V. T. & Lombardo, D. (2015). How self-assembly of amphiphilic molecules can generate complexity in the nanoscale. Colloids and Surfaces. A, Physicochemical and Engineering Aspects, 484, 164–183.

[164] Wang, Y., Song, S., Zhang, S. & Zhang, H. (2019). Stimuli-responsive nano theranostics based on lanthanide-doped upconversion nanoparticles for cancer imaging and therapy: current advances and future challenges. Nano Today, 25, 38–67.

[165] Jin, H., Jiao, F., Daily, M. D., Chen, Y., Yan, F., Ding, Y. H. . . . Chen, C. L. (2016). Highly stable and self-repairing membrane-mimetic 2D nanomaterials assembled from lipid-like peptoids. Nature Communications, 7(1), 1–8.

[166] Song, H., Rioux, R. M., Hoefelmeyer, J. D., Komor, R., Niesz, K., Grass, M., Yang, P. & Somorjai, G. A. (2006). Hydrothermal growth of mesoporous SBA-15 silica in the presence of PVP-stabilized Pt nanoparticles: synthesis, characterization, and catalytic properties. Journal of the American Chemical Society, 128(9), 3027–3037.

[167] Collins, S. D., Ran, N. A., Heiber, M. C. & Nguyen, T. Q. (2017). Small is powerful: recent progress in solution-processed small-molecule solar cells. Advanced Energy Materials, 7(10), 1602242.

[168] Zhang, J., Li, Y., Zhang, X. & Yang, B. (2010). Colloidal self-assembly meets nanofabrication: from two-dimensional colloidal crystals to nanostructure arrays. Advanced Materials, 22(38), 4249–4269.

[169] Xu, C., Lin, Y., Wang, J., Wu, L., Wei, W., Ren, J. & Qu, X. (2013). Nanoceria-Triggered Synergetic Drug Release Based on CeO2-Capped Mesoporous Silica Host-Guest Interactions and Switchable Enzymatic Activity and Cellular Effects of CeO2. Advanced Healthcare Materials, 2(12), 1591–1599.

[170] Carne, A., Carbonell, C., Imaz, I. & Maspoch, D. (2011). Nanoscale metal-organic materials. Chemical Society Reviews, 40(1), 291–305.

[171] Stock, N. & Biswas, S. (2012). Synthesis of metal-organic frameworks (MOFs): routes to various MOF topologies, morphologies, and composites. Chemical Reviews, 112(2), 933–969.

[172] Rubio-Martinez, M., Avci-Camur, C., Thornton, A. W., Imaz, I., Maspoch, D. & Hill, M. R. (2017). New synthetic routes towards MOF production at scale. Chemical Society Reviews, 46(11), 3453–3480.

[173] Vallé, K., Belleville, P., Pereira, F. & Sanchez, C. (2006). Hierarchically structured transparent hybrid membranes by in situ growth of mesostructured organosilica in host polymer. Nature Materials, 5(2), 107–111.

[174] Kaur, P., Hupp, J. T. & Nguyen, S. T. (2011). Porous organic polymers in catalysis: opportunities and challenges. ACS Catalysis, 1(7), 819–835.

[175] Hu, J., Deng, W. & Chen, D. (2017). Ceria hollow spheres as an adsorbent for efficient removal of acid dye. ACS Sustainable Chemistry & Engineering, 5(4), 3570–3582.

[176] Sun, J., Wang, C., Zeng, L., Xu, P., Yang, X., Chen, J., Xing, X., Jin, Q. & Yu, R. (2016). Controllable assembly of CeO2 micro/nanospheres with adjustable size and their application in Cr (VI) adsorption. Materials Research Bulletin, 75, 110–114.

[177] Hu, J., Deng, W. & Chen, D. (2017). Ceria hollow spheres as an adsorbent for efficient removal of acid dye. ACS Sustainable Chemistry & Engineering, 5(4), 3570–3582.

[178] Gogoi, A. & Sarma, K. C. (2017). Synthesis of the novel β-cyclodextrin supported CeO2 nanoparticles for the catalytic degradation of methylene blue in aqueous suspension. Materials Chemistry and Physics, 194, 327–336.

[179] Velusamy, P. & Lakshmi, G. (2017). The enhanced photocatalytic performance of (ZnO/CeO 2)-β-CD system for the effective decolorization of Rhodamine B under UV light irradiation. Applied Water Science, 7(7), 4025–4036.

[180] Khan, N. A., Hasan, Z. & Jhung, S. H. (2013). Adsorptive removal of hazardous materials using metal-organic frameworks (MOFs): a review. Journal of Hazardous Materials, 244, 444–456.

[181] Lee, H. A., Park, E. & Lee, H. (2020). Polydopamine and Its Derivative Surface Chemistry in Material Science: a Focused Review for Studies at KAIST. Advanced Materials, 1907505.

[182] Kobielska, P. A., Howarth, A. J., Farha, O. K. & Nayak, S. (2018). Metal-organic frameworks for heavy metal removal from water. Coordination Chemistry Reviews, 358, 92–107.

[183] Long, J. R. & Yaghi, O. M. (2009). The pervasive chemistry of metal-organic frameworks. Chemical Society Reviews, 38(5), 1213–1214.

[184] An, J., Farha, O. K., Hupp, J. T., Pohl, E., Yeh, J. I. & Rosi, N. L. (2012). Metal-adenine vertices for the construction of an exceptionally porous metal-organic framework. Nature Communications, 3(1), 1–6.

[185] Koh, K., Wong-Foy, A. G. & Matzger, A. J. (2008). A crystalline mesoporous coordination copolymer with high microporosity. Angewandte Chemie International Edition, 47(4), 677–680.

[186] Lu, G., Li, S., Guo, Z., Farha, O. K., Hauser, B. G., Qi, X., Wang, Y., Wang, X., Han, S., Liu, X. & DuChene, J. S. (2012). Imparting functionality to a metal-organic framework material by controlled nanoparticle encapsulation. Nature Chemistry, 4(4), 310–316.

[187] Zhuang, W., Ma, S., Wang, X. S., Yuan, D., Li, J. R., Zhao, D. & Zhou, H. C. (2010). Introduction of cavities up to 4 nm into a hierarchically-assembled metal-organic framework using an angular, tetratopic ligand. Chemical Communications, 46(29), 5223–5225.

[188] Eddaoudi, M., Li, H. & Yaghi, O. M. (2000). Highly porous and stable metal-organic frameworks: structure design and sorption properties. Journal of the American Chemical Society, 122(7), 1391–1397.

[189] Zhao, D., Timmons, D. J., Yuan, D. & Zhou, H. C. (2011). Tuning the topology and functionality of metal-organic frameworks by ligand design. Accounts of Chemical Research, 44(2), 123–133.

[190] Johnsen, A. R., Wick, L. Y. & Harms, H. (2005). Principles of microbial PAH-degradation in soil. Environmental Pollution, Jan 1 133(1), 71–84.

[191] Haritash, A. K. & Kaushik, C. P. (2009). Biodegradation aspects of polycyclic aromatic hydrocarbons (PAHs): a review. Journal of Hazardous Materials, 169(1–3), 1–15.

[192] An, J., Geib, S. J. & Rosi, N. L. (2009). Cation-triggered drug release from the porous zinc-adenine metal-organic framework. Journal of the American Chemical Society, 131(24), 8376–8377.

[193] USEPA (2004). Cleaning up the Nation's Waste Sites: markets and Technology Trends. In: US Environmental Protection Agency, Washington, DC, EPA 542-R-04-015.

[194] Chen, M., Xu, P., Zeng, G., Yang, C., Huang, D. & Zhang, J. (2015). Bioremediation of soils contaminated with polycyclic aromatic hydrocarbons, petroleum, pesticides, chlorophenols, and heavy metals by composting: applications, microbes, and future research needs. Biotechnology Advances, 33, 745–755.

[195] Li, H., Qu, R., Li, C., Guo, W., Han, X., He, F., Ma, Y. & Xing, B. (2014). Selective removal of polycyclic aromatic hydrocarbons (PAHs) from soil washing effluents using biochars produced at different pyrolytic temperatures. Bioresource Technology, 163, 193–198.

[196] Chen, Y., Cheng, J. J. & Creamer, K. S. (2008). Inhibition of anaerobic digestion process: a review. Bioresource Technology, 99, 4044–4064.

[197] Godoy, P., Reina, R., Calderón, A., Wittich, R.-M., García-Romera, I. & Aranda, E. (2016). Exploring the potential of fungi isolated from PAH-polluted soil as a source of xenobiotics-degrading fungi. Environmental Science and Pollution Research, 23, 20985–20996.

[198] Pathak, B., Gupta, S. & Verma, R. (2018). Biosorption and Biodegradation of Polycyclic Aromatic Hydrocarbons (PAHs) by Microalgae, Crini, G. & Lichtfouse, E. eds., Green Adsorbents for Pollutant Removal: fundamentals and Design (pp. 215–247), Cham.,London UK, Springer International Publishing.

[199] Basak, G., Das, D. & Das, N. (2014). Dual role of acidic diacetate sophorolipid as bio stabilizer for ZnO nanoparticle synthesis and biofunctionalized agent against Salmonella enterica and Candida albicans. Journal of Microbiology and Biotechnology, 24, 87–96.

[200] Basak, G., Hazra, C. & Sen, R. (2020). Biofunctionalized nanomaterials for in situ clean-up of hydrocarbon contamination: a quantum jump in global bioremediation research. Journal of Environmental Management, 256, 109913.

[201] Patil, S. S., Shedbalkar, U. U., Truskewycz, A., Chopade, B. A. & Ball, A. S. (2016). Nanoparticles for environmental clean-up: a review of potential risks and emerging solutions. Environmental Technology & Innovation, 5, 10–21.

[202] Patel, V., Jain, S. & Madamwar, D. (2012). Naphthalene degradation by a bacterial consortium (DV-AL) developed from Alang-Sosiya ship-breaking yard, Gujarat, India. Bioresource Technology, 107, 122–130.

[203] Lu, M., Zhang, -Z.-Z., Wang, J.-X., Zhang, M., Xu, Y.-X. & Wu, X.-J. (2014). Interaction of Heavy Metals and Pyrene on Their Fates in Soil and Tall Fescue (Festuca arundinacea). Environmental Science & Technology, 48, 1158–1165.

[204] Gauthier, P. T., Norwood, W. P., Prepas, E. E. & Pyle, G. G. (2014). Metal–PAH mixtures in the aquatic environment: a review of co-toxic mechanisms leading to more-than-additive outcomes. Aquatic Toxicology, 154, 253–269.

[205] Brito, E. M., De la Cruz Barrón, M., Caretta, C. A., Goñi-Urriza, M., Andrade, L. H., Cuevas-Rodríguez, G. . . . Guyoneaud, R. (2015). Impact of hydrocarbons, PCBs and heavy metals on bacterial communities in Lerma River, Salamanca, Mexico: investigation of hydrocarbon degradation potential. Science of the Total Environment, 521, 1–10.

[206] Zhang, Z., Guo, H., Sun, J. & Wang, H. (2020). Investigation of anaerobic phenanthrene biodegradation by a highly enriched co-culture, PheN9, with nitrate as an electron acceptor. Journal of Hazardous Materials, 383, 121191.

[207] Zhang, D., Vahala, R., Wang, Y. & Smets, B. F. (2016). International Bio-deterioration & Biodegradation Microbes in biological processes for municipal landfill leachate treatment:

community, function, and interaction. International Biodeterioration & Biodegradation, 113, 88–96.

[208] Giovanella, P., Vieira, G. A. L., Ramos Otero, I. V., Pais Pellizzer, E., de Jesus Fontes, B. & Sette, L. D. (2020). Metal and organic pollutants bioremediation by extremophile microorganisms. Journal of Hazardous Materials, 382, 121024.

[209] Gupta, A., Joia, J., Sood, A., Sood, R., Sidhu, C. & Kaur, G. (2016). Microbes as a potential tool for remediation of heavy metals: a review. Journal of Microbial & Biochemical Technology, 8(4), 364–372.

[210] Cecchin, I., Reddy, K. R., Thomé, A., Tessaro, E. F. & Schnaid, F. (2017). Nanobioremediation: integration of nanoparticles and bioremediation for sustainable remediation of chlorinated organic contaminants in soils. International Biodeterioration & Biodegradation, 119, 419–428.

[211] Cole, A. J., Yang, V. C. & David, A. E. (2011). Cancer theranostics: the rise of targeted magnetic nanoparticles. Trends in Biotechnology, 29(7), 323–332.

[212] Yang, J. & Yang, Y. W. (2020). Metal-organic framework-based cancer theranostic nanoplatforms. View, 1(2), e20.

[213] Wang, X. Q., Wang, W., Peng, M. & Zhang, X. Z. (2021). Free radicals for cancer theranostics. Biomaterials, 266, 120474.

[214] Teng, L., Song, G., Liu, Y., Han, X., Li, Z., Wang, Y. . . . Tan, W. (2019). Nitric oxide-activated "dual-key–one-lock" nanoprobe for in vivo molecular imaging and high-specificity cancer therapy. Journal of the American Chemical Society, 141(34), 13572–13581.

[215] Wang, K., Zhang, F., Wei, Y., Wei, W., Jiang, L., Liu, Z. & Liu, S. (2021). In situ imaging of cellular reactive oxygen species and caspase-3 activity using a multifunctional theranostic probe for cancer diagnosis and therapy. Analytical Chemistry, 93(22), 7870–7878.

[216] Dubey, M., Nagarkoti, S., Awasthi, D., Singh, A. K., Chandra, T., Kumaravelu, J. . . . Dikshit, M. (2016). Nitric oxide-mediated apoptosis of neutrophils through caspase-8 and caspase-3-dependent mechanism. Cell Death & Disease, 7(9), e2348–e2348.

[217] Ball, K. A., Castello, P. R. & Poyton, R. O. (2011). Low-intensity light stimulates nitrite-dependent nitric oxide synthesis but not oxygen consumption by cytochrome c oxidase: implications for phototherapy. Journal of Photochemistry and Photobiology. B, Biology, 102(3), 182–191.

[218] Karu, T. I., Pyatibrat, L. V., Kolyakov, S. F. & Afanasyeva, N. I. (2005). Absorption measurements of a cell monolayer relevant to phototherapy: reduction of cytochrome c oxidase under near IR radiation. Journal of Photochemistry and Photobiology. B, Biology, 81(2), 98–106.

[219] Ball, K. A., Castello, P. R. & Poyton, R. O. (2011). Low-intensity light stimulates nitrite-dependent nitric oxide synthesis but not oxygen consumption by cytochrome c oxidase: implications for phototherapy. Journal of Photochemistry and Photobiology. B, Biology, 102(3), 182–191.

[220] To, T. L., Fadul, M. J. & Shu, X. (2014). Singlet oxygen triplet energy transfer-based imaging technology for mapping protein-protein proximity in intact cells. Nature Communications, 5(1), 1–9.

[221] Mahalingam, S. M., Ordaz, J. D. & Low, P. S. (2018). Targeting a photosensitizer to the mitochondrion enhances the potency of photodynamic therapy. ACS Omega, 3(6), 6066–6074.

[222] Zhang, C., Hong, S., Liu, M. D., Yu, W. Y., Zhang, M. K., Zhang, L. . . . Zhang, X. Z. (2020). pH-sensitive MOF integrated with glucose oxidase for glucose-responsive insulin delivery. Journal of Controlled Release, 320, 159–167.

[223] Maganti, M. (2013). Functional metal-organic frameworks are applied in cancer diagnosis and therapy, Texas A&M University-Kingsville.

[224] Zhou, Z., Vázquez-González, M. & Willner, I. (2021). Stimuli-responsive metal-organic framework nanoparticles for controlled drug delivery and medical applications. Chemical Society Reviews, 50(7), 4541–4563.

[225] Liang, J. & Liang, K. (2020). Multi-enzyme Cascade Reactions in Metal-organic Frameworks. The Chemical Record, 20(10), 1100–1116.

[226] Khan, N. A., Hasan, Z. & Jhung, S. H. (2013). Adsorption and removal of sulfur or nitrogen-containing compounds with metal-organic frameworks (MOFs). Advanced Porous Materials, 1(1), 91–102.

[227] Wang, F., Ullah, A., Fan, X., Xu, Z., Zong, R., Wang, X. & Chen, G. (2021). Delivery of nanoparticle antigens to antigen-presenting cells: from extracellular specific targeting to intracellular responsive presentation. Journal of Controlled Release, 333, 107–128.

[228] Mahnashi, M. H., Mahmoud, A. M., Alhazzani, K., Alanazi, A. Z., Alaseem, A. M., Alghatani, M. M. & El-Wekil, M. M. (2021). Ultrasensitive and selective molecularly imprinted electrochemical oxaliplatin sensor based on a novel nitrogen-doped carbon nanotubes/Ag@cu MOF as a signal enhancer and reporter nanohybrid. Microchimica Acta, 188(4), 1–12.

[229] Wang, L., Qu, X., Zhao, Y., Weng, Y., Waterhouse, G. I., Yan, H. . . . Zhou, S. (2019). Exploiting single-atom iron centers in a porphyrin-like MOF for efficient cancer phototherapy. ACS Applied Materials & Interfaces, 11(38), 35228–35237.

[230] Zhuang, W., Yuan, D., Li, J. R., Luo, Z., Zhou, H. C., Bashir, S. & Liu, J. (2012). Highly potent bactericidal activity of porous metal-organic frameworks. Advanced Healthcare Materials, 1(2), 225–238.

[231] Bruck, S. D. & Mueller, E. P. (1988). Materials and biological aspects of synthetic polymers in controlled drug release systems: problems and challenges. Critical Reviews in Therapeutic Drug Carrier Systems, 5(3), 171–187.

[232] He, L., Liu, Y., Lau, J., Fan, W., Li, Q., Zhang, C. . . . Chen, X. (2019). Recent progress in nanoscale metal-organic frameworks for drug release and cancer therapy. Nanomedicine, 14(10), 1343–1365.

[233] Gao, X., Cui, R., Zhang, M. & Liu, Z. (2017). Metal-organic framework nanosheets that exhibit pH-controlled drug release. Materials Letters, 197, 217–220.

[234] Chen, X., Tong, R., Shi, Z., Yang, B., Liu, H., Ding, S. . . . Fang, W. (2018). MOF nanoparticles with encapsulated autophagy inhibitor in controlled drug delivery system for antitumor. ACS Applied Materials & Interfaces, 10(3), 2328–2337.

[235] Jensen, S., Tan, K., Feng, L., Li, J., Zhou, H. C. & Thonhauser, T. (2020). Porous Ti-MOF-74 framework as a strong-binding nitric oxide scavenger. Journal of the American Chemical Society, 142(39), 16562–16568.

[236] Lu, K., He, C., Guo, N., Chan, C., Ni, K., Weichselbaum, R. R. & Lin, W. (2016). Chlorin-based nanoscale metal-organic framework systemically rejects colorectal cancers via synergistic photodynamic therapy and checkpoint blockade immunotherapy. Journal of the American Chemical Society, 138(38), 12502–12510.

[237] Savino, W., Mendes-da-cruz, D. A., Lepletier, A. & Dardenne, M. (2016). Hormonal control of T-cell development in health and disease. Nature Reviews. Endocrinology, 12(2), 77–89.

[238] Tan, G., Zhong, Y., Yang, L., Jiang, Y., Liu, J. & Ren, F. (2020). A multifunctional MOF-based nanohybrid as an injectable implant platform for drug synergistic oral cancer therapy. Chemical Engineering Journal, 390, 124446.

[239] Nazio, F., Bordi, M., Cianfanelli, V., Locatelli, F. & Cecconi, F. (2019). Autophagy and cancer stem cells: molecular mechanisms and therapeutic applications. Cell Death and Differentiation, 26(4), 690–702.

[240] Ni, W., Zhang, L., Zhang, H., Zhang, C., Jiang, K. & Cao, X. (2022). Hierarchical MOF-on-MOF Architecture for pH/GSH-Controlled Drug Delivery and Fe-Based Chemodynamic Therapy. Inorganic Chemistry, 61(7), 3281–3287.

[241] Malone-Povolny, M. J., Maloney, S. E. & Schoenfisch, M. H. (2019). Nitric oxide therapy for diabetic wound healing. Advanced Healthcare Materials, 8(12), 1801210.

[242] Hamblin, M. R. (2018). Mechanisms and mitochondrial redox signaling in photobiomodulation. Photochemistry and Photobiology, 94(2), 199–212.

[243] Amaroli, A., Ravera, S., Baldini, F., Benedicenti, S., Panfoli, I. & Vergani, L. (2019). Photobiomodulation with 808-nm diode laser light promotes wound healing of human endothelial cells through increased reactive oxygen species production, stimulating mitochondrial oxidative phosphorylation. Lasers in Medical Science, 34(3), 495–504.

[244] Gao, X., Cui, R., Ji, G. & Liu, Z. (2018). Size and surface controllable metal-organic frameworks (MOFs) for fluorescence imaging and cancer therapy. Nanoscale, 10(13), 6205–6211.

[245] Wang, L., Li, J., Cheng, L., Song, Y., Zeng, P. & Wen, X. (2021). Application of hard and soft acid-base theory to uncover the destructiveness of Lewis bases to UiO-66 type metal-organic frameworks in aqueous solutions. Journal of Materials Chemistry A, 9(26), 14868–14876.

[246] Prasad, A., Sedlářová, M. & Pospíšil, P. (2018). Singlet oxygen imaging using fluorescent probe Singlet Oxygen Sensor Green in photosynthetic organisms. Scientific Reports, 8(1), 1–13.

[247] Su, Y., Song, H. & Lv, Y. (2019). Recent advances in chemiluminescence for reactive oxygen species sensing and imaging analysis. Microchemical Journal, 146, 83–97.

[248] Lee, B., Hwang, J. S. & Lee, D. G. (2019). Induction of apoptosis-like death by periplanetasin-2 in Escherichia coli and contribution of SOS genes. Applied Microbiology and Biotechnology, 103(3), 1417–1427.

[249] Wei, Z., Maganti, M., Martinez, B., Vangara, K. K., Palakurthi, S., Luo, Z. . . . . Liu, J. (2016). A biological study of metal-organic frameworks towards human ovarian cancer cell lines. Canadian Journal of Chemistry, 94(4), 380–385.

[250] Deria, P., Bury, W., Hupp, J. T. & Farha, O. K. (2014). Versatile functionalization of the NU-1000 platform by solvent-assisted ligand incorporation. Chemical Communications, 50(16), 1965–1968.

[251] Guardia, P., Di Corato, R., Lartigue, L., Wilhelm, C., Espinosa, A., Garcia-Hernandez, M. . . . . Pellegrino, T. (2012). Water-soluble iron oxide nanocubes with high values of specific absorption rate for cancer cell hyperthermia treatment. ACS Nano, 6(4), 3080–3091.

[252] Kelkar, S. S. & Reineke, T. M. (2011). Theranostics: combining imaging and therapy. Bioconjugate Chemistry, 22(10), 1879–1903.

[253] Ali, A., Ovais, M., Zhou, H., Rui, Y. & Chen, C. (2021). Tailoring metal-organic framework-based nanozymes for bacterial theranostics. Biomaterials, 275, 120951.

[254] Zhao, H., Serre, C., Dumas, E. & Steunou, N. (2020). Functional MOFs as theranostics. In: Metal-Organic Frameworks for Biomedical Applications (pp. 397–423), Woodhead Publishing.

[255] Alahri, M. B., Arshadizadeh, R., Raeisi, M., Khatami, M., Sajadi, M. S., Abdelbasset, W. K. . . . Iravani, S. (2021). Theranostic applications of metal-organic frameworks (MOFs)-based materials in brain disorders: recent advances and challenges. Inorganic Chemistry Communications, 134, 108997.

[256] Zhao, S., Yu, X., Qian, Y., Chen, W. & Shen, J. (2020). Multifunctional magnetic iron oxide nanoparticles: an advanced platform for cancer theranostics. Theranostics, 10(14), 6278.

[257] Jessel, S., Sawyer, S. & Hernández, D. (2019). Energy, poverty, and health in climate change: a comprehensive review of emerging literature. Frontiers in Public Health, 357.

[258] Ackerman, F. & Stanton, E. (2013). Climate economics: the state of the art (pp. 43–58), New York: Routledge Publication.

[259] Wei, Z., Maganti, M., Martinez, B., Vangara, K. K., Palakurthi, S., Luo, Z. . . . Liu, J. (2016). A biological study of metal-organic frameworks towards human ovarian cancer cell lines. Canadian Journal of Chemistry, 94(4), 380–385.

[260] Champod, C., Lennard, C., Margot, P. & Stoilovic, M. (2004). Chapter 1: friction Ridge Skin 1–14 Fingerprints and Other Skin Ridge Impressions, Champod, C., Lennard, C., Margot, P. & Stoilovic, M. eds., New York: CRC Press, 2005.

[261] Leung, W. F., Leung, S. H., Lau, W. H. & Luk, A. (1991, September). Fingerprint recognition using a neural network. In Neural Networks for Signal Processing [1991]., Proceedings of the 1991 IEEE Workshop (pp. 226–235).

[262] Jones, N. E., Davies, L. M., Russell, C. A., Brennan, J. S. & Bramble, S. K. (2001). A systematic approach to latent fingerprint sample preparation for comparative chemical studies. Journal of Forensic Identification, 51(5), 504.

[263] Jelly, R., Patton, E. L., Lennard, C., Lewis, S. W. & Lim, K. F. (2009). The detection of latent fingermarks on porous surfaces using amino acid-sensitive reagents: a review. Analytica Chimica Acta, 652(1), 128–142.

[264] Kruse, F. A., Lefkoff, A. B., Boardman, J. W., Heidebrecht, K. B., Shapiro, A. T., Barloon, P. J. & Goetz, A. F. H. (1993). The spectral image processing system (SIPS) – interactive visualization and analysis of imaging spectrometer data. Remote Sensing of Environment, 44(2–3), 145–163.

[265] Valenta, J. (2014). Determination of absolute quantum yields of luminescing nanomaterials over a broad spectral range: from the integrating sphere theory to the correct methodology. Nanoscience Methods, 3(1), 11–27.

[266] Stryer, L. (1965). The interaction of a naphthalene dye with apomyoglobin and apohemoglobin: a fluorescent probe of non-polar binding sites. Journal of Molecular Biology, 13(2), 482–495.

[267] Yaghi, O. M., Matzger, A. J., Benin, A. & Côte, A. P. (2009). U.S. Patent No. 7,582,798. Washington, DC: U.S. Patent and Trademark Office.

[268] Ricci, C. & Kazarian, S. G. (2010). Collection and detection of latent fingermarks contaminated with cosmetics on nonporous and porous surfaces. Surface and Interface Analysis, 42(5), 386–392.

[269] Fryer, P. (1991). Civilianization within the police force – the scope for further development. Local Government Studies, 17(1), 73–83.

[270] Nagaoka, T., Nakamura, A., Okutani, H., Kiyohara, Y. & Sota, T. (2012). A possible melanoma discrimination index based on hyperspectral data: a pilot study. Skin Research and Technology, 18(3), 301–310.

[271] Castelló, A., Francés, F. & Verdú, F. (2013). Solving underwater crimes: development of latent prints made on submerged objects. Science & Justice, 53(3), 328–331.

[272] Khoobehi, B., Beach, J. M. & Kawano, H. (2004). Hyperspectral imaging for measurement of oxygen saturation in the optic nerve head. Investigative Ophthalmology & Visual Science, 45(5), 1464–1472.

[273] International Fingerprint Research Group. (2014). Guidelines for the assessment of fingermark detection techniques. Journal of Forensic Identification, 64(2), 174–200.

[274] Arthur, A. M. (1972). A new method for taking fingerprints using photographic film. American Journal of Physical Anthropology, 36(3), 441–442.

[275] Dominick, A. J., NicDaeid, N. & Bleay, S. M. (2010). The recoverability of fingerprints on paper exposed to elevated temperatures-Part 1: comparison of enhancement techniques. Journal of Forensic Identification, 59(3), 325–339.

[276] Olsen, R. D. Sr. & Lee, H. R. (2001). Chapter2: identification of Latent Prints 41–63 In Advances in Fingerprint Technology, Lee, H. C. & Ramotowski, R. eds., Second Edition, New York: CRC Press, 2001.

[277] Spreadborough, J. & Christian, J. W. (1959). High-temperature X-ray diffractometer. Journal of Scientific Instruments, 36(3), 116.

[278] Peng, Y., Ben, T., Xu, J., Xue, M., Jing, X., Deng, F. & Zhu, G. (2011). A covalently-linked microporous organic-inorganic hybrid framework contains polyhedral oligomeric silsesquioxane moieties. Dalton Transactions, 40(12), 2720–2724.

[279] Zou, Y., Park, M., Hong, S. & Lah, M. S. (2008). A designed metal-organic framework based on a metal-organic polyhedron. Chemical Communications, (20), 2340–2342.

[280] Tozawa, T., Jones, J. T., Swamy, S. I., Jiang, S., Adams, D. J., Shakespeare, S. & Tang, C. (2009). Porous organic cages. Nature Materials, 8(12), 973–978.

[281] Vangara, K. K., Liu, J. L. & Palakurthi, S. (2013). Hyaluronic acid-decorated PLGA-PEG nanoparticles for targeted delivery of SN-38 to ovarian cancer. Anticancer Research, 33(6), 2425–2434.

[282] Medina-Ramirez, I., Liu, J., Gonzalez-Garcia, M. & Palakurthi, S. (2012). Application of nanometals fabricated using green synthesis in cancer diagnosis and therapy 33–62 Green Chemistry -Environmentally Benign Approaches, Kidwai, M. Dr. ed., ISBN: 978-953-51-0334-9. Rijeka, Croatia: INTECH Publisher.

[283] Wei, Z., Maganti, M., Martinez, B., Vangara, K. K., Palakurthi, S., Luo, Z. & Liu, J. (2015). The biological study of metal-organic frameworks towards human ovarian cancer cell lines. Canadian Journal of Chemistry, 94(4), 380–385.

[284] Ertaş, E., Özer, H. & Alasalvar, C. (2007). A rapid HPLC method for determination of Sudan dyes and Para Red in red chili pepper. Food Chemistry, 105(2), 756–760.

[285] Feng, X., Pisula, W. & Müllen, K. (2009). Large polycyclic aromatic hydrocarbons: synthesis and discotic organization. Pure and Applied Chemistry, 81(12), 2203–2224.

[286] Reichardt, C. (1994). Solvatochromic dyes as solvent polarity indicators. Chemical Reviews, 94(8), 2319–2358.

[287] Zhuang, W., Yuan, D., Li, J. R., Luo, Z., Zhou, H. C., Bashir, S. & Liu, J. (2012). Highly Potent Bactericidal Activity of Porous Metal-Organic Frameworks. Advanced Healthcare Materials, 1(2), 225–238.

[288] Gordon, G. W., Berry, G., Liang, X. H., Levine, B. & Herman, B. (1998). Quantitative fluorescence resonance energy transfer measurements using fluorescence microscopy. Biophysical Journal, 74(5), 2702–2713.

[289] Lock, N., Wu, Y., Christensen, M., Cameron, L. J., Peterson, V. K., Bridgeman, A. J. . . . Iversen, B. B. (2010). Elucidating negative thermal expansion in MOF-5. The Journal of Physical Chemistry C, 114(39), 16181–16186.

[290] Furukawa, H., Kim, J., Ockwig, N. W., O'Keeffe, M. & Yaghi, O. M. (2008). Control of vertex geometry, structure dimensionality, functionality, and pore metrics in the reticular synthesis of crystalline metal-organic frameworks and Polyhedra. Journal of the American Chemical Society, 130(35), 11650–11661.

[291] Fox, C. (1973). U.S. Patent No. 3,722,750. Washington, DC: U.S. Patent and Trademark Office.

[292] Boissiere, C., Grosso, D., Chaumonnot, A., Nicole, L. & Sanchez, C. (2011). The aerosol route to functional nanostructured inorganic and hybrid porous materials. Advanced Materials, 23(5), 599–623.

[293] Han, C. C., Cheng, H. L., Lin, C. L. & Fan, K. C. (2003). Personal authentication using palmprint features. Pattern Recognition, 36(2), 371–381.

[294] Merkel, R., Gruhn, S., Dittmann, J., Vielhauer, C. & Brautigam, A. (2012). On non-invasive 2D and 3D Chromatic White Light image sensors for age determination of latent fingerprints. Forensic Science International, 222, 52–70.

[295] Menzel, E. R. & Duff, J. M. (1979). Laser detection of latent fingerprints – treatment with fluorescents. Journal of Forensic Science, 24(1), 96–100.

[296] Egmont-Petersen, M., de Ridder, D. & Handels, H. (2002). Image processing with neural networks – a review. Pattern Recognition, 35(10), 2279–2301.

[297] Gantz, D., Gantz, D., Walch, M., Roberts, M. & Buscaglia, J. (2014). A novel approach for latent print identification using accurate overlays to prioritize reference prints. Forensic Science International, 245, 162–170.

[298] Jolivot, R., Vabres, P. & Marzani, F. (2011). Reconstruction of hyperspectral cutaneous data from an artificial neural network-based multispectral imaging system. Computerized Medical Imaging and Graphics, 35(2), 85–88.

[299] Hrechak, A. K. & McHugh, J. A. (1990). Automated fingerprint recognition using structural matching. Pattern Recognition, 23(8), 893–904.

[300] Fang, L. & Gossard, D. C. (1995). Multidimensional curve fitting to unorganized data points by nonlinear minimization. Computer-Aided Design, 27(1), 48–58.

[301] Walch, M. A., Gantz, D. T. & Gantz, D. T. (2012). U.S. Patent Application No. 13/367, 153–154.

[302] Walch, M. A. (2009). U.S. Patent Application No. 12/611,893.

[303] Schwarz, L. & Klenke, I. (2007). Enhancement of Ninhydrin-or DFO-Treated Latent Fingerprints on Thermal Paper. Journal of Forensic Sciences, 52(3), 649–655.

Xuan Wang, Jingbo Louise Liu

Chapter 2B
Bottom-up synthesis of nanomaterials

A bottom-up fabrication process is a wet-chemistry approach, the opposite of top-down synthesis, which constructs nanomaterials from building blocks of atoms, molecules, and clusters. Those small units self-assemble to form monolayers on the surface of the substrate in the presence of physical or chemical triggers. The key part is to modulate the triggers that will only provide nanosized materials. Therefore, optimization of the process is a must. In this case, both the thermodynamic and kinetic aspects of the self-assembling need to be considered. The following sections present the chemistry and synthesis of nanomaterials through bottom-up approaches, such as vapor-phase deposition, coprecipitation, sol–gel synthesis, and hydrothermal and solvothermal methods.

2B.1 Vapor-phase deposition

Vapor-phase deposition is a widely used technique to coat substrate surfaces. This coating method could be further divided into physical and chemical phase deposition depending on how the vapor behaves. The physical vapor deposition (PVD) starts from condensed materials to a gas phase, allowing for the coating of the material into a thin film on top of the substrate. The most common PVD process includes thermal vaporization and sputtering. On the other hand, volatile chemicals react or decompose to produce the desired species on the surface of a substrate in chemical vapor deposition (CVD).

2B.1.1 Physical vapor deposition

The basic PVD processes involve condensing vapor-phase species onto a surface to create a nanosized thin film under a vacuum. The conventional PVD processes are further categorized into sputtering and evaporation [4].

Xuan Wang, Texas A&M Higher Education Center at McAllen, McAllen, TX, USA, e-mail: xuan.wang@science.tamu.edu
Jingbo Louise Liu, The Department of Chemistry, Texas A&M University-Kingsville, MSC161, 700 University Blvd., Kingsville, TX, USA; Texas A&M Energy Institute, 1617 Research Parkway, Suite 308, College Station, TX 77843-3372 USA, e-mail: jingbo.liu@tamuk.edu and jingbo.liu@tamu.edu

https://doi.org/10.1515/9783110739879-004

2B.1.1.1 Sputtering deposition

One widely used sputtering deposition method is done through the planar diode glow discharge (Fig. 2B.1). The basic setup is a galvanic cell consisting of the target serving as the cathode side and an anode with the subtract on the surface facing each other in a vacuum chamber. Once a high negative voltage is applied between the target and substrate, plasma is created by ionizing the working gas molecules, usually the argon, which bombards the target and sputters off-target species for deposition. The sputtered target atoms fly off in all directions, including toward the substrate, to start coating. Therefore, carefully adjusting power density at the target surface, the size of erosion surface area, the pressure of noble gas, the distance between cathode and substrate, and so on, could precisely control the deposition rate. The simplicity and ease of setup allow a broad material fabrication spectrum with the planar diode glow discharge method. The limitations are the relatively small surface area, slow deposition rate (~60 nm/min), and rotation requirement for coating uniformity. In addition, the substrate heating caused by the bombardment of higher energy may lead to cracking, sublimation, or melting, which ultimately limits the amount of power supply.

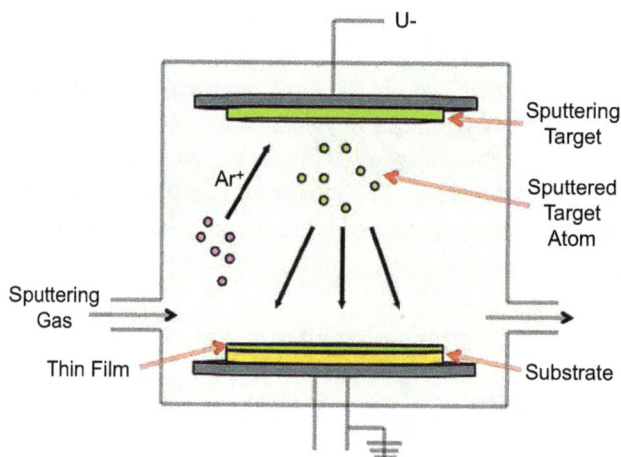

Fig. 2B.1: Representation of physical vapor deposition. Reproduced from ref. [5]. Copyright @ Sigma Aldrich 2018.

With improved sputter yield, the magnetron sputtering PVD method utilizes a magnetic field to concentrate high-energy ions on the target surface to eject more target atoms. A stronger outer field leads to unbalanced magnetic sources and more electrons escaping from cathodes in a given field line. Meanwhile, the neutral atoms could advance across the field through electron–gas collisions. As a result, the loss of electrons and ions (without collisions) is minimized. Common configurations of

PVD sputtering sources include cylindrical magnetron, planar magnetron, and S-gun magnetron sources. A long cathode is employed in the cylindrical magnetron, allowing uniform coating over a relatively large surface area. Examples of depositing metallic and dielectric films were reported with the planar magnetron sputtering sources. S-gun magnetron sputtering allows good isolation from the plasma; therefore, it could be used to deposit thermally sensitive films. Overall, the sputtering process is more suitable for ceramics and refractory metals. A simplified experimental setup is shown in Fig. 2B.2.

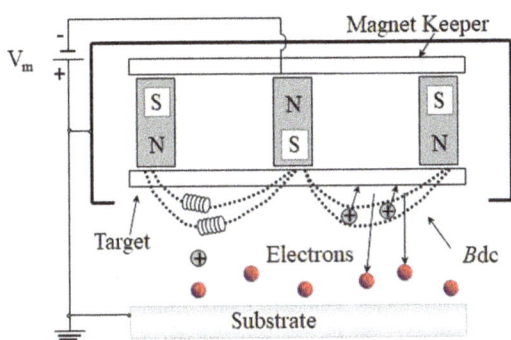

Fig. 2B.2: Representation of magnetron sputtering apparatus.

2B.1.1.2 Evaporation

In the PVD method, the ultrapure source (target) could vaporize from raising the temperature and then travel in the chamber and strike the substrate surface. Finally, those source atoms condense on the cooler substrate surface. Those evaporation methods are commonly employed to create thin metallic films. The heat to generate the vapor could be generated from serval sources - resistance, induction heating, cathodic, and anodic molecular beam sources, which can generate ions or an electron beam, using lasers to excite and generate the appropriate species. Thermal evaporation generates the right amount of heat to reach the melting point of metals from a direct current or radio frequency power supply. A tungsten boat holds and heats the metal and conducts the electricity without consuming itself. Instead of the high voltage, electron beam evaporation requires an electron gun to produce a high-energy beam of electrons. The electron beam is confined through magnetic field lines among the filament to target the metal. The source in a crucible made of tungsten, copper, or ceramic is heated up to evaporate. Benefiting from targeting only small areas in the crucible, it allows the rotating hearth of copper to hold multiple crucibles and different source materials. A water-cooling system is applied in the hearth to prevent the melting and mixing of source materials. As a result, the electron beam evaporating could coat single-component metal films and mixed layers of metals. The flash evaporation method is more applicable for targets with high vapor pressure, such as alloy, while maintaining the composition. Unlike the electron beams, the arc source evaporation processes operate at the same pressure of deposition and use a supersonic stream of fully ionized plasma from nonstationary electrodes [6]. Cathodic arc evaporation evaporates a small cathode volume, while the anodic arc evaporates the anode materials. PVD has been used in all inorganic materials and some selective organic materials compared to other synthetic methods. The entire process is still environmentally friendly. However, the pitfalls of PVD include the high cost but low yield and unpredictable coating shape [7].

2B.1.2 Chemical vapor deposition

2B.1.2.1 Fundamentals about CVD

CVD is a chemical process that produces high purity and higher performance. It involves thermal decomposition or other chemical reactions of gas-phase species (see Section 1.2.4). Unlike the PVD, the substrate is exposed to more than one volatile species on the surface in a reaction chamber. A bulk gas flow of precursors is first delivered into the reaction and then transformed into intermediates or gas by-products in the reaction zone. Another difference between PVD and CVD is the traveling path. The travel of target atoms is directional in PVD but multidirectional in CVD, so the target atoms are adsorbed on the surface of subtracts. Once more atoms diffuse to the growth area, they will start to nucleate and expand, leading the film formation. After achieving the desired size of the film, the remaining fragments will be desorbed and removed from the reaction zone [8]. Based on the surface chemistry, the growth rate could be adjusted by the temperature and concentrations of gaseous precursors, the temperature on the substrate, and so on.

Most of the chemical reactions in CVD are thermodynamically endothermic. Therefore, it requires an initiator of heat to overcome the activation energy. This energy input could be from various methods, such as directing resistance heating (thermal), thermal and photo radiation, and rf induction. Therefore, one way to categorize the CVD processes is based on the condition used to promote the reactions, pressure, heat, high-frequency radiation, or the stimulus used within the reactors, like the plasma-enhanced or plasma-assisted, oxygen-assisted, water-assisted, and photo-assisted CVDs. Researchers could carefully choose the various methods for desired features. The thermal CVD allows the creation of thin film at higher temperatures, so temperature-sensitive materials must use alternative forms of energy inputs, which allows the fabrication at low temperatures or even ambient temperatures.

Herein, the electrical energy is such a great substitution that the plasma-assisted CVD utilizes the plasma bombardment to turn the precursors into active species. The plasma-assisted CVD has been used in various precursors of inorganic

and organic materials and polymers. Another possible energy input could be provided through photons generated from UV radiation in the photon-assisted CVD procedure. A precursor is suitable for CVD if it has Aquent volatility at low and moderate
evaporation temperature, good thermal stability with high decomposition temperature, high chemical purity, low cost, and long shelf life [8]. Another way to group
CVD is based on the starting materials. To unify the way we present the exciting
research finding, we will discuss those methods based on the condition and the
stimuli of CVD for the rest of this chapter.

2B.1.2.2 Nanomaterial's synthesis

The current challenge of nanotechnology is to synthesize a large scale of nanomaterials. CVD has been approved as one of the most employed methods to synthesize
nanomaterials – graphene, nanotubes, nanofibers, fullerene, and other nanoscale
polymers [9]. We will present some examples of the CVD synthesis of graphene,
nanotubes, nanofibers, and fullerenes.

Graphene
Graphene, one single layer of graphite, is one novel 2D building block of many
other nanomaterials for its striking features of outstanding mechanical, thermal,
and electrical properties. It receives considerable attention in energy harvesting
and storage, stain sensing, and the steel industry [10]. The first successful isolation
of a single layer of graphene was observed in 2004, in which the cleaved graphite
was pressed on the oxidized SiO_2 surface [11]. After this report, research efforts focused on improving the quality of graphene sheets via electrostatic voltage-assisted
exfoliation or diffusion of graphene through solutions [12]. The current synthesis of
high-quality graphene on large-scale graphene is done through the CVD method
through two routes [13]. One route involves the deposition of a carbon-containing
gas onto the surface of a metal catalyst (Fig. 2B.3). For example, preheating gas

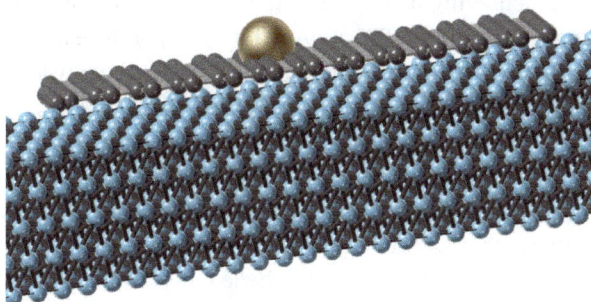

Fig. 2B.3: The schematic diagram of Ni-supported graphene using Au single atom on top as a catalyst.

mixture of H_2 and CH_4 to 1,000 °C will facilitate the growth of graphene layers onto the nickel surface before saturation with graphite formations [14]. The other route requires the carbon separation on the surface of metastable carbon–metal solid solution with the examples from transition metal of Ni, Pt, Re, Pd, Ru, and Ir [8, 15].

The carbon impurities are separated and diffused to the surface of the bulk metal during the annealing and cooling stage [15c]. In this case, the solubility of carbon in the metal–carbon solution and temperature, and other growth conditions are the controllable factors for the morphology and thickness of graphene films. Li and coworkers reported the growth of high-quality and uniform graphene films in a large area [16]. They carefully controlled the amount of carbon in a copper–carbon solution to limit the growth rate of the film and ultimately synthesized a centimeter-scale single-layer graphene film on the copper foil. Continuous attempts to utilize copper foil's flexibility successfully produce 30-inch long graphene sheets through a high-speed roll-to-roll transfer method [17].

Besides copper, nickel, another relatively inexpensive polycrystalline substrate, attracts significant attention in large-scale film deposition [15d, 15e, 16, 18]. One of the early studies synthesized 1–12 layers of graphene film on the Ni film via ambient pressure CVD [15e]. Another study also reported that single and few-layered graphene over only a few ten microns of nickel surface, [15d] which resulted from the rapid segregation of carbon from nickel carbide upon cooling and the heterogeneous grain boundaries.

Carbon isotope labeling studies reveal two metal-dependent mechanisms in the graphene growth on the metal substrates [19]. After the carbon-containing gaseous precursors decompose at high temperatures, the carbon atoms dissolve into the Ni catalyst, mix to almost saturation, then segregate and grow on the surface. Conversely, the copper catalyst has a low solubility of carbon at elevated temperatures. As a result, the dissolving step is omitted, and the carbon atoms directly segregate and grow on the copper surface [19]. Notably, the monolayer of graphene film is self-medicated due to the blockage on the surface of the copper catalyst [20]. The solubility of carbon in nickel is temperature-dependent, and it will precipitate on the surface during cooling [15e]. Therefore, cooling control is the key to regulating graphene film's growth rate and achieving a single layer. The optimal attempt of CVD synthesis at lower temperatures is around 550 °C. Any high temperature above 600 °C will induce the dissolving carbon into the metal, and a competing surface carbide phase impedes the graphene formation at lower than 500 °C [21].

Since copper has lower solubility than carbon, a synergetic combination of different solubilities between copper and nickel leads to exploring a Cu–Ni binary alloy in CVD synthesis [22]. As the percentage of Ni in the alloy increase, the thickness and uniformity of graphene layers increase. Careful change in the Ni percentage of the alloy produced over 95 % single layer and 91 % bilayer graphene films [22a].

Nanotubes

Carbon nanotubes (CNTs) are one allotrope of carbon, which results from the seamless rolling up of graphene sheets into a cylinder-like structure through van der Walls interaction [23]. Based on the number of layers, CNT could be grouped into the single-walled CNTs and the multiwalled CNTs [24], as shown in Fig. 2B.4. The multiwalled CNTs consist of spacing of ~0.34 nm between graphene layers. The single-walled CNTs could grow from a few nanometers up to 20 cm in length [25] and have a diameter between 0.4 and 3 nM [26], while the multiwalled CNTs could reach a diameter of 100 nm [27]. The CNT could also be categorized into zigzag, armchair, and chair based on the rolling angle [27b] (Fig. 2B.5).

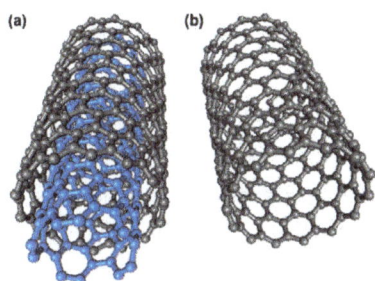

Fig. 2B.4: Structure representation of (a) single-walled and (b) multiwalled carbon nanotubes. Reproduced with permission from ref. [24]. Copyright @ American Chemical society 2009.

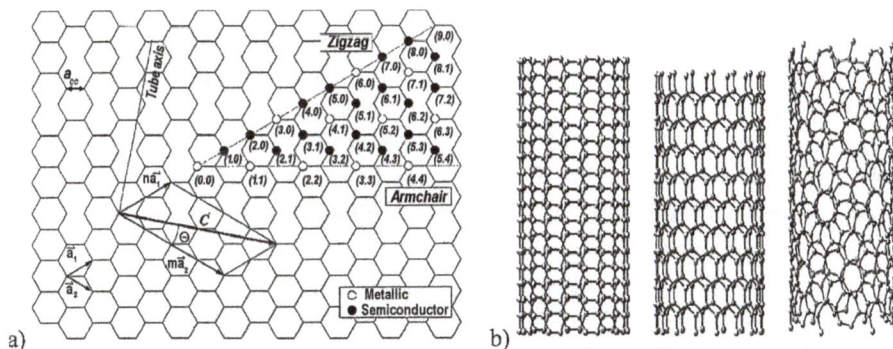

Fig. 2B.5: Use the (a) chair vector to obtain the (b) three configurations of zigzag (left), armchair (middle), and chair (right) nanotubes. Reproduced from ref. [28]. Copyright @ American Chemical Society 2007.

Besides the laser ablation and electric arc discharge, the CVD (Fig. 2B.6) synthesis of CNTs has been approved as an effective method for large-scale synthesis [29]. The high yield of CNTs requires a relatively lower temperature (500–1,000 °C). CNT's morphology and structure can be controlled by changing the alignment angle, the number of walls, the diameter, and the length [20]. Vast choices of catalysts, such as Fe, Mo, Co, Ni, and other metal-containing moieties, have been studied in the CVD synthesis of CNTs [28–30].

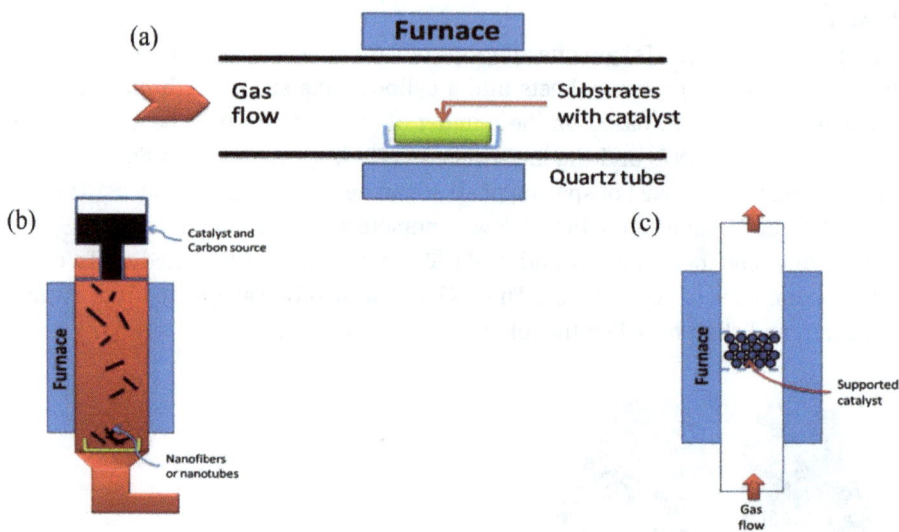

Fig. 2B.6: Three CVD methods to synthesize CNTs using (a) horizontal and (b) vertical furnaces of a fluidized bed reactor. Reproduced from ref. [31]. Copyright @ Multidisciplinary Digital Publishing Institute 2010.

The most common CVD methods are thermal CVD, plasma-assisted CVD, aerosol-assisted CVD, and catalytic pyrolysis of hydration. Early studies developed the floating catalyst CVD to synthesize the CNTs for continuous production. This method introduced the catalyst precursors and carbon feedstock gas simultaneously into the reactor with the aid of a carrier [29d, 32]. In this design, continuous production of CNT was reported by the in-situ decomposition ferrocene of the fluidized Fe/MgO, which facilitated the discharging of CNTs with the formation of small iron nanoparticles [33]. Another study also reported the utilization of small cobalt nanoparticles as catalysts to produce highly pure single-walled CNTs [34]. In this study, a small oxygen flow was introduced to the reactor to remove unreacted catalysts, leading to high-purity single-walled CNTs without serve defects. Lehman and other researchers grew the multiwalled CNTs on a pyroelectric detector in the hot wire CVD with a nickel film as the catalyst [35]. The reactor is enclosed in a clamshell furnace to achieve CNTs with good resistance to a higher temperature.

Multiple groups have also reported aerosol-assisted CVD synthesis of CNTs due to the effective dissolving of liquid hydrocarbons into the metal catalysts [31, 36]. The aerosol-assisted CVD produces free-standing vertical aligned multiwalled CNTs with precursors of benzene derivatives and other metallocenes as catalysts, which could finely tune the morphology and properties of CNTs. Similar efforts have been made by injecting a solution of ferrocene into toluene to control the catalyst to carbon ratio [37]. In this study, researchers found that the length of multiwalled CNTs increased, but the growth rate decreased with the injection time. A detailed investigation was done on synthesizing multiwalled CNTs from vast carbon sources with cobaltocene and nickelocene as the catalysts [38]. The combination of ferrocene and nickelocene yielded the maximum amount of CNTs. In addition, Hayashi and coworkers further improved the floating reactant methods using zeolite particles as the floating catalyst to synthesize the smallest free-standing single-walled CNTs [39]. Those CNTs were so small, with a diameter of ~0.43 nm, that they could be treated as a 1D material.

Carbon nanofibers

Carbon nanofibers (CNF) are cylindrical structures resulting from stacking one or double layers of graphene sheets into cones, cups, or plates. CNFs are another allotrope of carbon with a 50–200 nm diameter [40]. Compared to CNT, CNFs have low density and better mechanical properties with more defects [41]. More importantly, CNFs are cheaper to synthesize, simpler, and relatively more available. CNFs have attracted tremendous interest in fabricating composite materials [40].

The catalytic CVD growth of CNF requires converting small hydrocarbons, such as methane, carbon monoxide, ethylene, ethane, propane, acetylene, or benzene, with the help of metal catalysts such as Fe, Co, Ni, or metal alloys [42]. Like in the synthesis of CNT, CVD also demonstrates its advantages in tailoring the diameter and orientation of rolling by adjusting the synthetic conditions of CNFs. The carbon conversion from carbon feedstocks occurs at 700–1,200 K [43]. In this process, the hydrocarbon decomposes and releases carbon atoms. Then carbon atoms either dissolve in the catalyst or form metal carbides, then precipitate in the form of graphite on the metal surface. Besides the synthetic conditions, the structure of CNFs also depends on the shape of metal nanoparticles.

Fullerene

Fullerene, another allotrope of carbon, is a closed or partially closed cage connected to carbon–carbon single and double bonds. It consists of fused pentagonal and hexagonal rings [44]. The structure could be a hollow sphere, ellipsoid, tube shape, and many other forms [45]. Fullerene has demonstrated appealing structural tunability, hydrophobicity, low toxicity, and photodynamic properties [46]. Those unique features spur the studies of fullerene to focus on the antioxidizing and photosensitizing activities, drug and gene delivery, and biomarking for diagnosis [46].

The commercialization of fullerenes in medical applications suffers from the mass production of materials. Hence, the ongoing projects continuously aim to refine and improve synthetic methods and ultimately develop cost-efficiently with a high fullerene yield. Compared to the other synthetic methods via laser ablation and vaporization of graphite, CVD requires a relatively low temperature. Chow et al. obtained fullerene as a by-product of hot filament CVD synthesis of diamond [47]. The co-production of diamond and fullerene was later reported again by Keckley and coworkers [48]. They believed the formation of fullerene provided good nucleation sites, which facilitated the growth of diamonds. The same study also demonstrated fullerene synthesis through the microwave-assisted CVD method. During the synthesis, they observed the appearance of corannulene, which could be responsible for the formation of fullerene [48].

Besides the examples mentioned above about CVD synthesis of graphene, CNTs, CNFs, and fullerenes, many ongoing efforts also explore the discovery of other nanomaterials, such as nano onions, nanosheets, and nanowires [4, 20, 49]. However, mass production of high quality materials is still the major challenge to enabling the commercial use of those nanomaterials.

2B.1.3 Atomic layer deposition

So far, we have discussed the manual control of the growth of nanomaterial film on catalyst surfaces in the CVD section. Now we are looking at a more precious control of the reaction at an atomic level. Atomic layer deposition (ALD), a variant of CVD, differs in using the self-limiting chemical reaction for the film growth. Atomic layer deposition selectively introduces the precursors alternatively into the reactor until one layer of the precursors is fully saturated on the surface of subtracts [50]. Then the growth stops upon saturation. Researchers could introduce another precursor after purging the noble gas or evacuation and repeat the deposition process. The precursors do not decompose in this process but react separately on the substrate. Since the growth terminates upon saturation, there is no need to control the dose of precursors [50]. In addition, the precursors do not necessarily need to be very reactive in ALD, while CVD requires the reactive precursors as they need to react simultaneously. For the sake of high purity of film, the pulse time of the precursor needs to be long enough to reach saturation. Meanwhile, the purge time needs to be long enough to remove the excess number of previous precursors and any impurity. The overall process takes a longer time in comparison to that of CVD. Moreover, the production is highly dependent on the surface, which may lead to a low deposition rate.

ALD is a four-step process – pulse of one precursor, purge, another pulse of another precursor, and purge again. This stepwise growth allows the precious control of the thickness of each layer [51]. Early studies of ALD methods focused on preparing dense and uniform films of metal oxides, nitrides, sulfides, and some elemental films [51, 52]. The new advances lie in applying ALD as the post-synthetic modification method to change the surface properties of transition metal dichalcogenides (TMDC) and metal-organic frameworks (MOFs) [53]. This part will highlight significant achievements in the ALD-derived nanomaterials of oxides, TMDC, and MOFs.

The innovative design of ALD has been developed to overcome the issue of low deposition rate, for example, lift-off ALD, [54] spatial ALD, and [55] radical-enhanced ALD. However, the conventional ALD also requires a higher temperature, which could cause damage to the resistance layer, possible outgassing, and the shape change of the substrate [54]. As a consequence, the growth of film is disrupted. Biercuk et al. presented the lift-off method for the low-temperature ALD to overcome this issue. In this method, a resist pattern was designed on the surface before the growth of films. After the deposition, the resist layer could be removed along with that material region on the resist. Moreover, the spatial ALD method physically separates the half-reaction instead of through the use of purge steps [55b]. The spatial separation could be fulfilled using a modular rotating cylinder reactor with precursors and vacuum pump region division or a gas source with multiple slits [55b]. As a result, these spatial ALD methods shorten the reaction time, increase the deposition rate, and allow high throughput processing.

2B.1.3.1 Nanomaterials of oxides

Among all the allotropes of carbon, nanomembranes are light-weighted films with a thickness of up to 100 nm, demonstrating a large surface area, great flexibility, and appealing quantum confinement efforts [56]. Therefore, it has been applied in catalysis, sensing, biomedicine, and energy harvesting and conversion [56, 57]. In the lift-off ALD synthesis of nanomembranes, the inorganic film is deposited onto a sacrificial polymer substrate, which could be later removed by dissolving or chemical etching. The key to etching is the high selectivity between the sacrificial layer and the nanomembranes [58]. Though the limited selection of material sets makes the release of free-standing nanomembranes difficult, coupling the lift-off ALD with rolling-up techniques provides possibilities to transform the 2D nanomembranes into other 3D architectures.

In 2008, Mei and coworkers developed a versatile approach to tubular engineering structures by precisely releasing and simultaneously rolling nanomembranes on polymers [59]. The trick works with the strain gradient caused by the thermal expansion under different temperatures. During the deposition stage, the top layer of the substrate was kept at 80 °C and room temperature for the top layer. Then the researchers used acetone to dissolve the polymer, releasing the film. After rolling up those films, TiO_2, ZnO, and Al_2O_3 microtubes were successfully obtained.

Lee and coworkers chose polyvinyl alcohol as the sacrificial substrate to prepare free-floating nanosheets of TiO_2, ZnO, and Al_2O_3 [60]. In another study, polyvinyl alcohol and poly acyclic acid were used to produce TiO_2 nanomembrane [60b]. Those sacrificial layers were removed by dissolving the hot water, leaving the free-standing nanomembranes. The later study controlled the thickness of films by increasing the ALD cycles.

These two studies reveal that the sacrificial layers influence nanomembranes' size, thickness, and shape. Considering that the available surface area of planer polymers is limited, only a small amount of nanomembrane could be generated. Then increasing the surface area of polymer substrate leads to a high yield of nanomembranes. Porous sponge [61], porous carbon, [62] and other templates [63] have been reported in the ALD synthesis of nanosheets, nanomembranes, nanopillars, and nanotubes.

2B.1.3.2 Transitional metal dichalcogenides nanomaterials

TMDCs are 2D nanosheets developed for their unique photothermal conversion in the semiconductor. The transition metal could be Mo, W, and so on, while the chalcogen atom is among S, Se, or Te [64]. TMDC demonstrates high biocompatibility and bio-safety and high photothermal-conversion efficiency, so this family has been extensively studied in energy conversion, drug delivery, and photo cancer therapy [65].

Benefiting from the self-limiting feature of ALD, researchers could fabricate controllable thickness of those nanosheets, control the number of TMDC layers, manipulate the growth region, and manage the size of those TMDCs to tailor the desired properties [66]. In preparing TMDCs, one of the basic requirements for biomedical uses is the reliable synthesis of ultrathin materials. Herein, ALD is superior to conventional CVD for its precious control at an atomic level with fewer defects.

Using transition metal oxide as the predeposited thin film, researchers observed the thickness of TMDCs film increase along with the number of ALD cycles on a wafer scale [53b, 67]. In addition, successful fabrication of MoS_2, MoO_3, WO_3, Al_2O_3, TiO_2, and ZnO nanomaterials and their applications have been studied [60a, 68]. Besides optimizing ALDs and application exploration of TMDC layers, another new area of research is post-synthetic engineering to synthesize TMDC (Fig. 2B.7).

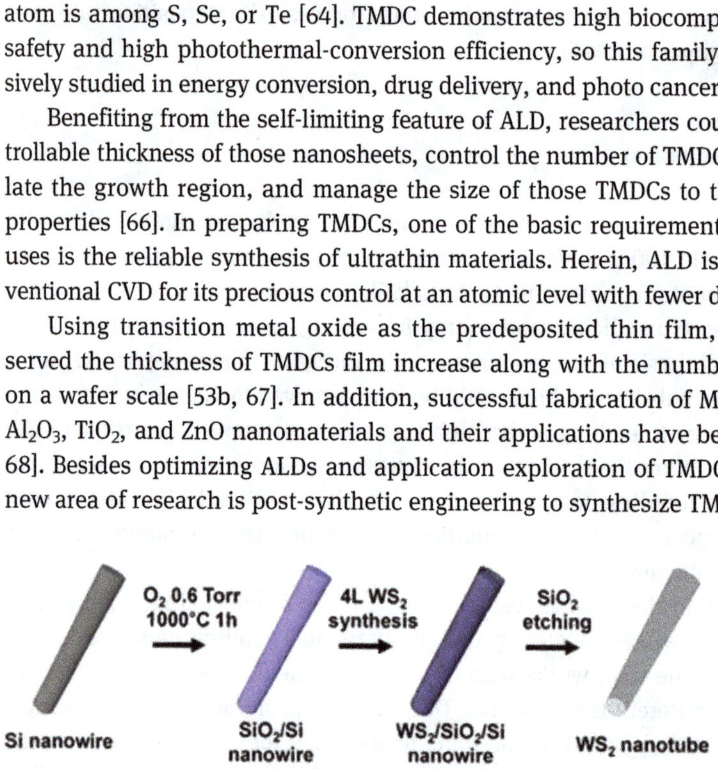

Fig. 2B.7: Synthetic scheme for the WS_2 nanotubes through ALD-guided growth. Reproduced from ref. [53b]. Copyright @ American Chemistry Society 2013.

After preparing MoO_3 film via ALD methods, Dai and coworkers induced the growth of $MoSe_2$ by following a selenization process in the furnace [69]. This ALD-induced growth of TMDC requires the preparation of one TMDC as a guide and followed by a modified synthesis through liquid or vapor post-treatment to produce another TMDC. In another study, the ALD deposited WO_3 film reacted with the gas generated from the sublimation of sulfur powder in the mix of H_2S and Ar [53b]. The resultant WS_2 nanotubes displayed a visually uniform outcome. After the applicable approval of ALD-guided growth of TMDC, other nanomaterials of MoS_2, Bi_2S_3, and $MoSe_2$ were successfully obtained using a similar strategy [53b, 68d, 68i, 69–70].

Process control in this ALD-guided growth reveals some significant findings in the growth mechanism. First, the replacement of oxides with sulfides starts around 500–700 °C, then nucleation of metal sulfide grows and completes around 900–1,000 °C [68d, 71]. The sulfur atom initiates the conversion, as evident by the decrease of ionic bonds between oxides and metals with increasing temperature when the metal–sulfur bonds start to form [72]. During the growth stage, water is critical to enhancing the transportation of metal oxides. Based on the advances in nanomaterials we surveyed above, ALD displays great appeal in the post-processing of TMDC. The resultant materials demonstrate their potential in energy storage, conservation, sensing, and catalysis, as shown in Fig. 2B.8.

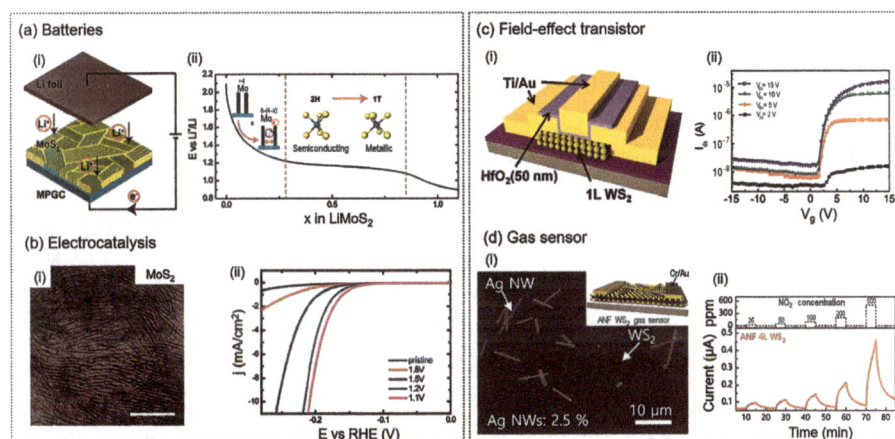

Fig. 2B.8: The application of ALD-induced growth of nanomaterials in battery, electrocatalysis, field-effect transmitter, and gas sensors. Reproduced from ref. [51]. Copyright @ American Chemical Society 2020.

2B.1.3.3 Metal-organic frameworks

Like what we discussed in the case of TMDCs, the flexible integration of layered materials through ALD has also been observed in the 3D architecture of MOFs films. Conventional synthesis of MOFs includes the solvothermal and hydrothermal reactions, which we will discuss in Section 3.4. Those synthetic methods produce three-dimensional crystals, so we need alternative ways to make MOF membranes.

The Karppinene group developed an ALD synthesis of MOF-2 thin film on various surfaces [73]. This study introduced us to metal and organic precursors in the gas phase. The thickness of MOF films also depends on the number of ALD cycles. The powder XRD patterns revealed the match of MOF films with the pristine MOF crystals.

Most MOFs take hours and days and require higher temperatures, hindering massive industrial applications. Zhao and his colleagues demonstrated a facile and ultrafast MOF synthesis at room temperature by taking advantage of the ALD method [53a]. The ALD-deposited ZnO membrane transformed into the (Zn, Cu) hydroxy double salt intermediate, which is also layered sheets connected by inorganic and organic anions [74]. The cationic nature allows a fast anion exchange in the linker solution, enabling the rapid growth of MOF HKUST-1 in the liquid within a minute [53a]. They obtained MOFs of Cu-BDC, ZIF-8, and IRMOF-3 by following a similar design.

The downside of solvent-based synthesis of MOFs is corrosion and contamination, and other surface-tension-related problems [75]. Therefore, solvent-free methods serve as an alternative to manufacturing and processing MOFs. Many recent reports discovered ways – dry synthesis or fluid-assisted synthesis – to prepare some prototypical MOFs in the presence of trace amounts of solvents and additives, or even not [75]. Those discoveries support the vapor-phase fabrication of MOFs. The first vapor-phase deposition was to synthesize ZIF-8 with the ALD-deposited ZnO film as the induce layer [76]. The vaporized 2-methyl imidazole ligand reacted with the ZnO film in the closed reactor. Though the material was partially converted, the MOF membranes turned on the selectivity of propylene from propane [77].

Those pioneered works open a new revenue of research in fabricating MOF-based nanomaterials. Other recent advances in MOFs include UiO-66, UiO-66-NH2, ZIF-67, Al-PMOF, Cu-TCPP, MAF-6, Fe-BDC, Fe-BTC, Al-BTC, and Al-BDC [78]. Those CVD and ALD-deposited MOF nanomaterials permit numerous opportunities to turn the nanostructure, composition, and associated optical and electrical features. However, the ALD synthesis methods still suffer from a low effective deposition rate, inevitable impurity in the films, and limited selections of precursors. Continuous efforts need to be made to optimize film quality, understand the growth mechanism, evolute the ALD technology, and develop complementary applications.

2B.2 Microemulsion methods–coprecipitation method

Emulsion describes a colloid mixture of two immiscible liquids, with live examples of milk, ice cream, and some cosmetic products as described in Section 1.2.2. Typically, we will see a mix of hydrocarbons, aqueous solutions, surfactants, and cosurfactants. Microemulsions are homogenous solutions consisting of a polar phase, a nonpolar phase, and a surfactant. The surfactant forms an interfacial film to separate the polar and nonpolar domains [3]. The resultant film structures include three phases – the oil droplets in the water phase (O/W), a bicontinuous sponge phase, and the water droplets in the oil phase (W/O). In general, the surfactant is the key to stabilizing the emulsion. The O/W microemulsion will form a traditional micelle, while the W/O microemulsion will form a reverse micelle. As shown below in Fig. 2B.9, the reverse micelle is a structure with entrapped aqueous phase in the core surrounded by the hydrophobic tails, which could be used as media for templates to fabricate nanoparticles.

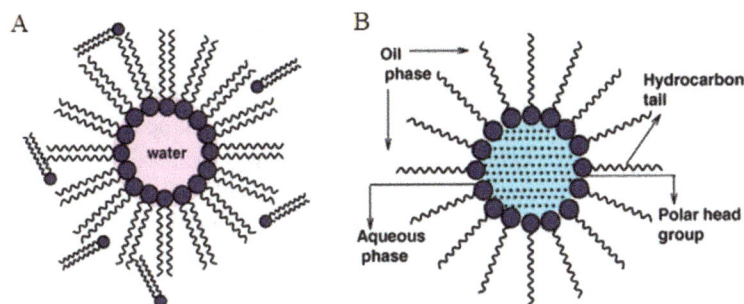

Fig. 2B.9: Microemulsion approached (A.) water-in-oil microemulsion and (B) formation of the structure of a reverse micelle. Reproduced from ref. [3]. Copyright @ Elsevier 2010.

Three involved stages in the formation of nanoparticles are the chemical reaction in the water droplets, nucleation, and growth of particles, as shown in Fig. 2B.10 [3, 79]. In the first step, reactants dissolved in the water droplets continuously collide, exchange the reactant, and form a microemulsion system. Then, the reactants in the microemulsion core react and form a rapid increase in the local concentration of clusters, which will dissolve and reform as the nucleation centers. The concentration gradually increases from saturation to supersaturation, and particles start to precipitate afterward. Those small nuclei aggregate into bigger particles and finally into the bulk-phase nanoparticles.

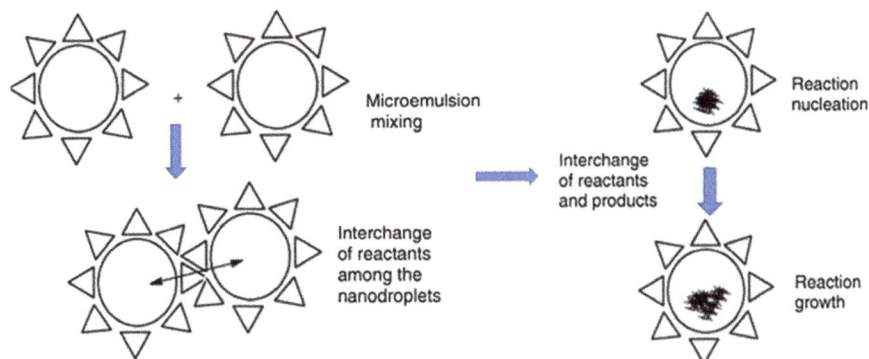

Fig. 2B.10: Growth mechanism of nanoparticles in microemulsions. Reproduced from ref. [3]. Copyright @ Elsevier 2010.

The microemulsion assisted the synthesis of both inorganic and organic nanomaterials. Inorganic nanomaterials include metals [80], metal sulfide [81], metal salts [82], and metal oxides [81a, 83], as well the integration with organic components to form nanocomposites [84]. For example, Pt, Pd, Rh, and Ir nanoparticles were prepared by reducing metal salts in the water pools of microemulsion with hydrogen and hydrazine [85]. They used two microemulsion systems of water/cetrimonium bromide (CTAB)/octanol and water/pentaethleneglycol dodecyl ether/hexane.

From the synthetic point of view, the concentration of precursors affects the extent of saturation and, ultimately, nucleation. From the perspective of microemulsion, surfactants are the best agent to shape nanomaterials by adsorbing the nucleation centers on the surface of surfactants [1]. The surfactant is a compound including a hydrophilic head and hydrophobic tail; therefore, it is an amphiphile. The hydrophilic head could be cationic, anionic, or polar, while the hydrophobic is usually a long hydrocarbon. The spectrum of surfactants ranges from proteins, carbohydrates, lipids, or surface-active polymers. The most common surfactants used in nanomaterials preparation are CTAB and sodium dodecyl sulfates. Surfactants utilize unique interfacial adsorption to construct desired architectures, evidenced by nanowires, nanoribbons, nanobelts, nanorods, nanoprisms, nanofilms, nanoflowers, and nanoleaves (Fig. 2B.11) [84b, 86]. Those surfactants provide precise shape control to obtain the desired morpholines of sustainable materials that could further be used in biomedical and pharmaceutical applications. The increasing industry demand continuously motivates future interest in microemulsion in large-scale realization.

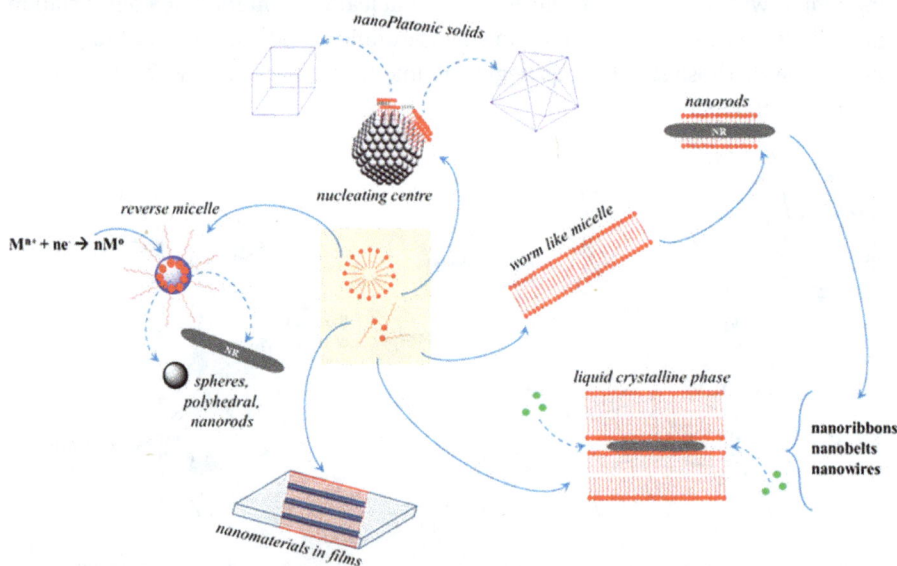

Fig. 2B.11: Surfactants shape nanostructures. Reproduced from ref. [1]. Copyright @ American Chemical Society 2015.

2B.3 Sol–gel methods

Sol–gel synthesis has been investigated for hundreds of years. This solution-based synthesis method could also shape the product with controllable particle size at low temperatures in the form of monoliths, fibers, films, and powder [87]. As demonstrated in Fig. 2B.12, there are steps – hydrolysis, condensation, and drying – to prepare nanomaterials. First, the precursors dissolve in a liquid and well-disperse in the mixture to produce *sol*. Those precursors are typically metal salts, which hydrolyze to produce metal hydroxide in the presence of water. After removing the solvent, the colloid materials become an interconnected and rigid network with pores, remnant water, and solvents, known as a *gel*. In the third stage, the drying treatment dictates the gel type obtained. Quick evaporation of solvent causes extreme shrinkages leading to a xerogel. At the same time, supercritical drying conditions induce less shrinkage with aerogel production. During the transformation, factors of the experimental condition, such as temperature, type of solvent, concentration, and pH of the solution, could be modified to tailor the desired particle size and morphology. Both the aerogel and xerogel are porous materials with nanopores, which allow foreign species to reside.

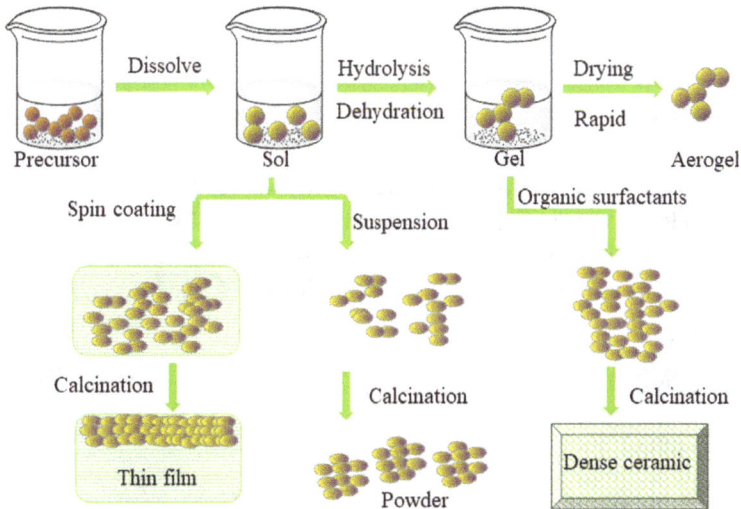

Fig. 2B.12: Scheme representation of the sol–gel synthesis of nanomaterials in different forms. Reproduced from ref. [2]. Copyright @AIP Publishing LL.C. 2015.

Using metal alkoxide as the precursor material for the sol–gel synthesis method is feasible and effective for producing various nanostructured materials. The metal alkoxide materials dissolved in organic solvent tend to form homogeneous solutions, especially when the chelating agents are introduced. Three key steps of sol–gel chemistry – hydrolysis, condensation, and gelation – are discussed in the following sections.

2B.3.1 Hydrolysis

The negatively charged hydroxyl ions become attached to the metal atoms during hydrolysis. The hydrolysis process can be expressed in the general equation (eq. (2B.1)), $M^{n+} + nH_2O \rightarrow M(OH)_n + nH^+$. The sol is formed by the following equation (eq. (2B.2)) when the metal alkoxide precursors are used, $M(OR)_4 + nH_2O \rightarrow M(OR)_{(4-n)}(OH)_n + nHOR$ [43]

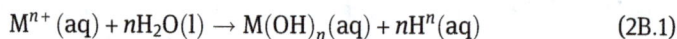

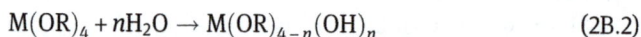

$$M^{n+}(aq) + nH_2O(l) \rightarrow M(OH)_n(aq) + nH^n(aq) \tag{2B.1}$$

$$M(OR)_4 + nH_2O \rightarrow M(OR)_{4-n}(OH)_n \tag{2B.2}$$

The metal cations can vary from alkaline earth metals to rare-earth metals, $M = Sr^{2+}$, La^{3+}, and Ti^{4+}, which can be hydrated in an aqueous solution. When a basic solution is added into the system, the reaction will be shifted forward, allowing hydrolysis completion. Based on Le Châtelier's principle, the addition of hydroxide will neutralize the proton, resulting in a forward shift of eq. (2B.1).

2B.3.2 Condensation

The polycondensation follows the hydrolysis, when the acid or base, as a homogeneous catalyst, is added to the sol system, as shown in eq. (2B.3). This condensation process allows the building up of larger molecules via oxoanion bridging and liberating small molecules (water or alcohol). The large molecules can be formed through polymerization formation of dimer, chain, and ring (as shown in eqs. (2B.4)–(2B.6)) [44]

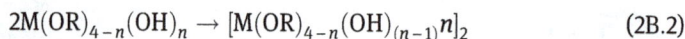

$$2M(OR)_{4-n}(OH)_n \rightarrow [M(OR)_{4-n}(OH)_{(n-1)}n]_2 \tag{2B.2}$$

Dimer formation eq. (2B.4)

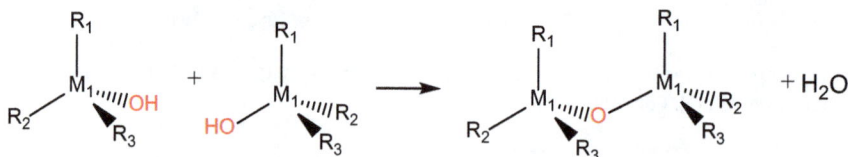

Chain formation: eq. (2B.5)

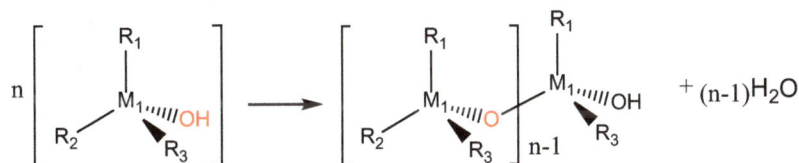

$$\left[\begin{array}{c} R_1 \\ | \\ M_{1\cdots\!/\!/\!/OH} \\ {}_{R_2}\!\!\!\nearrow\!\!\!\searrow_{R_3} \end{array}\right]_n \longrightarrow \left[\begin{array}{cc} R_1 & R_1 \\ | & | \\ M_{1\cdots\!/\!/\!/}\!-\!O\!-\!M_{1\cdots\!/\!/\!/OH} \\ {}_{R_2}\!\!\nearrow\!\!\searrow_{R_3} & {}^{R_3} \end{array}\right]_{n-1} {}^+(n\text{-}1)H_2O$$

Ring formation eq. (2B.5)

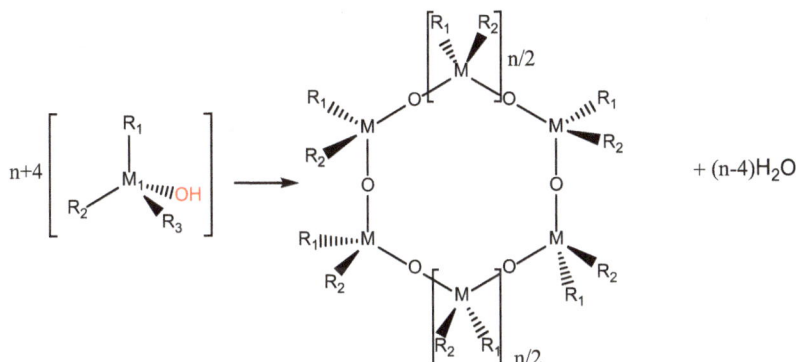

$$\left[\begin{array}{c} R_1 \\ | \\ M_{1\cdots\!/\!/\!/OH} \\ {}_{R_2}\!\!\nearrow\!\!\searrow_{R_3} \end{array}\right]_{n+4} \longrightarrow \qquad\qquad\qquad + (n\text{-}4)H_2O$$

2B.3.3 Gelation

The gelation is a spontaneous process due to the coagulation of sol precursors when the pH is increased. The water is removed through dehydration if gelation occurs under alkaline conditions. This gel reaches macroscopic dimensions by extending its networking frames throughout the solution due to the formation of giant molecules. This polymerized giant molecule is composed of a continuous skeleton in a continuous liquid phase. This continuity of solid gel structure ensures an improved elasticity of nanomaterials derived from a sol–gel method [45]. The formation of the M–O–M oxo bridges within the gel network begins and grows in three dimensions (eqs. (2B.4) and (2B.5)). The covalent bond or van der Waals forces are the major driving forces to form gelation. As a result, the chemical bonds and intermolecular forces in the gel skeleton can be permanent or reversible depending on the strength of the attractive forces. It was found that the system's viscosity increased exponentially over the gel formation and reached the maximum when the solid structure formed. Our previous experimental study on $La_xSr_{1-x}MnO_3$ (LSM) perovskite materials indicated that the unit cells' homogeneity of Sr^{2+} to La^{3+} distribution could be obtained using the sol–gel method. This homogeneity is crucial for the efficient conduction of electrons when LSM is used as a cathodic catalyst.

The versatility and low-cost nature of sol–gel methods enable a wide fabrication of metallic, organic, inorganic, and hybrid nanocomposites. Since nanoporous silica was prepared in 1995 [88], it has become one of the favorite host or templating materials for many other purposes. Later, the sol–gel-derived porous silica was developed using templating agents, such as surfactants or polymers [89]. The inherited porosity of porous silica permits the adsorption of functional organic species, giving rise to a new class of composites, the ormosils (organically modified silicate) [90]. For example, the polydimethylsiloxane and dimethyl-phenylmethyl siloxane copolymer could be incorporated into porous silica, generating the corresponding ormosils.

Ormosils could serve as the alternative matrix of the inorganic gel to develop nanomaterials. Many optically active components, like metals, oxides, ionics, and semiconductors have been prepared to prepare photo-active composites [90a, 91]. The sol–gel preparation of ferroelectric nanocrystals of LiNbO$_3$ could be achieved in both amorphous silica [92] and the ormosil matrix modified with triethoxy silypropyl dinitrophenylamine [93]. Similar fabrication has also been observed in metallic nanoparticle synthesis, Ag, Pt, Pd, and others [94]. Additionally, the nanomaterials of PbS, SbSI, CdTe, and CdS also successfully grow after the sequential heat or reduction treatment of oxides gel [95]. Besides porous silica, other porous ZrO$_2$, Al$_2$O$_3$, and SiC have also been produced via the sol–gel routes [96]. Since the development of sol–gel synthesis is relatively mature, more recent studies aim at developing innovative applications of those nanomaterials in the field of batteries [97], sensing and extraction [96b–d], photo-electrochemical conversion [97, 98], and fuel cells [99].

Overall, the sol–gel route is a versatile technique for fabricating vast numbers of nanomaterials due to the low-temperature crystallization. The sol–gel synthesis's flexibility, tunability, feasibility, and low energy consumption guarantee continuous interest in industrial realizations. The ongoing research emphasizes the advanced applications of sol–gel-derived nanomaterials, especially for sustainable energy. We will survey more cutting-edge research on energy storage and conversion, biomedical applications, and so on.

2B.4 Hydrothermal and solvothermal methods

Hydrothermal and solvothermal synthesis (as discussed in Section 1.2.3) has been highly present since its application in the mining industry in the nineteenth century [100]. They define the hydrothermal and solvothermal methods as solution-based chemical reactions in a sealed vessel. The temperature and pressures of the solvent are brought to their critical point through heating [100, 101]. Specifically, water is used as the solvent for a hydrothermal reaction, while organic solvents are the solvents for a solvothermal process, such as alcohol, ether, and other organic solvents. Water is an abundant and environment-friendly solvent with a high dielectric constant; thus, it has been the most widely used solvent in manufacturing. Organic solvents allow the reactions to occur at relatively low temperatures and pressures compared to water.

Moreover, it gives opportunities to those water-sensitive precursors to react. Such reactions may also include the precursors for the reaction, mineralizers, and any additive for growth control. Examples of mineralizers, such as HNO_3, HCl, $HCOOH$, H_2SO_4, acetic acid, $NaOH$, and KOH [101], used to control the pH of the solution and ultimately affect the phase and morphology of nanomaterials. From the safety point of view, reactions under high pressure and high temperature could be potential hazards so that they could be carried in an autoclave, a pressure vessel, or a bomb.

Like other solution-based synthetic methods, the nanomaterials are fabricated through nucleation and growth stages. When the solute exceeds the solubility limit in the solvent, it precipitates out into clusters [102]. Those clusters incorporate more units and grow into bulk crystals. Therefore, concentrations of reactants, additives, temperature, reaction time, the filling ratio, and solvents could all lead to fluctuations of solubility and ultimately shape the growth of crystals [101]. The hydrothermal and solvothermal processes emphasize the properties of solvent and heat. Properties of solvents, like the dielectric constant and density, are important. In the following, we will highlight some representative examples in the preparation of nanoparticles.

Metal oxide nanoparticles have unique optical, mechanical, thermal, conductive, and biological appeals that make them one of the most used nanomaterials in industry and research [103]. Nanoparticles of TiO_2, MnO_2, CeO_2, ZnO, ZrO_2, HfO_2, Al_2O_3, Fe_2O_3, $KNbO_3$, and many others have been fabricated through hydrothermal and solvothermal methods [104]. Moreover, hydrothermal and solvothermal reactions could generate semiconducting nanomaterials consisting of Group II to Group VI elements. The successful synthesis of ZnS, CdS, HgS, $ZnSe$, $ZnTe$, and $CdTe$ nanoparticles has also been reported [105]. The nanoparticles of nitrides, arsenides, and phosphides of elements from group III–V have also been generated for optoelectronic devices [105d, 106]. The nanoparticles of transition metal Pt and Ag synthesized through solvothermal methods displayed excellent performance in sensing, catalysis, and memory devices [107]. More recently, there have been a blooming number of researches on synthesizing advanced materials, such as MOFs, through hydrothermal and solvothermal methods [108]. Innovative interests target the enrichment of MOFs' functionality through post-synthetic modification via solvothermal processes [109].

One of the advantages of hydrothermal and solvothermal synthesis is the large-scale fabrication of nanomaterials. However, the thermal stability of organic chemicals constrains the temperature range of those syntheses. However, the increasing awareness of environmental safety and urgent need for energy utilization spur the further development of the sustainable synthetic method, like the hydrothermal and solvothermal processes. As a mature synthetic method, hydrothermal and solvothermal synthesis has established its profound status in nanomaterial science.

2B.5 Summary and outlook

The bottom-up approaches usually prepare ultrafine nanomaterials. In this case, the size of nanomaterials could be precisely controlled. Nanoparticles, nanoshells, nanotubes, and nanowires could be prejudicially designed with desire. Though those approaches are typically cheaper than the top-down approaches, it is still difficult for large-scale production. During the self-assembling process, there are many possibilities of linkage between substrates. Therefore, it is necessary to purify those nanomaterials to obtain pure nanomaterials. A continuously growing landscape of nanomaterials has been established to fulfill the rapid increase in energy demand. The selection of a suitable technique to generate a particular nanomaterial needs to be made under the consideration of all existing factors. Joint ventures on top-down and bottom-up means may eventually lead to the mass production of nanomaterials and the realization of industrial uses.

References

[1]   Bakshi, M. S. (2016). How surfactants control crystal growth of nanomaterials. Crystal Growth & Design, 16(2), 1104–1133.
[2]   Tripathi, S. K., Kaur, R., Kaur, H., Rani, M., Kaur, J., & Kaur, H. (2015, May). Fabrication and electrical characterization of memristor with TiO2 as an active layer. In AIP Conference Proceedings (Vol. 1661, No. 1, p. 110027–1 to 110027–3). AIP Publishing LLC, Melville, NY, USA.
[3]   Malik, M. A., Wani, M. Y., & Hashim, M. A. (2012). Microemulsion method: A novel route to synthesize organic and inorganic nanomaterials: 1st Nano Update. Arabian journal of Chemistry, 5(4), 397–417.
[4]   Wahl, G., Davies, P. B., Bunshah, R. F., Joyce, B. A., Bain, C. D., Wegner, G., & Edler, K. J. Ullmann's Encyclopedia of Industrial Chemistry, 2000, pp. 1–75, Wiley-VCH Verlag GmbH & Co. KGaA, Berlin, Germany.
[5]   Solution & Vapor Deposition Precursors https://www.sigmaaldrich.com/US/en/products/materials-science/energy-materials/solution-and-vapor-deposition-precursors (accessed on March 01, 2022, Sigma-Aldrich Corporation, Burlington, MA, USA).
[6]   Wang, Q. J., & Chung, Y. W. (2013). Encyclopedia of tribology, 2013th edition, pp. 1–4190, Springer, Boston, MA, USA.
[7]   Makhlouf, A. S. H., & Gajarla, Y. (2020). Advances in smart coatings for magnesium alloys and their applications in industry. In Advances in Smart Coatings and Thin Films for Future Industrial and Biomedical Engineering Applications (pp. 245–261, Editors: Abdel Salam Hamdy Makhlouf, Nedal Yusuf Abu-Thabit). Elsevier, Oxford, UK.
[8]   Luo, L., Kuzminykh, Y., Catalano, M. R., Malandrino, G., & Hoffmann, P. (2009). Optimization of Calcium Precursor Transport for High Vacuum Chemical Vapor Deposition (HVCVD). ECS Transactions, 25(8), 173–1 to 173–7.
[9]   Eatemadi, A., Daraee, H., Karimkhanloo, H., Kouhi, M., Zarghami, N., Akbarzadeh, A., Abasi, M., Hanifehpour, Y. & Joo, S. W. (2014). Carbon nanotubes: properties, synthesis, purification, and medical applications. Nanoscale research letters, 9(1), 1–13.

[10] Hatta, F. F., Mohammad Haniff, M. A. S., & Mohamed, M. A. (2022). A review on applications of graphene in triboelectric nanogenerators. International Journal of Energy Research, 46(2), 544–576.

[11] Novoselov, K. S., Geim, A. K., Morozov, S. V., Jiang, D. E., Zhang, Y., Dubonos, S. V., & Firsov, A. A. (2004). Electric field effect in atomically thin carbon films. science, 306(5696), 666–669.

[12] Liang, X., Fu, Z., & Chou, S. Y. (2007). Graphene transistors fabricated via transfer-printing in device active-areas on large wafer. Nano letters, 7(12), 3840–3844.

[13] Avouris, P., & Dimitrakopoulos, C. (2012). Graphene: synthesis and applications. Materials today, 15(3), 86–97.

[14] Angermann, H. H., & Hörz, G. (1993). Influence of sulfur on surface carbon monolayer formation and graphite growth on nickel. Applied surface science, 70, 163–168.

[15] Coraux, J., N 'Diaye, A. T., Busse, C., & Michely, T. (2008). Structural coherency of graphene on Ir (111). Nano letters, 8(2), 565–570.

[16] Li, X., Cai, W., An, J., Kim, S., Nah, J., Yang, D., & Ruoff, R. S. (2009). Large-area synthesis of high-quality and uniform graphene films on copper foils. science, 324(5932), 1312–1314.

[17] Ahn, S. H., & Guo, L. J. (2008). High-speed roll-to-roll nanoimprint lithography on flexible plastic substrates. Advanced materials, 20(11), 2044–2049.

[18] Kim, K. S., Zhao, Y., Jang, H., Lee, S. Y., Kim, J. M., Kim, K. S., & Hong, B. H. (2009). Large-scale pattern growth of graphene films for stretchable transparent electrodes. nature, 457(7230), 706–710.

[19] Li, X., Cai, W., Colombo, L., & Ruoff, R. S. (2009). Evolution of graphene growth on Ni and Cu by carbon isotope labeling. Nano letters, 9(12), 4268–4272.

[20] Manawi, Y. M., Samara, A., Al-Ansari, T., & Atieh, M. A. (2018). A review of carbon nanomaterials' synthesis via the chemical vapor deposition (CVD) method. Materials, 11(5), 822–1 to 822–36.

[21] Addou, R., Dahal, A., Sutter, P., & Batzill, M. (2012). Monolayer graphene growth on Ni (111) by low temperature chemical vapor deposition. Applied Physics Letters, 100(2), 021601–1 to 021601–3.

[22] Liu, X., Fu, L., Liu, N., Gao, T., Zhang, Y., Liao, L., & Liu, Z. (2011). Segregation growth of graphene on Cu–Ni alloy for precise layer control. The Journal of Physical Chemistry C, 115(24), 11976–11982.

[23] Liu, X., Wang, M., Zhang, S., & Pan, B. (2013). Application potential of carbon nanotubes in water treatment: a review. Journal of Environmental Sciences, 25(7), 1263–1280.

[24] Zhao, Y. L., & Stoddart, J. F. (2009). Noncovalent functionalization of single-walled carbon nanotubes. Accounts of chemical research, 42(8), 1161–1171.

[25] Zhu, H. W., Xu, C. L., Wu, D. H., Wei, B. Q., Vajtai, R., & Ajayan, P. M. (2002). Direct synthesis of long single-walled carbon nanotube strands. Science, 296(5569), 884–886.

[26] Ng, K. W., Lam, W. H., & Pichiah, S. (2013). A review on potential applications of carbon nanotubes in marine current turbines. Renewable and Sustainable Energy Reviews, 28, 331–339.

[27] Liu, W. W., Chai, S. P., Mohamed, A. R., & Hashim, U. (2014). Synthesis and characterization of graphene and carbon nanotubes: A review on the past and recent developments. Journal of Industrial and Engineering Chemistry, 20(4), 1171–1185.

[28] See, C. H., & Harris, A. T. (2007). A review of carbon nanotube synthesis via fluidized-bed chemical vapor deposition. Industrial & engineering chemistry research, 46(4), 997–1012.

[29] Andrews, R., Jacques, D., Rao, A. M., Derbyshire, F., Qian, D., Fan, X., & Chen, J. (1999). Continuous production of aligned carbon nanotubes: a step closer to commercial realization. Chemical physics letters, 303(5–6), 467–474.

[30] Peigney, A., Coquay, P., Flahaut, E., Vandenberghe, R. E., De Grave, E., & Laurent, C. (2001). A study of the formation of single-and double-walled carbon nanotubes by a CVD method. The Journal of Physical Chemistry B, 105(40), 9699–9710.

[31] Szabó, A., Perri, C., Csató, A., Giordano, G., Vuono, D., & Nagy, J. B. (2010). Synthesis methods of carbon nanotubes and related materials. Materials, 3(5), 3092–3140.

[32] Mora, E., Tokune, T., & Harutyunyan, A. R. (2007). Continuous production of single-walled carbon nanotubes using a supported floating catalyst. Carbon, 45(5), 971–977.

[33] Maghsoodi, S., Khodadadi, A., & Mortazavi, Y. (2010). A novel continuous process for synthesis of carbon nanotubes using iron floating catalyst and MgO particles for CVD of methane in a fluidized bed reactor. Applied Surface Science, 256(9), 2769–2774.

[34] Byon, H. R., Lim, H. S., Song, H. J., & Choi, H. C. (2007). A synthesis of high purity single-walled carbon nanotubes from small diameters of cobalt nanoparticles by using oxygen-assisted chemical vapor deposition process. Bulletin of the Korean Chemical Society, 28(11), 2056–2060.

[35] Lehman, J. H., Deshpande, R., Rice, P., To, B., & Dillon, A. C. (2006). Carbon multi-walled nanotubes grown by HWCVD on a pyroelectric detector. Infrared physics & technology, 47(3), 246–250.

[36] Meysami, S. S., Koos, A. A., Dillon, F., & Grobert, N. (2013). Aerosol-assisted chemical vapour deposition synthesis of multi-wall carbon nanotubes: II. An analytical study. Carbon, 58, 159–169.

[37] Singh, C., Shaffer, M., Kinloch, I., & Windle, A. (2002). Production of aligned carbon nanotubes by the CVD injection method. Physica B: Condensed Matter, 323(1-4), 339–340.

[38] Horváth, Z. E., Kertész, K., Pethő, L., Koós, A. A., Tapasztó, L., Vértesy, Z., & Biró, L. P. (2006). Inexpensive, upscalable nanotube growth methods. Current Applied Physics, 6(2), 135–140.

[39] Hayashi, T., Kim, Y. A., Matoba, T., Esaka, M., Nishimura, K., Tsukada, T., Endo, M. & Dresselhaus, M. S. (2003). Smallest freestanding single-walled carbon nanotube. Nano letters, 3(7), 887–889.

[40] Feng, L., Xie, N., & Zhong, J. (2014). Carbon nanofibers and their composites: a review of synthesizing, properties and applications. Materials, 7(5), 3919–3945.

[41] Al-Saleh, M. H., & Sundararaj, U. (2011). Review of the mechanical properties of carbon nanofiber/polymer composites. Composites Part A: Applied Science and Manufacturing, 42(12), 2126–2142.

[42] Hammel, E., Tang, X., Trampert, M., Schmitt, T., Mauthner, K., Eder, A., & Pötschke, P. (2004). Carbon nanofibers for composite applications. Carbon, 42(5–6), 1153–1158.

[43] De Jong, K. P., & Geus, J. W. (2000). Carbon nanofibers: catalytic synthesis and applications. Catalysis Reviews, 42(4), 481–510.

[44] Kroto, H. W., Heath, J. R., O'Brien, S. C., Curl, R. F., & Smalley, R. E. (1985). C60: Buckminsterfullerene. nature, 318(6042), 162–163.

[45] Yadav, B. C., & Kumar, R. (2008). Structure, properties and applications of fullerenes. International Journal of Nanotechnology and Applications, 2(1), 15–24.

[46] Bakry, R., Vallant, R. M., Najam-ul-Haq, M., Rainer, M., Szabo, Z., Huck, C. W., & Bonn, G. K. (2007). Medicinal applications of fullerenes. International journal of nanomedicine, 2(4), 639–649.

[47] Chow, L., Wang, H., Kleckley, S., Daly, T. K., & Buseck, P. R. (1995). Fullerene formation during production of chemical vapor deposited diamond. Applied physics letters, 66(4), 430–432.

[48] Kleckley, S., Wang, H., Oladeji, I., Chow, L., Daly, T. K., Buseck, P. R., & Marshall, A. Fullerenes and polymers produced by the chemical vapor deposition method (pp. 51–60)

In Synthesis and Characterization of Advanced Materials (Editors: Michael A. Serio, Dieter
M. Gruen, and Ripudaman Malhotra, 1998, Vol. 681, American Chemical Society Book
Series, Washington, DC, USA)

[49]   Kuznetsov, V. L., Chuvilin, A. L., Moroz, E. M., Kolomiichuk, V. N., Shaikhutdinov, S. K.,
Butenko, Y. V., & Mal'kov, I. Y. (1994). Effect of explosion conditions on the structure of
detonation soots: Ultradisperse diamond and onion carbon. Carbon, 32(5), 873–882.

[50]   Ritala, M., Niinisto, J (2008) Atomic Layer Deposition In Chemical vapour deposition:
precursors, processes and applications (pp. 158–206, Editors: Anthony C. Jones, and Michael
L. Hitchman), Royal Society of Chemistry, Cambridge, UK.

[51]   Zhang, Z., Zhao, Y., Zhao, Z., Huang, G., & Mei, Y. (2020). Atomic layer deposition-derived
nanomaterials: oxides, transition metal dichalcogenides, and metal–organic frameworks.
Chemistry of Materials, 32(21), 9056–9077.

[52]   Meng, X., Byun, Y. C., Kim, H. S., Lee, J. S., Lucero, A. T., Cheng, L., & Kim, J. (2016). Atomic
layer deposition of silicon nitride thin films: a review of recent progress, challenges, and
outlooks. Materials, 9(12), 1007–1 to 1007–20.

[53a] Zhao, J., Nunn, W. T., Lemaire, P. C., Lin, Y., Dickey, M. D., Oldham, C. J., & Parsons, G. N.
(2015). Facile conversion of hydroxy double salts to metal–organic frameworks using metal
oxide particles and atomic layer deposition thin-film templates. Journal of the American
Chemical Society, 137(43), 13756–13759.

[53b] Song, J. G., Park, J., Lee, W., Choi, T., Jung, H., Lee, C. W., Lee, Hwang, S.-H., Myoung, J. M.,
Jung, J.-H., Kim, S.-H., Lansalot-Matras, C. & Kim, H. (2013). Layer-controlled, wafer-scale,
and conformal synthesis of tungsten disulfide nanosheets using atomic layer deposition.
ACS nano, 7(12), 11333–11340.

[54]   Biercuk, M. J., Monsma, D. J., Marcus, C. M., Becker, J. S., & Gordon, R. G. (2003). Low-
temperature atomic-layer-deposition lift-off method for microelectronic and nanoelectronic
applications. Applied Physics Letters, 83(12), 2405–2407.

[55]   Sharma, K., Hall, R. A., & George, S. M. (2015). Spatial atomic layer deposition on flexible
substrates using a modular rotating cylinder reactor. Journal of Vacuum Science &
Technology A: Vacuum, Surfaces, and Films, 33(1), 01A132–1 to 01A132–8.

[56]   Rogers, J. A., Lagally, M. G., & Nuzzo, R. G. (2011). Synthesis, assembly and applications of
semiconductor nanomembranes. Nature, 477(7362), 45–53.

[57]   Khan, N. A., Khan, S. U., Ahmed, S., Farooqi, I. H., Dhingra, A., Hussain, A., & Changani,
F. (2019). Applications of nanotechnology in water and wastewater treatment: a review.
Asian Journal of Water, Environment and Pollution, 16(4), 81–86.

[58]   Huang, G., & Mei, Y. (2012). Thinning and shaping solid films into functional and integrative
nanomembranes. Advanced Materials, 24(19), 2517–2546.

[59]   Mei, Y., Huang, G., Solovev, A. A., Ureña, E. B., Mönch, I., Ding, F., Reindl, T., Fu, R. K. Y.,
Chu, P. K., & Schmidt, O. G. (2008). Versatile approach for integrative and functionalized
tubes by strain engineering of nanomembranes on polymers. Advanced Materials, 20(21),
4085–4090.

[60]   Lee, K., Kim, D. H., & Parsons, G. N. (2014). Free-floating synthetic nanosheets by atomic
layer deposition. ACS Applied Materials & Interfaces, 6(14), 10981–10985.

[61]   Edy, R., Huang, G., Zhao, Y., Zhang, J., Mei, Y., & Shi, J. (2016). Atomic layer deposition of
TiO2-nanomembrane-based photocatalysts with enhanced performance. AIP Advances, 6(11),
115113–1 to 115113–9.

[62]   Zhao, Y., Huang, G., Wang, D., Ma, Y., Fan, Z., Bao, Z., & Mei, Y. (2018). Sandwiched porous
C/ZnO/porous C nanosheet battery anodes with a stable solid-electrolyte interphase for fast
and long cycling. Journal of Materials Chemistry A, 6(45), 22870–22878.

[63] Ku, S. J., Jo, G. C., Bak, C. H., Kim, S. M., Shin, Y. R., Kim, K. H., Kwon, S. H., & Kim, J. B. (2013). Highly ordered freestanding titanium oxide nanotube arrays using Si-containing block copolymer lithography and atomic layer deposition. Nanotechnology, 24(8), 085301-1 to 085301-9.

[64] Xu, M., Liang, T., Shi, M., & Chen, H. (2013). Graphene-like two-dimensional materials. Chemical reviews, 113(5), 3766-3798.

[65] Ganatra, R., & Zhang, Q. (2014). Few-layer MoS2: a promising layered semiconductor. ACS nano, 8(5), 4074-4099.

[66] Hao, W., Marichy, C., & Journet, C. (2018). Atomic layer deposition of stable 2D materials. 2D Materials, 6(1), 012001-1 to 012001-45.

[67] Shi, M. L., Chen, L., Zhang, T. B., Xu, J., Zhu, H., Sun, Q. Q., & Zhang, D. W. (2017). Top-Down Integration of Molybdenum Disulfide Transistors with Wafer-Scale Uniformity and Layer Controllability. Small, 13(35), 1603157-1 to 1603157-7.

[68] Feng, Z., Kim, C. Y., Elam, J. W., Ma, Q., Zhang, Z., & Bedzyk, M. J. (2009). Direct Atomic-Scale Observation of Redox-Induced Cation Dynamics in an Oxide-Supported Monolayer Catalyst: WO x/α-Fe2O3 (0001). Journal of the American Chemical Society, 131(51), 18200-18201.

[69] Dai, T. J., Fan, X. D., Ren, Y. X., Hou, S., Zhang, Y. Y., Qian, L. X., Li, Y.-R. & Liu, X. Z. (2018). Layer-controlled synthesis of wafer-scale MoSe2 nanosheets for photodetector arrays. Journal of materials science, 53(11), 8436-8444.

[70] Liu, H. F., Antwi, K. A., Wang, Y. D., Ong, L. T., Chua, S. J., & Chi, D. Z. (2014). Atomic layer deposition of crystalline Bi 2 O 3 thin films and their conversion into Bi 2 S 3 by thermal vapor sulfurization. RSC Advances, 4(102), 58724-58731.

[71] Romanov, R. I., Kozodaev, M. G., Myakota, D. I., Chernikova, A. G., Novikov, S. M., Volkov, V. S., Slavich, A. S., Zarubin, S. S., Chizhov, P. S., Khakimov, R. R., Chouprik, A. A., Hwang, C. S. & Markeev, A. M. (2019). Synthesis of large area two-dimensional MoS2 films by sulfurization of atomic layer deposited MoO3 thin film for nanoelectronic applications. ACS Applied Nano Materials, 2(12), 7521-7531.

[72] Kastl, C., Chen, C. T., Kuykendall, T., Shevitski, B., Darlington, T. P., Borys, N. J., Krayev, A., Schuck, P. J., Aloni, S. & Schwartzberg, A. M. (2017). The important role of water in growth of monolayer transition metal dichalcogenides. 2D Materials, 4(2), 021024-1 to 021024-23.

[73] Ahvenniemi, E., & Karppinen, M. (2016). Atomic/molecular layer deposition: a direct gas-phase route to crystalline metal–organic framework thin films. Chemical Communications, 52(6), 1139-1142.

[74] Meyn, M., Beneke, K., & Lagaly, G. (1993). Anion-exchange reactions of hydroxy double salts. Inorganic Chemistry, 32(7), 1209-1215.

[75] Stassen, I., De Vos, D., & Ameloot, R. (2016). Vapor-Phase Deposition and Modification of Metal–Organic Frameworks: State-of-the-Art and Future Directions. Chemistry–A European Journal, 22(41), 14452-14460.

[76] Stassen, I., Styles, M., Grenci, G., Gorp, H. V., Vanderlinden, W., Feyter, S. D., Falcaro, P., Vos, D. D., Vereecken, P. & Ameloot, R. (2016). Chemical vapour deposition of zeolitic imidazolate framework thin films. Nature materials, 15(3), 304-310.

[77] Ma, X., Kumar, P., Mittal, N., Khlyustova, A., Daoutidis, P., Mkhoyan, K. A., & Tsapatsis, M. (2018). Zeolitic imidazolate framework membranes made by ligand-induced permselectivation. Science, 361(6406), 1008-1011.

[78] Lausund, K. B., Petrovic, V., & Nilsen, O. (2017). All-gas-phase synthesis of amino-functionalized UiO-66 thin films. Dalton Transactions, 46(48), 16983-16992.

[79] Thanh, N. T., Maclean, N., & Mahiddine, S. (2014). Mechanisms of nucleation and growth of nanoparticles in solution. Chemical reviews, 114(15), 7610-7630.

[80] Barnickel, P., & Wokaun, A. (1990). Synthesis of metal colloids in inverse microemulsions. Molecular Physics, 69(1), 1–9.

[81] Herron, N., Wang, Y., & Eckert, H. (1990). Synthesis and characterization of surface-capped, size-quantized cadmium sulfide clusters. Chemical control of cluster size. Journal of the American Chemical Society, 112(4), 1322–1326.

[82] Chew, C. H., Can, L. M., & Shah, D. O. (1990). The effect of alkanes on the formation of ultrafine silver bromide particles in ionic w/o microemulsions. Journal of Dispersion Science and Technology, 11(6), 593–609.

[83] Mozaffari, M., Hadadian, Y., Aftabi, A., & Moakhar, M. O. (2014). The effect of cobalt substitution on magnetic hardening of magnetite. Journal of Magnetism and Magnetic Materials, 354, 119–124.

[84] Chen, D. H., & Chen, C. J. (2002). Formation and characterization of Au–Ag bimetallic nanoparticles in water-in-oil microemulsions. Journal of Materials Chemistry, 12(5), 1557–1562.

[85] Boutonnet, M., Kizling, J., Stenius, P., & Maire, G. (1982). The preparation of monodisperse colloidal metal particles from microemulsions. Colloids and surfaces, 5(3), 209–225.

[86] Almora-Barrios, N., Novell-Leruth, G., Whiting, P., Liz-Marzan, L. M., & Lopez, N. (2014). Theoretical description of the role of halides, silver, and surfactants on the structure of gold nanorods. Nano letters, 14(2), 871–875.

[87] Hench, L. L., & West, J. K. (1990). The sol-gel process. Chemical reviews, 90(1), 33–72.

[88] Klein, L. C., & Woodmann, R. H. (1996). Porous silica by the sol-gel process. In Key Engineering Materials (Vol. 115, pp. 109–124). Trans Tech Publications Ltd.

[89] Baskaran, S., Liu, J., Li, X., Fryxell, G. E., Kohler, N., Coyle, C. A., Bimbaum, J. & Dunham, G. (2001). Molecular Templated Sol-Gel Synthesis of Nanoporous Dielectric Films. Ceramic Transactions, 123, 39–48.

[90] Mackenzie, J. D., Chung, Y. J., & Hu, Y. (1992). Rubbery ormosils and their applications. Journal of non-crystalline solids, 147, 271–279.

[91] Li, C. Y., Tseng, J. Y., Morita, K., Lechner, C. L., Hu, Y., & Mackenzie, J. D. (1992, December). ORMOSILS as matrices in inorganic-organic nanocomposites for various optical applications. In Sol-Gel Optics II (Vol. 1758, pp. 410–419). SPIE, Bellingham, WA USA.

[92] Mackenzie, J. D., & Bescher, E. P. (2007). Chemical routes in the synthesis of nanomaterials using the sol–gel process. Accounts of chemical research, 40(9), 810–818.

[93] Bescher, E. P. (1997). Ferroelectric-glass nanocomposites and related organically modified glass-ceramics. Ph.D Thesis (pp. 1–168). University of California, Los Angeles, CA, USA.

[94] Innocenzi, P., & Kozuka, H. (1994). Methyltriethoxysilane-derived sol-gel coatings doped with silver metal particles. Journal of Sol-Gel Science and Technology, 3(3), 229–233.

[95] Chia, C., Kao, Y. H., Xu, Y., & Mackenzie, J. D. (1997, October). Cadmium telluride quantum-dot-doped glass by the sol-gel technique. In Sol-Gel Optics IV (Vol. 3136, pp. 337–347). SPIE, Bellingham, WA USA.

[96] Carstens, S., & Enke, D. (2019). Investigation of the formation process of highly porous α-Al2O3 via citric acid-assisted sol-gel synthesis. Journal of the European Ceramic Society, 39(7), 2493–2502.

[97] Kim, S. W., Nam, K. W., Seo, D. H., Hong, J., Kim, H., Gwon, H., & Kang, K. (2012). Energy storage in composites of a redox couple host and a lithium ion host. Nano Today, 7(3), 168–173.

[98] Roy, P., Berger, S., & Schmuki, P. (2011). TiO2 nanotubes: synthesis and applications. Angewandte Chemie International Edition, 50(13), 2904–2939.

[99] Daiko, Y., Sakamoto, H., Katagiri, K., Muto, H., Sakai, M., & Matsuda, A. (2008). Deposition of ultrathin Nafion layers on sol–gel-derived phenylsilsesquioxane particles via layer-by-layer assembly. Journal of The Electrochemical Society, 155(5), B479-1 to B479-4.

[100] Byrappa, K., & Yoshimura, M. (2012). Handbook of hydrothermal technology, pp. 1–763, Elsvier, Oxford, UK.

[101] Li, J., Wu, Q., & Wu, J. (2016). Synthesis of nanoparticles via solvothermal and hydrothermal methods. In Handbook of nanoparticles (pp. 295–328, Editor: Mahmood Aliofkhazraei). Springer, Cham., Berlin, Germany.

[102] Kashchiev, D. (1982). On the relation between nucleation work, nucleus size, and nucleation rate. The Journal of Chemical Physics, 76(10), 5098–5102.

[103] Karak, N. (2019). Fundamentals of nanomaterials and polymer nanocomposites. In Nanomaterials and polymer nanocomposites (pp. 1–45, Editor: Niranjan Karak). Elsevier, Oxford, UK.

[104] Li, Y., Duan, X., Liao, H., & Qian, Y. (1998). Self-regulation synthesis of nanocrystalline ZnGa2O4 by hydrothermal reaction. Chemistry of materials, 10(1), 17–18.

[105] Biswas, S., Kar, S., & Chaudhuri, S. (2005). Optical and magnetic properties of manganese-incorporated zinc sulfide nanorods synthesized by a solvothermal process. The Journal of Physical Chemistry B, 109(37), 17526–17530.

[106] Biswas, K., Sardar, K., & Rao, C. N. R. (2006). Ferromagnetism in Mn-doped GaN nanocrystals prepared solvothermally at low temperatures. Applied Physics Letters, 89(13), 132503.

[107] Ji, W., Qi, W., Tang, S., Peng, H., & Li, S. (2015). Hydrothermal synthesis of ultrasmall Pt nanoparticles as highly active electrocatalysts for methanol oxidation. Nanomaterials, 5(4), 2203–2211.

[108] Lan, G., Ni, K., & Lin, W. (2019). Nanoscale metal–organic frameworks for phototherapy of cancer. Coordination chemistry reviews, 379, 65–81.

[109] Kaur, M., Kumar, S., Younis, S. A., Yusuf, M., Lee, J., Weon, S., Kim, K.-H. & Malik, A. K. (2021). Post-Synthesis modification of metal-organic frameworks using Schiff base complexes for various catalytic applications. Chemical Engineering Journal, 423, 130230–1 to 130230–43.

Telli Alia, Darem Sabrine, Sushesh Srivatsa Palakurthi,
Jingbo Louise Liu
Chapter 2C
Green pathways to synthesize nanomaterials

2C.1 Introduction

The increasing energy consumption, depletion of fossil energy sources, and pollution search for new and renewable energy sources with zero carbon emissions. The research for alternative energy sources required new and clean technologies allowing the generation, harvesting, conversion, and storage of energy. Nanotechnology is among the new technologies used, considered one of the paramount forefronts in science over the last decade. Green nanotechnology is a branch of green technology that utilizes the concepts of green chemistry and green engineering. Nanomaterials synthesis is performed using less hazardous, low-cost, eco-friendly approaches and biogenic formation through living organisms (plants, algae, bacteria, yeast, and fungi) or biomolecules as capping and reducing agents (phenolic compounds, polysaccharides, protein). This chapter will discuss different green synthesis methods adopted and the mechanisms of nanomaterial fabrication used for renewable energy.

Overcoming the global energy crisis and environmental pollution caused by the increasing consumption of nonrenewable energy sources prompted researchers to explore alternative clean and sustainable energy sources using new and green technologies. The

Acknowledgments: This work was supported by the Robert Welch Foundation (Departmental Grant, AC-0006) to analyze the data and write this chapter. The technical support from Texas A&M University-Kingsville and Texas A&M Energy Institute is also duly acknowledged. The authors also gratefully acknowledge the helpful discussions with colleagues, comments, and reviewers' suggestions, which have improved the presentation.

Author contribution: Telli Alia and Darem Sabrine collectively wrote the first draft. J. Liu revised the first draft and added other figures and data from S. Palakurthi, a graduate student. The final version was edited by Dr. Bashir (as co-editor) and submitted for review.

Telli Alia, Laboratoire de protection des écosystèmes en zone aride and semi-aride, Université de KASDI Merbah,BP 511 la route de Ghardaïa, Ouargla 30000, Algérie, e-mail: telli.alia@univ-ouargla.dz
Darem Sabrine, Laboratoire de sol et développement durable, Université Badji Mokhtar, PB 12, Annaba 23000, Algérie
Sushesh Srivatsa Palakurthi, The Department of Pharmaceutical Sciences, Irma Lerma Rangel College of Pharmacy, 1010 W. Avenue B., Kingsville, TX 78363, USA
Jingbo Louise Liu, The Department of Chemistry, Texas A&M University-Kingsville, MSC161, 700 University Blvd., Kingsville, TX 78363, USA; Texas A&M Energy Institute, 1617 Research Parkway, Suite 308, College Station, TX 77843-3372, USA, e-mail: jingbo.liu@tamuk.edu; and jingbo.liu@tamu.edu

https://doi.org/10.1515/9783110739879-005

use of nanotechnology to develop a suite of sustainable energy production schemes is one of the most important scientific challenges of the twenty-first century [1–71]. Green chemistry has been developed as an alternative to the utilization of environmentally harmful procedures and products because of the serious consequences that the world is facing and the limited available time to find efficient solutions [26]. Green nanotechnology is a branch of green technology that uses the notions of green chemistry and green engineering. It minimizes energy and fuel utilization by using less material and renewable inputs [72–140]. There are two basic synthetic strategies and three different techniques employed in nanotechnological synthesis. Bottom-up approaches for synthesizing nanostructured materials include physical methods such as physical vapor deposition, chemical vapor deposition, molecular beam epitaxy, pulsed laser deposition, atomic layer deposition, ion implantation, spray pyrolysis [6], sputter deposition, and electric arc deposition [137]. Coprecipitation, sol–gel, chemical reduction, sono- and photochemical, electrochemical, microemulsion, solvothermal, and surface-derived methods are used in the bottom-up chemical approach [7, 100]. The natural way of a bottom-up approach, a green synthesis technique of nanoparticles (NPs), includes the synthesis from plants, bacteria, fungi, yeast, algae, biomolecules, and agricultural and industrial wastes. The NP synthesis method is crucial because the synthesis procedure influences the morphology, size, shape, and functionality of the obtained NPs. Additionally, other parameters need to be taken into account, particularly the consequences of the chosen method on the environment.

The conventional synthesis methods involve the use of hazardous chemicals, the generation of toxic products, consumption of energy, and less biocompatibility leading to damaging impacts on living organisms and the environment [47, 30, 129]. The negative effects of traditional methods are undeniable and need to be surmounted. "Green" synthesis has gained extensive attention as a reliable, sustainable, and eco-friendly protocol for synthesizing many materials/nanomaterials (NMs), including metal/metal oxides, NMs, hybrid materials, and bioinspired materials [126]. Nowadays, it has been proved that green synthesis of NPs is not harmful to the environment, is low cost, and is less toxic when compared to other conventional synthesis methods [10]. The capping, reducing, and reaction solvents are considered important for the NPs synthesis via the green approaches [30]. The green synthesis of nanoparticles (NPs) utilizing environmentally friendly and cost-effective reducing and stabilizing materials from plants, microbes, and other natural resources without using toxic chemicals reduces health and environmental risks at the source level [47]. The final goal of green design aims to reduce the utilization or generation of hazardous substances. Green synthesis has attracted attention due to its easy availability, wide distribution of plants, microorganisms, and biomolecules, great diversity, and safe use. The 12 principles "demonstrate the breadth of the concept of green chemistry," listed below (Fig. 2C.1). In this context, this chapter provides an overview of NM's green synthesis and its usage in renewable energy [91].

The 12 principles of green chemistry, as articulated by Kharissova et al. [62], encompass a circular economy principle, where no hazardous byproducts are introduced, and whatever by-products are generated are utilized. These principles (1) prevent waste (leave no waste to clean up), (2) maximize atom economy (waste few or no atoms), (3) design less hazardous chemical syntheses, (4) design safer chemicals and products, (5) use safer solvents and reaction conditions, (6) increase energy efficiency, (7) use renewable feedstocks, (8) avoid chemical derivatives, (9) use catalysts, not stoichiometric reagents, (10) design chemicals and products to degrade after use, (11) analyze in real-time to prevent pollution, and (12) minimize the potential for accidents.

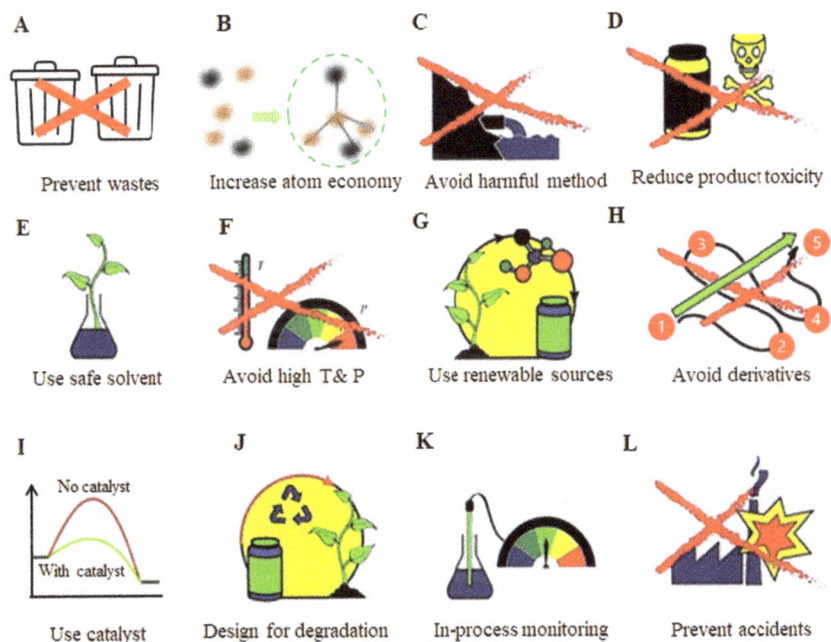

A Prevent wastes

B Increase atom economy

C Avoid harmful method

D Reduce product toxicity

E Use safe solvent

F Avoid high T& P

G Use renewable sources

H Avoid derivatives

I Use catalyst

J Design for degradation

K In-process monitoring

L Prevent accidents

Fig. 2C.1: The 12 principles in green synthesis guide the procedures and productions.

2C.2 Properties and importance of nanomaterials

Nanotechnology can measure, see, manipulate, and fabricate materials on the nanoscale (0–100 nm), at least from one dimension. These ultrafine products also have a large surface area-to-volume ratio, which is their most important characteristic responsible for the broad worldwide utilization of NMs in mechanics, optics, electronics, biotechnology, microbiology, environmental remediation, medicine, numerous engineering fields, and material science [118]. At this nanoscale, materials can exhibit unique, extraordinary properties absent or limited in their bulk (see Chapter 1). These properties include ultrahigh surface area, electro-optical properties, eco-friendliness, high mechanical strength, and advanced physical properties. The new features of nanomaterials intrigued the engineering and characterization of nanomaterials to explore their fine structure and further overcome the obstacles to their industrial adaption. The synthesis methods, characterization techniques, quantitation techniques, impact of NPs, different types, separation techniques, and applications of NPs are schematically depicted (Fig. 2C.2).

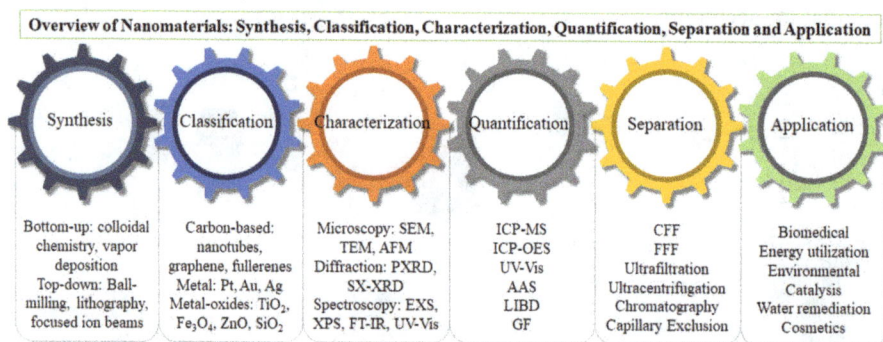

Overview of Nanomaterials: Synthesis, Classification, Characterization, Quantification, Separation and Application

Synthesis	Classification	Characterization	Quantification	Separation	Application
Bottom-up: colloidal chemistry, vapor deposition Top-down: Ball-milling, lithography, focused ion beams	Carbon-based: nanotubes, graphene, fullerenes Metal: Pt, Au, Ag Metal-oxides: TiO_2, Fe_3O_4, ZnO, SiO_2	Microscopy: SEM, TEM, AFM Diffraction: PXRD, SX-XRD Spectroscopy: EXS, XPS, FT-IR, UV-Vis	ICP-MS ICP-OES UV-Vis AAS LIBD GF	CFF FFF Ultrafiltration Ultracentrifugation Chromatography Capillary Exclusion	Biomedical Energy utilization Environmental Catalysis Water remediation Cosmetics

Fig. 2C.2: Overview of nanoparticles, synthesis, classification, characterization, quantification, separation, and application.

The nanomaterials are divided into organic, inorganic, ceramic, and carbon-based NPs with different crystallographic structures, summarized in Tab. 2C.1. The inorganic NPs are classified into metal (e.g., gold) and metal oxide (e.g., ZrO_2) NPs [11]. Similarly, carbon base NPs classified into fullerene, carbon nanotubes, graphene, carbon nanofiber, and carbon black [49]. Khan et al. [61] reported that NPs could broadly be divided into various categories depending on their morphology, size, and chemical properties on carbon-based NPs, metal NPs, ceramics NPs, semiconductor NPs, polymeric NPs, and lipid-based NPs. However, Pandey and Prajapati [98] reported that NPs are classified depending on their chemical composition into four major classes, carbon-based (nanotubes and nanofibers of carbon, etc.), metal and metal oxide-based (Ag or Cu), and bio-organic based (liposomes, micelles, etc.), and composite-based. In particular, Mr. Palarguthi systematically investigated NP synthesis from plants, bacteria, yeast, fungi, algae, and viruses. The plant extracts include *Pelargonium graveolens* (geranium leaves) extract used in the extracellular synthesis of silver NPs by reducing Ag^+ ions. Highly stable crystalline NPs of silver were synthesized using geranium leaf extract [106]. Terpenoids present in the plant extract were responsible for reducing Ag^+ ions, which FTIR determined. *Cymbopogon flexuosus* (lemongrass) plant extract was used to synthesize gold NPs, where reduction of $AuCl_4^-$ ion occurs in a single step at room temperature. The large surface area/volume ratio gives the NPs exceptional physicochemical properties (mechanically strong, optically active, and chemically reactive). It makes them powerful tools in various fields (medicine, food industry, cosmetics, energy generation, storage, etc.) [61].

Table 2C.1: Summary of representative nanostructured materials and their properties.

Carbon-based nanomaterials

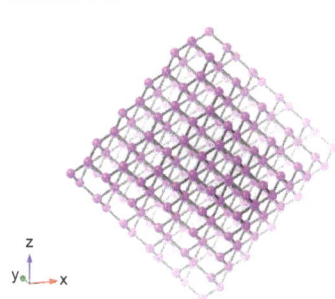

Diamond: the high-pressure form of carbon with an octahedral crystal shape [148]
Space group: *Fd3m* [origin 1]
Origin shift: ¼, ¼, ¼
Crystal system: cubic
a: 3.5597 Å
Cell volume: 45.107 Å^3
Asymmetric unit: 1 site
Unit cell: 8 sites/unit cell
Density: 3.5371 g/cm^3

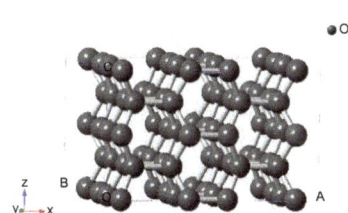

Carbon BCT: intermediate phase between graphite and diamond [154]
Space Group: *I4/mmm*
Crystal System: Tetragonal
a: 4.3220 Å
c: 2.4780 Å
Cell volume: 46.288 Å^3
Asymmetric unit: 1 site
Unit cell: 8 sites/unit cell
Density: 3.4468 g/cm^3

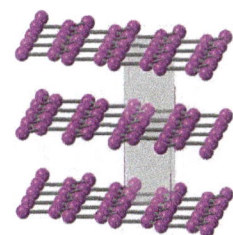

Graphite (the low-pressure/high-temperature polymorph). Three sp^2 bonds form strong networks in the basal (001) plane
[23]
Space group: *P3* (chirality; polar)
Crystal system: trigonal
a: 2.4560 Å
c: 6.6960 Å
Cell volume: 34.979 Å^3
Asymmetric unit: 4 sites
Unit cell: 4 sites/unit cell
Density: 2.2806 g/cm^3

Table 2C.1 (continued)

Carbon-based nanomaterials

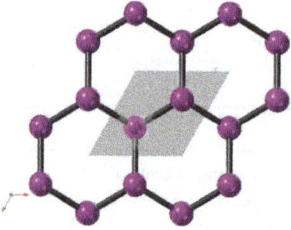

●C

Lonsdaleite – hexagonal carbon
Found in microscopic crystals [42]
Space Group: *P* 63/*mmc*
Crystal System: hexagonal
a: 2.5200 Å
c: 4.1200 Å
Cell volume: 22.658 Å³
Asymmetric unit: 1 site
Unit cell: 4 sites/unit cell
Density: 3.5207 g/cm³

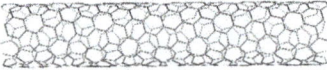

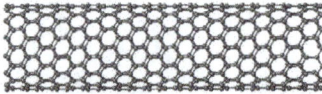

Carbon nanotube, an idealized structure
for buckytube [133].

●c

A buckminsterfullerene (C₆₀) molecule [67]

Metal

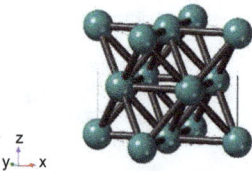

●Ni

Nickel [57]
Space group: *Fm* 3̄*m*
Crystal system: Cubic
a: 3.5168 Å
Cell volume: 43.495 Å³
Asymmetric unit: 1 site
Unit cell: 4 sites/unit cell
Density: 8.9657 g/cm³

Table 2C.1 (continued)

Carbon-based nanomaterials

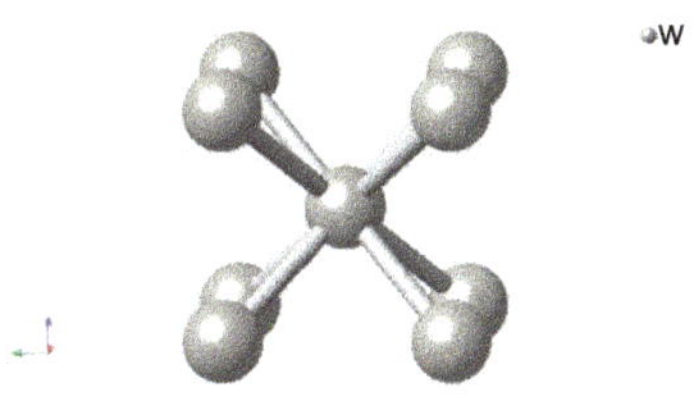

Tungsten (W) [83]
The cubic form of tungsten has a bond-centered cubic structure [147]
Space group: $I\,1$ (chirality; polar)
Crystal system: triclinic
a = b = c: 3.1647 Å
$\alpha = \beta = \gamma$: 90.000°
Cell volume: 31.695 Å^3
Asymmetric unit: 1 site
Unit cell: 2 sites/unit cell
Density: 19.2696 g/cm^3

Metal oxides

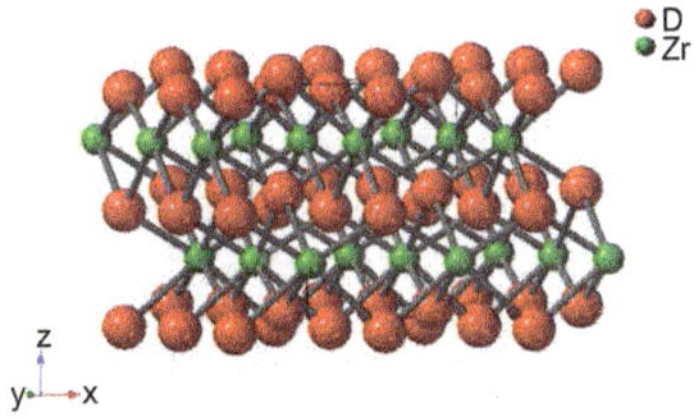

Tetragonal zirconia (ZrO_2)
Space group: $P\,4_2/nmc$ [origin 2]
Crystal system: tetragonal
a: 3.6008 Å
c: 5.1793 Å
Cell volume: 67.154 Å^3
Asymmetric unit: 2 sites
Unit cell: 6 sites/unit cell
Density: 6.0939 g/cm^3

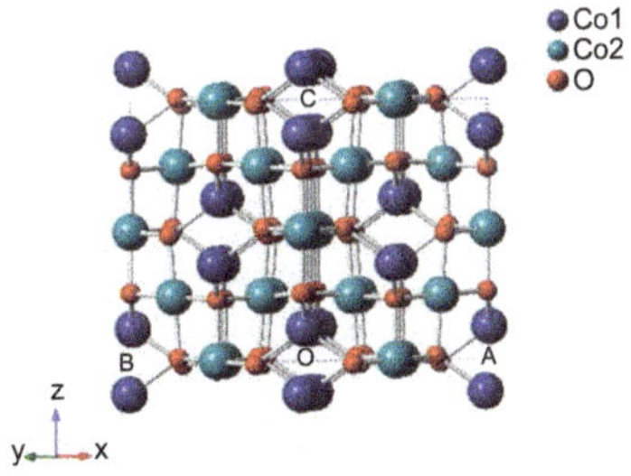

Cobalt oxide (Co_3O_4)
A normal spinel structure at room temperature (measured at 301 K) [72].
Space group: $Fd\,\bar{3}m$ [origin 2]
Crystal system: cubic
a: 8.0821 Å
Cell volume: 527.926 Å^3
Asymmetric unit: 3 sites
Unit cell: 56 sites/unit cell
Density: 6.0591 g/cm^3

Metal boride

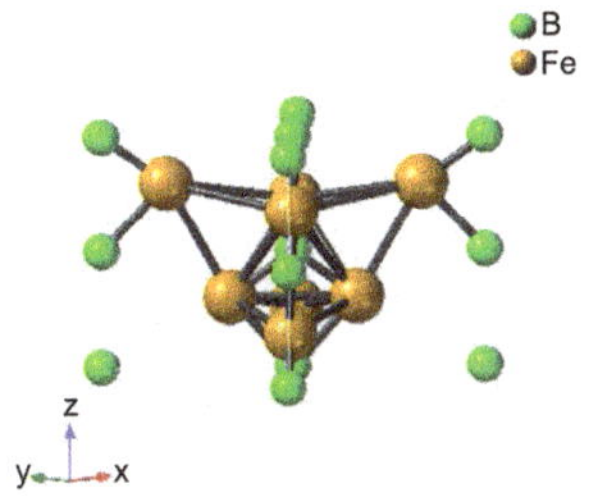

Iron boride (Fe_2B) [146].
Space group: $I\bar{4}2\,m$
Crystal system: tetragonal
a: 5.0780(50) Å
c: 4.2230(50) Å
Cell volume: 108.895 Å^3
Asymmetric unit: 3 sites
Unit cell: 12 sites/unit cell
Density: 7.4727 g/cm^3

Table 2C.1 (continued)

Carbon-based nanomaterials

Metal nitride

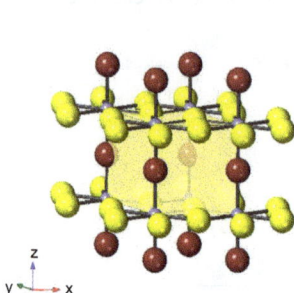

● Li1
● Li2
● N

Lithium nitride (Li_3N) [60]
N ions are [8] coordinated by Li, which occupies two sites in this structure.
Space group: *P6/mmm*
Crystal system: hexagonal
a: 3.6520(30) Å
c: 3.8660(20) Å
Cell volume: 44.653 Å^3
Asymmetric unit: 3 sites
Unit cell: 4 sites/unit cell
Density: 1.2954 g/cm^3

Silicate

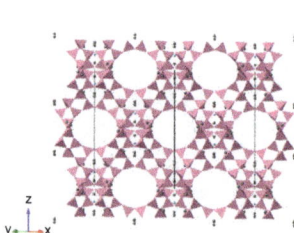

● Al
▲ Si
● Zn

Zeolite X ($Zn_{54.976}Al_{95.898}Si_{104.0}O_{416.06}$)
Based on the faujasite structure comprising a framework of corner-sharing SiO_4 tetrahedra, forming large, 12-fold rings along the [110] direction [16]
Space group: $Fd\bar{3}m$ [origin 2]
Crystal system: cubic
a: 24.7180 Å
Cell volume: 15102.192 Å^3
Asymmetric unit: 10 sites
Unit cell: 872 sites/unit cell
0.0444 atoms/Å^3
Density: 1.7329 g/cm^3

Metal hydroxide

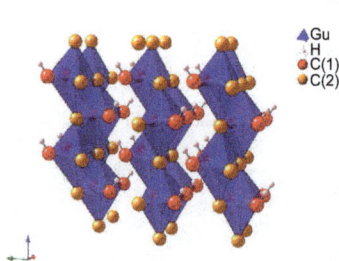

▲ Gu
· H
● C(1)
● C(2)

Copper hydroxide – $Cu(OH)_2$
Copper atoms are octahedrally coordinated by O and O–H groups. The octahedra share edges arranged in slabs parallel to the crystallographic *z*-axis [94]
Space group: $Cmc2_1$ (polar)
Crystal system: orthorhombic
a: 2.9470 Å
b: 10.5930 Å
c: 5.2560 Å
Cell volume: 164.080 Å^3
Asymmetric unit: 4 sites
Unit cell: 16 sites/unit cell
Density: 3.9088 g/cm^3

2C.3 Renewable energy and nanotechnology

Renewable energy utilizes energy sources continually replenished by nature, the sun, the wind, water, the Earth's heat, and plants [141–156]. Renewable energy technologies turn these fuels into usable forms of energy – most often, electricity, heat, chemicals, or mechanical power [157], [89]. The last decades have seen an increase in the use of NMs in the renewable energy sector due to their efficiency in converting and storing energy, reduced cost, and being environmentally friendly. Nanotechnology has been found to improve energy-conversion efficiency when used in photovoltaic cells and dye-sensitized solar cells. Wind energy harvesting can be enhanced using miniaturized nanogenerators composed of polyurethane and ZnO NPs. Biofuels were generated using nanocatalysts when fuel cell technology, carbon capture, and conversion significantly benefited from nanomaterials as catalysts. In the traditional oil and natural gas fields, nanomaterials have been used to inhibit damage formation and improve oil recovery. Figure 2C.3 summarized nanomaterial's application in energy systems, focusing on sustainable energy storage and conversion.

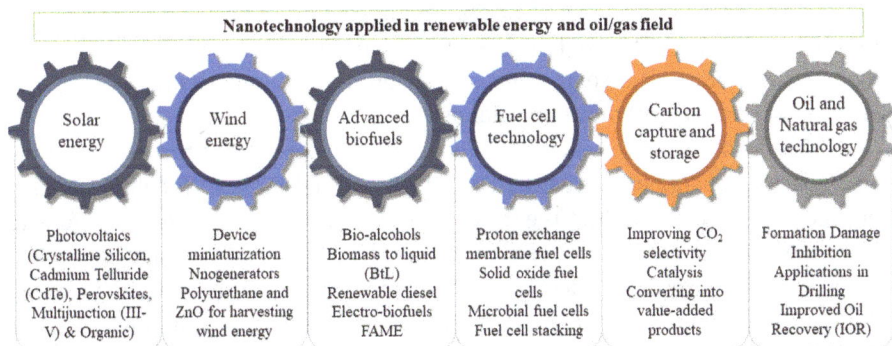

Nanotechnology applied in renewable energy and oil/gas field

Solar energy	Wind energy	Advanced biofuels	Fuel cell technology	Carbon capture and storage	Oil and Natural gas technology
Photovoltaics (Crystalline Silicon, Cadmium Telluride (CdTe), Perovskites, Multijunction (III-V) & Organic)	Device miniaturization Nnogenerators Polyurethane and ZnO for harvesting wind energy	Bio-alcohols Biomass to liquid (BtL) Renewable diesel Electro-biofuels FAME	Proton exchange membrane fuel cells Solid oxide fuel cells Microbial fuel cells Fuel cell stacking	Improving CO_2 selectivity Catalysis Converting into value-added products	Formation Damage Inhibition Applications in Drilling Improved Oil Recovery (IOR)

Fig. 2C.3: A summary of nanomaterial's application energy system focuses on sustainable energy conversion.

Solar energy is one of the best renewable energy sources available. The conversion of sunlight into electricity is done either by photovoltaics or concentrated solar power. A solar cell is a semiconductor electron that transforms sunlight energy directly into electricity through its photovoltaic influence. These conventional solar cells have two main problems: they can only achieve efficiencies of around 10 % and are expensive to manufacture [55]. However, solar technology's levelized cost is not yet economically competitive with conventional fossil fuel technologies without subsidies. They identify and develop practically efficient technologies to yield and store solar energy. Nanotechnology can be utilized as an alternative for producing, storing, and converting solar energy through the safest and cleanest means [97]. The application of nanotechnology in solar cells has opened the path to developing a new generation of high-performance products. Several works of two-dimensional (2D) layered materials are applied in Perovskite solar cells (PSCs) owing to their particular chemical and physical characteristics, like high carrier mobility and tunable bandgap, which highly determines the perovskite film growth kinetics, carrier transfer, and stability of PSCs [152]. The ever-increasing power conversion efficiency (PCE), low-cost materials constituents, and simple solution fabrication process ensured the future leading role of PSCs.

The nanomaterials possess desirable properties such as high catalytic activity, better stability in aqueous media, comparatively easier preparation techniques, and material economy. Still, NMs suffer from some drawbacks when utilized in photocatalytic and photoelectrochemical devices [2]. In their review, Moore and Wei [82] mentioned that PSCs based on metal halides (e.g., absorber material, methylammonium lead halides $CH_3NH_3PbX_3$) are rapidly emerging as the most promising and competing perovskite technology because of their high record power conversion efficiencies and potentially low production costs. Nanofluids exhibit improved heat absorbing, and transportability, credited to NPs suspended in base fluids. The main factor responsible for enhanced heat transfer and absorption ability of nanofluids is the multiplication of the surface-to-volume ratio of NPs [142]. Since the initiation of PSCs in 2009 with an efficiency of 3.8 %, PSCs have now achieved a PCE of 23.3 % on the lab scale [114]. This high efficiency rivaled the performance of commercial polycrystalline silicon (Si) solar cells, copper indium gallium selenide (CIGS), and cadmium telluride (CdTe) thin-film solar cells.

The US Department of Energy Solar Energy Technologies Office has supported the R&D projects, focusing on increasing the "efficiency and lifetime of hybrid organic-inorganic perovskite solar cells." The research and development suggested the efforts to address the challenges in stability, efficiency, manufacturing, validation, and bankability (Fig. 2C.4). The materials architecture, stability, and characterization are the starting points for future industry adaptation, in Fig. 2C.4A, methylammonium lead halides ($CH_3NH_3PbX_3$, MALHs) were chosen due to their high PCE (>19 %). This family of halides is solid compounds with perovskite structure, with a space group: $P1$ (allows chirality; polar), crystal system: triclinic; a: 17.6784 Å, b: 8.8392 Å, c: 12.6948 Å, $\alpha = \beta = \gamma$: 90.000°, cell volume: 1,983.726 Å^3. A simplified schematic of a tandem PSC solar cell is shown in Fig. 2C.4B. Figure 2C.4C shows the performance evaluation of "a V-shaped perovskite/silicon tandem device based on a bifacial heterojunction silicon cell." The plots in (a) and (b) are the current (J)-voltage (V) "J–V curves of the front side and rear side of the [intrinsic thin layer (HIT) silicon solar cell (SSC)] HIT SSC under illumination with different light intensities," respectively. Figure 2C.4C (c) shows the "wavelength-dependent external quantum efficiencies and integrated current densities of the HIT SSC." The VOC versus light intensity curves of the HIT SSC is measured in Fig.2C.4C (d). As of January 26, 2022, the efficiency records for perovskite PV cells, the current records were "25.7 % for single-junction perovskite devices," and "29.8 % for tandem perovskite-silicon devices," compared to other PV technologies as shown in Fig. 2C.4D (c) [136].

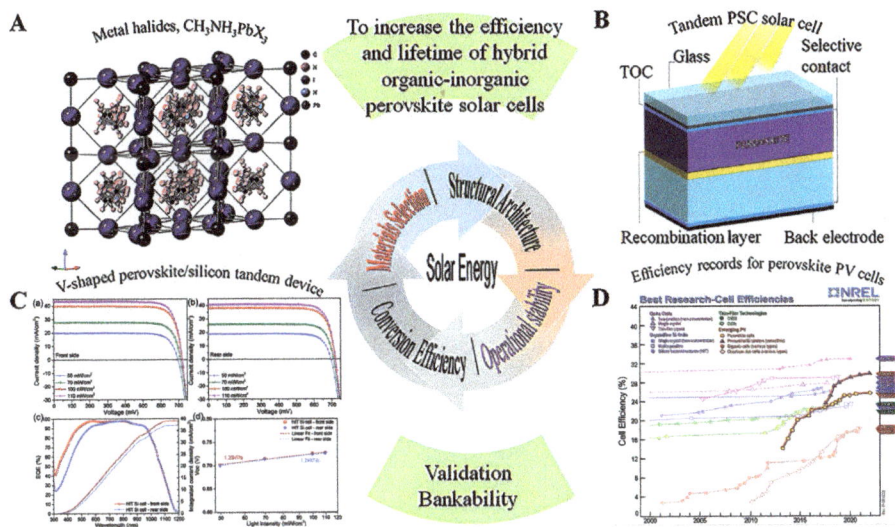

Fig. 2C.4: The research and development suggested the efforts to address the challenges in stability, efficiency, manufacturing, validation, and bankability. Plots from Fig. 2C.4C are adapted with permission from Zheng and Xuan [151], and Fig. 2C.4D from (NREL).

Wind energy is associated with the movement of air masses from areas of high atmospheric pressure to adjacent low-pressure areas, with velocities proportional to the pressure gradient. During the daytime, the air masses over the oceans, seas, and lakes remain cool compared to neighboring masses situated over land areas. Wind power, for now, is a trustworthy and founded technology that can generate electricity at a competitive cost. Alternative energy, such as nuclear, has been installed for approximately 10 years since 2005 [4]. The wind turbine generates electricity using the aerodynamic force from the rotor blades. The electricity will be further distributed to the end-users (Fig. 2C.5A). The main components of the turbine are the generator, tower, gearbox, and blades. Among all the components, the blade is considered the costliest component. Advanced nanotechnology has been employed in developing cost-effective materials with a higher strength-to-mass ratio [87]. This team found that the power eco-efficiency can be improved 2.8 times using a "cross axis wind turbine integrated with a 45° deflector" compared to the "vertical axis wind turbine." The research data also indicated that the rotor rotational speed could be increased by 70 %. To increase the wind energy harvesting efficiency, Li et al. proposed "triboelectric-electromagnetic flexible cooperation." This strategy successfully "combines the advantages of the triboelectric nanogenerator and electromagnetic generator " to adopt different wind speeds. The energy-harvesting capacity can be adjusted according to variables of wind speed, allowing for adaptation to the instable of natural wind.

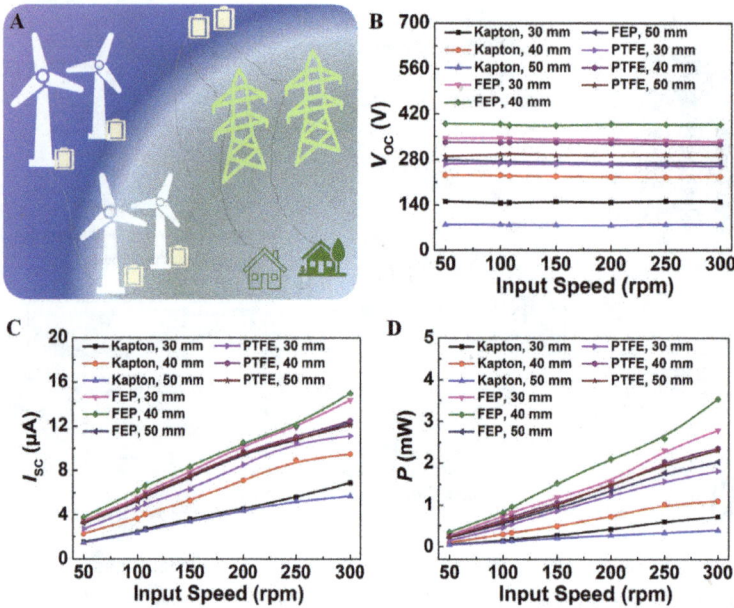

Fig. 2C.5: The nanotechnology used to enhance wind energy harvesting and output ability of the triboelectric nanogenerator (TENG) module with different materials and different lengths of dielectric layers (40 mm): (A) the electricity generation by a wind turbine and distribution of electricity, (B) the open-circuit voltage of TENG, (C) the short-circuit current, and (D) the average output power. Images (B)–(D) were adapted with permission from Li et al. [73].

Hydrogen (H_2) is an ideal alternative energy source from the point of view of ecology and the environment. Hydrogen is one of the most abundant elements in the universe (also see hydrogen production) [158], and it is stored in water (H_2O), hydrocarbons (e.g., methane, CH_4), and other organic matter. Since the 1990s, the United States has started to promote hydrogen fuel cell vehicles (FCV) by putting forward legislation [25]. In 2014, Toyota Mirai showcased the mid-size FCV in Los Angeles, which intrigued the global annual sales of FCV increases [149]. However, the hydrogen refueling station and availability (Fig. 2C.6) hindered the utilization of HFCV and its further commercialization. By May 2020, some "operating hydrogenation fueling stations" were established in several provinces of China, where hydrogen availability is foreseen to be increased shortly [76]. Hydrogen technology has been developed in other countries, including England, France, and Canada, to achieve carbon neutrality. The failure in nuclear energy management is a driving force for Japan to advance clean energy sources with an attempt at hydrogen techniques. The Japanese government issued the "Basic Hydrogen Energy Strategy" in December 2017, intending to promote hydrogen energy application [50]. This strategy includes ambitious goals to further spread about "800,000 FCVs by 2030, 160 hydrogen refueling stations by 2020, and 320 by 2025." Currently, Japan has become the first country to publish a "policy of subsidy for hydrogen vehicles" [135]. Chen et al. [29] applied multiple datasets for "data fusion and computation" to achieve two goals, "computation and analysis on the hotspot of hydrogen refueling station" and "location of hydrogen depot," as shown in Fig. 2C.7. 5. Separate datasets were used: (1) government policy, (2) massive GPS trajectory dataset, (3) geographical land use data, (4) geographical road network data, and (5) position of points-of-interest (POI) data [150].

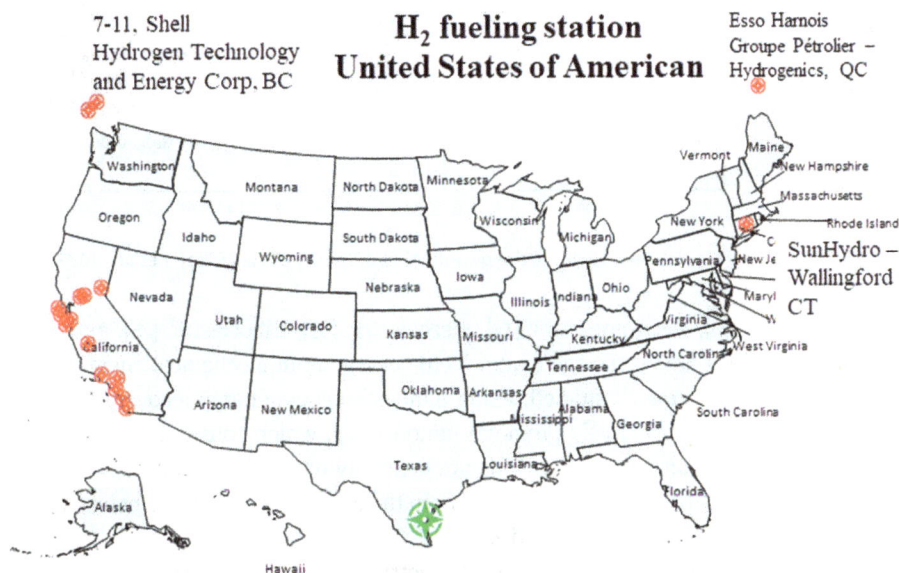

Fig. 2C.6: The hydrogen fueling stations in the United States and Canada [90].

Hydrogen becomes attractive due to the high energy density, in which heat produced per unit mass of hydrogen (heating value of 141.8 MJ/kg at 298 K) is three times more than gasoline. Moreover, it is a clean and renewable energy source with zero CO_2 emissions [145]. Around 120 million tons of hydrogen are produced each year (2/3 of the produced quantity is pure hydrogen). Around 95 % of all hydrogen is generated from natural gas and coal. Around 5 % is generated from chlorine production through electrolysis [52, 93]. A wide range of resources is available for hydrogen production, mainly fossil-based and renewable fuels [93]. The photoelectrochemical water splitting and photocatalytic hydrogen production, solid-state hydrogen storage, and proton exchange membrane fuel cells are fabricated based on different NMs [75].

The energy can also be extracted from the ambient environment. Human motion, water flow, wind, and so on are among various mechanical energy resources present in our surroundings, which are being wasted and can be efficiently converted into electrical energy to drive different practical and functional devices [111]. Numerous methods

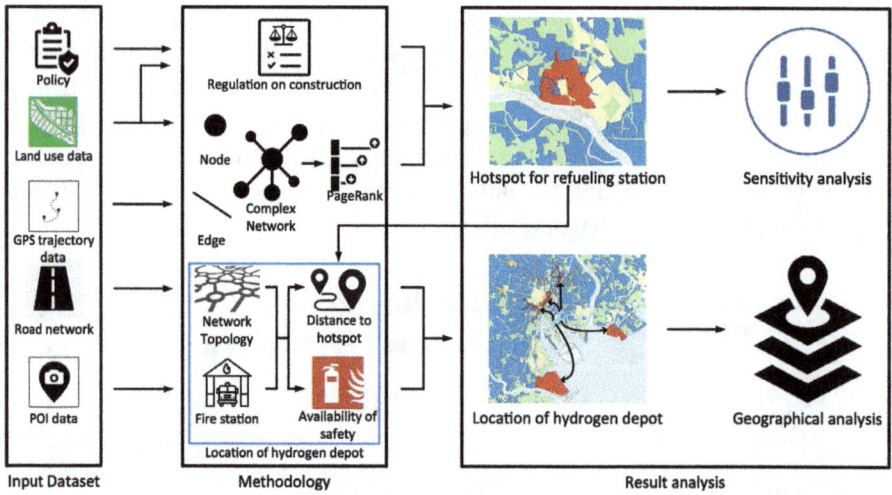

Fig. 2C.7: Five separate datasets to guide the priority of hydrogen facilities (Zhang et al., 2022).

like electrostatic [9], electromagnetic [13], piezoelectric [33], triboelectric [28], and pyro-electric [134] effects are used to generate electric energy from mechanical energy found in the environment are well studied. Using nanotechnology, energy recovery from the ambient environment effectively promotes nanodevices, which consume very little energy. The nanogenerators based on NMs are generally energy-harvesting devices that generate electricity from mechanical energy from the ambient to create self-powered systems [130], described in Chapters 1 and 2.

Microorganisms are used to produce electric energy without relying on noble metal catalysts. Microbial fuel cells (MFCs) have recently emerged as a green technology for direct electricity generation from polluted water loaded with organic and inorganic contaminants. The major setback for this technology's commercialization is its capacity to generate limited power. The efficiency loss during the scale-up procedure, higher cost of proton permeable membrane, and the internal resistances in a reactor are thought to be the hindering factors behind the marketing of this technology. Engineered NMs, which are sustainable and economical, can promote electron transfer and proton exchange in MFCs [12, 58]. It has been proven that the coating of electrode surfaces by NPs/NMs enhances power production in MFC [12]. Biofuel cells (BFCs) are devices made to transform the chemical energy of organic matter into electrical energy utilizing metabolic reactions occurring in microorganisms during the degradation of organic contaminants [157–159]. Bashir et al. recently presented their research data (Fig. 2C.8) on using the brewer yeast *Saccharomyces cerevisiae* as a biocatalyst. In this study, a novel proton-exchange membrane composed of polyvinyl alcohol as the matrix was prepared by a wet-chemistry sol–gel method, followed by crosslinking by tetraethoxysilane and sulfosuccinic acid. These Polyvinyl alcohol-tetraethoxysilane-sulfosuccinic acid (PTS) membranes displayed a greater solution and similar dry conductance to those values of the commercially available membrane, Nafion. The MFC devices were assembled and tested, demonstrating a 300 % increase in power density. The microbe *S. cerevisiae* was used as a biocatalyst to construct the MFC devices. The methyl blue (MB) "mediation," RL exposure of 2.88 J cm^{-2}, and graphene quantum dots anchoring at the anode collectively improved the MFC electrochemical performance. These results suggest that MFCs showed great promise as a platform for sustainable energy generation with lower costs and emissions than traditional energy production.

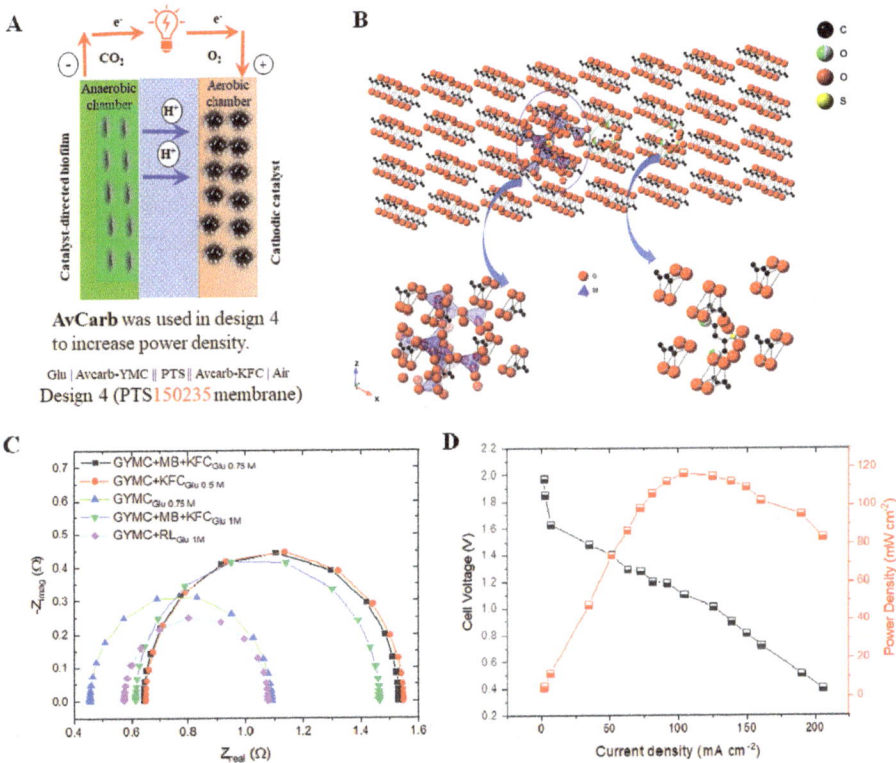

Fig. 2C.8: The performance of single microbial fuel cells (MFCs) with different catalysts using PTS as membrane: (A) the MFC configuration using AvCarb as the gas diffusion layer, (B) the PTS is composed of polyvinyl alcohol as the matrix, tetraethoxysilane, and sulfosuccinic acid as crosslinkers, (C) electrochemical impedance spectroscopic analysis, and (D) the cyclic voltammetry evaluation of the MFCs devices. Permission from Mulvaney et al. [20].

In Summary, the energy, food, and water resources are interdependent. The energy–food–water (EWF) nexus presents a new perspective on the inextricable link between these resources as actions on one side affect the others. NMs may help enhance the performance of the EWF nexus in regions with a disproportionate distribution of resources, especially arid regions [92].

2C.4 Green methods of synthesis

The two approaches for NPs synthesis are commonly known as the bottom-up approach, which uses physical processes, and the top-down approach, which is done with the help of chemical and biological processes [1]. Capping agents play a very crucial and versatile role in NP fabrication. NPs can be functionalized and stabilized using capping agents to give beneficial properties by controlling morphology and size and protecting the surface, preventing aggregation [123]. In addition, NMs synthesis can be subdivided into two main categories: traditional methods and green methods. Traditional synthesis methods usually affect the environment and human health due to contaminant reagents and the generation of byproducts. NMs, when manufactured using green synthesis methods, are especially desirable, being devoid of harsh operating conditions (high temperature and pressure), pH, hazardous chemicals, or external stabilizing or capping agents [39].

The synthesis procedure, experimental conditions, and interactions with reducing and capping affect the properties of NPs. The size, shape, and distribution of NPs can be controlled by changing the synthesis method and the types of reducing and stabilizing agents [38]. To improve the proprieties and control the morphology of NPs, sol–gel-based green synthesis can be implemented as a primary synthetic method assisted by other methods, particularly physical, like microwave, ultrasound, and pulsed laser ablation [38, 70, 107]. Several techniques are employed for characterizing produced NPs: light scattering, scanning electron microscope, energy dispersive spectroscopy, UV-Vis spectroscopy, X-ray diffraction, Fourier transform infrared spectroscopy, surface-enhanced remain spectroscopy, atomic force microscopy, high angle annular dark field, atomic absorption spectroscopy, and ray photoelectron spectroscopy [49].

2C.4.1 Physical and chemical methods

Some physical and chemical methods of NPs synthesis are improved to make them safer, less harmful, and economical. Microwaves are electromagnetic waves consisting of a kind of pure energy radiated in the form of a wave propagating at the speed of light. **Microwave propagation** usually happens at lower speeds within condensed matter than in air or vacuum, where the light speed is lesser due to refraction in a media with a higher refractive index. Adjusting power and time is necessary because they influence NPs synthesis by microwave. Microwaves can also assist with hydrothermal synthesis apparatus. The most commonly used microwave frequencies for NP synthesis range from 2–45 GHz, where all-dielectric parameters strongly depend on temperature [35]. Dr. Liu and her team produced metal-organic framework motifs using microwave-assisted solvothermal with high efficiency and well-distinctive structures (Fig. 2C.9A–C). *Pulsed laser ablation in liquids (PLAL)* allows the fabrication of NPs with no byproducts. In this method, irradiation of a target material with an ultra-short laser pulse leads to the formation of high-temperature plasma and removal of material, which has been termed pulsed laser ablation. With this clean and versatile technique, high-purity NP can be synthesized well suited for functionalization [122, 125]. Dell'Aglio et al. [36] reported the production of AgNPs by employing a nanosecond pulsed laser (Quantasystem, PILS-GIANT, 6 ns pulse duration) in the stainless still chamber. The operating conditions were controlled at the "second harmonic (532 nm), a repetition rate varying between 1 and 10 Hz and with an energy varying between 217 and 50 mJ, independence of the performed experiment." The deionized water (400 mL) was placed into the chamber as a solvent, where "the target was placed on the rotating holder." The axial movement was controlled from backward and forward. The laser was focused onto the target surface by a 4.0 cm lens inserted inside the spacer tube in the opposite position concerning the target (Fig. 2C.10A–C). The ultrasound process is also employed in the fabrication of NPs. The physical phenomena associated with ultrasound-assisted synthesis are cavitation (formation, growth, and implosive collapse of bubbles in a liquid) and nebulization (the creation of heated microdroplets like mist in a liquid) [104]. These microdroplets were considered reactors to facilitate nonaggregated NPs under mild conditions, with shorter reaction times with high yields [74].

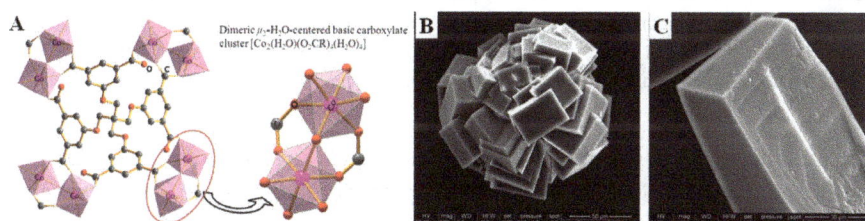

Fig. 2C.9: The microwave-assisted solvothermal chemistry to prepare Co-MOFs used as disinfectants, A. Tetragonal space group *P42₁m* with dimetric μ₂-H₂O-centered basic carboxylate cluster of [Co₂(H₂O) (O₂CR)₄(H₂O)₄], B. TEM image of Co-MOF motifs, and C. Zoomed-in TEM image of Co-MOF motifs [155].

2C.4.2 Biological methods

The biological entities are known to synthesize NPs both extra- and intracellularly. The ability of a living system to utilize its intrinsic organic chemistry processes in

A

Shadowgraph lamp

Laser induced
bubble

Laser pulse

Focusing
lens

Rotating
target

Water

y

x

z

To shadowgraph detection

B

Collapse

20 160 300 340
217

300 320
156

240 300
108

260 280
60

(mJ) Primary bubble evolution

C

$\lambda_{max} = 398$ nm

217
156
108

Laser energy
(mJ)

60

Absorbance (a.u.)

0.3
0.2
0.1
0.0

300   400   500   600

Wavelength (nm)

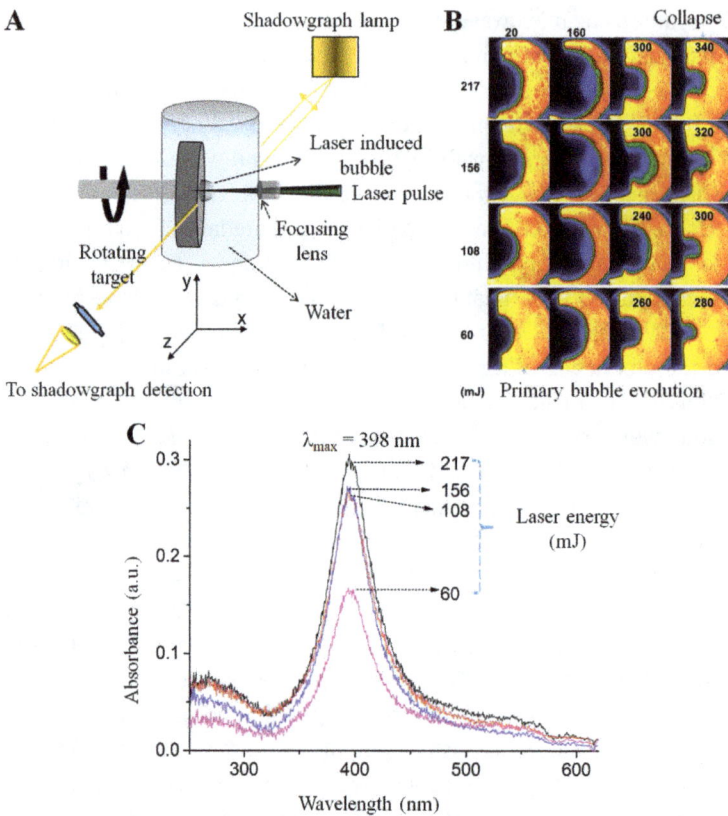

Fig. 2C.10: Pulsed laser ablation in liquids (PLAL) was used to produce nanostructured silver metal: (A) experimental setup, (B) temporal resolved shadowgraph images of the laser-induced bubble on Ag target immersed in water obtained at different laser energies, 217, 156, 108, and 60 mJ, respectively, and (C) Surface plasmon resonance spectra of AgNPs colloidal solutions produced with different laser energies.

remodeling inorganic metal ions into NPs has opened up an undiscovered area of biochemical analysis [150]. The biosynthesis procedure follows the bottom-up process and implicates either reduction or oxidation reactions [39]. The NPs production via the "green method" is a safe and non-noxious approach that uses various natural sources like enzymes, bacteria, fungi, and plants [64]. *Using plant extracts*: Plants are known as chemical factories of nature which are cost-efficient and need little maintenance. Plants have shown unique potential in heavy metal detoxification. The accumulation by which environmental pollutants can be controlled as very small traces of these heavy metals is too toxic even at very low concentrations [85]. Many plants used in the biosynthesis of NPs are reported in the literature. The use of plants as the production assembly of NPs has drawn attention because of its rapid, eco-friendly, non-pathogenic, economical protocol and providing a single-step technique for the biosynthetic processes [6]. Otherwise, bioactive compounds present in plant extracts: flavonoids, alkaloids, phenols, saponins, carbohydrates, proteins, quinine, glycosides, tannins, and steroids can help the biosynthesis of NPs in a single-step process [64]. The majority of active chemical constituents of the above natural products are tabulated in Tab. 2C.2.

In addition, the reaction rate is much faster as compared to that of the microorganism methods. The functional groups in these compounds are actively bonded or interact with the reactants of the targeted products. They also act as a surfactant, resulting in spatial hindrances to prevent particle growth into large clusters. More importantly, the richness of plant extracts by molecules is known for their antioxidant power, which participates in the process of reduction and stabilization of NPs during biosynthesis due to the presence of some functional groups (like $-O-H$, $-C-O-C$, $-C=O$, $-C=C$) [79, 141]. Different parts of plants are employed in the bioformation of NPs (leaf, stem, bark, flower, root, seed, and fruits) owing to the exceptional metabolites they produce. The leaves are the most used part as they are the site of photosynthesis and are rich in different metabolites [56, 119, 141]. The biochemical and molecular mechanisms of NPs biosynthesis remain poorly characterized and should be investigated to optimize further the process [31].

Tab. 2C.2: A summary of plant water extracts used to synthesize nanomaterials.

C, H, O Phenol, C_6H_5OH
Space group: $P112_1$ (chirality; polar)
Crystal system: monoclinic [8]
a: 6.0500 Å
b: 8.9250 Å
c: 14.5940 Å
β: 90.000°
Cell volume: 788.021 Å^3
Asymmetric unit: 39 sites
Unit cell: 78 sites/unit cell
Density: 1.1898 g/cm^3

C, H, N Nicotine (alkaloids) [63]
Being a naturally produced alkaloid in the nightshade family of plants;
constituting approximately 0.6–3.0% of the dry weight of tobacco;
Being highly addictive unless used in slow-release forms;
Using it as a tool for quitting smoking has a good safety history

Tab. 2C.2 (continued)

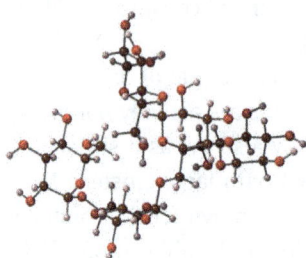	● C ● H ● O	Glycogen (a polysaccharide carbohydrate) the multibranched polysaccharide of glucose; functioning as one of two forms of energy reserves; stored in the body, particularly within the muscles and liver; stored in skeletal muscle serve as a form of energy storage for the muscle itself; and Being the analog of starch, a glucose polymer [24]
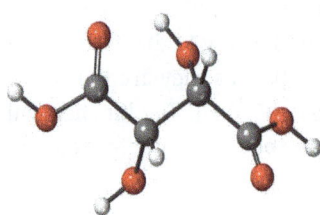	● C ● H ● O	Tartaric acid (2,3-dihydroxy-butanedioic acid, $C_4H_6O_6$) is a chiral molecule with two forms: L and D, of which the D form is the naturally occurring form and is illustrated here [131]
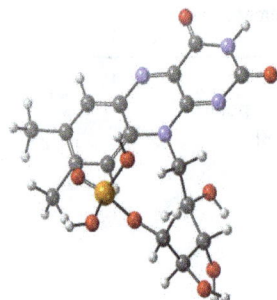	● C ● H ● N ● O ● P	Flavin mononucleotide (E101a, $C_{17}H_{21}N_4O_9P$) is a flavin mononucleotide that is an orange-red food color additive. It is found in many jams, milk, and sugar products. Other food products designed for small children and babies may also contain flavin mononucleotide [95].
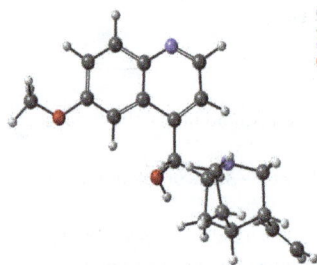	● C ● H ● N ● O	Quinine ($C_{20}H_{24}N_2O_2$) is a medication used to treat malaria. It is also used in Toni water to give its bitter taste [46].

Tab. 2C.2 (continued)

● C ◗ N ● O ◗ S	Lysozyme ($C_{613}N_{193}O_{277}S_{10}$) is found in egg-white lysozyme and tears and acts as an antimicrobial agent [68]
● C ◗ H ● O	Salicin ($C_{13}H_{16}O_7$) is produced in willow bark and acts as an anti-inflammatory Being the chemical from which aspirin was first developed [138].
● C ● F ◗ H ● O	Dexamethasone ($C_{22}H_{29}FO_5$) is a corticosteroid medication used to treat many conditions, including rheumatic problems, skin diseases, severe allergies, asthma, chronic obstructive lung disease, croup, and brain swelling [113].

The plant drying method and extraction method have a big influence on the anti-oxidant amount and, consequently, the rate of NPs synthesis [79]. Indeed, Farooqi et al. [41] found that the method of plant drying (fresh, sun-dried, and hot air oven-dried leaf) has a remarkable impact on the size of Ag NPs produced by leaf extract *Clerodendrum inerme*. These results showed that the methods used for drying plants are highly promising in developing Ag NPs of different sizes that could be tailored for specific applications. The NPs of the smallest size were obtained with the extract of fresh leaf, while the biggest one was observed in the extract of the sun-dried leaf. However, the processing conditions need to be optimized to extract spherical and monodispersed NPs [41]. Freeze-drying is one of the drying processes employed for conserving bioactive molecules in plants. Freeze-drying (lyophilization) is a drying process in which the solvent (usual water) or the suspension medium is crystallized at a low temperature and, after that, sublimated from the solid-state directly into the vapor phase [32]. Freeze-dried plant parts are the best method to preserve the active ingredients of the plant [81]. The main extraction methods for the synthesis of AuNPs and AgNPs are (a) solvent-based extraction, (b) microwave-assisted extraction, and (c) maceration extraction. The most utilized solvents for extraction are water, methanol, and ethanol. Water is preferred as a green solvent for extracting biomolecules of plant species [26].

Various parameters can affect the characteristic features, production rate, and quantity of the NPs synthesized with plant extracts like extract concentration, metal salt concentration, pH, temperature, time, and extraction solvent. These parameters need to be optimized to improve the NPs biosynthesis and make it competitive with traditional synthesis approaches [44, 141].

The concentration of plant extract and pH and the temperature of reaction affect the morphology, size, and shape of the obtained NPs [22, 34, 105]. Bezares et al. [22] showed that a higher concentration of extract implies a greater amount of reducing agent and stabilizing molecules in the reaction mixture, therefore, high reaction speed and smaller particle size. The study was done by Kredy [66] on the biosynthesis of silver NPs using *Lawsonia inermis* leaves extract proved an increase in the rate of silver NPs formation with increasing temperature. Monodispersive silver NPs were also obtained at a pH equal to 9. [102], also demonstrated the synthesis of Ag NPs from silver nitrate solution and seeds extracts of different vegetable seeds of the *Brassicaceae* family. The obtained results indicated that the optimal conditions of Ag NPs synthesis were concentration at 20 mL, temperature at 80 °C, and pH 8.5.

Many active ingredients in plant extracts may participate in reducing and stabilizing NPs. Primary metabolites (proteins and polysaccharides) and secondary metabolites in the extract (terpenoids, flavonoids, phenols, alkaloids, steroids, saponins, and tannins) operate as reducing agents [69, 88, 109, 143]. The essential oils can be utilized as reducing and capping agents; however, the study carried out by Dzimitrowicz et al. [40] on the green synthesis of AuNPs by using aqueous extracts and essential oils of *Eucalyptus globulus* and *Rosmarinus officinalis* exhibited that the aqueous extract of these two species led to obtaining small NPs in comparison with those synthesized by essential oils possibly because of the higher amounts of possible reducing and stabilizing agents in the crude extracts compared with the corresponding essential oils [40, 143]. The structure of salicin-coated cubic Cu metal was generated using CrystalMaker as a guide to preparing zero valence NPs (Fig. 2C.11).

Several works reported the utilization of phyto-assisted synthesis of NPs for manufacturing solar cells. Sutradhar and Saha [132] showed that the fabricated ZnO NPs using green tea leaf extract were further used to prepare ZnO/natural graphite (NG) composite material. The phenolic compound was used as a capping agent to form ZnO compounds preventing crystal growth (Fig. 2C.12). The current–voltage (I–V) characteristics of a ZnO/NG nanocomposite thin film were investigated. The ZnO/NG nanocomposite measured the short-circuit photocurrent, open-circuit photovoltage, fill factor, and solar cell efficiency was measured for ZnO/NG nanocomposite. Interestingly, the cell exhibited a good PCE of 3.54 % with high stability. *Aloe vera* aqueous extract is employed for TiO_2 NPs preparation. The prepared TiO_2 NPs fabricate solar

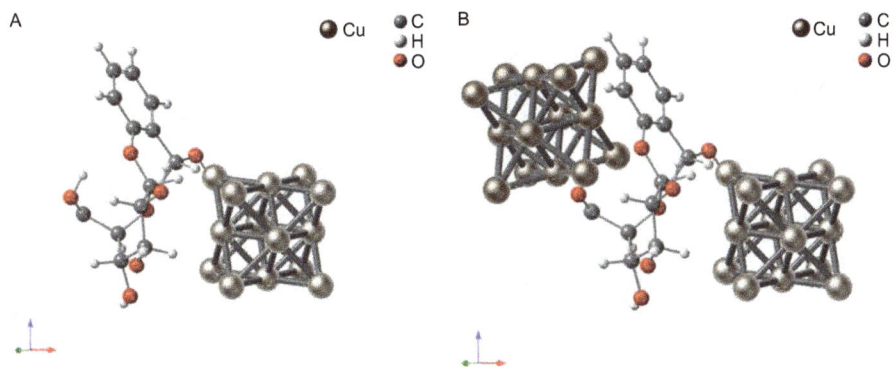

Fig. 2C.11: The simulated structure of salicin-coated cubic Cu metal to prepare zero valence nanoparticles: (A) the anticipated dipole ion intermolecular forces between Cu and O from salicin and (B) intermolecular forces between salicin with two Cu crystal units.

cells using indium tin oxide (ITO) glass, TiO_2, ruthenium, graphite, and potassium iodide [54]. Au-doped TiO_2 NPs were successfully synthesized using *Terminalia arjuna* bark extract. The energy conversion efficiency was more important for Au-doped TiO2 NPs than pure and Pt-doped TiO_2 NPs. The synthesized NPs were investigated for dye-sensitized solar cell applications. Au noble metal presents a TiO_2 matrix and an improved open-circuit voltage (V_{oc}) of dye-sensitized solar cells [45].

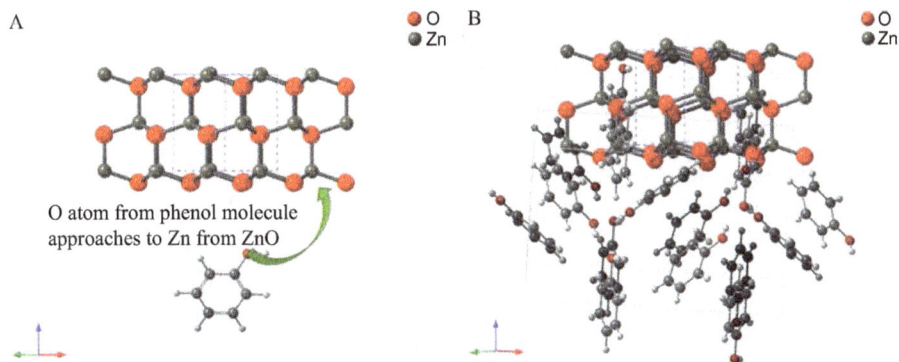

Fig. 2C.12: The simulated structure of zinc oxide (ZnO) nanoparticles coated by tea leaf extract: (A) the anticipated dipole ion intermolecular forces between Zn from ZnO and O from tea leaf extract (phenol as an example) and (B) intermolecular forces between phenolic compounds with ZnO crystal units inhibit the crystal growth into large clusters.

Many NPs are synthesized by green methods utilizing plants. Some papers are available on the applications of water splitting using these green NPs [19]. The leaf extract of *Ocimum sanctum* was used as a reducing agent for the synthesis of platinum NPs from an aqueous chloroplatinic acid ($H_2PtCl_{66}H_2O$). The reduced platinum showed similar hydrogen evolution potential and catalytic activity to pure platinum using linear scan voltammetry [128]. Silver deposited titanate nanotube array composite (Ag/TNA-c) was successfully produced using tea leaves and ground coffee as reducing agents. The main reducing agents were catechin and chlorogenic acid. The fabricated Ag/TNA-c showed the best photocatalytic (PC) performance in photocurrent response, electrochemical impedance spectrum, Ibuprofen degradation, and hydrogen generation [101].

Using microorganisms: Over the last few years, the biosynthesis of NPs using microbial agents such as bacteria, actinomycetes, fungi, yeast, viruses, and marine algae has received tremendous attention in green nanotechnology due to their biotransformation and bioaccumulation ability [43]. Microorganisms proved their potential in metal reduction and were deemed the potential biofactories for synthesizing metal NPs. They are very effective secretors of extracellular enzymes [74, 78]. At present, microbial methods in synthesizing nanomaterials of varying compositions are limited and confined to metals, some metal sulfide, and very low oxides. All these are restricted to microorganisms of terrestrial origin [14].

For NP formation, different biological agents react differently with different metal solutions. Numerous microbes produce inorganic materials, either extracellularly or intracellularly, and the mechanism differs from one organism to another, either intra or extracellularly [120]. In intracellular synthesis, metal ions are transported into the microbial cell, and NP forms enzymes inside the cell. In extracellular biosynthesis, it involves trapping metal ions on the surface of microbial cells and their reduction into NPs in the presence of various enzymes (like nitrate reductase) available at the surface [160]. Extracellular synthesis of NPs has an advantage over the intracellular method in terms of being less time-consuming since it does not need any downstream process to collect NPs from the organisms [124]. The precise mechanism for NPs synthesis employing biological agents has not been conceived yet. Different biological agents react differently with metal ions leading to the NPs formation [48].

Biological synthesis using microbes offers an advantage over plants since microbes are easily reproduced. Nonetheless, there are many drawbacks to isolating and screening potential microbes [80]. The microbial synthesis of NPs is characterized by the expensive cost of microorganism isolation [64]. In addition, the main problem includes the cost-effectiveness of the fabrication procedures as it is time consuming and implicates the utilization of chemicals for the growth medium [80]. Bacteria are readily available from the environment, can be cultivated quickly, and adapt to different conditions, making them great candidates for NP synthesis [84]. Some bacterial species resort to specific defense mechanisms to quell stresses like heavy metal ions or metals (e.g., *Pseudomonas stutzeri* and *Pseudomonas aeruginosa*). These mechanisms of resistance include efflux pumps, metal efflux systems, inactivation and complexation of metals, impermeability to metals and the lack of specific metal transport systems, alteration of solubility and toxicity by changes in the redox state of the metal ions, extracellular precipitation of metals, and volatilization of toxic metals by enzymatic reactions [21, 115]. Numerous bacterial species are applied in the biosynthesis of NPs, such as *Acinetobacter* sp., *Aeromonas* sp., *Bacillus* species, *Clostridium thermoaceticum*, *Corynebacterium* sp., *Desulfovibrio* species, the species of *Enterobacteriaceae* family (*E. coli*, *Klebsiella pneumonia*), *Geobacillus* sp., *Lactobacillus* sp., *Pseudomonas* sp., *Rhodobacter sphaeroides*, and *Rhodopseudomonas capsulata* [51, 65, 84]. Microorganisms' biomineralization of metal into NPs depends on environmental parameters, like pH, pO_2, pCO_2, redox potential, and temperature. Optimizing these physiological parameters requires synthesizing NPs with accurate size, morphology, and chemical compositions [3, 15]. Magnetotactic bacteria, known to synthesize magnetic nanocrystals with uniform shapes and sizes under physiological conditions, serve as an inspiration and source of some biological macromolecules used for the biomimetic synthesis of a variety of magnetic [108].

Fungi are recognized as eukaryotic organisms that reside in various ordinary lodgings, and they typically form decomposer organisms. From an anticipated sum of 1.5 million species of fungi on Earth, only about 70,000 species have been recognized [78]. As nano factories and extremely efficient secretors of extracellular enzymes, it is possible to obtain large-scale production of enzymes easily. Further advantages of using fungal-mediated green approaches for synthesizing metallic NPs include economic viability and ease in biomass handling [53]. According to Mukherjee et al. [86] trapping, bioreduction, and capping synthesizing gold NPs by the fungus *Verticillium* sp. Fungi have a high cell wall binding potential with metal ions and have a higher potential to tolerate metal concentrations; hence, fungi can yield more NPs than bacterial cells [65]. Several species of fungi like *F. oxysporum*, *Fusarium solani*, *Pleurotus sajorcaju*, *Fusarium semitectum*, *Alternaria alternata*, *Fusarium acuminatum*, *Penicillium* species, *Aspergillus* species have been utilized in the psychosynthesis of NPs [121].

Yeasts are eukaryotic unicellular microorganisms of the fungi kingdom, and generally, several strains play an important role in the food industry due to their ability to sugar fermentation. About 1,500 species are currently described in the family Saccharomycetaceae. Yeast production is easily controlled in laboratory conditions. The rapid growth of yeast strains and simple nutrients benefits the mass production of metal NPs [116, 127]. Some yeast species are employed in the biosynthesis of NPs: *Y. lipolytica*, *S. cerevisiae*, *Trichophyton* species, *Microsporum canis*, *Kluyveromyces marxianus*, and *Candida albicans*, *Candida glabrata*, *Schizosaccharomyces pombe*, *Rhodotorula mucilaginosa*, *Chlorococcum humicola*, *Aphanothece* sp., *Sargassum muticum*, and *Caulerpa racemosa* [17].

Algae are a diverse group of photoautotrophic, eukaryotic, aquatic, and unicellular/multicellular organisms. They have been classified based on the pigmentation they release into brown algae (phaeophytes), red algae (rhodophytes), and green algae (chlorophytes). In addition to their commercial and industrial utilization as food, feed, additives, cosmetics, pharmaceuticals, and fertilizer, the algae are used in NPs biofabrication. The secondary metabolites from algae reduce, cap, and stabilize the metal precursors to form NPs [27].

There is very little literature supporting the utilization of marine microorganisms in NPs fabrication. Among them, bacteria (*E. coli, Pseudomonas sp.*), cyanobacteria (*Spirulina platensis, Oscillatoria willei, Phormidium tenue*), yeasts (*Pichia capsulata, Rhodospiridium diobovatum*), fungi (*Thraustochytrium sp., Penicillium fellutanum, Aspergillus niger*), and algae (*Navicula atoms, Diadesmis gallica, Stauroneis* sp., *Sargassum wightii, Fucus vesiculosus*) are reported to produce inorganic NPs either inside or outside cells [14].

Using biomolecules: Several biomolecules employed to fabricate NMs as reducing, capping, and stabilizing agents are investigated. Many researchers have studied and proved the reducing capability of polysaccharides, peptides and proteins, amino acids, and phytochemicals. The methods adopted for biological NPs synthesis are two-step procedures based on emulsification and a one-step procedure involving nano-precipitation, desolvation, and gelation and drying methods [117].

Polysaccharides are natural biopolymers recognized as the most encouraging hosts for producing metallic nanoparticles (MNPs) because of their outstanding biocompatible and biodegradable properties [144]. Polysaccharides are natural polymers consisting of monosaccharide units linked by glycosidic bonds, chitosan, cellulose, starch, hyaluronic acid, and typical dextran examples [103]. Polysaccharides have hydroxyl groups, a hemiacetal reducing end, and other functionalities that can play important roles in reducing and stabilizing MNPs [99]. Dr. Liu and her team used starch to coat ceria-doped ZnO, which showed high potency in teeth whitening. The formation of ZnO was carried out using colloidal chemistry, which resulted in uniformly distributed NPs. The hexagonal, "wurtzite" structure of ZnO and triclinic fluorite structure of CeO_2 showed some discrepancy, which can be mitigated using the starch molecule to "glue" two crystalline phases together (Fig. 2C.13).

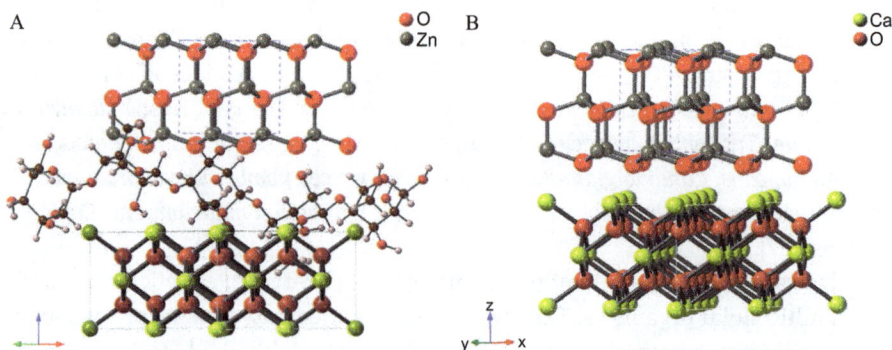

Fig. 2C.13: The simulated structure of ceria-doped zinc oxide (ZnO) nanoparticles coated by starch: (A) the anticipated dipole ion intermolecular forces between Zn from ZnO, Ce from CeO_2, and O from starch and (B) Intermolecular forces between phenolic compounds with ZnO crystal units inhibit the crystal growth into large clusters.

Protein is the predecessor of naturally occurring material used for the preparation of NP, attributed to its unique functionalities and defined primary structure [117]. Biological protein-based NPs such as silk, keratin, collagen, elastin, corn zein, and soy protein-based NPs are advantageous in having biodegradability, bioavailability, and relatively low cost [37]. Phytochemicals like phenolic compounds, terpenoids, essential oils, alkaloids, saponins, and so on are also used in NPs fabrication owing to their reducing power. These compounds may impart different biological activities to these NPs [5].

Using wastes: According to the Environment Program of the United Nations, nearly 11.2 billion tons of solid wastes are generated every year, which is a significant source of environmental degradation and negative health impacts [139]. Many agricultural and industrial food wastes considered renewable, sustainable, and nontoxic can be useful for synthesizing NPs. Peanut skin was easy to obtain in large quantities, relatively inexpensive, and environmentally friendly as an agricultural waste product. The mechanism for NPs synthesis involving the formation of intermediate complexes, an electron transfer reaction, and adsorption of non-reducing organic macromolecules at the solid–liquid interfaces was proposed [96]. In their work, Rajput et al. [110] synthesized silver NPs using a waste source, i.e., used tea leaves. The obtained results showed

Fig. 2C.14: The conversion of waste into value-added products with high porosity, used for water remediation: (A) Scanning electron microscopic images of zeolites, treated by 3 mol/L NaOH solution for different processing time (a) 0.5 h, (b) 1.0 h, and (B) adsorption curves of Cu^{2+} and Rh-B adsorbed by zeolites treated by 3 mol/L NaOH solution for different processing times: (a) 0.5 h, (b) 1.0 h. Permission from Zhou et al. [153] (Liu et al., 2020).

that the particles synthesized are in the nanoscale range and crystalline. Dr. F. Zheng and Dr. J. Liu have been collaborating on the convert metallurgy waste (coal gangue) into value-added products, such as porous zeolite, for photocatalysis (Fig. 2C.14).

2C.5 Advantages and disadvantages of green methods

The green synthesis of NMs using plant extract, microorganisms, biomolecules, and wastes is cheap, environmentally friendly, sustainable, and easy to operate. The parameters of plant extract-mediated NPs synthesis, microorganism cultivation parameters, and metal salt concentration still need to be optimized. Meanwhile, the relevant synthetic mechanisms need to be further elucidated. The lack of understanding of the chemical components involved in the synthesis and stabilization of NPs remains a great challenge for researchers [18]. Also, the cultivation of microorganisms and large-scale production remain problematic uptake. Nevertheless, the genetic manipulation to overexpress specific enzymes to intensify synthesis is much more difficult among eukaryotes [112]. The poor understanding of hazards associated with green-produced NPs is another drawback.

2C.6 Conclusion

Recently, nanotechnology has gained attention for confronting many challenges concerning economic and environmental issues. The traditional methods of synthesis can efficiently substitute the green synthesis of NPs. It reduces environmental pollutants and conserves natural resources without creating any environmental damage. The green and biosynthetized NPs are utilized in various fields, especially in the renewable energy sector and energy conversion. This greener route is inexpensive, easily scaled up, and has eco-friendliness. Despite their advantages, the green synthesis methods still need a few improvements that should be further looked into, particularly the control of the morphology, the size and shape of the produced NMs, and the large-scale manufacture of these NMs.

References

[1]   Aakash, L., Vinduja, B. S., Arunkumar, T., Sivakumar, T.,, & Gajalakshmi, D. (2020). A systematic review on green synthesis of nanoparticles and their medical applications. Plant Archives, 20(2), 6069–6076. http://www.researchgate.net/publication/343760340

[2]   Abdel-Mottaleb, M. S. A., Byrne, J. A.,, & Chakarov, D. (2011). Nanotechnology and solar
        energy. International Journal of Photoenergy, 2011, Article ID 194146. 10.1155/2011/194146
[3]   Abhilash, K. R.,, & Pandey, B. D. (2011). Microbial synthesis of iron-based nanomaterials- A
        review. Bulletin of Materials Science, 34(2), 191–198. https://www.ias.ac.in/article/fulltext/
        boms/034/02/0191-0198
[4]   Acosta-Silva, Y., de, J., Torres-Pacheco, I., Mastsumoto, Y., Toledano-Ayala, M., Soto-
        Zarazúa, G., Zelaya-Ángel, O., & Méndez-lópez, A. (2019). Applications of solar and wind
        renewable energy in agriculture: A review. Science Progress, 102(2), 127–140. 10.1177/
        0036850419832696
[5]   Afreen, A., Ahmed, R., Meboub, S., Tariq, M., Ahmed Alghamdi, H., Zahid, A. A., Ali, I., Malik,
        K., & Hasan, A. (2020). Phytochemical-assisted biosynthesis of silver nanoparticles from
        Ajuga bracteosa for biomedical applications. Materials Research Express, 7, 075404.
        https://doi.org/10.1088/2053-1591/aba5d0
[6]   Ahmed, S., Ahmed, M., Swami, B. L., & Ikram, S. (2016a). Green expertise is a review of plant
        extract-mediated synthesis of silver nanoparticles for antimicrobial applications. Journal of
        Advanced Research, 7, 17–28. https://doi.org/10.1016/j.jare.2015.02.007
[7]   Ahmed, S., Chaudhry, S. A., & Ikram, S. (2017). A review on biogenic synthesis of ZnO
        nanoparticles using plant extracts and microbes: A prospect towards green chemistry.
        Journal of Photochemistry and Photobiology. B, Biology, 166, 272–284.
[8]   Allan, D. R., Clark, S. J., Dawson, A., McGregor, P. A., & Parsons, S. (2002). Pressure-induced
        polymorphism in phenol. Acta Crystallographica. Section B, Structural Science, 58(6),
        1018–1024.
[9]   Aljaridi, R., Taha, L. Y., & Ivey, P. (2017). Electrostatic energy harvesting systems: A better
        understanding of their sustainability. Journal of Clean Energy Technologies, 5(5), 409–416.
        10.18178/jocet.2017.5.5.407
[10]   Ansari, M. A., Murali, M., Prasad, D., Alzohairy, M. A., Almatroudi, A., et al. (2020).
        Cinnamomum Verum bark extract mediated green synthesis of ZnO nanoparticles and their
        antibacterial potentiality. Biomolecules, 10, 336. 10.3390/biom10020336
[11]   Argyriou, D. N., & Howard, C. J. (1995). Re-investigation of Yttria–Tetragonal Zirconia
        Polycrystal (Y-TZP) by neutron powder diffraction – A Cautionary Tale. Journal of Applied
        Crystallography, 28(2), 206–208.
[12]   Arkatkar, A., Mungray, A. K., & Sharma, P. (2020). Conjugation of nanomaterials and
        bioanodes for energy production in microbial fuel cells. In Ledwani, L., & Sangwai,
        J. S. (eds.). Green energy and technology (pp. 169–184). Switzerland AG, Springer Nature.
        https://doi.org/10.1007/978-3-030-33774-2_7
[13]   Arroyo, E., & Badel, A. (2012). Electromagnetic vibration energy harvesting device
        optimization by synchronous energy extraction. Sensors and Actuators. A, Physical, 24.
        10.1016/j.sna.2011.06.024
[14]   Asmathunisha, N., & Kathiresan, K. (2013). A review on the biosynthesis of nanoparticles by
        marine organisms. Colloids and Surfaces. B, Biointerfaces, 103, 283–287. https://dx.doi.
        org/10.1016/j.colsurfb.2012.10.030
[15]   Bahrulolum, H., Nooraei, S., Javanshir, N., Tarrahimofrad, H., Mirbagheri, V. S., Easton,
        A. J., & Ahmadian, G. (2021). Green synthesis of metal nanoparticles using microorganisms
        and their application in the agrifood sector. Journal of Nanobiotechnology, 19, 86. https://
        doi.org/10.1186/s12951-021-00834-3
[16]   Bae, D., Zhen, S., & Seff, K. (1999). Structure of dehydrated Zn2+-exchanged zeolite
        X. Overexchange, framework dealumination, reorganization, stoichiometric retention of
        monomeric tetrahedral aluminate. The Journal of Physical Chemistry. B, 103(27), 5631–5636.

[17] Banejree, K., & Rai, V. R. (2021). A review on mycosynthesis, mechanism, and characterization of silver and gold nanoparticles. BioNanoScience, 8, 17–31. 10.1007/s12668-017-0437-8

[18] Bao, Y., He, J., Song, K., Guo, J., Zhou, X., & Liu, S. (2021). Plant-extract-mediated synthesis of metal nanoparticles. Journal of Chemistry, 2021, Article ID 6562687, 14. https://doi.org/10.1155/2021/6562687

[19] Basheer, A. A., & Ali, I. (2019). Water photo splitting for green hydrogen energy by green nanoparticles. International Journal of Hydrogen Energy. https://doi.org/10.1016/j.ijhydene.2019.03.040

[20] Bashir, S., Houf, W., Liu, J. L., & Mulvaney, S. P. (2021). 3D Conducting polymeric membrane and scaffold saccharomyces cerevisiae biofilms to enhance energy conversion in microbial fuel cells. ACS Applied Materials, & Interfaces. https://doi.org/10.1021/acsami.1c20445

[21] Beveridge, T. J., Hughes, M. N., Lee, H., Leung, K. T., Poole, R. K., Savvaidis, I., Silver, S., & Trevors, J. T. (1997). Metal-microbe interactions: Contemporary approaches. Advances in Microbial Physiology, 38, 177–243. 10.1016/s0065-2911(08)60158-7

[22] Bezares, B., Jaña, Y., Cottet, L., & Castillo, A. (2018). Mater. Express, 8(5), 450–456. 10.1166/mex.2018.1448

[23] Bundy, F. P., & Kasper, J. S. (1967). Hexagonal diamond – A new form of carbon. The Journal of Chemical Physics, 46(9), 3437–3446.

[24] Buschiazzo, A., Ugalde, J. E., Guerin, M. E., Shepard, W., Ugalde, R. A., & Alzari, P. M. (2004). Crystal structure of glycogen synthase: Homologous enzymes catalyze glycogen synthesis and degradation. The EMBO Journal, 23(16), 3196–3205.

[25] Cannon, J. S. (1994). Hydrogen vehicle programs in the USA. International Journal of Hydrogen Energy, 19(11), 905–909.

[26] Castillo-Henríquez, L., Alfaro-Aguilar, K., Uglade-Álvarez, J., et al. (2020). Green synthesis of gold and silver nanoparticles from plant extracts and possible antimicrobial agents in the agricultural area. Nanomaterials, 10, 1763. 10.3390/nano10091763

[27] Chaudhary, R., Nawaz, K., Khan, A. K., Hano, C., Abbasi, B. H., & Anjum, S. (2020). An overview of the algae-mediated biosynthesis of nanoparticles and their biomedical applications. Biomolecules, 10, 1498. 10.3390/biom10111498

[28] Chen, J., & Wang, Z. L. (2017). Reviving vibration energy harvesting and self-powered sensing by a triboelectric nanogenerator. Joule, 1, 480–521. http://doi.org/10.1016/j.joule.2017.09.004

[29] Chen, J., Zhang, Q., Xu, N., Li, W., Yao, Y., Li, P. . . . Zhang, H. (2022). Roadmap to hydrogen society of Tokyo: Locating priority of hydrogen facilities based on multiple big data fusion. Applied Energy, 313, 118688–1 to 118688–15.

[30] Cherian, T., Ali, K., Saquib, Q., Faisal, M., Wahab, R., & Musarrat, J. (2020). Cymbopogon citratus functionalized green synthesis of CuO-nanoparticles: Novel prospects as antibacterial and antibiofilm agents. Biomolecules, 10, 169. 10.3390/biom10020169

[31] Chung, I.-M., Park, I., Kim, S.-H., Thiruvengadam, M., & Rajkumar, G. (2016). Plant-mediated synthesis of silver nanoparticles: Their characteristic, properties, and therapeutic application. Nanoscale Research Letters, 11, 40. 10.1186/s11671-016-1257-4

[32] Ciurzyńska, A., & Lenart, A. (2011). Freeze-drying-application in food processing and biotechnology – A review. Polish Journal of Food and Nutrition Sciences, 61(3), 165–171. 10.2478/v10222-011-0017-5

[33] Cojocariu, B., Hill, A., Escudero, A., et al. (2012). Piezoelectric vibration energy harvester-Design and prototype. International Mechanical Engineering Congress, & Exposition, ASME 2012, 10. 10.1115/IMECE2012-85785

[34] Cruz, D., Falé, P. L., Mourato, A., Vaz, P. D., Serralheiro, M. L., & Lino, A. N. R. L. (2010). AG nanoparticles biosynthesized by Lippia citriodora (Lemon Verbena) are the preparation and physicochemical characterization. Colloids and Surfaces. B, Biointerfaces, 81, 67–73. 10.1016/j.colsurfb.2010.06.025

[35] Das, M., & Chatterjee, S. (2019). Green synthesis metal/metal oxide nanoparticles towards applications: Boon or bane. In: Ashutosh, K. S., & Siavash, I. (eds.). Green synthesis, characterization and applications of nanoparticles (pp. 265–301, Editors: Ashutosh Kumar Shukla, and Siavash Iravan). Elsevier, Oxford, UK. https://doi.org/10.1016/B978-0-08-102579-6.00011-3

[36] Dell'Aglio, M., & De Giacomo, A. (2020). Plasma charging effect on the nanoparticles releasing from the cavitation bubble to the solution during nanosecond Pulsed Laser Ablation in Liquid. Applied Surface Science, 515, 146031–1 to 146031–8.

[37] DeFrates, K., Markiewicz, T., Gallo, P., Rack, A., Weyhmiller, A., Jarmusik, B., & Hu, X. (2018). Protein polymer-based nanoparticles: Fabrication and medical applications. International Journal of Molecular Sciences, 19(6), 1717.

[38] Deshmukh, A. R., Gupta, A., & Kim, B. S. (2019). Ultrasound-assisted green synthesis of silver and iron oxide nanoparticles using fenugreek seed extract and their enhanced antibacterial and antioxidant activities. BioMed Research International, 2019, Article ID 1714358, 14. https://doi.org/10.1155/2019/1714358

[39] Dikshit, P. K., Kumar, J., Das, A. K., Sadhu, S., Sharma, S., Singh, S., Gupta, P. K., & Kim, B. S. (2021). Green synthesis of metallic nanoparticles: Application and limitations. Catalysts, 11, 902. http://doi.org/10.3390/catal11080902

[40] Dzimitrowicz, A., Berent, S., Motyka, A., Jamroz, P., Kurcbach, K., Sledz, W., & Pohl, P. (2019). Comparison of the characteristics of gold nanoparticles synthesized using aqueous plant extracts and natural plant essential oils of Eucalyptus globulus and Rosmarinus officinalis. The Arabian Journal of Chemistry, 12, 4795–4805. https://doi.org/10.1016/j.arabjc.2016.09.007

[41] Farooqi, Md.A., Chauhan, P. S., Moorthy, P. K., & Shaik, J. (2010). Extraction of silver nanoparticles from the leaf extracts of Clerodendrum inerme. Digest Journal of Nanomaterials and Biostructures, 5(1), 43–49. https://www.researchgate.net/publication/215568562

[42] Frondel, C., & Marvin, U. B. (1967). Lonsdaleite is a hexagonal polymorph of a diamond. Nature, 214(5088), 587–589.

[43] Gahlawat, G., & Choudhury, A. R. (2019). A review on the biosynthesis of metal and metal salt nanoparticles by microbes. RSC Advances, 9, 12944. 10.1039/c8ra10483b

[44] Godwin Christopher, J. S., Saswati, B., & Ezilrani, P. S. (2015). Optimization of parameters for the biosynthesis of silver nanoparticles using leaf extract of Aegle marmelos. Brazilian Archives of Biology and Technology, 58, n(5), 702–710 5. http://dx.doi.org/10.1590/S1516-89132015050106

[45] Gopinath, K., Kumaraguru, S., Bhakyaraj, K., Thirumal, S., & Arumugam, A. (2016). Eco-friendly synthesis of TiO2, Au and Pt doped TiO2 nanoparticles for dye-sensitized solar cell applications and evaluation of toxicity. Superlattices, and Microstructures. 10.1016/j.spmi.2016.02.012

[46] Hisaki, I., Hiraishi, E., Sasaki, T., Orita, H., Tsuzuki, S., Tohnai, N., & Miyata, M. (2012). Crystal structure of quinine: The effects of vinyl and methoxy groups on molecular assemblies of cinchona alkaloids cannot be ignored. Chemistry–An Asian Journal, 7(11), 2607–2614.

[47] Hossain, A., Abdallah, Y., Ali, Md.A., Masum, M. I. Md., Li, B., Sun, G., Meng, Y., Wang, Y., & An, Q. (2019). Lemon fruit-based green synthesis of zinc oxide nanoparticles and titanium dioxide nanoparticles against soft rot bacterial pathogen Dickeya dadantii. Biomolecules, 9, 863. 103390/biom9120863

[48] Hulkoti, N. I., & Taranath, T. C. (2014). Biosynthesis of nanoparticles using microbes- A review. Colloids and Surfaces. B, Biointerfaces, 121, 474–483. https://dx.doi.org/10.1016/j.col surfb.2014.05.027

[49] Ijaz, I., Gilani, E., Nazir, A., & Bukhari, A. (2020). A detailed review of chemical, physical and green synthesis, classification, characterization, and applications of nanoparticles. Green Chemistry Letters and Reviews, 13(3), 223–245. https://doi.org/10.1080/17518253.2020. 1802517

[50] Iida, S., & Sakata, K. (2019). Hydrogen technologies and developments in Japan. Clean Energy, 3(2), 105–113.

[51] Iravani, S. (2014). Bacteria in nanoparticle synthesis: Current status and future prospects. International Scholarly Research Notices, 2014, Article ID 359316, 18. https://doi.org/10.1155/ 2014/359316

[52] IRENA (2019). Hydrogen: A renewable energy perspective. International Renewable Energy Agency, Abu Dhabi. https://www.irena.org/-/media/Files/IRENA/Agency/Publication/2019/ Sep/IRENA_Hydrogen_2019.pdf

[53] Jacob, J. M., Lens, P. N. L., & Balakrishnan, R. M. (2015). Microbial synthesis of chalcogenide semiconductor nanoparticles: A review. Microbial Biotechnology, 9(1), 11–21. 10.1111/1751-7915.12297

[54] Jadhav, D. B., & Kokate, R. D., 2019. Characterization of TiO2 metal oxides nanoparticle was synthesized using plant extracts and solar cell fabrication using ITO glass, TiO2, Ruthenium, graphite, and potassium iodide. 3rd International Conference on "Advances in Power Generation from Renewable Energy Sources." SSRN-Elsevier, https://hq.ssrn.com/confer ence=2019-APGRES

[55] Jadhav, M. V., Todkar, A. S., Gambhire, V. R., & Sawat, S. Y. (2011). Nanotechnology for powerful solar energy. International Journal of Advanced Biotechnology and Research, 2(1), 208–212. http://www.bipublication.com

[56] Jadoun, S., Arif, R., Jangid, N. K., & Meena, R. K. (2020). Green synthesis of nanoparticles using plant extracts: A review. Environmental Chemistry Letters. https://doi.org/10.1007/ s10311-020-01074-x

[57] Jette, E. R., & Foote, F. (1935). Precision determination of lattice constants. The Journal of Chemical Physics, 3(10), 605–616.

[58] Kamali, M., Aminabhavi, T. M., Abbassi, R., Dewil, R., & Appels, L. (2021). Engineered nanomaterials in microbial fuel cells-recent developments, sustainability aspects, and future outlook. Fuel, 310, Part B, 122347. https://doi.org/10.1016/j.fuel.2021.122347

[59] Kapoor, R. T., Salvadori, M. R., Rafatullah, M., Siddiqui, M. R., Khan, M. A., & Alshareef, S. A. (2021). Exploration of microbial factories for synthesis of nanoparticles – A sustainable approach for bioremediation of environmental contaminants. Frontiers in Microbiology, 12, 658294. 10.3389/fmicb.2021.658294

[60] Kawada, I., Isobe, M., Okamura, F. P., Watanabe, H., Ohsumi, K., Horiuchi, H. . . . Ishii, T. (1986). Time-of-flight neutron diffraction study of Li3N at high temperature. Mineralogical Journal, 13(1), 28–33.

[61] Khan, I., Saeed, K., & Khan, I. (2019). Nanoparticles: Properties, applications, and toxicities. Arabian Journal of Chemistry, 12, 908–931. https://doi.org/10.1016/j.arabjc.2017.05.011

[62] Kharissova, O. V., Kharisov, B. I., Oliva González, C. M., Méndez, Y. P., & López, I. (2019). Greener synthesis of chemical compounds and materials. Royal Society Open Science, 6(11), 191378–1 to 191378–41.

[63] Koo, C. H., & Kim, H. S. (1965). The crystal structure of nicotine dihydrocodeine. Journal of the Korean Chemical Society, 9(3), 134–141.

[64] Korde, P., Ghotekar, S., Pagar, T., Pansambal, S., Oza, R., & Mane, D. (2020). Plant extract assisted eco-benevolent synthesis of selenium nanoparticles-A review on plant parts involved, characterization, and their recent applications. Journal of Chemical Reviews, 2(3), 157–168. 10.33945/SAMI/JCR.2020.3.3

[65] Koul, B., Poonia, A. K., Yadav, D., & Jin, J.-O. (2021). Microbe-mediated biosynthesis of nanoparticles: Applications and future prospects. Biomolecules, 11, 886. https://doi.org/10. 3390/biom11060886

[66] Kredy, H. M. (2018). The effect of pH, the temperature on the green synthesis, and biochemical activities of silver nanoparticles from Lawsonia inermis extract. Journal of Pharmaceutical Sciences and Research, 10(8), 2022–2026. https://www.researchgate.net/ publication/327571058

[67] Kroto, H. W., Heath, J. R., O'Brien, S. C., Curl, R. F., & Smalley, R. E. (1985). C60: Buckminsterfullerene. nature, 318(6042), 162–163.

[68] Kundrot, C. E., & Richards, F. M. (1987). Crystal structure of hen egg-white lysozyme at a hydrostatic pressure of 1000 atmospheres. Journal of Molecular Biology, 193(1), 157–170.

[69] Kuppusamy, P., Yusoff, M. M., Maniam, G. P., & Govindan, N. (2016). Biosynthesis of metallic nanoparticles using plant derivatives and their new avenues in pharmacological applications: An updated report. Saudi Pharmaceutical Journal, 24, 473–484. https://doi. org/10.1016/j.jsps.2014.11.013

[70] Le, L., Nguyen, T., & Nguyen, D. (2018). Microwave-assisted green synthesis of silver nanoparticles using mulberry leaves extract and silver nitrate solution. Technologies, 7(1), 7–1 to 7–9. 10.20944/preprints201811.0012.v1

[71] Liu, C.-J., Burghaus, U., Besenbacher, F., & Wang, Z. L. (2010). Preparation and characterization of nanomaterials for sustainable energy production. ACS NANO, 4(10), 5517–5526. 10.1021/nn102420c

[72] Liu, X., & Prewitt, C. T. (1990). High-temperature X-ray diffraction study of Co3O4: Transition from normal to disordered spinel. Physics and Chemistry of Minerals, 17(2), 168–172.

[73] Li, X., Gao, Q., Cao, Y., Yang, Y., Liu, S., Wang, Z. L., & Cheng, T. (2022). Optimization strategy of wind energy harvesting via triboelectric-electromagnetic flexible cooperation. Applied Energy, 307, 118311–1 to 118311–10.

[74] Manjamadha, V. P., & Muthukumar, K. (2016). Ultrasound-assisted green synthesis of silver nanoparticles using weed plant. The Journal Bioprocess and Biosystems Engineering. 10.1007/s00449-015-1523-3

[75] Mao, S. S., Shen, S., & Guo, L. (2012). Nanomaterials for renewable hydrogen production, storage, and utilization. Progress in Natural Science: Materials International, 22(6), 522–534. https://doi.org/10.1016/j.pnsc.2012.12.003

[76] Meng, X., Gu, A., Wu, X., Zhou, L., Zhou, J., Liu, B., & Mao, Z. (2021). Status quo of China's hydrogen strategy in transportation and international comparisons. International Journal of Hydrogen Energy, 46(57), 28887–28899.

[77] Mishra, A., Bhatt, R., Bajpai, J., & Bajpai, A. K. (2021). Nanomaterials based biofuel cells: A review. International Journal of Hydrogen Energy, 46(36), 19085–19105. https://doi.org/10. 1016/j.ijhydene.2021.03.024

[78]   Moghaddam, A. B., Namvar, F., Moniri, M., Md. Tahir, P., Azizi, S., & Mohamad, R. (2015). Nanoparticles biosynthesized by fungi and yeast: A review of their preparation, properties, and medical applications. Molecules, 20, 16540–16565. 10.3390/molecules200916540

[79]   Mohamed, N. A. N., Arham, N. A., Jai, J., & Hadi, A. (2014). Plant extract as a reducing agent in the synthesis of metallic nanoparticles: A review. Advanced Materials Research, 832, 350–355. 10.4028/www.scientific.net/AMR.832.350

[80]   Mohd Yusof, H., Mohamed, R., Zaidan, U. H., & Abdul Rahman, A. (2019). Microbial synthesis of zinc oxide nanoparticles and their potential application as an antimicrobial agent and a feed supplement in the animal industry: A review. Journal of Animal Science and Biotechnology, 10, 57. https://doi.org/10.1186/s40104-019-0368-z

[81]   Mohd Zainol, M. K., Abdul-Hamid, A., Abu Bakar, F., & Park Dek, S. (2009). Effect of different drying methods on the degradation of selected flavonoids in Centella Asiatica. International Food Research Journal, 16, 531–537. https://www.researchgate.net/profile/Khairi-Zainol/publication/270512422

[82]   Moore, K., & Wei, W. (2021). Applications of carbon nanomaterials in perovskite solar cells for solar energy conversion. NanoMaterials Science, 3, 276–290. http://doi.org/10.1016/j.nanoms.2021.03.005

[83]   Movchan, 7. A., & Demchishin, A. V. (1969). Structure and properties of thick condensates of nickel, titanium, tungsten, aluminum oxides, and zirconium dioxide in vacuum.. Fiz Metal Metalloved, 28, 653–660. Oct 1969.

[84]   Mughal, B., Zaidi, S. Z. J., Zhang, X., & Hassan, S. U. (2021). Biogenic nanoparticles: Synthesis, characterization, and applications. Applied Sciences, 11, 2598. https://doi.org/10.3390/app11062598

[85]   Shahid, M., Dumat, C., Khalid, S., Schreck, E., Xiong, T., & Niazi, N. K. (2016). Foliar heavy metal uptake, toxicity and detoxification in plants: A comparison of foliar and root metal uptake. Journal of Hazardous Materials. http://dx.doi.org/10.1016/j.jhazmat.2016.11.063

[86]   Mukherjee, P., Ahmad, A., Mandal, D., Senapati, S., Sainkar, S. R., Khan, M. I., Ramani, R. R., Parischa, R., Ajayakumar, P. V., Alam, M., Sastry, M., & Kumar, R. (2001). Bioreduction of AuCl4- ions by the fungus Verticillium sp. Moreover, surface trapping of the gold nanoparticles formed. Chemie International Edition, 40(19), 3586–3588. 10.1002/1521-3773 (20011001)40:19<3585::aid-anie3585>3.0.co;2-k

[87]   Muzammil, W. K., Rahman, Md.M., Fazlizan, A., Ismail, M. A., Phang, H. K., & Elias, M. A. (2019). Nanotechnology in renewable energy: Critical reviews for wind energy. In Siddiquee, S., et al. (ed.). Nanotechnology: Applications in energy, drug and food (pp. 49–71). Cham: Springer. https://doi.org/10.1007/978-3-319-99602-8_3

[88]   Naikoo, G. A., Mustaqeem, M., Hassan, I. U., Awan, T., Arshad, F., Salim, H., & Qurashi, A. (2021). Bioinspired and green synthesis of nanoparticles from plant extracts with antiviral and antimicrobial properties. Journal of Saudi Chemical Society, 25, 101304. https://doi.org/10.1016/j.jscs.2021.101304

[89]   National Renewable Energy Laboratory (NREL) Energy Basics (2022) https://www.nrel.gov/research/learning.html. Accessed on March 01 2022.

[90]   National Renewable Energy Laboratory (NREL) National Renewable Energy Laboratory Hydrogen Resource Data, Tools, and Maps (2006)https://www.nrel.gov/gis/hydrogen.html?print

[91]   Oh, Y. K., Hwang, K. R., Kim, C., Kim, J. R., & Lee, J. S. (2018). Recent developments and key barriers to advanced biofuels: A short review. Bioresource Technology, 257, 320–333.

[92]   Okonkwo, E. C., Abdullatif, Y. M., & AL-Ansari, T. (2021). A nanomaterial integrated technology approach to enhance the energy-water-food nexus. Renewable and Sustainable Energy Reviews, 145, 111118. https://doi.org/10.1016/j.rser.2021.111118

[93] Osman, A. I., Mehta, N., Elgarahy, A. M., Hefny, M., Al-Hinai, A., Al-Muhtaseb,, & Rooney, D. W. (2021). Hydrogen production, storage, utilization, and environmental impacts: A review. Environmental Chemistry Letters. https://doi.org/10.1007/s10311-021-01322-8

[94] Oswald, H. R., Reller, A., Schmalle, H. W., & Dubler, E. (1990). Structure of copper (II) hydroxide, Cu (OH) 2. Acta Crystallographica Section C: Crystal Structure Communications, 46(12), 2279–2284.

[95] Park, F., Gajiwala, K., Noland, B., Wu, L., He, D., Molinari, J. . . . Buchanan, S. (2004). The 1.59 A resolution crystal structure of TM0096, a flavin mononucleotide binding protein from Thermotoga Maritima. Proteins, 55(3), 772–774.

[96] Pan, Z., Lin, Y., Sarkar, B., Owens, G., & Chen, Z. (2020). Green synthesis of iron nanoparticles using red peanut skin extract: Synthesis mechanism, characterization, and effect of conditions on chromium removal. Journal of Colloid and Interface Science, 558, 106–114. https://doi.org/10.1016/j.jcis.2019.09.106

[97] Pandey, G. (2018). Nanotechnology for achieving a green economy through sustainable energy. Rasayäyan Journal of Chemistry, 11(3), 942–950. http://dx.doi.org/10.31788/RJC.2018.1133031

[98] Pandey, R. K., & Prajapati, V. K. (2018). Molecular and immunological toxic effects of nanoparticles. International Journal of Biological Macromolecules, 107, 1278–1293. https://doi.org/10.1016/j.ijbiomac.2017.09.110

[99] Park, Y., Hong, Y. N., Weyers, A., Kim, Y. S., & Linhardt, R. J. (2011). Polysaccharides and phytochemicals: A natural reservoir for the green synthesis of gold and silver nanoparticles. IET Nanotechnology, 5(3), 69–78. 10.1049/iet-nbt.2010.0033

[100] Patil, N., Bhaskar, R., Vyavhare, V., Dhadge, R., Khaire, V., & Patil, Y. (2021). Overview of methods of synthesis of nanoparticles. International Journal of Current Pharmaceutical Research, 13(2), 11–16. https://dx.doi.org/10.22159/ijcpr.2021v13i2.41556

[101] Peng, Y.-P., Liu, -C.-C., Chen, K.-F., Huang, C.-P., & Chen, C.-H. (2021). Green synthesis of nano-silver-titanium nanotube array (Ag/TNA) composite for concurrent ibuprofen degradation and hydrogen generation. Chemosphere, 264, 128407. https://doi.org/10.1016/j.chemosphere.2020.128407

[102] Perveen, R., Shujaat, S., Naz, M., Qureshi, M. Z., Nawaz, S., Shahzad, K., & Ikram, M. (2021). Meter. Res. Express, 8, 055007. https://doi.org/10.1088/2053-1591/ac006b

[103] Plucinski, A., Lyu, Z., & Schmidt, B. V. K. J. (2021). Polysaccharide nanoparticles: From fabrication to application. Journal of Materials Chemistry B, **9**, 7030–7062. 10.1039/D1TB00628B

[104] Popov, V., Hinkov, I., Diankov, S., Karsheva, M., & Handzhiyski, Y. (2015). Ultrasound-assisted green synthesis of silver nanoparticles and their incorporation in antibacterial cellulose packaging. Green Processing and Synthesis, 4, 125–131. https://doi.org/10.1515/gps-2014-0085

[105] Pourmortazavi, S. M., Taghdiri, M., Makari, V., & Rahimi-Nasrabadi, M. (2014). Procedure optimization for green synthesis of silver nanoparticles by aqueous extract of Eucalyptus oleosa, Spectrochimica Acta. Part A: Molecular and Biomolecular Spectroscopy. http://dx.doi.org/10.1016/j.saa.2014.10.010

[106] Prakash, P., Gnanaprakasam, P., Emmanuel, R., Arokiyaraj, S., & Saravanan, M. (2013). Green synthesis of silver nanoparticles from leaf extract of Mimusops elengi, Linn. for enhanced antibacterial activity against multi-drug resistant clinical isolates. Colloids and Surfaces. B, Biointerfaces, 108, 255–259.

[107] Priya, A. K., Krishna, R. S., Kumar, Y. G., Rohini, P., & Dillibabu, S., 2021. Green synthesis of silver nanoparticles by pulsed laser ablation using Citrus limetta juice extract for clad-

modified fiber optic gas sensing application. Proceedings Volume 11802, Nanoengineering: Fabrication, Properties, Optics, Thin Films, and Devices XVIII; 1180214 (2021) https://doi.org/ 10.1117/12.2597642. Event: SPIE Nanoscience + Engineering, 2021, San Diego, California, United States

[108] Prozorov, T., Bazylinski, D. A., Mallapragada, S. K., & Prozorov, R. (2013). Novel magnetic nanomaterials inspired by magnetotactic bacteria: Topical review. Materials Science and Engineering: R, 74(5), 133–172. https://dx.doi.org/10.1016/j.mser.2013.04.002

[109] Qamer, S., Romli, M. H., Che-Hamzah, F., Misni, N., Joseph, N. M. S., AL-Haj, N. A., & Amin-Nordin, S. (2021). Systematic review on biosynthesis of silver nanoparticles and antibacterial activities: Application and theoretical perspectives. Molecules, 26, 5057. https://doi.org/10. 3390/molecules26165057

[110] Rajput, D., Paul, S., & Gupta, A. D. (2020). Green synthesis of silver nanoparticles using waste tea leaves. Advanced Nano Research, 3(1), 1–14. https://doi.org/10.21467/anr.3.1.1-14

[111] Rathore, S., Sharma, S., Swain, B. P., & Ghadai, R. Kr., 2018. A critical review on triboelectric nanogenerator. International conference on mechanical, Materials, and Renewable Energy, IOP Conf. Series: Materials Science and Engineering, 377:012186. Doi:10.1088/1757-899X/ 377/012186

[112] Rauwel, P., Küünal, S., Ferdov, S., & Rau wal, E. (2015). A review on the Green synthesis of silver nanoparticles and their morphologies studied via TEM. Advances in Materials Science and Engineering, 2015, Article ID 682749, 9. http://dx.doi.org/10.1155/2015/682749

[113] Raynor, J. W., Minor, W., & Chruszcz, M. (2007). Dexamethasone at 119 K. Acta Crystallographica Section E: Structure Reports Online, 63(6), o2791–o2793.

[114] Rong, Y., Hu, Y., Mei, A., Tan, H., Saidaminov, M. I., Seok, S. I. . . . Han, H. (2018). Challenges for commercializing perovskite solar cells. Science, 361(6408), 1–7.

[115] Rouch, D. A., Lee, B. T. O., & Morby, A. P. (1995). Understanding cellular responses to toxic agents: A model for mechanism-choice in bacterial metal resistance. Journal of Industrial Microbiology, 14, 1326141. https://doi.org/10.1007/BF01569895

[116] Roychoudhury, A. (2020). Yeast-mediated green synthesis of nanoparticles for biological applications. Indian Journal of Pharmaceutical and Biological Research, 8(3), 26–31. https:// doi.org/10.30750/ijpbr.8.3.4

[117] Saallah, S., & Lenggoro, I. W. (2018). Nanoparticles carrying biological molecules: Recent advances and applications. KONA Powder and Particle Journal, 35, 89–111. 10.14356/ kona2018015

[118] Saif, S., Tahir, A., & Chen, Y. (2016). Green synthesis of iron nanoparticles and their environmental applications and implications. Nanomaterials, 6, 209. 10.3390/nano6110209

[119] Santhoshkumar, J., Rajeshkumar, S., & Venkat Kumar, S. (2017). Phyto-assisted synthesis, characterization, and applications of gold nanoparticles- A review. Biochemistry and Biophysics Reports, 11, 46–57. http://dx.doi.org/10.1016/j.bbrep.2017.06.004

[120] Saravanan, A., Kumar, P. S., Karishma, S., Vo, D.-V. N., Jeevanantham, S., Yaashikaa, P. R., & George, C. S. (2021). A review on the biosynthesis of metal nanoparticles and their environmental applications. Chemosphere, 264, 128580. https://doi.org/10.1016/j.chemo sphere.2020.128580

[121] Saxena, J., Sharma, M. M., Gupta, S., & Singh, A. (2014). The emerging role of fungi in nanoparticle synthesis and their applications. World Journal of Pharmacy and Pharmaceutical Sciences, 3(9), 1586–1613. https://www.researchgate.net/publication/273382953

[122] Shabalina, A. V., Svetlichny, V. A., & Kulinich, S. A. (2021). Green laser ablation-based synthesis of functional nanomaterials for hydrogen generation, storage, and detection. Current Opinion in Green and Sustainable Chemistry, 100566. https://doi.org/10.1016/j.cogsc.2021.100566

[123] Sharma, D., Kanchi, S., & Bisetty, K. (2019). Biogenic synthesis of nanoparticles: A review. Arabian Journal of Chemistry, 12, 3576–3600. https://doi.org/10.1016/j.arabjc.2015.11.002

[124] Singh, A., Gautam, P. K., Verma, A., Singh, V., Shivapriya, P. M., Shivalkar, S., Sahoo, A. K., & Samanta, S. K. (2020). Green synthesis of metallic nanoparticles as effective alternatives to treat antibiotics resistant bacterial infections: A review. Biotechnology Reports, 25, e00427. https://doi.org/10.1016/j.btre.2020.e00427

[125] Singh, A., Vihinen, J., Frankberg, E., Hyvärinen, L., Honkanen, M., & Levänen, E. (2016). Pulsed laser ablation-induced green synthesis of TiO2 nanoparticles and application of novel small angle X-ray scattering technique for nanoparticle size and size distribution analysis. Nanoscale research letters, 11(1), 1–9.

[126] Singh, J., Dutta, T., Kim, K.-H., Samddar, P., & Kumar, P. (2018). "Green" synthesis of metals and their oxide nanoparticles: Applications for environmental remediation. Journal of Nanotechnology, 16, 884. https://doi.org/10.1186/s12951-018-0408-4

[127] Skalickova, S., Baron, M., & Sochor, J. (2017). Nanoparticles biosynthesized by yeast: A review of their application. Kvasny Prumysl, 63(6), 290–292. 10.18832/kp201727

[128] Soundarrajan, C., Sankari, A., Dhandapani, P., Maruthamuthu, S., Ravichandran, S., Sozhan, G., & Palaniswamy, N. (2012). Rapid biological synthesis of platinum nanoparticles using Ocimum sanctum for water electrolysis applications. The Journal Bioprocess and Biosystems Engineering, 35, 827–833. 10.1007/s00449-011-0666-0

[129] Srihasam, S., Thyagarajan, K., Korivi, M., Reddy Lebaka, V., & Reddy Mallem, S. P. (2020). Phytogenic generation of NiO nanoparticles using Stevia leaf extract and evaluation of their in vitro antioxidant and antimicrobial properties. Biomolecules, 10, 89. 10.3390/biom10010089

[130] Sripadmanabhan Indira, S., Aravind Vaithilingam, C., Oruganti, K. S. P., et al. (2019). Nanogenerators as a sustainable power source: State of art, applications, and challenges. Nanomaterials, 9(5), 773. https://doi.org/10.3390/nano9050773

[131] Stern, F. T., & Beevers, C. A. (1950). The crystal structure of tartaric acid. Acta Crystallographica, 3(5), 341–346.

[132] Sutradhar, P., & Saha, M. (2015). Synthesis of zinc oxide nanoparticles using tea leaf extract and its application for solar cells. Bulletin of Materials Science, 38(3), 653–657. 10.1007/s12034-015-0895-y

[133] Tersoff, J., & Ruoff, R. S. (1994). Structural properties of a carbon-nanotube crystal. Physical Review Letters, 73(5), 676.

[134] Thakre, A., Kumar, A., Song, H.-C., et al. (2019). Pyroelectric energy conversion and its applications-flexible energy harvesters and sensors. Sensors, 19, 2170. 10.3390/s19092170

[135] Thananusak, T., Punnakitikashem, P., Tanthasith, S., & Kongarchapatara, B. (2020). The development of electric vehicle charging stations in Thailand: Policies, players, and key issues (2015–2020). World Electric Vehicle Journal, 12(1), 2.

[136] The Department of Energy Solar Energy Technologies Office (2022) https://www.energy.gov/eere/solar/solar-energy-technologies-office. Accessed on March 01 2022.

[137] Thunugunta, T., Reddy, A. C., & Lakshmana Reddy, D. C. (2015). Green synthesis of nanoparticles: Current prospectus. The Nanotechnology Reviews, 4(4), 303–323. https://www.researchgate.net/publication/277944432

[138] Ueno, K. (1984). Structure of salicin, C13H18O7. Acta Crystallographica Section C: Crystal Structure Communications, 40(10), 1726–1728.

[139] UNEP (2020). Solid Waste Management. UNEP-UN Environment Programme. Available online at: https://www.unenvironment.org/explore-topics/resourceefficiency/what-we-do/cities/solid-waste-management (accessed November 3, 2021).

[140] Verma, A., Gautam, S. P., Bansal, K. K., Prabhakar, N., & Rosenholm, J. M. (2019). Green nanotechnology: Advancement in phytoformulation research. Medicines, 6, 39. 10.3390/medicines6010039

[141] Verma, M. L., Dhanya, B. S., Thakur, M., Jeslin, J., & Jana, A. K. (2021). Plant-derived nanoparticles and their biotechnological applications. Comprehensive Analytical Chemistry, 331-362. https://doi.org/10.1016/bs.coac.2021.01.011

[142] Verma, S. K., & Tiwari, A. K. (2015). Application of nanoparticles in solar collectors: A review. Materials Today: Proceedings, 2(4-5), 3638–3647. 10.1016/j.matpr.2015.07.121

[143] Villaseñor-Basulto, D. L., Pedavoah, -M.-M., & Bandala, E. R. (2019). Plant materials for the synthesis of nanomaterials: Greener Sources In Handbook of Ecomaterials (pp. 105–121, Editors: Leticia Myriam Torres Martínez, Oxana Vasilievna Kharissova, Boris Ildusovich Kharisov) Springer, Cham. Berlin, Germany (pp. 105–121). 10.1007/978-3-319-68255-6_88

[144] Wang, C., Gao, X., Chen, Z., Chen, Y., & Chen, H. (2017). Preparation, characterization, and application of polysaccharide-based metallic nanoparticles: A review. Polymers, 9, 689. 10.3390/polym9120689

[145] Wang, M., Wang, G., Sun, Z., Zhang, Y., & Xu, D. (2019). Review of renewable energy-based hydrogen production processes for sustainable energy innovation. Global Energy Interconnection, 2(5), 436–443. 10.1016/j.gloei.2019.11.019

[146] Wever, F., & Müller, A. (1930). Über das Zweistoffsystem Eisen-Bor und über die Struktur des Eisenborides Fe4B2. Zeitschrift Für Anorganische Und Allgemeine Chemie, 192(1), 317–336.

[147] Wyckoff, R. W. G. (1960). Crystal structures: Chapt (Vol. 2, pp. 34–36). Interscience Publishers, Geneva, Switzerland.

[148] Wyckoff, R. W. G., & Wyckoff, R. W. (1963). Crystal structures (Vol. 1, pp. 26–27). New York: Interscience publishers.

[149] Yoshida, T., & Kojima, K. (2015). Toyota MIRAI fuel cell vehicle and progress toward a future hydrogen society. The Electrochemical Society Interface, 24(2), 45.

[150] Zhang, D., Ma, X.-L., Gu, Y., Huang, H., & Zhang, G.-W. (2020). Green synthesis of metallic nanoparticles and their potential applications to treat cancer. Frontiers in Chemistry, 8, 799. 10.3389/fchem.2020.00799

[151] Zheng, L., & Xuan, Y. (2021). Performance estimation of a V-shaped perovskite/silicon tandem device: A case study based on a bifacial heterojunction silicon cell. Applied Energy, 301, 117496–1 to 117496–12.

[152] Zhou, Q., Duan, J., Duan, Y., & Tang, Q. (2021). Review on engineering two-dimensional nanomaterials for promoting efficiency and stability of perovskite solar cells. Journal of Energy Chemistry, in Press, Journal Pre-proof. https://doi.org/10.1016/j.jechem.2021.09.017

[153] Zhou, J., Zheng, F., Li, H., Wang, J., Bu, N., Hu, P. . . . Liu, J. L. (2020). Optimization of post-treatment variables to produce hierarchical porous zeolites from coal gangue to enhance adsorption performance. Chemical Engineering Journal, 381, 122698–1 to 122698–15.

[154] Zhou, X. F., Qian, G. R., Dong, X., Zhang, L., Tian, Y., & Wang, H. T. (2010). Ab initio study of the formation of transparent carbon under pressure. Physical Review B, 82(13), 134126–1 to 134126–5.

[155] Zhuang, W., Yuan, D., Li, J. R., Luo, Z., Zhou, H. C., Bashir, S., & Liu, J. (2012). Highly potent bactericidal activity of porous metal-organic frameworks. Advanced Healthcare Materials, 1(2), 225–238.

[156] Teng, T. J., Arip, M. N. M., Sudesh, K., Nemoikina, A., Jalaludin, Z., Ng, E. P., & Lee, H. L. (2018). Conventional technology and nanotechnology in wood preservation: A review. BioResources, 13(4), 9220–9252.

[157] US Department of Energy–Energy Information Administration. (2001). Annual Energy Outlook with Projections to 2020.

[158] Zhang, L., Wang, Z., & Qiu, J. (2022). Energy-Saving Hydrogen Production by Seawater Electrolysis Coupling Sulfion Degradation. Advanced Materials, 34(16), 2109321.

[159] Kaur, M., Kaur, M., Singh, D., Oliveira, A. C., Garg, V. K., & Sharma, V. K. (2021). Synthesis of CaFe2O4-NGO nanocomposite for effective removal of heavy metal ion and photocatalytic degradation of organic pollutants. Nanomaterials, 11(6), 1471–1 to 1471–24.

[160] Ali, J., Hameed, A., Ahmed, S., Ali, M. I., Zainab, S., & Ali, N. (2016). Role of catalytic protein and stabilising agents in the transformation of Ag ions to nanoparticles by Pseudomonas aeruginosa. IET nanobiotechnology, 10(5), 295–300.

Saurabh Vyas, Vivek Anand, Roli Mishra

Chapter 2D
Synthesis and stabilization of metallic nanoparticles

Abstract: The development of varied methodologies for the synthesis of MNPs (metal nanoparticles) has gained tremendous boast because of the unique properties of the nanoparticles as compared to their bulk counterparts. The quantization of electronic effects results in special optical and magnetic properties. Moreover, a high surface-to-volume ratio promotes exceptional mechanical, thermal, and catalytic properties. Therefore, the synthesis and stabilization of metal nanoparticles are of paramount importance. Herein, we have highlighted the synthesis, stabilization, and applications of transition metallic nanoparticles like Pd, Ag, Au, Pt, Cu, and Co, using different methods, which include the sol–gel method, chemical reduction method, green method (natural resources), and colloidal injection method. Furthermore, we have elaborately discussed the different types of stabilizers used to stabilize the metallic nanoparticle. Moreover, the role of stabilizers and the effect on the size of nanoparticles have also been discussed.

Keywords: metallic nanoparticls, stabilization, nanotechnology, transition metals

2D.1 Introduction

Any particle with a diameter ranging between 1 and 100 nm in at least one dimension is defined as a nanoparticle [1]. Moreover, these materials can be zero-, one-, two-, or three-dimensional, depending on the shape. The properties of nanomaterials are unique in comparison with bulk materials. This is because of the better

Acknowledgments: RM and SV would like to thank IAR for supporting the research work. SV would like to recognize the Gujarat government for the Shodh research fellowship for carrying out the research. VA acknowledges Chandigarh University for its support.

Author contribution: The manuscript was written through the contributions of all the authors. All authorshave approved the final version of the manuscript.

Saurabh Vyas, Department of Engineering and Physical Sciences, Institute of Advanced Research, Gandhinagar, Gujarat 382426, India, e-mail: saurabhvyas.phd2020@iar.ac.in
Vivek Anand, Department of Chemistry, University Institute of Science, Chandigarh University, Gharuan, Mohali 140413, Punjab, India, e-mail: vivekanandac88@gmail.com
Roli Mishra, Department of Engineering and Physical Sciences, Institute of Advanced Research, Gandhinagar, Gujarat 382426, India, e-mail: rolimishra2@gmail.com

https://doi.org/10.1515/9783110739879-006

surface-to-volume ratio present in nanomaterials [2]. Because of their unique physical and chemical properties, the synthesis of metal nanoparticles exhibiting desired size, shape, morphology, surface area, and so on is significantly important. No surprise, nanomaterials have found their application in various fields such as catalysis [3], drug delivery [4], biosensors [5] paints and coatings [6], agriculture [7], and wastewater treatment [8].

Silver and gold have grabbed special attention among the noble metals because of their unparallel applications in catalysis, anticancer and antimicrobial properties. These metals in the nanoscale are excellent in exhibiting Gram-negative and Gram-positive bacteria. Nanoparticles can be synthesized by different methods like laser ablation [9], electro-explosion [10], chemical vapor deposition [11], and sol–gel synthesis [12]. The nanoparticles formed through chemical methods are often toxic to the environment. Therefore, biological methods involving living organisms such as bacteria, fungi, or plant extract are applied for environment-friendly and less toxic ways of obtaining nanoparticles.

Generally, there are two methods for the synthesis of nanoparticles: Top-down and bottom-up methods for the synthesis of metal nanoparticles [13]. In the top-down approach, bulk materials are broken down into nanosized materials. On the other hand, the bottom-up approach includes the synthesis of the nanoparticles atom by atom to nanosized particles with desired properties. The former is easy to implement and hence more frequently applied. The latter one is environmentally benign and less costly. Thus, both techniques have their pros and cons hence are used according to the requirements. MNPs are very reactive due to their high surface-to-volume ratio. To effectively utilize the properties of MNPs, stabilizers are used so that the agglomeration of MNPs can be restricted [14]. Metallic nanoparticles easily lose their size and shape when subjected to thermal and chemical changes. Therefore, stabilization is very important to maintain the shape and size and gain the maximum output from metallic nanoparticles [15]. Metal nanoparticles are stabilized by different stabilizers, such as steric stabilization, encapsulation of metal nanoparticles, and electrostatic stabilization via alloying [16]. The stability of nanoparticles is a challenge owing to their large surface energy. The utilization of stabilizing agents can check the coagulation of the particles to form bulk materials. These agents, through steric repulsion and electrostatic interaction, restrict coagulation. Examples include surfactants, polymers, and ligands.

The metal nanoparticles are characterized by the phenomenon of surface plasmon resonance (SPR), which is the vibration of the conducting electrons on interaction with the electromagnetic radiations. The noble metal nanoparticles show surface plasmon phenomenon due to their closely located conduction and valence band and free electron, which can move freely in these bands. Distinguished color and signature UV-visible peaks characterize the SPR. Changes in particle size, shape, and surrounding environment can affect this surface plasmon phenomenon [17].

2D.2 Synthesis and stabilization of transition metal nanoparticles

Transition metal nanoparticles are a vital class of MNPs with excellent properties, which include organic catalysis, electrocatalysis [18], energy storage, biological activity (antimicrobial, antibacterial), and optoelectronics [19]. The bimetallic nanoparticles are obtained when the transition metal is doped with noble metal nanoparticles to enhance its properties. Thus, these bimetallic nanoparticles have found applications in various fields. The synthesis of important metal nanoparticles like Cu, Ag, Au, Co, Pd, and Pt is discussed below.

2D.2.1 Synthesis and stabilization of palladium nanoparticles

Although an expensive metal, palladium finds significant importance in organic synthesis. Palladium nanoparticles (PdNPs) have excellent catalytic properties; hence, they are used in reactions like hydrogenation of alkynes or alkenes, petroleum cracking, and carbon-carbon cross-coupling reaction (Suzuki coupling and Heck coupling reaction). PdNPs can be prepared from different synthetic methods such as sonochemical, chemical reduction, and electrochemical [20]. The most commonly used method is the synthetic method, in which the reduction of palladium salt is accomplished using various reducing agents. The reducing agents used are hydrazine, sodium borohydride, ascorbic acid, and alcohol. Because of the high surface area of the PdNPs, they are very reactive and hence need stabilizers such as polymers, dendrimers, micelles, citrate salts, and toluene [21]. Some important synthetic methodologies applied to synthesize palladium nanoparticles are discussed below.

Athwale et al. synthesized the PdNPs by reflux and γ-radiolysis, while aniline was used as a nanoparticle stabilizer. To synthesize the PdNPs, the mixture of monomers, metal salt, and the solvent was bubbled with nitrogen gas to remove oxygen. Then, the mixture was refluxed for a few hours at a fixed temperature. Aniline was the best stabilizer among the other monomers like *N*-ethyl aniline, *N*-methyl aniline, *o*-anisidine, and *o*-toluidine (Fig. 2D.1). PdNPs showed polydispersity when synthesized by the γ-irradiation method [22].

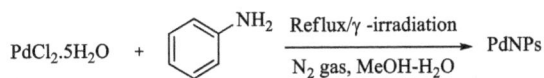

$$PdCl_2.5H_2O \quad + \quad \overset{NH_2}{\bigcirc} \quad \xrightarrow[\text{N}_2 \text{ gas, MeOH-H}_2\text{O}]{\text{Reflux/}\gamma \text{ -irradiation}} \quad PdNPs$$

Fig. 2D.1: Synthesis of PdNPs using aniline as a stabilizer.

K. Mallick et al. reported the synthetic route for the metal-polymer uniform composite material, in which metal nanoparticles were dispersed in the host polymer. The o-methoxyaniline solution was prepared by dissolving it in toluene with stirring. Then, palladium acetate was added drop by drop to obtain the precipitates with a deep yellow color (Fig. 2D.2). The PdNPs were stabilized in a polymer matrix under the reducing condition [23].

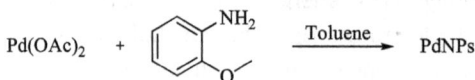

Pd(OAc)$_2$ + [o-methoxyaniline structure with NH$_2$ and O groups] $\xrightarrow{\text{Toluene}}$ PdNPs

Fig. 2D.2: Synthesis of PdNPs using o-methoxyaniline as a stabilizer.

Chen et al. [24] reported the synthesis of PdNPs. The H$_2$PdCl$_4$ and sodium citrate solution were mixed, stirred, and then H$_2$O$_2$ was added. After the complete dissolution, the gold seed was added. The color of the solution changed from yellow to gray. After the color transition, the solution was centrifuged, and suspensions were washed with Milli Q water and dried (Fig. 2D.3). The size of PdNPs was between 33 and 110 nm [24].

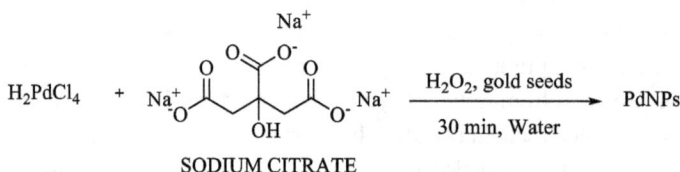

H$_2$PdCl$_4$ + Na$^+$ [sodium citrate structure] Na$^+$ $\xrightarrow[\text{30 min, Water}]{\text{H}_2\text{O}_2,\text{ gold seeds}}$ PdNPs

SODIUM CITRATE

Fig. 2D.3: Synthesis of PdNPs using sodium citrate as a stabilizer.

2D.2.2 Synthesis and stabilization of gold and silver nanoparticles

Gold nanoparticles (AuNPs) are most commonly obtained by the chemical reduction method. Gold salts are reduced in the presence of a reducing agent, such as reducing gold and chloroauric acid in sodium citrate. AuNPs are also predominantly obtained by using thiol ligands. This method (Burst-Schifrin method) involves two steps. First is the transfer of gold to the organic phase by tetrabutylammonium bromide salt of gold, then the reduction by the thiol ligand. Finally, the strong reducing agent NaBH$_4$ is added to give AuNPs protection by the thiolate group. During the synthesis of AuNPs, size control, thermal stability, and ease of preparation need to be optimized [25].

One of the most common metal nanoparticles synthesized vibrantly is the silver nanoparticles (AgNPs). This is because of many applications of AgNPs in catalysis, antimicrobial, anticancer activity, sensors, and so on. One of the popular methods is the reduction of silver salts by citrate anion. Citrate anion plays the dual role of stabilizer and a reducer for Ag$^+$ to Ag^0. Moreover, the concentration of citrate anion controls the size and shape of the AgNPs. Furthermore, at room temperature, the reduction of

Ag^+ to Ag^0 is obtained by gallic acid, having a reduction potential of 0.5 V. The $-OH$ groups present in the molecular structure of gallic acid are mainly involved in the reduction procedure [26].

Apart from the above-discussed methods, physical and biological methods are used to form AgNPs. For instance, UV irradiation, followed by the addition of PVP (capping agent) and ethylene glycol (reducing agent), leads to the formation of AgNPs. Nowadays, plant extracts are frequently used as reducing agents for synthesis. Moreover, many bacteria and fungi are also used for this purpose. Example includes eukaryotic fungus and prokaryotic bacteria. However, the reaction rates are slower than chemical methods [27].

Moreover, the size, shape, and ease of formation of AgNPs significantly depend on the concentration of the plant extract. Nevertheless, biological methods are economical and eco-friendly. Major synthetic methods used to prepare Ag and Au-NPs are discussed below.

Prabhu Charan et al. highlighted the synthesis of MNPs (MNP = gold and silver nanoparticles) by using polymeric ionic liquid (PIL) as a stabilizer. PIL was dissolved in water, and the aqueous solution of metal salt was added dropwise with continuous stirring. The reaction was kept on stirring at room temperature under dark conditions for 12 h. Here, sodium borohydride was used as a reducing agent and added to the reaction mixture with heating at 40 °C for 12 h (Fig. 2D.4). The mixture was centrifuged and collected [28].

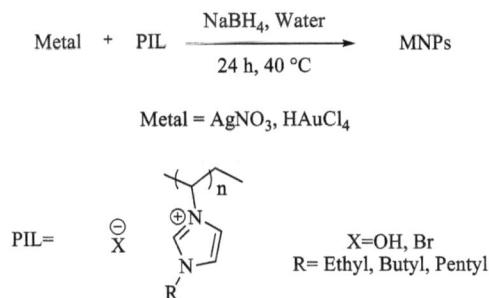

$$\text{Metal} \ + \ \text{PIL} \ \xrightarrow[\text{24 h, 40 °C}]{\text{NaBH}_4, \text{ Water}} \ \text{MNPs}$$

$$\text{Metal} = \text{AgNO}_3, \text{HAuCl}_4$$

PIL= $\overset{\ominus}{X}$ $X=OH, Br$
   $R=$ Ethyl, Butyl, Pentyl

Fig. 2D.4: Poly ionic liquid stabilized metal nanoparticles.

Sun et al. [29] proposed the synthesis of AgNPs using tea leaves extract. Solution of tea leaves extract was used as a reducing agent and capping agent. Different amount of the stock solution of tea leaves extract and silver nitrate solution was stirred (700 rpm) for 120 min. AgNPs were formed and purified by ultrafiltration and washed with Milli-Q-water (Fig. 2D.5). The maximum AgNPs production efficiency was achieved with 5 % (v/v) tea extract [29].

$$AgNO_3 \xrightarrow[\text{120 min, 55 °C}]{\text{Tea leaves extract, Water}} AgNPs$$

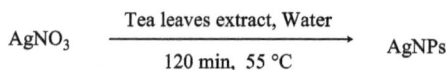

Fig. 2D.5: Synthesis of AgNPs using tea leaves extract.

Navaladian et al. prepared the solution of silver oxalate by mixing 50 mL of 0.5 M $AgNO_3$ solution with 30 mL of 0.5 M oxalic acid. Then, a convenient amount of polyvinyl alcohol (PVA) was added to 40 mL of water with stirring. Eventually, 0.05 g of $Ag_2C_2O_4$ was added with stirring, and after 10 min, the mixture was purged with N_2 gas. Then, the mixture was refluxed with a continuous nitrogen gas flow at 100 °C for 3 h in an oil bath till the mixture turned into a yellow-colored colloidal solution. The reaction mixture was then cooled to room temperature under the N_2 atmosphere. Finally, the solution was centrifuged to yield Ag nanoparticles (Fig. 2D.6). Polyvinyl alcohol was the capping agent, and it was also involved in the reduction process [30].

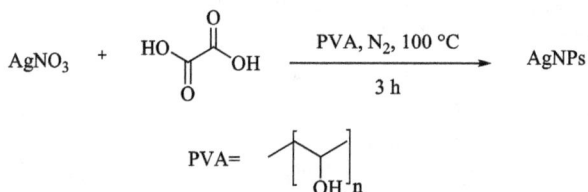

Fig. 2D.6: Synthesis of AgNPs using polyvinyl alcohol as a stabilizer.

Wei et al. reported the synthesis of Ag and Au NPs using chitosan as a mediator agent. To synthesize Ag/Au NPs, a metal salt solution was added to the chitosan solution under stirring conditions at room temperature. The reaction mixture was put in the water bath at 45–95 °C for 6– 12 h for reduction (Fig. 2D.7). The metal nanoparticles were characterized by transmission electron microscope (TEM) and FTIR [31].

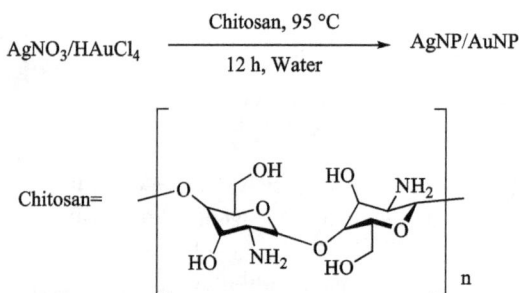

Fig. 2D.7: Synthesis of Ag/AuNP using chitosan as a stabilizer.

1-(3-(Acetylation)propyl)pyrazin-1-ium (APP) solution was prepared by adding APP into a 10 % mixture of methanol and deionized water by Anwar et al. This solution was added to the aqueous solution of tertachloroauric acid and stirred for 10 min. A freshly prepared $NaBH_4$ solution was added to reduce the reagent (Fig. 2D.8). The final reaction mixture was stirred for 2 h at room temperature to reduce the gold salt (if any). The solution was centrifuged to obtain the solid AuNPs, and the suspensions were washed with distilled water and dried [32].

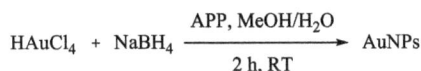

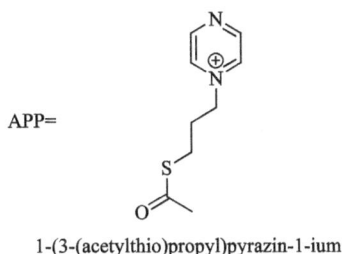

$$HAuCl_4 + NaBH_4 \xrightarrow[\text{2 h, RT}]{\text{APP, MeOH/H}_2\text{O}} AuNPs$$

APP=

1-(3-(acetylthio)propyl)pyrazin-1-ium

Fig. 2D.8: Synthesis and stabilization of AuNPs using APP.

2D.2.3 Synthesis and stabilization of copper nanoparticles

The chemical reduction method is the most popular one for synthesizing copper nanoparticles. Other widely used methods are microwave-assisted green synthesis, sol–gel auto combustion method, mechanochemical synthesis, sonochemical method, and colloidal injection method [33]. CuNPs exhibit unique properties such as very high electrical conductivity, low cost, and low electrochemical migration behavior. To stabilize CuNPs, OLA (oleylamine), TOP (tri-*n*-octyl phosphine), chitosan, and Tween-80 have been frequently used. Functional groups, such as amine and alcohol, present in the stabilizers above are mainly responsible for stabilizing CuNPs. CuNPs show antimicrobial and anti-tumor activity, and it has been reported that amino acid chelated CuNPs give good antimicrobial properties [34]. Moreover, CuNPs are also frequently used for the catalytic reduction of nitroaromatic compounds [35], in bio-sensing [36], and as a disinfectant for wastewater. The different synthetic methodologies adopted for the preparation of CuNPs are discussed below.

The aqueous solution of copper(II) sulfate pentahydrate ($CuSO_4 \cdot 5H_2O$) and orange juice extract were taken in 1:2 ratios and kept in a microwave system stirring for 15 min. After 15 min, a dark brown colloidal mixture was obtained (Fig. 2D.9) and the resultant copper nanoparticles [37].

$$CuSO_4.5H_2O \xrightarrow[\text{MW, 15 min.}]{\text{Water, Orange Juice Extract}} CuNPs$$

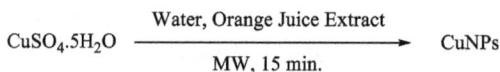

Fig. 2D.9: Synthesis of CuNPs using orange juice extract.

The copper (III) formate was dissolved in the mixture of oleylamine and dodecane at 60 °C by Oliva-Puigdomènech et al. The temperature was increased to 140 °C at the heating rate of 10 °C/min. Copper nanocrystals (CuNCs) were extracted from the reaction mixture and purified by repetitive precipitation/resuspensions cycles (Fig. 2D.10) using toluene as solvent and ethanol or methanol as non-solvents [38].

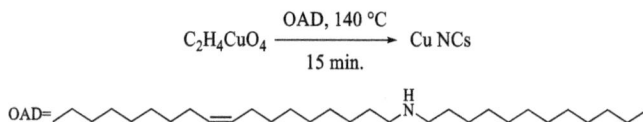

$$C_2H_4CuO_4 \xrightarrow[\text{15 min.}]{\text{OAD, 140 °C}} Cu\ NCs$$

OAD=

Fig. 2D.10: Synthesis of CuNCs using OAD as a stabilizer.

Shu et al. [39] reported the green synthesis of C/Cu-NPs by the sol–gel auto combustion method. Aqueous solutions of $Cu(NO_3)_2 \cdot 3H_2O$ and Tween-80 were mixed, and the pH of the resulting solution was maintained at 7.0 by using NaOH. Then, the g-PGA solution was added to the mixture and heated in a water bath at 60 °C so that water evaporated and the solution became thick. The viscous product obtained was heated to 150 °C in an oven. After that, it was heated at 400 °C in a muffle furnace for 10 min. Then, the reaction mixture was cooled to room temperature, washed with ethanol and distilled water, and finally dried at 100 °C. Thus, the biochar-conjugated copper NPs were obtained (Fig. 2D.11). In this method, γ-polyglutamic acid and Tween-80 are used as carbon sources. Tween-80 is a nonionic hydrophilic surfactant, and γ-PGA was used as a dispersant [39].

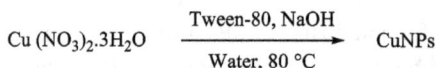

$$Cu\ (NO_3)_2.3H_2O \xrightarrow[\text{Water, 80 °C}]{\text{Tween-80, NaOH}} CuNPs$$

Fig. 2D.11: Synthesis of CuNPs using Tween-80 as a stabilizer.

N G et al. proposed the synthesis of copper oxide, copper sulfide, and copper selenide nanoparticles (Fig. 2D.12) by the colloidal hot injection method [40]. Zhou et al. reported the synthesis of CuNPs by using chitosan as a stabilizer and hydrazine as a reducing agent. To produce the CuNPs, chitosan solution was added to $CuCl_2$ solution with continuous stirring while the pH of the solution was maintained basic (9.0) using concentrated ammonia solution. After 10 min of stirring at room temperature, hydrazine hydrate was added to the mixture as a mild reducing agent. The reaction mixture was kept on stirring for 5 h. The color change of the solution to reddish-brown confirms the formation of copper nanoparticles (Fig. 2D.13). To stabilize the CuNPs, chitosan was used. Chitosan contains NH_2 and OH groups in its structure which help in stabilizing the nanoparticles by adsorbing on the surface of NPs [41].

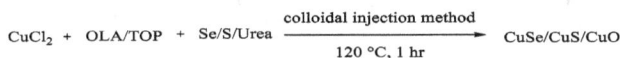

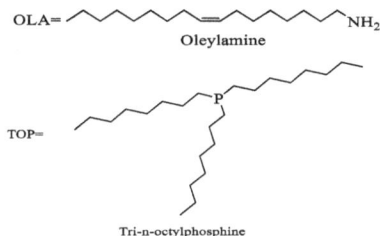

$$CuCl_2 + OLA/TOP + Se/S/Urea \xrightarrow[\text{120 °C, 1 hr}]{\text{colloidal injection method}} CuSe/CuS/CuO$$

OLA=

Oleylamine

TOP=

Tri-n-octylphosphine

Fig. 2D.12: Synthesis of CuONPs using OLA, TOP as a stabilizer.

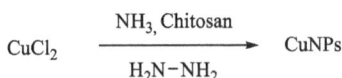

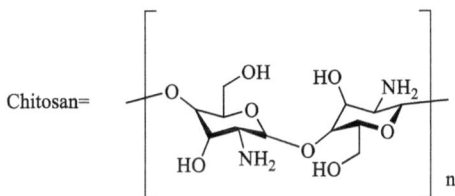

$$CuCl_2 \xrightarrow[\text{H}_2\text{N}-\text{NH}_2]{\text{NH}_3, \text{Chitosan}} CuNPs$$

Chitosan=

Fig. 2D.13: Chitosan stabilized CuNPs.

2D.2.4 Synthesis and stabilization of cobalt nanoparticles

Cobalt nanoparticles are gaining considerable research interest because of their biomedical and catalytic applications. Other major application includes the asymmetric catalytic hydrogenation of ketone [42]. CoNPs show antimicrobial, anticancer, antifungal, antioxidant, and enzyme inhibition properties in the biomedical field, and it is also used as biomedical sensors [43]. To prepare the CoNPs, chemical reduction, thermal decomposition, and precipitation methods are commonly used.

Shao et al. reported the synthesis of CoNPs by using cobalt acetate tetrahydrate through the thermal decomposition method. Trioctylphosphine (TOP), oleylamine, and oleic acid were used as stabilizers to control the size of CoNPs. To synthesize the CoNPs, the cobalt acetate solution was prepared by dissolving it in 2-pyrrolidinone and add mixture of oleylamine, oleic acid, and TOP. Then, the mixture was heated to 260 °C for 90 min under an N_2 gas atmosphere to prevent oxidation. The precipitation of CoNPs was achieved using ethanol. After frequent washing with ethanol, CoNPs were dried in a vacuum and characterized by TEM, X-ray diffractometer, and a vibrating sample magnetometer. The synthesized CoNPs were stabilized by the mixture of TOP, oleic acid, and oleylamine (Fig. 2D.14). With the addition of the stabilizer, the average particle size was reduced from 200 to 8 nm. The TOP as a stabilizer provides significant steric hindrance owing to the long-chain structure. Hence, the particles were well dispersed with no trace of agglomeration. Oleic acid binds tightly with the metal NP's surface. Hence, the use of the mixture of TOP, OA, and oleylamine enhances the stability of nanoparticles [44].

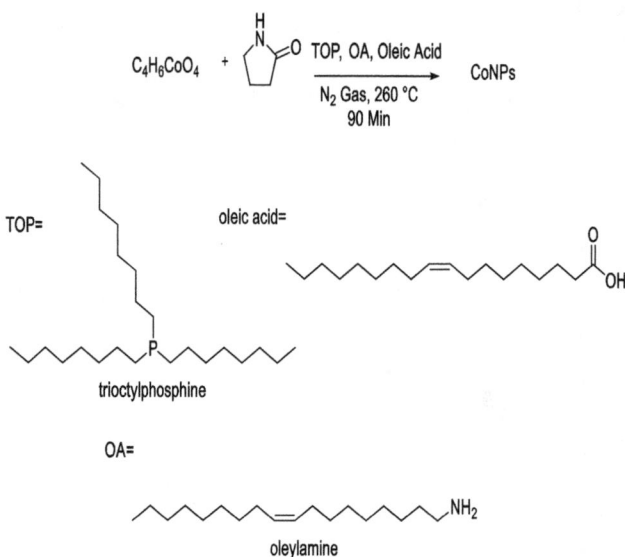

Fig. 2D.14: Synthesis of CoNPs using TOP, OA, and oleic acid as a stabilizer.

Santos et al. reported the synthesis of cobalt hydroxide nanoparticles using the precipitation method. For this method, an aqueous solution of $Co(NO_3)_2 \cdot 6H_2O$ was mixed with NaOH solution at room temperature with continuous stirring (2,400 rpm). As a result, black ppt was formed (Fig. 2D.15), which was extracted by centrifugation and washed with a sufficient amount of distilled water, and dried at 70 °C [45].

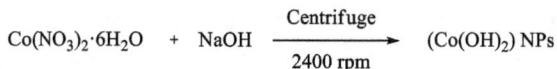

Fig. 2D.15: Synthesis of CoNPs of oleic acid in the presence of aniline.

In the following method, the KOH solution was added dropwise into the aqueous solution of CoCl$_2$.6H$_2$O with continuous stirring. Hydrazine monohydrate and oleic acid (OA) were added with vigorous stirring for 20 min. The mixture was kept in the autoclave. After the complete dissolution of the solid in solution, it was added to a Teflon-lined autoclave and was placed in an electric oven at 160 °C for 1 day (Fig. 2D.16). Then, the mixture was cooled at room temperature. To obtain the CoNPs, the solution was centrifuged, and the suspensions were washed with distilled water and dried [46].

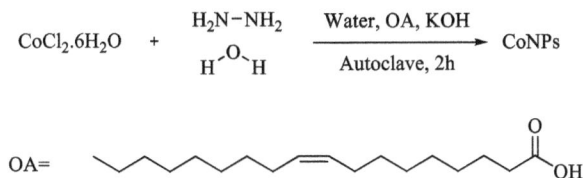

Fig. 2D.16: Synthesis of CoNPs using water OA and KOH.

In another study by Ansari et al., the mixture of solutions of sodium citrate dihydrate and aqueous cobalt sulfate was deaerated by bubbling argon gas through the solution. The solution was then heated for around 60–120 min at a fixed temperature (Fig. 2D.17). To obtain the CoNPs, the solution was centrifuged, and suspensions were washed with distilled water and dried [47].

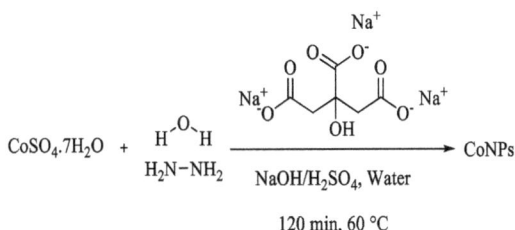

Fig. 2D.17: Synthesis of CoNPs using sodium citrate.

Metal salt and glycine (green fuel) were taken in a fixed ratio (1:4.4), and ammonia was used to maintain the basic pH of the solution (Fig. 2D.18). The final product was sintered at 600 °C for 5 h [48].

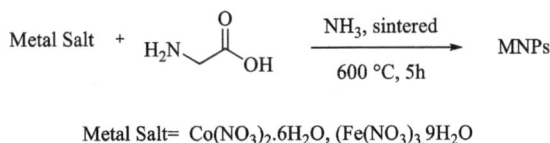

Metal Salt= Co(NO$_3$)$_2$.6H$_2$O, (Fe(NO$_3$)$_3$.9H$_2$O

Fig. 2D.18: Synthesis of CoNPs, Fe(NO$_3$)$_3$ NPs.

2D.2.5 Synthesis and stabilization of platinum nanoparticles

In the growing field of nanoscience, nanoparticles of noble metal platinum have attracted the attention of researchers because of their excellent applications and properties. For instance, platinum nanoparticles (PtNPs) show excellent catalytic properties for crucial organic transformations, antioxidant capabilities, low toxicity, anticancer, and antimicrobial properties [49]. PtNPs have also found crucial applications in the biomedical field. PtNPs can be synthesized by chemical reduction of Pt salt, green synthesis using natural bioactive compounds from different leaves extract [50] and water-in-oil micro-emulsion method. Metal nanoparticles are very sensitive to reaction conditions. Thus, the reaction time, the concentration of reducing or capping agent, and temperature can affect the size and shape of nanoparticles. To stabilize the PtNPs, different stabilizers such as thiol-containing polymer, natural leaves extract, and chlorogenic acid (CGA) are frequently used. The various synthetic procedures used for the preparation of PtNPs are discussed below.

Rong Chen et al. reported the synthesis of PtNPs using H_2PtCl_6 as a metal salt precursor and CGA as a reducing agent and stabilizer (Fig. 2D.19). The synthesis of platinum nanoparticles was accomplished using a single reaction flask containing the metal salt. A CGA solution was mixed with deionized water at 100 °C , and the reaction continued for 3 h. After that, a solution of H_2PtCl_6 was added to the mixture and incubated at 95 °C in a water bath for 2.5 h. Here, CGA was added as a reducing and capping agent for the preparation of PtNPs. Moreover, for higher yield, reaction time was increased [51].

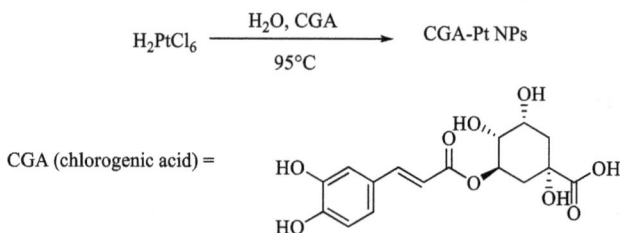

Fig. 2D.19: Synthesis of CGA capped PtNPs.

Roberto A et al. proposed the synthesis of PtNPs by the water-in-oil micro-emulsion method and investigated the effect of HCl on their surface structure. In this method, the nanoparticles were produced at room temperature, but this method allows only slight modification in the surface structure of the nanoparticles. For preparing PtNPs, H_2PtCl_6 was taken as a metal salt, and sodium borohydride was used to reduce the metal (Fig. 2D.20). As a stabilizer, polyethylene glycol dodecyl ether was used. The role of HCl was to control the size and shape of Pt nanoparticles. The optimal percentage was found to be 15–25 % of HCl [57].

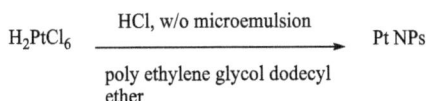

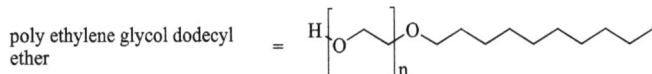

Fig. 2D.20: Synthesis of PtNPs using polyethylene glycol dodecyl ether as a stabilizer.

Raj Karthik et al. were the first to report the synthesis of ethylcellulose-supported platinum nanoparticles (Fig. 2D.21). The role of ethyl cellulose was to act as a reducing agent and stabilizer to prevent the aggregation of metal nanoparticles. The ion–dipole interaction in cellulose can disperse the metal nanoparticles on the entire surface of ethyl cellulose. For the synthesis of PtNPs, a solution of H_2PtCl_6 was added to the solution of ethyl cellulose in ethanol. The mixture was sonicated for 60 min. Then, the reducing agent ($NaBH_4$) was prepared freshly and added to the mixture till the color of the solution was completely converted to black [52].

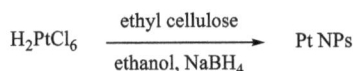

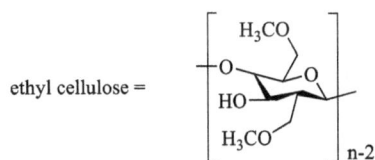

Fig. 2D.21: Synthesis of ethylcellulose-supported PtNPs.

Wei Li et al. reported the synthesis of PtNPs and their application in sensing. For the synthesis of PtNPs, dimethylamine borane (DMAB) was used as a reducing agent, and K_2PtCl_4 was used as a metal salt precursor. To stabilize the PtNPs, glutathione (GSH) was used as a stabilizer. Initially, the GSH and PBS buffer (pH 5.0) mixture was incubated with K_2PtCl_4 for 2 h at room temperature. Then, the reducing agent DMAB was freshly prepared and added to the mixture in a ratio ($DMAB/K_2PtCl_4$) of 5. The PtNPs were produced after 12 h of reduction at room temperature (Fig. 2D.22). GSH-capped PtNPs were utilized to sense heavy metal ions, selective towards Hg^{2+}. The limit of detection was found to be 0.25 nm.

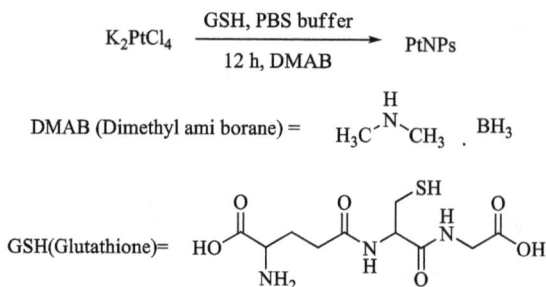

Fig. 2D.22: Synthesis of glutathione stabilized PtNPs.

2D.2.6 Synthesis of other important metal nanoparticles

To synthesize cerium nanoparticles (CeNPs), cerium (III) nitrate hexahydrate was used as a metal source and added to NaOH. In contrast, choline serinate ionic liquid was used as a stabilizer. The resulting mixture was stirred for 45 min and transferred to a glass tube. Then it was irradiated with an ultrasound bath for 12 h at room temperature (Fig. 2D.23). After the centrifugation, the product was separated and washed with ethanol and distilled water and then kept for drying at 80 °C overnight [53].

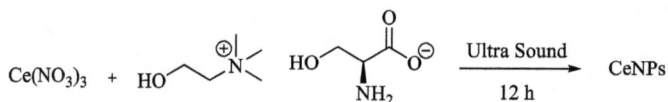

Fig. 2D.23: Synthesis of CeNPs using ionic liquid as a stabilizer.

To synthesize nickel nanoparticles (NiNPs), Wu and Chen took $NiCl_2$ as a Ni source and dissolved it in ethylene glycol. To the solution, hydrazine was added and followed by NaOH at 60 °C with stirring (Fig. 2D.24). The reaction mixture was kept on the same condition for 60 min, and finally, NiNPs were formed [54].

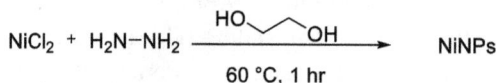

Fig. 2D.24: Synthesis of Ni NPs in the presence of ethylene glycol.

Hanying et al. reported the synthesis of semiconductor nanoparticles in salt-induced micelles (Fig. 2D.25). Polystyrene-*block*-2-vinyl pyridine (PS-*b*-P2VP) was taken and dissolved in THF with continuous stirring for about 1 h. Solution of cadmium acetate in the mixture of methanol and THF was added into the solution with continuous stirring at room temperature. The blue-colored solution showed the formation of polymer micelles. H_2S gas was bubbled through this solution for 10 min, and yellow cadmium sulfide nanoparticles (CdSNPs) were formed [55].

$$CdOAc \ + \ Polymer \ \xrightarrow[\text{H}_2\text{S Gas}]{\text{MeOH,THF}} \ CdSNPs$$

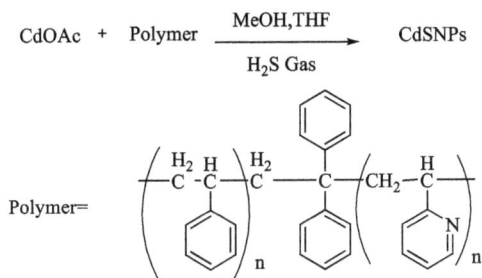

Polymer=

Fig. 2D.25: Synthesis and stabilization of CdNPs using (PS-b-P2VP).

2D.3 Conclusion

Due to their potential applications in vivid fields, metal nanoparticles have garnered special attention throughout the scientific communities working in chemical, physical, biological, and engineering domains. The noble metal nanoparticles have optical, magnetic, chemical, and biological properties ideal for vivid applications. Among the various methods for the preparation of nanoparticles, the chemical synthesis method has achieved special attention. Moreover, the radiolytic and photolytic methods are also important because of the advantages of low polydispersity and uniform size. Electron pulse radiolysis and laser techniques for nanoparticle synthesis help understand the nanoparticles' growth kinetics. This chapter discusses the various chemical methods used to synthesize transition metal nanoparticles of metals such as Pd, Ag, Au, Pt, Cu, and Co [56–57].

Moreover, the role of stabilizers and capping agents in the synthetic process has also been exhaustively discussed. More research needs to be done to synthesize at an industrial scale, and the applications may also be significantly enhanced. Therefore, the accumulation of many methodologies in the synthesis of NPs will help scientists design, engineer, and plan nobler approaches. This chapter is a humble contribution in this direction.

References

[1] Whitesides, G. M. (2003). The' right' size in nanobiotechnology. Nature Biotechnology, 21, 1161–1165.

[2] Rane, A. V., Kanny, K., Abitha, V. K. & Thomas, S. (2018). Methods for synthesis of nanoparticles and fabrication of nanocomposites. In Synthesis of inorganic nanomaterials (pp. 121–139, Editors: Sneha Mohan Bhagyaraj, Oluwatobi Samuel Oluwafemi, Nandakumar Kalarikkal, and Sabu Thomas), Woodhead publishing, Sawston, Cambridge.

[3] Taher, A., Nandi, D., Choudhary, M. & Mallick, K. (2015). Suzuki coupling reaction in the presence of polymer immobilized palladium nanoparticles: a heterogeneous catalytic pathway. New Journal of Chemistry, 39, 5589–5596. https://doi.org/10.1039/c5nj00969c.

[4] Kaphle, A., Navya, P. N., Umapathi, A. & Daima, H. K. (2018). Nanomaterials for agriculture, food, and environment: applications, toxicity, and regulation. Environmental Chemistry Letters, 16, 43–58. https://doi.org/10.1007/s10311-017-0662-y.

[5] Li, L. & Yang, Q.. Part II COATING MATERIALS NANOTECHNOLOGY 2019:175–233.

[6] Masoudi Asil, S., Ahlawat, J., Guillama Barroso, G. & Narayan, M. (2020). Nanomaterial based drug delivery systems for treating neurodegenerative diseases. Biomaterials Science, 8, 4088–4107. https://doi.org/10.1039/d0bm00809e.

[7] Willner, M. R. & Vikesland, P. J. (2018). Nanomaterial-enabled sensors for environmental contaminants Prof Ueli Aebi, Prof Peter Gehr. Journal of Nanobiotechnology, 16, 1–16. https://doi.org/10.1186/s12951-018-0419-1.

[8] Thekkudan, V. N., Vaidyanathan, V. K., Ponnusamy, S. K., Charles, C., Sundar, S. L., Vishnu, D. et al., (2017). Review on nano adsorbents: a solution for heavy metal removal from wastewater. IET Nanobiotechnology, 11, 213–224, https://doi.org/10.1049/iet-nbt.2015.0114.

[9] Lin, Z., Yue, J., Liang, L., Tang, B., Liu, B., Ren, L. et al., (2020). Rapid synthesis of metallic and alloy micro/nanoparticles by laser ablation towards the water. Applied Surface Science, 504, 144461, https://doi.org/10.1016/j.apsusc.2019.144461.

[10] Siwach, O. P. & Sen, P. (2008). Fluorescence properties of Fe nanoparticles prepared by electro-explosion of wires. Materials Science & Engineering B: Solid-State Materials for Advanced Technology, 149, 99–104. https://doi.org/10.1016/j.mseb.2007.12.007.

[11] Iravani, S., Korbekandi, H., Mirmohammadi, S. V. & Zolfaghari, B. Synthesis of silver nanoparticles: chemical, physicIravani, S., Korbekandi, H., Mirmohammadi, S. V, & Zolfaghari, B. (2014) Synthesis of silver nanoparticles: chemical, physical and biological methods. Research in Pharmaceutical Sciences, 9(6), 385–406. Res Pharm Sci 2014;9:385–406.

[12] Hasnidawani, J. N., Azlina, H. N., Norita, H., Bonnia, N. N., Ratim, S. & Ali, E. S. (2016). Synthesis of ZnO Nanostructures Using Sol-Gel Method. Procedia Chem, 19, 211–216. https://doi.org/10.1016/j.proche.2016.03.095.

[13] Saravanan, A., Kumar, P. S., Karishma, S., Vo, D. V. N., Jeevanantham, S., Yaashikaa, P. R. et al., (2021). A review on the biosynthesis of metal nanoparticles and its environmental applications. Chemosphere, 264, 128580, https://doi.org/10.1016/j.chemosphere.2020.128580.

[14] Pachón, L. D. & Rothenberg, G. (2008). Transition-metal nanoparticles: synthesis, stability and the leaching issue. Applied Organometallic Chemistry, 22, 288–299. https://doi.org/10.1002/aoc.1382.

[15] Feng, J., Gao, C. & Yin, Y. (2018). Stabilization of noble metal nanostructures for catalysis and sensing. Nanoscale, 10, 20492–20504. https://doi.org/10.1039/c8nr06757k.

[16] Cao, A., Lu, R. & Veser, G. (2010). Stabilizing metal nanoparticles for heterogeneous catalysis. Physical Chemistry Chemical Physics: PCCP, 12, 13499–13510. https://doi.org/10.1039/c0cp00729c.

[17] Baruwati, B. & Varma, R. S. (2009). High-value products from waste: grape pomace extract-a three-in-one package for synthesizing metal nanoparticles. ChemSusChem, 2, 1041–1044. https://doi.org/10.1002/cssc.200900220.

[18] Wang, C., Li, C., Liu, J. & Guo, C. (2021). Engineering transition metal-based nanomaterials for high-performance electrocatalysis. Mater Reports Energy, 1, 100006, https://doi.org/10.1016/j.matre.2021.01.001.

[19] Cid, A. & Simal-Gandara, J. (2020). Synthesis, Characterization, and Potential Applications of Transition Metal Nanoparticles. Journal of Inorganic and Organometallic Polymers and Materials, 30, 1011–1032. https://doi.org/10.1007/s10904-019-01331-9.

[20] Yang, X., Li, Q., Wang, H., Huang, J., Lin, L., Wang, W. et al., (2010). Green synthesis of palladium nanoparticles using broth of Cinnamomum camphora leaf. Journal of Nanoparticle Research, 12, 1589–1598, https://doi.org/10.1007/s11051-009-9675-1.

[21] Saldan, I., Semenyuk, Y., Marchuk, I. & Reshetnyak, O. (2015). Chemical synthesis and application of palladium nanoparticles. Journal of Materials Science, 50, 2337–2354. https://doi.org/10.1007/s10853-014-8802-2.

[22] Athawale, A. A., Bhagwat, S. V., Katre, P. P., Chandwadkar, A. J. & Karandikar, P. (2003). Aniline as a stabilizer for metal nanoparticles. Materials Letters, 57, 3889–3894. https://doi.org/10.1016/S0167-577X(03)00235-0.

[23] Mallick, K., Witcomb, M. J. & Scurrell, M. S. (2006). Formation of palladium nanoparticles in poly (o-methoxyaniline) macromolecule fibers: an in-situ chemical synthesis method. European Physical Journal E, 19, 149–154. https://doi.org/10.1140/epje/e2006-00027-2.

[24] Chen, H., Wei, G., Ispas, A., Hickey, S. G. & Eychmüller, A. (2010). Synthesis of palladium nanoparticles and their applications for surface-enhanced Raman scattering and electrocatalysis. The Journal of Physical Chemistry C, 114, 21976–21981. https://doi.org/10.1021/jp106623y.

[25] Venkatesh, N. (2018). Metallic Nanoparticle: a Review. Biomedical Journal of Scientific & Technical Research, 4, 3765–3775. https://doi.org/10.26717/bjstr.2018.04.0001011.

[26] Ibrahim, K., Saeed, K. & Idrees, K. (2019). Nanoparticles: properties, applications, and toxicities. Arabian Journal of Chemistry, 12, 908–931. https://doi.org/10.1016/j.arabjc.2017.05.011.

[27] Li, X., Xu, H., Chen, Z. S. & Chen, G. (2011). Biosynthesis of nanoparticles by microorganisms and their applications. Journal of Nanomaterials, 2011, https://doi.org/10.1155/2011/270974.

[28] Prabhu Charan, K. T., Pothanagandhi, N., Vijayakrishna, K., Sivaramakrishna, A., Mecerreyes, D. & Sreedhar, B. (2014). Poly(ionic liquids) as "smart" stabilizers for the metal nanoparticles. European Polymer Journal, 60, 114–122. https://doi.org/10.1016/j.eurpolymj.2014.09.004

[29] Sun, Q., Cai, X., Li, J., Zheng, M., Chen, Z. & Yu, C. P. (2014). Green synthesis of silver nanoparticles using tea leaf extract and evaluating their stability and antibacterial activity. Colloids Surface A Physicochem Eng Asp, 444, 226–231. https://doi.org/10.1016/j.colsurfa.2013.12.065.

[30] Navaladian, S., Viswanathan, B., Viswanath, R. P. & Varadarajan, T. K. (2007). Thermal decomposition as a route for silver nanoparticles. Nanoscale Research Letters, 2, 44–48. https://doi.org/10.1007/s11671-006-9028-2.

[31] Wei, D. & Qian, W. (2008). Facile synthesis of Ag and Au's nanoparticles utilizing chitosan as a mediator agent. Colloids Surfaces B Biointerfaces, 62, 136–142. https://doi.org/10.1016/j.colsurfb.2007.09.030.

[32] Anwar, A., Minhaz, A., Khan, N. A., Kalantari, K., Afifi, A. B. M. & Shah, M. R. (2018). Synthesis of gold nanoparticles stabilized by a pyrazinium thioacetate ligand: a new colorimetric nanosensor for detecting heavy metal Pd(II). Sensors Actuators, B Chem, 257, 875–881. https://doi.org/10.1016/j.snb.2017.11.040.

[33] Gawande, M. B., Goswami, A., Asefa, T., Huang, X., Silva, R., Zou, X. et al., (2016). Cu, and Cu-Based Nanoparticles: synthesis and Applications in Catalysis. Chemical Reviews, 116(6), 3722–3811. https://doi.org/10.1021/acs.chemrev.5b00482.

[34] Dealba-Montero, I., Guajardo-Pacheco, J., Morales-Sánchez, E., Araujo-Martínez, R., Loredo-Becerra, G. M., Martínez-Castañón, G. A. et al., (2017). Antimicrobial Properties of Copper Nanoparticles and Amino Acid Chelated Copper Nanoparticles Produced by Using a Soya Extract. Bioinorganic Chemistry and Applications, 2017, 15–17, https://doi.org/10.1155/2017/1064918.

[35] Kaur, R., Giordano, C., Gradzielski, M. & Mehta, S. K. (2014). Synthesis of highly stable, water-dispersible copper nanoparticles as catalysts for nitrobenzene reduction. Chemistry: An Asian Journal, 9, 189–198. https://doi.org/10.1002/asia.201300809.

[36] Din, M. I. & Rehan, R. (2017). Synthesis, Characterization, and Applications of Copper Nano-particles. Analytical Letters, 50, 50–62. https://doi.org/10.1080/00032719.2016.1172081.

[37] Jahan, I., Erci, F. & Isildak, I. (2021). Facile microwave-mediated green synthesis of non-toxic copper nanoparticles using Citrus sinensis aqueous fruit extract and their antibacterial potentials. Journal of Drug Delivery Science and Technology, 61, 102172, https://doi.org/10.1016/j.jddst.2020.102172.

[38] Oliva-Puigdomènech, A., De Roo, J., Van Avermaet, H., De Buysser, K. & Hens, Z. (2020). Scalable Approaches to Copper Nanocrystal Synthesis under Ambient Conditions for Printed Electronics. ACS Appl Nano Mater, 3, 3523–3531. https://doi.org/10.1021/acsanm.0c00242.

[39] Shu, X., Feng, J., Liao, J., Zhang, D., Peng, R., Shi, Q. et al., (2020). Amorphous carbon-coated nano-copper particles: novel synthesis by Sol-Gel and carbothermal reduction method and extensive characterization. Journal of Alloys and Compounds, 848, 156556, https://doi.org/10.1016/j.jallcom.2020.156556.

[40] Mbewana-Ntshanka, N. G., Moloto, M. J. & Mubiayi, P. K. (2020). Role of the amine and phosphine groups in oleylamine and trioctylphosphine in the synthesis of copper chalcogenide nanoparticles. Heliyon, 6, e05130, https://doi.org/10.1016/j.heliyon.2020.e05130.

[41] Hongfeng, Z., El-Kott, A., Ezzat Ahmed, A. & Khames, A. (2021). Synthesis of chitosan-stabilized copper nanoparticles (CS-Cu NPs): its catalytic activity for C-N and C-O cross-coupling reactions and treatment of bladder cancer. Arabian Journal of Chemistry, 14, 103259, https://doi.org/10.1016/j.arabjc.2021.103259.

[42] Michalek, F., Lagunas, A., Jimeno, C. & Pericàs, M. A. (2008). Synthesis of functional cobalt nanoparticles for catalytic applications. Use in asymmetric transfer hydrogenation of ketones. Journal of Materials Chemistry, 18, 4692–4697. https://doi.org/10.1039/b808383e

[43] Iravani, S. & Varma, R. S. (2020). Sustainable synthesis of cobalt and cobalt oxide nanoparticles and their catalytic and biomedical applications. Green Chemistry: An International Journal and Green Chemistry Resource: GC, 22, 2643–2661. https://doi.org/10.1039/d0gc00885k.

[44] Shao, H., Huang, Y., Lee, H. S., Suh, Y. J. & Kim, C. O. (2006). Cobalt nano-particles synthesis from Co(CH3COO)2 by thermal decomposition. Journal of Magnetism and Magnetic Materials, 304, 28–30. https://doi.org/10.1016/j.jmmm.2006.02.032.

[45] Santos, G. A., Santos, C. M. B., da Silva, S. W., Urquieta-González, E. A. & Sartoratto, P. P. C. (2012). Sol-gel synthesis of silica-cobalt composites by employing Co3O4 colloidal

dispersions. Colloids Surface A Physicochem Eng Asp, 395, 217–224. https://doi.org/10.1016/j.colsurfa.2011.12.033.

[46] Salman, S. A., Usami, T., Kuroda, K. & Okido, M. (2014). Synthesis and characterization of cobalt nanoparticles using hydrazine and citric acid. Journal of Nanotechnology, 2014, https://doi.org/10.1155/2014/525193.

[47] Ansari, S. M., Bhor, R. D., Pai, K. R., Sen, D., Mazumder, S., Ghosh, K. et al., (2017). Cobalt nanoparticles for biomedical applications: facile synthesis, physicochemical characterization, cytotoxicity behavior, and biocompatibility. Applied Surface Science, 414, 171–187, https://doi.org/10.1016/j.apsusc.2017.03.002.

[48] Borade, R. M., Kale, S. B., Tekale, S. U., Jadhav, K. M. & Pawar, R. P. (2021). Cobalt ferrite magnetic nanoparticles as highly efficient catalysts for the mechanochemical synthesis of 2-aryl benzimidazoles. Catalysis Communications, 159, 106349, https://doi.org/10.1016/j.catcom.2021.106349.

[49] Shiny, P. J., Mukherjee, A. & Chandrasekaran, N. (2014). Haemocompatibility assessment of synthesized platinum nanoparticles and its implication in biology. Bioprocess and Biosystems Engineering, 37, 991–997. https://doi.org/10.1007/s00449-013-1069-1.

[50] Zheng, B., Kong, T., Jing, X., Odoom-Wubah, T., Li, X., Sun, D. et al., (2013). Plant-mediated synthesis of platinum nanoparticles and its bioreductive mechanism. Journal of Colloid and Interface Science, 396, 138–145, https://doi.org/10.1016/j.jcis.2013.01.021.

[51] Chen, R., Wu, S. & Meng, C. (2021). Size-tunable green synthesis of platinum nanoparticles using chlorogenic acid. Res Chem Intermed, 47, 1775–1787. https://doi.org/10.1007/s11164-020-04377-4.

[52] Karthik, R., Karikalan, N. & Chen, S. M. (2017). Rapid synthesis of ethylcellulose supported platinum nanoparticles for the non-enzymatic determination of H2O2. Carbohydrate Polymers, 164, 102–108. https://doi.org/10.1016/j.carbpol.2017.01.077.

[53] Inbasekar, C. & Fathima, N. N. (2020). Collagen stabilization using ionic liquid functionalized cerium oxide nanoparticles. International Journal of Biological Macromolecules, 147, 24–28. https://doi.org/10.1016/j.ijbiomac.2019.12.271.

[54] Wu, S. H. & Chen, D. H. (2003). Synthesis and characterization of nickel nanoparticles by hydrazine reduction in ethylene glycol. Journal of Colloid and Interface Science, 259, 282–286. https://doi.org/10.1016/S0021-9797(02)00135-2.

[55] Zhao, H., Douglas, E. P., Harrison, B. S. & Schanze, K. S. (2001). Preparation of CdS nanoparticles in salt-induced block copolymer micelles. Langmuir, 17, 8428–8433. https://doi.org/10.1021/la011348q.

[56] Li, W., Zhang, H., Zhang, J. & Fu, Y. (2015). Synthesis and sensing application of glutathione-capped platinum nanoparticles. Analytical Methods: Advancing Methods and Applications, 7, 4464–4471. https://doi.org/10.1039/c5ay00365b.

[57] Roacho-Pérez, J. A., Ruiz-Hernandez, F. G., Chapa-Gonzalez, C., Martínez-Rodríguez, H. G., Flores-Urquizo, I. A., Pedroza-Montoya, F. E., . . . & Sánchez-Domínguez, C. N. (2020). Magnetite nanoparticles coated with PEG 3350-Tween 80: In vitro characterization using primary cell cultures. Polymers, 12(2), 300–1 to 300–321.

[58] Marínez-Rodríguez, R. A., Vidal-Iglesias, F. J., Solla-Gullon, J., Cabrera, C. R., & Feliu, J. M. (2014). Synthesis of Pt nanoparticles in water-in-oil microemulsion: effect of HCl on their surface structure. Journal of the American Chemical Society, 136(4), 1280–1283.

Section 3: **Focus on characterization methods**

Hao Zhang, Bingbao Mei, Zheng Jiang

Chapter 3A
Advances in understanding electrochemical reaction mechanisms of highly dispersed metal sites using X-ray absorption spectroscopy

Abstract: During the past decade, highly dispersed metal sites (HDMSs) have attracted considerable attention in electrocatalysis because of their intriguing catalytic performance and maximum efficiency of atomic utilization. These sites usually appear in single/dual-atom catalysts or subnanometric clusters, whose long-range structures are disordered and catalytic performances are intimately correlated with coordination environments and interactions with support. When applied in reaction, these unsaturated coordination sites within HDMSs strongly interact with the environment (support, electrolyte, ligands, adsorbates, reaction products, and intermediates), leading their structures to change with the reaction conditions. In this regard, clarifying the actual structure of HDMSs is of great importance for understanding the reaction mechanism and for further catalytic optimization. X-ray absorption spectroscopy (XAS) is an indispensable technique for probing the electronic and geometric structures of HDMSs. This chapter discusses the fundamental principles of the XAS method, introduces the experimental paradigm of data collection in the transmission and fluorescence models, and describes the data analysis approaches undertaken for deciphering X-ray absorption

Acknowledgments: This work was supported by the National Natural Science Foundation of China (grant no. 12105201), the Joint Fund of the National Natural Science Foundation of China (grant no. U1732267), Suzhou Key Laboratory of Functional Nano & Soft Materials, Collaborative Innovation Center of Suzhou Nano Science and Technology, the 111 Project, Joint International Research Laboratory of Carbon-Based Functional Materials and Devices, and Soochow University-Western University Centre for Synchrotron Radiation Research. The authors also gratefully acknowledge the helpful comments and suggestions of the reviewers, which have improved the presentation.

Authors' contribution H. Zhang completed the initial draft. B.B. Mei supplemented precious discussions in atomically dispersed metal sites and conclusion sections. Z. Jiang directed the structure and contents of the chapter and supervised the chapter preparation.

Hao Zhang, Institute of Functional Nano & Soft Materials (FUNSOM), Jiangsu Key Laboratory for Carbon-Based Functional Materials & Devices, Soochow University, 199 Ren'ai Road, Suzhou 215123, Jiangsu, PR China
Bingbao Mei, Zheng Jiang, Shanghai Synchrotron Radiation Facility, Zhangjiang Lab, Shanghai Advanced Research Institute, Chinese Academy of Sciences, Shanghai 201210, PR China, e-mail: jiangzheng@sinap.ac.cn

https://doi.org/10.1515/9783110739879-007

near the edge and extended X-ray absorption fine structure spectra. Moreover, we will illustrate the XAS studies of highly dispersed metal catalysts in a wide range of electrochemical reactions and highlight the application of in situ and operando XAS for revealing the nature of the active sites and establishing links between the structural motifs in HDMSs, local electronic structures, and catalytic properties.

Keywords: highly dispersed metal sites, electrocatalysis, XAS, operando characterization

3A.1 Introduction

As the contradiction between increasing energy demands and impending climate change gradually intensifies, developing sustainable energy to replace fossil fuels in the future energy framework has attracted global attention [1]. However, the intermittent nature of sustainable energy (e.g., solar, wind, and tides) limits its direct utilization in powering human life and manufacturing activities, a critical challenge. Electrochemical conversion processes provide a promising pathway to harvest the electricity generated by sustainable energy, which converts molecules in the atmosphere (e.g., water, carbon dioxide, and nitrogen) into higher-value products (e.g., hydrogen, hydrocarbons, oxygenates, and ammonia). Meanwhile, carbon dioxide emissions are reduced, mitigating climate change. Electrocatalysts play critical roles in these processes as their properties are closely related to the reaction rate, catalytic efficiency, and product selectivity of corresponding chemical transformations. However, the current electrocatalysts are inadequate. Greater effort is needed to develop efficient, robust, and low-cost electrocatalysts to promote the widespread penetration of electrochemical conversion technologies.

In recent years, highly dispersed metal sites (HDMSs) have become a hot topic in many electrocatalytic reactions, such as the hydrogen evolution reaction (HER), oxygen evolution reaction (OER), and oxygen reduction reaction (ORR), and carbon dioxide reduction reaction (CO2RR) [2]. Bottom-up or up-down strategies can construct the catalysts with HDMSs, elaborately reviewed in the previous works [3–6]. Generally, HDMSs present with a subnanometric size or low-dimensional shape often observed in single/dual-atom catalysts and subnanometric clusters. The most obvious characteristic of HDMSs is that almost every metal site is exposed to the environment, which makes HDMSs participate in reactions with an atomic-utilization efficiency of nearly 100 %. This characteristic is especially beneficial for noble-metal-based HDMSs because they show better performance but use a lower amount of noble metals than their bulk counterparts.

Nevertheless, this characteristic endows HDMSs with complicated and disordered long-range structures, making corresponding structural characterization challenging. Moreover, as these exposed sites possess high surface energy and strongly interact with the adsorbates, the HDMSs structure will change dynamically under reaction conditions.

Therefore, the huge obstacles to HDMS studies include accurately identifying the real structure of active sites and understanding their structure-activity relationships under realistic reaction conditions.

X-ray absorption spectroscopy (XAS) is a powerful characterization method that can provide catalysts electronic and geometric structures in a single experiment [7, 8]. As XAS originates from the core electron transitions of the selected element, it is an element-specific method. It enables one to extract the structural information of active sites from a multielement system. In addition, the XAS method shows less dependence on the form of catalysis, which allows it to detect a wide range of materials (ordered, disordered, nanostructured, liquids, etc.). Most importantly, comprehensive XAS analysis can provide very diverse information about a given sample: oxidation states, type of coordination atoms, coordination numbers, the interatomic distance, structure disorder, and so on, which are helpful to decipher the structure of the catalyst; if XAS measurements are performed under reaction conditions, one can investigate the catalysts structures in real-time and even probe the interactions between the active sites and adsorbates. Therefore, the XAS method is very suitable for HDMS studies [7–16].

Given that most research involving XAS studies of HDMSs comes from chemical or material fields, this chapter will start with the basic rules of XAS in a simplified way. Next, a brief overview of important experimental aspects and data analysis of the XAS method is presented; subsequently, the electrochemical instrumentations required for operando spectroscopic studies are introduced. Finally, we will give several examples to elaborate the studies of the XAS method on HDMSs in some typical electrocatalytic reactions (HER, OER, ORR, and CO2RR) and demonstrate the application of in situ operando XAS for electrochemical studies (Fig. 3A.1).

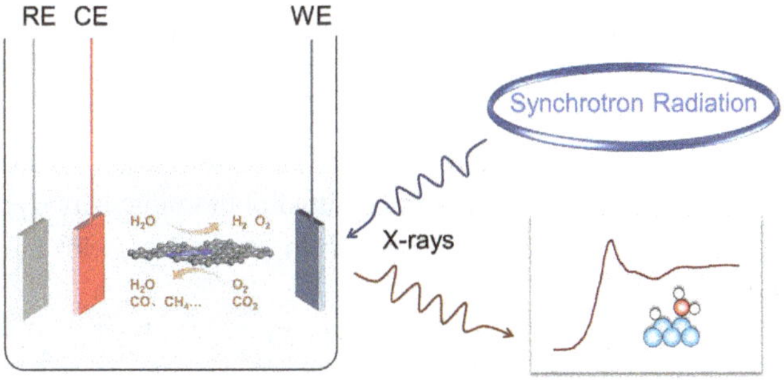

Fig. 3A.1: Schematic illustration of the XAS studies in electrochemical reactions.

3A.2 X-ray absorption spectroscopy

When an X-ray beam irradiates a material (Fig. 3A.2a), the intensity of the photons will be attenuated because of their complicated interactions with the atoms of the

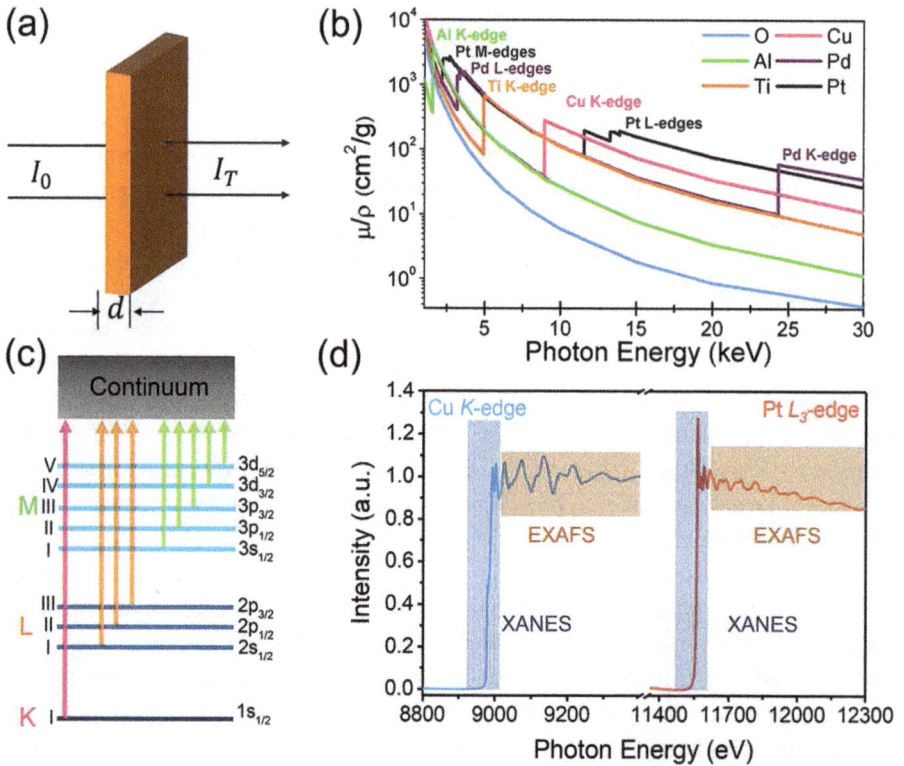

Fig. 3A.2: Fundamentals of X-ray absorption: (a) schematic of the incident and transmitted X-ray beam; (b) general behavior of the X-ray absorption coefficient as a function of the incident X-ray energy; (c) X-ray absorption edges nomenclature; (d) two representative XAS spectra at Cu K-edge from Cu foil and Pt L_3-edge from Pt foil (images of (b) and (c) were adapted from ref. [8] with permission).

material. Among these interactions, the absorption of X-ray can be described by the Beer–Lambert law in eq. (3A.1), which states that the transmitted photon intensity (I_T) decreases exponentially with increasing sample thickness (d) in comparison with the incident photon intensity (I_0):

$$I_T = I_0 \exp(-\mu d) \qquad (3A.1)$$

where μ is the so-called absorption coefficient, closely related to the incident photon energy (E). The empirical relation between $\mu(E)$ and the nature of the materials can be found as follows:

$$\mu(E) \sim \frac{\rho Z^4}{AE^3} \qquad (3A.2)$$

where ρ is the density of the material, Z is the atom's atomic number in the periodic table of elements, and A is the relative atomic mass. In a wide X-ray energy range, $\mu(E)$ decreases smoothly as the photon energy increases, as shown in Fig. 3A.2b. However, when the photon energy matches the binding energy of the core-level electrons, the intensity of $\mu(E)$ presents a sharp increase. In this case, the incident photons excite the core-level electrons and then produce photoelectrons to occupy the available unoccupied states in the valence band, resulting in a jump in the intensity of $\mu(E)$. Such a jump is defined as an absorption edge or absorption jump. As the photon energy increases, core-level electrons are excited to be free photoelectrons with a certain kinetic energy. These emitted electrons will interact with the neighboring atoms. Therefore, the intensity and features of the absorption edge depend on the electronic structures of the materials as well as their atomic structures. Thus, XAS is sensitive to the electronic and atomic structures of the absorbing atom and applies to both ordered and disordered samples in the gas, liquid, or solid phase [8, 17].

According to the excited core-level electrons (Fig. 3A.2c), the absorption edge can be classified as K, L, M, N, and so on. For example, Fig. 3A.2d shows a K-shell absorption edge at ~9.0 keV due to the absorption of X-ray photons by $1s$ electrons of Co atoms in Cu foil, which is referred to as the Cu K-edge. For comparison, an L-shell absorption at ~11.5 keV results from the absorption of X-ray photons by $2p_{3/2}$ electrons of Pt atoms in Pt foil referred to as the Pt L_3-edge. The position of the absorption edge (E_0) is defined at the maximum of the first derivative or position at the half-height of the jump. Before and after the absorption, the oscillated curves are identified as the X-ray absorption fine structure (XAFS). XAFS features at energies below and around the absorption edge are X-ray absorption near edge structures (XANES). The region approximately 20–30 eV above the absorption edge mainly results from the scattering of the photoelectrons and is referred to as the extended X-ray absorption fine structure (EXAFS). The fundamental theories of these two regions and corresponding structural information of the materials they can provide are discussed below.

3A.2.1 XANES basics

The absorption coefficient of X-rays corresponds to the transition probability between the initial state ($|i\rangle$) and final state ($|f\rangle$), which Fermi's golden rule describes:

$$\mu(E) \approx \sum_f |f|\hat{T}|i|^2 \delta\left(\epsilon_f - \epsilon_i - E\right) \tag{3A.3}$$

where ϵ_f is the final state energy, ϵ_i is the initial state energy, and E is the energy of the incident photon. The initial state is a two-particle state, containing one core-level electron and one incident photon; the final state is a mono-particle state that only contains a photoelectron. In the XANES regions, the core-level electron is excited out of the core-level orbital and into an unoccupied localized or delocalized state at a higher level. This process ensures the conservation of energy, as required by the Dirac function $\delta(\in_f - \in_i - E)$ in eq. (3A.3). The operator $\hat{T}$ deals with the interactions between electron and photon in the transition process, which involves a common dipole or more complex quadrupole approximation. It shows that the intensity and shape of the XANES spectrum are determined by both the number of unoccupied states and the selection rule of the transition process. In most cases, the dipole approximation with the selection rule of $l \rightarrow l+1$ can provide a good interpretation of the spectrum, in which transitions from $1s$ core-level are related to the p-DOS (e.g., K- and L_1-edges) and transitions from $2p$ core-level are related to the s- and d-DOSs (e.g., $L_{2,3}$-edges), whereas some K-edge XANES spectra show obvious pre-edge features, the quadrupole approximation with the selection rule of $\Delta l = 0, \pm 2$, for example, must be considered for a better explanation, where transitions from $1s$ states are related to the s- and d-DOSs.

3A.2.2 EXAFS basics

In the EXAFS region, the incident photon's energy is sufficient to make the core-level electron a free photoelectron. The photoelectron is then backscattered by the neighboring atoms and interferes with the outgoing photoelectron. This process will slightly perturb the absorption probability and result in the oscillatory structure's absorption coefficient $\mu(E)$. These oscillatory structures were related to the sample form (e.g., in the gas or solid phase), crystalline phase, temperature, and so on. The intrinsic relationships between the EXAFS oscillations and the local structure of the absorbing atoms became explicit thanks to Sayers, Stern, and Lytle [18]. Since then, EXAFS spectroscopy has become a powerful characterization tool. Their work has been considered a milestone for the development of EXAFS spectroscopy. Such transformation is based on the fact that the EXAFS signal mainly originates from the interactions between the excited photoelectron and the neighboring atoms. As shown in Fig. 3A.3a, to separate the EXAFS signal related to the neighboring atoms, one should remove the absorption probability $\mu_0(E)$ that belongs to the absorption of a bare atom. The normalization of the EXAFS component $\chi(E)$ is thus given by

$$\chi(E) = \frac{\mu(E) - \mu_0(E)}{\Delta \mu_0(E_0)} \tag{3A.4}$$

where $\mu(E)$ is the XAFS spectrum measured in the experiment, $\mu_0(E)$ is the background function, $\Delta \mu_0(E_0)$ is the absorption jump value, and E_0 represents the position of the absorption edge discussed in the previous section. Next, the EXAFS signal is converted from energy space to photoelectron wavenumber (k-) space by using the relation (Fig. 3A.3b)

$$k = \sqrt{\frac{2m_e}{\hbar^2}(E - E_0)} \qquad (3A.5)$$

where $\hbar$ is Planck's constant, and m_e is the electron mass.

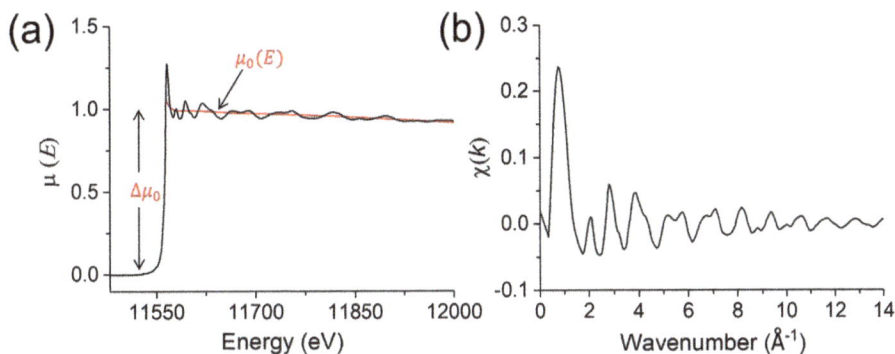

(a)

(b)

Fig. 3A.3: Transformation of raw XAS spectra from energy space to k-space: (a) Normalized Pt L_3-edge XAFS spectra of Pt foil; (b) Corresponding k-space spectra.

In the frame of the single-scattering (SS) approach, the k-weighted EXAFS signal can be expressed as a sum of contributions of different possible photoelectron scattering paths:

$$k\chi(k) = S_0^2 \sum_i \frac{N_i A_i}{r_i^2} e^{(-2r_i/\lambda)} e^{(-2\sigma_i^2 k^2)} \sin[2kr_i + 2\phi_i(k)] \qquad (3A.6)$$

where S_0^2 is the so-called amplitude reduction factor, λ is the photoelectron mean-free path, the sum contains the different coordination shells around the absorbing atom, A_i is the element-specific amplitude function of the neighboring atom, $\phi_i(k)$ is the phase function of the couple absorber/scatter, N_i is the number of scattering atoms in the ith shell, r_i is the corresponding interatomic distance, and σ_i is the relative Debye–Waller factor. It should be noted that the contributions of SS paths with close radial distance are grouped into a single coordination shell. The standard EXAFS formula in eq. (3A.6) is simplified to consider only the SS contributions. Multiple-scattering (MS) paths with more than two atoms involved can also contribute to the interference phenomenon and thus to the EXAFS signal. Compared to SS paths, MS paths often contribute less to the overall EXAFS signal because the low free mean path of the photoelectron penalizes longer paths. As the MS paths produce an obvious effect on the EXAFS signal, the dependence of the A_i and $\phi_i(k)$ functions on the relative positions of the absorbing and scattering atoms involved (e.g., bonding angles) must be considered.

3A.3 XAS experiment and data analysis

3A.3.1 XAS measurements in transmission and fluorescence modes

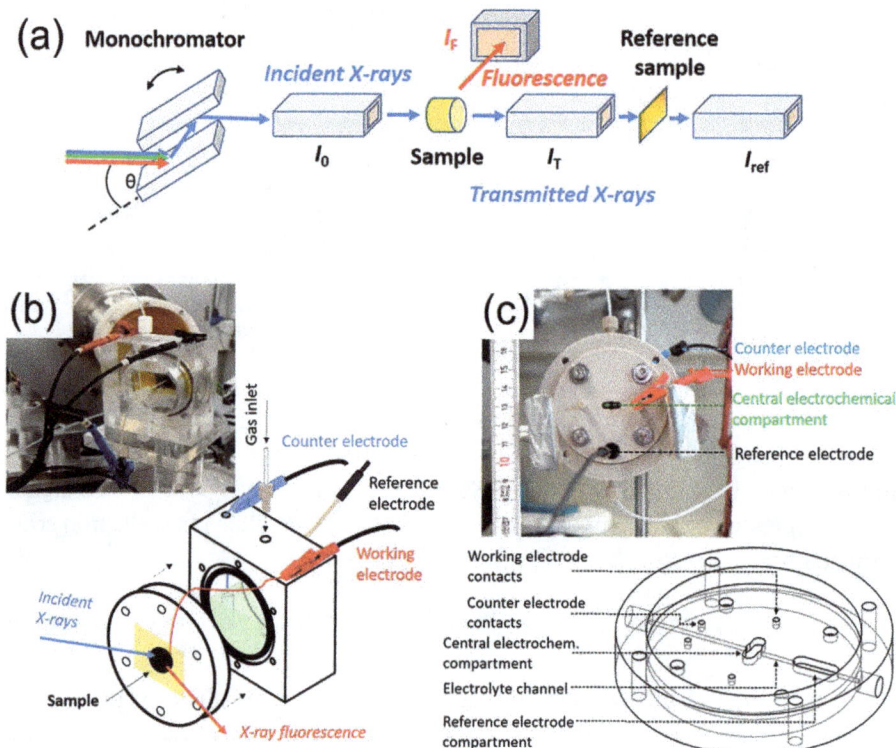

Fig. 3A.4: Schematics of common setups for XAS measurements: (a) XAS measurements in transmission and fluorescence modes with double crystal monochromator; electrochemical cell for XAS measurements (b) with large volume and (c) with continuous electrolyte (images were adapted from ref. [13] with permission).

In XAS experiments, the absorption coefficients of X-rays are recorded as functions of the photon energy. To acquire an eligible XAS spectrum, the X-ray sources should be guaranteed to have at least the following three factors: a broad X-ray energy range to cover the XANES and EXAFS regions simultaneously; an energy resolution up to a fraction of an eV to separate the XAS features accurately; and the high-intensity incident photons to ensure a signal-to-noise ratio better than 1,000 [13]. The properties of the investigated sample also determine the signal-to-noise ratio; commonly, the higher the concentration of the absorbing atoms, the stronger the signal-to-noise ratio. Some significant progress has been made in lab-based

XAS setups in the past decades; however, the X-ray sources used in these setups hardly meet the demands for XAS studies involving extremely dilute atoms or those conducted under operando conditions. Most XAS studies are still using X-ray sources from synchrotron radiation facilities.

There are two basic modes for XAS signal collection at synchrotron radiation facilities: transmission and fluorescence modes. Figure 3A.4a shows the common setup for the two XAS measurement modes. The X-ray beam produced by synchrotron radiation first reaches the monochromator. By rotating the monochromator crystals, the X-rays with wavelength λ are selected that fulfill the Bragg law $n\lambda = 2d_m \sin\theta$, where d_m is the crystal lattice spacing, and θ is the incident angle of the X-ray beam relative to the monochromator crystal. In transmission mode, the monochromatized X-rays pass through a series of detectors and the studied sample, where the intensities of the X-rays before and after the sample are measured and labeled as I_0 and I_T, respectively. According to the Beer–Lambert law in eq. (3A.1), the XAS signal of the selected element in the sample can be calculated. Considering that the obtained XAS signal will be normalized in a standard way, it is not necessary to know the sample thickness d. In fluorescence mode, the X-ray absorption intensity is often quantified by the emitted fluorescence intensity because of the intrinsic link between the absorption and fluorescence. The fluorescence detector is usually located at 90° concerning the X-ray beam, which can minimize the elastic scattering of the sample and eventually reduce the background signal. To record the intensity of the fluorescence photons (I_f), one can exploit different detectors, including classical Lytle detectors, photodiode detectors, multichannel silicon detectors, and silicon drift detectors. Once the intensity of the fluorescence photons has been recorded, the fluorescence XAS signal can be expressed as $\mu d = \frac{I_f}{I_0}$.

3A.3.2 In situ/operando XAS setup for electrochemical studies

The conventional XAS measurements conducted before or after electrochemical reaction are valuable to investigate the electronic and atomic structures of the specific metal sites. However, the catalysts, especially those containing HDMSs, will strongly interact with the reaction environment (support, electrolyte, adsorbates, intermediates, etc.), causing the structure of the active sites to change dynamically with the reaction conditions. In situ, XAS experiments can study the realistic structure of catalysts while the electrochemical reaction is proceeding and contribute to a better understanding of the structure-activity relationships. The most challenging aspect of the operando XAS measurements is to design a suitable electrochemical cell, which is required to achieve a satisfying compromise between electrochemical performance evaluation and X-ray detection. Figure 3A.4b and 3A.4c shows two single-compartment cells commonly used for operando XAS measurements [13, 19, 20]. These two cells adopt the same geometry with a three-electrode setup: a reference electrode, a counter electrode, and the sample as a working electrode. The catalysts are often deposited on conducting carbon-based electrodes (e.g., carbon paper and cloth), not interfering with data collection. The cell with a large volume (Fig. 3A.4b) is designed for fluorescence XAS measurements. The working electrode with the catalysts facing the electrolyte is mounted on the front panel of the cell. This geometry enabled the incident X-rays and emitted fluorescence to pass through the back of the sample, which can reduce the effect of the reaction bubble on the XAS signals. Another cell shown in Fig. 3A.4c can realize the operando XAS measurements in both transmission and fluorescence modes. A syringe pump can ensure the electrolyte pass through the cell continuously.

3A.3.3 Qualitative and semiquantitative analysis of XANES spectra

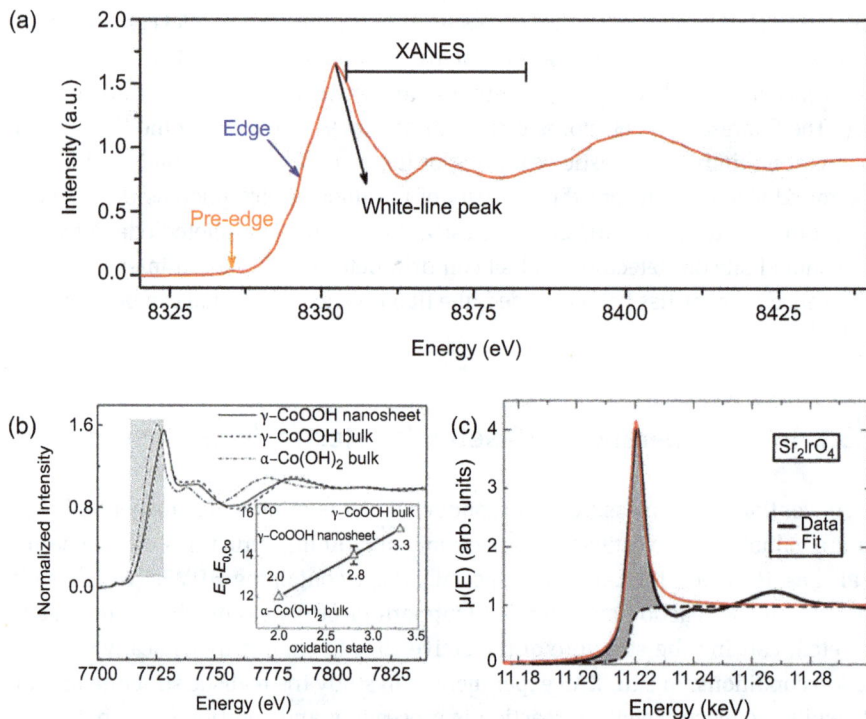

Fig. 3A.5: Details in XANES spectra: (a) XANES spectrum of NiO; (b) XANES spectra of Co-based materials showing how to fit oxidation state of Co in the sample quantitatively; (c) experimental determination of the white-line intensity at the Ir L_3-edge in Sr_2IrO_4 (image of (b) was adapted from ref. [21] with permission; image of (c) was adapted from ref. [22] with permission).

As discussed above, the spectral features in the XANES region mainly result from the transitions of excited core-level electrons to unoccupied localized or delocalized states, which can be used to probe the atomic and electronic structure of the investigated samples. The structural information provided by these features is closely related to their positions relative to the absorption edge. Such features can be separated into three components: pre-edge, edge, and XANES as shown in Fig. 3A.5a. Regarding these features, qualitative or semiquantitative analyses are often adopted to extract valuable structural information about the samples. More specifically, qualitative analysis of XANES spectra is conducted to identify the structural origins of the observed features by comparison with the spectra of reference materials with well-known structures; semiquantitative analysis of XANES spectra mostly relies on the positions and intensities of these features.

Pre-edge features are mainly caused by the electronic transitions to empty bound states. For the K-edge (s-to-p transition) of pure metal, the s-to-d transition is dipole-forbidden, and the allowed quadrupole is quite weak, eventually making their pre-edge features insignificant. When the metal atoms form chemical bonds with non-metallic elements (e.g., O atoms), the vacant d-orbital may assume the character of O $2p$-type states due to the hybridization effect. The corresponding transition occurs and leads to a sharp pre-edge peak. Commonly, the hybridization depends on the symmetry of the absorbing atoms and determines the intensity of the pre-edge: the pre-edge peak intensity for T_d symmetry is often larger than that for O_h symmetry. Thus, one can verify the symmetry of the absorbing atoms in a studied sample by comparing its pre-edge peak with a reference with well-known symmetry.

The absorption edge position represents the threshold of ionization of the core-level electrons to the continuum states. It is very sensitive to the oxidation state of the absorbing atom. In general, the edge position can shift to an energy several eVs higher when the valence state of the absorbing increases by one electron. As shown in Fig. 3A.5b, the linear relationship between the valence state and the position of the absorption edge is fitted and further exploited to calculate the valence state of the atoms in the studied sample [21]. For the L-edge of transition metals, the integration of the area under the white-line peak is intimately related to the density of unoccupied d-states, which can be used to probe the valence states of the absorbing atoms (Fig. 3A.5c) [22]. Note that the integration should subtract an arctangent function centered on the inflection point of the absorption edge.

XANES part is dominated by the MS resonances of the excited photoelectrons with low kinetic energy and thus is correlated with the atomic structure of the absorbing atoms. In other words, the features observed in the XANES regions may correspond to a definite atomic structure. For example, linear combination fitting (LCF) analysis is often employed to quantify the ratio of different species in a multi-component sample. In this case, the experimental spectrum of the studied sample can be fitted using XANES spectra of possible references as standards. For HDMSs applied in electrocatalysis, surface reconstruction and interactions with adsorbates make the changes of corresponding operando XANES spectra insignificant, which can hardly be resolved by LCF methods due to the lack of available reference spectra.

Nevertheless, these subtle variations in the XANES part can be distinguished by differential analysis (or Δ-XANES). The variations are emphasized by subtracting the spectra of the catalysts in the initial states, and one can analyze the difference spectrum. On the other hand, XANES modeling based on appropriate approximation, such as FMS, time-dependent density functional theory, and the Bethe–Salpeter method, provides the chance to match the theoretically simulated spectra to corresponding experimental data by modifying the structural parameters of the proposed structures. Such modeling provides a comprehensive understanding of the atomic structures of the catalysts and their evolutions as a function of the reaction conditions.

3A.3.4 Qualitative analysis of EXAFS data

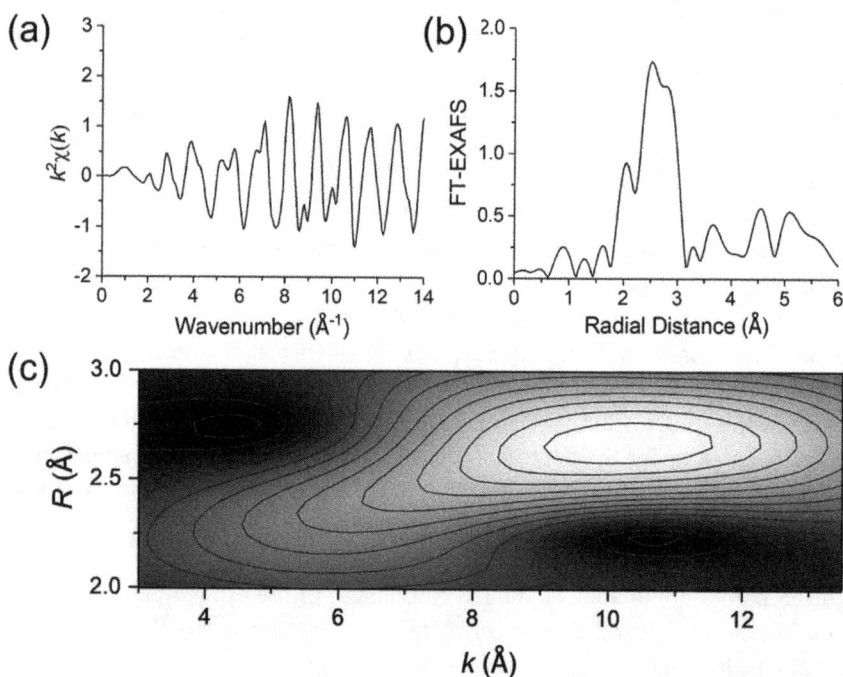

Fig. 3A.6: Schematic of FT and WT analysis of EXAFS spectra: (a) k^2-weighted k-space spectrum of Pt foil; corresponding FT-EXAFS spectrum (b) and WT-EXAFS spectrum.

For EXAFS data analysis, the raw data are usually reduced to a k-weighted $\chi(k)$ spectrum (Fig. 3A.6a), where the experimentally measured E-space parameters (E and $\mu(E)$) are converted into k-space based on relations in eqs. (3A.4, 3A.5). As noted in eq. (3A.6), the theoretical $\chi(k)$ function consists of a sum of sine waves, whose amplitudes depend on the type of atoms and their distributions for the absorbing atoms. To get the structural parameters of neighboring atoms qualitatively, $\chi(k)$ can be directly analyzed in k-space or Fourier transformed into an r-space radial distribution function (RDF) (Fig. 3A.6b). In k-space, although the wave functions from different scattering shells and atoms are summed up in one spectrum, the significant features related to the oscillation frequency or amplitude contain information about the coordination atoms. Commonly, the oscillation amplitudes of low-Z atoms decay more quickly than those of high-Z atoms; rapid oscillation in the $\chi(k)$ spectrum corresponds to a long radial distance; a large oscillation amplitude indicates high coordination or well-ordered structure. This information gives a preliminary insight into the scattering atoms. Moreover, Fourier transformed (FT) EXAFS analysis can provide more detailed information about the local atomic structure. The first peak in the FT-EXAFS with a distance of 1.5–2.0 Å typically corresponds to the bonding between the absorbing atom and a low-Z element. For comparison, metal-metal bonds usually contribute to the FT-EXAFS at a distance of 2.0–2.5 Å due to the larger ionic size of the reduced ions.

It should be noted that the position of the peak in the FT-EXAFS spectra is 0.3–0.5 Å shorter than the realistic length of the corresponding coordination. The detailed atom types, interatomic distance, and structural disorder can be obtained by fitting the EXAFS curves based on the intimate association between the scattering functions and neighboring atoms. However, when two or more neighboring atoms are close in the periodic tables, it is difficult for FT-EXAFS analysis to discriminate their respective contributions. Facing this challenge, wavelet transformed (WT) EXASF analysis provides an effective solution by presenting a two-dimensional representation of the analyzed spectrum in k- and r-space (Fig. 3Ac). The WT-EXAFS spectrum simultaneously shows the oscillation frequency and contribution part in the RDF, enabling one to distinguish between different neighbors. The combination of WT-EXAFS analysis and FT-EXAFS fitting has great potential for accurately probing the local atomic structures in heterogeneous catalysis.

3A.4 Applications of XAS in electrocatalytic reactions over HDMSs

HDMSs have recently attracted great interest in various reactions because they maximize atom efficiency and possess unique catalytic properties. In electrochemical reactions, HDMSs that are often constructed in the forms of atomically dispersed metal catalysts and subnanometric clusters have been widely exploited in the HER, OER, ORR, and

CO2RR. Therefore, it is critical to probe the electronic and atomic structures of HDMSs and establish links between structural motifs in HDMSs, local electronic structures, and catalytic properties. HDMSs mostly feature highly dispersed metal sites, in which almost all metal sites are exposed and are available to participate in the reactions. This feature introduces great challenges to the structural characterization of the corresponding catalysts, as HDMSs are often deposited on other substrates with low-dimensional and disordered structures. During the past decades, the visualization of HDMSs deposited even on strong support has been achieved thanks to advanced microscopic imaging technology. However, only small parts of HDMSs can be observed because of the limited area of the detection window, and the atomic structures of the observed HDMSs remain elusive. Referring to the previous introductions, it is well known that XAS can provide the bulk-average atomic and electronic structures of HDMSs. This section will introduce the studies of XAS electrochemical reactions, especially those catalyzed by HDMSs.

3A.4.1 Atomically dispersed metal catalysts

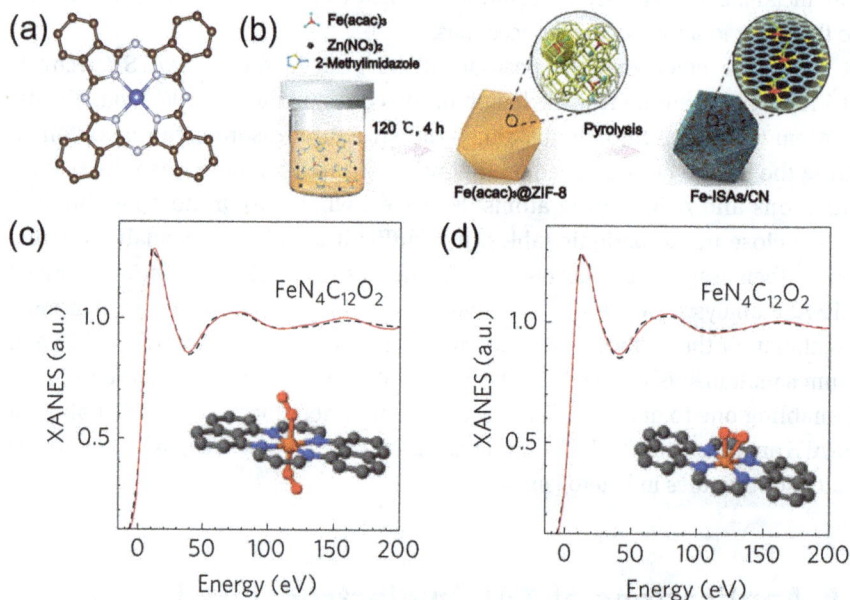

Fig. 3A.7: Identification of the structures of atomically dispersed metal sites: (a) Structure of cobalt phthalocyanine; (b) Schematic illustration of the formation of atomically dispersed Fe sites; comparison between the K-edge XANES experimental spectrum and the theoretical spectrum calculated with the FeN_4C_{12} with two O_2 molecules adsorbed in end-on mode (c) and with the FeN_4C_{12} moiety with one O_2 molecule adsorbed in side-on mode (d). (Image of Fig. 3A.7b was adapted from ref. [27] with permission; images of (c) and (d) were adapted from ref. [28] with permission).

In electrochemical reactions catalyzed by atomically dispersed metal catalysts, metal-nitrogen-carbon (M-N-C) catalysts are critical because their distinct active sites possess unique electrocatalytic properties and provide a good platform for understanding structure-activity relationships at the atomic scale [4–6], [23–26]. As shown in Fig. 3A.7a, macrocyclic compounds (e.g., phthalocyanine) composed of active center, nitrogen ligands, and carbon substrates were initially exploited for ORR, which initiated the application of M-N-C catalysts in electrocatalytic reactions. Now, metal-organic framework (MOF) materials have become the most commonly used precursors for constructing M-N-C catalysts. Their periodical metal nodes with organic linkers in MOF are close to M-N-C catalysts. In addition, the microporous nature of MOF materials is beneficial for improving the density of M-N-C catalysts. Fig. 3A.7b illustrates how to construct Fe-N-C catalysts based on MOF materials. Fe ions partially replace Zn ions in tetrahedral $Zn-N_4$ complexes and form chemical bonds with imidazolate ligands [27]. The subsequent high-temperature pyrolysis converts these $Fe-N_4$ complexes into atomically dispersed FeN_4 sites, which show significantly enhanced ORR activity in acidic media. EXAFS analysis that is only sensitive to the local structure of the absorbing atoms reveals that the Fe sites in both tetrahedral $Fe-N_4$ complexes and pyrolyzed FeN_4 sites coordinate with four N atoms. The detailed structure of atomically dispersed FeN_4 sites remains elusive.

Regarding this issue, Zitolo et al. prepared NH_3-pyrolyzed Fe-N-C materials as catalytic sites for oxygen reduction [28]. Then, they conducted a more sensitive XANES simulation and identified the active structure of FeN_4 sites having a porphyrin-like architecture with a FeN_4C_{12} core and dioxygen adsorbed in the side-on or end-on configuration (Fig. 3A.7c and 3A.7d). Under realistic ORR conditions, Jia et al. conducted operando XAFS measurements to probe the structural evolution of these atomically dispersed Fe sites and then unravel corresponding structure-activity relationships [29]. As shown in Fig. 3A.8, operando FT-EXAFS spectra indicate that the Fe-N/O peak increased in intensity and shifted to a longer radial distance. The applied potential increased from 0.1 to 0.9 V (vs. RHE); the Fe-N/O peak was restored to its initial state when decreasing the potential. To match these in situ changes related to the Fe-N peak, the authors modified the structure of FeN_4 sites by shifting the Fe center out of the original plane toward the adsorbed intermediates. Interestingly, the theoretically simulated XANES spectra based on modified FeN_4 structures reproduced those subtle changes observed in experimental XANES spectra.

On the contrary, the less-active $Fe-N_4$ complexes without being pyrolyzed at high temperatures show insignificant changes in corresponding XAFS spectra as the potential changes. Therefore, the ORR activities of Fe-N-C catalysts are essentially governed by their Fe-N switching behaviors. These examples demonstrate that comprehensive XAS studies of M-N-C catalysts provide a chance to identify the active structure at the atomic scale and uncover the related structure-activity relationships under realistic reaction conditions [30–33].

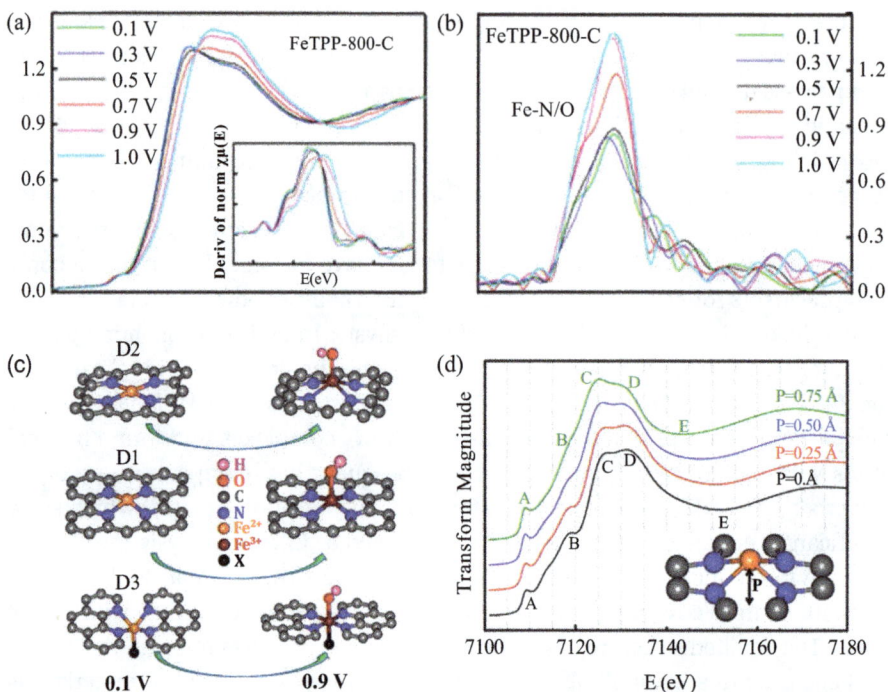

Fig. 3A.8: Operando XAS spectra of atomically dispersed Fe sites during ORR: (a) XANES spectra at the Fe K-edge and (b) the corresponding FT-EXAFS spectra as a function of applied potentials; (c) Schematic illustration of the Fe-N switching behavior during ORR; (d) XANES spectra calculated by FEFF9 based on the Fe-N_4-C_8 model with various central Fe displacements (images were adapted from ref. [29] with permission).

To expand the applications of M-N-C catalysts in electrochemical reactions, tremendous efforts have been devoted to enabling critical reaction intermediates to adsorb appropriately on M-N-C catalysts by modulating the local structures of metal sites. Theoretical calculations guide most related works, and the screened M-N-C catalysts are then prepared to verify the predictions. XAS plays a critical role in identifying the designed structure, and developing a new type of metal center has greatly broadened the applicable electrocatalytic reactions of M-N-C catalysts. Currently, the M-N-C family covers almost all 3d metals and has been further extended to 4d and 5d metals, which show superior performance in the HER, OER, and CO2RR [34–42]. In these studies, XANES and EXAFS characterization methods are essential to revealing the electronic and local atomic structure of newly developed M-N-C catalysts. For example, Wang et al. employed operando XAFS measurements to reveal the dynamic structural changes of a copper (II) phthalocyanine (CuPc) molecular model catalyst with a typical CuN_4C_8 configuration CO2RR [43]. As shown in

Fig. 3A.9a, the Cu (II) peak at 8985 eV attributed to the $1s \rightarrow 3d$ quadrupole transition decreases with the applied potential shifting to -0.66 V (vs. RHE).

Meanwhile, the Cu(I) feature was formed, as evidenced by the peak appearing at 8981 eV. The XANES is dominated by the Cu (0) feature with more negative potentials, indicating that metallic Cu clusters are formed. Combined with the FT-EXAFS analysis (Fig. 3A.9b) results, the quantitative changes of the local coordination environment clarified that the Cu-N/O coordination was gradually reduced, whereas that of Cu-Cu increased, indicating that the restructuring of CuPc was in favor of CH_4 production.

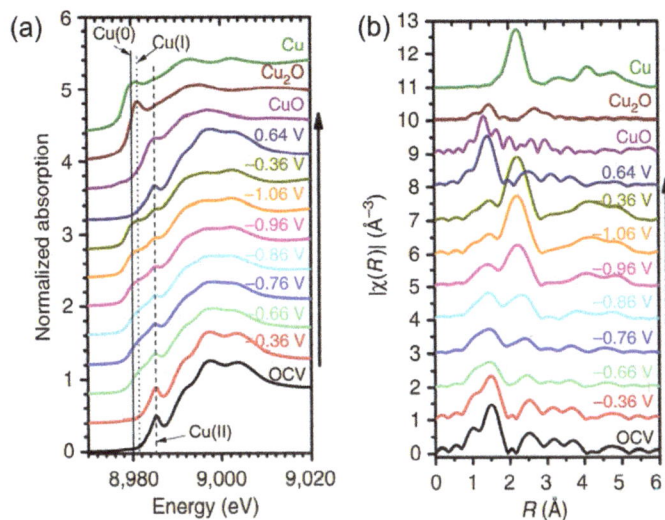

Fig. 3A.9: Operando XAS spectra under electrocatalytic reaction conditions: (a) XANES spectra at the Cu K-edge and (b) the corresponding FT-EXAFS spectra as a function of applied potentials (images were adapted from ref. [43] with permission).

Modulating the number and species of ligands surrounding the metal center is another promising pathway to improve the intrinsic activities of M-N-C catalysts. According to the EXAFS basis described in eq. (3A.6), ligand number variations will only influence the intensity of the corresponding shell in the FT-EXAFS spectrum. For example, Yin et al. investigated the effect of the pyrolysis temperature on the electrochemical activity of Co-N-C catalysts toward the ORR. They found that the Co-N-C catalysts pyrolyzed under higher temperatures (900 °C) exhibited superior ORR performance [44]. The FT-EXAFS spectrum showed that a high-temperature temperature led the Co-N peak to decrease; the corresponding coordination number was estimated to be 2.0, smaller than conventional Co-N_4 sites. According to the obtained parameters, the theoretical calculation further revealed that Co-N_2 species show stronger interactions with peroxide than Co-N_4 and then accelerated the ORR process. In the case of modulating the ligand species, the types of introduced heteroatoms determine the difficulty of precisely quantifying the atomic structures of the M-N-C catalysts. As described by Shang et al., the unsymmetrically arranged Cu-S_1N_3 derived from MOF materials exhibits excellent ORR activity [45]. The atomic structure of Cu-S_1N_3 catalysts can be easily decoded by FT-EXAFS analysis, as the introduced S ion with a larger radius (ca. 1.0 Å) than that of N (ca. 0.7 Å) will shift the first-shell coordination peak into the higher R region. Suppose it is necessary to distinguish two light atoms (e.g., C and N). In that case, WT-EXAFS analysis is suggested to be employed carefully, as such atoms show subtle differences in FT-EXAFS spectra. As shown in Fig. 3A.10a and 10b, Fei et al. distinguished Co-N and Co-C paths in their HER catalysts, as the two paths showed different maximum positions in WT contour plots [24]. Importantly, constructing binuclear sites has been an efficient means of activating O_2 molecules during the ORR. As displayed in Fig. 3A.10c, Wu et al. prepared a novel electrocatalyst with Fe-Co dual sites, exhibiting superior ORR performance [46]. Except for the prominent peak belonging to the first shell of the Co/Fe-N path, one high-shell Fe-Co coordination peak can be observed, which is missing from conventional M-N-C catalysts. Based on the combination of XANES simulation and EXAFS fitting results, the local atomic structure of binuclear sites was determined to be (Fe, Co)-N_6-C moieties. Many similar works have been carried out and have demonstrated the intriguing electrochemical properties of binuclear sites [47–49]. In summary, the adopted strategies in these modulations can be roughly classified in the following aspects: the development of a new type of metal center, modification of the number of nearest coordination atoms, replacement of the nearest coordination atoms with heteroatoms, and construction of binuclear sites.

Aside from microporous carbon materials, metal oxides and other metal compounds (e.g., sulfides and carbides) can support atomically dispersed metal catalysts. In these cases, the strong metal-support interactions can provide rich opportunities to regulate the properties of the active metal sites; conversely, the deposited metal atoms affect the electrochemical properties of supports to a certain extent. Nevertheless, the latter is insignificant because of the relatively low loading of the deposited metal atoms. These active metal atoms can be positioned at completely different sites in crystalline supports, such as adsorption sites on the support surface, defective sites, and substitutional sites. Therefore, identifying the location of active sites within the support using structure-sensitive XAFS characterization is essential to understanding the underlying metal-support interactions. Compared to carbon materials as supports, atomically dispersed metal sites on metal compounds display more pronounced high-shell EXAFS features, as ordered metal atoms in crystalline supports possess stronger scattering capacities toward ionized photoelectrons. Such high-shell EXAFS features combined with the first shell can be used to discriminate the locations of the atomically dispersed metal sites. For example, to explore the support effect in single-atom platinum catalysts for ORR toward H_2O_2, Yang et al. prepared a single-atom platinum catalyst on two different supports: titanium carbide (Pt/TiC) and titanium nitride (Pt/TiN) [50]. The pronounced FT-EXAFS peaks of the two

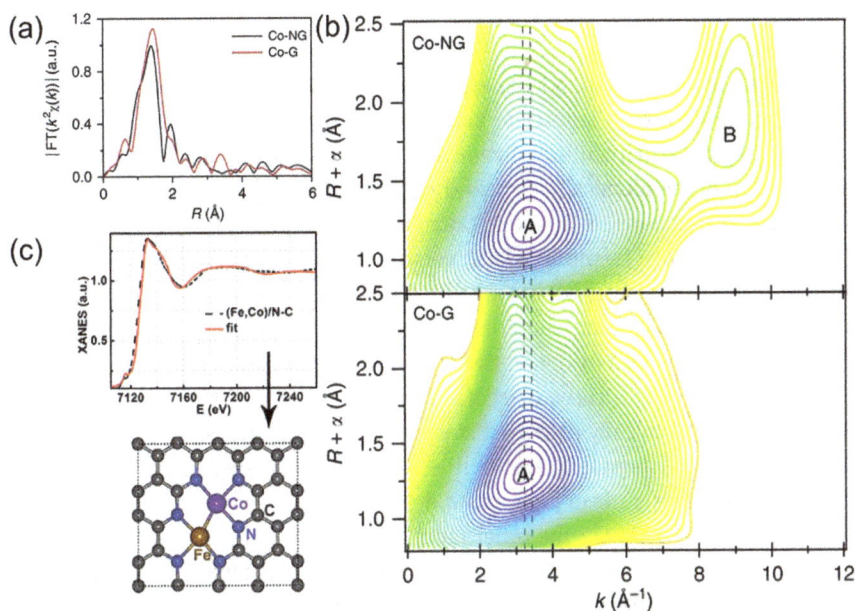

Fig. 3A.10: Spectroscopic identification of the structure of the atomically dispersed metal sites: (a) FT-EXAFS spectra of Co-NG and Co-G; (b) the corresponding WT-EXAFS spectra; (c) comparison between the Fe K-edge XANES experimental spectrum and theoretical spectrum calculated with depicted structure (images of (a) and (b) were adapted from ref. [24] with permission; image of (c) was adapted from ref. [46] with permission).

catalysts are all located in the low-R region (1–2 Å), indicating that the Pt atoms mainly coordinate Cl and Ti atoms on the support surface. Meanwhile, the extremely low Pt-Pt coordination reveals that only a small fraction of Pt exists as nanoparticles. Therefore, the Pt atoms in the two catalysts are predominately dispersed as single-atom catalysts. It can be concluded that the higher $2e^-$ of ORR activity in Pt/TiC can be ascribed to the positive effect of TiC in preserving O–O bonds. The Pd-MoS$_2$ prepared by Luo et al. showed highly efficient performance in the HER, in which the Pd K-edge FT-EXAFS spectrum showed the Pd-S path rather than the Pd-Mo path in the nearest shell, and no high-shell feature (e.g., Pd-S-Mo) could be observed [51]. It was suggested that Pd atoms mainly replace the interfacial Mo vacant sites. If introduced metal atoms are located at the substitutional sites, the corresponding FT-EXAFS spectrum will share features close to those of the metal atoms from the support. Duan et al. found that doping Co ions into monolayer MoS$_2$ could significantly improve their HER activity compared to pristine MoS$_2$ [52]. The Co K-edge FT-EXAFS spectrum of the sample presents two prominent coordination peaks at 1.9 Å (Co-S scattering path) and 2.8 Å (Co-Mo scattering path). These two features are consistent with those in the Mo K-edge FT-EXAFS spectrum, which suggests the substitutional doping of Co into MoS$_2$.

3A.4.2 Subnanometric clusters

Particles with diameters less than 1 nm are denoted as subnanometric clusters [53–57]. The unique property of high surface-to-bulk ratios in subnanometric clusters enables each metal atom to participate in the catalytic processes, exhibiting the same atomic utilization as that of the atomically dispersed metal catalysts (nearly 100 %) [53, 55–57]. One single cluster that consists of several atoms possesses more abundant active sites, which is a great challenge for atomically dispersed metal catalysts. Additionally, various surface sites (e.g., metal–metal bonds) are formed by ensembles of metal atoms in subnanometric clusters, catalyzing some important reactions requiring metal ensemble sites, such as hydrogenolysis oxidation [58–60], isomerization [61, 62], cracking [63, 64], and hydrogenolysis [65, 66]. These reactions are usually catalyzed by nanoparticles but cannot be driven by atomically dispersed metal sites. It shows that subnanometric clusters catalysts partially integrate the properties of atomically dispersed metal sites and nanoparticles, providing a novel platform to design effective and efficient catalysts for certain chemical processes. On the other hand, the support effects in subnanometric clusters are likely to be more prominent than those of typical nanoparticle catalysts, as a considerable fraction of atoms are located at the cluster-support interface. Accordingly, support effects are more often and efficiently used to modulate subnanometric clusters' structural and electronic characteristics.

Numerous works have demonstrated that subnanometric Pt clusters catalyze the HER with comparable or superior performance to Pt nanoparticles and atomically dispersed Pt sites [67]. XAS studies of these catalysts' electronic and local atomic structures play a critical role in understanding the structure-activity relationships involved. For example, Sun et al. prepared distinct Pt-based catalysts on N-doped graphene/graphitic-C_3N_4 for the HER. They found that the catalysts containing more atomic clusters (AC Pt-NG/C) show better catalytic performance than those composed of atomically dispersed Pt sites [68]. The XANES spectra at the Pt-L_3 edge show that the WL-peak of AC Pt-NG/C is located between those of Pt foil and PtO_2, suggesting that the metal-support interaction modulates the electronic structures of Pt atoms and makes them positively charged. This interaction results from the interfacial bonding between the Pt clusters and N-doped carbon support, demonstrated by the Pt-N/O/C coordination observed in the corresponding FT-EXAFS spectra. In addition to carbon-based materials selected as supports, Cheng et al. investigated the effects of TiO_2 support on the subnanometric Pt clusters for the HER by conducting XAS measurements at the Pt L_3-edge [69]. The XANES spectra indicated the high oxidation state of Pt atoms in Pt clusters; the Pt-O (II) (at 2.1 Å) and Pt-O-Ti (at 2.7 Å) contributions in the corresponding EXAFS spectra, respectively, represent the Pt-O interactions between the Pt clusters and TiO_2 support and the Pt-Ti bonds bridging the nearest O atoms. Therefore, the studied catalysts' excellent HER activity and stability are ascribed to the unique electronic structure of the oxidized Pt clusters and the strong metal-support interactions.

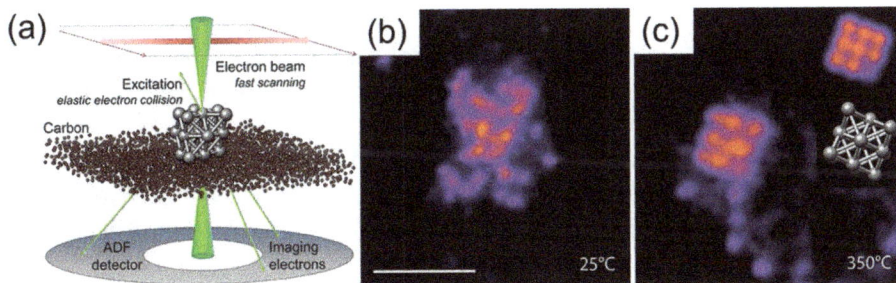

Fig. 3A.11: Fast STEM imaging of Pt clusters at variable temperatures: (a) Schematic of the experimental system; (b) A typical amorphous Pt cluster at room temperature; (c) A crystalline Pt_{17} fcc cluster at high temperature (images were adapted from ref. [70] with permission).

The property of a high surface-to-bulk ratio endows subnanometric clusters with unique properties and interaction with support materials. However, it simultaneously reduces their structural stability because of the dramatically increased proportion of under-coordinated atoms. When these clusters are applied in a specific reaction, the structural properties and interaction with support will evolve with the reaction conditions, resulting in a dynamic structure-activity relationship. Therefore, operando characterization of the structures of subnanometric clusters under realistic conditions is essential to reveal the catalytic mechanism of the corresponding reaction. As shown in Fig. 3A.11, Henninen et al. employed fast dynamic scanning transmission electron microscopy to investigate the structure and stability of Pt clusters deposited on carbon [70]. The environment temperature plays a critical role in determining the structure of Pt_n clusters: dynamic amorphous 2D structures are stabilized at room temperature; as the temperature rises above 300℃, the clusters transform into a crystalline state with an fcc structure. This result reflects that in realistic catalytic processes, the reaction energy barrier and the structure of the subnanometric clusters will be influenced by the reaction temperature.

Furthermore, Gorczyca et al. investigated the effect of absorbed hydrogen (H_{ads}) on the morphology of the nanometric Pt cluster on γ-Al_2O_3 by conducting in situ XANES measurements recorded in high energy resolution fluorescence mode [71]. The authors modulated the reaction temperature and H_2 pressure to control the H_{ads} coverage on the Pt surface. By comparing the theoretically simulated XANES spectra with experimental results, it can be known that the morphology of Pt clusters on the (110) face of γ-Al_2O_3 transforms from a biplanar structure to a 3D-like structure with increasing H_{ads}. In electrochemical reactions, the interaction between those under-coordinated atoms and the environment (e.g., support, electrolyte, and adsorbates) becomes more complex due to a solid-liquid-gas interface. To this issue, Zhang et al. prepared highly dispersed PtO_x clusters on carbon nanotube (PtO_x/CNT) as HER catalysts. They then conducted in situ XAS characterization to reveal the potential-dependent structural evolution of PtO_x/CNT during HER [72]. As shown in Fig. 3A.12, the XANES spectra showed that the WL peak of the sample significantly decreases in low locations when applying the HER potential. Considering the observed metallic Pt-Pt bonds in the corresponding EXAFS spectra, it was suggested that PtO_x clusters were reduced to metallic Pt clusters as the actual active sites for the HER. As the potential decreases, the shift of the WL-peak can be attributed to the increase in H coverage; simultaneously, the increased Pt-Pt coordination number and decreased interfacial Pt-O/C bonds indicate that the morphologies of the resulting Pt clusters evolved from a 2D shape toward a more 3D-like structure.

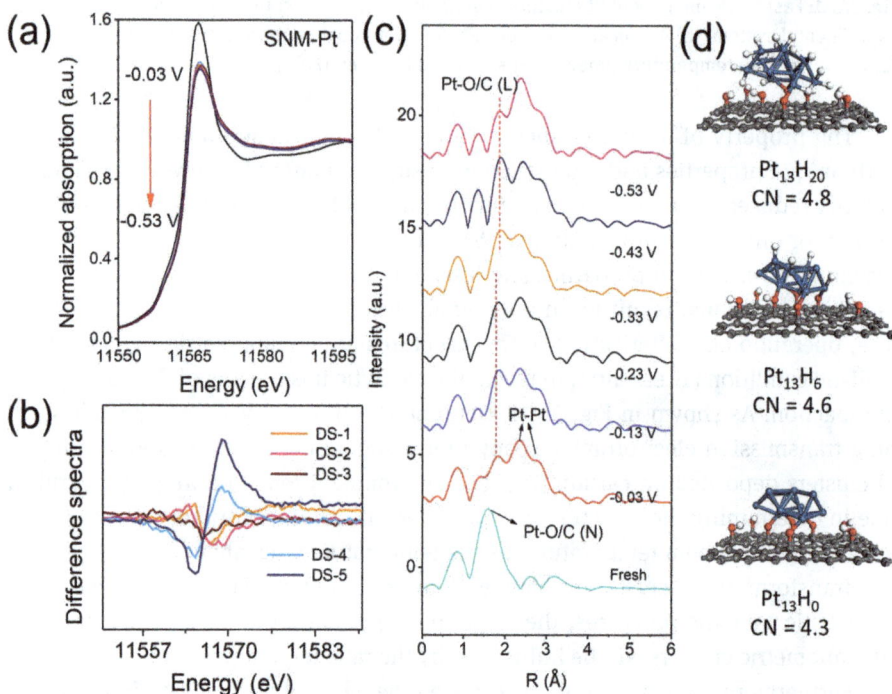

Fig. 3A.12: Operando XAS spectra of PtO_x clusters during HER: (a) Normalized Pt L_3-edge XANES spectra as a function of applied potential; (b) The corresponding difference spectra collected at adjacent potentials; (c) The corresponding FT-EXAFS spectra; (d) Theoretically optimized Pt_{13} cluster structure with different hydrogen coverages (images were adapted from ref. [72] with permission).

3A.5 Conclusion

This chapter focused on applying XAS characterization methods to understand electrochemical reactions' structure-activity relationships over HDMSs. First, Section 3A.2 carefully reviewed the physical basics of XAS characterization methods. Then,

Section 3A.3 introduced how to conduct synchrotron-based XAS measurements, especially those under electrochemical reactions, in terms of the in situ cell design and critical technical skills for improving the quality of the collected data. This section also discussed how to rationally analyze the obtained XAFS data and obtain useful and reliable information. Section 3A.4 reviewed recent progress in XAS studies of HDMSs in electrochemical reactions, in which atomically dispersed metal sites and subnanometric metal clusters were taken as examples, respectively.

It can be determined that advanced XAS characterizations can lead to a comprehensive understanding of the electrochemical processes. In the future, improved XAS characterizations that can achieve high energy resolution and spatial resolution are crucial to understanding the fundamental aspects of reaction systems, including active centers, support interactions, and processes occurring at interfaces. The development and popularization of high-energy-resolution X-ray spectroscopy methods will render the fine electronic structure that determines reaction activity uncovered directly. Ultrafast time-resolved XAS techniques such as energy-dispersed XAS and pump-probe XAS will make adsorption and desorption monitored in real-time. Furthermore, X-ray spectroscopy coupled with other characterization techniques has been a research trend, as it provides multiple dimensions to identify the activity origin. More importantly, it is highly recommended to introduce advanced computational methods (e.g., theoretical spectra simulation and machine learning-based spectra analysis) into XAS analysis, which have shown great potential in the making up for the resolution limitations of detection instruments.

References

[1]   Seh, Z. W., Kibsgaard, J., Dickens, C. F., Chorkendorff, I. B., Norskov, J. K. & Jaramillo, T. F. (2017). Combining theory and experiment in electrocatalysis: insights into materials design. Science, 355, 6321.

[2]   Zhao, G. X., Liu, H. M. & Ye, J. H. (2018). Constructing and controlling highly dispersed metallic sites for catalysis. Nano Today, 19, 108–125.

[3]   Lai, W. H., Wang, H., Zheng, L. R., Jiang, Q., Yan, Z. C., Wang, L., Yoshikawa, H., Matsumura, D., Sun, Q., Wang, Y. X., Gu, Q. F., Wang, J. Z., Liu, H. K., Chou, S. L. & Dou, S. X. (2020). General synthesis of single-atom catalysts for hydrogen evolution reactions and room-temperature Na-S batteries. Angewandte Chemie International Edition, 59(49), 22171–22178.

[4]   Yang, X. F., Wang, A. Q., Qiao, B. T., Li, J., Liu, J. Y. & Zhang, T. (2013). Single-Atom catalysts: a new frontier in heterogeneous catalysis. Accounts of Chemical Research, 46(8), 1740–1748.

[5]   Chen, Y. J., Ji, S. F., Chen, C., Peng, Q., Wang, D. S. & Li, Y. D. (2018). Single-Atom Catalysts: synthetic strategies and electrochemical applications. Joule, 2(7), 1242–1264.

[6]   Zhu, C. Z., Fu, S. F., Shi, Q. R., Du, D. & Lin, Y. H. (2017). Single-Atom electrocatalysts. Angewandte Chemie International Edition, 56(45), 13944–13960.

[7]   Zhang, H., Li, X. P. & Jiang, Z. (2019). Probe active sites of heterogeneous electrocatalysts by X-ray absorption spectroscopy: from the single atom to complex multielement composites. Current Opinion in Electrochemistry, 14, 7–15.

[8]   Bordiga, S., Groppo, E., Agostini, G., van Bokhoven, J. A. & Lamberti, C. (2013). Reactivity of surface species in heterogeneous catalysts probed by in situ X-ray absorption techniques. Chemical Reviews, 113(3), 1736–1850.

[9]   Li, Y. Y. & Frenkel, A. I. (2021). Deciphering the local environment of single-atom catalysts with X-ray absorption spectroscopy. Accounts of Chemical Research, 54(11), 2660–2669.

[10] Zhu, Y. P., Kuo, T. R., Li, Y. H., Qi, M. Y., Chen, G., Wang, J. L., Xu, Y. J. & Chen, H. M. (2021). Emerging dynamic structure of electrocatalysts unveiled by in situ X-ray diffraction/ absorption spectroscopy. Energy & Environmental Science, 14(4), 1928–1958.

[11] van Oversteeg, C. H. M., Doan, H. Q., de Groot, F. M. F. & Cuk, T. (2017). In situ X-ray absorption spectroscopy of transition metal-based water oxidation catalysts. Chemical Society Reviews, 46(1), 102–125.

[12] Wang, M. Y., Arnadottir, L., Xu, Z. C. J. & Feng, Z. X. (2019). In situ X-ray absorption spectroscopy studies of nanoscale electrocatalysts. Nano-Micro Letters, 11(1), 1–18.

[13] Timoshenko, J. & Cuenya, B. R. (2021). In situ/operando electrocatalyst characterization by X-ray absorption spectroscopy. Chemical Reviews, 121(2), 882–961.

[14] Fabbri, E., Abbott, D. F., Nachtegaal, M. & Schmidt, T. J. (2017). Operando X-ray absorption spectroscopy: a powerful tool toward water-splitting catalyst development. Current Opinion in Electrochemistry, 5(1), 20–26.

[15] Fang, L. Z., Seifert, S., Winans, R. E. & Li, T. (2021). Operando XAS/SAXS: guiding design of single-atom and subnanocluster catalysts. Small Methods, 5(5), 2001194–1 to 2001194–13.

[16] Jia, Q. Y., Liu, E. S., Jiao, L., Pann, S. & Mukerjee, S. (2019). X-Ray absorption spectroscopy characterizations on PGM-Free electrocatalysts: justification, advantages, and limitations. Advanced Materials (Deerfield Beach, Fla.), 31(31), 1805157–1 to 1805157–8.

[17] Billinge, S. J. L. & Levin, I. (2007). The problem with determining atomic structure at the nanoscale. Science, 316(5824), 561–565.

[18] Sayers, D. E., Stern, E. A. & Lytle, F. W. (1971). New technique for investigating noncrystalline structures – fourier analysis of extended x-ray – absorption fine structure. Physical Review Letters, 27(18), 1204-&.

[19] Timoshenko, J., Jeon, H. S., Sinev, I., Haase, F. T., Herzog, A. & Roldan Cuenya, B. (2020). Linking the evolution of catalytic properties and structural changes in copper-zinc nanocatalysts using operando EXAFS and neural-networks. Chemical Science, 11(14), 3727–3736.

[20] Sasaki, K., Wang, J. X., Naohara, H., Marinkovic, N., More, K., Inada, H. & Adzic, R. R. (2010). Recent advances in platinum monolayer electrocatalysts for oxygen reduction reaction: scale-up synthesis, structure and activity of Pt shells on Pd cores. Electrochimica Acta, 55(8), 2645–2652.

[21] Huang, J. H., Chen, J. T., Yao, T., He, J. F., Jiang, S., Sun, Z. H., Liu, Q. H., Cheng, W. R., Hu, F. C., Jiang, Y., Pan, Z. Y. & Wei, S. Q. (2015). CoOOH nanosheets with high mass activity for water oxidation. Angewandte Chemie International Edition, 54(30), 8722–8727.

[22] Clancy, J. P., Chen, N., Kim, C. Y., Chen, W. F., Plumb, K. W., Jeon, B. C., Noh, T. W. & Kim, Y. J. (2012). Spin-orbit coupling in iridium-based 5d compounds probed by x-ray absorption spectroscopy. Physical Review B, 86(19), 195131–1 to 195131–9.

[23] Wang, X. X., Swihart, M. T. & Wu, G. (2019). Achievements, challenges, and perspectives on cathode catalysts in proton exchange membrane fuel cells for transportation. Nature Catalysis, 2(7), 578–589.

[24] Fei, H. L., Dong, J. C., Arellano-Jimenez, M. J., Ye, G. L., Kim, N. D., Samuel, E. L. G., Peng, Z. W., Zhu, Z., Qin, F., Bao, J. M., Yacaman, M. J., Ajayan, P. M., Chen, D. L. & Tour, J. M. (2015). Atomic cobalt on nitrogen-doped graphene for hydrogen generation. Nature Communications, 6(1), 1–8.

[25] Liu, M. M., Wang, L. L., Zhao, K. N., Shi, S. S., Shao, Q. S., Zhang, L., Sun, X. L., Zhao, Y. F. & Zhang, J. J. (2019). Atomically dispersed metal catalysts for the oxygen reduction reaction: synthesis, characterization, reaction mechanisms, and electrochemical energy applications. Energy & Environmental Science, 12(10), 2890–2923.

[26] He, Y. H., Liu, S. W., Priest, C., Shi, Q. R. & Wu, G. (2020). Atomically dispersed metal-nitrogen-carbon catalysts for fuel cells: advances in catalyst design, electrode performance, and durability improvement. Chemical Society Reviews, 49(11), 3484–3524.

[27] Chen, Y. J., Ji, S. F., Wang, Y. G., Dong, J. C., Chen, W. X., Li, Z., Shen, R. A., Zheng, L. R., Zhuang, Z. B., Wang, D. S. & Li, Y. D. (2017). Isolated Single Iron Atoms Anchored on N-doped Porous Carbon as an Efficient Electrocatalyst for the Oxygen Reduction Reaction. Angewandte Chemie International Edition, 56(24), 6937–6941.

[28] Zitolo, A., Goellner, V., Armel, V., Sougrati, M. T., Mineva, T., Stievano, L., Fonda, E. & Jaouen, F. (2015). Identification of catalytic sites for oxygen reduction in iron- and nitrogen-doped graphene materials. Nature Materials, 14(9), 937-+.

[29] Jia, Q., Ramaswamy, N., Hafiz, H., Tylus, U., Strickland, K., Wu, G., Barbiellini, B., Bansil, A., Holby, E. F., Zelenay, P. & Mukerjee, S. (2015). Experimental observation of redox-induced Fe-N switching behavior as a determinant role for oxygen reduction activity. Acs Nano, 9(12), 12496–12505.

[30] Cao, L. L., Luo, Q. Q., Chen, J. J., Wang, L., Lin, Y., Wang, H. J., Liu, X. K., Shen, X. Y., Zhang, W., Liu, W., Qi, Z. M., Jiang, Z., Yang, J. L. & Yao, T. (2019). Dynamic oxygen adsorption on single-atomic Ruthenium catalyst with high performance for acidic oxygen evolution reaction. Nature Communications, 10(1), 1–9.

[31] Cao, L. L., Luo, Q. Q., Liu, W., Lin, Y. K., Liu, X. K., Cao, Y. J., Zhang, W., Wu, Y., Yang, J. L., Yao, T. & Wei, S. Q. (2019). Identification of single-atom active sites in carbon-based cobalt catalysts during electrocatalytic hydrogen evolution. Nature Catalysis, 2(2), 134–141.

[32] Su, H., Zhou, W. L., Zhou, W., Li, Y. L., Zheng, L. R., Zhang, H., Liu, M. H., Zhang, X. X., Sun, X., Xu, Y. Z., Hu, F. C., Zhang, J., Hu, T. D., Liu, Q. H. & Wei, S. Q. (2021). In-situ spectroscopic observation of dynamic-coupling oxygen on atomically dispersed iridium electrocatalyst for acidic water oxidation. Nature Communications, 12(1), 1–9.

[33] Fang, S., Zhu, X. R., Liu, X. K., Gu, J., Liu, W., Wang, D. H., Zhang, W., Lin, Y., Lu, J. L., Wei, S. Q., Li, Y. F. & Yao, T. (2020). Uncovering near-free platinum single-atom dynamics during electrochemical hydrogen evolution reaction. Nature Communications, 11(1), 1–8.

[34] Xiao, M. L., Zhu, J. B., Li, S., Li, G. R., Liu, W. W., Deng, Y. P., Bai, Z. Y., Ma, L., Feng, M., Wu, T. P., Su, D., Lu, J., Yu, A. P. & Chen, Z. W. (2021). 3d-Orbital Occupancy Regulated Ir-Co Atomic Pair Toward Superior Bifunctional Oxygen Electrocatalysis. ACS Catalysis, 11(14), 8837–8846.

[35] Yang, H. B., Hung, S. F., Liu, S., Yuan, K. D., Miao, S., Zhang, L. P., Huang, X., Wang, H. Y., Cai, W. Z., Chen, R., Gao, J. J., Yang, X. F., Chen, W., Huang, Y. Q., Chen, H. M., Li, C. M., Zhang, T. & Liu, B. (2018). Atomically dispersed Ni(i) as the active site for electrochemical CO2 reduction. Nature Energy, 3(2), 140–147.

[36] Xiao, M. L., Gao, L. Q., Wang, Y., Wang, X., Zhu, J. B., Jin, Z., Liu, C. P., Chen, H. Q., Li, G. R., Ge, J. J., He, Q. G., Wu, Z. J., Chen, Z. W. & Xing, W. (2019). Engineering Energy Level of Metal Center: ru Single-Atom Site for Efficient and Durable Oxygen Reduction Catalysis. Journal of the American Chemical Society, 141(50), 19800–19806.

[37] Liu, J., Jiao, M. G., Lu, L. L., Barkholtz, H. M., Li, Y. P., Wang, Y., Jiang, L. H., Wu, Z. J., Liu, D. J., Zhuang, L., Ma, C., Zeng, J., Zhang, B. S., Su, D. S., Song, P., Xing, W., Xu, W. L., Wang, Y., Jiang, Z. & Sun, G. Q. (2017). High-performance platinum single atom electrocatalyst for the oxygen reduction reaction. Nature Communications, 8(1), 1–10.

[38] Chen, W. X., Pei, J. J., He, C. T., Wan, J. W., Ren, H. L., Zhu, Y. Q., Wang, Y., Dong, J. C., Tian, S. B., Cheong, W. C., Lu, S. Q., Zheng, L. R., Zheng, X. S., Yan, W. S., Zhuang, Z. B., Chen, C., Peng, Q., Wang, D. S. & Li, Y. D. (2017). Rational design of single molybdenum atoms anchored on N-Doped carbon for effective hydrogen evolution reaction. Angewandte Chemie International Edition, 56(50), 16086–16090.

[39] Chen, W. X., Pei, J. J., He, C. T., Wan, J. W., Ren, H. L., Wang, Y., Dong, J. C., Wu, K. L., Cheong, W. C., Mao, J. J., Zheng, X. S., Yan, W. S., Zhuang, Z. B., Chen, C., Peng, Q., Wang, D. S. & Li, Y. D. (2018). Single Tungsten Atoms Supported on MOF-Derived N-Doped Carbon for Robust Electrochemical Hydrogen Evolution. Advanced Materials (Deerfield Beach, Fla.), 30(30), 1800396–1 to 1800396–6.

[40] Luo, E. G., Zhang, H., Wang, X., Gao, L. Q., Gong, L. Y., Zhao, T., Jin, Z., Ge, J. J., Jiang, Z., Liu, C. P. & Xing, W. (2019). Single-Atom Cr-N-4 sites designed for durable oxygen reduction catalysis in acid media. Angewandte Chemie International Edition, 58(36), 12469–12475.

[41] Xiao, M. L., Zhu, J. B., Li, G. R., Li, N., Li, S., Cano, Z. P., Ma, L., Cui, P. X., Xu, P., Jiang, G. P., Jin, H. L., Wang, S., Wu, T. P., Lu, J., Yu, A. P., Su, D. & Chen, Z. W. (2019). A single-atom iridium heterogeneous catalyst in oxygen reduction reaction. Angewandte Chemie International Edition, 58(28), 9640–9645.

[42] Zhang, C. H., Sha, J. W., Fei, H. L., Liu, M. J., Yazdi, S., Zhang, J. B., Zhong, Q. F., Zou, X. L., Zhao, N. Q., Yu, H. S., Jiang, Z., Ringe, E., Yakobson, B. I., Dong, J. C., Chen, D. L. & Tour, J. M. (2017). Single-Atomic ruthenium catalytic sites on nitrogen-doped graphene for oxygen reduction reaction in acidic medium. Acs Nano, 11(7), 6930–6941.

[43] Weng, Z., Wu, Y. S., Wang, M. Y., Jiang, J. B., Yang, K., Huo, S. J., Wang, X. F., Ma, Q., Brudvig, G. W., Batista, V. S., Liang, Y. Y., Feng, Z. X. & Wang, H. L. (2018). Active sites of copper-complex catalytic materials for electrochemical carbon dioxide reduction. Nature Communications, 9(1), 1–9.

[44] Yin, P. Q., Yao, T., Wu, Y., Zheng, L. R., Lin, Y., Liu, W., Ju, H. X., Zhu, J. F., Hong, X., Deng, Z. X., Zhou, G., Wei, S. Q. & Li, Y. D. (2016). Single cobalt atoms with precise n-coordination as superior oxygen reduction reaction catalysts. Angewandte Chemie International Edition, 55(36), 10800–10805.

[45] Shang, H. S., Zhou, X. Y., Dong, J. C., Li, A., Zhao, X., Liu, Q. H., Lin, Y., Pei, J. J., Li, Z., Jiang, Z. L., Zhou, D. N., Zheng, L. R., Wang, Y., Zhou, J., Yang, Z. K., Cao, R., Sarangi, R., Sun, T. T., Yang, X., Zheng, X. S., Yan, W. S., Zhuang, Z. B., Li, J., Chen, W. X., Wang, D. S., Zhang, J. T. & Li, Y. D. (2020). Engineering unsymmetrically coordinated Cu-S1N3 single-atom sites with enhanced oxygen reduction activity. Nature Communications, 11(1), 1–11.

[46] Wang, J., Huang, Z. Q., Liu, W., Chang, C. R., Tang, H. L., Li, Z. J., Chen, W. X., Jia, C. J., Yao, T., Wei, S. Q., Wu, Y. & Lie, Y. D. (2017). Design of N-coordinated dual-metal sites: a stable and active pt-free catalyst for acidic oxygen reduction reaction. Journal of the American Chemical Society, 139(48), 17281–17284.

[47] Xiao, M. L., Zhang, H., Chen, Y. T., Zhu, J. B., Gao, L. Q., Jin, Z., Ge, J. J., Jiang, Z., Chen, S. L., Liu, C. P. & Xing, W. (2018). Identification of binuclear Co2N5 active sites for oxygen reduction reaction with more than one magnitude higher activity than single atom CoN4 site. Nano Energy, 46, 396–403.

[48] Kong, F. P., Si, R. T., Chen, N., Wang, Q., Li, J. J., Yin, G. P., Gu, M., Wang, J. J., Liu, L. M. & Sun, X. L. (2022). Origin of hetero-nuclear Au-Co dual atoms for efficient acidic oxygen reduction. Applied Catalysis B: Environmental, 301, 120782–1 to 120782–10.

[49] Sun, M. Z., Wong, H. H., Wu, T., Dougherty, A. W. & Huang, B. L. (2021). Stepping out of transition metals: activating the dual atomic catalyst through main group elements. Advanced Energy Materials, 11(30), 2101404–1 to 2101404–13.

[50] Yang, S., Tak, Y. J., Kim, J., Soon, A. & Lee, H. (2017). Support effects in single-atom platinum catalysts for electrochemical oxygen reduction. ACS Catalysis, 7(2), 1301–1307.

[51] Luo, Z. Y., Ouyang, Y. X., Zhang, H., Xiao, M. L., Ge, J. J., Jiang, Z., Wang, J. L., Tang, D. M., Cao, X. Z., Liu, C. P. & Xing, W. (2018). Chemically activating MoS2 via spontaneous atomic palladium interfacial doping towards efficient hydrogen evolution. Nature Communications, 9(1), 1–8.

[52] Duan, H. L., Wang, C., Li, G. N., Tan, H., Hu, W., Cai, L., Liu, W., Li, N., Ji, Q. Q., Wang, Y., Lu, Y., Yan, W. S., Hu, F. C., Zhang, W. H., Sun, Z. H., Qi, Z. M., Song, L. & Wei, S. Q. (2021). Single-atom-layer catalysis in a mos2 monolayer activated by long-range ferromagnetism for the hydrogen evolution reaction: beyond single-atom catalysis. Angewandte Chemie International Edition, 60(13), 7251–7258.

[53] Tyo, E. C. & Vajda, S. (2015). Catalysis by clusters with precise numbers of atoms. Nature Nanotechnology, 10(7), 577–588.

[54] Ni, B. & Wang, X. (2016). Chemistry, and properties at a sub-nanometer scale. Chemical Science, 7(7), 3978–3991.

[55] Peng, M., Dong, C. Y., Gao, R., Xiao, D. Q., Liu, H. Y. & Ma, D. (2021). Fully Exposed Cluster Catalyst (FECC): toward Rich Surface Sites and Full Atom Utilization Efficiency. Acs Central Science, 7(2), 262–273.

[56] Ni, B., Shi, Y. & Wang, X. (2018). The sub-nanometer scale as a new focus in nanoscience. Advanced Materials (Deerfield Beach, Fla.), 30(43), 1802031–1 to 1802031–24.

[57] Liu, L. C., Lopez-Haro, M., Calvino, J. J. & Corma, A. (2021). Tutorial: structural characterization of isolated metal atoms and subnanometric metal clusters in zeolites. Nature Protocols, 16(4), 1871-+.

[58] Bae, J., Kim, J., Jeong, H. & Lee, H. (2018). CO oxidation on SnO2 surfaces enhanced by metal doping. Catalysis Science & Technology, 8(3), 782–789.

[59] Jeong, H., Lee, G., Kim, B. S., Bae, J., Han, J. W. & Lee, H. (2018). Fully dispersed Rh ensemble catalyst to enhance low-temperature activity. Journal of the American Chemical Society, 140(30), 9558–9565.

[60] Gabelnick, A. M., Capitano, A. T., Kane, S. M., Gland, J. L. & Fischer, D. A. (2000). Propylene oxidation mechanisms and intermediates using in situ soft X-ray fluorescence methods on the Pt(111) surface. Journal of the American Chemical Society, 122(1), 143–149.

[61] An, K., Alayoglu, S., Musselwhite, N., Na, K. & Somorjai, G. A. (2014). Designed catalysts from Pt nanoparticles supported on macroporous oxides for selective isomerization of n-Hexane. Journal of the American Chemical Society, 136(19), 6830–6833.

[62] Lee, I., Delbecq, F., Morales, R., Albiter, M. A. & Zaera, F. (2009). Tuning selectivity in catalysis by controlling particle shape. Nature Materials, 8(2), 132–138.

[63] Kliewer, C. J., Aliaga, C., Bieri, M., Huang, W. Y., Tsung, C. K., Wood, J. B., Komvopoulos, K. & Somorjai, G. A. (2010). Furan Hydrogenation over Pt(111) and Pt(100) Single-Crystal Surfaces and Pt Nanoparticles from 1 to 7 nm: a Kinetic and Sum Frequency Generation Vibrational Spectroscopy Study. Journal of the American Chemical Society, 132(37), 13088–13095.

[64] An, K. & Somorjai, G. A. (2012). Size and shape control of metal nanoparticles for reaction selectivity in catalysis. Chemcatchem, 4(10), 1512–1524.

[65] Wang, G. H., Hilgert, J., Richter, F. H., Wang, F., Bongard, H. J., Spliethoff, B., Weidenthaler, C. & Schuth, F. (2014). Platinum-cobalt bimetallic nanoparticles in hollow carbon nanospheres for hydrogenolysis of 5-hydroxymethylfurfural. Nature Materials, 13(3), 294–301.

[66] Chia, M., Pagan-Torres, Y. J., Hibbitts, D., Tan, Q. H., Pham, H. N., Datye, A. K., Neurock, M., Davis, R. J. & Dumesic, J. A. (2011). Selective Hydrogenolysis of Polyols and Cyclic Ethers over Bifunctional Surface Sites on Rhodium-Rhenium Catalysts. Journal of the American Chemical Society, 133(32), 12675–12689.

[67] Wan, X. K., Wu, H. B., Guan, B. Y., Luan, D. Y. & Lou, X. W. (2020). Confining Sub-Nanometer Pt Clusters in Hollow Mesoporous Carbon Spheres for Boosting Hydrogen Evolution Activity. Advanced Materials (Deerfield Beach, Fla.), 32(7), 1901349–1 to 1901349–8.

[68] Sun, M. H., Ji, J. P., Hu, M. Y., Weng, M. Y., Zhang, Y. P., Yu, H. S., Tang, J. J., Zheng, J. C., Jiang, Z., Pan, F., Liang, C. D. & Lin, Z. (2019). Overwhelming the Performance of Single Atoms with Atomic Clusters for Platinum-Catalyzed Hydrogen Evolution. ACS Catalysis, 9(9), 8213–8223.

[69] Cheng, X., Li, Y. H., Zheng, L. R., Yan, Y., Zhang, Y. F., Chen, G., Sun, S. R. & Zhang, J. J. (2017). Highly active, stable oxidized platinum clusters as electrocatalysts for the hydrogen evolution reaction. Energy & Environmental Science, 10(11), 2450–2458.

[70] Henninen, T. R., Bon, M., Wang, F., Passerone, D. & Erni, R. (2020). The Structure of Sub-nm Platinum Clusters at Elevated Temperatures. Angewandte Chemie International Edition, 59(2), 839–845.

[71] Gorczyca, A., Moizan, V., Chizallet, C., Proux, O., Del Net, W., Lahera, E., Hazemann, J. L., Raybaud, P. & Joly, Y. (2014). Monitoring Morphology and Hydrogen Coverage of Nanometric Pt/gamma-Al2O3 Particles by In Situ HERFD-XANES and Quantum Simulations. Angewandte Chemie International Edition, 53(46), 12426–12429.

[72] Zhang, H., Cao, L. N., Wang, Y. L., Gan, Z. D., Sun, F. F., Xiao, M. L., Yang, Y. Q., Mei, B. B., Wu, D. S., Lu, J. L., He, H. Y. & Jiang, Z. (2021). Interfacial Proton Transfer for Hydrogen Evolution at the Sub-Nanometric Platinum/Electrolyte Interface. ACS Applied Materials & Interfaces, 13(39), 47252–47261.

Dan Kong

Chapter 3B
In situ spectroscopic studies of the electrochemistry

Abstract: Most electrocatalysts would undergo a structural reconstruction during the electrochemical reactions, which involves the electron transfer between two-phase interfaces (solid/liquid, liquid/liquid, solid/solid; in most cases, it is the electrode/electrolyte interface). In situ spectroscopic studies on the electrode/electrolyte interface can facilitate the real-time examinations of the structure and composition, which help the researchers better understand the mechanism of electrocatalysis. In this perspective, the historical development and recent applications of in situ spectroscopic techniques in tracking the structural reconstruction of electrocatalysis are thoroughly summarized. In particular, the electrochemical in situ Raman spectroscopy, attenuated total reflection Fourier-transform infrared spectroscopy, scanning probing microscopy, and mass spectrometer will be introduced in detail. Based on each in situ technique's unique capabilities and limitations, an "in situ spectroscopic map" is established for characterizing electrode/electrolyte interfaces in a dynamic scene, which offers guidelines for developing the next-generation efficient electrocatalysts.

Keywords: in situ, online, spectroscopy, microscopy, electrocatalysis, electrochemistry

3B.1 Introduction

3B.1.1 A brief history of electrochemistry

The origin of electrochemistry can be traced back to the phenomenon of "animal electricity" discovered by Luigi Galvani in 1780, which reveals the profound connection between biology and electrochemistry. In 1800, Alessandro Volta invented the first electrochemical cells, a voltaic pile formed by alternating copper and zinc

Acknowledgments: The author gratefully acknowledges the support of the Fundamental Research Funds for the Central Universities of China University of Mining and Technology (2021QN1114), Startup Foundation for Talented Scholars of China University of Mining and Technology, and 2021 Double Innovation Doctor of Jiangsu Province (JSSCBS20211205).

Dan Kong, Department of Materials Science and Physics, China University of Mining and Technology, No.1, Daxue Road, Xuzhou City, Jiangsu Province, China,
e-mail: dan.kong.14@ucl.ac.uk

https://doi.org/10.1515/9783110739879-008

sheets on both sides of felt soaked in an acid solution. In 1833, Michael Faraday discovered Faraday's laws of electrolysis, making electrochemistry a quantitative science. In addition, Faraday created a series of terms of electrochemistry, such as electrolysis, electrolyte, electrode, anode, cathode, ion, anion, and cation. In 1889, Walther Nernst formulated the Nernst equation, revealing the relationship between the electrode potential and solution concentration and temperature. However, the Nernst equation is only applicable to the standard-state thermodynamics.

In contrast, it is not accurate to predict reaction trends for dynamic conditions where the ion concentration or temperature changes. In 1905, Julius Tafel deduced the Tafel equation experimentally, which is an equation relating the current density through an electrode to the overpotential and was a pioneer in studying electrochemical kinetics. However, until the late 1950s, electrochemical kinetics received due attention and rapid development [1]. For example, Alexander Frumkin and his team discovered that the structure of the interface between electrode and solution could influence the rate of electron transfer and introduced the framework of contemporary physical electron transfer models [2, 3], which became the basis for the scientists to understand the structure of the electrode interface.

At the same time, a variety of electrochemical methods and experimental techniques emerged in the 1930s. For example, Heinz Gerischer created the electronic potentiostat and monitored fast electrochemical processes by potential double-step and AC modulation methods. This work laid the foundation for a mechanistic interpretation of electrode reactions and vastly impacted our present understanding of electrode kinetics. Another internationally known scientist, Paul Delahay, coinvented cyclic voltammetry (CV) and profoundly clarified electrochemistry processes, electrode kinetics, and electroanalytical methods. Moreover, Fred C. Anson and his group systematically illustrated the behavior of reactants absorbed or otherwise attached to the surface of electrodes and summarized the basic law of the absorption behavior of the measurable substance on the electrode. In the 1980s, Jean Clavilier invented a technique to solve the problem of surface cleaning and pollution-free transferring of the metal single-crystal electrode. He and his team treated the metal single-crystal electrode with an oxyhydrogen flame to remove the impurities and restore the atomic arrangement at the electrode surface. They then transferred the electrode to the electrolyte under the protection of ultrapure water [4]. This technique explored the electrochemical research on the atomic arrangement level of the electrode surface structure firstly.

The above methods mainly rely on the phenomenological study of the measurement and analysis of the electrochemical parameters, such as current, potential, capacitance, and quantity. The measured data limits the understanding of electrode interface structure and reaction process from the molecular level. The greatest improvement in electrochemistry occurred in the last three decades of the twentieth century, which combined spectroscopic techniques with electrochemical measurements in the electrolytic equipment so that in situ spectroscopic observation of the dynamic process at the electrode/electrolyte interface can be carried out. At the same time, the electrochemical reaction is proceeding, which realizes the understanding of electrochemical phenomena and mechanisms at the molecular level. In the in situ spectroscopic observation, the electrochemical methods can easily control the reaction process by adjusting the electrode/electrolyte interface; spectroscopy is beneficial in identifying the reaction species, especially intermediates and transient species. Together with the fast development of spectroscopic technology from the 1980s, in situ spectroscopic research has received increasing attention and has been widely applied in the research of electrochemical kinetics.

3B.1.2 The development of in situ spectroscopic electrochemistry

Since the 1960s, a series of in situ spectroscopic techniques in electrochemistry have been set up along with the rapid development of spectroscopic technology [5]. For example, electron spin resonance [6], UV–visible absorption spectroscopy [7], infrared spectroscopy [8, 9], electric reflectance [10], and Raman spectroscopy [11] have been used to characterize the formation of the film at the electrode surface. Table 3B.1 lists brief information on the major technological inventions of in situ spectroscopic electrochemistry.

Tab. 3B.1: Introduction of the invented years and inventors of the in situ spectroscopic techniques in electrochemistry.

Year	Spectroscopic technique in electrochemistry	Inventor	Reference
1959	Electron spin resonance	August H. Maki	[6]
1963	Ellipsometry	J.O'M. Bockris	[12]
1964	UV–vis absorption spectroscopy	Theodore Kuwana	[7]
1966	Electric reflectance	D.F.A. Koch; J. Feinleib	[10, 13]
1966	Infrared spectroscopy based on frustrated multiple internal reflectance electrolysis technique	H.B. Mark	[8]
1968	Fluorescence spectroscopy	Charles N. Reilley	[14]
1970	Mossbauer spectroscopy	J.O'M Bockris	[15]
1973	Mass spectroscopy	Stanley Bruckenstein	[16]
1973	Raman spectroscopy	Martin Fleischmann	[17]
1975	Nuclear magnetic resonance	Jeffrey A. Richards	[18]

Tab. 3B.1 (continued)

Year	Spectroscopic technique in electrochemistry	Inventor	Reference
1980	Modulated specular reflectance spectroscopy	B. S. Pons	[19]
1983	X-ray diffraction spectroscopy	M. Fleischmann	[20]
1983	Extended X-ray absorption fine structure spectrometry	M. Kuriyama	[21]
1986	Rutherford backscattering spectroscopy	S. Stucki	[22]
1988	Scanning tunneling microscopy	D.M. Kolb; Kingo Itaya; Allen J. Bard; P. Lustenberger	[23–26]
1990	Sum frequency spectroscopy	A. Tadjeddine	[27]

The first-time application of the in situ spectroscopic technique in electrochemistry is electron spin resonance, as shown in Tab. 3B.1, which observed transient chlorine-containing radicals in the electrolysis directly by electron spin resonance spectrum [6]. And then, J.O'M. Bockris et al. reported a new "chronoellipsometry," which controlled the constant current by the conventional chronopotentiometry. They recorded the intensity change of optical response by ellipsometry simultaneously [12]. The experimental diagram is shown in Fig. 3B.1. However, the information from the chronoellipsometry at that time was very limited because of the high requirements of ellipsometry on the smoothness of the electrode surface. Only some simple characterization of thin-film structure with the strong signal can be performed, whereas the angular resolution information cannot be obtained.

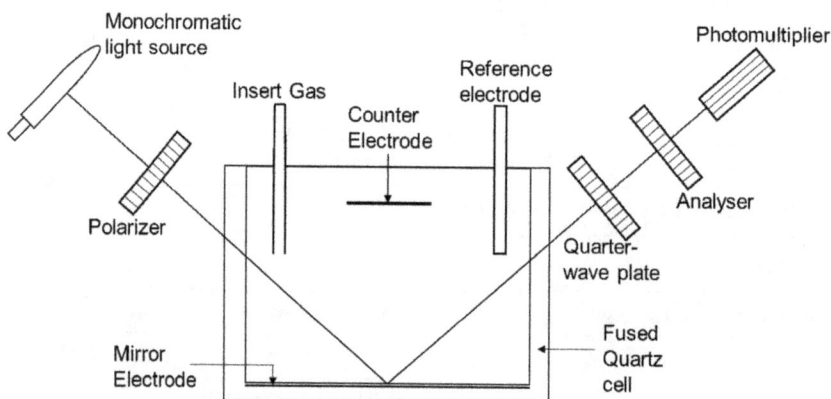

Fig. 3B.1: Synthetic quartz cells are used for chronoellipsometry [12].

In 1964, T. Kuwana applied another feasible spectroscopic technique, UV–vis spectroscopy, to the electrochemical system for the first time [7]. Their research initially focused on the redox couple of potassium ferrocyanide and potassium ferricyanide, which have different colors. When voltage was applied to the cathode or anode, the reduction or oxidation reaction production was verified by the color change of the solution phase near the electrode, which the spectrometer could easily detect. This is the first report of spectroscopic application in researching electrochemical reaction products in the solution phase, whereas the information obtained from this technique was very limited. Even with the naked eyes, the color change of the solution near the electrode during the reaction could be observed. Because the reactants and products need to have obvious color differences, very few molecular systems are suitable for this technique.

Both J. Feinleib and D. F. A. Koch reported their electroreflectance studies of metal electrodes in 1966 [10, 13]. Based on that, the dipole layer formed in the electrolyte at the interface of the electrode/electrolyte could cause a significant change in the reflectivity of the metal electrode surface. Researchers recorded the original reflection spectra of the metal electrode surface at different potentials and then compared them with the standard sample at the starting potential. Therefore, researchers obtained electroreflectance with the applied voltage in the etching process of the gold electrode surface in the KCN solution and the oxide film formation on the platinum electrode under the high potential. The configuration of the electrochemical cell is illustrated in Fig. 3B.2.

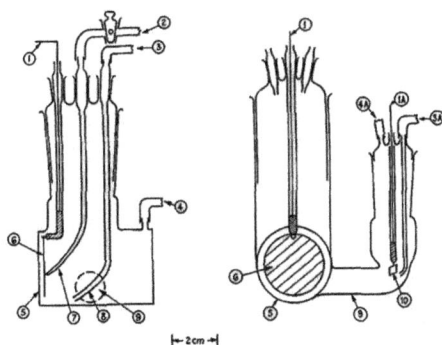

Fig. 3B.2: Electrochemical cell for reflectance measurements: 1 and 1A, leads to potentiostat; 2, H_2SO_4-filled bridge to saturated calomel reference electrode; 3 and 3A, gas inlets; 4 and 4A, gas outlets; 5, silica window; 6, working electrode; 7, Luggin capillary; 8, gas bubbler; 9, sidearm connecting to a compartment with a counter electrode, gas inlet, and outlet; 10, counter electrode [10].

Applying the above three types of spectroscopies was a reasonable attempt, demonstrating the feasibility of in situ spectroscopic electrochemistry. Ellipsometry and electroreflectance have a high requirement of a smooth electrode surface, but the detection sensitivity of the instruments was limited. Thus, the research was limited to a certain thickness of the electrode surface film. The signal of UV–vis spectroscopy was only from the solution species with relatively higher concentrations. These early reports of in situ spectroscopy technology did not reflect the advantages of spectroscopy and electrochemistry but just found the direction of spectroscopic electrochemistry.

The intense absorption dominates the infrared spectrum of the aqueous solution due to the fundamental O–H stretching vibration. Thus, the propagation distance of infrared light is very short. The signal will be significantly attenuated in a water layer of tens of microns, which causes great difficulties in studying the in situ infrared spectroscopic electrochemistry of the electrode/aqueous solution system. H. B. Mark et al. investigate infrared spectra of two organic molecules under electrolysis at germanium frustrated multiple internal reflectance (FMIR) electrode, which prevents infrared light from entering the aqueous solution and being absorbed by the water [8]. As shown in Fig. 3B.3, the infrared light only spreads in the germanium FMIR electrode. Firstly, the infrared light was introduced from one side of the cell, performed multiple internal reflections, came out from the opposite side of the cell, and was finally collected by IR spectrometer. For the first time, this research reported the in situ molecular vibrational spectroscopy in electrochemistry. This FMIR-IR technique could be applied to detect nonabsorbent or colorless molecular systems in the visible light region, which has a wider range of applications than the in situ UV–vis transparent spectroscopy electroreflectance and ellipsometry in electrochemistry. More importantly, IR spectroscopy belongs to molecular vibrational spectroscopy, which can reveal the fingerprint information of molecules.

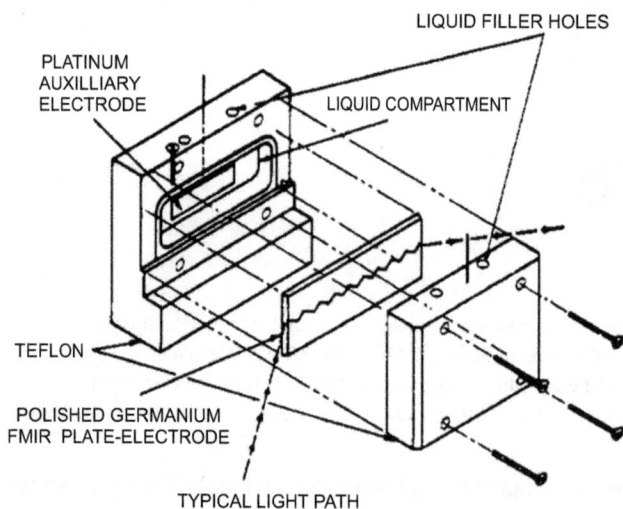

Fig. 3B.3: Details of Teflon electrolysis cell-FMIR plate holder [8].

However, it was very difficult to design and process FMIR electrolysis cells with germanium materials and manipulate invisible IR light at that time. Thus, there is no more development of the in situ IR spectroscopy in electrochemistry. After the 1990s, with more metal materials as IR light-conducting materials in thin films, the FMIR technology revived.

In 1968, C.N. Reilley et al. firstly reported the research on the electrochemistry–fluorescence performance of rubrene molecules [14]. This new method reduced the concentration magnitude of measured materials to a micromolar level, which was lower than that detectable by UV–vis absorption spectroscopy of aromatic hydrocarbons.

In 1973, M. Fleischmann et al. employed the in situ Raman spectroscopy for the first time to detect and characterize the small quantities of species present at the electrode surface [17]. Due to the advent of laser sources, the incident light would not be absorbed and interfered with by the aqueous sol. This technique could detect the species at the electrode/electrolyte interface. The huge difference between this work and previous spectroscopy–electrochemistry reports was using the external collection mode of reflectance for the first time (as shown in Fig. 3B.4). In this work, researchers chose the mercury halides and oxides as the application objects, which had a strong Raman signal and benefited the improvement and optimization of the spectroscopic, electrochemical cell. In their later research, the same team distinguished two types of pyridine absorption at a silver electrode by the in situ Raman spectra in their later research [11], which obtained the Raman signal of molecules absorbed on a monolayer.

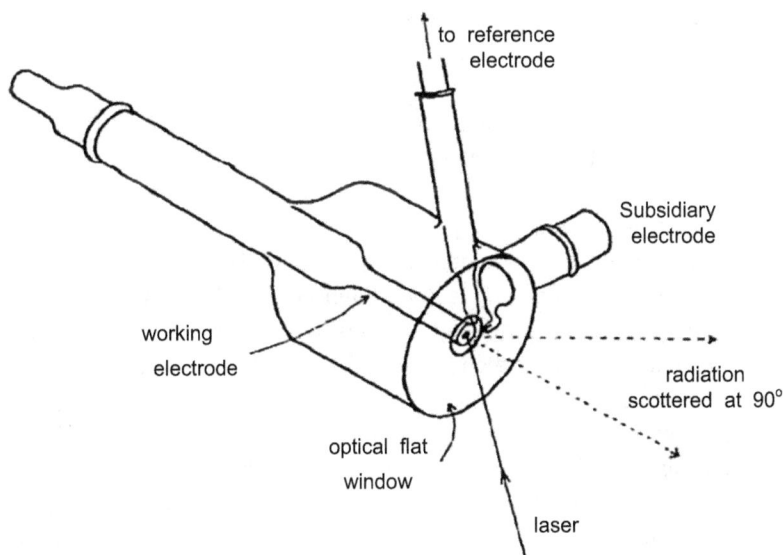

Fig. 3B.4: The electrochemical cell was used for the Raman experiments [11].

In summary, from the 1960s to the 1980s, J.O'M. Bockris' group from the University of Pennsylvania in the USA [12, 15], Theodore Kuwana's group from the University of California in the USA [7], and M. Fleischmann's group from the University of Southampton in the UK [11, 17, 20] made a key contribution to the initial development of the in situ spectroscopic electrochemistry. Their research has an important role in understanding the physical and chemical processes and mechanisms of the electrochemical interface.

3B.1.3 The classification of the in situ spectroscopic electrochemistry

Based on the excitation modes, the in situ spectroscopic electrochemistry can be classified by four branches, as illustrated in Fig. 3B.5. As the photon can pass through most media, the incident light can penetrate the solution layer of the electrode/electrolyte interface or the solid electrode material with optical transparency easily, which can reflect various optical, electrical, thermal, and acoustic signals of the structure and properties of the electrochemical interface. Therefore, using light as the means of excitation and detection has been the pioneer in developing in situ spectroscopic electrochemistry since 1963. With the rapid development of laser technology, in situ spectroscopic technology has received increasing attention and has become one of the important branches of in situ spectroscopic electrochemistry. Especially molecular vibrational spectroscopy technology (IR, Raman, and sum-frequency generation spectroscopy) has received exceptional attention from researchers because of its ability to obtain the fingerprint information of interface species at the molecular level.

Another important branch of in situ spectroscopic electrochemistry is the in situ scanning probe microscopic technique, which uses an extremely sharp probe tip of metal, alloy, or semiconductor materials to approach and scan the surface of the solid electrode from the liquid side. A simplified sketch of in situ scanning probe microscopy in the electrochemical cell was demonstrated in Fig. 3B.6. The tip can localize the surface at a very small scope and is not influenced by optical diffraction. The spatial resolution of this kind of technology can easily reach tens to several nanometers. It can even achieve atomic-scale resolution by measuring the tuning current following the most protruding atom of the conductive tip and the electrode surface. The surface imaging with ultra-high-space resolution far exceeding the optical diffraction limit can be obtained through two-dimensional scanning.

Since the first scanning tunneling microscope (STM) was invented in 1981, researchers have realized the great potential of STM in electrochemical research and carried out a large number of exploratory works. The inventors Gerd Binning and Heinrich Rohrer were thus awarded the Nobel Prize in Physics in 1986. The STM experiment in a solution environment was first reported in 1986 [28, 29]. The researchers

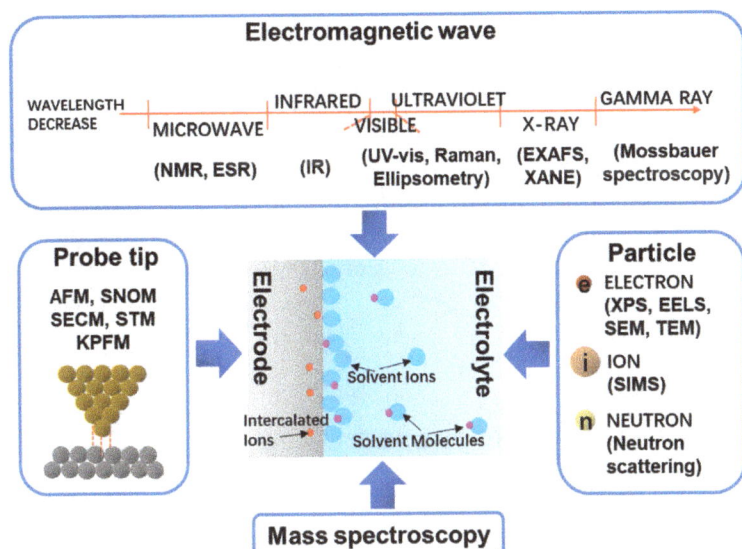

Fig. 3B.5: There are three branches of in situ spectroscopic electrochemistry based on the excitation sources, including electromagnetic wave, probe, and particle sources; mass spectroscopy detects electrochemical species, thus being attributed to the fourth branch (NMR, nuclear magnetic resonance; ESR, electron spin resonance; EXAFS, extended X-ray absorption fine structure; XANE, X-ray absorption near the edge; AFM, atomic force microscope; SNOM, scanning near-field optical microscopy; SECM, scanning electrochemical microscopy; STM, scanning tunneling spectroscopy; KPFM, Kelvin probe force microscope; XPS, X-ray photoelectron spectroscopy; EELS, electron energy loss spectroscopy; SEM, scanning electron microscope; TEM, transmission electron microscope; SIMS, secondary-ion mass spectrometry).

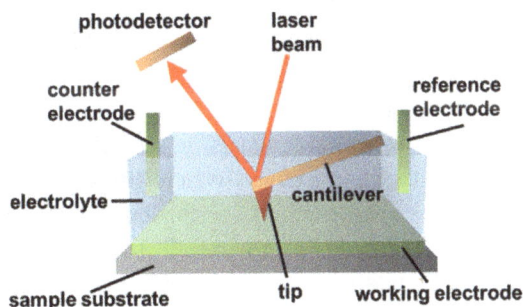

Fig. 3B.6: A schematic diagram of an electrochemical cell for in situ scanning probe microscopic observation.

initially overcame the difficulties caused by the electrochemical system by partially covering the tunneling tip with thin glass, avoiding the serious interference problem of the tunneling current caused by the electrochemical current generated on the tip. Thus, an image of a graphite surface immersed in deionized water was obtained with features less than 3 Å apart, resolved by the in situ STM [29]. Since then, the electrochemical STM (ECSTM) has been successfully applied to the in situ characterization and structure research of various electrode/electrolyte interfaces. ECSTM and electrochemical atomic force microscopy have become the most common in situ techniques in electrochemical scanning probe microscopy (ECSPM). In addition to providing the fine surface topography, ECSPM can further obtain richer interface information when combined with tunneling spectroscopy, capacitance/voltage spectroscopy, force spectroscopy, or other spectroscopic techniques. In short, ECSPM belongs to one of the four significant branches of in situ spectroscopic electrochemistry.

3B.1.4 The resolution of the in situ spectroscopic electrochemistry

There are two main indicators for the various techniques when characterizing the electrochemistry system's surface or interface structure and process. One is how to improve the sensitivity of detecting sublayers or monolayers. The other is how to improve the energy, space, and time resolution. This section mainly evaluates these three kinds of resolution of in situ spectroscopic electrochemistry technology to help researchers have a brief view of how to choose a suitable in situ technique.

The molecule structure will change slightly when molecules are adsorbed on the electrode surface. The technology with high energy resolution can distinguish the tiny energy difference between the same molecules on the electrode surface and in the solution phase. In this way, the structure of the absorbed or redox molecule can be observed clearly, and the absorption process and reaction process and mechanism can be analyzed in detail. That is why energy resolution is very important for in situ spectroscopic techniques.

Electrochemical technology can only investigate the phase change process on the surface of a single-crystal electrode with an unambiguous structure. However, the electrodes are polycrystal or powder for most surface electrochemistry systems. When two or more reactions or electrochemical absorptions are carried out in the system, it is hard to distinguish by the electrochemical techniques if their reaction or absorption potentials are too close. Fortunately, the spectroscopic techniques can easily reach a wavenumber resolution. Especially for the Raman spectroscopy with a visible laser as the detection source, the energy resolution is equivalent to 0.1 meV. Compared with conventional electrochemical technology, the energy resolution of in situ spectroscopic, electrochemical technology has been improved by at least two orders of magnitude.

Compared with spectroscopic technology, microwave technology has a higher energy resolution by 2–4 orders. Among them, NMR technology is the most representative and has become a powerful tool for investigating the local structure of nanoscale and amorphous materials. The in situ NMR technology in electrochemistry can also provide the structure information of the ions in the electrolyte and surrounding electrons on the electrode. Therefore, in situ NMR technology has been important in researching electrode/electrolyte materials. For example, the in situ NMR electrochemical technology is applied in observing the tiny structural changes of Li-ion/Na-ion battery materials in the electrochemical reaction. Thus, this is a very detailed and systematic mechanism of the related electrochemical reaction [30].

In terms of spatial resolution, electrochemical technology also has its shortcomings. The measured current, voltage, or electrical impedance signals are derived from the overall contribution of the local electrochemical signals. For example, even if the active sites on the electrode surface are very uneven, the electrical current generated everywhere can only be collected by a wire and then averaged by the electrochemical software, so the average value of the electrochemical signals cannot fully reflect the details of the electrode/electrolyte interface.

Since the 1980s, the scanning probe microscope (SPM) series with ultra-high-space resolution, represented by the STM), enables the researchers to observe the electronic behavior of the individual atoms on the surface for the first time in real time. More importantly, the SPM can be used for in situ electrochemical research, that is, to characterize and analyze the electrochemical interface without changing the solution or electrochemical conditions. Therefore, SPM has become the most important technique for investigating the structure and process of the electrode/electrolyte interface with a high spatial resolution. However, the beginning of the in situ SPM in electrochemistry was also quite difficult. The tunnel bias applied between the tip and sample in the electrolyte was also equivalent to the potential applied in a two-electrode electrochemical system. Thus, the additional Faraday current was created between the tip and sample. Since the tunneling current value is on the order of nano-amperes, the typical Faraday current value is on the order of microamperes, which will cause serious interference to the tunneling current and damage the geometry stability of the tip. Therefore, it is necessary to encapsulate most of the tip with the insulating materials to suppress the Faraday current to obtain an STM image with an atomic resolution under the electrochemical conditions.

Thus, in situ STM electrochemical technology has the most advantages in terms of spatial resolution. Its spatial resolution can easily reach the order of nanometers or even as high as angstroms, so it is expected to distinguish the fine topographic structure on the electrode surface. If the potential between the tip and the sample is controlled to a constant value, the tunneling current is recorded while changing the distance between the tip and the sample. This method is called distance tunneling spectroscopy (DTS). Through this method, the information of the structure perpendicular to the interface can be obtained by STM to disclose the double-layer structure in the interface (see Chapter 4 for specific examples).

The weakness of the STM technology lies in energy resolution. Although this technology can achieve high energy resolution equivalent to vibrational spectroscopy, it often needs to work at ultra-low temperatures. Because the working temperature of the electrochemical system is generally higher than 0 ℃, it is extremely difficult to obtain high-quality tunneling spectra. Researchers have been trying to combine the scanning probe technology with high spatial resolution and the spectroscopy technology with high energy resolution. Several research groups simultaneously established tip-enhanced Raman spectroscopy (TERS) [31–34]. The key is to change the tip of the conventional tungsten or platinum–iridium alloy materials instead of silver or gold materials, generating the localized surface plasmon resonance (LSPR) effect under laser conditions. Thus, Raman signals with a resolution of tens of nanometers or even a few nanometers can be obtained. Although TERS technology has high spatial resolution and high energy resolution, it is not easy to apply TERS to electrochemical (EC) systems. The various substances with different reflective indexes, such as electrolytic cell windows and electrolytes, can cause optical path distortion, seriously reducing the sensitivity of TERS detection. Although several research groups in the world have been working hard to develop EC-TERS instruments for more than 10 years, it was not until 2015 that Ren Bin et al. made a breakthrough and reported the EC-TERS setup for the first time [35]. Figure 3B.7a illustrates a schematic diagram of EC-TERS, which involves the combination of SPM, plasmon-enhanced Raman spectroscopy, and electrochemistry in a setup. In this mode, the potentials of the tip (WE1) and the substrate (WE2) are controlled by a bipotentiostat relative to a Pt wire. The laser is introduced horizontally via a long working distance microscope objective to the electrochemical cell and is illuminated in the gap of the tip and Au(111) surface. The light has to pass through the air, the electrolyte of different refractive indexes in a tilted way, and the optical path will be severely distorted (see Fig. 3B.7b). Therefore, the researchers modified the EC-STM cell by tilting the single crystal to about 10° so that the laser can properly focus on the tip and the Raman signal can be collected with high efficiency. In this configuration, the tip can still approach vertically downward. Compared with titled illumination, the optical path will not change even if electrolyte evaporation during the EC-TERS measurement allows long-time measurement.

The time resolution of electrochemical technology is also limited by its deficiencies. It is inevitable to encounter the problem of electrochemical double-layer charging when using the fast-scan CV or potential step method with time resolution capability. Therefore, the time resolution is often milliseconds or microseconds for the conventional electrode system if the microelectrode system is used. Suppose that the pulse laser is used as the light source of the spectroscopy. In that case, the time resolution can reach the order of picoseconds or even femtoseconds, so the in situ spectroscopy technology has obvious advantages in the time resolution. At the same time, the time resolution of STM is poor. Depending on the size of the STM's imaging areas, its time resolution may range from a few seconds to a few milliseconds. In recent years, various types of spectral imaging technologies have developed rapidly due to the significant improvement of the optical detection methods and the performance of the charge-coupled device (CCD) detectors. When the space resolution of two-dimensional imaging changes from submicron to micro-meters, the time resolution can also reach several milliseconds.

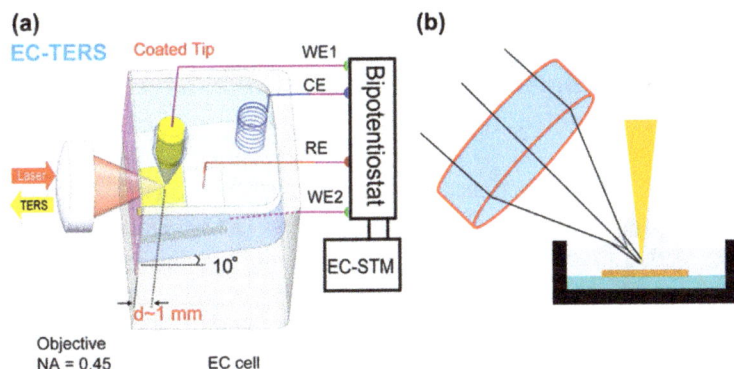

Fig. 3B.7: Schematic illustration of the EC-TERS setup (a) and the optical distortion in the side-illustration mode (b) [35].

In short, it is necessary to select the most appropriate technology in a targeted manner based on the energy, space, or time resolution of the electrochemical systems. The sensitivity and resolution of the in situ technologies of electrochemistry (CV), spectroscopy (Raman), and scanning probe microscopy (STM) are compared in Tab. 3B.2. It can be seen that no one type of technology can be fully dominant over the others. It is impossible to increase the three kinds of resolutions simultaneously because the three are interrelated. For example, to improve the energy resolution, reducing the width of the inlet and outlet slits of the spectrometer reduces the number of photons collected by the spectrometer's detector. Therefore, it will be inevitable to increase the time to collect the signal and reduce the time resolution ability to maintain the quality of the spectrum. When designing and customizing an in situ electrochemical system, it is

Tab. 3B.2: A brief comparison of the three in situ technologies of electrochemistry, Raman spectroscopy, and scanning probe microscopy.

Item	Electrochemistry (CV)	Spectroscopy (Raman)	Scanning probe microscopy (STM)
Sensitivity	Excellent (monolayer molecule)	General (10^2–10^4-layer molecule)	Excellent (monolayer molecule)
Energy resolution	General (10^{-2} V)	Excellent (10^{-4} V)	Poor (10^{-2} V, low temperature)
Space resolution	General (1 μm)	Good (0.5 μm)	Excellent (0.1 nm)
Time resolution	Good (20^{-6} s)	Excellent (10^{-9} s)	General (0.1 s)

necessary to consider which information is the most critical for the research system and then decide which one of the energy, space, or time resolution is the main objective.

3B.1.5 The quantum chemical calculations of the in situ spectroscopic electrochemistry

The electrochemical theory and spectroscopic theory provide a theoretical basis for the in situ spectroscopic electrochemistry to understand the electrochemical intermediate species and reaction process. The theories of the in situ spectroscopic electrochemistry are helpful for the in-depth analysis of the adsorption and co-adsorption of the complex molecules on the electrode surface.

The expansion of the in situ spectroscopic electrochemistry research from electrode surface adsorption to interfacial electrochemical reaction has been a key development. Certain molecules' solid or liquid phases can be measured first for the surface adsorption system, and their spectrum will be used as references. The spectrum obtained by the in situ characterizations will be compared with these references. The interaction between the structure and performance of the molecules can be deduced after the comparison. However, there is no sample or reference spectrum for the relatively unstable reaction intermediate products, so the characterization and analysis of these intermediates are much more difficult than the known species. Therefore, combining the two-dimensional or multidimensional spectroscopic techniques with electrochemical theoretical calculations is necessary to analyze the intermediate products and the reaction mechanism.

With the rapid improvement of computing technology, the simulation and theoretical calculations of the electrochemical models can match the experimental data well. The role of prediction and guidance of the simulation and calculation results is also enhanced and recognized by more and more experimental workers. At present, there are mainly two types of quantum chemical calculation methods. One is to process the molecule and cluster models as molecular systems. Their nuclear motion is calculated as a harmonic oscillator model to obtain harmonic vibration frequency and corresponding spectral peak intensity information, such as infrared and Raman intensity. This is very important for in situ spectroscopic electrochemistry research. The peak frequency needs to be combined with the peak intensity to comprehensively reflect the interaction's strength between the molecules and the surface. For vibration modes involving strong anharmonic characteristics, such as a torsional vibration mode with a periodic potential energy function, or a flip vibration mode with a double potential energy function, it will cause large calculation errors if the harmonic approximation is still applied to describe these vibration modes. In these cases, it is necessary to consider whether the nonadiabatic coupling, the anharmonicity of the vibrator, and the vibration–rotation coupling effect are ignored in the above theoretical calculation process. Due to the influence of the anharmonic effect on the vibration energy level, the potential energy function must be corrected by the quartic term to obtain a more accurate theoretical vibration frequency and wave function.

The other quantum chemical calculation method is a theoretical simulation method based on the periodicity of the solid/liquid interface, which considers the solid's quantum chemical effects and the liquid's statistical mechanical properties. The theoretical simulation methods based on periodicity have developed rapidly in these years, which can reasonably describe the electrode surface's ordered structure or the phase transition phenomenon of the molecular layer. This method can obtain accurate peak frequencies, but it cannot provide the peak intensity. Therefore, the ideal approach combines the cluster model with the periodic model in the same system, which is beneficial for comprehensively analyzing the interactive spectroscopy and electrochemical data and then revealing the images of the complex interface structure.

In the in situ spectroscopic electrochemistry theoretical research, the calculations involving surface enhancement spectroscopy are particularly difficult. For example, the surface plasmon resonance (SPR) created in the surface-enhanced Raman spectroscopy (SERS) can enhance the surface molecules' spectroscopic signals. However, it may also directly induce the chemical reaction of the absorbed species and get new substances. The latter should be avoided in the in situ spectroscopic electrochemistry research. Many experimenters do not realize the spectra from the new substances, which leads to misinterpretation of the experimental data. Special attention should be paid to the metal nanoparticle systems, such as gold, silver, and copper nanoparticles, with high activity of SERS. The SPR induces new charge transfer (CT) excited states at the metal nanoparticle/molecular interface. Thus, the generated electron/hole pairs provide new photon excitation and reaction channel. In addition, the nonradiative relaxation of SPR will be converted into heat eventually. This photothermal effect will provide the necessary activation energy for the surface chemical reactions. If there is a significant change between the spectra observed in the experiment and that of certain molecules, such as the appearance of some new peaks, the new substances have likely emerged. Generally, it is necessary to systematically change the power density and wavelength of the incident light, analyze a series of the spectra obtained, and pay attention to the correlation of the change with the new peaks. Therefore, theoretical calculations and related predictions are particularly important. Suppose that the production of new substances in the experiment cannot be avoided; in that case, it is a better choice to take advantage of it and to study the SPR-induced photochemical reaction.

So far, the theory of in situ spectroscopic electrochemistry is still in the preliminary stage of development. Combining quantum chemical calculations with surface-enhanced infrared absorption spectroscopy (SEIRAS) and SERS is common in studying the electrochemical interface adsorption systems. The adsorption state of the adsorbed molecules changes with the potential, which leads to the variation of the spectrum of the research system with the potential and time. Infrared spectroscopy usually only involves the properties of the surface electronic ground state, while Raman spectroscopy involves the properties of the excited surface state. On the surface of the metal electrode, the dynamic properties of the vibrationally excited state and the electronically excited state of molecules are different from those in the solid or liquid phase. Due to the continuous energy band of metals, the lifetime of the excited state of the absorbed molecules or the formed complex is significantly shortened. Thus, the timescale of its dynamic process is shortened. It is necessary to develop a time-resolved in situ spectroscopic technique under the guidance of the theory.

Although the different surface-enhanced spectroscopy has been applied in the research of the electrochemical interfaces, the related surface selection law of the quantum chemical calculation cannot be established accurately and quantitatively due to the complex enhancement mechanism (EM) and the uncertainty of the surface nanostructures. On the one hand, the electromagnetic field EM focuses on the interaction between the photons and the nanoparticles on the surface from the perspective of the photoelectric effect. The electromagnetic field enhancement was recognized as having nothing to do with the types and properties of the absorbed molecules, which is probably one-sided. On the other hand, the chemical EM considers the interaction of photons with molecules and metal surfaces from the perspective of excited-state energy levels. The related research has not yet introduced the strong surface photoelectric field caused by SPR, nor has it introduced the possible CT process of the light drive. The current problem is how to integrate the electromagnetic field EM with the chemical EM and establish a unified theoretical mode that considers the contributions of these two mechanisms. The ideal prospect is to combine the theoretical methods of molecular spectroscopy and plasmon photonics to establish a system that can study the interaction of photons, molecules, and metal nanostructures simultaneously and thus develop quantitative calculation methods and theories of electrochemical surface enhancement spectroscopy.

3B.1.6 Outlook

In short, in situ spectroscopic electrochemistry has gone from the initial to a mature stage after half a century's development. Various techniques and materials can be used, and the energy, space, and time resolutions have been greatly improved. With the development of various new spectroscopic technologies, excitation sources, detection sampling and analysis technologies, and even nanotechnology, it is foreseeable that the in situ spectroscopic electrochemistry will further innovate in theory and experiments and develop into a more universal and targeted technology. The following aspects of the application and basic research will be discussed.

In terms of application, the main direction of electrochemistry is electrochemical energy, along with global energy and environmental issues. All kinds of energy conversion and storage systems are facing strong demands of high quality, high efficiency, and low pollution and the challenges of various new systems. At the same time, they put forward unprecedented high requirements for in situ spectroscopic electrochemistry technology. The electrochemical reactions and energy conversion and storage processes are all physical and chemical processes on the interface. The various interface structures of the actual energy electrochemical system are very complex. The active sites with higher activity often have specific structures (from a single atom to clusters).

The key is to obtain reliable and useful information on various electrochemical interfaces under online conditions to solve the practical problems in electrochemical science and technology. The in situ spectroscopic electrochemistry strives to ensure the high detection sensitivity under the actual or close to actual working conditions of the electrochemical energy system so that the detection resolution (energy, time, and space resolution) can be further improved, and the dynamic and structural characterization of interface trace species can be achieved. Therefore, the focus of the in situ spectroscopic electrochemistry will gradually shift from in situ to online, from adsorption to reaction, and from steady state to dynamic state.

Scientific instruments such as spectroscopy, wave spectroscopy, and energy spectroscopy are often strongly dependent on the increase in the intensity of the incident light source and the detector's sensitivity. For example, the phenomenon of Raman scattering was discovered in 1928, but it was not rapidly developed and widely used until the invention of the laser in 1960. Laser technology is indispensable to the development and popularization of various spectroscopies. To conduct the online research, the laser's power, efficiency, and wavelength-tuning range are more demanding, and the performance of the laser needs to be improved by the researchers in new ways. Among them, the free-electron laser (FEL) has a series of advantages beyond the reach of the existing laser light sources. In 1976, L. R. Elias et al. realized the far-infrared FEL at Stanford University for the first time [36]. Infrared FELs can provide continuously tunable lasers with wavelengths ranging from a few to hundreds of microns. It has high power, continuously tunable broadband, and short pulse width, which is suitable for infrared spectroscopy research with the wavelength ranging from mid-infrared to far-infrared. The brightness of the infrared region is about six orders of magnitude higher than that of blackbody radiation and synchrotron radiation sources. The FELs mainly work in the region between far-infrared and ultraviolet at present. With the continuous development of laser technology, especially the advancement of accelerator technology, FEL will continue to push forward the application in the shortwave (vacuum ultraviolet, soft X-ray, and hard X-ray).

Since the development and maturity of FEL will take time, the synchrotron radiation light source is a better choice at present. For example, in situ X-ray Raman scattering electrochemical technology can be developed. The biggest feature is using high-energy hard X-rays to obtain low-energy soft X-ray absorption spectra. On the one hand, the changes in the electronic properties of the research object under potential modulation can be obtained. On the other hand, the potential difference spectroscopic technique can capture the surface electronic structure information of the research object and the potential follow-up relationship. In addition, the strong penetrability of X-rays makes it possible to obtain the X-ray Raman scattering signals in a forward irradiation-backward collection mode, which effectively improves the signal collection capability. X-ray source based on synchrotron radiation has great potential in electrochemical energy and storage applications. For example, these technologies can characterize the electronic and molecular structure of the interface and the dynamic changes of the electrode materials during the charging and discharging process of various batteries, which will provide key information for the optimization of the electrode materials and the improvement of the battery system.

So far, there are few examples of the electrochemical system of solid/liquid/gas three-phase coexistence, whereas it is closer to the practical fuel cell electrocatalytic system. Suppose that it is possible to establish an online real-time system to characterize the behavior of the same adsorbed species under different interface phases. In that case, it is necessary to design the relevant detection technology of the in situ spectroscopic electrochemistry and make a multifunctional electrolytic cell with variable temperature, pressure, and solution in a three-phase system. Considering the complexity and criticality of the solid/liquid/gas three-phase interface, the research in this area has the dual significance of application and basic research.

In terms of basic research, if the in situ spectroscopic electrochemistry is combined with other instrument hardware (especially for the detectors) and software, it will surely promote the engagement in major basic research issues that have been difficult to develop so far, such as characterizing the details of the electric double-layer structure and the weak adsorption system on the surface.

Electrochemical SPM with high spatial resolution will continue to play an important role in in situ spectroscopic electrochemistry. For example, the common spectroscopic methods often obtain the average information of a two-dimensional surface by observing solid electrolyte interface (SEI) film. It is difficult to understand SEI film's formation process and microscopic mechanism, whereas SPM can observe and measure the uneven structure of SEI in situ and measure its mechanical properties. Therefore, the high-resolution observation and research of these surface structure details under online conditions are of great significance, and it is also a new challenge for electrochemical SPM technology. In addition, studying the different concentration distributions and changes of reactants and products in the solution during the electrochemical reaction requires a longitudinal space resolution of several nanometers to several angstroms. For example, applying various probe force spectroscopy and electron microscopy techniques based on electron or ion beams can increase the resolution of the film system to angstroms in the Z-direction. However, it is extremely difficult to characterize the structure and detect the potential distribution of the interface on the more common and important liquid side (the tight and the dispersed layers). It is necessary to develop together with the tip-enhanced spectroscopic technology and the theoretical calculation and data analysis methods.

In short, the current situation in in situ spectroscopic electrochemistry is facing continuous breakthroughs. It will impact the progress of electrochemistry, surface and interface science, and even materials science and technology. Due to space limitations, this chapter covers only electrochemical in situ Raman spectroscopy, attenuated total reflection (ATR) Fourier-transform (FT) infrared (FTIR) spectroscopy, scanning probing microscopy, and mass spectrometer technologies in more detail in the following sections.

3B.2 Electrochemical Raman spectroscopy

Raman spectroscopy is a kind of scattering spectrum. It uses a laser with good monochromaticity as the light source to perform vibrational spectral analysis of samples in solid, liquid, and gas phases to obtain fine molecular structure information. However, the detection sensitivity of Raman spectroscopy is very low, which is particularly prominent in electrochemical research. The discovery of surface-enhanced Raman scattering (SERS) has greatly improved the detection sensitivity of surface Raman spectroscopy. SERS has been applied to the in situ electrochemistry for the in-depth characterization of various surface structures and processes at the molecular level and identification of the species' bonding, configuration, and orientation on the surface. In the past decade, the electrochemical SERS field has achieved remarkable development. Important breakthroughs have been made in researching electrochemical interface structures, electrocatalysis, corrosion, and the surface of single-crystal electrodes [37]. This section introduces the basic theories of Raman spectroscopy and SERS, the current research status of electrochemical SERS, and the problems and challenges in electrochemical SERS.

3B.2.1 Fundamentals of Raman spectroscopy

Figure 3B.8 compares the infrared adsorption and fluorescence molecular scattering (Raman and Rayleigh scattering) processes with the Jablonski energy diagram. Infrared adsorption is a process of photon absorption. A broadband light source is usually applied to illuminate the sample. The energy matching the vibrational energy level of the infrared molecular activity will be absorbed to excite the molecule to a certain vibrational excited state. The light source in the UV–visible range can excite electrons from the ground state to the excited state in the fluorescence process. The electrons in the excited state may relax to the vibrational ground state (usually within 10^{-9} s) through vibration. They will emit fluorescence during the transition back to the electronic ground state simultaneously. In the Raman scattering process, a beam of monochromatic light with a frequency of $h\upsilon_0$ is introduced to interact with molecules to deform an electron cloud around the molecular core into a short-lived, unstable state (virtual state). The electrons re-emit the scattering light with a frequency of $h(\upsilon_0 - \upsilon)$ quickly (usually within 10^{-12} s) without vibrational relaxation.

Although the diagram shows two successive processes, the scattering process is an instantaneous process that occurs almost together with re-emission. Suppose that the scattering process only involves the deformation of the electron cloud, and the change in the energy of the scattered photons is extremely small. In that case, this corresponds to the elastic scattering process, namely Rayleigh scattering. Suppose that the scattering process involves the vibration of the atomic nucleus, causing the change of the surrounding electron cloud (the change of polarizability). In that case, the process of the energy transferring from incident light to the molecules or from the molecules to the scattering light corresponds to the inelastic scattering process, namely Raman scattering. In this situation, the frequency of the scattering light is different from that of the incident light. The frequency difference υ between the incident light and the scattering light corresponds to the characteristic vibration frequency of the molecule and does not change with the incident light frequency. The Raman frequency shift υ of the horizontal axis in the Raman spectrum is the frequency shift of the scattered light relative to the incident light. In Raman scattering where the frequency of the scattering light is lower than that of the incident light is called Stokes Raman scattering, which corresponds to the process of transferring from the vibrational ground state to the vibrationally excited state. On the opposite side, in Raman scattering where the frequency of the scattering light is higher than that of the incident light is called anti-Stokes Raman scattering, which corresponds to the transition from the vibrationally excited state back to the vibrational ground state, as shown in Fig. 3B.8. The absolute value of the frequency of these two Raman processes is the same. According to the Boltzmann distribution, the population number of molecules in the vibrational ground state is much higher than that in the vibrationally excited state. Therefore, the intensity of Stokes Raman scattering is much higher than that of the anti-Stokes Raman scattering. Especially, Fig. 3B.8 only shows the energy diagram of one vibration mode. The characteristic vibration energy level diagram can be constructed for different vibration modes.

Raman spectrum is usually plotted based on the intensity of the scattering light and the Raman shift v in cm^{-1} units (energy difference between the scattering light and incident light). Therefore, in the Raman spectrum, the Stokes and anti-Stokes lines are distributed symmetrically on both sides of the Rayleigh line. No matter

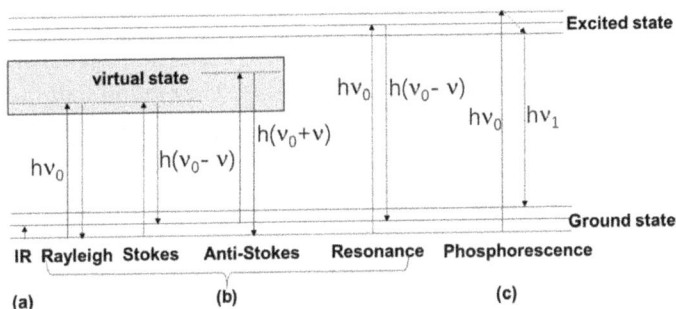

Fig. 3B.8: The diagram of three kinds of luminescence processes: (a) IR absorption, (b) Raman scattering, and (c) phosphorescence. v_0 is the frequency of incident light; v is the Raman frequency shift of a certain molecule vibration; v_1 is the frequency of phosphorescence.

which wavelength of the excitation light is, the vibrational frequency of the Raman spectrum will not change with the frequency of the excitation light if the laser's energy does not cause a significant change in the molecular structure and system temperature.

If the incident laser frequency v_0 is very closed to or within the electron absorption peak range of the molecule, the probability of Raman transition will be greatly increased, which can increase the Raman scattering cross section of certain vibration modes up to 10^6 times. This phenomenon is called the resonance Raman effect [38], as shown in Fig. 3B.8. The closer the energy of the excitation light is to the electronic resonance condition of the system, the stronger the intensity of the Raman peak is. The detection of sublayer molecules can be achieved using the resonance Raman enhancement effect. Resonance Raman spectroscopy is commonly used in the research of biomolecules containing chromophores. Since the chromophores mainly contribute to the Raman signal, the Raman peaks of chromophores can be separated from the peaks of other groups. It should be pointed out that the resonance Raman spectroscopy requires a laser source with a tunable wavelength. Although today with the rapid development of laser technology, the price of lasers with tunable wavelengths is still very high. More importantly, resonance Raman spectroscopy is not a surface-specific effect. The same species in the solution will cause serious interference to the detection signal, limiting its surface and interface research application.

3B.2.2 Electrochemical surface-enhanced Raman spectroscopy technology

In 1974, Fleischman et al. obtained high-quality Raman spectroscopy signals of single-layer absorbed pyridine molecules with potential changes on the roughened Ag electrode surface [11]. They believed that a highly rough surface could adsorb more molecules, giving a stronger Raman signal. Van Duyne et al. repeated the experiment with different tip materials and electrode treatment methods. They found that the Raman signal of pyridine adsorbed on the rough Ag surface was enhanced by about six orders of magnitude than that of the same amount of pyridine in the solution [39]. This phenomenon was deemed a huge enhancement effect related to rough surfaces at that time. This effect was later called the surface-enhanced Raman scattering effect. The related technology or the obtained spectrum was called SERS.

The SERS phenomenon was first discovered in the electrochemical system, and after that, electrochemical surface-enhanced Raman scattering spectroscopy (EC-SERS) played an irreplaceable role in SERS development. Through long-term experimental research, eight important characteristics of SERS are summed up.

(1) Depending on the optical properties of the metal and wavelength of the excitation light, Au, Ag, or Cu nanoparticles with particle sizes of 10–200 nm are usually required to obtain strong SERS activity. Aggregation nanoparticles or structures with nanogap can produce SERS effects far stronger than single particles.

(2) Various methods for obtaining nanoscale particles and surfaces can prepare SERS substrates, including electrochemical oxidation–reduction cycle(s) (ORC), chemical etching, physical deposition, and photolithography. Besides, as the substrates of EC-SERS, their surfaces must have good conductivity.

(3) The surface enhancement factor (SEF) of Ag, Cu, and Ag rough electrode or nanoparticle system is usually up to 10^6, and the SEF of other transition metals (such as Pt, Rh, Pd, Ni, Co, and Fe) is usually around 10^3. An enhancement effect of more than 10^9 is usually required to obtain a single-molecule SERS signal.

(4) Different substrates require lasers of different wavelengths to create SERS. Ag substrate can be excited by lasers with the wavelength from the visible to infrared region. In contrast, Au and Cu substrates usually use lasers from the red to infrared region, and most transition metals can use lasers from the UV to infrared region.

(5) SERS has a long-range effect, which means the enhancement effect decays exponentially with the distance of the nanostructure. Generally, it can still provide an enhancement effect within the distance of 5–10 nm.

(6) SERS is a kind of surface (interface)-sensitive technology, and the first layer of molecules absorbed on the surface gets the greatest enhancement. The smaller surface structures (such as surface complexes and absorbed atoms or clusters) play a very important role in the chemical EM, usually called active sites.

(7) Many molecules absorbed on the metal surface or nearby can produce SERS, whereas their enhancement effectivity differs. For example, the Raman scattering cross sections of CO and N_2 are equivalent, but their SERS intensity differs by 200 times under the same experimental conditions. Generally, the enhancement of molecules physically absorbed on the surface is small.

(8) In an electrochemical environment, the frequency and intensity of the SERS peaks of the surface species usually change with the electrode potential, and different vibration modes may have different relevance to the potential. For example, when a very negative potential is applied on the surface, the SERS activity of electrodes such as Ag or Cu may disappear irreversibly. However, if these electrodes are subjected to ORC treatment, their surfaces will have SERS activity again.

Various SERS mechanism proposed so far cannot fully explain all the SERS phenomenon that has been reported. Therefore, the research of the SERS mechanism is still an important research field. It is generally believed that two mechanisms mainly contribute to SERS enhancement.

One is the electromagnetic EM. EM refers to the enhancement of the molecular Raman signals caused by the enhancement of the localized photoelectric field of the substrate, which is not only related to the optical properties and geometric morphology of the surface materials and the excitation light frequency but also related to the local geometry of the measured molecules. As the dipole field of a metal particle is inversely proportional to the third power of the distance to the particle's center, it is essentially a long-range effect. The disadvantage of this mechanism is that it only considers the role of incident light and the metal surface while ignoring the molecule's role in it. EM plays a leading role in SERS enhancement technologies such as Au, Ag, and Cu.

The other one is the CT mechanism. This mechanism mainly describes the interaction between molecules, surface, and incident photons, including chemical adsorption, photoinduced CT between the adsorbate and the substrate, and coupling between electron–hole pairs and absorbed molecules. The most important contribution is the CT mechanism. The CT enhancement effect depends on the absorption site, the configuration of the bonding, the electronic structure of the electrode and the absorbed molecule, and the wavelength of the excitation light, which only contributes to SERS in the range of molecular scale. Essentially, it is a short-range effect.

The following formula can simply express the SERS intensity of an absorbed species:

$$I_{\text{SERS}} \propto G_{\text{EM}} \sum_{\rho,\sigma} \left| (\alpha_{\rho\sigma})_{nm} \right|^2 \tag{3B.1}$$

In formula (3B.1), G_{EM} represents the EM enhancement effect related to incident light and scattering light. The sum term of $\alpha_{\rho\sigma}$ represents the optical response caused by the interaction between the molecule and surface, representing the CT enhancement effect. From the formula, it can be seen that the two effects of EM and CT are the product relationship, not the additive relationship in the total SERS effect. Therefore, even if the contribution of the CT mechanism is much smaller than that of the EM mechanism, it can significantly change the frequency and relative intensity of the spectral peaks.

3B.2.3 Comparison of SERS and surface infrared spectroscopy technology

SERS and SEIRAS are the two main experimental methods for determining the changes in molecular vibrational energy levels. However, the mechanisms for these two spectra are completely different, and their surface selection rules are also different. Thus, the relative peak intensities of Raman and infrared spectra will differ for the same sample. The following summarizes the characteristics of these two technologies, illustrated in Tab. 3B.3.

Tab. 3B.3: Comparison of the characteristics of infrared and Raman technology.

Classification	Raman	Infrared
Optical phenomenon	Scattering	Absorption
Photon process	Two photons	Single-photon
Sensitivity	Low[1]	High
Spectral frequency range/cm^{-1}	Usually 50–4,000	Usually 900–4,000[2]
Spectral peak shape	Sharp	Relatively wide
Relationship between intensity and concentration	Linear relationship	Exponential relationship
Surface enhancement	Yes	Yes
SEF	10^2–10^6[3]	10–80
Surface selection rule	Loose and complicated	Strict and simple
Surface molecular orientation determination	Difficult and complex	Clear and simple
Typical sampling method	Backscattering or 90° collection	Reflection
Spatial resolution	About 1 μm	About 30 μm
Disturbance from the surrounding environment	Few	Strong interference from water and CO_2
Typical measurement method	No special requirements (absolute spectrum)	Modulation of electrode potential or polarization (subtract spectrum)
Structure of spectral electrolytic cell	No special requirements	Thin-layer electrolytic cell (about 10 μm)
Typical surfaces	Rough[4]	Smooth mirror

Tab. 3B.3 (continued)

Classification	Raman	Infrared
Typical light source	Laser	Silicon carbide or Nernst filament
Destructive to the surface	Local heating or decomposition	Nondestructive

①The probability of producing Raman scattering is much smaller for the molecules than that of infrared absorption. However, the development of Raman instruments and laser technology has narrowed the gap between them. Moreover, Raman scattering efficiency can be improved by using surface enhancement and the Raman resonance effect greatly.
②Infrared technology is usually difficult to study in the frequency range below 900 cm^{-1}, which may be overcome if a synchrotron radiation source is used as a high-intensity infrared light source.
③The SEF of single-molecule SERS can be as high as 10^{14}.
④A certain surface roughness is required to obtain a relatively high SERS signal.

3B.2.4 Electrochemical-SERS (EC-SERS) experiment

Figure 3B.9 displays the experimental setup for in situ EC-SERS. The experimental instruments include laser (excitation source), Raman spectrometer, computer (control wave function generator, acquire and process data), potentiostat (control and detect the potential and current of the electrochemical system), and spectroelectrochemical cell. The Raman spectrometer comprises a sample stage, incident light path, collection light path, monochromator, and detector. Before the laser enters the Raman spectrometer, it is usually necessary to use a band-pass filter to obtain pure monochromatic light. The signal collected from the sample contains excitation light, Rayleigh scattered light, and Raman scattered light. Notch filters, long-pass edge filters, and volume Bragg gratings can eliminate the excitation or Rayleigh scattered light. Then a good Raman scattering signal can be obtained through the monochromatic spectroscope.

Conventional Raman spectrometers are divided into single-channel and multi-channel Raman spectrometers. The photomultiplier tube (PMT) is the most commonly used high-gain laser detector in single-channel spectrometers at the early stages. Because of their high quantum efficiency and low dark current, they are very suitable for weak signal detection. However, for detecting extremely weak light signals, a long integration time is generally required. Thus, special attention needs to be paid to avoid the saturation and damage of the detector due to the accumulation of dark current. Avalanche photodiode (APD) is gradually replacing the PMT detector due to its advantages, such as high detection sensitivity and ease of maintenance. PMT and APD have important applications in developing high time-resolved Raman spectroscopy.

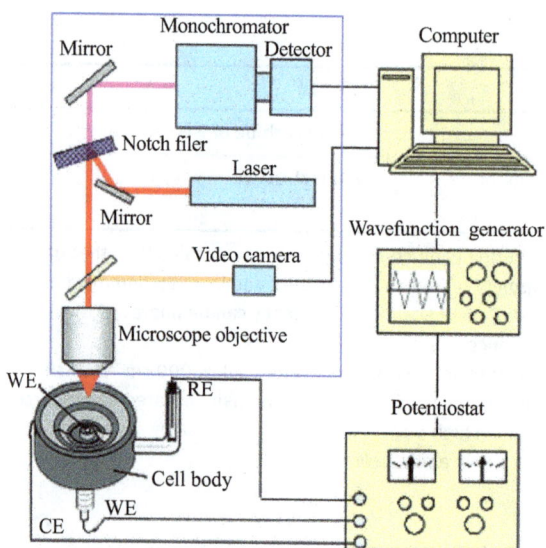

Fig. 3B.9: A diagram of the experimental setup for EC-SERS includes a Raman spectrometer (in the light blue block), potentiostat, wave function generator, and an EC-SERS cell. WE, working electrode; CE, counter electrode; RE, reference electrode [37].

A multichannel Raman spectrometer is mainly composed of a monochromator and a multichannel detector. In the experiment, the position of the grating (the number of center waves) is selected in advance, and the Raman spectrum with a wide frequency range can be recorded at one time. Therefore, the multichannel Raman spectrometer has gradually become the most important instrument in the Raman laboratory. The most commonly used detector for multichannel spectrometers is the CCD detector, which has high quantum efficiency and high detection sensitivity. The most prominent feature of CCD is that its dark noise is very small, which is particularly important for detecting weak signal systems that require a long-time integration to obtain a satisfying signal-to-noise ratio. However, the shortcoming of CCD is that the readout speed is very slow. If there is no pulsed laser, optical switch, or other special technologies, it is not easy to use it for time-resolved Raman spectroscopy research. The latest electron multiplied CCD has the characteristics of conventional CCDs, such as low noise, wideband, ease of manipulation, and high detection sensitivity and readout speed. Under the condition of electron multiplication, it has outstanding advantages in the short-time detection of strong signal systems (such as time-resolved spectroscopy), which can significantly improve the signal-to-noise ratio of the spectrum. If a higher time resolved is needed for the Raman spectroscopy, the time shutter can be used with intensified CCD.

At present, there are many kinds of lasers used in the research of Raman spectroscopy, and the wavelength range covers the ultraviolet region to the near-infrared region, which is usually divided into continuous laser and pulsed lasers. Pulsed lasers have high peak power density, which can easily damage samples. Therefore, it is only necessary to use pulsed lasers in fast time-resolved or nonlinear Raman spectroscopy. The most commonly used Raman light sources are small helium, neon gas lasers (632.8 nm), 532 nm solid-state diode-pumped lasers, and 785 nm semiconductor lasers. The blue (488.0 nm) and green (514.5 nm) wavelengths are most commonly used in argon particle lasers. The most commonly used laser source in the UV region is the helium–cadmium laser (325 nm). In the deep UV region, argon particle lasers can be frequency doubled to obtain wavelengths of 244 and 257 nm, and the solid lasers can be frequency quadrupled to obtain a wavelength of 266 nm. The near-infrared excitation light can reduce the fluorescence interference when the sample has a strong fluorescence signal. In this situation, FT-Raman spectrometers with a wavelength of 1,064 nm are usually used, eliminating the interference of fluorescence effectively and having advantages in studying fluorescent thin-film electrodes, with the multiplexing advantage of converting the frequency domain representation to a function of space or time using a Fourier Transformation [Transformation de Fourier] which is the mathematical operation that associates the frequency domain representation.

Under the same spectral resolution, the throughput of the FT-Raman spectrometer is better than that of the conventional Raman spectrometer. However, since the excitation source of the FT-Raman spectrometer is a near-infrared laser and the near-infrared detector generally has high noise and low quantum efficiency, the total detection sensitivity of the FT-Raman spectrometer is still lower than that of conventional Raman spectrometers. In addition, the aqueous solution also has strong absorption of incident light and Raman signals in the near-infrared region, which can make it very difficult to research the electrode/solution interface on an FT-Raman spectrometer.

3B.2.5 Application of Raman spectroscopy in electrochemistry

The discovery of SERS promotes the research and understanding of electrochemical interfaces. Driven by nanotechnology, the sensitivity of SERS detection has greatly improved, which has expanded EC-SERS from the previous limitation on Ag, Au, and Cu surfaces to transition metals such as the more important Pt and Fe groups in the electrochemical system. The research objects have been extended to electrocatalysis, corrosion, and energy systems. Some extremely complicated systems can also be studied, such as interfacial water, adsorbed hydrogen, and the adsorption on the surface of single crystals, which is difficult for the traditional electrochemical methods and electrochemical infrared and frequency spectroscopic techniques with the bonding information. Here are some typical examples to illustrate the application of Raman spectroscopy in electrochemical research.

The orientation and structure of water molecules will affect the electric field distribution and electrochemical reaction at the electrode/solution interface. The Raman scattering cross section of water itself is very small. However, SERS can solve the influence of the Raman signal from the solution body water on the signal from interface water. In order to obtain a strong SERS signal of the interface water, firstly, Au nanoparticles of about 55 nm are synthesized as the core that provides the enhancement effect and then wrapped up with the desired transition metal shell, such as Pt and Pd, by chemical synthesis. These nanoparticles are spread uniformly on the surface of the conductive electrode to form a uniform multilayer. Then the surface impurities are removed through a negative potential, used for EC-SERS water research. Figure 3B.10(a) shows the SERS signals of water from the surface of Pt, Pd, and Au at different potentials [40]. The peaks at around 1,615 and 3,400 cm^{-1} are water molecules' bending vibration and stretching vibration. In conventional Raman spectroscopy, the intensity of stretching vibration of water is about 20 times larger than that of bending vibration. However, in the SERS spectrum, the intensity of bending vibration is almost equal to that of stretching vibration. Figure 3B.10(b) was obtained by plotting the frequency of the Raman peak in Fig. 3B.10(a) against the potential [40]. It is found that the vibration frequency of the water on three metals is blueshifted with the positive shift of the potential obviously, especially for the stretching vibration. In addition, the frequency shift varies for these three metals under the same condition. Since the compositions of the solution and the morphology of the nanoparticles in three systems are almost the same, the main reason for such a big difference of frequencies may be the difference in the reaction of water molecules and interface. The intensity of SERS spectra of water absorbed on the surface of Au electrode is the highest among that of three metals, illustrating the highest electrochemical vibrational Stark effect. These results provide important information for us to understand the electrochemical interface.

Raman spectroscopy can identify the absorption of the same substance at different positions on the same electrode surface. In previous IR studies, it has been found that the stretching vibration frequency of carbon–oxygen (v_{CO}) is very sensitive to the chemical and physical properties of the surface, and the adsorption configuration largely depends on the properties of the surface. If the metal substrate is changed, the absorption signal of CO will also change considerably. Figure 3B.11 shows the Raman spectra of CO saturated adsorption on the different metal surfaces. The peaks of CO and M–CO are very sensitive to the metal substrates. The frequency and the corresponding intensity of the Raman signals v_{CO} of the linear and bridged CO at the top position (located at 2,040–2,080 and 1,870–1,960 cm^{-1}, respectively) change with the change of the metal substrates. On the Ir surface, CO adsorption is mainly linear adsorption, while it is mainly bridge adsorption for the Pd surface. For Pt and Rh surfaces, both bridged and linear adsorption exist. Moreover, the CO adsorption of the same geometric configuration on different metal surfaces has different vibration frequencies. For example, the v_{CO} order of the top absorbed CO on the metal surface is as follows: Pt > Ir > Pd > Rh; and the corresponding v_{M-CO} order is: Ir > Pt > Rh > Pd. From the peak position and relative peak intensity of the bridge and linear absorption of CO, we can use the Raman spectrum of CO adsorption to identify different metals.

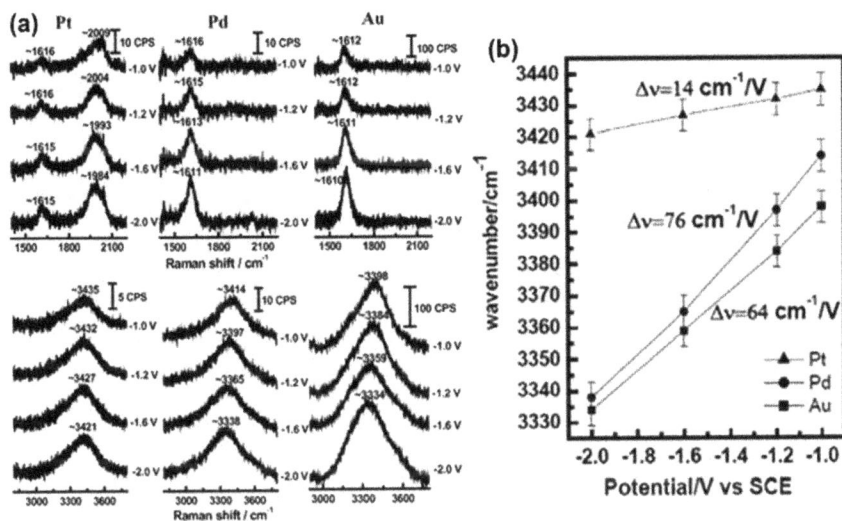

Fig. 3B.10: (a) SERS of water adsorbed on Pt, Pd, and Au at different potentials in 0.1 M NaClO$_4$ by excitation at 632.8 nm; (b) plots of stretching frequencies versus applied potentials for the O–H bonds of water absorbed on Pt, Pd, and Au electrodes in 0.1 M NaClO$_4$ [40].

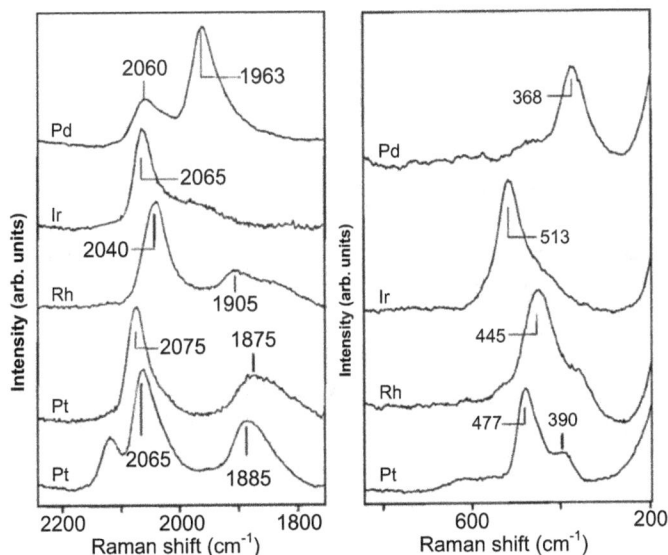

Fig. 3B.11: SERS in the (a) high-frequency spectral region for C–O (v_{CO}) stretching vibration and (b) low-frequency spectral region for metal–carbon ($v_{M–CO}$) stretching vibration [41].

Lithium-ion batteries are used more and more widely in daily life due to their high energy density and power density and low self-discharge performance. However, the performance of current lithium-ion batteries is much lower than the theoretical capacity.

Therefore, the design of suitable cathode and anode materials, good compatibility of electrolyte and separator materials, and the production of SEI with excellent performance are hotspots in the field of lithium-ion batteries currently. In situ research methods can avoid structural changes caused by ex situ research methods during the transfer and characterization process, which can reflect the true state of the system more accurately. Therefore, in situ characterizations of electrode materials and SEI layer structures during charging and discharging will help promote the development of this field. Among them, Raman spectroscopy can be used not only to characterize the composition and structure of electrodes and electrolytes but also to in situ characterize the cathodes and anode species and solution composition of the solution during the charging and discharging process of lithium-ion batteries, such as lithium insertion/extraction processes on the positive and negative electrodes [42, 43]. Stancovski and Badilescu gave a good overview of the history and application of Raman spectroscopy in lithium-ion battery research [44]. The research of SEI film has always been a difficult point in this field, mainly because the film thickness is thin, the Raman signal is weak, and it is easily interfered by the body materials. Recently, with the help of the shell-isolated nanoparticle-enhanced Raman spectroscopy (SHINERS) technology with the high surface sensitivity and nonchemical interference characteristics, Hwang et al. studied the structural characteristics of the SEI film in lithium-ion batteries and observed the deoxidation activation process of layered lithium-rich materials and the formation of LiO_2 at the interface successfully. EC-Raman technology has also been used to track the valence state changes of the electrode materials during the charging and discharging process of the lithium-sulfur battery [45].

As the SPR effect cannot be excited effectively, single metal crystals' surface electrochemistry is difficult to study with the conventional SERS technology. The combination of SHINRS and TERS can explore the influence of different crystal faces on the electrochemical adsorption behavior of the interface from the perspective of high sensitivity and high spatial resolution, respectively.

Figure 3B.12 shows the EC-SHINERS sequence obtained on the surface of three single-crystal electrodes of Au in the solution of 0.1 mol/L $NaClO_4$ and 1 mmol/L pyridine (Py) [46]. Among them, the strong peaks at about 1,010 and 1,035 cm^{-1} are attributed to the v_1 ring breathing vibration and the v_{12} symmetrical triangular ring deformation vibration of Py; the peaks at about 630, 1,210, and 1,600 cm^{-1} come from v_{6a}, v_{19a}, and v_{8a} vibrations, respectively. The blueshift Stark effect with the positive-shift v_1 frequency was detected on these three crystal planes, and their slopes were closed (about 5.8 $cm^{-1}\cdot V^{-1}$). However, the onset potential produced by the Stark effect is significantly different due to the difference in the crystal plane structure: Au(111) (0.24 V) > Au(100) (−0.04 V) > Au(110) (−0.27 V). This trend is consistent with the potential (E_{pzc}) changes of zero charge of the three crystal planes, which shows that the surface charge of the electrode determines the electrochemical adsorption behavior of Py on different crystal faces. At the same time, according to the peak intensity scale shown in Fig. 3B.12, it

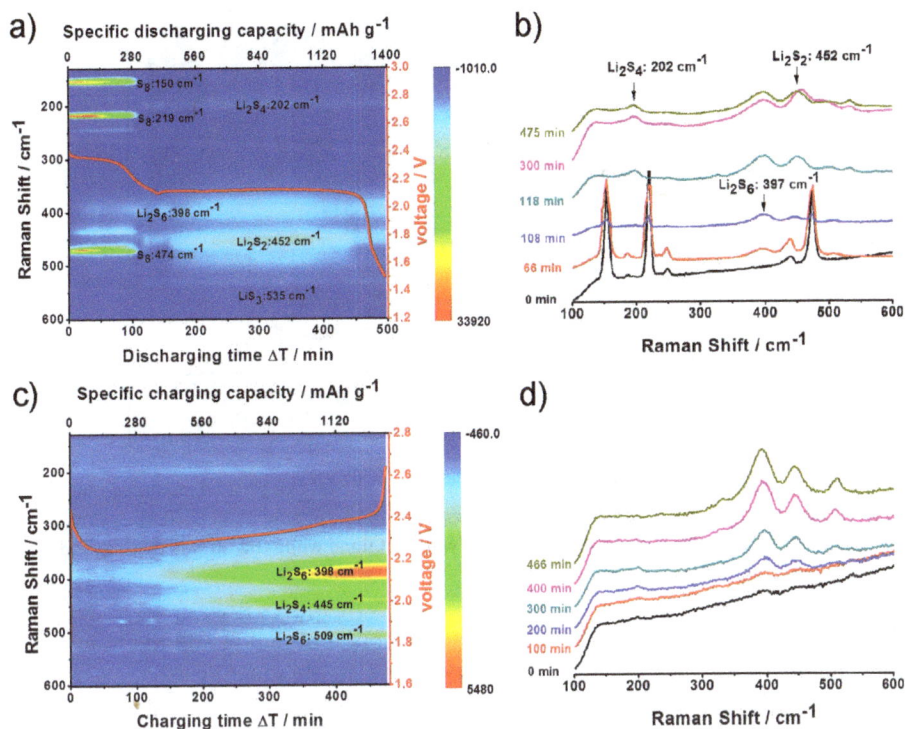

Fig. 3B.12: SHINERS of pyridine (Py) absorption on Au(111) (a), Au(110) (b), and Au(100) (c) in 0.1 mol/L NaClO$_4$ + 1 mmol/L Py [46].

can be observed that there are significant differences in the SERS peak intensity of Py absorbed on the three crystal planes. For example, the peak intensity of ν_1 increases in the order of Au(111) < Au(100) < Au(110), and the potential of the strongest peak values of these peaks is different, which in turn are located at 0.6, 0.3, and 0.2 V. The SHINERS intensity related to this crystal plane may come from the dielectric properties of the single-crystal surface, which leads to the difference in electromagnetic field coupling performance between SHINERS nanoparticles and different crystal planes.

In 1985, based on STM and combined with SERS, Wassel et al. proposed a similar concept to the existing TERS and called it surface-enhanced optical microscopy [47]. Extending TERS to electrochemical systems is a very challenging direction. The key is overcoming the optical path distortion caused by multiple refractive indexes of the different materials, eliminating the system's optical aberration, and improving the collection efficiency of EC-TERS. After nearly 10 years of effort, Bin Ren et al. developed the EC-TERS instrument device for the first time globally, as shown in Fig. 3B.13(a) [35]. Figure 3B.13(b) is an SEM image of the gold tip encapsulated with melt adhesive. It can be seen that only the top of the tip is exposed and still sharp. Figure 3B.13(c) is an optical imaging diagram of the tip and the substrate. The

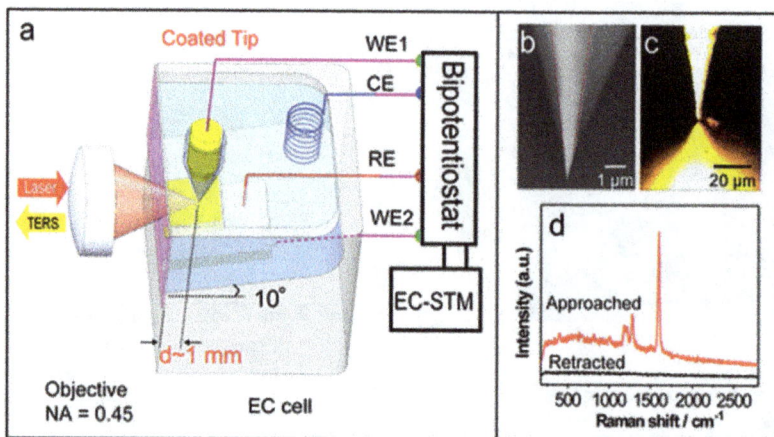

Fig. 3B.13: (a) Schematic illustration of the EC-TERS setup. (b) SEM image of an insulated gold tip. (c) Microscopic image of the tip, single-crystal substrate, and laser spot in an EC-TERS system. (d) TERS of 4-PBT adsorbed on the Au(111) surface were obtained while the tip was approached (top) and retracted (bottom).

clear microscopic image makes the coupling operation of the tip and the laser more convenient. Figure 3B.13(d) is the EC-TERS spectrum obtained by this device. The electrochemical absorption behavior of 1-methhylenetiol-4-pyridinebiphenyl molecules on the Au(111) crystal surface was observed in the system. This case demonstrates the advantages of the high spatial resolution of EC-TERS, which can obtain more information than that of EC-SERS. It is also necessary to design and develop a new model of EC-TERS instrument, which gets rid of the space constraints of the commercial SPM instruments completely and further improves the detection sensitivity.

3B.2.6 Development prospects of EC-Raman

The basic principles, instrumentation, and some important application developments of EC-Raman have been introduced systematically above. Special attention must be paid to the following issues in the EC-SERS research:

① When using nanoparticles as an enhancement substrate, the most matching excitation light wavelength is generally selected according to the substrate's LSPR. However, the common error operation is to determine the excitation light wavelength according to the peak position of the extinction spectrum obtained in the absorption spectrum test of the nanosol. When spreading the nanoparticles on a substrate, it is usually required to have a good coupling between the particles or between the particles and substrate to increase the signal strength. This coupling effect will make a significant difference between the SPR of the system and the single particle. The optimal SERS signal will not be obtained if a single-particle SPR excitation experimental system is used. Therefore, it is necessary to use the reflection (or transmission, for transparent samples) spectrum to obtain the actual absorption spectrum of the electrode to assist in selecting the excitation light wavelength. In addition, it is difficult to achieve the same coupling state of every point on the surface in a conventional system. Therefore, the absorption peak is usually very wide, and the allowable light wavelength can be wider, which will affect the ultimate sensitivity inevitably.

② In Raman microscopy, the power density at the focus spot is very high. Usually, the laser focus spot area is about $1\ \mu m^2$. Assuming that the laser power is 1 mW, the power density at the focus spot can be as high as $108\ mW/cm^2$. Although the signal intensity can be increased by increasing the laser power, the molecules absorbed on the surface, especially the nonchemically absorbed species on the surface of the single crystal, will be decomposed easily due to the high power density, which leads to thermal desorption and even photochemical reactions. Therefore, it is necessary to pay attention to whether the signal is stable during the experiments. If the signal is unstable, reducing the power density by defocusing or using the line focus method is better.

③ The development of microscopy technology has brought about a significant increase in signal collection efficiency. However, at the same time, it has also led to the inability to change the incident and collection angles of the laser. Thus, it is difficult to achieve the optimal excitation of surface processes and the most efficient collection of Raman signals. Therefore, there is still much space to improve the detection sensitivity of the surface Raman spectroscopy.

④ Synthesizing single-crystal nanoparticles with various sizes and shapes can provide enhanced electromagnetic field.

⑤ The space and time resolutions obtained by EC-Raman instruments are still relatively low, and the high resolutions cannot be obtained simultaneously. Solving this problem will help research the electrochemical interface and provide information for understanding the electrochemical process. The preliminary results of EC-TERS show that introducing the EC-TERS system is expected to utilize TERS' high sensitivity and space resolution to promote the development of electrochemical interface research.

⑥ If the spectroscopy research can be carried out simultaneously as the transient electrochemical experiment, the direct correlation between the electrochemical signal and the Raman signal will be realized. Electrochemically active species with overlapping reaction potential ranges cannot distinguish their contributions directly. However, through the fingerprint information of Raman spectroscopy, the information of these species can be easily distinguished in the same potential ranges.

⑦ In situ X-ray Raman scattering technology should be developed in electrochemistry. Like conventional Raman scattering, the biggest feature of X-ray Raman scattering is using high-energy hard X-rays to obtain low-energy soft X-ray absorption spectra. The difference is that conventional Raman scattering provides molecular vibrational information about the research object, while X-ray Raman scattering is nuclear-electron excitation spectroscopy. The latter can obtain electronic information on bulk materials. If X-ray Raman scattering is applied to an in situ electrochemical system, on the one hand, the changes in bulk electronic properties under potential modulation can be obtained; on the other hand, the relativity of surface electronic structure information and the corresponding potential can be captured with the potential difference spectra technology. In addition, the strong penetrability of X-rays enables the forward irradiation-backward collection mode applied in the X-ray Raman scattering, which effectively improves the signal collection capability. A potential application area of this technology is to study the charging and discharging processes of various batteries. Examining the dynamic changes in the electronic structure of the body and surface of the electrode materials (such as SEI films) provides a reference for optimizing and improving the electrode materials.

In short, SERS has been widely used in electrochemistry and has provided important and beneficial data for the study of electrochemical systems after more than 30 years of development. Combining SERS with the new experimental methods and nanoscience technologies will provide richer information and a deeper understanding of the electrochemical interfaces and promote the in-depth development of electrochemical disciplines.

3B.3 Electrochemical attenuation total reflection surface-enhanced infrared absorption spectroscopy (ATR-SEIRAS)

In situ spectroscopic electrochemistry is one of the most active fields in electrochemical research today. It mainly refers to applying UV–visible, infrared spectroscopy, Raman spectroscopy, and frequency generation spectroscopic techniques to obtain the structure of intermediate end products in electrochemistry. Among them, electrochemical SEIRAS [48] and SERS [41] have received extensive simple attention because of their high surface sensitivity, simple operation, and ability to provide detailed molecular structure information on the electrode surface and interface. Compared with electrochemical SERS, electrochemical SEIRAS has disadvantages such as difficulty in low wavelength measurement. However, its advantages are that the SEIRA effect is less affected by the type of metal. The surface selection law is simple, sensitive to the small polar molecules, reversible spectral signals with potential changes, and so on.

At present, the electrochemical surface infrared spectroscopy method based on the FTIR spectrometer can be divided into two working modes, namely, the external reflection mode (infrared absorption spectroscopy (IRAS)) and the ATR (or internal reflection) mode (ATR-SEIRAS). The former is usually used to study the surface of smooth metal electrodes. Although the sensitivity of IRAS is sufficient for detecting species absorbed by a single layer, it is necessary to collect hundreds of interference patterns and overlaps and average them to improve the signal-to-noise ratio [49, 50]. In contrast, ART-SEIRAS is usually used for the surface research of nanofilm electrodes, and its surface sensitivity is dozens of times higher than that of IRAS [51]. A few interferograms are needed to obtain a spectrum with the same signal-to-noise ratio as IRAS [51]. At the same time, ATR-SEIRAS also has the advantages of low background signal interference and unimpeded mass transfer. Therefore, electrochemical ATR-SEIRAS is more conducive to obtaining absorption structure and reaction dynamic information of electrode/electrolyte interface. After years of practice, electrochemical ATR-SEIRAS has been successfully applied to the research of small organic molecular configuration [52–54], molecular configuration [55], and interface coordination reaction [56, 57] and has a remarkable achievement.

This chapter introduces the electrochemical ATR-SEIRS technology from EM, optical path system, and practical application.

3B.3.1 Surface-enhanced infrared absorption (SEIRA) effect

ATR-SEIRAS is based on the SEIRA effect. The so-called SEIRA effect refers to the infrared absorption intensity of species absorbed on the island surface of the metal, which is one to three orders of magnitude higher than that on a smooth metal film or infrared window. In 1980, Hartstein et al. discovered this effect for the first time when studying p-nitrobenzoic acid (PNBA) absorbed on Ag and Au island films [58]. Subsequent studies have illustrated that coin group metals with island-like structures exhibit strong SEIRA effects. Electromagnetic field EM theory and chemical enhancement mechanism (CM) theory are widely used to explain this enhancement [59, 60], which is similar to the EM of SERS [61].

Generally, the infrared absorption intensity A can be expressed by the following formula:

$$A \propto \left|\frac{\partial \mu}{\partial Q} \cdot E\right|^2 = \left|\frac{\partial \mu}{\partial Q}\right|^2 |E|^2 \cos^2 \alpha \qquad (3B.2)$$

where $\partial\mu/\partial Q$ is the differential value of the molecular dipole moment concerning the orthogonal molecular vibration; E is the electric field that excites the molecular vibration; α is the angle between $\partial\mu/\partial Q$ and E; $|E|^2$ is the electric field. The intensity E generally refers to the mean square electric field intensity inside the metal film; and $|\partial\mu/\partial Q|^2$ is the absorption coefficient. Theoretically, the EM mechanism is considered to be the enhancement of $|E|^2$, leading to the enhancement of infrared absorption intensity, while the CM mechanism is mainly explained by enhancing the absorption coefficient $|\partial\mu/\partial Q|^2$.

The EM theory is that under the excitation of the incident light, the freely moving charges in the metal or highly doped semiconductor resonate collectively to form a localized surface plasmon, which couples and oscillates with the incident electromagnetic field. The surface-localized electric field is enhanced. Figure 3B.14 illustrates the polarization of metal islands by the incident IR radiation and the electric field around the islands produced by the polarization [62]. When the substrate surface is plated with a very thin metal film (about 10 nm), these films are composed of a layer of ellipsoid metal islands. When the incident IR light is irradiated vertically, the incident photoelectric field is parallel to the long axis of the island (i.e., parallel to the metal surface), which causes the free electrons on the surface of the metal island to oscillate, forming a plasma wave, which is perpendicular to the surface of the island. A strong electromagnetic field is generated, as shown in Fig. 3B.14, and its intensity decays exponentially. Since the island dimension, a and b, is much smaller than that of the wavelength λ, that is, in the Rayleigh limit $2\pi d \ll \lambda$, the light field in the island is highly localized, and the corresponding island is also polarized. The dipole moment p induced at the center of an island can be written as

$$p = \alpha V E_i \tag{3B.3}$$

where α and V are the polarizability and the volume of the metal island, respectively, and E_i is the intensity of the incident electric field. This dipole produces an electric field around the island and absorbs molecules. The islands are modeled by rotating ellipsoids with a dielectric function of ε_m. A thin layer models the absorbed molecule with dielectric function ε_d, covering the ellipsoids. The vibrations of absorbed molecules excited by the enhanced local field also polarize the metal island. When the oscillation frequency of the surface plasmon is close to the vibration frequency of the molecule, the two will take a couple resonantly so that the infrared vibration signal of the molecule is significantly enhanced. The local electromagnetic field causes this enhancement effect.

The CM theory involves CT between the energy level of the molecule and the Fermi level of the metal. Under the excitation of the incident light, the electrons are transferred from the Femi level of the metal to the excited state electronic orbital of the absorbed molecule, or vice versa from the highest occupied orbital of the molecule to the Fermi level of the metal. When the incident photon's energy is equal to the difference between the Femi and molecular energy levels, the CT process is

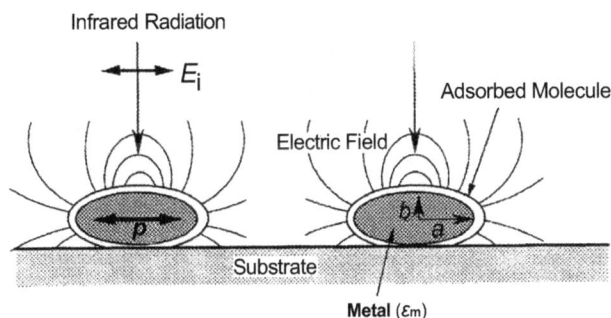

Fig. 3B.14: Schematic diagram of EM of metal islands by the incident IR radiation [62].

aggravated, increasing the effective dipole moment of molecular vibration and the infrared signal. The CT enhancement has been observed for tetracyanoethylene anion radical [63] and thiocyanate [64] absorbed on the metal electrode surfaces.

The selection rule of SEIRAS is that only those molecular vibrations normal to the surface can be effectively detected [62, 65]. Wan et al. reported that the absorption of thiophenol on the Au(111) surface is mutually corroborated with the results of the high-resolution STM [66], and they also confirmed that the surface selection rule of ATR-SEIRAS is similar to that of conventional external reflectance infrared spectroscopy (IRAS) [67]. The vibration mode of the absorbed molecule will have infrared absorption only when the component of its dipole changes normal to the surface that is not zero.

3.B.3.2 Electrochemical cells and optical arrangements of ART-SEIRAS

ATR-SEIRAS borrows a practical electrolytic cell design and is equipped with a corresponding optical path system, which can be conveniently applied to the in situ characterization of the adsorption and reaction of the electrode surface (especially the solid/liquid interface). The classic ATR-SEIRAS electrolytic cell is shown in Fig. 3B.15 [51]. The commonly used ATR crystals are Si, Ge, ZnSe, and so on. Si is widely used due to its stable properties and mature surface film techniques. However, there are also disadvantages such as poor applicability of alkaline conditions and the ineffective acquisition of fingerprint information below 1,200 cm^{-1} wavenumber. In contrast, although Ge and ZnSe have a wider infrared window, they are unstable electrolytes, limiting their application range.

From Fig. 3B.15, the working principle of the electrochemical ART-SEIRAS can be obtained: the infrared laser is the first incident from the curved surface of the infrared window to the reflection plane coated with the nanomembrane electrode,

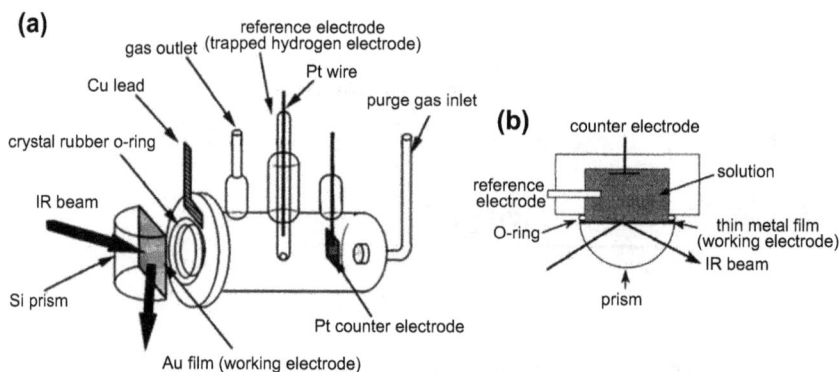

Fig. 3B.15: Electrochemical cell (a) [51] and optical arrangement (b) [62] of ART-SEIRAS.

where the plane infrared light is reflected. The generated evanescent wave penetrates the nanomembrane, enters into the electrode/electrolyte interface, and is then absorbed by the species on the electrode surface. The unabsorbed infrared light is reflected multiple times in the optical path outside the cell and enters into the mercury cadmium telluride detector. Then the infrared spectrum is generated by the computer FT.

3.B.3.3 Application of electrochemical ATR-SEIRAS

The electrochemical ATR-SEIRAS has high sensitivity, good signal-to-noise ratio, high time synchronization, and weak background signal. Moreover, the mass or CT is undisturbed. Thus, it is very suitable for dynamic and real-time research of the surface reaction (adsorption). This technology is mainly used to research the structure of the surface absorption molecules and the electrocatalytic mechanism of small organic molecules. This section explains these two important applications of electrochemical ATR-SEIRAS to illustrate the important role of this in situ technology in the research of surface electrocatalysis and electrosorption.

The interaction between aromatic molecules and metal surfaces has always been one of the hot topics of surface electrochemistry. The recent interest in molecular devices requires a deep understanding of the absorption configuration of aromatic molecules on the electrode surface. In particular, molecules containing benzene ring or pyridine (Py) ring and mercapto groups, cyano groups, or carboxyl groups [55, 68], which show strong interactions with metal surfaces, have become the main molecules of molecular electronic devices. In the following content, PNBA and Py molecules will be studied as model molecules for their adsorption configuration on metal electrodes through electrochemical ATR-SEIRAS technology.

3.B.3.3.1 Adsorption of p-nitrobenzoic acid (PNBA) on the electrode *surface*

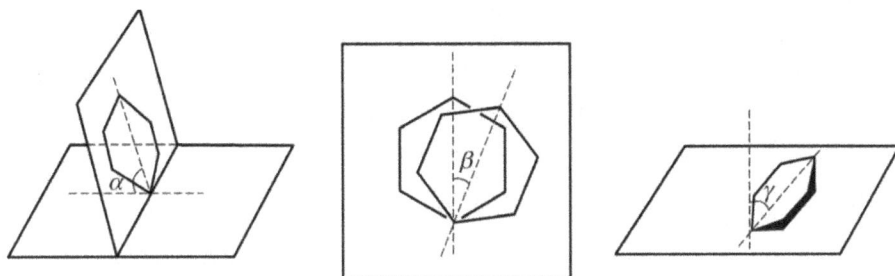

Fig. 3B.16: Schematic diagram of the adsorption configuration.

PNBA contains nitro and carbonyl groups with strong infrared adsorption and has simple C_{2v} symmetry after deprotonation. PNBA is a model molecule for studying the adsorption configuration on the electrode surface. When molecules with simple C_{2v} symmetry are adsorbed on the surface of the electrode, the angular of Fig. 3B.16 can be used to illustrate the adsorption configuration. In Fig. 3B.16, α represents the angle between the molecular plane and the electrode surface, and β represents the angle between the C_2 axis before and after the molecule's rotation in the same molecular plane. These two angles can be estimated by formulas (3B.3) and (3B.4), where I^0 and I represent the peak intensity of the bulk molecule and surface molecule, respectively; γ represents the angle between the normal line of the electrode surface and the C_2 axis of the molecule, which can be calculated by formula (3B.5). The dipole moments of the in-plane vibration modes a_1 and b_2 change along the z-axis and y-axis, respectively. The dipole moments of the out-of-plane vibration mode b_1 change perpendicular to the molecular plane (along the x-axis). The calculation formula demonstrates that detecting the out-of-plane vibration mode b_1 is very important for obtaining the dihedral angle α and other related adsorption parameters:

$$\tan^2 \alpha = \frac{I^0(a_1)}{I^0(b_1)} \times \frac{I(b_1)}{I(a_1)} \tag{3B.3}$$

$$\tan^2 \alpha = \frac{I^0(a_1)}{I^0(b_2)} \times \frac{I(b_2)}{I(a_1)} \tag{3B.4}$$

$$\cos = \cos \sin \tag{3B.5}$$

The ATR-SEIRA spectrum of the PNBA molecule absorbed on the Pt electrode surface is shown in Fig. 3B.17. The strongest peaks at 1,357 and 1,398 cm^{-1} correspond to the symmetric vibration modes of nitro and carboxyl groups, namely V_s (ONO) and V_s (OCO), respectively. Both of them belong to the a_1 mode. The peaks at 802 and 881 cm^{-1} (b_1 mode) were not detected, indicating that two oxygen atoms of the carboxyl group are vertically absorbed on the electrode surface after the deprotonation of PNBA. Thus, $I(b_1) \approx 0$ and $\alpha \approx 90°$. The peak intensities of V_s (ONO) (a_1 mode) and V_{as} (ONO) (b_2 mode) can be obtained from the bulk p-nitrobenzoate and PNBA absorbed on the surface of the Pt electrode. The bulk p-nitrobenzoate V_s (ONO) and V_{as} (ONO) are almost equal. The ratio of V_s (ONO) and V_{as} (ONO) of the PNBA absorbed on the Pt electrode surface is about 15, so by substituting to the above formula we get $\gamma \approx \beta \approx 14°$.

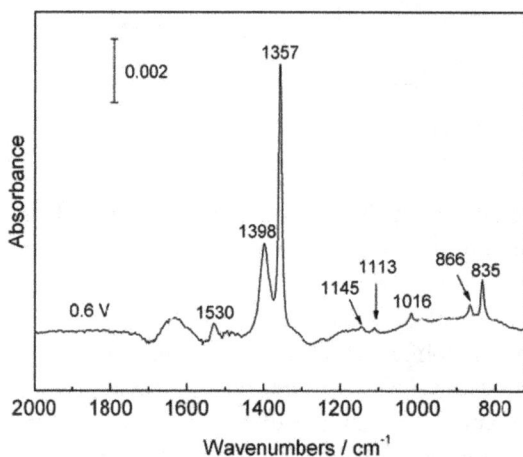

Fig. 3B.17: SEIRA spectrum for a Pt electrode in 0.1 M HClO$_4$ saturated with PNBA at 0.6 V with the reference spectrum later taken at −0.1 V [69].

3.B.3.3.2 Adsorption configuration of Py molecules on the electrode surface

As the most basic structural unit, Py has a highly symmetrical molecular structure and a clear bond formation, a typical molecular model for studying the absorption of aromatic molecules on the electrode surface. In terms of molecular symmetry, the gas-phase Py molecule belongs to the C_{2v} group and has 27 vibration modes, including $10A_1$, $3A_2$, $9B_1$, and $5B_2$. A_2 only has Raman activity, A_1 and B_1 belong to the in-plane vibration mode, and A_2 and B_2 belong to the out-of-plane vibration mode. In terms of electronic structure, in addition to a large π-bond, the Py molecule also has a pair of unbonded lone pair of electrons on the N atom. Therefore, when Py is absorbed on the metal surface, there is a competition between the large π-bond and the lone pair of electrons on the N atom to form bonds with the metal surface atoms. According to different factors such as the coverage of Py adsorption on the surface, temperature, metal surface structure (different metal or crystal surface), and interface potential, the main adsorption configuration of Py on various metal surfaces in the literature [70–74] is illustrated in Fig. 3B.18: (a) flat-on absorption;

(b) the lone pair of electrons of the nitrogen atom and the π-electrons of the ring form a bond with the metal surface at the same time, which is titled adsorption; (c) Py bonds with the metal surface through the lone pair of electrons of the nitrogen atom to form end-on adsorption; (d) Py bonds with the metal surface through the lone pair of electrons of the nitrogen atom and the *ortho*-position α-carbon atom after removing the H atom (i.e., α-pyridyl), which is almost edge-on adsorption. ATR-SEIRAS can easily analyze the various vibrational modes of Py molecules, which is very suitable for studying the surface adsorption configuration.

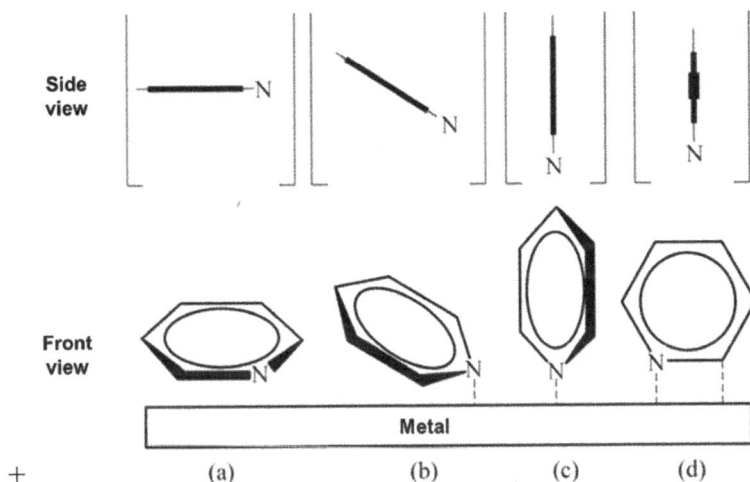

Fig. 3B.18: Absorption models of pyridine on the electrode surface [74].

Cai et al. combined ATR-SEIRAS with STM to study the adsorption configuration of Py molecules on the surface of a highly ordered Au(111) electrode [51]. The ATR-SEIRAS spectrum is shown in Fig. 3B.19, which demonstrates that the Py molecules began to absorb on the electrode surface when the electrode potential was higher than −0.6 V. When the electrode potential is higher than −0.3 V, the in-plane vibration (A_1 mode) of Py ring can be observed. However, the antisymmetric in-plane vibration (B_1 mode) can hardly be detected. According to the SEIRAS selection rule, the A_1 mode vibration is easier to be detected than the B_1 model, indicating that the C_2 symmetry axis of the Py molecule is either normal or parallel to the electrode surface. Whereas the possibility of the N and C atoms used as the adsorption site mode at the same time, that is, the existence of α-pyridyl species (Fig. 3B.18(d)), is very slight because the intensity of the vibration peaks corresponding to the A_1 mode and the B_1 mode in this situation (Fig. 3B.18(d)) is basically the same [75]. In addition, the symmetrical ring breathing vibration (v_1 mode) of Py molecules absorbed in SEIRAS is located at 1,012 cm^{-1}, which has a significant blueshift relative to the 992 cm^{-1} of liquid bulk Py, which also indicates that Py molecules are absorbed on the electrode surface through N atoms [61]. At the same time, this research found that the adsorption configuration of Py molecules on the electrode surface is closely related to the electrode potential. Under a relatively negative potential, Py molecules were absorbed on the surface of the electrode in a flat-on absorption (Fig. 3B.18(a)). With the positive shift of the electrode potential and the increase of the surface coverage, Py molecules gradually became titled absorption (Fig. 3B.18(b)). They finally even became end-on adsorption (Fig. 3B.18(c)).

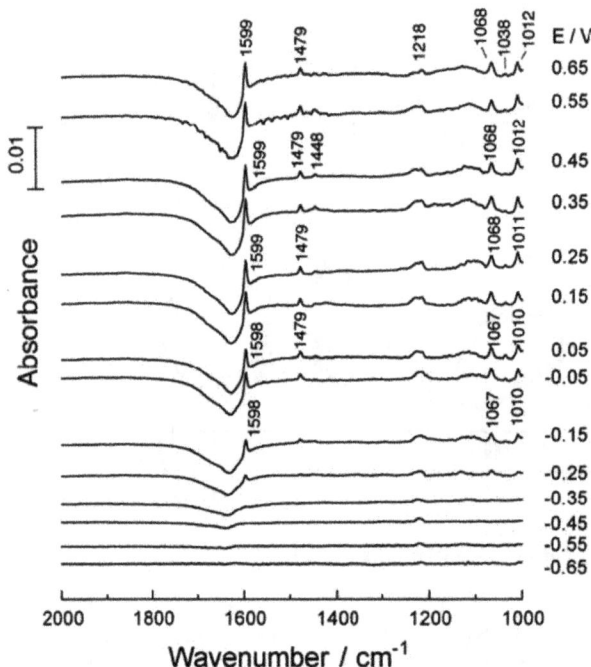

Fig. 3B.19: Infrared spectra of pyridine adsorbed on a highly ordered Au(111) surface as a function of the applied potential. The reference potential was −0.7 V versus SCE [65].

In addition, the adsorption configuration of Py molecules on the surface of other transition metal electrodes has also been studied. Figure 3B.20 compares the spectral characteristics of Py molecules on Au [74], Ag [76], Cu [70], and Cd [77] metal electrodes. First of all, there is no sign peak of α-pyridyl species on these spectra, indicating that Py is absorbed on the surface of these metals in a complete molecular form. Secondly, almost all the strong peaks are observed, namely 1,603, 1,485, 1,068, 1,042, and 1,013 cm^{-1}, belonging to the in-plane vibration of A_1 mode. The strongest peak (v_{19b}, B_1 mode) measured in the Py solution becomes very weak on these electrodes, and its position is blueshifted from 1,437 to 1,447 cm^{-1}, reflecting that the absorption form of Py is end-on adsorption. Although it is not certain whether the entire molecular plane is titled, there must be no tilt of the edge shape.

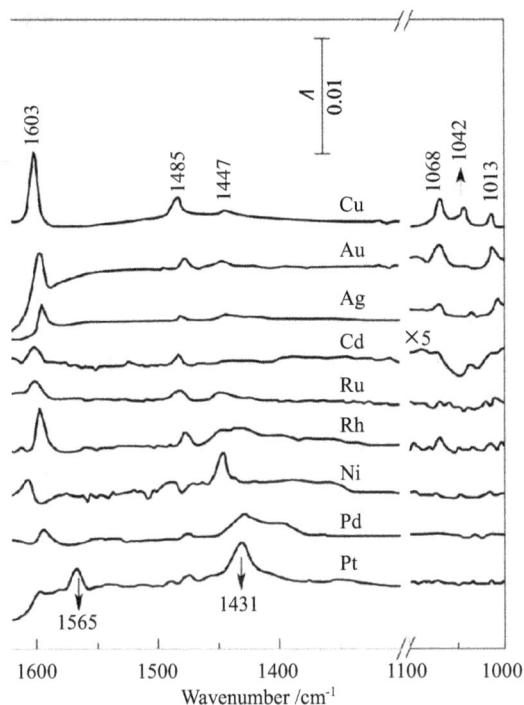

Fig. 3B.20: Typical infrared spectra of Py molecules are absorbed on different metal electrodes [77].

In addition, for the Py solution, the blueshift of the ring breathing vibration (v_1) of the absorbed Py molecule also means that Py is absorbed on the electrode surface with an N-terminal single bond.

From the adsorption spectrum of Py on the Ni electrode, the strongest peak is the peak with the B_1 (v_{19b}) vibration mode at the wavenumber of 1,448 cm^{-1}, while the peak intensity of the relative A_1 mode is weaker, such as 1,072, 1,045, and 1,016 cm^{-1}. In addition, a signature peak of α-pyridyl species appeared on the Ni electrode, a weak adsorption peak around 1,550 cm^{-1}. The simultaneous appearance of the vibration peaks of the A_1 and B_1 modes indicates that the Py molecule tilts along the C_2 axis on the surface of the Ni electrode, forming an edge configuration. Combined with the adsorption behavior of Py on Rh and Ru electrodes [77], similar peaks are also observed on the spectrum: 1,600, 1,482, 1,448, 1,070, 1,037, and 1,003 cm^{-1}. The difference is that the v_{19b} mode vibration peaks of Py on the Ru and Rh electrodes are much weaker than that of Ni electrodes, indicating that there is only a small amount of edge-tilted adsorption on the Rh and Ru electrodes, while Py on the Ni electrode is mainly absorbed in an edge-tilted configuration. At the same time, the spectroscopic study also found that most Py is absorbed on the Pt electrode in the α-pyridyl configuration, while on the Pd electrode, the α-pyridyl configuration's edge-tilted configuration of a small number of intact Py molecules appears at the same time.

In summary, the preferential adsorption orientation rule of Py on the transition metal electrode surface is from Cd (Cu), Ag (Au), Ru (Rh), Ni, Pd to Pt electrode, which tends to form end-on, edge-tilted, and α-pyridyl configuration.

3B.3.4 Summary

In this chapter, the electrochemical ATR-SEIRAS technology is introduced. The infrared EM, the design of the optical path system, and the application in surface electrochemistry research are described. This technology can provide detailed vibration information about the molecules on the electrode surface and the possibility of understanding the adsorption structure of the electrode/electrolyte surface and clarifying the reaction mechanism. If this technology is combined with external reflectance infrared spectroscopy, highly sensitive online differential electrochemical mass spectrometry (DEMS), or high-performance liquid chromatography (HPLC), this powerful method of electrochemical research will provide more comprehensive research of the surface adsorption species and interface reaction products.

3B.4 Electrochemical scanning tunneling microscopy (EC-STM)

EC-STM technology can obtain atomic-level structural information on the electrode surface in the electrolyte under the condition of controlling the electrode potential, which is a powerful in situ characterization with high spatial resolution. STM provides the three-dimensional distribution of the surface, which is very important for characterizing the surface roughness, observing surface defects, and obtaining the behavior of molecules and their aggregates. It can be used to study the atomic arrangement state of the metal or semiconductor surface, the adsorbed species on the surface, and the properties related to surface electronic behavior. More importantly, because the working principle of STM is based on the quantum tunneling of electrons at a short distance, STM has no special requirements on the working environment. It can work in a vacuum and atmosphere and work in solutions. This advantage provides the possibility for the application of STM in electrochemistry.

In 1986, Sonnenfeld and Hansma obtained the atom-resolved STM image of highly oriented pyrolytic graphite in the electrolyte [29], indicating that STM can be measured in the electrolyte for the study of electrochemical solid/liquid interface structure. After that, STM was gradually used in electrochemical research. After the 1990s, commercial electrochemical (EC)-STM instruments became mature, which strongly promoted the application of EC-STM, enabling people to study a variety of electrochemical processes, including metal and semiconductor surface structure, ion, and molecular adsorption, metal deposition, and corrosion. In recent years, based on the unique physical and chemical properties of ionic liquids at room temperature, such as nonvolatility, high stability, designability, high conductivity, and wide electrochemical window, have been widely used in electrochemistry [78]. Ionic liquids have almost no vapor pressure and are stable to air and water, making EC-STM technology an effective method for studying the electrochemistry of ionic liquids. Important processes have also been made in studying electrode/ionic liquid interface structure, adsorption in ionic liquids, electrodeposition, surface architecture, and conductivity measurement.

This chapter intends to briefly introduce the basic principles of STM technology, focusing on the experimental methods of STM in electrochemical research. In terms of application, this chapter introduces the progress of EC-STM in the study of electrode surface structure, underpotential deposition (UPD), surface architecture, and conductivity measurements in ionic liquids.

3B.4.1 Basic principles of STM

STM works are based on the quantum tunneling effect. The tunneling effect is a manifestation of the volatility of microscopic particles. The tunneling effect is a manifestation of the volatility of microscopic particles. According to the principles of quantum mechanics, even if the height of the potential barrier is greater than the particle's energy, the probability of a particle passing through the barrier and appearing on the other side of the barrier is not zero. In this case, the current formed by electrons penetrating the barrier between two conductors is called tunnel current. The greater the barrier height, the lower the tunneling probability.

The quantum tunneling effect in the metal/vacuum/metal system can be described as follows: the distance between two pieces of metal M_1 and M_2 is S when the distance between M_1 and M_2 is small enough (usually need to be <1 nm), the electron wave functions of the two metals overlap, the electrons in metal M_1 can tunnel to M_2 and vice versa, a net directional tunnel current will be formed and can be detected if a lower bias voltage (2 mV to 2 V) is applied between two metals.

In the design of the STM instrument, M_1 is used as a metal tip, M_2 is replaced by the sample (as shown in Fig. 3B.21), and a bias voltage is applied between these two. When the distance is less than 1 nm, a detectable tunnel current will be formed, and the simple form of its expression is as follows:

$$I \approx V_b \rho_s (0, E_F) \exp\left(-A\Phi^{\frac{1}{2}}S\right) \tag{3B.6}$$

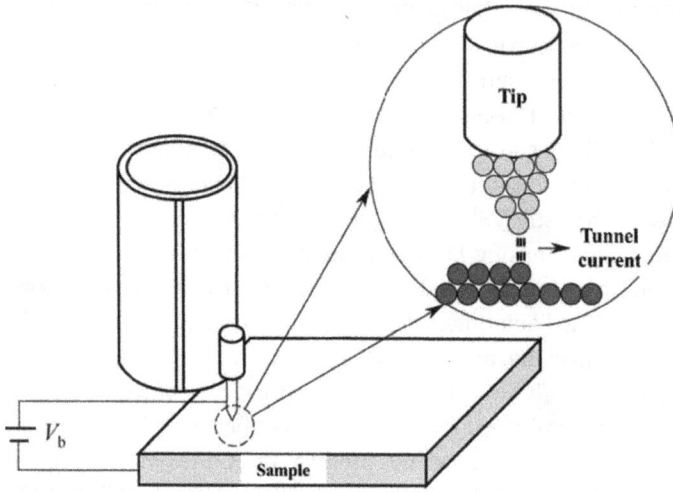

Fig. 3B.21: Schematic diagram of STM imaging setup.

where V_b is the bias voltage; ρ_s (0, E_F) is the local density of states (LDOS) near the Fermi level of the sample surface; Φ is the effective barrier (usually approximated by the average work function of the two metals), eV; S is the distance from the tip to the sample; A is the scale factor, which is 1.025 $(eV)^{-1/2} \cdot Å^{-1}$.

It can be seen from the above formula that the tunnel current has an exponential relationship with the distance between the tip and the sample. When Φ takes a typical value of 4 eV, the current will attenuate by about e^2 each increase of 1 Å in the distance, that is, about an order of magnitude. This sensitive change of current to distance brings the atomic-level spatial resolution capability of STM.

Because the tunnel current has an exponential dependence on the distance from the tip to the sample and the square root of the effective barrier, both the geometric morphology of the sample surface and the electronic factors will cause the change in the tunnel current. If the types of atoms on the sample surface are different, STM cannot distinguish whether the change is caused by surface topography changes or surface work function changes. At this time, the STM image corresponds to the fluctuations of the surface atoms of the sample and is the combined effect of the geometric shape factors and the electronic factors. Generally, the surface-localized state density of a single metal is consistent with the location of atoms, whose STM image can reflect the geometry of the surface. For semiconductors, the directionality of covalent bonds plays an important role. Thus, the image is not completely equivalent to the position of atoms, and the imaging mechanism of adsorbed molecules on the surface is more complicated, which is related to the front-line trajectory of the absorbed molecules, surface conditions, bias voltage, tunnel current, and other factors.

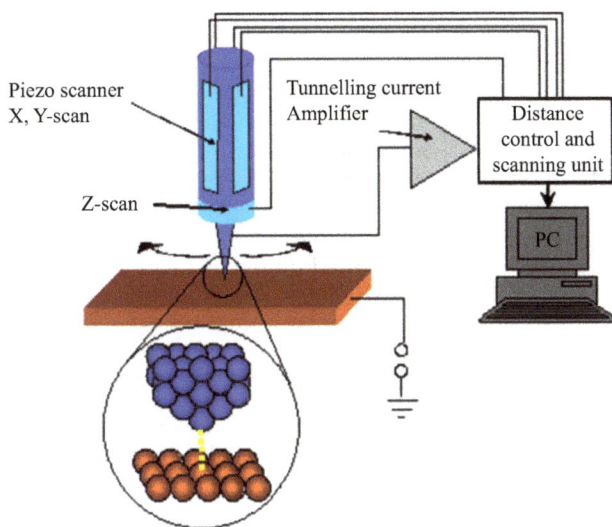

Fig. 3B.22: Schematic diagram of the main components of STM.

As shown in Fig. 3B.22, the STM instrument comprises the microscope probe, electronic circuit control, and computer control parts. EC-STM system needs to add the electrochemical control part. In addition to the sample and the tip, the core of the microscope probe is a piezoelectric ceramic scanner, which is a cylindrical piezoelectric ceramic tube coated with a thin and uniform metal on the inside and outside common. The metal electrode on the outer cylindrical surface is equally divided into four areas insulated along the Z-axial direction, called X and $-X$, Y, and $-Y$. The metal electrode on the inner cylindrical surface is a continuous metal plating layer. Suppose a bias voltage is applied between the external electrodes (X, $-X$, Y, and $-Y$). In that case, the piezoelectric ceramic tube will be deflected, thereby achieving a scanning image in the XY-plane. In addition, applying a bias to the internal electrode Z will cause the entire ceramic tube to expand and contract, thereby achieving height control in the Z-direction. The computer control part sets scan parameters, including XY scan parameters, and collects the information in the Z-direction to generate STM images.

The imaging modes of the STM instrument can be divided into constant current mode and constant height mode, as shown in Fig. 3B.23. A set voltage is applied to the scanner in the constant current mode, driving the probe's scanning in the XY-direction. The tunnel current between the tip and the sample is amplified and compared with the preset tunnel current to form a different signal. The difference signal is fed back to the high-voltage amplifier that controls the Z-direction of the scanner. The change of the Z-direction voltage on the scanner can adjust the distance between the tip and the sample so that the tunnel current follows the preset value. During the scanning process, the zoom information of the scanner in the Z direction is recorded by the computer to

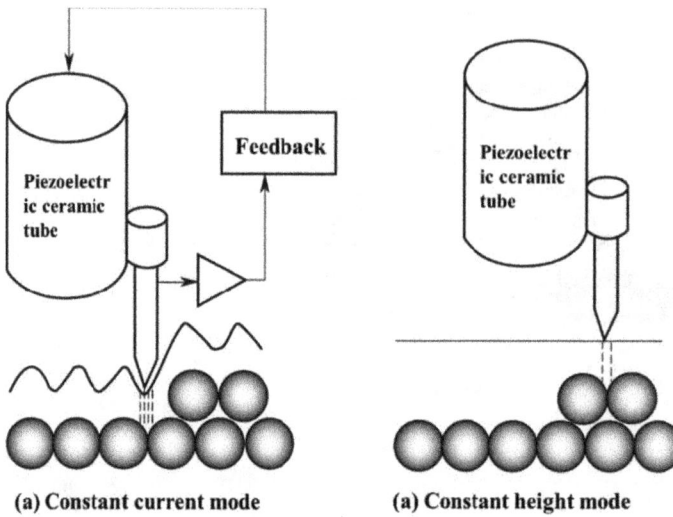

Fig. 3B.23: STM imaging mode.

form an STM image, which is related to the three-dimensional tomography of the sample surface at the atomic scale, as shown in Fig. 3B.23(a). The feedback loop does not work in the constant height mode, and the response is very slow. Therefore, the probe cannot follow the fluctuations of the sample surface. It can be considered that the voltage applied to the scanner in the Z-direction is unchanged, and the absolute height of the tip is constant. When the scanner deforms and drives the probe to scan in the XY-direction, the different distances between the tip and the sample at different positions will cause the change in the tunnel current, and the surface topography information can be obtained by recording the current tunnel change as shown in Fig. 3B.23(b). The constant height mode is suitable for small-scale atomic resolution imaging and can use faster scanning speeds to avoid image distortion caused by thermal drift. However, it is not suitable for large-scale imaging because the height of the tip is the same during scanning, and large surface fluctuations may cause the tip to collide with the sample surface. As the most commonly used mode of STM, the constant current mode can be used for large-scale imaging and observation of samples with large surface fluctuations. However, due to the limitation of the response speed of the feedback loop, a relatively low scanning speed is required for the constant current mode.

The fundamentals of both techniques are essentially the same. However, two different elements must be introduced in a conventional STM setup to operate as an ECSTM: an electrochemical cell consists of two working electrodes with a bipotentiostat approach and suitable ECSTM probes [79], as shown in Fig. 3B.24. The main purpose of the bipotentiostat is to control the potential of both the tip and the metal substrate concerning a current-less reference electrode to control electrochemical

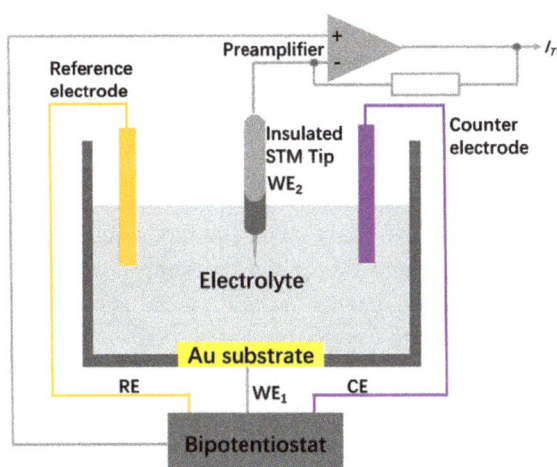

Fig. 3B.24: EC-STM configuration, the bipotentiostat, controls the potential of the tip and sample concerning the reference electrode.

reactions taking place at its surface. Generally, the tip is virtually grounded, and the tunneling current is measured by a high-gain current follower or fed to the STM control unit through a preamplifier. The counter electrode in the EC-STM setup is obtained by an Au or Pt wire with enough stability. A good reference electrode is obtained when a metal is used in electrochemical equilibrium with the corresponding metal cation in solution. Thus, EC-STM has advantages over classical STM, which allows precise and independent control of both tip potential and sample electrode potential through a bipotentiostat and a quasi-reference electrode [80].

3B.4.2 Application of ECSTM in aqueous solution

This section only introduces some research results from three applications of EC-STM : surface imaging, nanostructure, and spectroscopy measurement so that readers have a basic understanding of EC-STM applications. The selection of some examples compares the research results in the next section on ionic liquids.

Au(111), Au(100), and Au(110) of the low-index crystal planes of Au single crystals can undergo different forms of reconstruction to form a denser structure [81]. For the unreconstructed Au(111) surface, it can be considered that the surface atoms maintain their position in the bulk phase so that the bulk structure can be inferred directly from the surface structure, which is called the (1×1) structure. However, for the reconstructed Au(111) surface, the atoms no longer maintain their position in the bulk phase, and the arrangement has undergone significant changes, forming a so-called $(\sqrt{3} \times 22)$ structure. The Au(111) surface reconstruction depends on the applied electrode potential. When the electrode potential is slightly more positive than the zero-charge potential of the Au(111) reconstruction surface, the ($\sqrt{3}$×22) structure will be removed, and the surface will return to the same (1 × 1) structure of the bulk phase. Since the density of atoms on the reconstructed surface is higher than that of the unreconstructed surface, the excess atoms will be squeezed out after the reconstruction is removed, thus forming some Au islands.

In the electrochemical environment, processes other than the tunneling mechanism may occur when the distance between the STM tip and the sample surface is closed enough. For example, when the cohesive energy of the tip metal is less than the cohesive energy of the base metal, a jump-to-contact process of atoms on the tip to the surface of the base may occur. Using this process, Kolb's research at Ulm University in Germany developed a method of using EC-STM to construct metal nanoclusters on the surface of the metals and semiconductors [82]. The construction process is shown in Fig. 3B.25. Firstly, the EC-STM tip is covered by the deposited metal. Then a certain magnitude of pulse voltage in the piezoelectric ceramic Z-direction is applied to drive the tip close to the sample. A jump-to-contact process will occur when the distance between the tip and the sample is close enough. That is, the metal deposited on the tip will transfer to the sample's surface. When the pulse ends, the tip leaves the sample, and the metal transferred to the sample's surface forms a nanostructure. The metal ions in the solution can be continuously deposited on the surface of the tip to replenish the metal atoms of the tip consumed by the construction. Therefore, an array or pattern of nanostructures can be created by repeating the above steps. By adjusting the applied pulse, the height of the nanoclusters can be controlled within a certain range. This method is the highest resolution nanostructure method in the solution currently, and it has constructed Cu, Pd, Ag, and Cd nanostructures in an aqueous solution

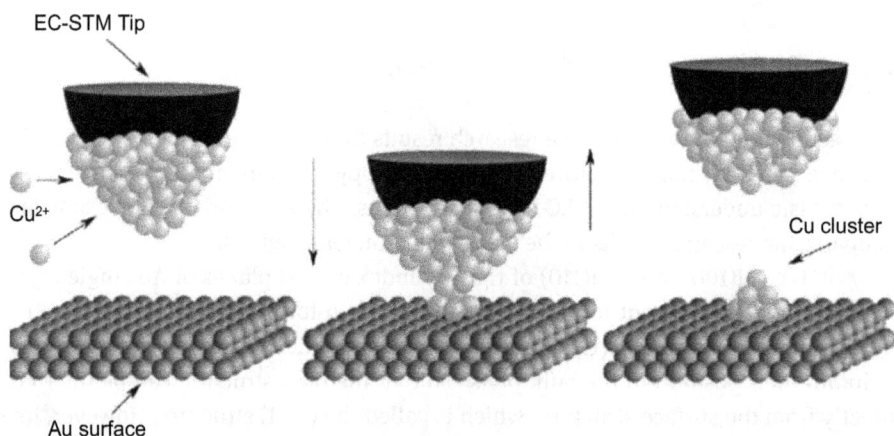

Fig. 3B.25: Schematic diagram of EC-STM using the jump-to-contact process for surface nanostructure [82].

successfully [82–85]. If the construction is treated in an aqueous solution, it should be pointed out. The disposition of some highly active metals is often affected by hydrogen evolution, and EC-STM cannot normally work, thus limiting the scope of application of this method.

EC-STM can observe the structure of the adsorbed species on the electrode surface at the atomic resolution level and obtain structural information perpendicular to the electrochemical interface by scanning the tunneling spectrum. Among them, the method of controlling the potential between the tip and the sample to a constant value, changing the distance between the tip and the sample, and recording the tunnel current is called DTS, which can be applied to calculate the effective energy barrier height. At the electrochemical interface, the surface state of the electrode and the interface structure will affect the relationship between the tunneling current and the distance. Schindler et al. reported that at the Au(111)/HClO$_4$ solution interface, the effective energy barrier height changes in a fluctuating manner with the distance from the Au electrode surface, which is also related to the structure of the interface water [86]. The effective energy barrier height measured by DTS reflects the charge density distribution. The density functional theory (DFT) calculation can calculate the charge density distribution of the absorbed species on the surface. Jacob et al. combined the DTS and DFT calculation to correlate the effective energy barrier height data with the spatial distribution of ions and water and obtained experimental and theoretical evidence for the structure of the electric double layer in the direction perpendicular to the electrode surface for the first time [87]. A detailed model of the electrochemical interface structure of Au(111) in H$_2$SO$_4$ solution was obtained in this research. This work also showed that experimental measurement and theoretical calculation could provide a distance calibration in the direction normal to the electrode surface without the assumption of the STM tip and the substrate point contact resistance.

3B.4.3 Application of EC-STM in ionic liquids

Room-temperature ionic liquid (referred to as ionic liquid for short) is composed of cations and anions, and their melting point is close to or lower than the room temperature. Due to their special physical and chemical properties, such as nonvolatility, high stability, designability, high conductivity, and wide electrochemical window, ionic liquids have become a new electrochemical solvent [88]. This section introduces the application of EC-STM in the study of electrode surface adsorption, UPD, surface construction, and conductivity measurement in ionic liquids.

EC-STM can accurately track the adsorption behavior of the anions and cations on the electrode surface and provide a direct experimental basis for judging the point of zero charge (PZC) of the electrode/ionic liquid interface and understanding the structure of the electric double layer. However, there are differences in the literature on how to judge PZC based on the differential capacitance curve. Freyland

and Pan observed the ordered adsorption structure formed on the Au(111) surface and its phase transition process with potential [89]. However, the Au(111) electrode is not suitable for observing the adsorption of cations. By choosing Au(100) single-crystal electrode, BF_4 forms an ordered structure with a rhombus unit cell in the more positive potential range, and BMI^+ is observed to form a striped structure in the more negative potential range as illustrated in Fig. 3B.26. In this structure, the butyl side chains of BMI^+ nearby interact through van der Waals forces to form a stable adsorption structure. The above-observed anion/cation adsorption and conversion process on the electrode surface occurs near the potential corresponding to the maximum value of the differential capacitance in the differential capacitance curve, which indicates that the PZC of the system should be near this potential. The above work combines the electrode surface process observed on the single-crystal electrode with the measurement result of the differential capacitance in the ionic liquid for the first time and then infers the PZC of the system.

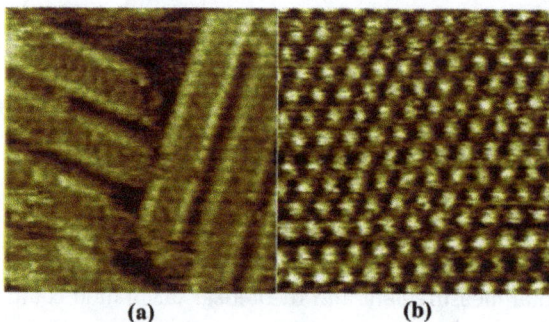

Fig. 3B.26: High-resolution STM image of (a) BMI^+ strip structure on Au(100), 8×8 nm^2; (b) BF_4^- ordered structure, 5×5 nm^2 [90].

In the initial metal electrodeposition process, if the interaction between the deposited atoms and the heterogeneous substrate atoms is stronger than the deposited atoms, the metal ions can be reduced and form a (sub)monoatomic layer on the substrate when the equilibrium potential is positive. This phenomenon is called UPD. Here the UPD process of Sb on Au(111) and Au(100) signal crystal electrodes will be introduced in a BMIBF$_4$ ionic liquid [91]. Before UPD occurs, a reversible adsorption process occurred at 0 V with an initially small quantity of scattered species that gradually formed domains and grew more rapidly at decreasing potential. Eventually, a nearly full-coverage layer was reached at 0 V, as shown in Fig. 3B.25a. The periodicity and sixfold symmetry of the cluster array suggest an Au(111) ($\sqrt{31} \times \sqrt{31}$) R8.9° adlayer structure, as depicted in Fig. 3B.25b. A structural transformation of the clusters distinguishes the initiation of UPD as the potential is decreased to −0.2 V, where the clusters and strips coexist.

The transformation can be accelerated when the potential is decreased to −0.3 V, where all the clusters are transformed into strips within a short period. The atomic strips are more densely arranged, and in some areas, gaps between atomic strips are not distinguishable when the potential is decreased to −0.4 V (Fig. 3B.25c). It reveals a nearly complete Sb(III) to Sb(0) reduction. Figure 3B.27d illustrates the structural model of the atomic strips in registration with the Au(111) substrate, where every three Sb atoms occupy four Au atoms along the strip direction. Unlike on Au(111), where a unique cluster structure is formed prior to the deposition processes, Sb UPD on Au (100) takes place directly from the solution species of Sb(III). The high-resolution image in Fig. 3B.27e illustrates the structure details of the strips, which are composed of double lines, and has a large gap between each other. The defects in the strips with missing atoms in the double lines are also observed, as indicated by the white circles. This research illustrates that the UPD behavior of Sb in ionic liquids is completely different from that in an aqueous solution. The anion and cation composition of the solvent and the high ionic forces form the special structure and the electrochemical interface, leading to the adsorption and self-organization of the precursor. With the influence of the reactants' mass transfer and CT behavior in ionic liquids, the UPD of Sb exhibits new characteristics. In-depth research on the interaction of ionic liquid-electrode surface and ionic liquid-metal ions is of great significance in revealing the nature of UPD.

In the aqueous environment, the special chemical nature of Fe makes its electrochemical research face greater challenges, and its electrodeposition process is also severely disturbed by the hydrogen evolution process, which limits the research of in situ STM. Fortunately, the characteristics of room temperature ionic liquids enable its research to be carried out. At the same time, ionic liquids also provide special solvation and high ionic strength environments, which uniquely impact the mass transfer and discharge processes of reactant and intermediate species. These characteristics may lead to the discovery of new phenomena and reactions in the electrodeposition of metals (especially for ferromagnetic metals). In addition, STM technology provides a powerful tool for studying the electrodeposition process with its ultra-high spatial resolution and manipulation capabilities. It is also a method for achieving the highest surface nanostructure efficiency and resolution in a liquid environment. Figure 3B.28 shows a ring composed of 48 Fe nanoclusters constructed by the electrochemical STM-BJ (break junction) method based on "jump contact" using ionic liquid as a solvent. The height of the Fe clusters is about 0.6 nm, and the ring diameter is 120 nm.

Microcrystalline Ge is usually a direct bandgap semiconductor. Compared with Si, Ge has a narrow bandgap of 0.67 eV, high carrier mobility, a low effective mass of electrons and holes, and a low dielectric constant, making it more suitable for light-emitting devices and storage devices, and photodetector devices. In addition to preparing various Ge nanostructures and studying their morphology, structure, optical, and electrical properties, the construction of Ge atomic wires and the study of their quantum transport properties are also important research contents in the

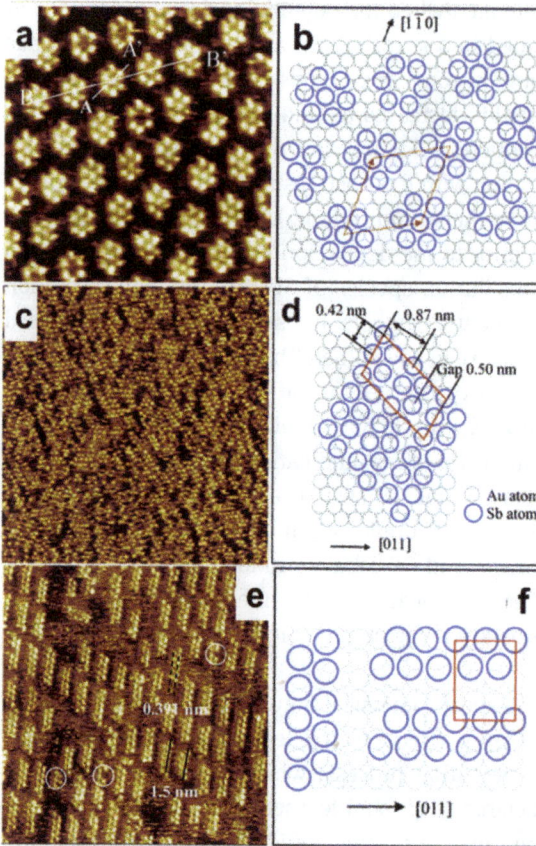

Fig. 3B.27: High-resolution in situ STM image showing the detailed arrangement of the Sb monolayer cluster array (a) on bare Au(111) surface at 0 V, (c) on Au(111) substrate at −0.4 V, (e) on Au(100) substrate at −0.2 V, and the corresponding proposed model of the atomic strips (b), (d), (f), respectively. Each blue circle represents an Sb atom [91].

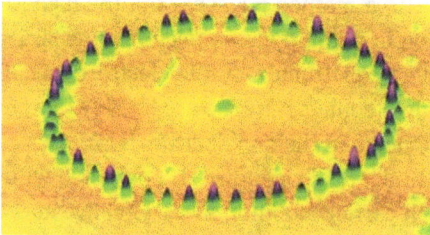

Fig. 3B.28: STM images of Au(111) surfaces decorated with 48 Fe clusters [92].

field of nanoelectronics. The electrochemical STM-BJ method based on jump-to-contact can ensure that homogeneous fracture occurs whenever it is pulled, which can overcome the problem of complex fracture conditions [93]. Fig. 3B.29 shows the complete (3 × 3) lattice obtained when the tip potential is in the underpotential zone. The cluster heights in the lattice are 0.4–1.1 nm, and the diameters are 3–8 nm. During this nanostructuring process, Ge atom lines are formed between the tip and the substrate. In constructing the Ge atomic wire, recording the tip current can obtain the tip I–t curve (pull curve), and the conductance curve can be obtained after appropriate mathematical conversion to reflect the quantum transport properties of the Ge atomic wire.

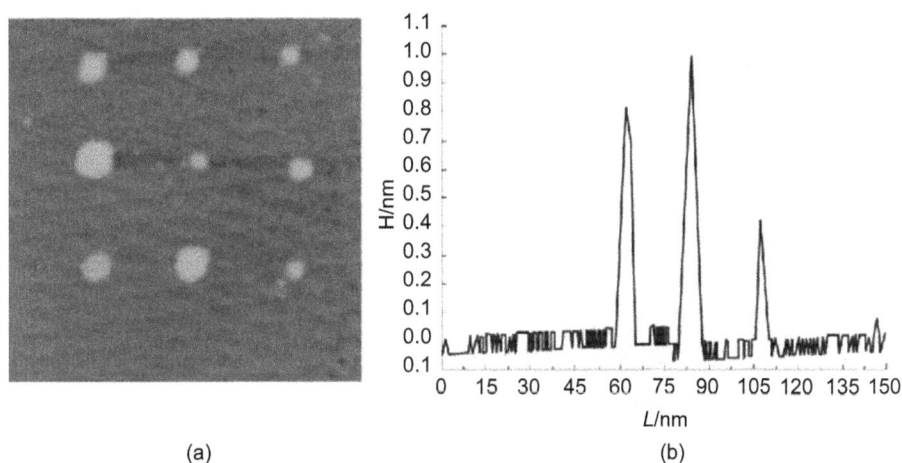

(a)   (b)

Fig. 3B.29: (3 × 3) Ge nanolattice constructed on Au(111) surface [93].

Experiments have found that the pull curve does not necessarily have a conductivity step when Ge clusters are left. The reason might be that the mechanical stretchability of Ge is not good enough. The atomic junction was broken before its diameter is equivalent to the Fermi wavelength (2.1 nm) when retracting the tip to construct the atomic wire. Therefore, the quantum conductance of Ge can only be obtained through a large number of experiments and statistical analysis of the stepped curve from the pulling curve. The typical conductance curve (Fig. 3B.30) has steps at 0.25 G_0 and 0.05 G_0. The conductance statistics show that the quantum conductance of Ge is concentrated between 0.02 G_0 and 0.15 G_0, and a relatively pronounced conductivity peak appears at 0.025 G_0 and 0.05 G_0.

In general, the success rate of Ge nanostructures is much lower than that of metal systems that have been nanostructured by this method. The size uniformity of nanoclusters is not as good as that of metals. The possible reason is that the mechanical pull ability of Ge and its electrodeposition behavior are not as good as metals. Although the electrodeposition behavior of Ge on Au(111) and Pt(111) surfaces is

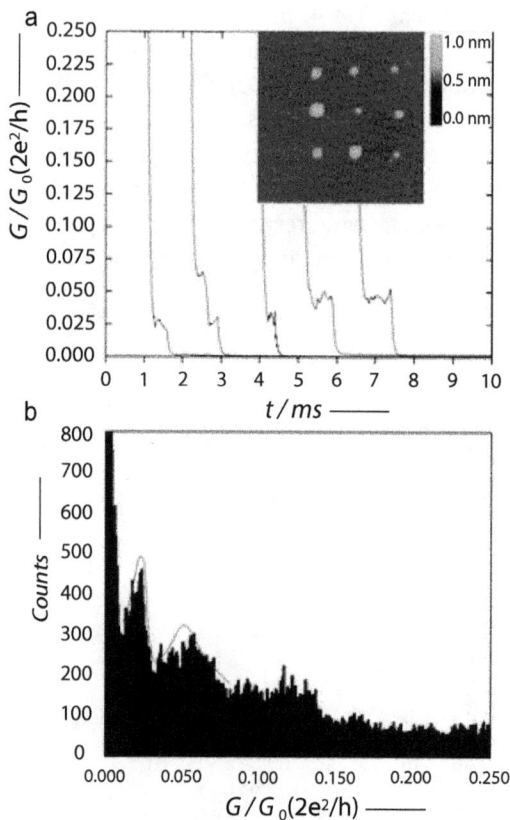

Fig. 3B.30: Typical conductance traces (a) and a conductance histogram (b) of Ge nanowires fabricated between Pt/Ir tips and Au(111) surfaces. Inset: (3 × 3) Ge array on Au(111) surfaces, scan size: 150 × 150 nm².

slightly different and the conditions for its nanostructure and quantum conductance measurement are not the same, the quantum conductance measurement results obtained on the surface of these two electrodes are consistent. This illustrates that the homogeneous fracture of Ge occurs when the Ge atomic wire is constructed by this experimental method. That means the quantum transport properties of Ge are not affected by the substrate material.

3B.4.4 Summary

After more than 20 years of development, EC-SPM has not been limited to the characterization of electrode surface structure and the spectrometric measurement of electrode/electrolyte interface. Driven by the demand and the development of nanotechnology, EC-SPM has become an important medium to use the advantages of electrochemistry to conduct nanostructure preparation and nanoelectronics research in an electrochemical environment. The emergence of new research fields will provide new opportunities for developing the theory and experimental technology of EC-SPM, which will also play an important role in understanding new fields. Early mass spectrometry (MS) researches on electrochemical reactions were offline until 1971.

3B.5 Electrochemical mass spectrometry

The electrochemical in situ infrared spectroscopy, SERS , and scanning probe microscopy can provide information on the chemical type, adsorption configuration, and atomic configuration, which have been described in detail earlier. In addition to the above information, obtaining information such as the quality and quantity of reaction products and by-products is also indispensable for an accurate understanding of the mechanism and kinetics of the electrocatalytic process. MS is an effective method for qualitative and quantitative analysis of reactants and products. Early MS researches on electrochemical reactions were offline. It was not until 1973 that Bruckenstein et al. built an electrochemical mass spectrometer for online analysis of volatile reaction products successfully [16, 94]. Subsequently, Heitbaum et al. [95] combined an electrochemical cell, a two-stage turbomolecular pump, and a quadrupole mass spectrometer to establish a DEMS technology. Using an MS monitor, they realized real-time, in situ tracking of electrocatalytic reaction products. DEMS can qualitatively identify volatile species in the solution and time-resolved (millisecond level) to give the concentration or absolute amount of the species quantitatively. Its detection sensitivity can be 10^{-6} magnitude [96]. However, DEMS can only give information about the volatile substances in the solution. The combined technology of electrochemical in situ infrared spectroscopy and DEMS was realized at Ulm University in Germany [97]. This technology can obtain multidirectional information simultaneously, such as the chemical properties of the reaction intermediates, adsorption configuration, the type and output of the reactants and products, the Faraday current, and the reaction overpotential of the reaction. This part will mainly introduce the basic principles of MS and DEMS and the corresponding experimental techniques. The combined technology of electrochemical in situ infrared spectroscopy and DEMS is introduced, and the applications are illustrated.

Finally, the future development prospects of this combined technology will be described briefly.

3B.5.1 Fundamentals of mass spectrometry technology

MS is a technique for qualitative and quantitative analyses of the analyte by measuring the ratio of the mass to the charge (m/z or m/e) of the ionized species and the signal intensity. The composition of the mass spectrometer is illustrated in Fig. 3B.31, which contains an ionizer, accelerator, deflector, detector, computer control system, and vacuum system. The analytes entering the mass spectrometer's vacuum chamber will first volatilize under low pressure and then collide with the low-energy electron beam and lose one or more electrons to become positive ions. Then these positive ions are focused and accelerated into the analyzer under the action of the electric field. There is usually a magnetic field or alternating electric field in the analyzer. In the case of a magnetic field with a magnetic flux B, the positive ions will deflect under the action of the magnetic force and make Lorentz motion along an arc of radius r. If the voltage of the accelerating electric field is U, the radius of the arc channel is

$$r = \tfrac{1}{B} \sqrt{\tfrac{2mU}{Z_0}} \tag{3B.7}$$

If the channel radius of the analyzer is RC, then

$$\tfrac{m}{Z_0} = \tfrac{B^2 r_c^2}{2U} \tag{3B.8}$$

Only ions with a suitable mass-to-charge ratio can pass through the analyzer to the detector. The ions with a higher or lower mass-to-charge ratio will collide with the inner wall of the vacuum chamber and be neutralized by the stainless-steel electrodes, and finally be pumped out of the vacuum. This is the basic principle of mass-to-charge ratio selection for MS. According to equations (3B.7) and (3B.8), we can change the accelerating electric field voltage U or the deflection magnetic field strength B. The ions with a specific mass-to-charge ratio can reach the detector through the mass analyzer.

The last step of the MS characterization program is the detection of specific ion signals. Different detectors can be selected for the mass spectrometer according to the characteristics of the species to be tested (such as structure, relative molecular mass, and quantity). For a general quadrupole mass spectrometer, one of the simplest detectors is a Faraday cup. It is a metal cup placed in the ion flow path. When the ions hit the wall of the metal cup, they are neutralized by the charge in the metal. The electrometer will measure the corresponding ion current intensity. Depending on the measurement of the integration time (time constant, from several seconds to 100 ms), the detection limit is between 10^{-16} and 10^{-14} A. The Faraday cup has the advantages of working under relatively high pressure (10^{-5}–10^{-4} mbar), good long-term stability, high heat resistance, and a simple design. Its disadvantage is that the signal-to-noise ratio of the detection is low, and the sensitivity is not good. In addition to Faraday cups, other commonly used detectors include channel multiplier tube detectors (Daly detectors), secondary electron multipliers, and microchannel plate detectors.

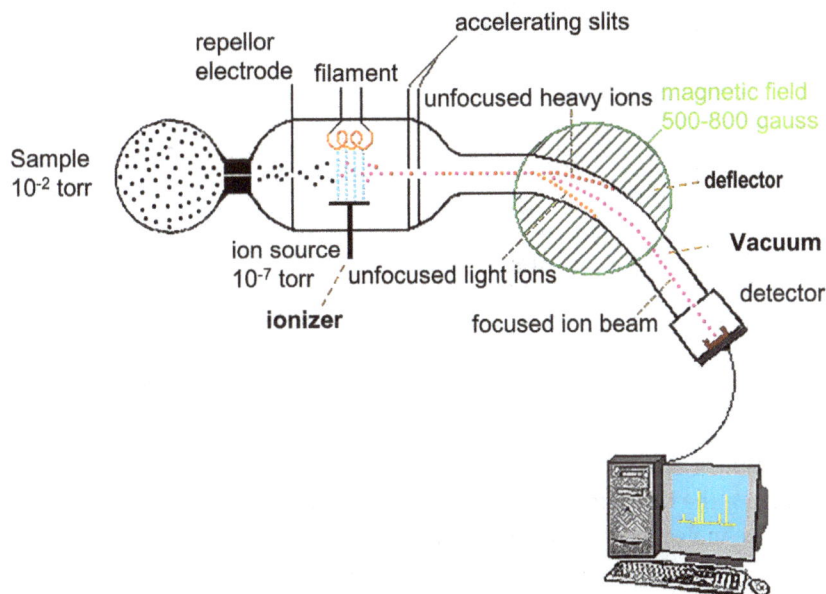

Fig. 3B.31: Schematic diagram of the composition of the mass spectrometer system.

3B.5.2 Differential electrochemical mass spectrometry (DEMS)

Figure 3B.32 shows the electrochemical MS device reported by Bruckenstein in 1971 [94] and 1973 [16]. This electrochemical MS and the DEMS can be combined with traditional electrochemical current and voltage measuring instruments to perform CV, potential step, constant current, and other experiments. The design of this mass spectrometer is different from DEMS in that the signal value of electrochemical MS is the cumulative value of the signal corresponding to the species to be measured over the reaction time. Based on this data, the electrochemical reaction rate can be calculated as equal to the derivative value (or tangent slope). After all, electrochemical in situ mass spectroscopy has already made a breakthrough in experimental electrochemical technology in the real-time detection of electrochemical reactions.

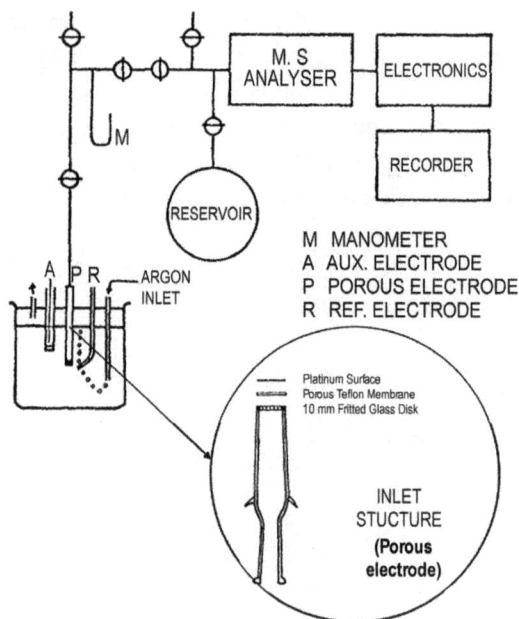

M MANOMETER
A AUX. ELECTRODE
P POROUS ELECTRODE
R REF. ELECTRODE

Platinum Surface
Porous Teflon Membrane
10 mm Fritted Glass Disk

INLET
STUCTURE
(Porous
electrode)

Fig. 3B.32: The design of electrochemical mass spectrometry by Bruckenstein et al. [16].

Electrochemical in situ MS technology was developed rapidly in the following decades. In 1984. the DEMS technology was developed by combining an electrolytic cell, a two-stage molecular pump, and a quadrupole mass spectrometer for the first time [95]. DEMS uses a vacuum pump to remove the injected species during the detection process quickly. The detected signal value can be directly proportional to the products' reaction rate or generation rate. In addition, DEMS is quite different from the former in electrolytic cells, vacuum systems, and mass spectrometer systems.

Figure 3B.33 shows a schematic diagram of the DEMS system. In order to improve the efficiency and the detection resolution of DEMS, the mass spectrometer part needs: ① choose a smaller pre-vacuum chamber and install the quadrupole mass analyzer and detector reasonably and compactly so that the smaller power pump can maintain the vacuum conditions satisfying the working environment required by the ion source; ② the ion source should be as close as possible to the analysis room to allow as much as ionized species to be analyzed and detected, and the detector can be installed in an orientation of 90° perpendicular to the quadrupole to prevent the cathode rays of the filament from reaching the multiplier; ③ the pre-vacuum chamber and the analysis chamber are connected with two molecular pumps, respectively, to ensure the high degree of vacuum. Requirements for external accessories: ① the area of the electrolytic cell/vacuum interface and the pore size and porosity of the porous hydrophobic material layer are determined according to the suction speed of the pump; ② the electrolytic cell is designed according to the research system and experimental conditions, which will be introduced below. The entire system, including DEMS and electrochemical signal acquisition, should be controlled by the same computer effectively and simultaneously.

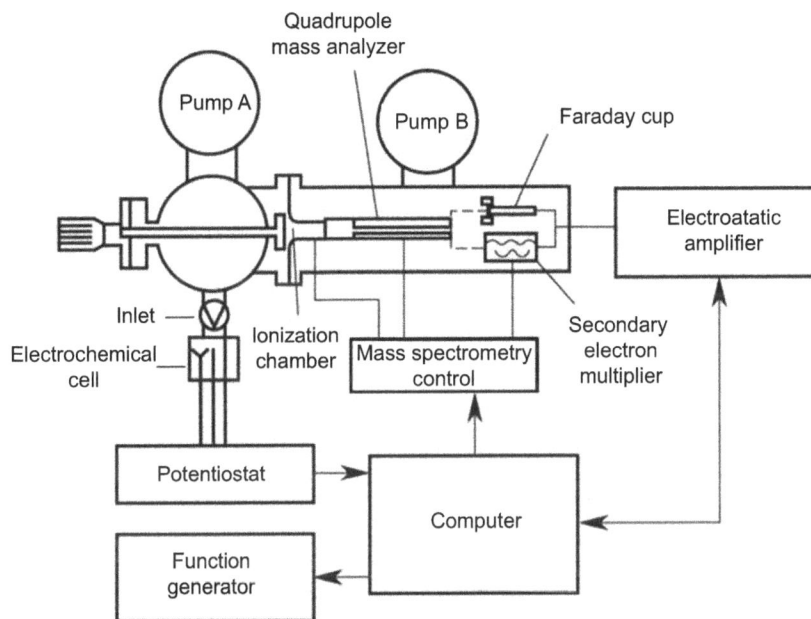

Fig. 3B.33: The general design of the DEMS system.

The electrolytic cell is one of the cores of the entire DEMS system as the reactor for the electrochemical reaction and the sampling interface of the mass spectrometer. It is necessary to adopt the corresponding designs according to the research reaction system and the working conditions (such as an electrode, temperature, and pressure). There are two mainstream sampling methods: membrane sampling and capillary sampling. The relevant electrolytic cell design will be introduced from these two aspects.

The classic DEMS electrolytic cell design is shown in Fig. 3B.34. It includes ① working electrode, reference electrode, auxiliary electrode, and conductive wiring; ② sample injection membrane and electrolyte/MS injection interface; ③ electrolyte solution channel; and ④ insulation support and sealing part. In this design, the working electrode is close to the sample injection membrane to ensure the high sensitivity of MS signal detection. However, due to the position of the working electrode being between the electrolyte and the MS injection membrane, the space is very limited. Thus, only thin-film electrodes can be used instead of single-crystal electrodes or larger rod-shaped electrodes. At the same time, the design of the upper opening cannot be completely sealed. Therefore, there will be the loss of the solutes or highly volatile substances in the reaction due to vaporization. Although the working electrode is quite restrictive, loading working electrode or catalysts on the MS injection membrane can be combined directly with gas diffusion electrode technology well, which can benefit the study of the effect of catalyst loading or diffusion layer thickness on reaction kinetics by changing the catalyst loading.

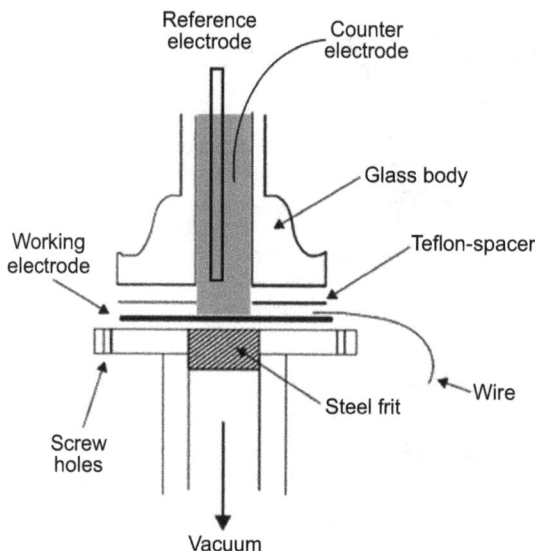

Fig. 3B.34: DEMS electrolytic cell in the early stage [98].

In the thin-layer electrolysis cell design, the catalyst loading on the working electrode is much lower than that of the fuel cell. As a result, the thickness of the catalyst is very small, and there is little or almost no gas diffusion and microporous diffusion in the catalyst layer, resulting in the contact between the reactants and the catalyst being very short. However, these parameters may greatly influence the reaction kinetics of the fuel cell. In order to make the experimental results closer to the working conditions of real fuel cells and to guide the synthesis and preparation of fuel cell catalysts, T. Seiler et al. [99] designed a new type of DEMS thin-layer flow electrolytic cell for direct methanol fuel cells, as shown in Fig. 3B.35. This kind of flow electrolytic cell suitable for actual fuel cell measurement takes full advantage of the high sensitivity of the thin-layer flow electrolytic cell in the sampling part of the MS. At the same time, as the working electrode refers to the electrode structure of the real fuel cell, that is, the design of the layered separator supporting the catalyst, the experimental conditions are closer to the real fuel cell conditions and can even apply to the membrane electrode test of the current commercial fuel cells directly with a certain degree of improvement.

   In addition to membrane injection, DEMS can also use capillary injection, which researchers have used at earlier stages [16, 94]. Compared with membrane injection, the capillary design does not require the complicated electrolytic cell design and only needs to be modified slightly based on the conventional electrochemical cell for usage. When the pore size meets the requirements, the capillary nozzle does not need to be covered with a microporous membrane to ensure a higher subsequent vacuum. However, PTFE membranes are used to reduce the entry of water molecules and improve the signal-to-noise ratio in general. Since the membrane sampling generally requires sealing and electrolyte flow design, the electrode needs to be fixed by mechanical force, which is unsuitable for sensitive and fragile single-crystal electrodes. Thus, capillary injection is more suitable for studying electrochemical reactions on single electrodes than membrane injection [100, 101].

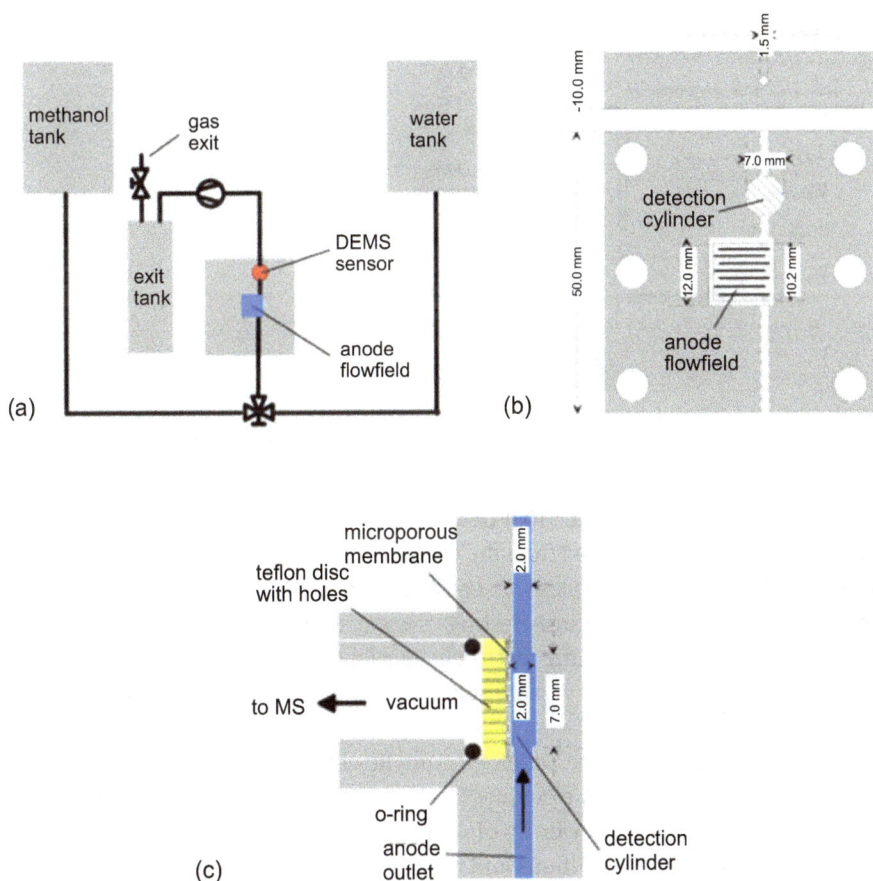

Fig. 3B.35: Scheme of the experimental setup: (a) flow system, (b) fuel cell anode, and (c) DEMS sensor at the outlet of the fuel cell [99].

Capillary sampling is also suitable for nonaqueous systems, as shown in Fig. 3B.36, suitable for the DEMS flow electrolytic cell of the lithium battery system [102]. Its advantages are: ① the gaseous substances produced by the reaction are generally brought into the capillary by the inert carrier gas and then enter into the MS for online detection. Under the conditions of good sealing performance, it is easy to realize the quantitative analysis of various trace gases; ② the material of the main body of the electrolytic cell has good heat and pressure resistance, which can be used in high-temperature environments; ③ the measurement of the full battery reaction can be achieved by replacing the counter electrode material.

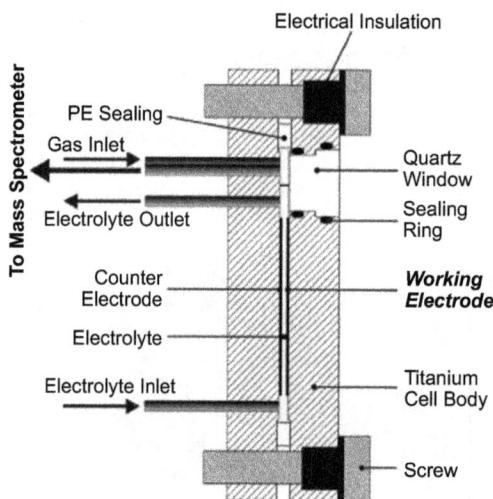

Fig. 3B.36: The capillary injection of the DEMS flow electrolytic cell in the lithium battery system [102].

Recently, Richard N. Zare et al. [103] developed a new type of electrochemical MS (Fig. 3B.37), which was borrowed from the design of desorption electrospray ionization MS (DESI-MS), and used the "waterwheel design" creatively. This design combined the high-speed spray technology, three-electrode system, and capillary injection MS, realizing the real-time tracking of the electrochemical reactions. Compared with DEMS, this type of electrochemical MS is more suitable for the electrochemical reaction of organic macromolecules.

It should be noted that the DEMS electrolytic cell has continued development and improvement. There is no electrolytic cell suitable for all research systems and experimental conditions. It is worthy of designing and processing electrolytic cells according to the research conditions and requirements.

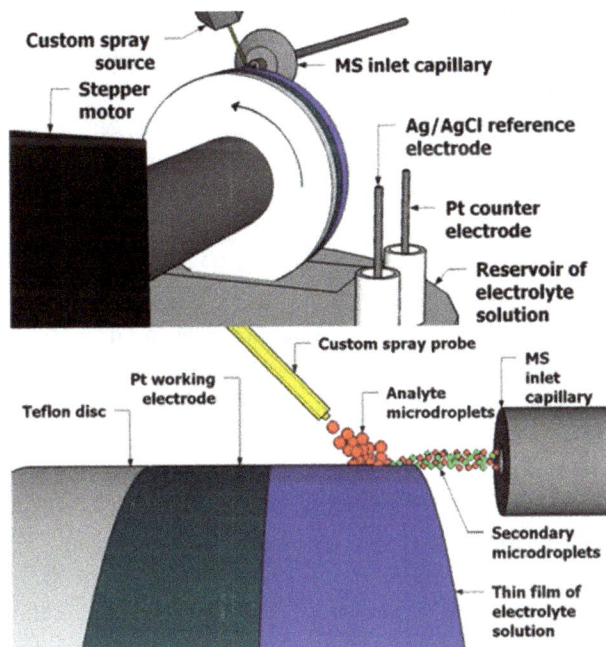

Fig. 3B.37: (Top) Apparatus used to detect electrochemical reaction intermediates, employing a rotating waterwheel electrode. The arrow indicates the direction of rotation. (Bottom) Zoomed-in area showing sample ionization [103].

3B.5.3 Applications of differential electrochemical mass spectrometry

Since the invention of the DEMS technology more than 30 years ago, it was initially used for the related electrocatalytic research of highly dispersed, high-loaded fuel cell nanocatalysts. The investigated reactions are mainly the oxidation of small organic molecules, hydrogen precipitation, and hydroxide reaction. With the development of various electrolytic cells, people have realized the combination of a rotating disk electrode system and a vacuum system, making the mass transfer of reactants between the solution phase and electrode surface more rapid and regular [104, 105]. At the same time, the detection sensitivity of the mass spectrometer signal is greatly developed with the improvement of the mass transfer of the electrolytic cell and the performance of the various components of the mass spectrometer so that it can directly detect the small number of reaction products from a small area of the smooth electrode or even a single-crystal electrode. The invention of the dual thin-layer electrolytic flow cell effectively separates the electrode surface area where the electrochemical reaction is carried out from the MS injection area so that DEMS has enough space to realize the combination with other traditional or modern electrochemical measurement technologies, such as its real-time connection with quartz microcrystal balance and electrochemical in situ infrared spectroscopy.

Recently, DEMS has also been gradually extended to the study of oxygen reduction mechanism [106–109], the charge and discharge processes of lithium-ion or lithium-air batteries [110–112], the enrichment and quantitative analysis of trace organic matter in solutions [96, 113], and the electrochemical reduction of carbon dioxide [102, 114, 115]. The application examples in these areas will be introduced.

Most alcohol fuel cells produce some by-products at the anode during the reaction process. These by-products reduce energy conversion efficiency and may cause damage to the membrane electrodes, bipolar plates, and other components of the fuel cell. Generally, these by-products have good volatility and high stability. Thus, DEMS is very suitable for real-time detection of them. DEMS is also suitable for studying the structure–activity relationship between the surface structure of electrocatalysts and the oxidation reaction activity of alcohol molecules. Baltruschat et al. explored the influence of a series of factors such as the surface structure of the electrode and the introduction of Ru atoms on the methanol oxidation reaction activity on polycrystalline and single-crystal Pt electrodes [116]. They found that the higher the step density of the Pt single-crystal electrode, the better methanol oxidation activity.

In contrast, the carbon dioxide current efficiency has no obvious difference in polycrystalline Pt, Pt(111), and Pt(332). In the low potential range (below 0.65 V), Ru atoms can increase methanol oxidation activity through the carbon monoxide pathway. This process occurs near the Ru atom islands, while the reaction pathway of the soluble intermediates occurs at the step or defect site. Koper et al. used the online electrochemical MS to study the activity of methanol oxidation reaction on a series of single-crystal Pt electrodes [101] and found that the reaction activity on the Pt crystal surface increased in the following order: Pt(111) < Pt(110) < Pt(100). According to the results of the DFT calculation, the highest activity of the Pt(100) plane is mainly since the crystal plane of (100) is beneficial for molecular adsorption and bond breaking.

In the research on lithium batteries, lithium-ion batteries, and lithium-air batteries, DEMS is also a very important tool for studying the electrode charge and discharge reactions and electrolyte stability [102, 117, 118]. In the case of lithium-ion batteries, a crucial prerequisite for good charge–discharge cycle stability is that the negative electrode must form a complete and stable passivation layer during the initial charge–discharge process, that is, the SEI. During the formation of SEI, a certain amount of gas may be generated due to the reductive decomposition of the EC solvent. At the same time, the oxidative decomposition or reduction of the electrolyte and the electrode itself will also generate gas. These gas products can be detected by real-time in situ DEMS when using DEMS to monitor the reaction behavior of the graphite anode during the charge–discharge process in the standard electrolyte of

1 mol/L LiPF$_6$ and ethylene carbonate/dimethyl carbonate, it was found that the formation of CO$_2$ was observed when the electrode potential (relative to Li/Li$^+$) was slightly higher than 1 V. When the potential is more negative, C$_2$H$_4$ and H$_2$ are also detected. At the same time, the production of H$_2$ can always be observed when Li^{1+} is embedded in the graphite during the subsequent charge and discharge process, indicating that even after the initial SEI is formed, the reduction and decomposition of the electrolyte can still occur [119].

When Luntz et al. researched the reaction mechanism of lithium-air batteries and the stability of solvents, they combined DEMS with ex situ XRD and Raman spectroscopy. They found that carbonate-based solvents would irreversibly decompose during battery discharge [120]. In a lithium-air battery using dimethoxyethane (DME) as a solvent, Li$_2$O$_2$ is mainly generated during discharge and will liberate O$_2$ and oxidize DME at a high potential during charging. In addition to the high potential that may oxidize the solvent, the electrode carbon material will also corrode in the presence of oxygen. In this report, the author also used the isotope calibration method and found that the ^{13}CO$_2$ generated from the carbon black cathode calibrated by ^{13}C accounted for about 40 % of the total CO$_2$, which strongly proved that the cathode is accompanied by a decomposition process during the whole charge–discharge process, as shown in Fig. 3B.38. Isotope calibration has also played an important role in the real-time tracking of specific reactants and products in this research.

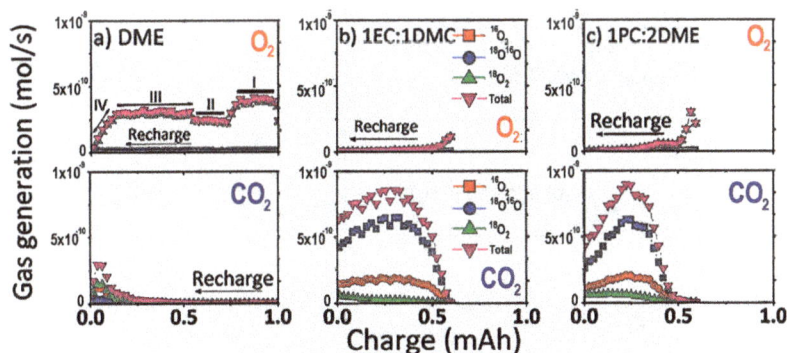

Fig. 3B.38: Isotopically labeled O$_2$ and CO$_2$ gas evolution rate during charging of (a) DME-based, (b) 1:1 (v:v) ethylene carbonate/dimethyl carbonate-based, and (c) 1:2 (v:v) PC/DME-based cells [120].

The products of electrochemical reduction of CO$_2$ to synthesize organic fuel molecules include CO, HCOOH/HCOO$^-$, HCHO, CH$_3$OH, and even C$_2$ products such as olefins, alcohols, or carboxylic acids [121]. Because these products are all volatile, DEMS should be a powerful tool for the in situ exploration of CO$_2$ reduction reactions. Brisard et al. used DEMS to explore the CO$_2$ reduction reaction on the Pt electrode. They found that CO$_2$ can be reduced to form formic acid, methanol, and a small amount of methane in the hydrogen evolution potential range [114]. The experimental results are shown in Fig. 3B.39. These correspond to the following products: formic acid ($m/z = 45$), methanol ($m/z = 32$), formaldehyde ($m/z = 30$), and methane ($m/z = 16$). The onset of formation of formic acid, methanol, and formaldehyde can be extrapolated to the beginning of hydrogen evolution. At the same time, methane is formed at potentials significantly more negative (between −0.1 and −0.2 V).

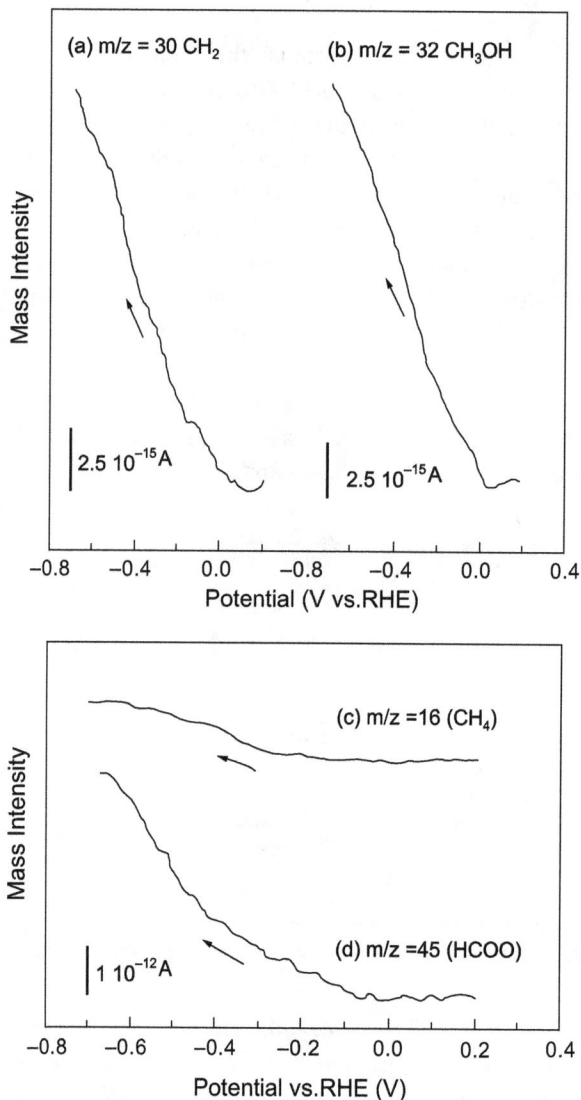

Fig. 3B.39: Mass signals during a potentiodynamic scan of Pt electrode in CO_2 saturated solution saturated at 0.1 M $HClO_4$ solution [114].

The effect of the adsorption of anions on the CO_2 reduction reaction on the Cu electrode was studied by Dubé et al. The formation rate of formaldehyde and methanol was significantly higher than that of ethylene and ethanol [122]. In addition, the adsorption of sulfate ions would promote the formation of formaldehyde, thereby inhibiting the formation of methanol. Plana et al. studied the effect of the thickness of the Pd shell layer on the Au core nanocatalysts on the reduction of CO_2. They found that when the thickness of the Pd shell layer was changed from 10 to 1 nm, the Faraday efficiency of CO_2 reduction was double [115]. These studies show that in situ MS can well track the products of the CO_2 reduction process and the generation quantity. With the population of this technology, it should play an irreplaceable role in this field in the future.

In addition to studying various electrochemical reaction mechanisms, DEMS can also perform quantitative analysis of organic matter in solution. Through the test of the dual thin-layer electrolytic cell of DEMS, as shown in Fig. 3B.40, the detection limit has reached 10^{-6}. This result proves that it can be used for trace species analysis, such as detecting the number of species consumed or generated on the electrode surface due to adsorption and desorption during the electrochemical reactions.

Research has shown that DEMS is also very useful in studying the hydrogeneration or oxidation reactions of small molecules such as ethylene and benzene sensitive to surface structures [123]. It is also very suitable for the trace characterization of macromolecule species [113]. The trace species in the solution itself can be detected by other means such as chromatography. However, the trace species consumed and generated in the adsorption process are difficult to achieve real-time online detection by these means. DEMS can analyze trace species in the solution and play a powerful role in real-time tracking and monitoring the amount of species produced or consumed in electrochemical reactions.

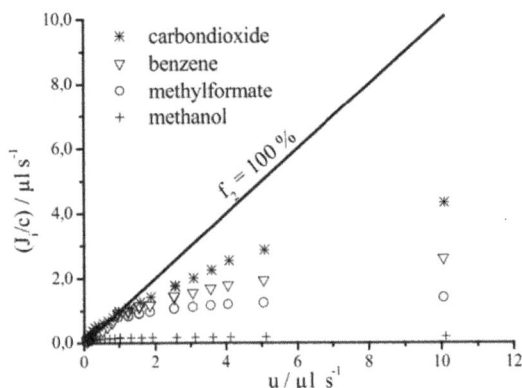

Fig. 3B.40: Performance of the dual thin-layer cell: ion currents converted to flux and normalized by the concentration (CO_2: 38 mM, benzene: 23 mM, formic acid methyl ester: 2 mM, methanol: 2 mM) [96].

3B.5.4 Real-time technology of electrochemical infrared and mass spectrometry

The combined technology of electrochemical in situ FTIR spectroscopy and DEMS can simultaneously obtain various information through a single experiment, such as the chemical properties of the reaction intermediates adsorbed on the electrode surface, the adsorption configuration, the type and output of the reaction product, as well as the Faraday current and the reaction overpotential. Using these experimental results in a single experiment will overcome the incomparability of the results obtained from different experimental techniques due to the difference in experimental conditions, shorten the experimental time greatly, and realize the real-time analysis and diagnosis of some electrode systems with poor stability and reproducibility. This section introduces the combined technology of electrochemical in situ FTIR and DEMS and several examples to demonstrate its advantages and potential in the electrocatalytic field.

Figure 3B.41 shows the structure diagram of the dual thin-layer electrolytic flow cell developed by Heinen et al. at Ulm University in Germany to combine in situ ATR-FTIR spectroscopy (FTIRS) and online DEMS technology [97]. As the key to the combined technology, its specific structure will be introduced here. The body of the electrolytic cell is cylindrical. The flow of the electrolyte is all carried out through the capillary tubes, where the capillary in the middle of the electrolytic cell is used as the inlet, and the six peripheral capillaries normal to the surface of the working electrode connect to the second chamber, which is close to the perimeter of the cell. This geometry enabled a radial laminar electrolyte flow over the working electrode surface, driven by the hydrostatic pressure in the electrolyte supply bottles. An external flow rate pump can achieve more precise control. The electrolyte's inlet and outlet are equipped with suitable three-way or four-way connections. The triplet at the entrance connects two bottles with different electrolytes and a reference electrode or auxiliary electrode.

In contrast, the triplet interface at the exit connects the electrolyte outlet and the reference electrode or auxiliary electrode. The infrared window is a semicylindrical silicon prism. The thin metal film or conductive substrate deposited on the reflective plane of the silicon prims is used as the working electrode (the SEM image of a platinum metal film with a thickness of about 50 nm deposited by a chemical method as shown in the middle part of Fig. 3B.41) and is then pressed against the cylindrical electrolytic cell through a ring gasket.

In the second thin-layer chamber, the electrolyte was flowing over a porous membrane from the six connecting capillaries to the outlet capillary positioned in the center. While passing through the second chamber, the gaseous species could evaporate via the porous membrane into the mass spectrometer and be detected online. There will be a delay of ca. 1 s between the Faradaic current signal and the mass spectrometric response due to the electrolyte flow time to the second chamber.

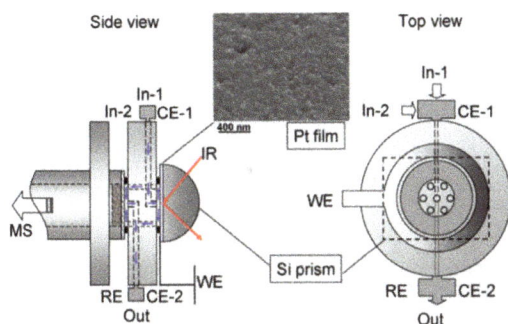

Fig. 3B.41: Schematic diagram of the thin-layer flow cell for simultaneous online DEMS and in situ ATR-FTIRS measurements and an SEM image (center) of the P thin film deposited on the Si prism. In-1 and In-2: two inlet ports for electrolyte; CE-1 and CE-2: ports for the connecting two counter electrodes; WE and RE: ports for connecting to the working and reference electrodes; Out: electrolyte outlet port [97].

Heinen et al. explored the mechanisms and kinetics of CO oxidation on a Pt film electrode with the setup of the combination of DEMS and ATR-FTIRS [97]. For chemical CO oxidation under open-circuit conditions, the combined IR–DEMS data show that (i) the rapid change in the open-circuit potential is indeed due to CO_2 formation, (ii) the CO_2 formation rate is initially slow, increases with PtO removal, and passes through a maximum at about 0.5 mL PtO coverage, and (iii) the buildup of a CO adlayer starts only after much of the PtO has been reductively removed at an oxide coverage of about 0.5 $\theta_{PtO, sat}$ ($\theta_{PtO, sat}$: PtO saturation coverage), as shown in Fig. 3B.42. The CO_{ad} (preabsorbed CO) surface diffusion and reaction are sufficiently fast to prevent the buildup of a CO adlayer in the earlier stages of the reaction, at oxide coverage >0.5 $\theta_{PtO, sat}$ and potential >0.8 V. When CO bulk oxidation underenforces mass transport conditions, the onset of CO_{ad} oxidation and the potential range of constant CO_{ad} coverage can be precisely defined. Over this potential range, the frequency of linearly adsorbed CO shows a Stark-tuning slope of 29 cm^{-1} V^{-1}, in good agreement with theoretical predictions. Furthermore, the absence of any detectable CO_{ad} bands in the PtO region suggests a negligible CO_{ad} steady-state coverage under these conditions due to fast mass transport-limited CO oxidation.

In the later research, Heinen and Behm et al. demonstrated the CO adsorption kinetics and adlayer buildup in the same setup [124]. They found that the intensity of linearly (CO_L) and multiply bonded (CO_M) CO_{ad}, at a constant CO_{ad} coverage, depends strongly on the potential. The intensity of CO_L is linearly correlated with the CO_{ad} coverage in the coverage regime 0.2 < $\theta_{CO, rel}$ < 0.8. At higher coverages, depolarization effects caused by (static) adsorbate–adsorbate interactions and dynamic dipole–dipole coupling between neighboring CO_{ad} molecules lead to deviations from the linear relation. At low coverages, preferential adsorption in a multiply bonded configuration also causes deviations from the linear relation. The COM adsorption configuration for CO is

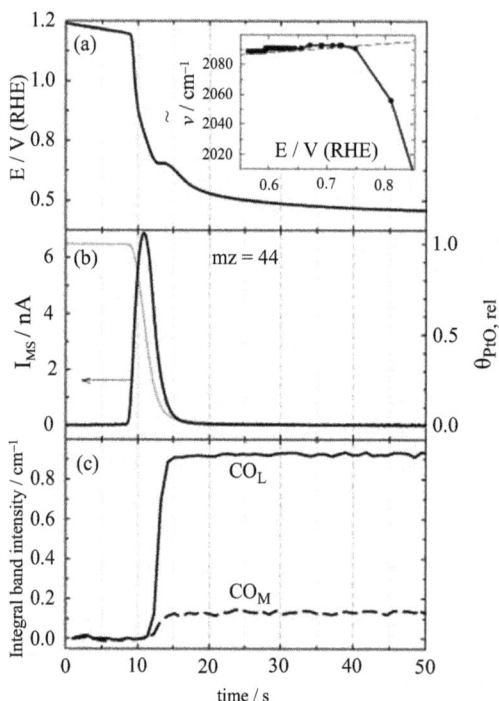

Fig. 3B.42: Simultaneously recorded chronopotentiometry (a) and chronoamperometric ($m/z = 44$ ion current), (b) transients and time-dependent IR band intensity profiles for COL and COM, (c) measured upon interaction of a preformed PtO layer (preformed at 1.2 V for 1 min) with CO-saturated solution (admission stars at ca. 8 s) under open-circuit conditions ($t = 0$ s corresponds to open-circuiting, reference potential: 1.2 V). Inset in (a): COL frequency as a function of the open-circuit potential (dashed line indicates a ~29 $cm^{-1} V^{-1}$ Stark-tuning slope) [97].

preferred at low potentials and low coverages. With increasing coverage, the intensity of linearly adsorbed CO_L increases much stronger than the intensity for multiply adsorbed CO_M, which was attributed to the preferential adsorption of CO in linear configuration (at higher CO_{ad} coverage) and intensity transfer from the lower (CO_M) to the higher (CO_L) frequency component due to dynamic coupling. Significant new insight into the CO adsorption kinetics, the CO adlayer buildup, and the potential dependence of these features was gained in this study.

The examples given above are only results on platinum thin-film electrodes. The same research can also be carried out on electrodes with other compositions and structures (such as PtRu electrodes) [125] and other experimental conditions (reaction temperature, potential, electrolyte flow rate, etc.) [126, 127], or it can be used to study the electro-oxidation mechanism and kinetics of other more complex organic small molecules [128–130]. Although for this more complex reaction system, the simple use of electrochemical infrared and MS technology cannot give all the information about the reaction mechanism and kinetic behavior, it can help deter-mine the key factors that affect the reaction kinetics. This information plays a sig-nificant role in guiding the rational design of efficient anode electrocatalysts.

3B.5.5 Summary and outlook

This section systematically introduces the working principle of DEMS, the com-bined technology of in situ FTIRS and DEMS, and the applications of these technol-ogies in electrocatalysis, electrochemical energy conversion, enrichment of trace organic matter in solution, and the CO_2 carbon cycle.

DEMS has a high sensitivity to the volatile substances in the detected solution. There are no harsh requirements for the electrolytic cell design and the structure and composition of the electrode. This technology can quantify the species gener-ated or desorbed during the reaction through reasonable and appropriate calibra-tion. It is the preferred technology for online quantitative analysis of reaction products and by-products in the electrochemical field. Compared with various chro-matographic or fluorescence techniques, it has incomparable advantages in mea-suring species' instantaneous production and consumption rate. In addition to the current important applications in fuel cell electrocatalysis, lithium batteries and the CO_2 carbon cycle will greatly contribute to photoelectric catalysis, electro-organic synthesis, and bioelectrochemistry. DEMS technology will be a promising experi-mental detection method for electrochemical researchers, whether in a basic theo-retical research or practical application development.

After proper calibration, electrochemical in situ ATR-FTIRS and DEMS are ap-plied to study the electrode reaction at the interface and provide information about the reaction products absorbed on the electrode surface and the intermediates' chemical nature, the adsorption configuration, and the formation or consumption rate. The real-time combination of these techniques has formed a powerful tool for characterizing the mechanism and dynamics of electrode processes. As the energy and environmental issues become increasingly prominent, these technologies will play an important role in electrochemical energy conversion.

References

[1] Bard, A. J. & Faulkner, L. R. (2001). Fundamentals and applications. Electrochemical Methods, 2(482), 580–632.

[2] Frumkin, A., Petrii, O. & Damaskin, B. (1980). Potentials of zero charge. In: Comprehensive treatise of electrochemistry (pp. 221–289), Springer, Boston, MA, USA.

[3] Frumkin, A. N., et al., *Kinetics of electrode processes.* 1967, FOREIGN TECHNOLOGY DIV WRIGHT-PATTERSON AFB OH.

[4]   Clavilier, J., Armand, D. & Wu, B. L. (1982). Electrochemical study of the initial surface condition of platinum surfaces with (100) and (111) orientations. Journal of Electroanalytical Chemistry and Interfacial Electrochemistry, 135(1), 159–166.

[5]   Heineman, W. R. & Jensen, W. B., Spectroelectrochemistry using transparent electrodes: an anecdotal history of the early years. 1989.

[6]   Maki, A. H. & Geske, D. H. (1959). Detection of electrolytically generated transient free radicals by electron spin resonance. The Journal of Chemical Physics, 30(5), 1356–1357.

[7]   Kuwana, T., Darlington, R. & Leedy, D. (1964). Electrochemical studies using conducting glass indicator electrodes. Analytical Chemistry, 36(10), 2023–2025.

[8]   Mark, H. & Pons, B. S. (1966). An in situ spectrophotometric method for observing the infrared spectra of species at the electrode surface during electrolysis. Analytical Chemistry, 38(1), 119–121.

[9]   Bewick, A. et al., (1984). Electrochemically modulated infrared spectroscopy (EMIRS): experimental details. Journal of Electroanalytical Chemistry and Interfacial Electrochemistry, 160(1–2), 47–61.

[10] Koch, D. & Scaife, D. (1966). Measurement of the reflectance of an electrode surface. Journal of the Electrochemical Society, 113(3), 302.

[11] Fleischmann, M., Hendra, P. J. & McQuillan, A. J. (1974). Raman spectra of pyridine adsorbed at a silver electrode. Chemical Physics Letters, 26(2), 163–166.

[12] Reddy, A., Devanathan, M. & Bockris, J. M. (1959). Chronoellipsometry: a new technique for studying anodic processes of the dissolution-precipitation type. Journal of Electroanalytical Chemistry, 1963, 6(1), 61–67.

[13] Feinleib, J. (1966). Electroreflectance in metals. Physical Review Letters, 16(26), 1200.

[14] Yildiz, A., Kissinger, P. T. & Reilley, C. N. (1968). Evaluation of an improved thin-layer electrode. Analytical Chemistry, 40(7), 1018–1024.

[15] O'Grady, W. & Bockris, J. M. (1970). Observing thin films of iron oxides with Mossbauer spectroscopy. Chemical Physics Letters, 5(2), 116.

[16] Bruckenstein, S. & Comeau, J. (1973). Electrochemical mass spectrometry. Part 1. – Preliminary studies of propane oxidation on platinum. Faraday Discussions of the Chemical Society, 56, 285–292.

[17] Fleischmann, M., Hendra, P. J. & McQuillan, A. J. (1973). Raman spectra from electrode surfaces. Journal of the Chemical Society. Chemical Communications, 1(3), 80–81.

[18] Richards, J. A. & Evans, D. H. (1975). Flow cell for electrolysis within a nuclear magnetic resonance spectrometer probe. Analytical Chemistry, 47(6), 964–966.

[19] Bewick, A., Mellor, J. & Pons, B. (1980). Distinction between ECE and disproportionation mechanisms in the anodic oxidation of methyl benzenes using spectroelectrochemical methods. Electrochimica Acta, 25(7), 931–941.

[20] Fleischmann, M. et al., (1983). Raman spectroscopic and x-ray diffraction studies of electrode-solution interfaces. Journal of Electroanalytical Chemistry and Interfacial Electrochemistry, 150(1–2), 33–42.

[21] Long, G. et al., (1983). Structure of passive films on iron using a new surface-EXAFS technique. Journal of Electroanalytical Chemistry and Interfacial Electrochemistry, 150(1–2), 603–610.

[22] Kötz, R. et al., (1986). In situ Rutherford backscattering spectroscopy for electrochemical interphase analysis. Electrochimica Acta, 31(2), 169–172.

[23] Wiechers, J. et al., (1988). An in-situ scanning tunneling microscopy study of Au (111) with atomic-scale resolution. Journal of Electroanalytical Chemistry and Interfacial Electrochemistry, 248(2), 451–460.

[24]   Itaya, K. & Tomita, E. (1988). Scanning tunneling microscope for an electrochemistry-a new concept for the in situ scanning tunneling microscope in electrolyte solutions. Surface Science, 201(3), L507–L512.

[25]   Lev, O., Fan, F. R. & Bard, A. J. (1988). The application of scanning tunneling microscopy to in situ studies of nickel electrodes under potential control. Journal of the Electrochemical Society, 135, 783–784.

[26]   Lustenberger, P. et al., (1988). Scanning tunneling microscopy at potential controlled electrode surfaces in the electrolytic environment. Journal of Electroanalytical Chemistry and Interfacial Electrochemistry, 243(1), 225–235.

[27]   Guyot-Sionnest, P. & Tadjeddine, A. (1990). Spectroscopic investigations of adsorbates at the metal – electrolyte interface using sum-frequency generation. Chemical Physics Letters, 172(5), 341–345.

[28]   Liu, H. Y. et al., (1986). Scanning electrochemical and tunneling ultramicroelectrode microscope for high-resolution examination of electrode surfaces in solution. Journal of the American Chemical Society, 108(13), 3838–3839.

[29]   Sonnenfeld, R. & Hansma, P. K. (1986). atomic-resolution microscopy in water. Science, 232(4747), 211–213.

[30]   Chevallier, F. et al., (2003). In situ 7Li-nuclear magnetic resonance observation of reversible lithium insertion into disordered carbons. Electrochemical and Solid-State Letters, 6(11), A225.

[31]   Anderson, M. S. (2000). Locally enhanced Raman spectroscopy with an atomic force microscope. Applied Physics Letters, 76(21), 3130–3132.

[32]   Hayazawa, N. et al., (2000). Metallized tip amplification of near-field Raman scattering. Optics Communications, 183(1–4), 333–336.

[33]   Pettinger, B. et al., (2000). Surface-enhanced Raman spectroscopy: towards single-molecule spectroscopy. Electrochemistry, 68(12), 942–949.

[34]   Stöckle, R. M. et al., (2000). Nanoscale chemical analysis by tip-enhanced Raman spectroscopy. Chemical Physics Letters, 318(1–3), 131–136.

[35]   Zeng, Z.-C. et al., (2015). Electrochemical tip-enhanced Raman spectroscopy. Journal of the American Chemical Society, 137(37), 11928–11931.

[36]   Elias, L. R. et al., (1976). Observation of stimulated emission of radiation by relativistic electrons in a spatially periodic transverse magnetic field. Physical Review Letters, 36(13), 717.

[37]   Wu, D.-Y. et al., (2008). Electrochemical surface-enhanced Raman spectroscopy of nanostructures. Chemical Society Reviews, 37(5), 1025–1041.

[38]   Smith, E. & Dent, G. (2019). Modern Raman spectroscopy: a practical approach, pp. 1–233. John Wiley & Sons, Hoboken, NJ, USA.

[39]   Jeanmaire, D. L. & Van Duyne, R. P. (1977). Surface Raman spectroelectrochemistry: part I. Heterocyclic, aromatic, and aliphatic amines adsorbed on the anodized silver electrode. Journal of Electroanalytical Chemistry and Interfacial Electrochemistry, 84(1), 1–20.

[40]   Jiang, Y.-X. (2007). Surface water characterization on Au core Pt-group metal shell nanoparticles coated electrodes by surface-enhanced Raman spectroscopy. Chemical Communications, 1(44), 4608–4610.

[41]   Zou, S. & Weaver, M. J. (1998). Surface-enhanced Raman scattering on uniform transition-metal films: toward a versatile adsorbate vibrational strategy for solid-nonvacuum interfaces?. Analytical Chemistry, 70(11), 2387–2395.

[42]   Endo, M. et al., (1998). In situ Raman study of PPP-based disordered carbon as an anode in a Li-ion battery. Synthetic Metals, 98(1), 17–24.

[43]   Dokko, K. et al., (2002). In situ Raman spectroscopic studies of LiNi x Mn 2– x O 4 thin film cathode materials for lithium-ion secondary batteries. Journal of Materials Chemistry, 12(12), 3688–3693.

[44] Stancovski, V. & Badilescu, S. (2014). In situ Raman spectroscopic–electrochemical studies of lithium-ion battery materials: a historical overview. Journal of Applied Electrochemistry, 44(1), 23–43.

[45] Chen, -J.-J. et al., (2015). Conductive Lewis base matrix to recover the missing link of Li2S8 during the sulfur redox cycle in Li–S battery. Chemistry of Materials, 27(6), 2048–2055.

[46] Li, J.-F. et al., (2015). Electrochemical shell-isolated nanoparticle-enhanced Raman spectroscopy: correlating structural information and adsorption processes of pyridine at the Au(hkl) single crystal/solution interface. Journal of the American Chemical Society, 137(6), 2400–2408.

[47] Wessel, J. (1985). Surface-enhanced optical microscopy. JOSA B, 2(9), 1538–1541.

[48] Griffiths, P. R. (2006). Introduction to the theory and instrumentation for vibrational spectroscopy. Applications of Vibrational Spectroscopy in Food Science, Part One: Introduction and Basic Concepts, 1, 31–46.

[49] Suetaka, W., Suëtaka, W. & Yates, J. T. Jr (1995). Surface infrared and Raman spectroscopy: methods and applications (Vol. 3, pp. 1–284, Editors: W. Suetaka, and J.T. Yates, Jr.), Springer Science & Business Media.

[50] Hoffmann, F. M. (1983). Infrared reflection-absorption spectroscopy of adsorbed molecules. Surface Science Reports, 3(2–3), 107–192.

[51] Ataka, K.-I., Yotsuyanagi, T. & Osawa, M. (1996). Potential-dependent reorientation of water molecules at an electrode/electrolyte interface studied by surface-enhanced infrared absorption spectroscopy. The Journal of Physical Chemistry, 100(25), 10664–10672.

[52] Wang, J.-Y. et al., (2011). From HCOOH to CO at Pd electrodes: a surface-enhanced infrared spectroscopy study. Journal of the American Chemical Society, 133(38), 14876–14879.

[53] Osawa, M. et al., (2011). The role of bridge-bonded adsorbed formate in the electrocatalytic oxidation of formic acid on platinum. Angewandte Chemie International Edition, 50(5), 1159–1163.

[54] Joo, J. et al., (2013). Importance of acid-base equilibrium in the electrocatalytic oxidation of formic acid on platinum. Journal of the American Chemical Society, 135(27), 9991–9994.

[55] Diao, Y.-X. et al., (2006). Adsorbed structures of 4, 4'-bipyridine on Cu (111) in acid studied by STM and IR. Langmuir, 22(8), 3640–3646.

[56] Gnanaprakash, G. et al., (2007). Effect of digestion time and alkali addition rate on physical properties of magnetite nanoparticles. The Journal of Physical Chemistry. B, 111(28), 7978–7986.

[57] Jiang, X., Ataka, K. & Heberle, J. (2008). Influence of the molecular structure of carboxyl-terminated self-assembled monolayer on the electron transfer of cytochrome c adsorbed on an Au electrode: in situ observation by surface-enhanced infrared absorption spectroscopy. The Journal of Physical Chemistry C, 112(3), 813–819.

[58] Hartstein, A., Kirtley, J. & Tsang, J. (1980). Enhancement of the infrared absorption from molecular monolayers with thin metal overlayers. Physical Review Letters, 45(3), 201.

[59] Osawa, M. & Ikeda, M. (1991). Surface-enhanced infrared absorption of p-nitrobenzoic acid deposited on silver island films: contributions of electromagnetic and chemical mechanisms. The Journal of Physical Chemistry, 95(24), 9914–9919.

[60] Merklin, G. T. & Griffiths, P. R. (1997). Influence of chemical interactions on the surface-enhanced infrared absorption spectrometry of nitrophenols on copper and silver films. Langmuir, 13(23), 6159–6163.

[61] Otto, A. et al., (1992). Surface-enhanced Raman scattering. Journal of Physics: Condensed Matter, 4(5), 1143.

[62] Osawa, M. (1997). Dynamic processes in electrochemical reactions studied by surface-enhanced infrared absorption spectroscopy (SEIRAS). Bulletin of the Chemical Society of Japan, 70(12), 2861–2880.

[63] Pons, S. et al., (1984). Infrared spectroelectrochemical study of adsorbed tetracyanoethylene anions at a platinum electrode. The Journal of Physical Chemistry, 88(16), 3575–3578.

[64] Wadayama, T. et al., (1988). Charge-transfer enhancement in infrared absorption of thiocyanate ions adsorbed on a gold electrode in the Kretschmann ATR configuration. Surface Science, 198(3), L359–L364.

[65] Nishikawa, Y., Fujiwara, K. & Shima, T. (1991). A study on the qualitative and quantitative analysis of nanogram samples by infrared transmission spectroscopy with the use of silver island films. Applied Spectroscopy, 45(5), 747–751.

[66] Wan, L.-J. et al., (2000). Molecular orientation and ordered structure of benzenethiol adsorbed on gold(111). The Journal of Physical Chemistry. B, 104(15), 3563–3569.

[67] Greenler, R. G. (1966). Infrared study of adsorbed molecules on metal surfaces by reflection techniques. The Journal of Chemical Physics, 44(1), 310–315.

[68] Wandlowski, T., Ataka, K. & Mayer, D. (2002). In situ infrared study of 4, 4'-bipyridine adsorption on thin gold films. Langmuir, 18(11), 4331–4341.

[69] Xue, X.-K. et al., (2008). Practically modified attenuated total reflection surface-enhanced IR absorption spectroscopy for high-quality frequency-extended detection of surface species at electrodes. Analytical Chemistry, 80(1), 166–171.

[70] Bandy, B., Lloyd, D. R. & Richardson, N. V. (1979). Selection rules in photoemission from adsorbates: pyridine adsorbed on copper. Surface Science, 89(1–3), 344–353.

[71] DiNardo, N., Avouris, P. & Demuth, J. (1984). Chemisorbed pyridine on Ni (001): a high-resolution electron energy loss study of vibrational and electronic excitations. The Journal of Chemical Physics, 81(4), 2169–2180.

[72] Johnson, A. L. et al., (1985). Chemisorption geometry of pyridine on platinum (111) by NEXAFS. The Journal of Physical Chemistry, 89(19), 4071–4075.

[73] Ikezawa, Y. et al., (1998). In situ FTIR study of pyridine adsorbed on a polycrystalline gold electrode. Electrochimica Acta, 43(21–22), 3297–3301.

[74] Cai, W.-B. et al., (1998). Orientational phase transition in a pyridine adlayer on gold (111) in aqueous solution studied by in situ infrared spectroscopy and scanning tunneling microscopy. Langmuir, 14(24), 6992–6998.

[75] Bridge, M. E. et al., (1987). Electron spectroscopic studies of pyridine on metal surfaces. Spectrochimica Acta. Part A: Molecular Spectroscopy, 43(12), 1473–1478.

[76] Huo, S.-J. et al., (2006). Seeded-Growth Approach to Fabrication of Silver Nanoparticle Films on Silicon for Electrochemical ATR Surface-Enhanced IR Absorption Spectroscopy. The Journal of Physical Chemistry. B, 110(51), 25721–25728.

[77] Li, Q.-X. et al., (2007). Application of Surface-Enhanced Infrared Absorption Spectroscopy to Investigate Pyridine Adsorption on Platinum-Group Electrodes. Applied Spectroscopy, 61(12), 1328–1333.

[78] Liu, H., Liu, Y. & Li, J. (2010). Ionic liquids in surface electrochemistry. Physical Chemistry Chemical Physics, 12(8), 1685–1697.

[79] Dieluweit, S. & Giesen, M. (2002). Determination of step and kink energies on Au (100) electrodes in sulfuric acid solutions by island studies with electrochemical STM. Journal of Electroanalytical Chemistry, 524, 194–200.

[80] Albrecht, T. et al., (2006). Scanning tunneling spectroscopy in an ionic liquid. Journal of the American Chemical Society, 128(20), 6574–6575.

[81] Kolb, D. M. (1996). Reconstruction phenomena at metal-electrolyte interfaces. Progress in Surface Science, 51(2), 109–173.

[82]   Kolb, D. M., Ullmann, R. & Will, T. (1997). Nanofabrication of Small Copper Clusters on Gold (111) Electrodes by a Scanning Tunneling Microscope. Science, 275(5303), 1097–1099.

[83]   Engelmann, G., Ziegler, J. & Kolb, D. (1998). Nanofabrication of small palladium clusters on Au (111) electrodes with a scanning tunneling microscope. Journal of the Electrochemical Society, 145(3), L33.

[84]   Kolb, D., Engelmann, G. & Ziegler, J. (2000). Nanoscale decoration of electrode surfaces with an STM. Solid State Ionics, 131(1–2), 69–78.

[85]   Zhang, Y., Maupai, S. & Schmuki, P. (2004). EC-STM tip induced Cd nanostructures on Au (1 1 1). Surface Science, 551(1–2), L33–L39.

[86]   Hugelmann, M. & Schindler, W. (2004). In situ distance tunneling spectroscopy at Au (111)/ 0.02 M HClO4: from faradaic regime to quantized conductance channels. Journal of the Electrochemical Society, 151(3), E97.

[87]   Simeone, F. C. et al., (2007). The Au(111)/electrolyte interface: a tunnel-spectroscopic and DFT investigation. Angewandte Chemie International Edition, 46(46), 8903–8906.

[88]   Buzzeo, M. C., Evans, R. G. & Compton, R. G. (2004). Non-haloaluminate room-temperature ionic liquids in electrochemistry-a review. ChemPhysChem, 5(8), 1106–1120.

[89]   Pan, G.-B. & Freyland, W. (2006). 2D phase transition of PF6 adlayers at the electrified ionic liquid/Au (1 1 1) interface. Chemical Physics Letters, 427(1–3), 96–100.

[90]   Su, Y. Z. et al., (2009). Double layer of Au (100)/ionic liquid interface and its stability in imidazolium-based ionic liquids. Angewandte Chemie International Edition, 48(28), 5148–5151.

[91]   Fu, Y.-C. (2007). In situ STM studies on the underpotential deposition of antimony on Au (111) and Au (100) in a BMIBF4 ionic liquid. The Journal of Physical Chemistry C, 111(28), 10467–10477.

[92]   Wei, Y. M. et al., (2008). The Creation of Nanostructures on an Au (111) Electrode by Tip-Induced Iron Deposition from an Ionic Liquid. Small, 4(9), 1355–1358.

[93]   Xie, X. et al., (2013). Measurement of the quantum conductance of germanium by an electrochemical scanning tunneling microscope break junction based on a jump-to-contact mechanism. Chemistry–An Asian Journal, 8(10), 2401–2406.

[94]   Bruckenstein, S. & Gadde, R. R. (1971). Use of a porous electrode for in situ mass spectrometric determination of volatile electrode reaction products. Journal of the American Chemical Society, 93(3), 793–794.

[95]   Wolter, O. & Heitbaum, J. (1984). Differential electrochemical mass spectroscopy (DEMS) – a new method for the study of electrode processes. Berichte der Bunsengesellschaft Für Physikalische Chemie, 88(1), 2–6.

[96]   Baltruschat, H. (2004). Differential electrochemical mass spectrometry. Journal of the American Society for Mass Spectrometry, 15(12), 1693–1706.

[97]   Heinen, M. et al., (2007). In situ ATR-FTIRS coupled with on-line DEMS under controlled mass transport conditions – A novel tool for electrocatalytic reaction studies. Electrochimica Acta, 52(18), 5634–5643.

[98]   Cremers, C. & Bayer, D. (2012). Differential electrochemical mass spectrometry (DEMS) technique for direct alcohol fuel cell characterization. In: Polymer electrolyte membrane and direct methanol fuel cell technology, Hartnig, C. & Roth, C. eds., (pp. 65–86), Woodhead Publishing, Philadelphia, PA, USA.

[99]   Seiler, T. et al., (2004). Poisoning of PtRu/C catalysts in the anode of a direct methanol fuel cell: a DEMS study. Electrochimica Acta, 49(22), 3927–3936.

[100] Gao, Y. et al., (1994). New online mass spectrometer system designed for platinum-single crystal electrode and electroreduction of acetylene. Journal of Electroanalytical Chemistry, 372(1–2), 195–200.

[101] Housmans, T. H., Wonders, A. H. & Koper, M. T. (2006). Structure sensitivity of methanol electrooxidation pathways on platinum: an on-line electrochemical mass spectrometry study. The Journal of Physical Chemistry. B, 110(20), 10021–10031.
[102] Novák, P. et al., (2005). Advanced in situ characterization methods applied to carbonaceous materials. Journal of Power Sources, 146(1), 15–20.
[103] Brown, T. A., Chen, H. & Zare, R. N. (2015). Identification of fleeting electrochemical reaction intermediates using desorption electrospray ionization mass spectrometry. Journal of the American Chemical Society, 137(23), 7274–7277.
[104] Iwasita, T. (2002). Electrocatalysis of methanol oxidation. Electrochimica Acta, 47(22–23), 3663–3674.
[105] Lee, J. K., et al., *Microdroplet fusion mass spectrometry for fast reaction kinetics*. Proceedings of the National Academy of Sciences, 2015. **112**(13): p. 3898–3903.
[106] Amin, H. M. et al., (2015). A highly efficient bifunctional catalyst for alkaline air-electrodes based on an Ag and Co3O4 hybrid: RRDE and online DEMS insights. Electrochimica Acta, 151, 332–339.
[107] Jusys, Z. & Behm, R. (2004). Simultaneous oxygen reduction and methanol oxidation on a carbon-supported Pt catalyst and mixed potential formation-revisited. Electrochimica Acta, 49(22–23), 3891–3900.
[108] Gómez-Marín, A. et al., (2012). Interaction of hydrogen peroxide with a Pt (111) electrode. Electrochemistry Communications, 22, 153–156.
[109] Antolini, E. & Gonzalez, E. R. (2010). Tungsten-based materials for fuel cell applications. Applied Catalysis. B, Environmental, 96(3–4), 245–266.
[110] Wang, H. et al., (2014). CO2 and O2 evolution at high voltage cathode materials of Li-ion batteries: a differential electrochemical mass spectrometry study. Analytical Chemistry, 86(13), 6197–6201.
[111] Girishkumar, G. et al., (2010). Lithium-air battery: promise and challenges. The Journal of Physical Chemistry Letters, 1(14), 2193–2203.
[112] Novák, P. et al., (2005). Advanced in situ characterization methods applied to carbonaceous materials. Journal of Power Sources, 146(1–2), 15–20.
[113] Löffler, T. et al., (2003). Adsorption and desorption reactions of bicyclic aromatic compounds at polycrystalline and Pt (111) studied by DEMS. Journal of Electroanalytical Chemistry, 550, 81–92.
[114] Brisard, G. et al., (2001). On-line mass spectrometry investigation of the reduction of carbon dioxide in acidic media on polycrystalline Pt. Electrochemistry Communications, 3(11), 603–607.
[115] Plana, D. et al., (2013). Tuning CO2 electroreduction efficiency at Pd shells on Au nanocores. Chemical Communications, 49(93), 10962–10964.
[116] Wang, H., Löffler, T. & Baltruschat, H. (2001). Formation of intermediates during methanol oxidation: a quantitative DEMS study. Journal of Applied Electrochemistry, 31(7), 759–765.
[117] Zhong, J.-H. et al., (2014). Quantitative correlation between defect density and heterogeneous electron transfer rate of single-layer graphene. Journal of the American Chemical Society, 136(47), 16609–16617.
[118] Novák, P. et al., (2000). Advanced in situ methods for the characterization of practical electrodes in lithium-ion batteries. Journal of Power Sources, 90(1), 52–58.
[119] La Mantia, F. & Novák, P. (2008). Online detection of reductive CO2 development at graphite electrodes in the 1 M LiPF6, EC: DMC battery electrolyte. Electrochemical and Solid-State Letters, 11(5), A84.
[120] McCloskey, B. D. et al., (2011). Solvents' critical role in nonaqueous lithium-oxygen battery electrochemistry. The Journal of Physical Chemistry Letters, 2(10), 1161–1166.

[121] Gattrell, M., Gupta, N. & Co, A. (2006). A review of the aqueous electrochemical reduction of CO2 to hydrocarbons at copper. Journal of Electroanalytical Chemistry, 594(1), 1–19.

[122] Dubé, P. & Brisard, G. (2005). Influence of adsorption processes on the CO2 electroreduction: an electrochemical mass spectrometry study. Journal of Electroanalytical Chemistry, 582(1–2), 230–240.

[123] Abd-El-Latif, A. et al., (2010). Electrooxidation of ethanol at polycrystalline and platinum stepped single crystals: a study by differential electrochemical mass spectrometry. Electrochimica Acta, 55(27), 7951–7960.

[124] Heinen, M. et al., (2007). CO adsorption kinetics and adlayer build-up studied by combined ATR-FTIR spectroscopy and online DEMS under continuous flow conditions. Electrochimica Acta, 53(3), 1279–1289.

[125] Tao, Q. et al., (2014). Study on methanol oxidation at Pt and PtRu electrodes by combining in situ infrared spectroscopy and differential electrochemical mass spectrometry. Chinese Journal of Chemical Physics, 27(5), 541.

[126] Heinen, M., Jusys, Z. & Behm, R. J., Reaction pathway analysis and reaction intermediate detection via simultaneous differential electrochemical mass spectrometry (DEMS) and attenuated total reflection Fourier transform infrared spectroscopy (ATR-FTIRS), in Handbook of Fuel Cells. 2007.

[127] Schnaidt, J. et al., (2013). Oxidation of the partly oxidized ethylene glycol oxidation products glycolaldehyde, glyoxal, glycolic acid, glyoxylic acid, and oxalic acid on Pt electrodes: a combined ATR-FTIRS and DEMS spectroelectrochemical study. The Journal of Physical Chemistry C, 117(24), 12689–12701.

[128] Liao, L. W. et al., (2011). A method for kinetic study of methanol oxidation at Pt electrodes by electrochemical in situ infrared spectroscopy. Journal of Electroanalytical Chemistry, 650(2), 233–240.

[129] Schnaidt, J. et al., (2012). Electro-oxidation of ethylene glycol on a Pt-Film electrode studied by combined in situ infrared spectroscopy and online mass spectrometry. The Journal of Physical Chemistry C, 116(4), 2872–2883.

[130] Belén Molina Concha, M. et al., (2013). In situ Fourier transform infrared spectroscopy and on-line differential electrochemical mass spectrometry study of the NH3BH3 oxidation reaction on gold electrodes. Electrochimica Acta, 89, 607–615.

Sohaib Mohammed, Hassnain Asgar, Greeshma Gadikota
Chapter 3C
Integrated X-ray scattering and molecular-scale simulation approaches to probe the behavior of confined fluids for a sustainable energy future

Abstract: Anomalous fluid flow, thermodynamics, and reactivity of confined fluids compared to bulk fluids have significant implications for storing and recovering fluids in the subsurface and engineered materials for sustainable energy and the environment. In this context, the fate of CO_2 in nanoporous environments for subsurface CO_2 storage, stability of ice hydrates and carbon-bearing hydrates in response to a changing climate, and storage of excess renewable methane in porous materials are closely linked to their organization in confinement which is a function of the fluid–solid interactions. Developing quantifiable insights on the structure of confined fluids is now possible due to advances in advanced X-ray and neutron scattering measurements, which can be used to validate molecular-scale predictions. Furthermore, *operando* investigations enable us to resolve temporal and spatial effects associated with the organization of confined fluids. This chapter introduces recent advances in harnessing scattering measurements and molecular-scale simulations to probe the organization of confined fluids in porous materials. The basic principles of X-ray and neutron scattering measurements and molecular-scale simulations are introduced, and case studies that harness these approaches are discussed. This chapter sheds fundamental insights into experimental and simulation approaches to resolving mesoscale and molecular-scale phenomena associated with the organization of confined fluids.

Acknowledgment: This work was supported as part of the Multi-scale Fluid-Solid Interactions in Architected and Natural Materials (MUSE), an Energy Frontier Research Center funded by the US Department of Energy, Office of Science, Basic Energy Sciences under Award no. DE-SC0019285.

Note: All the authors contributed equally to this chapter.

Sohaib Mohammed, School of Civil and Environmental Engineering, Cornell University, Ithaca, NY 14850, USA, e-mail: sm2793@cornell.edu
Hassnain Asgar, School of Civil and Environmental Engineering, Cornell University, Ithaca, NY 14850, USA, e-mail: ha356@cornell.edu
Greeshma Gadikota, School of Civil and Environmental Engineering, Cornell University, Ithaca, NY 14850, USA; Smith School of Chemical and Biological Engineering, Cornell University, Ithaca, NY 14853, USA, e-mail: gg464@cornell.edu

https://doi.org/10.1515/9783110739879-009

Keywords: confined fluids, X-ray scattering, neutron scattering, MD simulations, nanopores

3C.1 Introduction

The organization of fluids confined in nanoporous environments and the underlying fluid–solid interactions is the fundamental basis for designing energy-efficient separation processes [1, 2], heterogeneous catalysis [3], tuning oil and gas interactions in subsurface environments, the presence of CO_2 for sustainable recovery and storage [4, 5], and engineering gas separation and storage behaviors grounded in molecular-scale transport [6, 7]. However, characterizing the organization of confined fluids has been limited by experimental approaches and our ability to develop experimentally validated molecular-scale models [8, 9] (see Fig. 3C.1). Resolving these knowledge gaps is crucial for revealing the structure, phase transitions, dynamics, reactivity, and rheology of fluids in nanoporous environments.

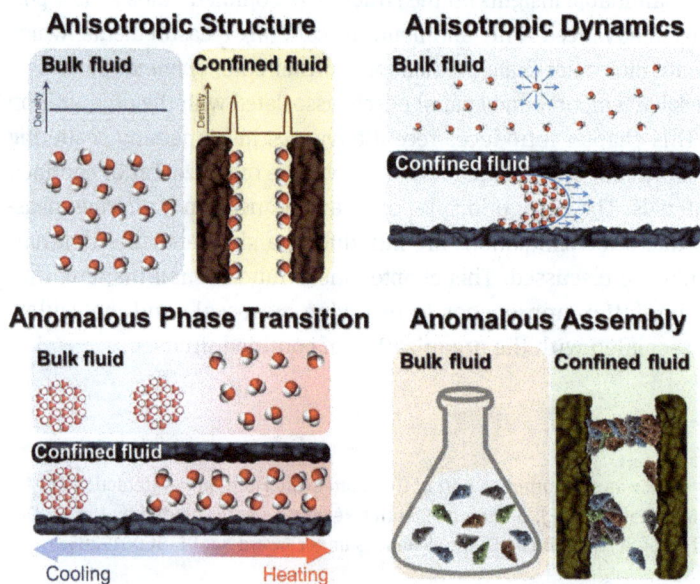

Anisotropic Structure

Bulk fluid   Confined fluid

Anisotropic Dynamics

Bulk fluid

Confined fluid

Anomalous Phase Transition

Bulk fluid

Confined fluid

Cooling   Heating

Anomalous Assembly

Bulk fluid   Confined fluid

Fig. 3C.1: Schematic representation of the structure, dynamics, phase transitions, and self-assembly of confined fluids compared to bulk fluids.

Fundamental insights into the dynamics and organization of confined fluids have been uncovered using X-ray scattering [10–12], neutron scattering [13–16], adsorption isotherms [17–19], and differential scanning calorimetry [20, 21]. Linking the organization of confined fluids to the pore geometry and physicochemical interactions with the solid interface to observed properties such as reactivity and thermodynamics (e.g., melting point and boiling point) provides the rational basis for tuning interactions to achieve the desired properties.

In this context, the ability to delineate the structure of confined fluids such as gases, liquids, and supercritical fluids under various thermal and mechanical conditions using small-angle neutron scattering (SANS) and small/wide-angle X-ray scattering (SAXS/WAXS) measurements enables us to develop unprecedented insights. SANS measurements have been used successfully to investigate the structure of fluids such as CO_2, CH_4, H_2, and water in siliceous, carbonaceous, and architected porous materials that are highly relevant for energy and environmental applications. For example, SANS measurements have been used to resolve the structure of CO_2 and CH_4 in MCM-41 and SBA-14 porous silica [22–25], supercritical CO_2 in silica aerogels [26, 27], H_2 storage in carbonaceous materials [28], water [11], electrolytes [29], and ionic liquids [30] in porous silica and carbon-based materials. Insights into the "core–shell" organization of confined fluids emerging from fluid adsorption on solid interfaces (which constitutes the shell) are delineated from SANS measurements.

Similarly, X-ray scattering measurements have also been used to explore the structural properties of liquids in confinement over spatial scales that range from few angstroms (10^{-10} m) to micrometers (10^{-6} m). SAXS and WAXS measurements have been used to explore the structural characteristics of water [31–33] and hydrocarbons [34, 35] in confinement. Rapid temporal evolution of the structures of confined fluids is possible through the fast measurements and high resolution of synchrotron SAXS/WAXS measurements. Advances in the scientific understanding of the phase transitions of confined fluids associated with the anomalous freezing of confined fluids and the crystallization of fluids and the resulting structures that differ from bulk fluids [5] have been made possible by SAXS and WAXS measurements.

Atomic structuring of fluid–fluid and fluid–solid interfaces is obtained from the distribution functions extracted from the X-ray and neutron scattering of confined fluids [36]. Various forms and normalizations of the partial distribution functions (PDF) have been adopted to describe the structural correlations of confined fluids, including the real-space radial distribution functions (RDFs). RDFs have been used to elucidate the structural correlations of hydrogen-bonded and nonhydrogen-bonded liquids confined in various porous materials, including water and methanol, in the silica-based porous materials [37].

In addition to probing the structural characteristics of "confined fluids," neutron scattering measurements have also been harnessed to investigate the dynamics of confined fluids using quasielastic neutron scattering (QENS) techniques.

The diffusivities of water, hydrocarbons, and gases confined in porous alumina [38], MCM-41 [39, 40], and a mixture of ethane–CO_2 in silica nanopores [41] are obtained from "QENS measurements." Evidence of the anisotropic diffusion of the confined fluids such that the interfacial molecules diffuse slower than the bulk molecules has been uncovered from QENS measurements.

Although significant advances in understanding the structure and dynamics of confined fluids in nanopores have been achieved by conducting X-ray and neutron scattering measurements, linking these advances to the insights obtained from molecular-scale simulations is far more powerful in elucidating the underlying energetic interactions of confined fluids. In this context, molecular dynamic (MD) simulations have been widely implemented to investigate the properties of gases [42, 43], water [44], hydrocarbons [45, 46], and their mixtures [47–49] in confinement. MD simulations can probe the fluid–solid interfaces from timescales ranging from picoseconds to microseconds and spatial scales ranging from angstroms to nanometers. The spatial scale of MD simulations is comparable to the micropores and mesopores in which confined fluids exhibit considerable differences in structures, phase transitions, and reactivity than counterpart bulk fluids.

Insights from MD simulations include the anisotropic structure [49–51], anisotropic dynamics [52], enhanced flow rates [53, 54], and depressed freezing and melting points in confinement [33]. Nonetheless, MD simulations are limited by the short timescale and dependence on the utilized force fields that define the intermolecular and intramolecular interactions of the involved atoms. To overcome the limitations of MD simulations, using experimental-computational approaches can elucidate the experimental observations and validate the force fields of MD simulations.

They integrate X-ray and neutron scattering experiments with MD simulations successfully to resolve the structure and dynamics of confined fluids. In this context, SAXS-MD, WAXS-MD, and SANS-MD have been used to explore the structure of confined fluids in nanoporous environments. In contrast, QENS-MD has been utilized to investigate the diffusion of confined fluids in a wide range of confined spaces. This chapter reports the recent advances in investigating the behavior of confined fluids using tandem experimental and computational approaches. The fundamentals of X-ray scattering and neutron scattering measurements and MD simulations used to study the properties of confined fluids, namely, SAXS, SANS, WAXS, QENS, and MD, are discussed. Case studies on harnessing the experimental and computational approaches discussed are described in detail in this chapter.

3C.2 Methods

3C.2.1 Experimental approaches to probe the organization of confined fluids

X-ray and neutron scattering measurements have been proposed and implemented to investigate the behavior of confined fluids in nanoscale porous environments owing to their advantages of being nondestructive and the ability to conduct in situ and *operando* investigations. X-rays are electromagnetic waves that interact with and are scattered by the electrons in the matter, while neutrons are scattered by the nuclei of atoms [55–57]. The X-ray scattering from each electron has the same scattering amplitude (b_x) that is proportional to the atomic number of the element (Z) such that

$$b_x = b_0 Z \tag{3C.1}$$

where b_0 is the Thompson scattering factor for one electron (0.282×10^{-12} cm). Similarly, for "neutron scattering," each atom has a characteristic scattering length (b) that determines the strength of neutron scattering. The neutron scattering length for heterogeneous materials is given as the sum of the scattering length (for both X-rays and neutrons) of its constituent atoms (b_i) as follows [55–58]:

$$b \ \text{ or } \ b_x = \sum_{i=1}^{n} b_i \tag{3C.2}$$

In this section, we discuss the elastically (small-angle scattering (SAS), wide-angle scattering (WAS), and PDF) and quasi-elastically (QENS) scattered X-rays or neutrons that have been adopted to understand the properties of "confined fluids."

3C.2.1.1 Small- and wide-angle scattering

Depending on the angle of scattered X-ray photons or neutrons, the scattering can be referred to as either SAS or WAS, with scattering at angles <10° or >10°, respectively [57]. The primary aspects of SAS and WAS experiments lie in determining the probability that an X-ray photon or neutron having wave vector k_i is scattered as wave vector k_f when incident on a sample. The magnitude of a wave vector can be given as its wave number as $|\mathbf{k}| = 2\pi/\lambda$, where λ is the wavelength. The difference between these wave vectors gives the intensity of the scattered radiation based on the momentum transfer $\mathbf{Q}$ as follows:

$$\hbar Q = \hbar(k_f - k_i) \tag{3C.3}$$

$\mathbf{Q}$ in eq. (3C.3) is called the scattering vector (Fig. 3C.2). During the elastic scattering process, $|\mathbf{k_i}| = |\mathbf{k_f}| = k$, thus the magnitude of scattering vector $\mathbf{Q}$, in terms of scattering angle "2θ," can be written as follows:

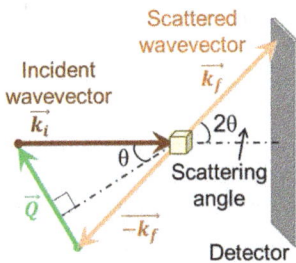

Fig. 3C.2: Schematic representation of scattering from an object. $\vec{Q}$ represents the scattering vector.

$$|Q| = Q = \frac{4\pi}{\lambda}\sin(\theta) \tag{3C.4}$$

Linking eq. (3C.4) to Bragg's law ($\lambda = 2d \sin \theta$) results in eq. (3C.4) as follows:

$$d = \frac{2\pi}{Q} \tag{3C.5}$$

where d is the distance between the porous materials' crystallographic planes, pores, or interlayers. This relation suggests that probing features at length scales of ~5–10 Å are resolved by scattering at larger Q values (or wider angles). In comparison, features at relatively larger length scales can be resolved by scattering at smaller Q values (or small angles) (see Fig. 3C.3).

The differential scattering cross section ($d\sigma/d\Omega$) in the basic quantity determined by SAS and WAS experiments is given by

$$\left(\frac{d\sigma}{d\Omega}\right) = \frac{I_{sc}}{\Phi_0 \, \Delta\Omega} \tag{3C.6}$$

In the expression above, Φ_0 is the number of photons or neutrons passing through a unit area per second, and I_{sc} is the number of scattered neutrons or photons recorded per second on the detector located at a distance M from the scatterer and subtended at the solid angle $\Delta\Omega$. The differential scattering cross sections are size-dependent. Therefore, normalization by the sample volume "V" is performed as follows:

$$\frac{d\Sigma}{d\Omega}(Q) = \frac{1}{V}\frac{d\sigma}{d\Omega}(Q) \tag{3C.7}$$

The differential scattering cross section per unit volume ($d\Sigma/d\Omega$) is interchangeably written as $I(Q)$ and has the units of cm^{-1}. $I(Q)$ is widely termed as the scattering intensity in the literature and is proportional to the contrast factor (X-rays or neutrons) of the scattering object and the surrounding medium:

$$I(Q) \sim (\Delta\rho^*)^2 = \left(\rho_1^* - \rho_2^*\right)^2 \tag{3C.8}$$

where ρ_1^* and ρ_2^* are the scattering length densities (SLDs) of the sample and the medium, respectively.

For SAS, eq. (3C.8) can be written in a general form, as follows:

$$I(Q) = NV_P^2 \, (\Delta\rho^*)^2 \, F(Q)S(Q) + B \tag{3C.9}$$

In this expression, N is the number density of scattering particles per unit volume, $F(Q)$ is the form factor which is a dimensionless quantity, $S(Q)$ is the structure factor, V_P is the particle volume, $(\Delta\rho^*)^2$ is the contrast factor obtained from SLDs, and B is the background signal [57]. The background signal that comes primarily from the air and the sample cell can be removed by subtracting the scattering intensity of the empty container.

The analytical expressions for form factors ($F(Q)$) for the most common shapes, such as spheres, disks, and rods [55, 57, 59], are established and can be used to provide useful information about the intrinsic physical characteristics of the sample such as the shape and size of the individual objects. $S(Q)$ provides information about the spatial arrangements of scatterers and is mathematically described as follows [57]:

$$S(Q) = 1 + 4\pi N \int (g(r) - 1)(r^2) \frac{\sin(Qr)}{Qr} dr \qquad (3C.10)$$

where $g(r)$ is the pair distribution function of the particles as a function of mean separation distance, r.

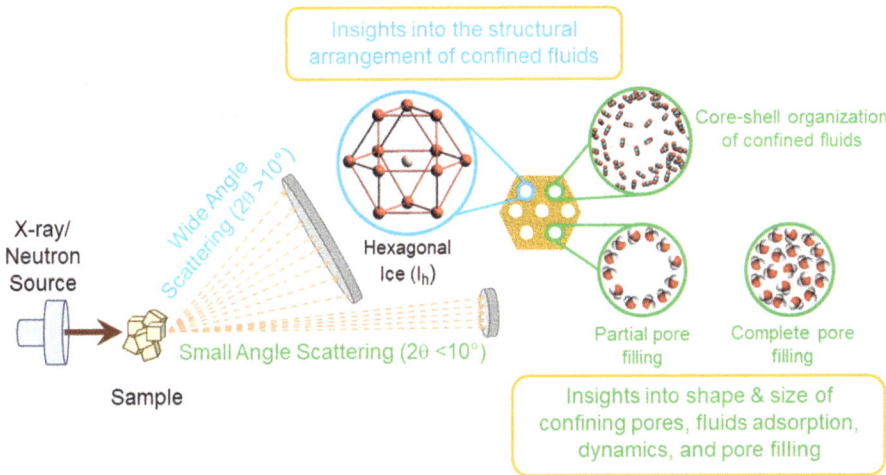

Fig. 3C.3: Illustration of small-angle and wide-angle scattering and the associated phenomena that can be probed from these measurements in "confined fluids".

3C.2.1.2 Pair distribution function

For a system of particles that produce isotropic scattering, the structure factor $S(Q)$ is related to the pair distribution function or RDF and can be given as $g(r)$, as in eq. (3C.10) [60–63]. This equation shows that the Fourier transform of $g(r) - 1$ gives $S(Q) - 1$, allowing the conversion between real ($g(r)$) and reciprocal ($S(Q)$) spaces. To obtain the pair distribution function from scattering data, the inverse Fourier transform of $S(Q)$ can be used and is written as follows:

$$g(r) = 1 + \frac{1}{2\pi^2 N} \int (S(Q) - 1)Q^2 \frac{\sin(Qr)}{Qr} \, dQ \tag{3C.11}$$

3C.2.1.3 Quasielastic neutron scattering (QENS)

For inelastic scattering experiments, such as "QENS," the energy of scattered neutrons is also considered in addition to the scattering angle and the double differential cross section, as represented by the following expression:

$$\frac{d^2\sigma}{d\Omega \, dE} = \frac{k_f}{k_i} \frac{N}{4\pi} \{\sigma_{coh} S_{coh}(Q, \omega) + \sigma_{inc} S_{inc}(Q, \omega)\} \tag{3C.12}$$

where $S_{coh}(Q, \omega)$ and $S_{inc}(Q, \omega)$ are scattering functions, which are double Fourier transform in space and time of the correlation functions for coherent scattering (in case of correlation function) and incoherent scattering (for self-correlation function), respectively [38, 64–66]. The coherent scattering provides information about the long-term average atomic structure and excitations. In contrast, the incoherent scattering describes the stochastic (diffusive) motions and is dominated by the hydrogen-containing species. In the incoherent approximation, assuming that all motions are independent for H-isotope-rich systems, the scattering function can be given as the convolution of the translational, rotational, and vibrational orientations:

$$S(Q, \omega) = S_{trans} \otimes S_{rot} \otimes S_{vib} \tag{3C.13}$$

For QENS experiments, the vibrational reorientations are diminishing due to the Debye–Waller factor [66], while the scattering functions S_{trans} and S_{rot} are given as follows:

$$S(Q, \hbar\omega)_{trans} = \frac{1}{\pi} \frac{\Gamma(Q)}{\hbar\omega^2 + \Gamma(Q)^2} \tag{3C.14}$$

$$S(Q, \hbar\omega)_{rot} = A_0(Q)\delta(\hbar\omega) + \sum A_i(Q) \frac{1}{\pi} \frac{\Gamma}{\hbar\omega^2 + \Gamma^2} \tag{3C.15}$$

where A_i is the spectral width and Γ_i is the half-width of the Lorentzian function. $\Gamma \sim 1/\tau$ and τ is the average residence time between successive jumps. Further, A_0 is termed the elastic incoherent scattering structure (EISF), and it provides information about the spatial extension of the rotational reorientations of interest. EISF can be described as the Fourier transform of the time average of the probability function of finding an atom at position **r**. When using a model-independent approach, EISF can be written as the ratio of elastic to elastic plus quasielastic scattering as follows:

$$EISF = A_0 = \frac{I^{elastic}}{I^{elastic} + I^{quasielastic}} \tag{3C.16}$$

The expression in eq. (3C.16) contains the elastic scattering component, which indicates that the estimated motion (reorientation) is localized and emerges from the confined fluids and is not caused by the host lattice.

3C.2.2 Molecular dynamic simulations

For a system composed of N interacting atoms, MD simulations solve Newton's equation of motion as follows:

$$m_i \frac{\partial^2 r_i}{\partial t^2} = F_i \tag{3C.17}$$

where m_i, r_i, and F_i are the mass, position, and force on atom i that is obtained from the potential function U as follows:

$$F_i = - \frac{\partial U}{\partial r_i} \tag{3C.18}$$

This equation is solved simultaneously for all atoms in the system in small time steps, and the updated coordinates are written to an output file at regular time intervals (trajectory). The trajectory can be used to calculate various structural and dynamical properties such that the property $\langle X \rangle$ can be calculated as follows:

$$\langle X \rangle = \frac{1}{t} \int_0^t X(t) dt \tag{3C.19}$$

Nonbonded and bonded interactions are accounted for in the total potential. The nonbonded interactions consist of van der Waals potential and electrostatic potential. Van der Waals and electrostatic interactions are calculated by Lennard-Jones (U_{LJ}) and Coulomb (U_C) models, respectively, as follows:

$$U_{nonbonded} = U_{LJ} + U_C = 4\varepsilon_{ij} \left[\left(\frac{\sigma_{ij}}{r_{ij}} \right)^{12} - \left(\frac{\sigma_{ij}}{r_{ij}} \right)^6 \right] + \kappa \frac{q_i q_j}{r_{ij}} \tag{3C.20}$$

where ε_{ij} is the depth of the potential well, σ_{ij} is the distance at which the potential energy U_{LJ} is zero, r_{ij} is the distance between the interacting atoms, k is the Coulomb constant, and q_α is the charge of atom α (where α is i and j). LJ parameters and the atomic charges are reported as a force field, and LJ parameters between different atom types are calculated according to Lorentz-Berthelot rules as follows:

$$\sigma_{ij} = \frac{\sigma_{ii} + \sigma_{jj}}{2} \tag{3C.21}$$

$$\varepsilon_{ij} = \sqrt{\varepsilon_{ii}\varepsilon_{jj}} \qquad (3C.22)$$

The bonded potentials were calculated for two-body (bond stretching) (U_b), three-body (angle) (U_a), and four-body (dihedral) (U_d) interactions.

The forces between the molecules in the initial system are extremely high. Therefore, energy minimization is essential before performing the equilibrium simulation to remove the inappropriate geometries and optimize the positions of the atoms in the system. Steepest descent is the most used algorithm to optimize the energy of the initial configurations of confined fluids simulations. In this method, the forces and potential for step $n + 1$ are evaluated based on those of step n as follows:

$$r_{n+1} = r_n + \frac{F_n}{\max(|F_n|)}h_n \qquad (3C.23)$$

where h_n is the maximum displacement and F_n is the force. The new position is accepted or rejected according to the following algorithm:

If $U_{n+1} < U_n$, the new position is accepted and set $h_{n+1} = 1.2\, h_n$.

If $U_{n+1} \geq U_n$, the new position is rejected and set $h_{n+1} = 0.2\, h_n$.

The algorithm stops when $max(|F_n|)$ is smaller than a specified value.

Different codes are available for performing MD simulations. However, the most widely used algorithms are LAMMPS [67] and GROMACS [68].

3C.3 Case studies of confined fluids

Harnessing SANS, SAXS, and WAXS measurements in concert with MD simulations has enabled advances in the fundamental understanding of the structure and interfacial organization of gases, water, and hydrocarbons in confinement. These insights provide the fundamental basis for observed properties of confined fluids in natural and architected porous materials. This section introduces and discusses case studies on the recent advances in utilizing X-ray and neutron scattering measurements and molecular-scale simulations to provide fundamental insights into the structure and dynamics of confined fluids.

3C.3.1 Structure of confined fluids

3C.3.1.1 Confined compressed gases

The combined SANS-MD approach is uniquely suited for linking the structure of confined fluids to various physicochemical parameters such as the applied pressure, the chemistry of the confining space, and the pore sizes (see Tab. 3C.1). For example, Mohammed and coworkers used an in situ SANS-MD approach to resolve the pressure-driven adsorption and organization of CO_2 and CD_4 on the interfaces of MCM-41 and SBA-15 pores with a diameter of 3.3 and 6.8 nm, respectively (see Fig. 3C.4) [23, 24]. These studies are performed at pressures ranging from vacuum to about 100 bars and a temperature of 25 °C. The study concluded that CO_2 and CD_4 form a core–shell structure in the nanopores, with the shell emerging from the adsorbed gas molecules on the pore surface. The shell thicknesses increase with gas pressure. However, the adsorption extent of CO_2 is greater than CD_4 on the pore surfaces under the same pressure due to the stronger intermolecular interactions between CO_2 and silica surfaces. The stronger intermolecular interactions of CO_2 with the pore surfaces are primarily driven by van der Waals interactions, electrostatic interactions, and hydrogen bonding.

In contrast, CD_4 interactions with the pore surface are mainly driven by van der Waals interactions. Close agreements were found between the shell thicknesses and core radii extracted from SANS measurements and MD simulations for CO_2 and CD_4 in confinement.

SANS-MD approach was recently used to elucidate the reduced methane recovery from shale pores at high pressures [70]. The study revealed that higher pressure led to the trapping of dense, liquid-like methane in micropores with a diameter of <2 nm that comprise 90 % of the used pore volume. The study attributed this trapping to the irreversible deformation of kerogen. In addition, the study suggested that the methane-trapping behavior in shale micropores might also be affected by other factors such as shale mineralogy, fracking fluid chemistry, reservoir temperature, and overburden stress.

Similarly, X-ray-based scattering and imaging techniques and MD simulations have been utilized to probe the structure of fluids in confinement. For instance, an integrated approach that harnesses scanning transmission X-ray microscopy measurements and MD simulations were used to explore the density of oxygen confined in surface nanobubbles in ambient water [71]. The results demonstrated that the density of oxygen inside the nanobubble is 1–2 orders of magnitude higher than that in atmospheric pressure. The study concluded that gas molecules within the confinement of bubbles in supersaturated liquid could maintain a dense state of aggregated gas molecules instead of the ideal gas state. This condensed state has significant implications for catalysis, hydrate formation, and hydrogen storage and transportation.

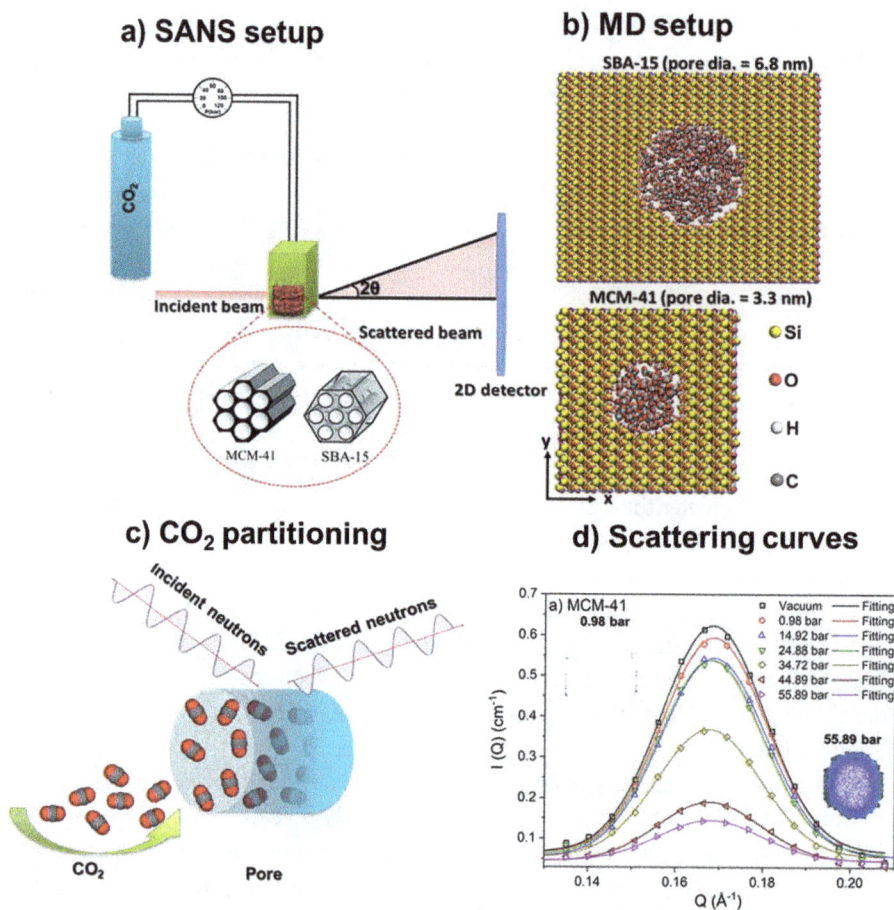

Fig. 3C.4: Schematic representation of the setup of SANS experiments, including the incident and scattered neutron beams, the experimental cell, and the structure of MCM-41 and SBA-15 particles, is shown in panel **(a)**. Snapshots of the initial configuration of the confined CO_2 molecules in silica nanopores with diameters of 3.3 and 6.8 nm are shown in panel **(b)**. A schematic representation of the partitioning of CO_2 into the nanopores is shown in panel **(c)**. The evolution of the small-angle diffraction peak corresponding to the mesopore filled with CO_2 is shown in panel **(d)**. This figure was reproduced with permission from the Royal Society of Chemistry [23].

3C.3.1.2 Confined liquids

The combination of X-ray spectroscopic techniques and MD simulations is a powerful systematic approach that has been utilized to harness the hydration structure of monovalent ions dissolved in confined aqueous phases. For example, the hydration of K^+ ions confined in hydrated montmorillonite pores was investigated using extended X-ray absorption fine structure (EXAFS) spectroscopy. MD simulations revealed significant differences in the number of ions in the hydration shell structure of ions [72]. The first coordination shell of K^+ in monohydrated montmorillonite comprises five water molecules instead of seven molecules in the aqueous bulk solution. Further, the study revealed that montmorillonite interlayer spacing undergoes a swelling effect as the number of confined water molecules increases in these micropores. Similarly, EXAFS-MD approaches showed that the number of water molecules in the first hydration shell of K^+, Na^+, and Ca^{2+} ions in 2.5 nm MCM-41 hydrated pores is higher than the ions solvated in bulk water solutions [73].

Hybrid reverse Monte Carlo simulations of X-ray diffraction (XRD) patterns showed that the number of water molecules in the first coordination shell of the Na^+ ion is 5.7 and 5.0 in confined water in carbon nanotubes and bulk water, respectively. Hydrophobic and confined environments stabilize the ion hydration shell despite weakening hydrogen bonds of water in confinement relative to bulk water [74]. These studies illustrate that the molecular basis for anomalous reactivity of confined fluids emerges from the differences in the ion hydration environment, which can be quantified using integrated experimental and simulation approaches.

3C.3.1.3 Confined polymers and surfactants

The assembly of polymers and surfactants in confined fluids differs from bulk environments due to surface interactions. Integrated experimental and simulation studies such as XRD-MD and SANS-MD studies have successfully revealed the organization of these confined components and the associated energetic interactions. One specific example is investigating the structure of hyperbranched polyamide polymer in natural sodium-montmorillonite clays using XRD-MD approaches [81]. The results show an intercalated structure for polymer chains in the interlayer spacing of the clays that are similar to the bulk structure of the polymer at temperatures below the T_g and a frozen behavior under confinement at temperatures above the T_g. Similarly, the effect of confinement on the self-assembly and temperature-induced liquid–liquid phase separation of surfactant solutions in 8.6-nm-sized SBA-15 mesopores was investigated using SANS-MD approaches [82]. Enhanced surfactant self-assembly in confinement was noted above the lower critical solution temperature. Further, an enhanced surfactant uptake and aggregate size in confinement are driven by the adsorbed surfactant molecules on the pore surface that act as anchor sites for subsequent adsorption.

3C.3.2 Dynamics of confined fluids

3C.3.2.1 Confined gases

A typical approach to probe the dynamics of confined fluids is QENS-MD, as it can resolve the mobility of fluids in a wide range of timescales. This powerful approach has been used to investigate the dynamics of various fluids confined in natural and architected porous materials. Investigations of the dynamics of propane confined in 1.5-nm-sized MCM-41-S nanopores in the temperature range of 230 and 250 K in the presence of water using QENS-MD studies showed that water hinders the diffusion of propane, and high water content in the nanopores reduces the translational and rotational diffusions of propane more significantly [39]. Further, water displaced propane from the pore surface toward the pore center. The displacement of propane molecules from the pore surface is countered by molecular crowding that suppresses propane mobility.

Similarly, the QENS-MD approach to investigate the diffusion of CO_2 in NaY and NaX zeolites showed that the diffusivity of confined CO_2 by QENS increased with CO_2 loading. In contrast, the simulated diffusivity decreased [75]. The diffusivity of captured CO_2 showed a strong dependence on CO_2 concentration and interactions with confining pores. Similarly, the QENS-MD approach has been used to understand the dynamics of light hydrocarbons in MOFs [76]. The diffusivity of confined hydrocarbons decreases with increasing the molecular size and confinement extent. These studies show that QENS-MD is a powerful combination for resolving the dynamics of gases confined in various porous environments.

3C.3.2.2 Confined liquids, polymer, and surfactants

QENS-MD approach has also been used to investigate the dynamics of confined liquids in various confining environments. For example, specific insights into the dynamics of water confined in montmorillonite interlayers were obtained using QENS-MD analyses [77]. The study demonstrated that the size of the clay interlayer and clay particles determines the diffusion of water confined in the clay interlayers. The oscillatory motion dominates short-time water dynamics in the cage of the first coordination shell. At the same time, long-time diffusion is highly affected by the interactions of water with the surface. Investigations into the motion of water in aerosol reverse micelles with varying water content using the QENS-MD approach revealed that translational diffusion dominates cases with low water content [78]. The rotational diffusion coefficient of confined water is smaller than that of bulk water and increases with the confined water content.

The fate and transport of hydrocarbon contaminants motivated investigations into the behavior of benzene confined in nanopores. QENS-MD approaches revealed that a combination of unhindered transactional and jump diffusions contributed to the lower diffusivity of confined benzene relative to the bulk fluid in MCM-41 nanopores at 300, 325, and 350 K [40].

Rising interest in harnessing ionic liquids for advanced separations motivated investigations into their dynamics in carbide-derived carbons and graphene nanoplatelets [79]. The study demonstrated that oxidized pore surfaces draw ions closer to pore surfaces and enhance potential-driven ion transport during electrosorption. Further, MD simulations showed that surface functional groups significantly influence ion's orientations, accumulation densities, and capacitance.

Tuning porous polymer composites for desired dynamics targeting CO_2 capture is possible using the QENS-MD approach. Nonmonotonic dynamics of confined polymer, poly(ethylenimine), is related to the structure of the polymer and is a function of the monomer concentration within an adsorbing cylindrical SBA-15 silica mesopore. Branched polymers, in contrast, do not exhibit nonmonotonic dynamics [83]. These studies illustrate the use of QENS-MD approaches to inform the design of materials for applications related to decarbonization.

3C.3.3 Phase transitions of confined fluids

3C.3.3.1 Confined gases

Phase transitions of confined fluids have significant implications for storing and recovering fluids from natural and engineered materials. Advances in X-ray and neutron scattering measurements combined with computational molecular-scale predictions have unlocked unprecedented insights into confined fluids' organization and phase transitions. SANS-MD approaches to probe the phase transitions of confined methane in silica mesopores revealed that the thickness of the adsorbed gases is influenced by the roughness of the pore surface and the pore size [25, 80]. The gas–liquid condensation temperature is sensitive to the pore size such that it shifts to higher values in smaller pores. The subsequent MD simulations showed that the densification of captured methane results from structural reorganization from the disordered amorphous phase to the two-dimensional hexagonal structure.

3C.3.3.2 Confined liquids

Due to a changing climate, the rising global temperature has significant implications for the stability of ice hydrates and gas hydrates, particularly those bearing CO_2 and CH_4. Furthermore, small temperature changes have significant implications for the melting and freezing behavior of ice in these environments. The freezing and melting behavior of water confined in nanopores is considerably different from that of bulk water. However, significant uncertainties regarding the structure of confined ice need to be resolved to establish the fundamental basis for the observed anomalous freezing and melting behavior of fluids.

To address this challenge of liquids or difficult to analyze substrates, WAXS and classical MD simulations were combined to investigate the freezing of water confined in cylindrical silica nanopores with a diameter of 4, 6, and 8 nm in the temperature range of 170–300 K [33]. Ice produced in 4 nm pores is a mixture of hexagonal and cubic structures, while that formed in 6- and 8-nm-sized pores has a hexagonal structure (Fig. 3C.5). Higher freezing point depressions are noted in smaller pores. MD simulations showed that the freezing of confined water undergoes first-order transitions. The diffusion coefficients of ice are four orders of magnitude lower than that of liquid water in confinement.

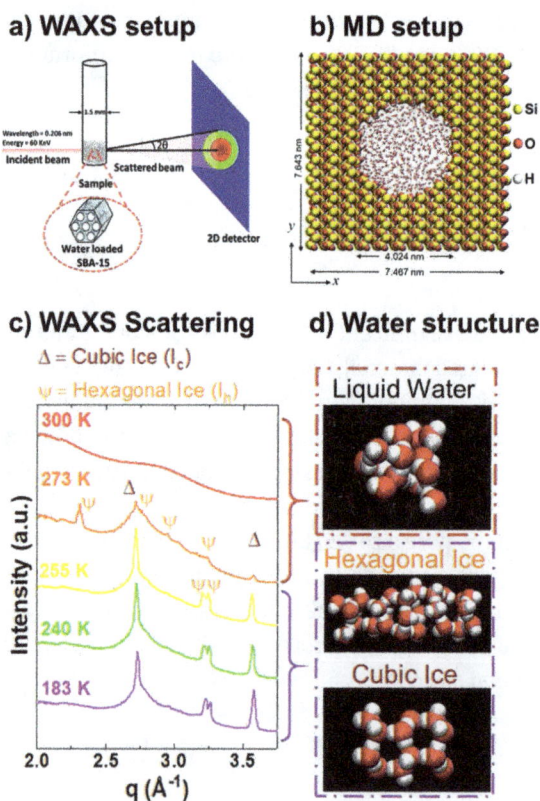

Fig. 3C.5: A schematic representation of the experimental setup for WAXS measurements from water loaded in 4-, 6-, and 8-nm-sized SBA-15 porous particles is shown in panel **(a)**. Snapshots of the initial configurations of water confined in 4-nm-sized silica pores used in MD simulations are shown in panel **(b)**. WAXS scattering patterns from water confined in 4-nm-sized pores as a function of the applied temperature are shown in panel **(c)**. Snapshots of cubic ice, hexagonal ice, and liquid water structures from MD simulations at different temperatures are shown in panel **(d)**. This figure was reproduced with permission from the Royal Society of Chemistry [33].

Tab. 3.1: Experimental-computational approaches are used to probe the structure, dynamics, and phase transitions of fluids confined in natural and architected porous materials.

Fluid	Solid interface	Approach	Key outcomes	Reference
Structure of confined fluids				
CO_2	MCM-41 and SBA-15	SANS-MD	CO_2 is organized in core–shell structures in cylindrical pores	[23]
CD_4	MCM-41 and SBA-15	SANS-MD	CD_4 is organized in core–shell structures in cylindrical pores	[24]
C-O-H fluids	Silica and alumina	SANS-MD	The structure and dynamics of confined CO_2 and hydrocarbons are different from their bulk counterparts	[69]
CH_4	Shale matrix	SANS-MD	High pressure results in the trapping of dense, liquid-like methane in nanopores with sub-2-nm pores	[70]
O_2	Nanobubbles	STXM-MD	Gas molecules within confinement could maintain a dense state instead of the ideal gas state if their surrounding liquid is supersaturated	[71]
Brine	Montmorillonite	EXAFS-MD	Solvated ions in hydrated montmorillonite interlayers are highly influenced by the water content and interlayer spacing	[72]
Brine	MCM-41	EXAFS-MD	The dynamic hydration number in mesopores is larger than that of bulk solutions.	[73]
Brine	CNT	XRD-MC	The structure of hydrated ions differs in confined and bulk fluids	[74]
Dynamics of confined fluids				
Propane	Silica nanopores	QENS-MD	The adsorbed molecules dominate the dynamics of confined propane at low pressures and result in lower diffusion coefficients	[45]
CO_2	NaY and NaX zeolites	QENS-MD	The experimentally measured diffusivity increases with the loading while the simulated diffusivity decreases	[75]
Light hydro carbons	UiO-66(Zr) MOF	QENS-MD	The self-diffusion coefficient decreases with increasing the hydrocarbon's molecular XE "molecular" size and increasing confinement	[76]

Tab. 3.1 (continued)

Fluid	Solid interface	Approach	Key outcomes	Reference
Dynamics of confined fluids				
Water	Montmorillonite	QENS-MD	The size of the clay interlayer and the clay particles determine the time-dependent behavior of the confined water diffusion	[77]
Water	Aerosol OT	QENS-MD	The water content highly influences the diffusion of confined water in confinement	[78]
Benzene	MCM-41	QENS-MD	The diffusion coefficient of benzene decreases on confinement. Confined benzene shows a different structure compared to bulk benzene	[40]
RTILs	Carbide-derived carbons and graphene nanoplatelets	QENS-MD	Surface functional groups influence ion orientations, accumulation densities, and capacitance	[79]
Phase transitions of confined fluids				
CD_4	MCM-41 and SBA-15	SANS-MC	Structural organization-induced densification of confined methane is noted	[25, 80]
Water	MCM-41	WAXS-MD	Hexagonal and cubic ice emerge in 4-nm-sized pores, while only hexagonal ice emerges in 6- and 8-nm-sized pores	[33]

SANS, small-angle neutron scattering; MD, molecular dynamics; STXM, scanning transmission X-ray microscopy; XRD, X-ray diffraction; QENS, quasielastic neutron scattering; WAXS, wide-angle X-ray scattering; CNT, carbon nanotubes; EXAFS, extended X-ray absorption fine structure.

3C.4 Conclusions

This chapter discusses approaches to unlock fundamental insights into the structure, dynamics, and phase transitions of gases, liquids, polymers, and surfactants confined in micropores and mesopores of natural and engineered materials using X-ray and neutron scattering measurements with MD simulations. The basic principles of the relevant X-ray and neutron scattering techniques and molecular simulations are discussed. Case studies that involve resolving the structure and dynamics of confined fluids using X-ray and neutron scattering and MD simulations are discussed. Advances in in situ and *operando* experimental methods have enabled us to resolve the spatial and temporal evolution of the structures and dynamics of confined fluids and experimentally validate computational molecular-scale models. The approaches described in this book chapter are essential to advance the scientific basis for emerging decarbonization technologies, including carbon capture, conversion, storage, low carbon hydrogen conversion and storage, and energy storage for sustainable energy, environmental, and climate future.

References

[1]   Lin, L. C., & Grossman, J. C. (2015). Atomistic understandings of reduced graphene oxide as an ultrathin-film nanoporous membrane for separations. Nature Communications, 6, 1–7.
[2]   Wang, Z., Knebel, A., Grosjean, S., Wagner, D., Bräse, S., Wöll, C., Caro, J., & Heinke, L. (2016). Tunable molecular separation by nanoporous membranes. Nature Communications, 7, 1–7.
[3]   Na, K., & Somorjai, G. A. (2015). Hierarchically nanoporous zeolites and their heterogeneous catalysis: Current status and future perspectives. Catalysis Letters, 145, 193–213.
[4]   Zheng, J., Wang, Z., Gong, W., Ju, Y., & Wang, M. (2017). Characterization of nanopore morphology of shale and its effects on gas permeability. Journal of Natural Gas Science and Engineering, 47, 83–90.
[5]   Mohammed, S., Asgar, H., Deo, M., & Gadikota, G. (2021b). Interfacial and confinement-mediated organization of gas hydrates, water, organic fluids, and nanoparticles for the utilization of subsurface energy and geological resources. Energy & Fuels : An American Chemical Society Journal, 35, 4687–4710.
[6]   Bhatia, S. K., Bonilla, M. R., & Nicholson, D. (2011). Molecular transport in nanopores: A theoretical perspective. Physical Chemistry Chemical Physics, 13, 15350–15383.
[7]   Keyser, U. F. (2011). Controlling molecular transport through nanopores. Journal of the Royal Society, Interface / the Royal Society, 8, 1369–1378.
[8]   Mansoori, G. A., & Rice, S. A. (2014). Confined fluids: Structure, properties, and phase behavior. Advances in Chemical Physics, 156, 1–97.
[9]   Zhang, K., Jia, N., Li, S., & Liu, L. (2018). Thermodynamic phase behavior and miscibility of confined fluids in nanopores. Chemical Engineering Journal, 351, 1115–1128.
[10] Takamuku, T., Maruyama, H., Kittaka, S., Takahara, S., & Yamaguchi, T. (2005). Structure of methanol confined in MCM-41 investigated by large-angle X-ray scattering technique. The Journal of Physical Chemistry. B, 109, 892–899.
[11] Erko, M., Wallacher, D., Hoell, A., Hauss, T., Zizak, I., & Paris, O. (2012). Density minimum of confined water at low temperatures: A combined study by small-angle scattering of X-rays and neutrons. Physical Chemistry Chemical Physics: PCCP, 14, 3852–3858.
[12] Paineau, E., Albouy, P. A., Rouzière, S., Orecchini, A., Rols, S., & Launois, P. (2013). X-ray scattering determination of the structure of water during carbon nanotube filling. Nano Letters, 13, 1751–1756.
[13] Crupi, V., Magazu, S., Majolino, D., Migliardo, P., Venuti, V., & Bellissent-Funel, M. C. (2000). Confinement influence in liquid water studied by Raman and neutron scattering. Journal of Physics: Condensed Matter, 12, 3625.
[14] Bordallo, H. N., Aldridge, L. P., Churchman, G. J., Gates, W. P., Telling, M. T., Kiefer, K., Fouquet, P., Seydel, T., & Kimber, S. A. (2008). Quasielastic neutron scattering studies on clay interlayer-space highlighting the effect of the cation in confined water dynamics. The Journal of Physical Chemistry C, 112, 13982–13991.

[15] Bertrand, C. E., Zhang, Y., & Chen, S. H. (2013). Deeply-cooled water under strong confinement: Neutron scattering investigations and the liquid-liquid critical point hypothesis. Physical Chemistry Chemical Physics, 15, 721–745.

[16] Xu, H. (2020). Probing nanopore structure and confined fluid behavior in shale matrix: A review on small-angle neutron scattering studies. International Journal of Coal Geology, 217, 103325–1 to 103325–9.

[17] Coasne, B., Di Renzo, F., Galarneau, A., & Pellenq, R. J. (2008). Adsorption of simple fluid on silica surface and nanopore: Effect of surface chemistry and pore shape. Langmuir, 24, 7285–7293.

[18] Zeng, K., Jiang, P., Lun, Z., & Xu, R. (2018). Molecular simulation of carbon dioxide and methane adsorption in shale organic nanopores. Energy & Fuels: An American Chemical Society Journal, 33, 1785–1796.

[19] Yang, G., Chai, D., Fan, Z., & Li, X. (2019). Capillary condensation of single-and multi-component fluids in nanopores. Industrial & Engineering Chemistry Research, 58, 19302–19315.

[20] Luo, S., Lutkenhaus, J. L., & Nasrabadi, H. (2018). Use of differential scanning calorimetry to study the phase behavior of hydrocarbon mixtures in nanoscale porous media. The Journal of Petroleum Science and Engineering, 163, 731–738.

[21] Qiu, X., Tan, S. P., Dejam, M., & Adidharma, H. (2019). Simple and accurate isochoric differential scanning calorimetry measurements: Phase transitions for pure fluids and mixtures in nanopores. Physical Chemistry Chemical Physics: PCCP, 21, 224–231.

[22] Chiang, W. S., Fratini, E., Baglioni, P., Georgi, D., Chen, J. H., & Liu, Y. (2016b). Methane adsorption in model mesoporous material, SBA-15, studied by small-angle neutron scattering.". The Journal of Physical Chemistry C, 120, 4354–4363.

[23] Mohammed, S., Liu, M., & Gadikota, G. (2021c). Resolving the organization of CO2 molecules confined in silica nanopores using in situ small-angle neutron scattering and molecular dynamics simulations. Environmental Science: Nano, 8, 2006–2018.

[24] Mohammed, S., Liu, M., Liu, Y., & Gadikota, G. (2020b). Probing the core-shell organization of nano-confined methane in cylindrical silica pores using in-situ small-angle neutron scattering and molecular dynamics simulations. Energy & Fuels : An American Chemical Society Journal, 34, 15246–15256.

[25] Chiang, W. S., Fratini, E., Baglioni, P., Chen, J. H., & Liu, Y. (2016a). Pore size effect on methane adsorption in mesoporous silica materials studied by small-angle neutron scattering.". Langmuir, 32, 8849–8857.

[26] Melnichenko, Y. B., & Wignall, G. D. (2009). Density and volume fraction of supercritical CO2 in pores of native and oxidized aerogels. International Journal of Thermophysics, 30, 1578–1590.

[27] Melnichenko, Y. B., Wignall, G. D., Cole, D. R., & Frielinghaus, H. (2006). Adsorption of supercritical CO2 in aerogels as studied by small-angle neutron scattering and neutron transmission techniques. The Journal of Chemical Physics, 124, 204711–1 to 204711–11.

[28] Gallego, N. C., He, L., Saha, D., Contescu, C. I., & Melnichenko, Y. B. (2011). Hydrogen confinement in carbon nanopores: Extreme densification at ambient temperature. Journal of the American Chemical Society, 133, 13794–13797.

[29] Boukhalfa, S., He, L., Melnichenko, Y. B., & Yushin, G. (2013). Small-angle neutron scattering for in situ probing of ion adsorption inside micropores. Angewandte Chemie International Edition, 52, 4618–4622.

[30] Stefanopoulos, K. L., Romanos, G. E., Vangeli, O. C., Mergia, K., Kanellopoulos, N. K., Koutsioubas, A., & Lairez, D. (2011). Investigation of confined ionic liquid in nanostructured

materials by a combination of SANS, contrast-matching SANS, and nitrogen adsorption. Langmuir, 27, 7980–7985.

[31] Bellissent-Funel, M. C. (2001). Structure of confined water. Journal of Physics: Condensed Matter, 13, 9165.

[32] Morishige, K., & Iwasaki, H. (2003). X-ray study of freezing and melting water confined within SBA-15. Langmuir, 19, 2808–2811.

[33] Mohammed, S., Asgar, H., Benmore, C. J., & Gadikota, G. (2021a). Structure of ice confined in silica nanopores. Physical Chemistry Chemical Physics: PCCP, 23, 12706–12717.

[34] Falkowska, M., Bowron, D. T., Manyar, H., Youngs, T. G., & Hardacre, C. (2018). Confinement effects on the benzene orientational structure. Angewandte Chemie, 130, 4655–4660.

[35] Wang, N., Zhi, Y., Wei, Y., Zhang, W., Liu, Z., Huang, J., Sun, T., Xu, S., Lin, S., He, Y., & Zheng, A. (2020). Molecular elucidating an unusual growth mechanism for polycyclic aromatic hydrocarbons in confined space. Nature Communications, 11, 1–12.

[36] Terban, M. W., & Billinge, S. J. (2021). Structural analysis of molecular materials using the pair distribution function. Chemical Reviews, 122(1), 1208–1272.

[37] Yamaguchi, T., Yoshida, K., Smirnov, P., Takamuku, T., Kittaka, S., Takahara, S., Kuroda, Y., & Bellissent-Funel, M. C. (2007). Structure and dynamic properties of liquids confined in MCM-41 mesopores. The European Physical Journal: Special Topics, 141, 19–27.

[38] Mitra, S., Mukhopadhyay, R., Tsukushi, I., & Ikeda, S. (2001). Dynamics of water in confined space (porous alumina): QENS study. Journal of Physics: Condensed Matter, 13, 8455.

[39] Gautam, S., Le, T. T. B., Rother, G., Jalarvo, N., Liu, T., Mamontov, E., Dai, S., Qiao, Z. A., Striolo, A., & Cole, D. (2019). Effects of water on the stochastic motions of propane confined in MCM-41-S pores. Physical Chemistry Chemical Physics: PCCP, 21, 25035–25046.

[40] Dervin, D., O'malley, A. J., Falkowska, M., Chansai, S., Silverwood, I. P., Hardacre, C., & Catlow, C. R. A. (2020). Probing the dynamics and structure of confined benzene in MCM-41 based catalysts. Physical Chemistry Chemical Physics: PCCP, 22, 11485–11489.

[41] Liu, T., Gautam, S., Cole, D. R., Patankar, S., Tomasko, D., Zhou, W., & Rother, G. (2020). Structure and dynamics of ethane confined in silica nanopores in the presence of CO2. The Journal of Chemical Physics, 152, 084707–1 to 084707–14.

[42] Le, T., Striolo, A., & Cole, D. R. (2015). CO2–C4H10 mixtures simulated in silica slit pores: Relation between structure and dynamics. The Journal of Physical Chemistry C, 119, 15274–15284.

[43] Mohammed, S., Sunkara, A. K., Walike, C. E., & Gadikota, G. (2021d). The role of surface hydrophobicity on the structure and dynamics of CO2 and CH4 confined in silica nanopores. Front Climate, 3, 713708–1 to 713708–14.

[44] Bourg, I. C., & Steefel, C. I. (2012). Molecular dynamics simulations of water structure and diffusion in silica nanopores. The Journal of Physical Chemistry C, 116, 11556–11564.

[45] Gautam, S., Le, T., Striolo, A., & Cole, D. (2017). Molecular dynamics simulations of propane in slit-shaped silica nanopores: Direct comparison with quasielastic neutron scattering experiments. Physical Chemistry Chemical Physics: PCCP, 19, 32320–32332.

[46] Mohammed, S., & Gadikota, G. (2019b). The role of calcite and silica interfaces on the aggregation and transport of asphaltenes in confinement. Journal of Molecular Liquids, 274, 792–800.

[47] Phan, A., Cole, D. R., Weiß, R. G., Dzubiella, J., & Striolo, A. (2016). Confined water determines transport properties of guest molecules in narrow pores. ACS Nano, 10, 7646–7656.

[48] Le, T. T. B., Striolo, A., Gautam, S. S., & Cole, D. R. (2017). Propane–water mixtures confined within cylindrical silica nanopores: Structural and dynamical properties probed by molecular dynamics. Langmuir, 33, 11310–11320.

[49]   Mohammed, S., & Gadikota, G. (2018). The effect of hydration on the structure and transport properties of confined carbon dioxide and methane in calcite nano-pores. Frontiers in Energy Research, 6, 86.

[50]   Mohammed, S., & Gadikota, G. (2019a). The influence of CO2 on the structure of confined asphaltenes in calcite nanopores. Fuel, 236, 769–777.

[51]   Mohammed, S., & Gadikota, G. (2020a). Exploring the role of inorganic and organic interfaces on CO2 and CH4 partitioning: Case study of silica, illite, calcite, and kerogen nano-pores on gas adsorption and nanoscale transport behaviors. Energy & Fuels : An American Chemical Society Journal, 34, 3578–3590.

[52]   Mosaddeghi, H., Alavi, S., Kowsari, M. H., & Najafi, B. (2012). Simulations of structural and dynamic anisotropy in nano-confined water between parallel graphite plates. The Journal of Chemical Physics, 137(18), 184703–1 to 184703–10.

[53]   Yasuoka, H., Takahama, R., Kaneda, M., & Suga, K. (2015). Confinement effects on liquid-flow characteristics in carbon nanotubes. Physical Review E, 92(6), 063001–1 to 063001–9.

[54]   He, J., Ju, Y., Kulasinski, K., Zheng, L., & Lammers, L. (2019). Molecular dynamics simulation of methane transport in confined organic nanopores with high relative roughness. Journal of Natural Gas Science and Engineering, 62, 202–213.

[55]   Glatter, O. K. (1982). Small angle X-ray scattering. London, New York: Academic Press.

[56]   Melnichenko, Y. B., & Gorge, D. W. (2007). Small-angle neutron scattering in materials science: Recent practical applications. Applied Physics Reviews, 102(2), 3–1 to 3–24.

[57]   Melnichenko, Y. B. (2016). Small-angle scattering from confined and interfacial fluids. Switzerland: Springer International Publishing.

[58]   Lovesey, S. W. (1984). Theory of neutron scattering from condensed matter. Oxford: Clarendon.

[59]   Li, T., Senesi, A. J., & Lee, B. (2016). Small angle X-ray scattering for nanoparticle research. Chemical Reviews, 116, 11128–11180.

[60]   Warren, B. E. (1990). X-ray diffraction. New York: Dover.

[61]   Billinge, S. J. L., & Kanatzidis, M. G. (2004). Beyond crystallography: The study of disorder, nanocrystallinity, and crystallographically challenged materials with pair distribution functions. ChemComm, 7, 749–760.

[62]   Farrow, C. L., & Billnge, S. J. (2009). Relationship between the atomic pair distribution function and small-angle scattering: Implications for modeling of nanoparticles. Acta Crystallographica, A65, 232–239.

[63]   Billinge, S. J. L. (2019). The rise of the X-ray atomic pair distribution function method: A series of fortunate events. Philosophical Transactions of the Royal Society A, 377(2147), 20180413–1 to 20180413–17.

[64]   Bee, M. (1988). Quasielastic neutron scattering. Principles and applications in solid state chemistry. Biology and materials science. Bristol, PA: Adam Hilger.

[65]   Embs, J. P., & Hempelmann, F. J. R. (2010). Introduction to quasielastic neutron scattering. Zeitschrift Für Physikalische Chemie, 224, 5–32.

[66]   Lohstroh, W., & Heere, M. (2020). Structure and dynamics of borohydrides studied by neutron scattering techniques: A review. Journal of the Physical Society of Japan, 89(5), 051011–1 to 051011–12.

[67]   Thompson, A. P., Aktulga, H. M., Berger, R., Bolintineanu, D. S., Brown, W. M., Crozier, P. S., In't Veld, P. J., Kohlmeyer, A., Moore, S. G., Nguyen, T. D., & Shan, R. (2021). LAMMPS-a flexible simulation tool for particle-based materials modeling at the atomic, meso, and continuum scales. Computer Physics Communications, 271, 108171–1 to 108171–34.

[68]   Abraham, M. J., Murtola, T., Schulz, R., Páll, S., Smith, J. C., Hess, B., & Lindahl, E. (2015). GROMACS: high-performance molecular simulations through multi-level parallelism from laptops to supercomputers. SoftwareX, 1, 19–25.

[69] Cole, D. R., Ok, S., Phan, A., Rother, G., Striolo, A., & Vlcek, L. (2013). Carbon-bearing fluids at nanoscale interfaces. Procedia Earth and Planetary Science, 7, 175–178.

[70] Neil, C. W., Mehana, M., Hjelm, R. P., Hawley, M. E., Watkins, E. B., Mao, Y., Viswanathan, H., Kang, Q., & Xu, H. (2020). Reduced methane recovery at high pressure due to methane trapping in shale nanopores. Communications Earth & Environment, 1, 1–10.

[71] Zhou, L., Wang, X., Shin, H. J., Wang, J., Tai, R., Zhang, X., Fang, H., Xiao, W., Wang, L., Wang, C., & Gao, X. (2020). Ultrahigh density of gas molecules confined in surface nanobubbles in ambient water. Journal of the American Chemical Society, 142, 5583–5593.

[72] Vao-Soongnern, V., Pipatpanukul, C., & Horpibulsuk, S. (2015). A combined X-ray absorption spectroscopy and molecular dynamics simulation to study the local potassium ion in hydrated montmorillonite. Journal of Materials Science, 50, 7126–7136.

[73] Ogura, R., & Ueda, T. (2019). The first evaluation of the dynamic hydration number of hydrated ions confined in mesoporous silica MCM-41. Adsorption, 25, 1057–1066.

[74] Ohba, T. (2014). Anomalously enhanced hydration of aqueous electrolyte solution in hydrophobic carbon nanotubes to maintain stability. ChemPhysChem, 15, 415–419.

[75] Plant, D., Jobic, H., Llewellyn, P., & Maurin, G. (2007). Diffusion of CO_2 in NaY and NaX Faujasite systems: Quasielastic neutron scattering experiments and molecular dynamics simulations. The European Physical Journal: Special Topics, 141, 127–132.

[76] Ramsahye, N. A., Gao, J., Jobic, H., Llewellyn, P. L., Yang, Q., Wiersum, A. D., Koza, M. M., Guillerm, V., Serre, C., Zhong, C. L., & Maurin, G. (2014). Adsorption and diffusion of light, hydrocarbons in UiO-66 (Zr): A combination of experimental and modeling tools. The Journal of Physical Chemistry C, 118, 27470–27482.

[77] Churakov, S. V., Gimmi, T., Unruh, T., Van Loon, L. R., & Juranyi, F. (2014). Resolving diffusion in clay minerals at different time scales: Combination of experimental and modeling approaches. Applied Clay Science, 96, 36–44.

[78] Harpham, M. R., Ladanyi, B. M., Levinger, N. E., & Herwig, K. W. (2004). Water motion in reverse micelles studied by quasielastic neutron scattering and molecular dynamics simulations. The Journal of Chemical Physics, 121, 7855–7868.

[79] Dyatkin, B., Zhang, Y., Mamontov, E., Kolesnikov, A. I., Cheng, Y., Meyer Iii, H. M., Cummings, P. T., & Gogotsi, Y. (2016). Influence of surface oxidation on ion dynamics and capacitance in porous and nonporous carbon electrodes. The Journal of Physical Chemistry C, 120, 8730–8741.

[80] Siderius, D. W., Krekelberg, W. P., Chiang, W. S., Shen, V. K., & Liu, Y. (2017). Quasi-two-dimensional phase transition of methane adsorbed in cylindrical silica mesopores. Langmuir, 33, 14252–14262.

[81] Chrissopoulou, K., Fotiadou, S., Androulaki, K., Tanis, I., Karatasos, K., Prevosto, D., Labardi, M., Frick, B., & Anastasiadis, S. H. (2014) Dynamics of dendritic polymers in bulk and under confinement. In AIP Conf Proc 1599:250–253.

[82] Wu, Y., Ma, Y., He, L., Rother, G., Shelton, W. A., & Bharti, B. (2019). Directed pore uptake and phase separation of surfactant solutions under confinement. The Journal of Physical Chemistry C, 123, 9957–9966.

[83] Carrillo, J. M. Y., Sakwa-Novak, M. A., Holewinski, A., Potter, M. E., Rother, G., Jones, C. W., & Sumpter, B. G. (2016). Unraveling the dynamics of amino polymer/silica composites. Langmuir, 32, 2617–2625.

Section 4: **Focus on select example applications of nanoscience in energy, environment, and health**

Yeshu Tan, Ivan P. Parkin, Guanjie He

Chapter 4A
Electrocatalytic hydrogen production

Abstract: Hydrogen is a clean energy alternative to adjust and interconnect the energy landscape as it is renewable, cost-effective, and relatively easily processed. The rapid development of advanced materials and deep mechanistic studies for electrocatalytic processes provides a good backdrop for developing more efficient hydrogen generation. This chapter summarizes the development of hydrogen energy, current research progress in electrocatalytic hydrogen evolution, and advanced materials in hydrogen production, storage, and utilization.

Keywords: hydrogen development, hydrogen evolution reaction, electrocatalyst, electrocatalytic water splitting, hydrogen storage

4A.1 Hydrogen energy and development

Fossil fuel consumption has led to emissions of polluting gases that have led to a global warming crisis. This has been driven by rapid economic development that has led to a doubling of global energy requirements over the last 30 years [1]. Greenhouse gases are mainly generated from carbon-containing sources and have accelerated climate change. Moreover, other pollution gases containing sulfur and nitrogen are hazardous to humans and the environment [2]. Thus, more and more countries have announced guidelines for developing clean energy systems, including solar energy, wind energy, tidal power, and geothermal energy [3]. These renewable energy sources rely on natural phenomena, such as solar and wind, which are intermittent. Nuclear power is also a suitable alternative to fossil fuels but risks widespread

Acknowledgments: The authors acknowledge the Engineering and Physical Sciences Research Council (EPSRC, EP/V027433/1, EP/L015862/1) and the Royal Society (RGS\R1\211080; IEC\NSFC\201261) for the funding support.

Yeshu Tan, Christopher Ingold Laboratory, Department of Chemistry, University College London, 20 Gordon Street, London WC1H 0AJ, United Kingdom, e-mail: yeshu.tan.19@ucl.ac.uk
Guanjie He, Christopher Ingold Laboratory, Department of Chemistry, University College London, 20 Gordon Street, London WC1H 0AJ, United Kingdom, e-mail: yeshu.tan.19@ucl.ac.uk; School of Chemistry, University of Lincoln, Joseph Banks Laboratories, Green Lane, Lincoln, LN6 7DL, United Kingdom, e-mail: g.he@ucl.ac.uk
Ivan P. Parkin, Christopher Ingold Laboratory, Department of Chemistry, University College London, 20 Gordon Street, London WC1H 0AJ, United Kingdom, e-mail: i.p.parkin@ucl.ac.uk

https://doi.org/10.1515/9783110739879-010

damaging leaks (Chernobyl, Fujishima). Further, nuclear waste disposal is a significant problem. It is perhaps not a safe alternative for long-term energy needs, but it has the short-term advantage of not producing climate-damaging gases [4].

Due to the different usage of electricity between days and nights, batteries, pump storage, or other energy vectors such as hydrogen could hold the key to widespread renewable energy usage. As clean energy, hydrogen is widely regarded as a promising, cost-effective, environment-friendly energy alternative to fossil fuels [5, 6]. Hydrogen can be produced from water, plants, and geothermal processes, which are fairly evenly distributed worldwide, unlike the distribution of fossil fuel sources. Developing clean energy can mitigate the energy shortage for those areas relying heavily on imported fossil fuels. Hydrogen has a high energy density (142 MJ/kg), and proton-exchange hydrogen fuel cells are promising devices for utilizing hydrogen, and the only by-product is water [7, 8]. Hydrogen can also be stored in tanks and transferred through pipes like natural gas to the energy area [9]. The increasing trends of interest in hydrogen energy are becoming more practical for the programs carried out in many countries [10]. The related utilization of hydrogen energy has been achieved, such as in fuel-cell cars [11]. The hydrogen system consists of production, storage, and application and will become more and more important for future energy systems [12].

Continuous and stable hydrogen production is required for the hydrogen economy. The main methods to produce hydrogen are steam reforming and electrochemical water splitting. The steam-methane reforming method is now the cheapest and the most mature way to produce hydrogen. However, the reactions occur at high temperatures and form by-product CO, making the process environmentally damaging [13]. A different reaction is needed to transfer CO to CO_2, increasing energy consumption and releasing greenhouse gases. The reaction of steam-methane reforming is carried out at around 700 °C, assisted by a catalyst:

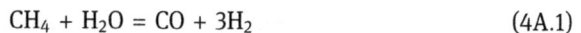

$$CH_4 + H_2O = CO + 3H_2 \tag{4A.1}$$

Carbon monoxide is a toxic by-product, which is often converted to carbon dioxide with further oxidation by an excess of steam:

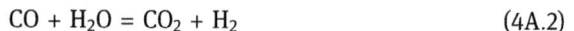

$$CO + H_2O = CO_2 + H_2 \tag{4A.2}$$

The whole reaction of steam-methane reforming is described as follows:

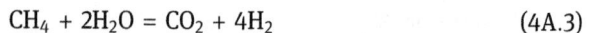

$$CH_4 + 2H_2O = CO_2 + 4H_2 \tag{4A.3}$$

The problems with the steam-methane process include catalyst poisoning by CO and the necessary purification of hydrogen from the gas mixture [14]. The poisoning is inevitable and leads to a decrease in production efficiency. The purification process is based on membranes with selective transport, which increases the cost. The whole process makes steam-methane reforming much more complicated than electrochemical water splitting. Besides the production efficiency, the safety issues are significant during steam-methane reforming.

The purity of hydrogen produced from electrochemical water splitting is comparatively high, with oxygen being formed at the other electrode, reflected by the equation:

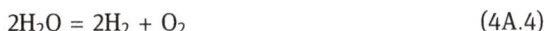

$$2H_2O = 2H_2 + O_2 \qquad (4A.4)$$

The schematic illustration of electrochemical water splitting is shown in Fig. 4A.1.

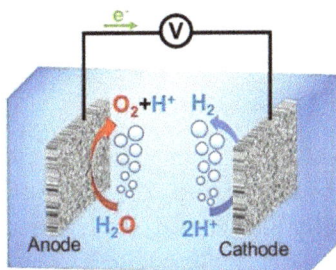

Fig. 4A.1: Schematic illustration of electrochemical water splitting.

The hydrogen can be directly collected from electrochemical water splitting. With the assistance of a membrane during the reaction, the purification process is quite simple. However, electrical consumption for water splitting is significantly higher than the estimated energy utilization of the produced hydrogen, which means that the hydrogen usage directly from water splitting is not that economical [15]. Electrochemical water splitting is an alternative to transfer excess power to hydrogens, such as electrical power from solar energy, tidal energy, and wind energy. These natural energy sources are unstable and intermittent, which is not optimal for long-term continuous output. Thus, water splitting can be an easy and excellent method to store the energy as hydrogen, which is environmental-friendly and mobile. To reduce the cost of electrochemical water splitting, photocatalytic water splitting is becoming popular, acquiring energy from the Sun. However, the transfer efficiency is relatively low, combined with electrocatalytic water splitting [16].

4A.2 Electrocatalytic hydrogen evolution reaction

Electrocatalysis is the foundation of hydrogen-based renewable and green energy conversion systems. Many devices and applications rely on hydrogen as an energy vector, including the petroleum refining industry, fuel cells, and the synthesis of ammonia [17, 18]. Therefore, the electrocatalytic hydrogen evolution reaction (HER) has increased attention over recent years.

Nowadays, electrocatalytic HER in the alkaline electrolyte is widely studied and applied commercially. It has relatively low cost and excellent gas generation efficiency [19]. The hydrogen can be easily collected with high purity using a simple proton-conduction membrane. The reaction mechanism of hydrogen evolution on the cathode includes three steps such as the Volmer step, the Heyrovsky step, and the Tafel step, involving hydrogen intermediates (*H) adsorption and desorption. When the reaction is performed in a neutral and basic environment, the first step is dissociating water molecules in the Volmer step, as described below. Then hydrogen is produced through either the Heyrovsky step or the Tafel step:

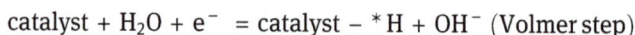

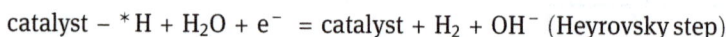

$$\text{catalyst} + H_2O + e^- = \text{catalyst} - {}^*H + OH^- \text{ (Volmer step)}$$

$$\text{catalyst} - {}^*H + H_2O + e^- = \text{catalyst} + H_2 + OH^- \text{ (Heyrovsky step)}$$

$$2\,\text{catalyst} - {}^*H \rightarrow 2\,\text{catalyst} + H_2 \text{ (Tafel step)}$$

When the hydrogen evolution happens in acidic media, the reaction pathway is similar, except the dissociation of water occurs in the Volmer step. The hydrogen intermediates are formed in the Volmer step, and then the pathway undergoes a Heyrovsky step, or Tafel step, to form hydrogen:

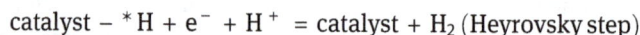

$$H^+ + e^- + \text{catalyst} = \text{catalyst} - {}^*H \text{ (Volmer step)}$$

$$\text{catalyst} - {}^*H + e^- + H^+ = \text{catalyst} + H_2 \text{ (Heyrovsky step)}$$

$$2\,\text{catalyst} - {}^*H \rightarrow 2\,\text{catalyst} + H_2 \text{ (Tafel step)}$$

The reaction steps are different because of the H^+ or OH^- domination in acidic or basic electrolytes, which also influences hydrogen production efficiency, as shown in the schematic illustration of HER steps in Fig. 4A.2. The adsorption and desorption of *H on the catalyst are significant for hydrogen evolution, which can be the rate-determining steps.

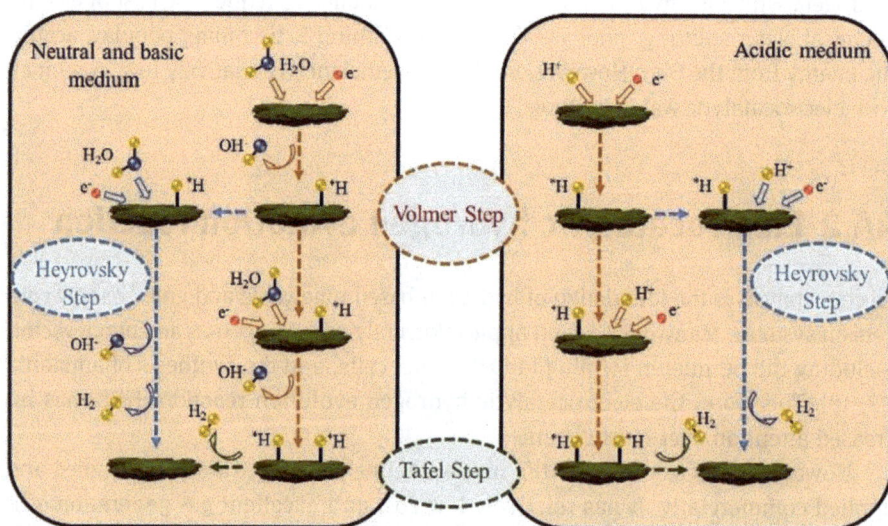

Fig. 4A.2: Schematic illustration of three steps of hydrogen evolution reaction in different media.

Suitable hydrogen bond formation can encourage an efficient HER process. H-bonding can limit the pathway to either weak or strong, making the useful description of hydrogen adsorption energy (ΔG_H) close to zero. The schemes shown in Fig. 4A.3 provide guidelines for the metal material design according to the changes in ΔG_H [20]. The volcano plot shows the inherent advantages of noble metal for the appropriate ability to promote HER catalysis.

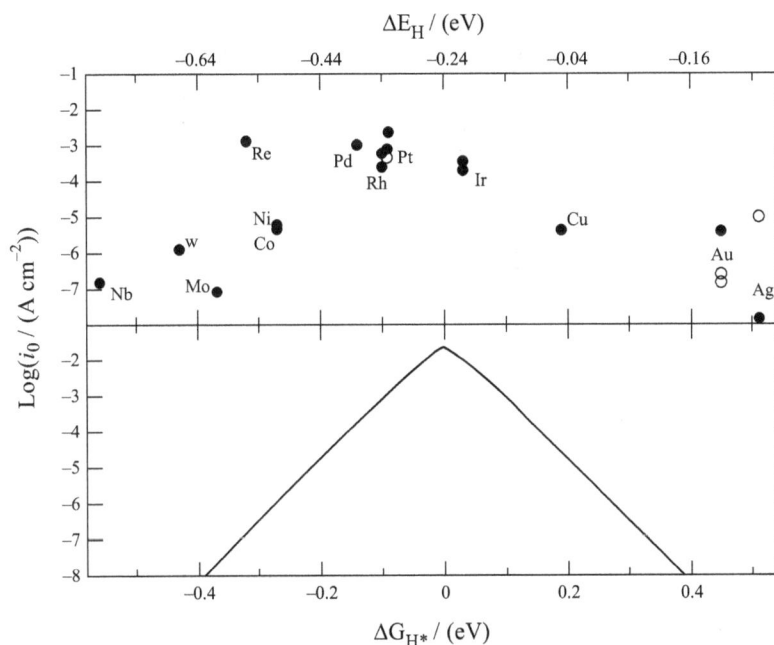

Fig. 4A.3: (Top) Experimentally measured exchange current, log(i_0), for hydrogen evolution over different metal surfaces plotted as a function of the calculated hydrogen chemisorption energy per atom, $\triangle E_H$ (top axis); (bottom) the result of the simple kinetic model is plotted as a function of the free energy for hydrogen adsorption [20] (Copyright 2005, Institute of Physics).

The increasing research interest in HER has accelerated the development of electrocatalysts. Electrocatalysts with lower overpotential and enhanced stability are desirable to promote applications and devices relying on hydrogen. This has promoted various synthesis methods.

4A.3 Advanced materials in hydrogen production

HER from water electrolysis is a promising strategy for producing clean hydrogen. Many challenges need to be overcome to lower the overpotential and accelerate the

HER kinetics in different media. Thus, for realizing efficient HER, numerous kinds of catalysts have been developed, including noble-metal, transition-metal-based, and metal-free electrocatalysts, which show promising performances and specific advantages for HER [21, 22].

4A.3.1 Noble-metal-based electrocatalyst

Noble metals, especially Pt, have attracted attention due to their intrinsic efficient HER properties. The inherent hydrogen adsorption energy which is close to zero makes it a promising candidate as an HER catalyst. However, the limitation and expense of precious metal sources restrict their use as electrocatalysts. Therefore, researchers have developed various kinds of methods to modify the materials. Figure 4A.4 illustrates the schematic of three main ways of structural regulation of noble-metal electrocatalysts, including shape modification, alloy composition, and single atom incorporation.

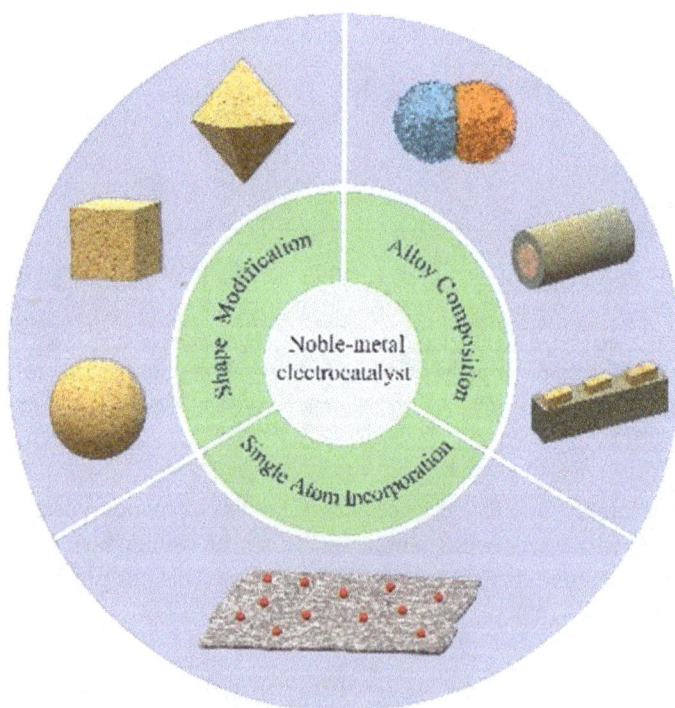

Fig. 4A.4: Schematic illustration of three ways of utilization of noble-metal electrocatalysts.

One possible way is to adjust the shape of the noble-metal nanoparticles to achieve a large specific surface area and high density of electrocatalytic sites. Among noble-metal electrocatalysts, platinum-based electrocatalysts are considered the best performance for HER. Different strategies have been developed to synthesize numerous catalyst shapes, including nanowires [23], hollow structures [24], dendrite-like shapes [25], nanocubes [26], and nanospheres [27].

The shape control of Pt nanoparticles is vital for improving HER performance. Different crystal planes are exposed on the surface to minimize the surface energy, which leads to different active planes. (111) and (100) facets are desirable for Pt nanoparticles [28]. Synthesis methods, including hydrothermal, solvothermal, sol–gel, and electrochemical deposition methods, are widely reported for HER electrocatalyst fabrication [29]. As shown by transmission electron microscopic (TEM) images in Fig. 4A.5, Pt nanoparticles with cuboctahedra, spheres, tetrapod, truncated cubs, stars, and dendrite-like shapes are presented [30]. The changing amounts of precursors, reaction temperature, surface ligands, and nucleation processes will produce different shapes due to the different nucleation rates and growth orientation. By tuning the shapes, the specific surface areas can be increased, along with electrocatalytic sites, increasing HER performances.

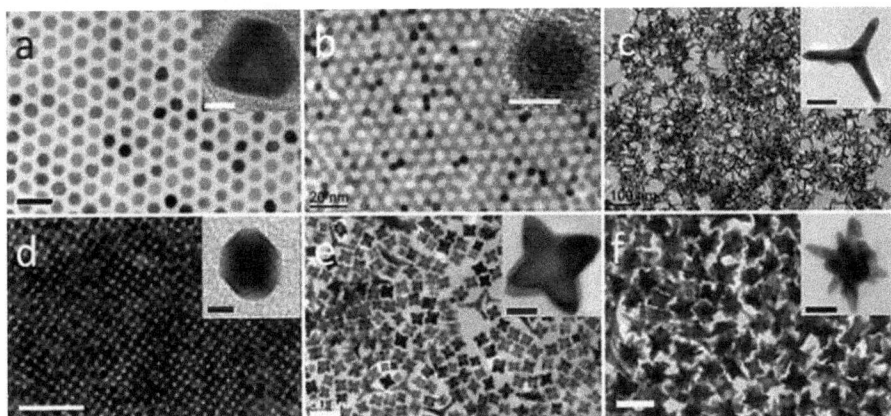

Fig. 4A.5: TEM images of (a) Pt cuboctahedra, (b) Pt spheres, (c) Pt tetrapods, (d) Pt truncated cubes, (e) Pt stars (octapods), (f) Pt multipods. Scale bars: (a, b) 20 nm, (c) 100 nm, (d, f) 50 nm; insets of (a, b, d) 5 nm; insets of (c, f) 20 nm; and inset of (e) 10 nm [30] (Copyright 2013, American Chemical Society).

Besides modifying shapes for noble nanoparticles, alloy nanoparticles are popular to achieve synergistic effects, which further improve the HER efficiency and stability of electrocatalysts compared with single-component noble-metal ones. The reaction is usually controlled by mixing two or several metal precursors to obtain alloy nanoparticles. Under different solvents, temperatures, and surfactants based on single-metal electrocatalysts, noble metals will be simultaneously reduced, and the ratio between metals can be controlled simply by adjusting the precursors.

Pt$_2$Pd alloy nanoparticles were synthesized through a one-step method [31]. Platinum phthalocyanine, palladium phthalocyanine, 2, 2-dipyridylacetyene, and graphene were added into a Teflon-lined autoclave and hydrothermally treated. Finally, the Pt$_2$Pd nanoparticles were embedded in N-doped carbon materials and shown to act as efficient HER electrocatalysts, as shown in Fig. 4A.6. The TEM image shows that individual Pt$_2$Pd alloy nanoparticles are hosted within the carbon materials. The high-resolution TEM (HRTEM) presents the lattice space of 0.225 nm, corresponding to the Pt(111) plane. The ratio between Pt and Pd is 2:1 according to the line-scanning image across an alloy nanoparticle. The energy-dispersive X-ray spectroscopy (EDS) mapping images confirm the chemical composition of the alloy nanoparticle.

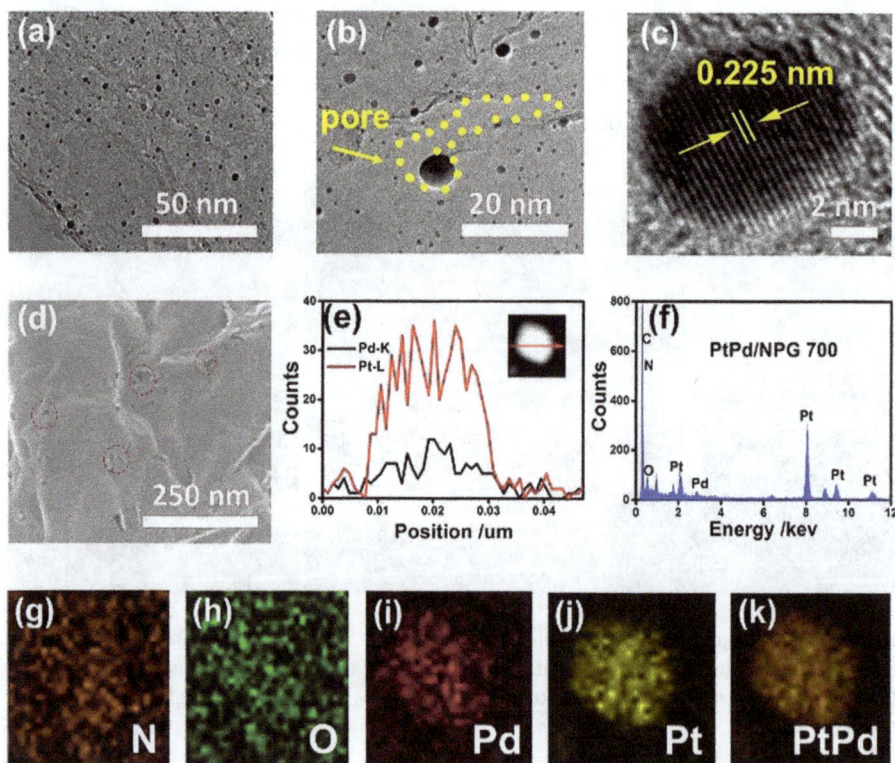

Fig. 4A.6: (a, b) TEM images of Pt$_2$Pd/NPG 700. (c) HRTEM image. (d) SEM image of Pt$_2$Pd/NPG 700. (e) Line-scanning profile across a Pt$_2$Pd nanoparticle, as indicated in the inset of (e). (f) EDX pattern of Pt$_2$Pd/NPG 700 catalyst. (g–k) EDS elemental mapping of N, O, Pd, Pt, and Pt$_2$Pd alloy, respectively [31] (Copyright 2017, Elsevier).

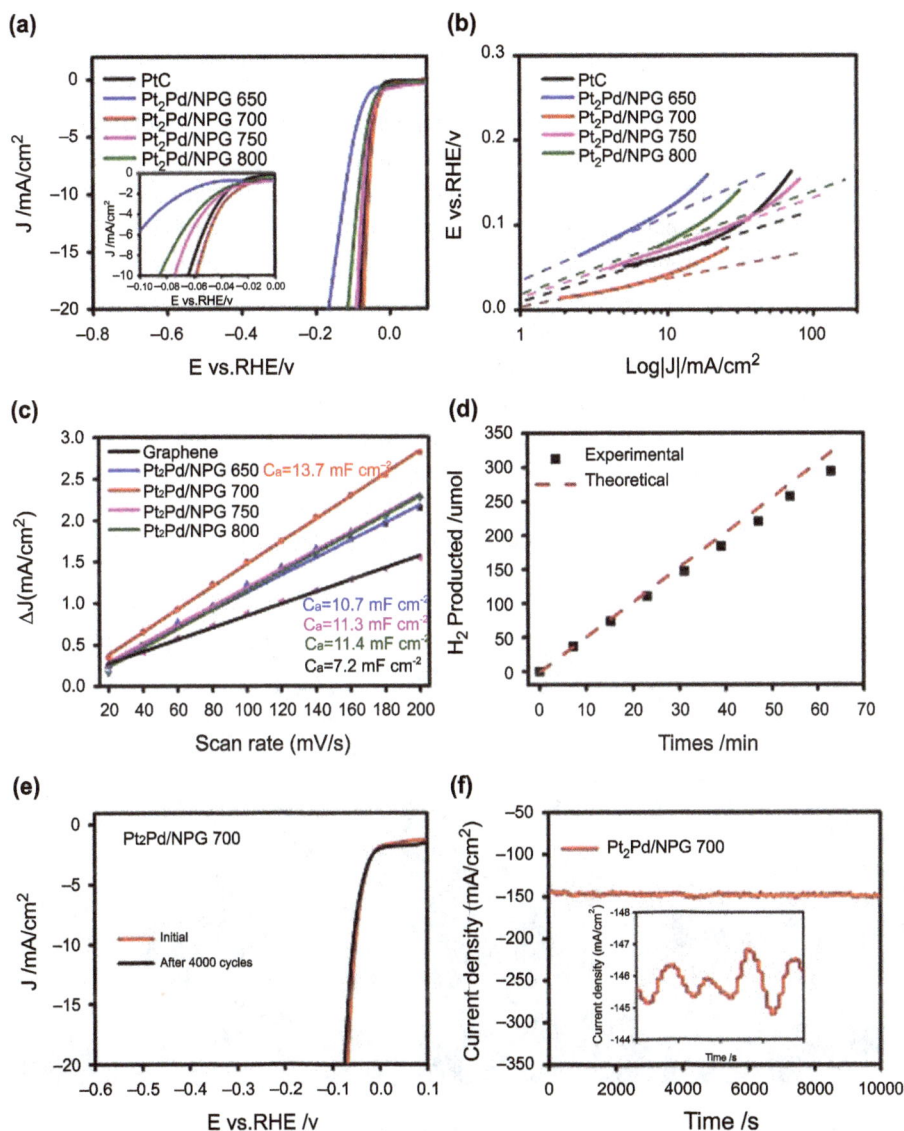

Fig. 4A.7: (a) HER performance in 0.5 M H_2SO_4. (b) Tafel plots and (c) double-layer capacitances (C_{dl}) for Pt_2Pd/NPG electrocatalysts treated with different temperatures. (d) Faradic efficiency of PtPd/NPG 700 catalysts. (e) Initial and 4,000th LSV curves of Pt_2Pd/NPG 700. (f) Time dependence of current density for Pt_2Pd/NPG 700 under a static overpotential of 140 mV. The inset shows an enlarged image [31] (Copyright 2017, Elsevier).

The HER performance was evaluated in 0.5 M H_2SO_4, as shown in Fig. 4A.7. Pt_2Pd nanoparticles annealed at different temperatures show remarkable performance, and Pt_2Pd/NPG 700 exhibits superior performance, even better than commercial Pt/C materials. The lowest is overpotential (η_{10}) at the current density of –10 mA/cm^2 for Pt$_2$Pd/NPG 700. The Tafel slopes and electrochemical double-layer capacitance (C_{dl}) are illustrated in Fig. 4A.7b and c, which reflect the electrochemical active surface area (ECSA). Hydrogen production is consistent with the theoretical simulation. Additionally, the durability was verified from 4,000 linear sweep voltammetry (LSV) cycles under long-term continuous applied potential. Pt$_2$Pd alloy nanoparticles embedded in graphene provide a synergistic effect on interfaces and facilitate the HER reaction.

Additionally, non-noble metals can be introduced to noble-metal electrocatalysts to simultaneously lower usage and tune the electrical property. The triple metal alloy nanowires were fabricated with S-doping through reduction methods. TEM and EDS elemental mapping images are shown in Fig. 4A.8. Firstly, AuPb nanowires were fabricated through reduction methods in an aqueous solution. Then the platinum source was added and reduced on AuPb nanowires to form AuPbPt alloy nanowires [32]. The lattice fringes in HRTEM are 0.235 and 0.234 nm, corresponding to Au's (111) plane and the intermetallic PbPt, respectively. The high-angle annular dark-field scanning TEM (HAADF-STEM) and the corresponding EDS mapping images illustrate the composition of Au, Pt, and Pb in the alloy nanowires.

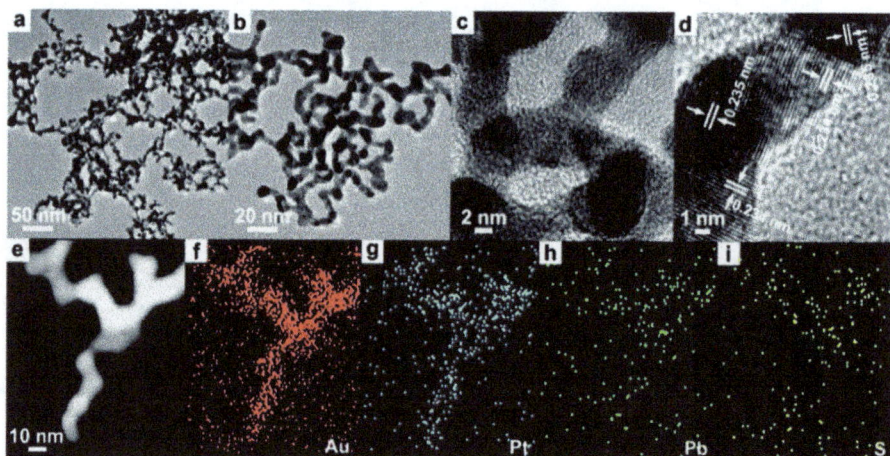

Fig. 4A.8: (a, b) TEM images and (c, d) HRTEM images of S-doped AuPbPt alloy NWs. (e) HAADF-STEM image and the corresponding EDS mapping images of (f) Au, (g) Pt, (h) Pb, and (i) S [32] (Copyright 2013, Royal Society of Chemistry).

The HER performance in acidic electrolytes is much better than commercial Pt/C electrocatalysts, as shown in Fig. 4A.9. The mass activity is 12.3 times (at –0.05 V vs. RHE) and 3.4 times (at –0.07 V vs. RHE) higher than the commercial Pt/C materials. The Tafel slope of the AuPbPt alloy is 17.7 mV/dec, lower than that of Pt/C, indicating that the Volmer–Tafel process is the determining step. After 5,000 cycles of cyclic voltammetry tests, the performance has almost no decrease, showing remarkable stability.

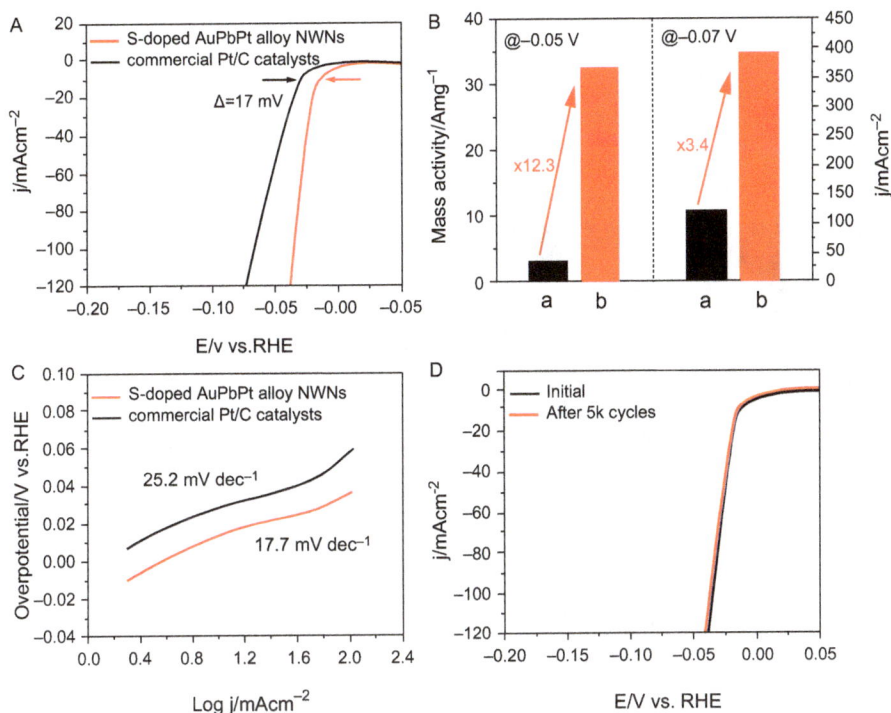

Fig. 4A.9: (a) HER performance of S-doped AuPbPt alloy NWs and commercial Pt/C catalysts in N₂-saturated 0.5 M H₂SO₄ solution. (b) Mass activity (at 0.05 V vs. RHE) and current density (at 0.07 V vs. RHE) of the S-doped AuPbPt alloy NWs and commercial Pt/C catalysts. (c) Tafel plots of S-doped AuPbPt alloy NWs and commercial Pt/C catalysts. (d) LSV curves of Si-doped AuPbPt alloy NWs before and after 5,000 cycles of the accelerated durability test in a 0.5 M H₂SO₄ solution [32] (Copyright 2013, Royal Society of Chemistry).

Single-atom catalysts (SACs) are realized to reach a maximum utilization rate of noble metals near 100 %. Methods that utilize the noble metal at an atomic level can substantially decrease the cost. Various methods, including photochemical reaction, atomic-layer deposition, and thermal reduction, have been developed [33]. Meantime, the HER performance is superior to noble-metal nanoparticles with the same noble-metal loading. Single atoms of noble metals anchored on substrates have been developed very quickly over recent years [34].

For SACs, the surface coordination status of the metal atom will greatly affect their electronic structures, which subsequently influence the catalytic activity. The substrates supporting single atoms are quite significant, such as graphene, C_3N_4, and MoS_2, and are widely explored [35].

A stable substrate and adjustable coordination conditions were studied to understand Pt single-atom electrocatalyst. Graphdiyne (GDY) was chosen as the substrate, and the atomically dispersed Pt atoms were loaded on GDY through a wet-chemical strategy. The as-prepared single-atom Pt can be observed from TEM images, as shown in Fig. 4A.10. The Pt-GDY1 and Pt-GDY2 samples are under different temperature treatments. The Pt-GDY1 was synthesized at room temperature while Pt-GDY2 was further treated at 200 °C from Pt-GDY1, making Pt in Pt-GDY2 a more positive valence state than in Pt-GDY1 [36].

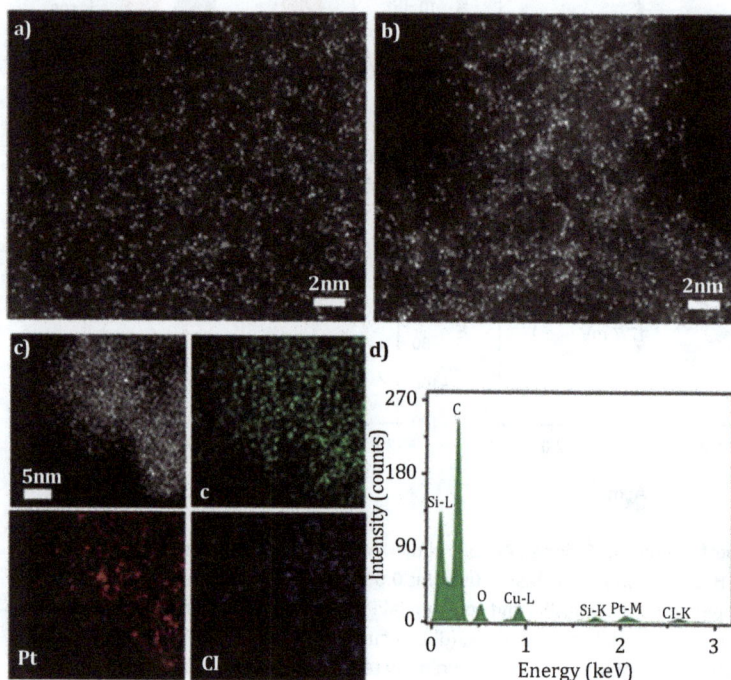

Fig. 4A.10: Atomic-resolution HAADF-STEM images of (a) Pt-GDY1 and (b) Pt-GDY2. (c) Elemental mapping of Pt-GDY2. (d) The EDS analysis of Pt-GDY2 [36] (Copyright 2018, John Wiley and Sons).

Figure 4A.10 shows the HAADF-STEM image of Pt single atom on GDY. The individual Pt atoms are observed, and the EDS mapping confirms the elemental composition. With the successful Pt loading on GDY, the HER performance of Pt-GDY2 shows much better HER performance than commercial Pt/C. The mass activity is 23.64 times that of Pt/C, illustrating the efficient Pt usage for electrocatalysis, as shown in Fig. 4A.11. The Tafel slope indicates that Volmer–Heyrovsky is the determining step. The 1,000 cycles of LSV tests confirm the excellent stability of SACs. The isolated Pt atoms were anchored on GDY by the coordination interactions of the C-Pt-Cl$_4$ group, and the higher unoccupied density of states of Pt 5d orbital makes Pt-GDY2 more efficient for HER.

In addition, Pt, Ru, Ir, and Pd were studied for HER with the development of SACs [37, 38]. With the development of noble-metal materials, from shape modification and alloy composition to single-atom incorporation, the HER performance increases with the decrease in the amount of noble-metal loading, making them promising for future

HER applications. Noble-metal-based electrocatalysts lead the best-in-class HER performance. The latest HER performance is listed in Tab. 4A.1.

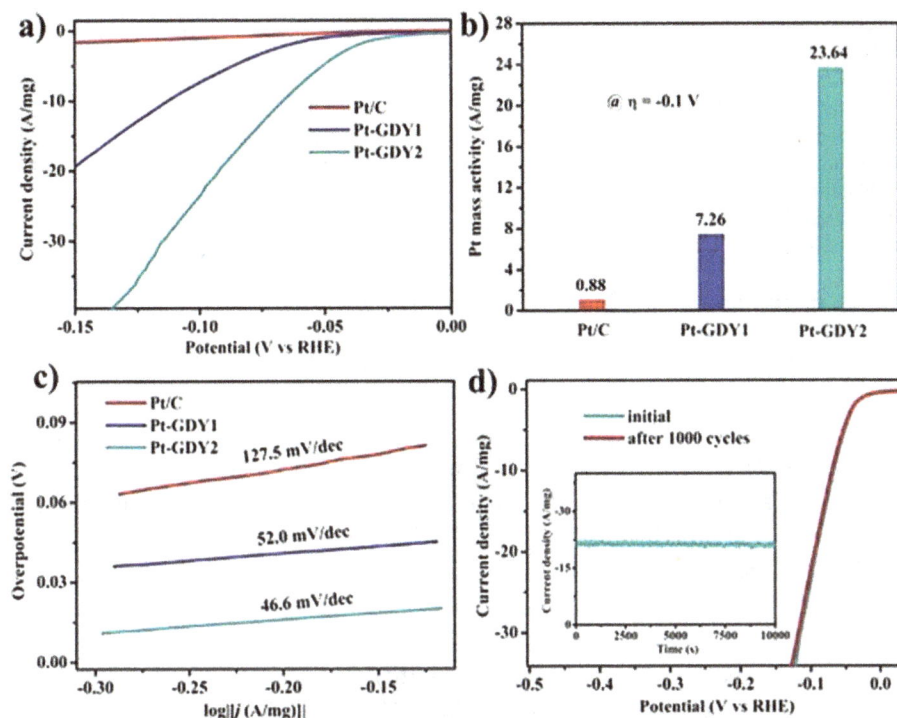

Fig. 4A.11: (a) LSV curves of Pt-GDY1, Pt-GDY2, and commercial Pt/C in 0.5 M H_2SO_4 solution. (b) The HER mass activity at $\eta = 0.1$ V for Pt/C, Pt-GDY1, and Pt-GDY2. (c) Tafel plots of Pt/C, Pt-GDY1, and Pt-GDY2. (d) The LSV curves of Pt-GDY2 at initial and after 1,000 cycles with a scan rate of 5 mV/s. The inset shows the time-dependent current density curve at 95 mV versus RHE [36] (Copyright 2018, John Wiley and Sons).

4A.3.2 Transition-metal-based electrocatalyst

Transition metals outside the Pt, Ir, Pd, and Ru groups are promising alternatives due to their abundance and occasional excellent HER performance. First-row transition metals have been introduced into noble-metal electrocatalysts to replace part of the noble metals and form alloy nanoparticles, Pt–Co [47] and Pt–Ni [48], that show impressive HER performances. Among transition-metal-based electrocatalysts, transition-metal phosphides (TMPs) have gained intensive attention due to their adjustable compositions and structures, tunable electronic properties, and remarkable electrical conductivities, which present comparable HER performance with commercial Pt/C [49].

Tab. 4A.1: Comparison of latest noble-metal-based electrocatalysts on HER performance.

Electrocatalyst	Overpotential η_{10} (mV)	Tafel slope (mV/dec)	Electrolyte	Reference
Pt dendrite	15	31	0.5 M H_2SO_4	[39]
Ru nanodot	14	32.5	1 M PBS	[40]
Pt cluster	35	61	1 M PBS	[25]
Pt/Ru nanocrystal	22	19	1 M KOH	[41]
Ru/Pd nanowire	11	50	1 M KOH	[42]
Pt/Ni nanowire	13	29	1 M KOH	[43]
Pt_{SA}-MWCNTS	43.9	30	0.5 M H_2SO_4	[44]
Ru_{SA}-N-$Ti_3C_2T_X$	27	29	1 M KOH	[45]
Pt_{SA}-CN	13	34	1 M $HClO_4$	[46]

More than 100 bimetallic phosphides have been studied [50], which provide large possibilities for tuning the ratios and compositions of TMPs and their electronic structures. The phosphorization process can successfully introduce P into transition-metal electrocatalysts, forming P-metal bonds. The comparatively higher electronegativity of P in the TMPs connected with metal atoms leads to a proton-acceptor process for promoting HER performance [50]. CoP [51], NiP [52], and bimetallic CoP/Ni_2P [53] and CoP/Co_2P [54] have been explored, and their HER performances are remarkable.

TMPs are often fabricated on substrates as self-standing electrodes. The vertical CoP nanoarray was fabricated through three steps. Firstly, the CoCH nanorods were formed from the hydrothermal method on Ti foil and then immersed in 2-methylimidazole to grow the ZIF-67 shell. Finally, CoP/NPC/TF was acquired through a phosphatization process at 400 °C, as shown in Fig. 4A.12.

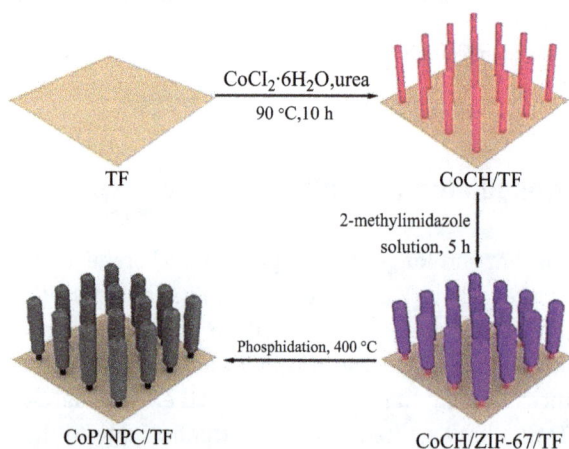

Fig. 4A.12: Schematic illustration of the synthetic process of the CoP/NPC/TF [55] (Copyright 2019, John Wiley and Sons).

The as-prepared CoP nanoarray performed in 0.5 M H_2SO_4 is shown in Fig. 4A.13. The optimized CoP/NPC/TF electrode shows comparable HER performance with commercial Pt/C. The Tafel slopes and overpotentials (η_{10}) are close to Pt/C; double-layer capacitance reflects the ECSA for the modified materials. The LSV cycling and long-term durability tests indicate its superb stability [55].

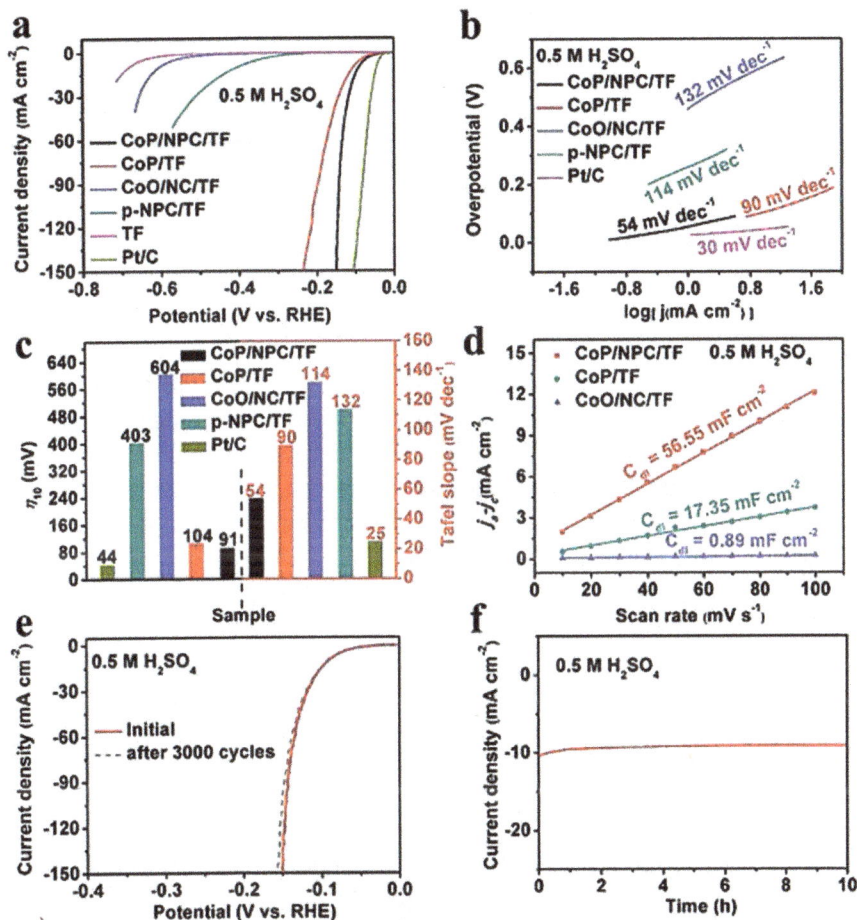

Fig. 4A.13: (a) HER performance and (b) Tafel plots of CoP/NPC/TF and other references with a scan rate of 2 mV/s in 0.5 M H_2SO_4. (c) The comparison of overpotential at 10 mA/cm^2 and Tafel slopes of CoP/NPC/TF and other references. (d) Double-layer capacitances (C_{dl}) at −0.15 V versus RHE as a function of scan rate for CoP/NPC/TF, CoP/TF, and CoO/NC/TF. (e) LSV curves of CoP/NPC/TF before and after 3,000 cycles. (f) Time-dependent current density curve of CoP/NPC/TF under a constant overpotential of 103 mV for 10 h [55] (Copyright 2019, John Wiley and Sons).

Transition-metal-based electrocatalysts have flourished, and some of the latest electrocatalysts are listed in Tab. 4A.2.

Tab. 4A.2: Comparison of latest transition-metal-based electrocatalyst on HER performance.

Electrocatalyst	Overpotential η_{10} (mV)	Tafel slope (mV/dec)	Electrolyte	Reference
Mn-Ni	80	68	1 M KOH	[56]
Co_{SA}/C	230	99	0.5 M H_2SO_4	[57]
$Ni_{SA}-MoS_2$	98	74	1 M KOH	[58]
B-CoP/CNT	79	80	1 M PBS	[59]
NiO-NiP	76	98	1 M KOH	[52]
CoP/Co_2P	103	61.2	1 M KOH	[60]
NiCoP/CoP	73	91.3	1 M KOH	[61]
CoP/CoMoP	34	33	1 M KOH	[62]

4A.3.3 Metal-free electrocatalyst

Noble-metal-based materials are regarded as the most efficient HER electrocatalysts. However, the limited abundance and high cost restrict future application, and thus developing electrocatalysts with earth-abundance materials is promising. Metal-free electrocatalysts are considered potentially efficient HER electrocatalysts, mostly carbon-based materials. However, the pristine carbon materials have limited and poor electrocatalytic activity. To increase the HER performance of carbon materials, a surface chemical environment is required by doping heteroatoms, such as N, S, and P [63]. The dopants in carbon materials lead to physical- and chemical-tuning properties, increasing the electrocatalytic HER performance.

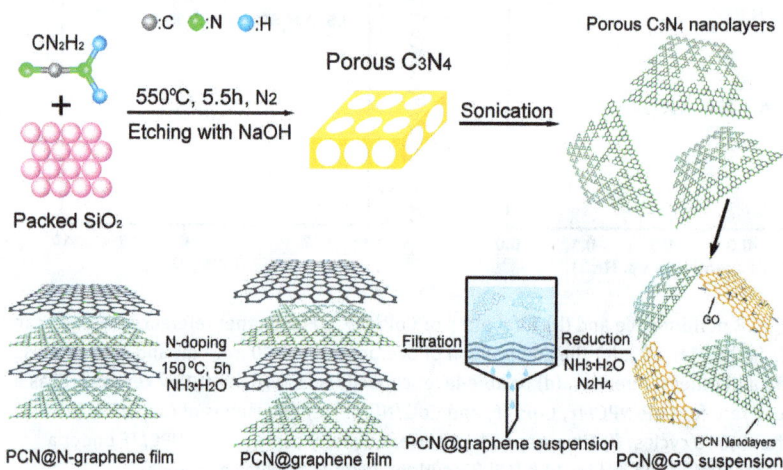

Fig. 4A.14: Schematic illustration of the fabrication process of PCN@N-graphene film [64] (Copyright 2015, American Chemical Society).

Nitrogen-doped carbon is a widely explored material in metal-free electrocatalysts. When nitrogen is introduced into the carbon, properties such as electronic structures and electrocatalytic activities are improved. The HER performance is based on the amount of N dopants. Moreover, the efficient N-doped carbon electrodes also depend on the surface structure. As shown in Fig. 4A.14, porous C_3N_4 is integrated with N-doped graphene film to form the catalyst without substrate [64]. The hierarchical structure, large specific surface area, and highly exposed electrocatalytic sites enable the electrode to achieve fabulous HER performance with a low onset overpotential and excellent stability, as shown in Fig. 4A.15.

The HER performance is much closer to Pt/C with N doping in graphene and annealing treatment, verified by the low Tafel slopes and overpotentials (η_{10}). The charge transfer resistance (R_{ct}) is indicated in Fig. 4A.15d, which is related to the kinetics. The lower value of R_{ct} presents the fast reaction during HER. The 5,000 cycles of LSV tests activate the as-prepared materials, resulting in better performance. Meantime, the stability is also remarkable for long-term tests.

However, due to the intrinsic properties of carbon materials, the efficient HER performance is still a challenge, which could not hinder its advantages of stability, low cost, and recycling properties. More importantly, double or triple hetero-atom doping was further explored, such as N and S and N and P doping in carbon material [63]. The metal-free electrocatalyst is another way to realize stable and efficient hydrogen production. The latest HER performance of metal-free-based electrocatalyst is listed in Tab. 4A.3.

4A.4 Hydrogen storage

The utilization of hydrogen directly from water splitting is limited, and thus hydrogen storage and transport are quite significant. Developing a safe, cost-effective method for hydrogen storage is still a challenge. Nowadays, compression, liquid, chemical, and physical storage are the four main storage methods that are widely explored [70].

4A.4.1 Compression storage

Hydrogen can be stored as a compressed gas in high-pressure tanks, and it can be transported through pipelines like natural gas. The compression storage method is simple and can increase the energy density for hydrogen but requires high-level storage equipment.

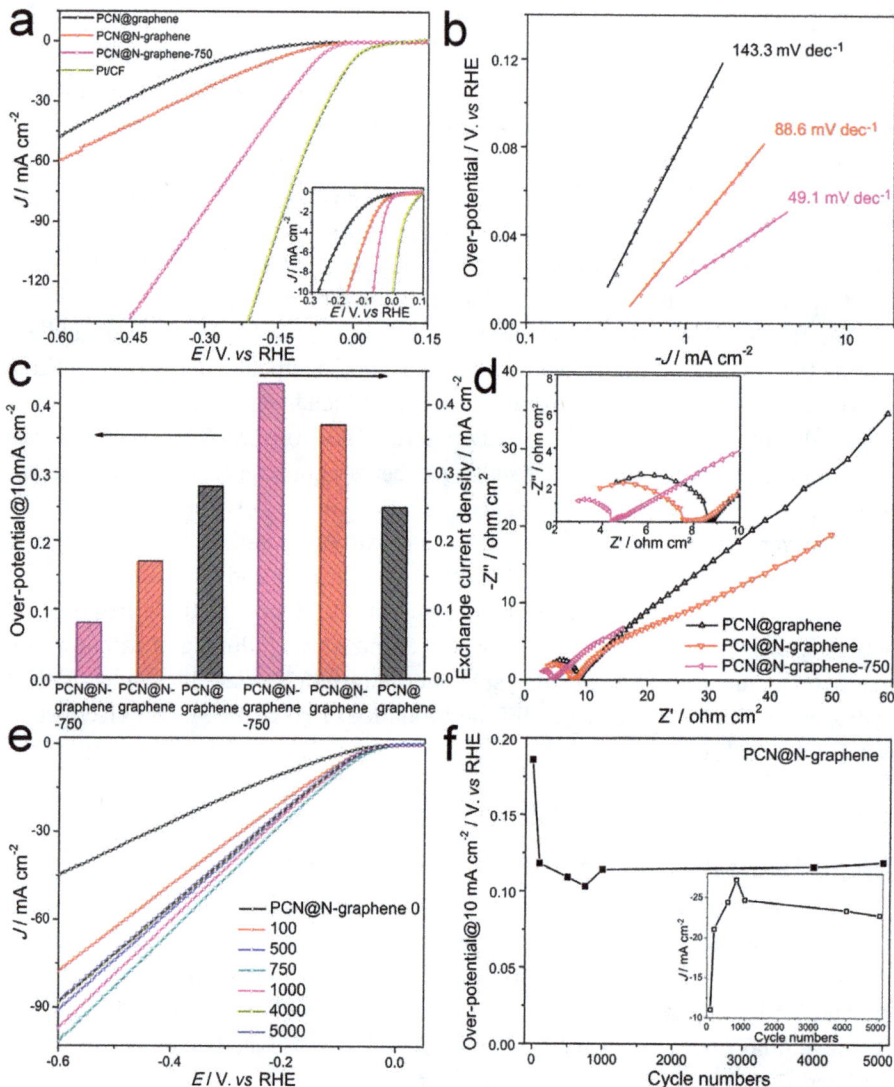

Fig. 4A.15: (a) LSV curves (inset shows LSV curves with the current density below 10 mA/cm^2). (b) Tafel plots and (c) overpotential@10 mA/cm^2 versus RHE (left) and exchange current density (right). (d) Electrochemical impedance spectra at 0.2 V versus RHE of PCN@graphene, PCN@N-graphene, and PCN@N-graphene-750 films. (e) The polarization curves after different cyclic voltammetry (CV) cycles. (f) Required overpotential@10 mA/cm^2 versus RHE plotted as CV cycle numbers of PCN@N-graphene film (inset is the current density@0.2 V versus RHE [64] (Copyright 2015, American Chemical Society).

Tab. 4A.3: Comparison of latest metal-free electrocatalyst on HER performance.

Electrocatalyst	Overpotential η_{10} (mV)	Tafel slope (mV/dec)	Electrolyte	Reference
g-C$_3$N$_4$/N-graphene	80	49.1	0.5 M H$_2$SO$_4$	[64]
N, S-graphene	130	80.5	0.5 M H$_2$SO$_4$	[65]
N, P-graphene	420	91	0.5 M H$_2$SO$_4$	[66]
N, P-carbon	204	58	0.5 M H$_2$SO$_4$	[67]
g-C$_3$N$_4$ nanoribbons/ graphene	207	54	0.5 M H$_2$SO$_4$	[68]
TpPAM	250	106	0.5 M H$_2$SO$_4$	[69]

4A.4.2 Liquid storage

Liquid hydrogen storage is a normal storage method, but the high pressure required for liquid hydrogen storage leads to many challenges for the fabrication of storage equipment. Meantime, the low temperature reaching 21 K is also quite strict. The efficient load and release processes are also difficult.

4A.4.3 Chemical storage

Nowadays, more and more researchers focus on hydrogen storage materials to realize the fast adsorption and release of hydrogen in a favorable environment. The chemical storage materials involve chemical bond forming and breaking. Under certain temperatures and pressures, hydrogen can be adsorbed and generated through a chemical reaction in many materials, including metal hydrides (LiBH$_4$, NaAlH$_4$, NaBH$_4$, etc.) [71, 72].

4A.4.4 Physical storage

Molecular hydrogen can get absorbed on the surface of some materials, which are called physical storage materials. Physisorption only involves physical adsorption and desorption, which requires high pressure for large amounts of storage, unlike chemical storage materials. The most studied materials are carbon materials (graphene and nanotubes) [73], organic metal frameworks [74], and covalent organic frameworks [75].

With the rapid development of electrocatalytic water splitting and advanced materials, hydrogen storage technology and materials are pursued as an essential foundation for renewable energy systems. The comparison among different storage methods is shown in Tab. 4A.4.

Tab. 4A.4: Comparison of four hydrogen storage methods.

	Chemical storage	Physical storage	Compressed storage	Liquid storage
Interaction	Chemical bonding	Van der Waals	Compression	High-pressure compression
Energy requirement	Large	Low	Large	Large
Transfer rate	Slow	Fast	Fast	Fast
Temperature (K)	400–600	Varies	273	21
Volumetric capacity (kg/m^3)	~150	~20	~30	~70
Advantage	High storage density Feature selectivity	Fast reversible process Low storage cost	Lightweight Easy storage technology	Long-term storage Highest energy density
Limitation	The slow activation process Impurity absorption High temperature Slow-release kinetic	Weak interaction High pressure Volumetrically inefficient	High-pressure tanks Volumetrically inefficient	High-pressure tanks Energy loss High pressure Low temperature

4A.5 Summary

With the fast development of hydrogen energy, future energy systems will be more adjustable, reducing pollution and optimizing the energy network caused by the uneven consumption and distribution of fossil fuels. Electrochemical water splitting has great potential due to its easy fabrication processes, pure hydrogen collection, and safety steps. Meanwhile, the studies of numerous electrocatalysts, working mechanisms, and hydrogen storage materials lead to more and more efficient hydrogen production and storage, which shows promising future applications in the whole clean energy system.

References

[1]   Barreto, L., Makihira, A. & Riahi, K. (2003). The hydrogen economy in the twenty-first century: a sustainable development scenario. International Journal of Hydrogen Energy, 28, 267–284.

[2]   Chen, G., Li, J., Li, K., Lin, F., Tian, W., Che, L., Yan, B., Ma, W. & Song, Y. (2020). Nitrogen, sulfur, and chlorine-containing pollutants are releasing characteristics during pyrolysis and combustion of oily sludge. Fuel, 273, 117772.

[3]   Gielen, D., Boshell, F., Saygin, D., Bazilian, M. D., Wagner, N. & Gorini, R. (2019). The role of renewable energy in the global energy transformation. Energy Strategy Reviews, 24, 38–50.

[4]   Neumann, A., Sorge, L., von Hirschhausen, C. & Wealer, B. (2020). Democratic quality, and nuclear power: reviewing the global determinants for introducing nuclear energy in 166 countries. Energy Research & Social Science, 63, 101389.

[5]   Esposito, D. V. (2017). Membraneless electrolyzers for low-cost hydrogen production in a renewable energy future. Joule, 1, 651–658.

[6]   Mayyas, A., Wei, M. & Levis, G. (2020). Hydrogen as a long-term, large-scale energy storage solution when coupled with renewable energy sources or grids with dynamic electricity pricing schemes. International Journal of Hydrogen Energy, 45, 16311–16325.

[7]   Song, Y., Zhang, C., Ling, C.-Y., Han, M., Yong, R.-Y., Sun, D. & Chen, J. (2020). Review on current research of materials, fabrication, and application for bipolar plate in proton exchange membrane fuel cell. International Journal of Hydrogen Energy, 45, 29832–29847.

[8]   Zhang, G. & Jiao, K. (2018). Multi-phase models for water and thermal management of proton exchange membrane fuel cell: a review. Journal of Power Sources, 391, 120–133.

[9]   Hassanpouryouzband, A., Joonaki, E., Edlmann, K., Heinemann, N. & Yang, J. (2020). Thermodynamic and transport properties of hydrogen-containing streams. Scientific Data, 7, 222.

[10] Xu, R., Chou, L.-C. & Zhang, W.-H. (2019). The effect of CO_2 emissions and economic performance on hydrogen-based renewable production in 35 European Countries. International Journal of Hydrogen Energy, 44, 29418–29425.

[11] Changizian, S., Ahmadi, P., Raeesi, M. & Javani, N. (2020). Performance optimization of hybrid hydrogen fuel cell electric vehicles in real driving cycles. International Journal of Hydrogen Energy, 45, 35180–35197.

[12] Sazali, N. (2020). Emerging technologies by hydrogen: a review. International Journal of Hydrogen Energy, 45, 18753–18771.

[13] Cai, L., He, T., Xiang, Y. & Guan, Y. (2020). Study on the reaction pathways of steam methane reforming for H_2 production. energy, 207, 118296.

[14] Ji, G., Zhao, M. & Wang, G. (2018). Computational fluid dynamic simulation of a sorption-enhanced palladium membrane reactor for enhancing hydrogen production from methane steam reforming. energy, 147, 884–895.

[15] Filippov, S. P. & Yaroslavtsev, A. B. (2021). Hydrogen energy: development prospects and materials. Russian Chemical Reviews, 90, 627–643.

[16] Ganguly, P., Harb, M., Cao, Z., Cavallo, L., Breen, A., Dervin, S., Dionysiou, D. D. & Pillai, S. C. (2019). 2D nanomaterials for photocatalytic hydrogen production. ACS Energy Letters, 4, 1687–1709.

[17] Soloveichik, G. (2019). Electrochemical synthesis of ammonia as a potential alternative to the Haber–Bosch process. Nature Catalysis, 2, 377–380.

[18] Chen, S., Li, M., Gao, M., Jin, J., van Spronsen, M. A., Salmeron, M. B. & Yang, P. (2020). High-performance Pt-Co nanoframes for fuel-cell electrocatalysis. Nano Letters, 20, 1974–1979.

[19] Ďurovič, M., Hnát, J. & Bouzek, K. (2021). Electrocatalysts for the hydrogen evolution reaction in alkaline and neutral media. A comparative review. Journal of Power Sources, 493, 229708.

[20] Nørskov, J. K., Bligaard, T., Logadottir, A., Kitchin, J. R., Chen, J. G., Pandelov, S. & Stimming, U. (2005). Trends in the exchange current for hydrogen evolution. Journal of the Electrochemical Society, 152, J23.

[21] Cai, J., Javed, R., Ye, D., Zhao, H. & Zhang, J. (2020). Recent progress in noble metal nanocluster and single-atom electrocatalysts for the hydrogen evolution reaction. Journal of Materials Chemistry A, 8, 22467–22487.

[22] Zou, X. & Zhang, Y. (2015). Noble metal-free hydrogen evolution catalysts for water splitting. Chemical Society Reviews, 44, 5148–5180.

[23] Li, H., Wu, X., Tao, X., Lu, Y. & Wang, Y. (2020). Direct synthesis of ultrathin Pt nanowire arrays as catalysts for methanol oxidation. Small, 16, 2001135.

[24] Ge, C., Wu, R., Chong, Y., Fang, G., Jiang, X., Pan, Y., Chen, C. & Yin, -J.-J. (2018). Synthesis of Pt hollow nanodendrites with enhanced peroxidase-like activity against bacterial infections: implication for wound healing. Advanced Functional Materials, 28, 1801484.

[25] Tan, Y., Xie, R., Zhao, S., Lu, X., Liu, L., Zhao, F., Li, C., Jiang, H., Chai, G., Brett, D. J. L., Shearing, P. R., He, G. & Parkin, I. P. (2021). Facile fabrication of robust hydrogen evolution electrodes under high current densities via Pt@Cu interactions. Advanced Functional Materials, 31, 2105579.

[26] Darmadi, I., Stolaś, A., Östergren, I., Berke, B., Nugroho, F. A. A., Minelli, M., Lerch, S., Tanyeli, I., Lund, A., Andersson, O., Zhdanov, V. P., Liebi, M., Moth-Poulsen, K., Müller, C. & Langhammer, C. (2020). Bulk-processed Pd nanocube–poly(methyl methacrylate) nanocomposites as plasmonic plastics for hydrogen sensing. ACS Applied Nano Materials, 3, 8438–8445.

[27] Su, D. W., Dou, S. X. & Wang, G. X. (2015). Hierarchical Ru nanospheres as highly effective cathode catalysts for Li-O_2 batteries. Journal of Materials Chemistry A, 3, 18384–18388.

[28] Duan, S., Du, Z., Fan, H. & Wang, R. (2018). Nanostructure optimization of platinum-based nanomaterials for catalytic applications. Nanomaterials, 8.

[29] Zhang, L., Doyle-Davis, K. & Sun, X. (2019). Pt-based electrocatalysts with high atom utilization efficiency: from nanostructures to single atoms. Energy & Environmental Science, 12, 492–517.

[30] Kang, Y., Pyo, J. B., Ye, X., Diaz, R. E., Gordon, T. R., Stach, E. A. & Murray, C. B. (2013). Shape-controlled synthesis of Pt nanocrystals: the role of metal carbonyls. ACS Nano, 7, 645–653.

[31] Zhong, X., Qin, Y., Chen, X., Xu, W., Zhuang, G., Li, X. & Wang, J. (2017). PtPd alloy embedded in nitrogen-rich graphene nanopores: high-performance bifunctional electrocatalysts for hydrogen evolution and oxygen reduction. carbon, 114, 740–748.

[32] Zhang, X., Wang, S., Wu, C., Li, H., Cao, Y., Li, S. & Xia, H. (2020). Synthesis of S-doped AuPbPt alloy nanowire-networks as superior catalysts towards the ORR and HER. Journal of Materials Chemistry A, 8, 23906–23918.

[33] Zhang, L., Ren, Y., Liu, W., Wang, A. & Zhang, T. (2018). Single-atom catalyst: a rising star for green synthesis of fine chemicals. National Science Review, 5, 653–672.

[34] Liu, Q. & Zhang, Z. (2019). Platinum single-atom catalysts: a comparative review towards effective characterization. Catalysis Science and Technology, 9, 4821–4834.

[35] Deng, J., Li, H., Xiao, J., Tu, Y., Deng, D., Yang, H., Tian, H., Li, J., Ren, P. & Bao, X. (2015). Triggering the electrocatalytic hydrogen evolution activity of the inert two-dimensional MoS_2 surface via single-atom metal doping. Energy & Environmental Science, 8, 1594–1601.

[36] Yin, X.-P., Wang, H.-J., Tang, S.-F., Lu, X.-L., Shu, M., Si, R. & Lu, T.-B. (2018). Engineering the coordination environment of single-atom platinum anchored on graphdiyne for optimizing electrocatalytic hydrogen evolution. Angewandte Chemie International Edition, 57, 9382–9386.

[37] Lai, W.-H., Zhang, L.-F., Hua, W.-B., Indris, S., Yan, Z.-C., Hu, Z., Zhang, B., Liu, Y., Wang, L., Liu, M., Liu, R., Wang, Y.-X., Wang, J.-Z., Hu, Z., Liu, H.-K., Chou, S.-L. & Dou, S.-X. (2019). General π-electron-assisted strategy for Ir, Pt, Ru, Pd, Fe, Ni single-atom electrocatalysts with

bifunctional active sites for highly efficient water splitting. Angewandte Chemie International Edition, 58, 11868–11873.

[38] Li, C. & Baek, J.-B. (2020). Recent advances in noble metal (Pt, Ru, and Ir)-based electrocatalysts for efficient hydrogen evolution reaction. ACS Omega, 5, 31–40.

[39] Lin, L., Sun, Z., Yuan, M., He, J., Long, R., Li, H., Nan, C., Sun, G. & Ma, S. (2018). Significant enhancement of hydrogen evolution reaction performance through a shape-controlled synthesis of hierarchical dendrite-like platinum. Journal of Materials Chemistry A, 6, 8068–8077.

[40] Xu, C., Ming, M., Wang, Q., Yang, C., Fan, G., Wang, Y., Gao, D., Bi, J. & Zhang, Y. (2018). Facile synthesis of effective Ru nanoparticles on carbon by adsorption-low temperature pyrolysis strategy for hydrogen evolution. Journal of Materials Chemistry A, 6, 14380–14386.

[41] Li, Y., Pei, W., He, J., Liu, K., Qi, W., Gao, X., Zhou, S., Xie, H., Yin, K., Gao, Y., He, J., Zhao, J., Hu, J., Chan, T.-S., Li, Z., Zhang, G. & Liu, M. (2019). Hybrids of PtRu nanoclusters and black phosphorus nanosheets for highly efficient alkaline hydrogen evolution reaction. ACS Catalysis, 9, 10870–10875.

[42] Kweon, Y., Noh, S. & Shim, J. H. (2021). Low content Ru-incorporated Pd nanowires for bifunctional electrocatalysis. RSC Advances, 11, 28775–28784.

[43] Xie, Y., Cai, J., Wu, Y., Zang, Y., Zheng, X., Ye, J., Cui, P., Niu, S., Liu, Y., Zhu, J., Liu, X., Wang, G. & Qian, Y. (2019). Boosting water dissociation kinetics on Pt–Ni nanowires by N-induced orbital tuning. Advanced Materials, 31, 1807780.

[44] Ji, J., Zhang, Y., Tang, L., Liu, C., Gao, X., Sun, M., Zheng, J., Ling, M., Liang, C. & Lin, Z. (2019). Platinum single-atom and cluster anchored on functionalized MWCNTs with ultrahigh mass efficiency for electrocatalytic hydrogen evolution. Nano Energy, 63, 103849.

[45] Liu, H., Hu, Z., Liu, Q., Sun, P., Wang, Y., Chou, S., Hu, Z. & Zhang, Z. (2020). Single-atom Ru anchored in nitrogen-doped MXene (Ti_3C_2Tx) as an efficient catalyst for the hydrogen evolution reaction at all pH values. Journal of Materials Chemistry A, 8, 24710–24717.

[46] Chen, S., Lv, C., Liu, L., Li, M., Liu, J., Ma, J., Hao, P., Wang, X., Ding, W., Xie, M. & Guo, X. (2021). High-temperature treatment to engineer the single-atom Pt coordination environment towards highly efficient hydrogen evolution. Journal of Energy Chemistry, 59, 212–219.

[47] Zhang, S. L., Lu, X. F., Wu, Z.-P., Luan, D. & Lou, X. W. (2021). Engineering Platinum–Cobalt Nano-alloys in Porous Nitrogen-Doped Carbon Nanotubes for Highly Efficient Electrocatalytic Hydrogen Evolution. Angewandte Chemie International Edition, 60, 19068–19073.

[48] Zhang, C., Chen, B., Mei, D. & Liang, X. (2019). The OH^--driven synthesis of Pt–Ni nanocatalysts with atomic segregation for alkaline hydrogen evolution reaction. Journal of Materials Chemistry A, 7, 5475–5481.

[49] Shi, Y. & Zhang, B. (2016). Recent advances in transition metal phosphide nanomaterials: synthesis and applications in hydrogen evolution reaction. Chemical Society Reviews, 45, 1529–1541.

[50] Weng, C.-C., Ren, J.-T. & Yuan, Z.-Y. (2020). Transition metal phosphide-based materials for efficient electrochemical hydrogen evolution: a critical review. ChemSusChem, 13, 3357–3375.

[51] Yang, X., Lu, A.-Y., Zhu, Y., Hedhili, M. N., Min, S., Huang, K.-W., Han, Y. & Li, L.-J. (2015). CoP nanosheet assembly, grew on carbon cloth: a highly efficient electrocatalyst for hydrogen generation. Nano Energy, 15, 634–641.

[52] Sun, C., Wang, H., Ren, J., Wang, X. & Wang, R. (2021). Inserting ultrafine NiO nanoparticles into amorphous NiP sheets by in situ phase reconstruction for high-stability of the HER catalysts. Nanoscale, 13, 13703–13708.

[53] Feng, T., Wang, F., Xu, Y., Chang, M., Jin, X., Yulin, Z., Piao, J. & Lei, J. (2021). CoP/Ni$_2$P heteronanoparticles integrated with atomic Co/Ni dual sites for enhanced electrocatalytic performance toward hydrogen evolution. International Journal of Hydrogen Energy, 46, 8431–8443.

[54] Chen, T., Ye, B., Dai, H., Qin, S., Zhang, Y. & Yang, Q. (2021). Ni-doped CoP/Co$_2$P nanospheres as highly efficient and stable hydrogen evolution catalysts in acidic and alkaline mediums. Journal of Solid State Chemistry, 301, 122299.

[55] Huang, X., Xu, X., Li, C., Wu, D., Cheng, D. & Cao, D. (2019). Vertical CoP Nanoarray Wrapped by N, P-Doped Carbon for Hydrogen Evolution Reaction in Both Acidic and Alkaline Conditions. Advanced Energy Materials, 9, 1803970.

[56] Shao, Q., Wang, Y., Yang, S., Lu, K., Zhang, Y., Tang, C., Song, J., Feng, Y., Xiong, L., Peng, Y., Li, Y., Xin, H. L. & Huang, X. (2018). Stabilizing and Activating Metastable Nickel Nanocrystals for Highly Efficient Hydrogen Evolution Electrocatalysis. ACS Nano, 12, 11625–11631.

[57] Hossain, M. D., Liu, Z., Zhuang, M., Yan, X., Xu, G.-L., Gadre, C. A., Tyagi, A., Abidi, I. H., Sun, C.-J., Wong, H., Guda, A., Hao, Y., Pan, X., Amine, K. & Luo, Z. (2019). Rational Design of Graphene-Supported Single-Atom Catalysts for Hydrogen Evolution Reaction. Advanced Energy Materials, 9, 1803689.

[58] Wang, Q., Zhao, Z. L., Dong, S., He, D., Lawrence, M. J., Han, S., Cai, C., Xiang, S., Rodriguez, P., Xiang, B., Wang, Z., Liang, Y. & Gu, M. (2018). Design of active nickel single-atom decorated MoS$_2$ as a pH-universal catalyst for hydrogen evolution reaction. Nano Energy, 53, 458–467.

[59] Cao, E., Chen, Z., Wu, H., Yu, P., Wang, Y., Xiao, F., Chen, S., Du, S., Xie, Y., Wu, Y. & Ren, Z. (2020). Boron-Induced Electronic-Structure Reformation of CoP Nanoparticles Drives Enhanced pH-Universal Hydrogen Evolution. Angewandte Chemie International Edition, 59, 4154–4160.

[60] Chen, L., Zhang, Y., Wang, H., Wang, Y., Li, D. & Duan, C. (2018). Cobalt-layered double hydroxides derived CoP/Co$_2$P hybrids for electrocatalytic overall water splitting. Nanoscale, 10, 21019–21024.

[61] Liu, H., Ma, X., Hu, H., Pan, Y., Zhao, W., Liu, J., Zhao, X., Wang, J., Yang, Z., Zhao, Q., Ning, H. & Wu, M. (2019). Robust NiCoP/CoP heterostructures for highly efficient hydrogen evolution electrocatalysis in alkaline solution. ACS Applied Materials & Interfaces, 11, 15528–15536.

[62] Huang, X., Xu, X., Luan, X. & Cheng, D. (2020). CoP nanowires coupled with CoMoP nanosheets as a highly efficient cooperative catalyst for hydrogen evolution reaction. Nano Energy, 68, 104332.

[63] Zhou, W., Jia, J., Lu, J., Yang, L., Hou, D., Li, G. & Chen, S. (2016). Recent developments of carbon-based electrocatalysts for hydrogen evolution reaction. Nano Energy, 28, 29–43.

[64] Duan, J., Chen, S., Jaroniec, M. & Qiao, S. Z. (2015). Porous C$_3$N$_4$ nanolayers@N-graphene films as catalyst electrodes for highly efficient hydrogen evolution. ACS Nano, 9, 931–940.

[65] Ito, Y., Cong, W., Fujita, T., Tang, Z. & Chen, M. (2015). High catalytic activity of nitrogen and sulfur co-doped nanoporous graphene in the hydrogen evolution reaction. Angewandte Chemie International Edition, 54, 2131–2136.

[66] Zheng, Y., Jiao, Y., Li, L. H., Xing, T., Chen, Y., Jaroniec, M. & Qiao, S. Z. (2014). Toward design of synergistically active carbon-based catalysts for electrocatalytic hydrogen evolution. ACS Nano, 8, 5290–5296.

[67] Wei, L., Karahan, H. E., Goh, K., Jiang, W., Yu, D., Birer, Ö., Jiang, R. & Chen, Y. (2015). A high-performance metal-free hydrogen-evolution reaction electrocatalyst from bacterium derived carbon. Journal of Materials Chemistry A, 3, 7210–7214.

[68] Zhao, Y., Zhao, F., Wang, X., Xu, C., Zhang, Z., Shi, G. & Qu, L. (2014). Graphitic carbon nitride nanoribbons: graphene-assisted formation and synergic function for highly efficient hydrogen evolution. Angewandte Chemie International Edition, 53, 13934–13939.
[69] Patra, B. C., Khilari, S., Manna, R. N., Mondal, S., Pradhan, D., Pradhan, A. & Bhaumik, A. (2017). A metal-free covalent organic polymer for electrocatalytic hydrogen evolution. ACS Catalysis, 7, 6120–6127.
[70] Niaz, S., Manzoor, T. & Pandith, A. H. (2015). Hydrogen storage: materials, methods, and perspectives. Renewable and Sustainable Energy Reviews, 50, 457–469.
[71] Ali, N. A. & Ismail, M. (2021). Modification of NaAlH4 properties using catalysts for solid-state hydrogen storage: a review. International Journal of Hydrogen Energy, 46, 766–782.
[72] Li, Z., Gao, M., Gu, J., Xian, K., Yao, Z., Shang, C., Liu, Y., Guo, Z. & Pan, H. (2020). In situ introduction of Li_3BO_3 and NbH leads to superior cyclic stability and kinetics of a LiBH4-based hydrogen storage system. ACS Applied Materials & Interfaces, 12, 893–903.
[73] Salehabadi, A., Salavati-Niasari, M. & Ghiyasiyan-Arani, M. (2018). Self-assembly of hydrogen storage materials based on multi-walled carbon nanotubes (MWCNTs) and Dy3Fe5O12 (DFO) nanoparticles. Journal of Alloys and Compounds, 745, 789–797.
[74] Suresh, K., Aulakh, D., Purewal, J., Siegel, D. J., Veenstra, M. & Matzger, A. J. (2021). Optimizing hydrogen storage in MOFs through engineering of crystal morphology and control of crystal size. Journal of the American Chemical Society, 143, 10727–10734.
[75] Tong, M., Zhu, W., Li, J., Long, Z., Zhao, S., Chen, G. & Lan, Y. (2020). An easy way to identify high performing covalent organic frameworks for hydrogen storage. ChemComm, 56, 6376–6379.

Wen Lei, Shaowei Zhang, Haijun Zhang

Chapter 4B
Nanostructured materials for electrocatalytic hydrogen evolution reaction

Abstract: Hydrogen (H$_2$) is one of the most importantly clean and renewable energy sources for future energy sustainability. Electrocatalytic hydrogen evolution reaction (HER) from water splitting is considered one of the most efficient ways to convert sustainable energy to the clean energy carrier, H$_2$. In recent years, nanomaterials have attracted much research interest for renewable energy applications due to their diversity and readily tunable electronic, optical, physical, and chemical properties. Nanomaterial-based catalysts have been extensively investigated as next-generation promising ones for HER. In this chapter, the fundamental mechanisms underpinning the electrocatalytic HER are highlighted. The strategies for improving the catalytic activity and long-term durability are introduced. The main challenges and prospects in developing electrodes for water electrolysis are discussed. It is expected that the content covered in this chapter would provide useful guidance for future exploration of novel low-cost nanostructured catalysts for electrochemical HER.

Keywords: nanomaterials, hydrogen evolution reaction, electrocatalysts, overpotential

4B.1 Introduction

The global energy demand has increased rapidly in the past decades due to rapid population growth and industrialization. Consequently, there has been an evident increase in the usage of traditional fossil fuels, causing severe environmental problems, particularly

Acknowledgments: This work was financially supported by the National Natural Science Foundation of China (grant nos. 52104309, 51872210, and 51702241), the Special Project of Central Government for Local Science and Technology Development of Hubei Province (no. 2019ZYYD076), and Open Foundation of State Key Laboratory of Advanced Refractories (no. SKLAR202002). We also acknowledge the fund of the Scientific Research Project of the Education Department of Hubei Province (D20201103), and the support of the "Macao Young Scholars Program," China (AM2020004).

Wen Lei, The State Key Laboratory of Refractories and Metallurgy, Wuhan University of Science and Technology, Wuhan 430081, China
Shaowei Zhang, College of Engineering, Mathematics and Physical Sciences, University of Exeter, Exeter EX4 4QF, UK
Haijun Zhang, The State Key Laboratory of Refractories and Metallurgy, Wuhan University of Science and Technology, Wuhan 430081, China, e-mail: zhanghaijun@wust.edu.cn

https://doi.org/10.1515/9783110739879-011

the greenhouse effect, air pollution, and water pollution [1]. It has become increasingly important and necessary to use renewable/clean energy sources such as solar and wind. However, most of them suffer from daily and seasonal intermittent and regional variability. To cope with these issues, exploring stable and readily storable clean energy is of realistic significance [2]. Thanks to its highest energy density and zero carbon emission, hydrogen energy is regarded as one of the most competitive candidate energies to replace traditional fossil fuels [3].

The success of the hydrogen economy is highly dependent on the availability of reliable routes to the large-scale production of hydrogen. Since water is a very stable compound composed of hydrogen and oxygen atoms, water splitting techniques including photocatalytic, photoelectrochemical, electrocatalytic, thermochemical, sonochemical water splitting, biophotolysis, gasification, and microbial fermentation have been extensively investigated. Using electric energy over an electrocatalyst, direct water splitting has proved highly promising because of its sustainability and efficiency [4].

4B.2 Fundamentals of hydrogen evolution reaction

4B.2.1 Electrocatalytic hydrogen evolution reaction

The entire electrocatalytic water splitting includes the hydrogen evolution reaction (HER) on the cathode side and the oxygen evolution reaction (OER) on the anode side [5].

4B.2.1.1 Basic principles

HER involved with a two-electron transfer is generally identified as the intermediate adsorption/desorption process [6]. In an acidic electrolyte (normally H_2SO_4), H^+ is initially adsorbed onto the surface of the electrocatalyst, followed by reduction, via gaining one electron, to the adsorbed intermediate H^* ($*$ stands for the adsorbed active site). This step is known as the Volmer step, as indicated in eq. (4B.1). Then, the adsorbed H^* combines with another H^+ and gains one electron, resulting in an H_2 molecule. This step is known as the Heyrovsky step, as indicated in eq. (4B.2). Simultaneously, the adsorbed H^* could alternatively combine with another H^* to form an H_2 molecule in the presence of high coverage of surrounding H^*. This step is known as the Tafel step, as indicated in eq. (4B.3):

$$H^+ + e^- \rightarrow H^* \tag{4B.1}$$

$$H^* + H^+ + e^- \rightarrow H_2 \tag{4B.2}$$

$$2H^* \rightarrow H_2 \tag{4B.3}$$

Regardless of the Volmer–Heyrovsky (eqs. (4B.1) and (4B.2)) or Volmer–Tafel (eqs. (4B.1) and (4B.3)) mechanism, the hydrogen adsorption free energy (ΔG_{H*}) is widely accepted as the criterion of HER activity [3]. Differently from the two-step process in an acidic electrolyte (normally H_2SO_4), the adsorbed H_2O molecule on the electrocatalyst surface in an alkaline electrolyte (normally KOH) starts to dissociate in the initial stage of HER due to the absence of proton, as indicated by eq. (4B.4). Accordingly, the H_2 molecule could be obtained via H^* with the H_2O molecule or coupling the intermediates (the Tafel step), as shown in eq. (4B.5):

$$H_2O + e^- \rightarrow H^* + OH^- \tag{4B.4}$$

$$H^* + H_2O + e^- \rightarrow H_2 + OH^- \tag{4B.5}$$

It is necessary to balance the desired water dissociation and usable hydrogen adsorption/desorption energy. Therefore, exploring a qualified HER electrocatalyst with an ideal trade-off between the binding and release of H atoms is extremely important for high-efficiency H_2 production.

4B.2.1.2 HER energetics and computational activity descriptors

Benefiting from the rapid development of computational techniques, density functional theory (DFT) calculations, a powerful research tool, have been widely used to provide an in-depth insight into the electrochemical reactions at the atomistic level. Various material properties are named catalytic descriptors, where the practical goal of linking material properties with the reaction rate is to reveal the material/activity relationships. Regardless of the path that HER takes, H^* is indispensably involved in the HER process. Therefore, the free energy of hydrogen adsorption (ΔG_{H*}) is always adopted as one of the key descriptors to evaluate the HER activity, which reflects the binding strength of H^* on the catalyst surface [7]. The linear relationship between exchange current and catalyst–hydrogen bond strength exhibited a typical "volcano" plot, as obtained by the experimental result and the mathematical model. The relationship perfectly follows the Sabatier principle, which qualitatively defines the prerequisite for a high reaction rate as "not too strong, nor too weak" binding of intermediates.

4B.2.2 Factors determining the electrocatalytic HER activity

The catalytic activity of an HER electrocatalyst relies highly on its intrinsic activity and the exposure of active sites (Fig. 4B.1 for main transition metals (a) and select metal-hydrogen bond strengths (b)). The former is closely related to its composition, crystal structure, defect concentration, electronic conductivity, and charge transfer capability, and for the latter, designing an electrocatalyst with a nanostructure, that

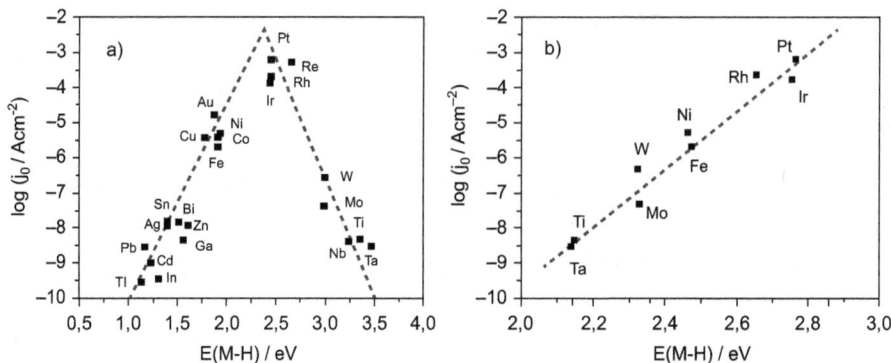

Fig. 4B.1: (a) Experimental "volcano" plot derived by Trasatti (replotted using the data from Ref. [8]). The exchange current is given as a function of the metal–hydrogen bond strength, and (b) the linear relationship between the exchange current and the metal–hydrogen bond strength is given by the same author (replotted using the data from Ref. [9]).

is, with significantly increased surface area, is efficient, which not only provides more active sites but also facilitates the diffusions of ions/electrolyte/H_2.

4B.2.2.1 Overpotential

Theoretically, electrocatalytic water splitting reactions occur upon the equilibrium potential reaching 1.23 V[3]. However, an overpotential (η) is inevitable to overcome the intrinsic activation barrier of electrocatalysis. Therefore, reducing the overpotential should be the primary task in the electrocatalyst's design, promoting superior efficiency in the electrocatalytic water splitting process. The electrochemical workstation can obtain overpotential due to linear scan voltammetry (LSV) curves measured with IR correction. Commonly, the overpotential at the current density of 10 mA/cm^2 is accepted as the benchmark to evaluate the performance of an electrocatalyst.

4B.2.2.2 Tafel slope and exchange current density

The relationship between overpotential and current density (j) could be obtained from the Tafel slope, elicited through the logarithmic transformation of LSV curves. The Tafel plot can be linearly fitted based on the following equation: $\eta = a + b \log j$ (where b is the Tafel slope, η is the overpotential, and a is a related constant electrode property, electrolyte, and temperature). Normally, a small Tafel slope value is desired because it signifies a more rapid current density and increased applied potential. The current exchange density (j_0) is obtained by extrapolating the Tafel plots at the equilibrium potential (overpotential is 0 V). It indicates the intrinsic charge transfer between electrocatalyst and reactants. The higher the j_0 value, the better the HER catalytic performance of the electrocatalyst.

4B.2.2.3 Faradaic efficiency

The faradaic efficiency is the efficiency of electrons used to catalyze the desired water reduction/oxidation reactions rather than unwanted reactions (e.g., electrode redox reactions). Specifically, it is quantified by the molar ratio of the hydrogen/oxygen produced to the hydrogen/oxygen produced theoretically.

4B.2.2.4 Stability

In practical applications, stability is a significant factor in evaluating electrocatalysts. It is generally measured by recording the relationship between the current change and reaction time under a fixed potential or performing a chronopotentiometry experiment upon the potential change. In addition, durability can be fully evaluated utilizing a multicurrent step chronopotentiometry technique with a measurement range from 10 to 10^3 mA/cm^2.

4B.3 Nanomaterial-based catalysts for electrocatalytic HER

In the case of an electrocatalyst for HER, the exposed active sites can be dramatically increased by constructing a high-surface-area structure, optimizing morphology, and creating particular nanostructures, highlighting the significance of nanostructure construction. On the other hand, the hybridization with other catalytic materials or conductive scaffolds may lead to improved electrocatalytic activity owing to the so-called synergistic effect.

The most effective ways to maximize the HER activity include adjusting surface and/or volume characteristics (by selecting cations and anions), optimizing morphology (by using advanced synthesis procedures), enhancing the charge transfer (by modifying the electronic structure of functional surfaces), and forming a composite or hybrid. These methods may generate more HER active sites and provide an ideal way to transport reactants and gaseous products (i.e., H_2). The strategy to improve electrocatalytic activity is not limited to oxides but can be applied in principle to other types of electrocatalysts. In addition, researchers often use several combination strategies to improve the efficiency of electrocatalysts in HER.

Comparing nanoparticles of different shapes could better understand the surface chemistry of metal (metal compounds) nanoparticles with different sizes and their effects on the HER kinetics. Indeed, nanoscale morphology regulation offers an effective strategy for controlling the ratio of crystal planes, the number of atoms in a corner, and edge positions, which might cause noteworthy differences in bond enthalpy, adsorption and desorption energy, and so on. As a result of the high surface-to-volume ratio, geometric surface effect, unique electronic properties, and quantum size effect, small particles always exhibit high activity (Fig. 4B.2). Besides, further reducing the particle size to nanoscale or even atomic scale to form nanocluster and single atoms could significantly impact the electronic properties of electrocatalysts, affecting their catalytic behaviors.

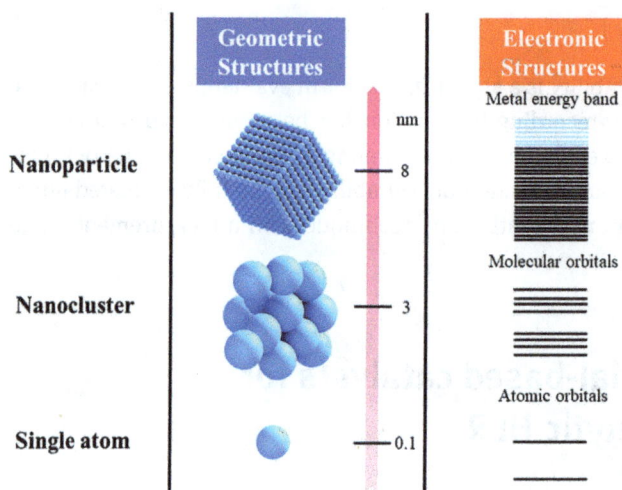

Fig. 4B.2: Geometric and electronic structures of a single atom, nanocluster, and nanoparticle [10] .

4B.3.1 Metals

Metals are the most common type of HER electrocatalyst, which can be further divided into two main subtypes, that is, noble metals and transition metals.

4B.3.1.1 Noble metals

Despite the high cost of noble metal electrocatalysts, which may need to be considered for their practical application in HER, the most representative noble metal, platinum (Pt), remains still the optimal choice for all ranges of pH, exhibiting a nearly zero overpotential ($\Delta G_{H*} \approx 0$), low Tafel slope, and excellent long-term stability. According to

Tymoczko et al. [11], the high cost of noble metal electrocatalysts is not the main issue that hinders electrocatalytic H_2 production on an industrial scale. The system inefficiency is majorly responsible for the low proportion (only 4 %) of the H_2 produced from the water electrolysis [12]. The electricity expense of an industrially applicable electrocatalyst is much higher than the cost of the noble metal electrocatalyst itself. Under ideal conditions, a Pt electrocatalyst's particle size and morphology significantly affect its HER performance. For this reason, great research efforts have been made in recent years to explore Pt nanostructures in various forms such as thin film [13], nanoparticle [14], nanocage [15], nanowire [16], and nanosheet [17].

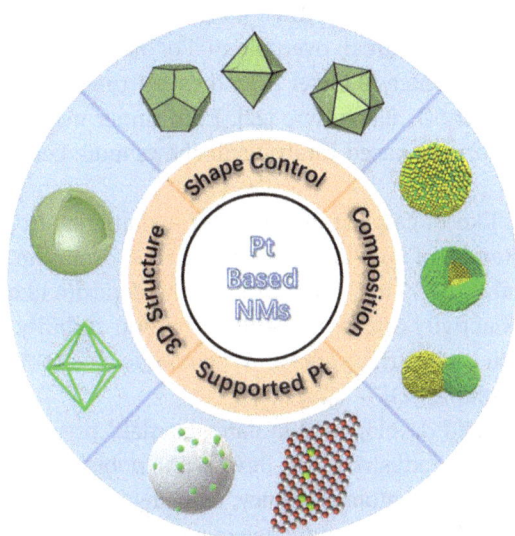

Fig. 4B.3: A schematic illustrating various approaches to the structural regulation of Pt-based nanomaterials [18] .

Given the scarcity and high cost of Pt, researchers are motivated to minimize its usage and enhance its electrocatalytic activity by constructing specific nanostructures of it (Fig. 4B.3). Apart from the above-mentioned physical structure design, the regulation of chemical structure (composition-engineered strategy) to form Pt-based alloys or intermetallic nanocrystals via introducing nonprecious metals is another effective strategy. The optimal introduction of nonprecious metals could generate favorable electronic and geometric effects. The catalytic performance of these bimetallic (or trimetallic) nanocrystals can be retained or even improved. A typical example is alloying Pt with 3d transition metals such as Fe, Co, and Ni [19]. According to the d-band center theory, strong adsorption will occur when the d-band center is close to the Fermi level, and weak adsorption will occur when the d-band center is far away from the Fermi level will occur. Therefore, the alloying can downshift the d-band center of Pt, leading to weaker adsorption energy of OH* on surface Pt atoms [20]. Compared to the metallic alloy nanocrystals, the intermetallic ones with special geometrical structures and local geometrical properties are more stable under critical catalytic conditions. Due to their highly ordered structures, Pt-based intermetallic nanocrystals are widely used as electrode materials for various electrocatalytic reactions in fuel cells. They show high stability against oxidation and etching [21].

Regarding the morphology-engineered strategy, the enhancement of electrocatalytic activity can be achieved via controlling crystal sizes and morphologies to increase the surface area and expose the active-referable crystal facets [19]. For example, there has always been a debate about the correlation between the size of Pt nanoparticles and their electrocatalytic activities. The ~1.5-nm-sized Pt shows significantly enhanced HER catalytic activity with an overpotential of only 13 mV to achieve a current density of 10 mA/cm^2, which is lower than those of the ~ 3-nm- and ~ 6-nm-sized Pt catalysts in 0.5 M H_2SO_4 electrolyte [22]. Even under neutral and alkaline conditions, the ~ 1.5-nm-sized Pt-based catalyst still shows more beneficial HER electrocatalytic activity.

For another example, a dendrite-like Pt catalyst prepared via a simple hydrothermal approach shows an onset potential of only 15 mV in a 0.5 M H_2SO_4 solution, which is much lower than commercial Pt/C (30 mV). Furthermore, the dendrite-like Pt catalyst also exhibits higher HER activity and remarkable long-term stability, owing to its hierarchical structure with high surface area and the unique electronic structure with a large percent of (220) planes [23].

The past decades have witnessed rapid development in nanomaterials' preparation technologies in academic research and industrial trial. Downsizing noble metals to clusters or single atoms effectively maximize atom efficiency, reducing production costs considerably. For example, isolated Pt atoms and clusters could be prepared using the atomic layer deposition technique [24], achieving a high HER activity of 16 mA/cm^2 (at the overpotential of 0.05 V) with a small Tafel slope of 31 mV/dec.

Apart from Pt, ruthenium (Ru) and iridium (Ir) have evoked increasing interest as the least expensive metals in the Pt group and active HER electrocatalysts, thanks to their similar metal–H bond strength to Pt. To date, some electrocatalysts based on Ru (Ir), including metallic Ru (Ir) and their alloys, Ru (Ir)-based compounds, and corresponding composites, have been investigated to boost the HER performance. Many of them exhibit performance comparable to or even better than the commercial Pt/C, especially in an alkaline solution [25].

4B.3.1.2 Transition metals

According to the volcano theory, the HER performance of transition metals is poor. By alloying two or more metals with different binding energies toward H, the HER activity of transition metals can be dramatically improved. In detail, two metals with different binding energies with hydrogen atoms can be combined to form alloy materials with different H adsorption capabilities, which is beneficial to the HER. It is generally accepted that H is adsorbed on metal (M) via the strong M–H bond, and the adsorbed H can diffuse and move on the surface of alloy, while the weak M–H bond is easily destroyed and is conducive to the separation of the formed H_2 molecules from the electrode surface. Therefore, the introduction of another metal of different sizes can result in a significant change in the original lattice structure and thus the properties of the metal, eventually contributing to more active sites required for the HER process. So far, numerous alloys such as Ni–Cu, Fe–Co–Ni, Ni-Cr-Mo-Cu, and Fe-Al-Ni-Mo-Cu have shown excellent HER performances [26–29].

The coupling between the carbon material and the encapsulated metal compound nanoparticles is critical as it affects the charge transfer between different components and the cyclic stability. However, conventional physical methods such as mechanical mixing or electrostatic adsorption cannot achieve such intimate contact. The metal (metal compounds) nanoparticles could be easily separated from the carbon support upon continuous release of H_2 during the HER process [30]. Therefore, catalysts formed via in situ anchoring nanoparticles of metal compounds on the transition metal and nitrogen co-doped carbon, namely M–N–C, have gained tremendous attention due to their unique structural, surface stability, and unexpectedly high electrocatalytic activity [31]. Several research groups have reported that some transition metals such as Co, Ni, and Fe nanoparticles encapsulated by an ultrathin nitrogen-doped carbon layer exhibited excellent HER activity [31, 32]. The carbon shell of metal-organic framework (MOF)-derived Co@NC nanocomposites with optimal structure delivered a remarkable HER activity with an overpotential of 95 mV to achieve the current density of 10 mA/cm^2 in 1.0 M KOH [33].

The HER properties of some representative metals and alloys are listed in Tab. 4B.1.

4B.3.2 Transition metal oxides (TMOs)

Transition metal oxides (TMOs) composed of low-cost and earth-abundant elements, including simple metal oxides (MnO_2, TiO_2, CoO, NiO, etc.), spinels (Co_3O_4, $NiCo_2O_4$, $CoMn_2O_4$, etc.), perovskites ($Ba_{0.5}Sr_{0.5}Co_{0.8}Fe_{0.2}O_{3-\delta}$, $SrNb_{0.1}Co_{0.7}Fe_{0.2}O_{3-\delta}$, etc.), are regarded as an important class of OER catalysts. In comparison with other types of metal compounds, the superiorities of metal oxides lie in their unique compositional and structural diversities, which guarantee the electronic and crystal structure flexibility with various desirable physical/chemical properties (Fig. 4B.4a). However, pristine metal oxides, especially the bulk materials, were supposed to exhibit poor catalytic activities toward HER in the past due to their poor electrical conductivity, limited hydrogen adsorption ability, as well as an insufficient catalytic-active site.

Tab. 4B.1: Compilation of HER performance based on various metals and alloys.

Material	Substrate	Electrolyte	Overpotential (mV)		Tafel slope (mV/dec)	Year	Ref.
			η_{10}	η_{100}			
Ni NSs	Carbon cloth	1.0 M KOH	80	150	70	2017	[34]
Fe/SWCNTs	GCE	0.5 M H_2SO_4	77	–	40	2015	[35]
Co-NRCNTs	GCE	0.5 M H_2SO_4	260	–	80	2015	[36]
Co/NG	GCE	0.5 M H_2SO_4	265	–	98	2015	[37]
Co/N-CNTs/GR	GCE	0.5 M H_2SO_4	87	~170	52	2018	[38]
Cu-NRCNT	Cu foam	1.0 M KOH	123	–	61	2017	[39]
SA-Ni/GR	GCE	0.5 M H_2SO_4	180	–	45	2015	[40]
SA-Mo/NC	GCE	0.1 M KOH	132	–	90	2017	[41]
SA-Co/NG	GCE	0.5 M H_2SO_4	147	–	82	2015	[42]
SA-Pt/NC	GCE	0.5 M H_2SO_4	19	~36	19	2020	[43]
NiW	Ni foam	1.0 M KOH	36	–	43	2018	[44]
MoNi$_4$	Ni foam	1.0 M KOH	15	–	30	2017	[45]
NiMo NSs	Ni foam	1.0 M KOH	30	–	86	2016	[46]
NiCu/C	Graphite plate	0.5 M H_2SO_4	48	160	63.2	2017	[47]
NiCo/NC	GCE	0.1 M H_2SO_4	142	–	105	2015	[48]
PtNi NWs	GCE	1.0 M KOH	39.7	–	/	2016	[49]
PtRh DNAs	GCE	0.5 M H_2SO_4	27	–	40	2020	[50]

NSs, nanosheets; SWCNTs, single-walled carbon nanotubes; GCE, glass carbon electrode; NRCNT, nitrogen-rich carbon nanotubes; NG, N-doped graphene; GR, graphene; SA, single atom; NC, N-doped porous carbon; NWs, nanowires; DNAs, dendritic nanoassemblies; GR, graphene.

Various techniques have been investigated to address these issues, including defect engineering, morphology engineering, valence regulation, and phase (crystallinity) engineering (Fig. 4B.4b).

Defect engineering, which normally includes vacancy control and doping, is an effective strategy to modulate the electronic properties of metal oxides. Optimal defect engineering gives favorable physical and chemical properties, breaking through the intrinsic bottleneck and likely boosting the electrocatalytic HER performance. Typical examples include stoichiometric TiO_2, MoO_2, MoO_3, and WO_3, which show inherently unfavorable acidic HER activity. Creating oxygen vacancies in stoichiometric

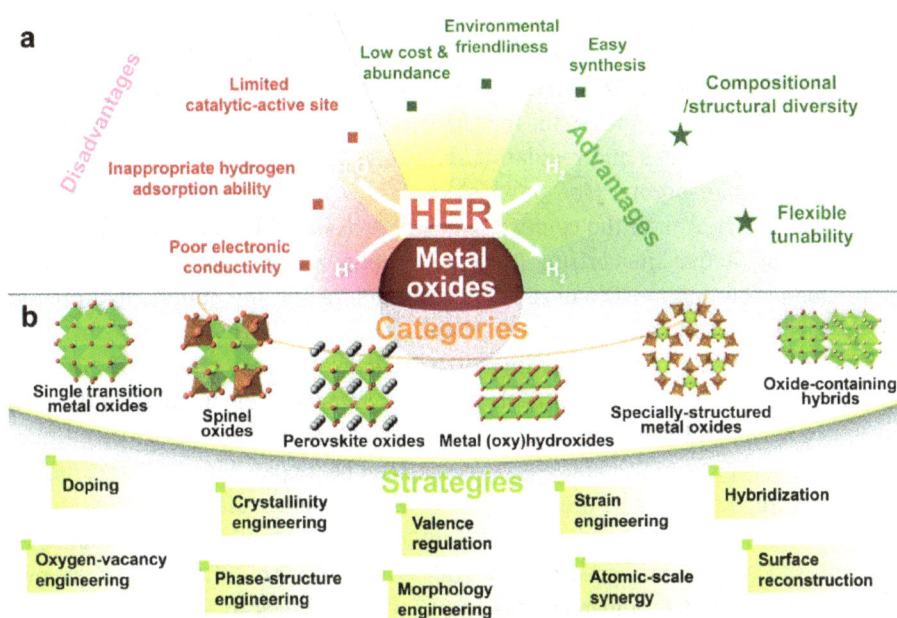

Fig. 4B.4: (a) Advantages and disadvantages of metal oxides as electrocatalysts for HER, and (b) summary of design strategies for the development of metal-oxide-based HER electrocatalysts [51].

oxides is one of the most common and efficient ways to boost intrinsic catalytic activity by modulating the electronic structure, conductivity, and hydrogen adsorption energies. For example, molybdenum oxides have been extensively investigated as electrocatalysts for HER due to their rich valence states. The electrocatalytic activities of molybdenum oxides strongly rely on the valence states of Mo, which can be significantly improved by introducing oxygen vacancies in the oxides. An electrodeposited MoO_{3-y} film electrode exhibits an overpotential of 201 mV (vs. RHE) to achieve the current density of 10 mA/cm^2. In contrast, the LSV curve of the MoO_{2+x} electrode is flat and cannot reach the current density of 10 mA/cm^2 (within 250 mV overpotential) [52].

In addition, hetero-nonmetal atom (e.g., N, P, and S) doping effectively enhances the HER performance of TMOs. Utilizing nonmetal atom doping, the electronic conductivity of TMOs could be improved, beneficial for promoting charge transfer. Meanwhile, the electronic structure of the electrocatalysts could be optimized, modulating the adsorption and dissociation of reactants. Furthermore, more importantly, nonmetal atom doping is generally straightforward and does not involve any complex operations, making it feasible for commercial applications. Apart from nonmetal atom doping, metal doping effectively improves the HER performance of TMOs. For example, Mo doping can significantly improve the HER activity of $W_{18}O_{49}$ [53]. Therefore, Mo doping can result in more active sites and optimal hydrogen adsorption based on DFT calculations,. Likewise, due to the synergistic electronic effect (energy-level matching), MoS_2 doped with the optimal amount of Zn exhibits superior electrochemical activity toward HER with the onset potential of -0.13 V versus RHE Tafel slope of 51 mV/dec [54].

HER performance improves the morphology engineering of metal oxides by producing more active sites and regulating the superoleophobic (hydrophilia) properties. For example, porous MoO_2 nanosheets show much higher electrocatalytic activity toward HER than the compact MoO_2 particles, owing to the higher surface area and more active sites arising from the morphology engineering [55]. Furthermore, due to the unique 3D nanosheet array structure and the optimal hydrophilia for gas evolution, the as-constructed $NiMoO_4$ nanosheet array catalyst showed a low HER overpotential of 117 mV with a high faradaic efficiency of up to 99 %.

Phase (crystallinity) engineering also plays an important role in affecting the HER activity of metal oxides. For example, three different Ti_2O_3 polymorphs (α-Ti_2O_3 (trigonal), o-Ti_2O_3 (orthorhombic), and γ-Ti_2O_3 (cubic)) can be prepared by regulating the epitaxial growth process, and the last one shows the highest HER activity with the smallest overpotential of 271 mV at 10 mA/cm^2 in 0.5 M H_2SO_4 electrolyte. For another example, DFT calculations reveal that the (110) crystal planes of $NiCo_2O_4$ show the lowest ΔG_H^* of 0.15 eV. In comparison, the (100) and (111) planes show the value of 0.62 and 0.36 eV, respectively, implying the higher HER activity of (110) surface [56].

Despite broad application prospects of metal oxides, their HER electrocatalytic activity is still not satisfactory. Very few of them can exhibit a similar activity to the "holy grail" Pt. An ideal metal-oxide-based HER electrocatalyst should meet several requirements, particularly high intrinsic conductivity, simple and low-cost manufacturing process, similar (or better) HER activity to (than) that of Pt, and desirable stability in both acidic and alkaline electrolytes upon long-term HER operation.

The HER properties of some representative TMOs are shown in Tab. 4B.2.

4B.3.3 Metal chalcogenides

The structures of the metal chalcogenides can be divided into two distinct classes: layered MX_2 (M = Mo, W, Sn; X = S, Se) and nonlayered M_xX_y (M = Fe, Co, Ni, Cu, Zn, etc.). A typical layered MX_2 exhibits a sandwich-like structure, in which one layer of metal is bonded to two layers of sulfur or selenium [68]. Each MX_2 unit cell stacks vertically on top of each other via the weak van der Waals forces, which allows the exfoliation of MX_2 to single layers. According to the bonding and configurations of MX_2, they can be classified into three main phases: 1 T, 2 H, and 3 R (the digit is the layer number, while T, H, and R are the abbreviations of tetragonal, hexagonal, and trigonal lattices, respectively), as shown in Fig. 4B.5.

Due to the rich tunable 2D surface chemistries and precisely tunable properties based on size, defects, strain, and doping, the ΔG_{H^*} of 2D TMDs can be adjusted to 0 eV, showing high activities similar to those of Pt-based electrocatalysts (Fig. 4B.6).

Tab. 4B.2: Comparison of HER performance of various TMOs.

Material	Substrate	Electrolyte	Overpotential (mV)		Tafel slope (mV dec^{-1})	Year	Ref.
			η_{10}	η_{100}			
WO$_3$ NSs	GCE	0.5 M H$_2$SO$_4$	66	–	34.8	2017	[57]
WO$_{3-x}$ NPLs	CNFs	0.5 M H$_2$SO$_4$	185	–	89	2016	[58]
NiO NRs	Carbon fiber	1.0 M KOH	110	280	100	2018	[59]
δ-MnO$_2$ NSs	Nickel foam	1.0 M KOH	197	~350	66	2017	[60]
N-MoO$_3$ SLs	GCE	0.5 M H$_2$SO$_4$	–	540	101	2016	[61]
CoNiO$_x$/CNFs	GCE	1.0 M KOH	145	~220	40	2020	[62]
TiO$_2$/ZrO$_2$ NPs	GCE	0.5 M H$_2$SO$_4$	160	–	87	2020	[63]
MgCoO$_2$	/	1.0 M KOH	400	–	174	2020	[64]
MoO$_2$/PC-GR	GCE	0.5 M H$_2$SO$_4$	64	–	41	2015	[65]
NiO/Ni-CNT	GCE	1.0 M KOH	80	–	82	2014	[66]

NPLs, nanoplates; NRs, nanorods; SLs, single layers; CNFs, carbon nanofibers; PC, polyoxometalate-based MOF-derived carbon; NPs, nanoparticles, GCE, glass carbon electrode.

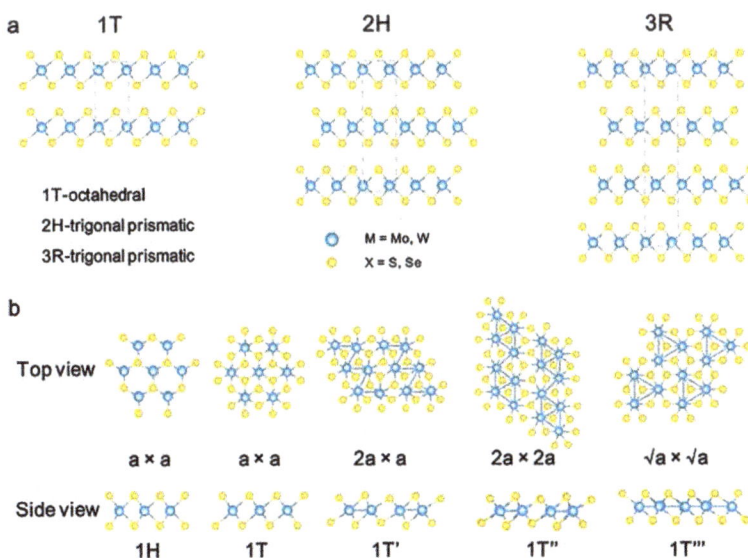

Fig. 4B.5: Molecular structure of 1 T, 2 H, and 3 R phases [67].

There are two kinds of surface sites in the case of single trilayer X–M–X, that is, the basal plane and the edge. The basal plane of MX$_2$ shows much lower surface energy and is typically inert compared to the highly active edge states. The HER catalytic activity of metallic and semiconducting sites of TMDs are different. Even though the basal plane of some 1 H/2 H-TMDs (e.g., MoS$_2$, WS$_2$, MoSe$_2$, and WSe$_2$) is HER inert with $\Delta G_{H*} \geq 2$ eV [69], the ΔG_{H*} at the metallic edge (the edge exhibits metallic features) in some cases is close to that of Pt, and is thermoneutral, as shown in Fig. 4B.6a [70]. For example, the ΔG_{H*} of the optimal edge configurations of MoS$_2$ and WS$_2$ are −0.45 and −0.06 eV, respectively, with 100 % S edges [69].

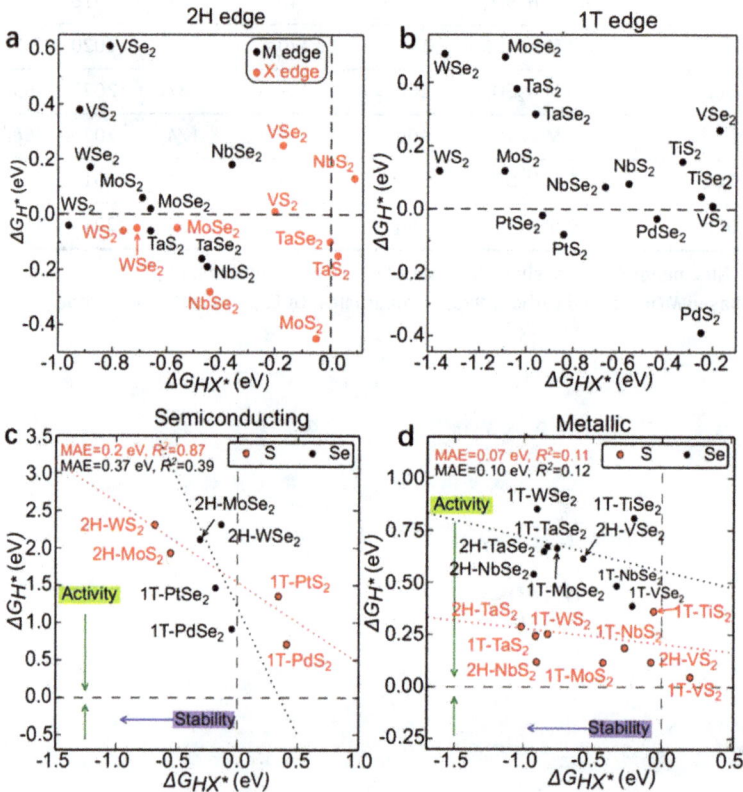

Fig. 4B.6: ΔG_{H*} as a function of the HX adsorption free energy (ΔG_{HX*}, X = S or Se) at (a) 2 H edge, (b) 1 T edge, (c) 2 H plane, and (d) 1 T plane. MAE, mean absolute error [71].

As illustrated in Fig. 4B.6a, b, we can also plot ΔG_{H*} of various TMDs at different edges against the absorption free energy of H–X (ΔG_{HX*}) based on the data reported in the literature. When $\Delta G_{HX*} > 0$, with the desorption of H–X groups, the reaction tends to form H$_2$X. Apart from MoS$_2$, several other TMDs such as WS$_2$, TaS$_2$, MoSe$_2$, and WSe$_2$ show highly promising electrocatalytic HER activities with 2 H or 1 T edges. Replacing the Se (electronegativity: 2.55) in TMDs with S (electronegativity: 2.58) can slightly regulate the ΔG_{H*} close to more thermoneutral states, but it could cause significantly deteriorated stability at the same time (Fig. 4B.6c). The stability of the semiconducting basal plane should be carefully considered during the activation for an efficient HER. Like the semiconducting basal plane, the ΔG_{H*} on the metallic plane is also inversely proportional to ΔG_{HX*} (Fig. 4B.6d). However, the metallic planes exhibit higher electroactivity toward HER with similar stabilities to the semiconducting basal plane. As shown in Fig. 4B.6d, ΔG_{H*} on the plane of 1 T-MoS_2 is 0.12 eV, while it is 1.92 eV on that of 1 H/2 H-MoS_2. Given this, improving the electronegativity of the X atom should be an effective approach to the improved HER activity concerning that the semiconducting TMDs (at the basal plane) and the basal planes of some metallic TMDs are suitable for the catalytic HER. For example, the 1 T-MoS_2 and 2 H-NbS_2 exhibit a similar ΔG_{H*} around 0.12 eV on the basal planes, which is close to that at the 2 H edges of some TMDs.

Based on the above analysis, the unsatisfactory HER performance of TMDs is considered to be mainly due to the lack of sufficient edge sites and the low electrical conductivity. For instance, the controlled growth of MoS_2 atomic film was used to prove the HER catalytic properties of MoS_2, and the results revealed that due to the poor electron hopping efficiency in multilayers, the HER catalytic activity of MoS_2 decreased by 4.47 times with each additional layer [72].

In the past few years, the fundamental understanding of the electrocatalytic active sites, reaction limiting factors, and the significance of optimal chemical/structural control of TMDs have provided new strategies for their HER performance enhancement, including phase engineering, defect creation (size control, edge enrichment, and strain engineering), chemical doping, and hybrid structure formation.

As early as the 1970s, there were reports on using MoS_2 as the HER electrocatalyst. However, bulk MoS_2 exhibited poor electrocatalytic performance due to its poor conductivity and limited active sites. Because of this, few researchers paid attention to MoS_2 until 2005, when Hinnemann et al. [70] found that the Mo boundary structure of MoS_2 is very similar to the structure of the HER active site of nitrogenase could potentially exhibit a high HER electrocatalytic activity. Their further theoretical investigations revealed that the ΔG_{H*} of the Mo edge of MoS_2 is equivalent to that of Pt. In addition, they successfully anchored MoS_2 nanoparticles on graphite and confirmed experimentally that MoS_2 does have active sites for HER.

Nevertheless, given the issues of lack of edge sites and inherent poor electrical conductivity, two main strategies are generally adopted to improve the activity of MoS_2: (1) increasing the number of active sites and activity, and (2) enhancing the transport of electrons between electrode and catalyst to ensure the effective reduction of protons. This finding is equally applicable to other 2D TMDs. Various TMD catalysts with different configurations and high electroactivity have been synthesized via different methods such as mechanical exfoliation, chemical vapor deposition (CVD), hydrothermal/solvothermal, liquid exfoliation, and electrochemical Li$^+$ intercalation exfoliation [73–75].

CVD shows many distinct advantages over the other methods in controlling the lateral size and thickness. In 2014, Yu et al. [72] used this method to precisely control the layer number of MoS$_2$ via controlling the amount of MoCl$_5$. Electrochemical tests revealed that with each additional layer, the current exchange density of MoS$_2$ decreased by 4.47 times. The main reason for the layer-dependent catalytic activity of MoS$_2$ is that the increase in the layer number reduces the efficiency of electron jumping between the layers.

Shi et al. [76] also used the CVD method to synthesize monolayered MoS$_2$ with uniform triangular morphology on a gold foil. By changing the growth temperature and the distance between the precursor and the substrate, the size of the resultant monolayer MoS$_2$ can be varied from nanometers to micrometers (Fig. 4B.7). The product shows high HER activity with a Tafel slope of 61 mV/dec and a high exchange current density of 3.8×10^{-2} mA/cm^2.

Wang et al. [77] prepared Fe$_{1-x}$Co$_x$S$_2$/CNT (carbon nanotube) composites by incorporating Co into FeS$_2$ nanosheets and compounding with CNTs to improve HER performance. Their electrochemical test results showed that the doping amount of Co in the material plays an important role in the electrocatalytic activity. At the same time, DFT calculations reveal that the S atoms located at the edge of the pyrite are the actual active sites for HER. Compared with the Pt(111) surface, Fe$_{1-x}$Co$_x$S$_2$/CNTs and FeS$_2$/CNTs exhibit higher H$^+$ adsorption energy and similar H$_2$ adsorption energy. On the other hand, compared with FeS$_2$/CNTs (1.62 eV), Fe$_{1-x}$Co$_x$S$_2$/CNTs show a smaller H adsorption energy barrier (1.23 eV). Co doping can activate the (110) crystal plane of Fe$_{1-x}$Co$_x$S$_2$/CNTs and weaken the S–H bond on the catalyst's surface, greatly promoting the adsorption of H atoms in the reactive sites and the formation of the H–H bond.

In the periodic table, both selenium and sulfur belong to the VIA group, located in the fourth and third periods. The metallicity of selenium is stronger than that of sulfur, indicating the better electrical conductivity of metal selenides. The atomic radius of sulfur is smaller than that of selenium, and its ionization energy is greater than that of selenium. Based on the above analysis, metal selenides may exhibit unique HER properties differently from metal sulfides. Among many metal selenides reported, CoSe$_2$ is considered the most powerful candidate electrocatalyst for the HER [78]. McCarthy et al. [79] used a thiol/amine solvent mixture to prepare a thin film of phase-pure marcasite-type CoSe$_2$, which exhibited an onset potential of 117 mV and Tafel slopes of 60 mV/dec when serving as the HER electrocatalyst. Sun et al. [80] used NiSe$_2$ as a bifunctional catalyst for complete water splitting. By controlling the doping ratio between Fe and Co, both the HER and OER activities of NiSe$_2$ could be significantly improved. At the current density of 10 mA/cm^2, the overpotentials were 92 mV (HER) and 251 mV (OER), respectively.

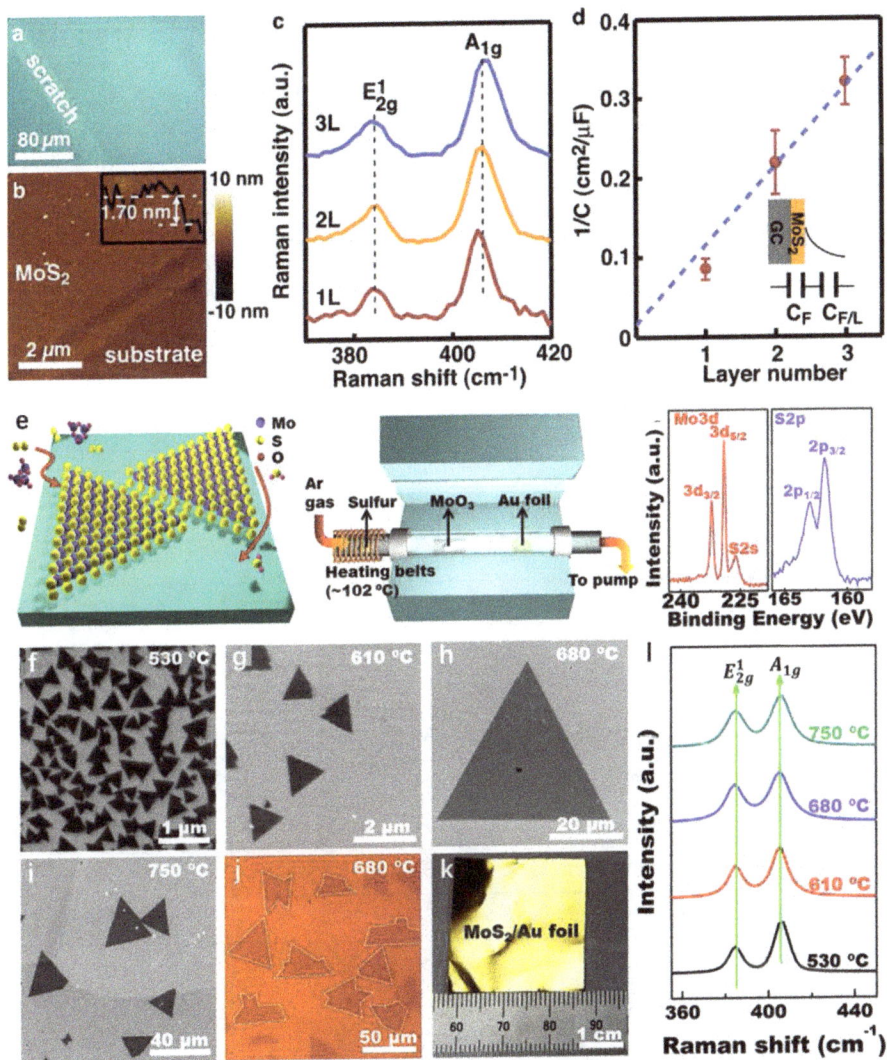

Fig. 4B.7: Optical image (a) and AFM image (b) of a bilayer MoS_2 film. (c) Raman spectra of MoS_2 films with monolayer (1 L), bilayer (2 L), and trilayer (3 L). (d) The reciprocal of measured capacitances of the films as a function of the layer number, and the inset is a schematic illustrating the two capacitors in the series model. (e) Schematic illustration of the surface growth of MoS_2 on Au foils and XPS spectra of as-grown MoS_2 on Au foil. (f–k) SEM images of MoS_2 triangular flakes grown under different growth temperatures displaying different domain sizes. (l) Raman spectra of MoS_2 flakes at different preparation temperatures [76].

To sum up, studies on metal sulfides for electrocatalytic HER applications are still early. The most extensively studied TMDs for HER are MoS_2 and WS_2. At the same time, other TMDs, including Sn-, Mn-, Sb-, Re-, V-, Ni-, Cu-, Nb-, and Ta-based mono-, bi-, or multimetal sulfides with precisely defined shapes, sizes, and compositions, have received limited attention. In addition, the stability of TMDs under a wide range of pH is a technical challenge that needs to be perfectly resolved.

The HER properties of some representative TMDs are shown in Tab. 4B.3.

Tab. 4B.3: Comparison of HER performance of various TMDs.

Material	Substrate	Electrolyte	Overpotential (mV)		Tafel slope (mV/dec)	Year	Ref.
			η_{10}	η_{100}			
Co_8S_9 HNs	Carbon paper	1.0 M KOH	193	~360	100	2018	[81]
MoS_2 film	Mo foil	0.5 M H_2SO_4	187	–	55	2016	[82]
WS_2 nanoflakes	GCE	0.5 M H_2SO_4	100	–	48	2014	[83]
WTe_2 nanoflakes	GCE	0.5 M H_2SO_4	119	–	79	2020	[84]
1 T-2 H $MoSe_2$ NSs	GCE	0.5 M H_2SO_4	155	–	60	2020	[85]
MoS_2 NSs/GR	GCE	0.5 M H_2SO_4	129	–	43.5	2017	[86]
Co-$MoS_{1.4}$ NSs	GCE	0.5 M H_2SO_4	56	–	32	2020	[87]
$Co_{0.75}Ni_{0.25}Se$ NRs	Ni foam	1.0 M KOH	106	~215	58	2019	[88]
Fe-WS_2 NSs	Carbon cloth	0.5 M H_2SO_4	166	~340	82.2	2021	[89]
N-Ni_3S_2 NSs	Ni foam	1.0 M KOH	155	–	113	2018	[90]
Mo-Ni_3S_2 NRs	Ni foam	1.0 M KOH	/	180	31.8	2018	[91]
WS_2/MoS_2/GR	Graphite sheet	0.5 M H_2SO_4	110	–	41	2020	[92]
Cu-$MoTe_2$/GR	Graphite electrode	0.5 M H_2SO_4	99.2	–	31	2020	[93]

HNs, hollow nanospheres; NS, nanosheets; GCE, glass carbon electrode; NG, N-doped graphene; GR, graphene.

4B.3.4 Transition metal phosphides (TMPs)

Transition metal phosphides (TMPs) have been investigated extensively as electrocatalysts due to their high abundance, HER activity, and favorable stability [94]. They have a trigonal prism structure with metallic or covalent bonds and metallic and semiconducting properties, rendering them high catalytic activity and stability.

Among the reported TMPs, the following six types, FeP_x, Ni_xP_y, Co_xP_y, Cu_3P, MoP_x, and WP_x, exhibit favorable HER activities (Fig. 4B.8). Their applications for HER are inspired by their proven electrocatalytic performance in hydrodesulfurization, which shares a catalytic mechanism analogous to HER [96]. The phosphorous and metal sites

Crystal Structures of Transition Metal Phosphides

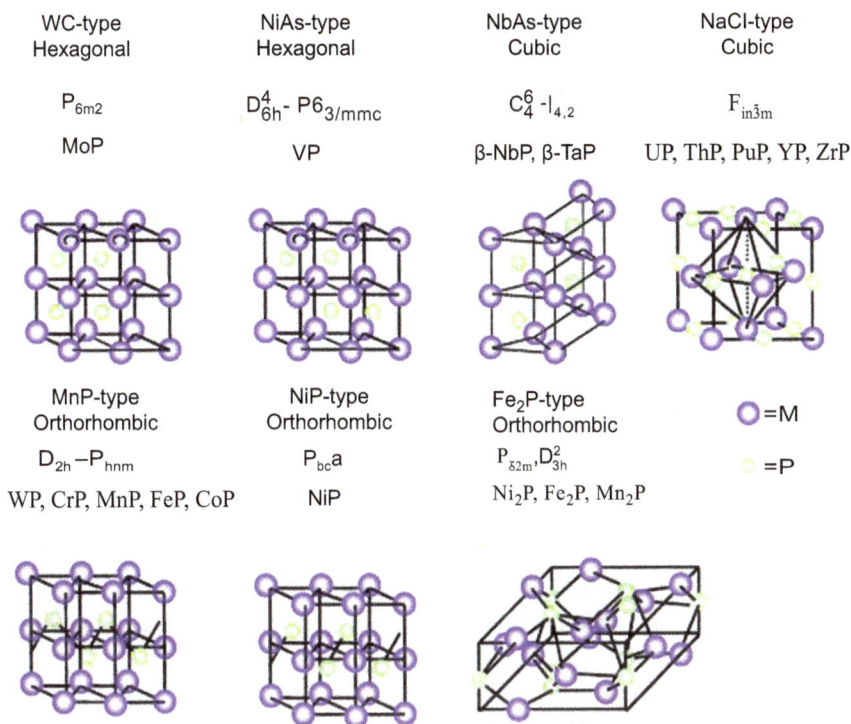

WC-type Hexagonal	NiAs-type Hexagonal	NbAs-type Cubic	NaCl-type Cubic
$P_{\bar{6}m2}$	D_{6h}^4- $P6_{3/mmc}$	C_4^6 -$I_{4,2}$	$F_{in\bar{3}m}$
MoP	VP	β-NbP, β-TaP	UP, ThP, PuP, YP, ZrP

MnP-type Orthorhombic	NiP-type Orthorhombic	Fe₂P-type Orthorhombic	
$D_{2h}-P_{hnm}$	$P_{bc}a$	$P_{\bar{6}2m}, D_{3h}^2$	◯ =M
WP, CrP, MnP, FeP, CoP	NiP	Ni_2P, Fe_2P, Mn_2P	◌ =P

Fig. 4B.8: Crystal structures of typical TMPs [95].

in TMPs act as proton-acceptor sites and hydride-acceptor sites, contributing to the HER process activity. This could be demonstrated with a typical TMPs – Ni_2P [97]. Specifically, the presence of P dilutes the concentrations of active Ni sites and provides new active sites, that is, Ni–P bonds can bond intermediates and/or products in a more moderate way and give rise to more favorable desorption of H_2 when compared to the case of using metal "Ni" that has too strong bonding with intermediates and/or products. In addition, Fe-, Co-, and Ni-based phosphides exhibit excellent OER performances. The real active sites in these phosphides are in situ formed surface oxy/hydroxides (M-OOH, M = Fe, Co, Ni, etc.) and phosphates during the OER process. In this case, phosphides contribute to the formation of active oxy/hydroxides or phosphates and facilitate the charge transfer due to the higher electrical conductivity of phosphides than their oxide counterparts.

In the past, TMPs were mainly prepared at a high temperature or high pressure, using flammable white phosphorus or highly toxic phosphine as the phosphorus source. Unfortunately, such methods suffer from several shortcomings, hindering

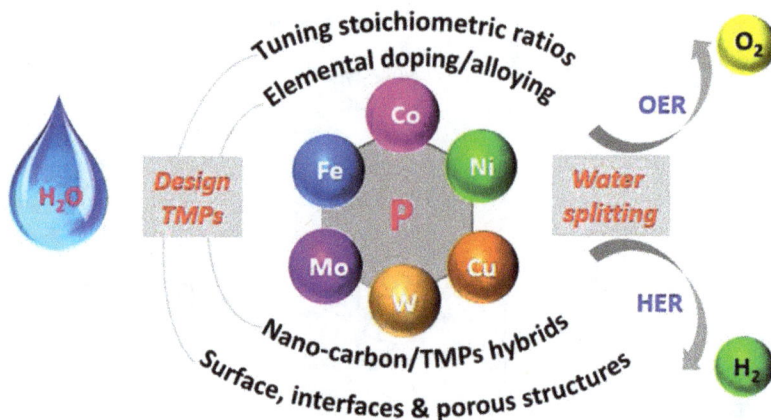

Fig. 4B.9: Designing strategies and six types of transition metal phosphides used for water splitting [98].

the development and application of TMPs. Some methods for preparing TMP catalysts have been developed recently, driven by the constant exploration of attractive physical/chemical properties of TMPs. The preparation methods of TMPs are closely related to the type of phosphorus source chosen. They can be divided into three main categories: (1) Liquid-phase synthesis using organic phosphorus (such as tri-*n*-octyl phosphine, TOP) as the phosphorus source, (2) gas–solid synthesis using inorganic phosphorus such as hypophosphite and orthophosphate as the phosphorus source, and (3) the general method using elemental phosphorus such as red phosphorus and white phosphorus as the phosphorus source.

Specifically, the liquid-phase synthesis of TMPs is achieved via the reaction of TOP with a metal-organic compound (e.g., metal acetylacetonate and metal carbonyl compounds) or elemental metal (e.g., bulk metals and metal nanoparticles) in an organic solvent at a temperature around 300 °C. In the gas–solid synthesis, a hypophosphite or orthophosphate (such as NaH_2PO_2 and $NH_4H_2PO_2$) is mainly used as the phosphorus source, which might decompose and produce gaseous PH_3 at a high temperature. The generated PH_3 can further react with the metal precursors such as metal oxides, hydroxides, MOFs, and other compounds, forming the desired TMPs. In addition, reducing metal orthophosphates directly by H_2 at a higher temperature (over 650 °C) is another gas–solid reaction route to prepare TMPs [99]. In the case of TMP preparation using red phosphorus and white phosphorus as the phosphorus source, the principle is that the elemental phosphorus can react with a metal salt under relatively mild hydrothermal conditions (about 180 ℃), during ball milling, or via other solid–gas reaction routes. Special attention should be paid to the synthesis temperature, given the low ignition point of white phosphorus.

Although TMPs show good catalytic activity and stability as electrocatalysts for HER, some critical issues such as high overpotential and poor charge/electron transfer conductivity limit their widespread application [100]. Because of these, various countermeasures have been suggested, including morphology control, compounding, doping, and interface engineering. By designing and tailoring the morphology of TMPs, their specific surface area could be increased, which is beneficial to the exposure of more active sites, thus improving the HER catalytic activity. For example, cobalt phosphides were prepared using a one-step gas–solid phosphatization using cobalt acetate tetrahydrate and cobalt acetylacetonate cobalt as the cobalt sources and NaH_2PO_2 as the phosphorus source. It was revealed that the effectiveness of the phosphatization reaction was associated with the state of the cobalt precursors. Specifically, a higher Co_xP yield with an alveolar sac-like nanostructure resulting from cobalt(II) acetate is related to its better stability. At the same time, the easily oxidizable salts (e.g., cobalt(II) acetylacetonate) tend to generate a mixture composed of mainly metallic cobalt and minor cobalt phosphides [101].

By introducing conductive materials or substrates, the catalytic activity and stability of TMPs could be significantly improved due to the synergistic effects between different components. Jiang et al. [102] used a two-step method to prepare P-rich NiP_2 nanosheet arrays supported on carbon cloth (NiP_2/CC). The first step was hydrothermal growth of $Ni(OH)_2$ nanosheet arrays on CC. The second step was a chemical conversion of the $Ni(OH)_2$/CC precursor into NiP_2/CC via the low-temperature phosphatization reaction. Such a unique 3D nanosheet array with a large surface area achieves low overpotentials of 75 and 204 mV at 10 and 100 mA/cm^2, respectively, showing a broad application prospect. The CNT-supported interconnected hollow cobalt phosphide nanospheres exhibited remarkable HER electrocatalytic performance with low onset (18 mV) and overpotential ($\eta_{10} = 73$ mV), small Tafel slope (54.6 mV/dec), and high turnover frequency (0.58 s^{-1} at $\eta = 73$ mV) [103].

On the other hand, some other strategies such as modulation of electronic structures (metallic element doping [104], nonmetallic element doping [99], vacancy defects [105]), engineering of multicomponent hybrids (constructing heterogeneous structures [106], coupling with carbon materials), tailoring of microstructures, and construction of working electrode interface have also been reported and proved effective in improving the HER performance.

Among these strategies, metallic element doping has received particular attention in the past decades. Owing to the ability to modify the electronic structure of the host elements and consequently enhance the intrinsic HER activity, many metal dopants such as Fe, Co, Ni, Cu, and Mn have been proven effective in acting as the electrocatalytic activity promoters of TMPs. A series of transition metals (i.e., Fe, Ni, Co, Mn, Cu, Cr, Mo, and V) have been investigated to clarify the impact of metal dopants on the HER activity of CoP in alkaline media [107]. A correlation was revealed between the activity and the dopant's portion of unoccupied d orbitals (P_{un}). DFT calculations indicate that the P_{un} of the undoped CoP is small, which means that a small number of unoccupied 3d orbitals of Co are available to accommodate the HOMO level lone pair electrons of the water molecules to initiate the water adsorption step (Fig. 4B.10a). Moreover, Co's high 3d electron population leads to strong electron–electron repulsion with the HOMO electrons of water and thus weakens the water adsorption and subsequent dissociation probability. The presence of a dopant with a higher P_{un} compared to Co can provide more unoccupied d orbitals to accommodate the lone pair electrons of water and thus strengthen the adsorption and initiate the dissociation step (Fig. 4B.10b). DFT results also indicate that Cr–CoP shows lower water adsorption and water dissociation barrier than CoP. Furthermore, they imply that due to more thermoneutral hydrogen adsorption free energy of Cr–CoP, Cr doping would enhance the water adsorption/dissociation and modulate the electronic structure simultaneously. The experimentally measured HER activity in 1.0 M KOH of the doped CoP electrocatalysts increased in the order V-CoP > Cr/Mo-CoP > Mn-CoP > Fe-CoP > CoP > Ni-CoP > Cu-CoP, which parallels the trend of the P_{un} values (Fig. 4B.10c). This comprehensive study emphasized that transition metal dopants providing the host with higher P_{un} can lead to oxophilic sites for enhanced water activation and modulate the electronic structure of the host phosphide to endow an optimized H adsorption energy.

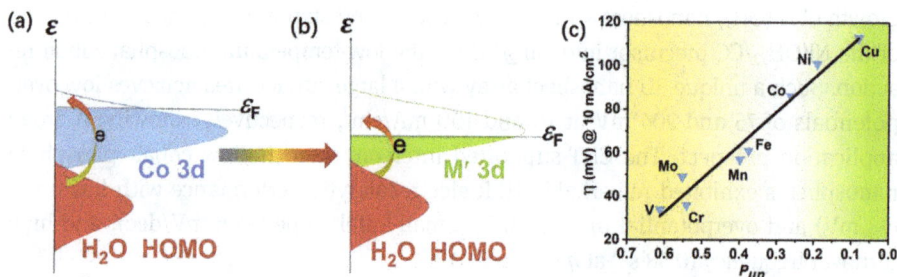

Fig. 4B.10: (a) Schematic energy bands of individual water HOMO and Co 3d orbital on undoped CoP (111) surface. (b) Schematic energy bands of individual water HOMO and M′ d orbital on M′-substituted CoP (111) surface. ε_F represents the Fermi level. (C) Trends in η@10 mA/cm^2 for alkaline HER are shown as a function of P_{un} [107]

Nevertheless, the HER activity of TMPs still needs to be further improved in comparison with the state-of-the-art Pt/C catalyst, and several challenges such as the surface oxidation of TMPs, in-depth understanding of catalytic behavior of TMPs, and corrosion resistance to the electrolyte (regardless of acidic or alkaline solution) need to be addressed properly.

The HER properties of some representative TMPs are summarized in Tab. 4B.4.

4B.3.5 Metal-organic frameworks (MOFs)

MOFs represent a new class of porous materials with unique electronic, optical, and catalytic properties. They can also be used as precursors to fabricate various metals,

Tab. 4B.4: Comparison of HER performances of various TMPs.

Material	Substrate	Electrolyte	Overpotential (mV)		Tafel slope (mV/dec)	Year	Ref.
			η_{10}	η_{100}			
NiP_2 NSs	Carbon cloth	1.0 M KOH	190	/	54	2014	[102]
Ni_5P_4	Ti foil	1.0 M H_2SO_4	23	62	66	2015	[108]
Ni_2P NPs	Ti foil	0.5 M H_2SO_4	120	180	46	2013	[109]
CoP NPs	Ti foil	0.5 M H_2SO_4	75	–	50	2014	[110]
CoP NWs	Carbon cloth	0.5 M H_2SO_4	67	204	51	2014	[111]
CoP NCs	GCE	0.5 M H_2SO_4	105	180	46	2015	[112]
Co_2P NPs	Ti foil	0.5 M H_2SO_4	95	180	45	2015	[113]
CoP_3 NS	Carbon cloth	1.0 M KOH	121	–	66	2017	[114]
Mo-CoP NA	Carbon cloth	1.0 M KOH	40	~130	65	2018	[115]
Co_3S_4/CoP	GCE	0.5 M H_2SO_4	34	–	45	2017	[116]
FeP NPs	Ti foil	0.5 M H_2SO_4	154	–	65	2016	[117]
FeP NA	Ti foil	0.5 M H_2SO_4	55	127	38	2014	[118]
Fe_2P	FTO	0.5 M H_2SO_4	83	163	66	2018	[119]
Fe_3P	FTO	0.5 M H_2SO_4	49	135	57	2018	[119]
Cu_3P NWs	Cu foil	0.5 M H_2SO_4	143	276	67	2014	[120]
MoP	GCE	0.5 M H_2SO_4	130	–	54	2014	[121]
MoP NPs	GCE	0.5 M H_2SO_4	125	200	54	2014	[122]
MoP/CNT	Carbon fiber	0.5 M H_2SO_4	83	142	60	2018	[123]
MoP/RGO	GCE	1.0 M KOH	70	–	58	2016	[124]
WP NPs	Ti foil	0.5 M H_2SO_4	120	–	54	2014	[125]
WP NWs	Carbon cloth	0.5 M H_2SO_4	109	190	56	2016	[126]
FeCoP	Carbon cloth	0.5 M H_2SO_4	37	98	30	2016	[127]
FeCoP	Cobalt foam	1.0 M KOH	44	~155	92	2021	[128]
NiCoP/NiOOH	Cobalt foam	1.0 M KOH	47	126	92	2020	[129]
MoWP	Carbon cloth	0.5 M H_2SO_4	–	138	52	2016	[130]
$NiCo_2P_x$	Carbon felt	1.0 M KOH	58	127	34	2016	[131]
W-NiCoP	Ni foam	1.0 M KOH	29.6	–	38	2019	[132]

NS, nanosheets; NPs, nanoparticles; NWs, nanowires; NA, nanoarrays; GCE, glassy carbon electrode; FTO, fluorine-doped tin oxide.

metal oxide–carbon composites, or pure carbon materials with rich morphological structures and versatile properties (Fig. 4B.11).

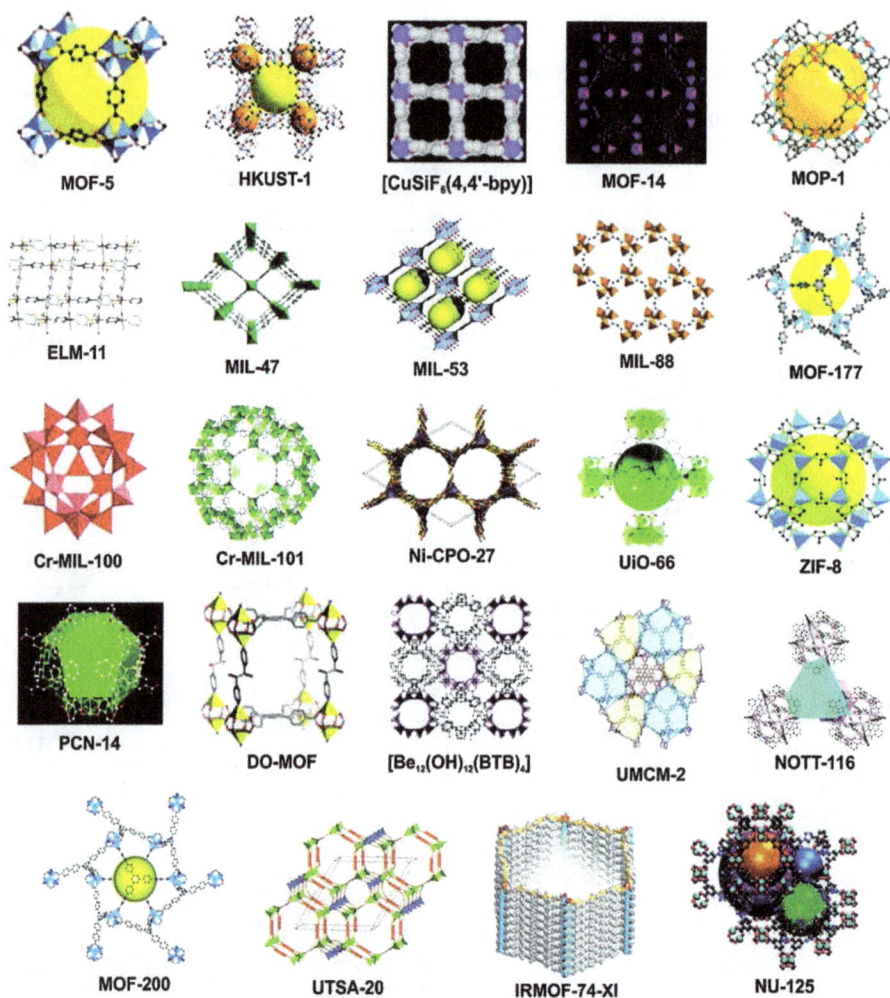

| MOF-5 | HKUST-1 | [CuSiF$_6$(4,4'-bpy)] | MOF-14 | MOP-1 |

| ELM-11 | MIL-47 | MIL-53 | MIL-88 | MOF-177 |

| Cr-MIL-100 | Cr-MIL-101 | Ni-CPO-27 | UiO-66 | ZIF-8 |

| PCN-14 | DO-MOF | [Be$_{12}$(OH)$_{12}$(BTB)$_4$] | UMCM-2 | NOTT-116 |

| MOF-200 | UTSA-20 | IRMOF-74-XI | NU-125 |

Fig. 4B.11: Schematic illustration of reported MOF structures [133].

When used as electrocatalysts or their precursors, MOFs offer several advantages, including high design flexibility, tunable pore channel, large surface-to-volume ratio, flexibility in functionalization with various ligands and metal centers, and rich compositions. The metal centers separated by organic linkers in MOFs can be considered quantum dots; consequently, short diffusion lengths of the charge carriers can be achieved during the electrocatalytic and photocatalytic reactions. The specific surface areas and bandgaps of MOFs can be tailored by tuning the organic ligands and/or metal centers so that their electrocatalytic and photocatalytic activities can be tailored to maximize their performance. Recently, MOF-based materials also have been proved to be particularly suitable for electrocatalytic water splitting, and the interest in this research field is projected to continue increasing.

Pristine MOF-based HER catalysts mostly depend on the metallic catalytic active sites [134]. For example, the highly conductive (electric conductivity: 10^3 S/cm) copper-dithiolene-based coordination polymer Cu-BHT (BHT stands for benzene-hexathiol) showed a good HER performance with an overpotential of 450 mV to achieve the current density of 10 mA/cm^2 [135]. Moreover, DFT calculations reveal that the "Cu edge" on the (100) surface is the most active site in Cu-BHT MOF. The first report on using pristine MOF as an HER catalyst can be dated to early 2011. In this work, polyoxometalate-based MOFs (POMOFs) were successfully synthesized with a unique 3D open framework built of molecular Keggin units connected by trim linkers. The HER performance of as-synthesized POMOFs was better than that of commercial Pt. Neither extensive porosity nor the presence of conjugated ligands connecting the POMs is necessary for the satisfactory HER catalytic behavior.

Later in 2015, another type of POMOFs [136] named [TBA]$_3$[ε-PMo$^V_8$Mo$^{VI}_4$O$_{36}$(OH)$_4$-Zn$_4$][BTB]$_{4/3}$ xGuest (x = NENU-500, BTB, benzenetribenzoate; TBA$^+$, tetrabutylammonium ion) was synthesized and proved to be an ultra-stable HER electrocatalyst in an acid electrolyte, showing an onset overpotential of 180 mV. An overpotential of 237 mV to achieve the current density of 10 mA/cm^2. For another example, the [Co(X$_4$-PTA)(bpy)(H$_2$O)$_2$]$_n$ (X = F, Cl, and Br) based MOFs could be hydrothermally synthesized from the dual ligand system of halogen-substituted phthalate and a dipyridyl molecule (bpy = 4,4′-dipyridyl), which showed the minimal overpotential of 283 mV at the current density of 10 mA/cm^2, low Tafel slope of 86 mV/dec, and long-term stability [137], suggesting that the introduction of halogen elements into MOFs is a promising approach to the enhancement of the electrocatalytic HER performance.

To further improve the HER performance of MOFs, the exposure of more electrocatalytic active sites via constructing various nanostructures is generally considered. For example, 2D BHT-Co and THT-Co (BHT, benzene-hexathiolate; THT, triphenylene-2,3,6,7,10,11-hexathiolate) showed low overpotentials of 340 and 530 mV, respectively, at the current density of 10 mA/cm. The free-standing 2D single-layer sheet of Ni-THT MOF with a thickness of 0.7–0.9 nm showed excellent electrocatalytic activity for HER with a small Tafel slope of 81 mV/dec and an overpotential of 333 mV at the current density of 10 mA/cm^2 in an acidic electrolyte [138]. In addition, the insulation nature of most MOFs is a critical obstacle that hinders their wide application as the HER catalysts, which could be overcome by forming a composite with a conductive material, for example, graphene, CNTs, and conductive substrates (Ni foam, Cu foam, and Ti foil) to accelerate the electron and charge transfer [139].

Thanks to the diversity of MOFs and the rapid development of nanotechnology, MOF-derived materials with controllable structures and compositions can be used to tackle some of the issues mentioned above in HER. MOFs' unique structural and

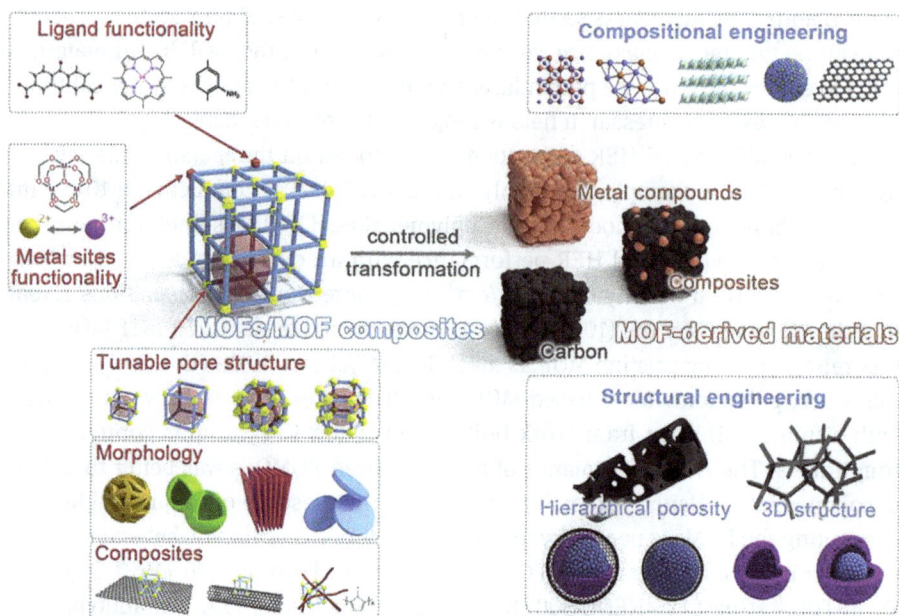

Fig. 4B.12: Schematic illustration of various MOF-derived materials [140].

compositional features make them ideal sacrificial templates for MOF-derived materials. Most of the MOFs are composed of the fourth-period transition metals (Fe, Co, Ni, Mn, C, and Cu) and organic ligands (mainly composed of C, H, O, and N) through coordination bonds. Some advanced strategies such as carbonization and etching, carbonization, sulfonating, phosphating, and nitridation have been developed to prepare various HER catalysts like porous carbon, metals, metal oxides, hydroxides, carbides, chalcogenides, phosphides, nitrides, phosphates, or their hybrids or carbon composites, which showed desirable performances (Fig. 4B.12) [141].

For example, a porous CoP HER electrocatalyst with a concave polyhedron structure was prepared by a topological conversion method using cubic-like ZIF-67 as the precursor [142]. The resultant CoP electrocatalyst showed an overpotential of 133 mV to achieve the current density of 10 mA/cm^2. For another example, bimetallic MOFs (MCo-MOFs, M = Zn, Ni, and Cu) were used as self-templates and precursors to synthesize bimetallic sulfides (M$_x$Co$_{3-x}$S$_4$) via solvothermal sulfidation and calcination treatment [143]. The resultant bimetallic sulfides, especially the Zn$_{0.30}$Co$_{2.70}$S$_4$, fulfilled HER performance over a wide pH range.

Compared to their solid counterparts, hollow nanostructures with larger surface area and extra void spaces can provide many electrochemically active sites and more sufficiently accessible contact interfaces between the electrode and the electrolyte, which is beneficial to the efficient mass transfer. Considerable efforts have been made to prepare MOF-derived electrocatalysts with hollow (or hierarchical) structures. Unlike the traditional hard-template strategy for preparing hollow (or hierarchical) structures, the MOF templates can serve as both morphology-determining scaffolds and essential components in the resulting hollow (or hierarchical) structures, greatly simplifying the preparation process and conferring open porous channels on the resulting hollow (or hierarchical) structures [144]. For instance, the porous carbon matrix with encapsulated Co nanoparticles could be readily prepared by annealing ZIF-67 at 650 °C in an inert atmosphere [145]. The resultant hollow HER catalyst only requires an overpotential of ca. 220 mV to achieve the current density of 10 mA/cm^2, indicating a better HER activity than in most cases using the reported solid Co-based HER electrocatalysts.

The MOF-derived materials can also be used as scaffolds for the nanoparticles with HER electrocatalytic activity. For example, MoS$_x$ nanoparticles were anchored on the porous Zr-MOF framework, showing a low Tafel slope of 59 mV/dec and a low onset potential of nearly 125 mV [146]. The excellent HER performance was ascribed to the favorable delivery of local protons within the porous Zr-MOF structure during the electrocatalytic reaction.

Because of their advantages, such as high crystallinity/porous structure/high surface area, atomically distributed catalytic metal sites, and tailorable ligands, MOFs and MOF-derived materials have been extensively investigated as HER electrocatalysts recently. However, several critical issues still need to be addressed properly. Their instability in electrolytes and poor electrical conductivity seriously hinder their wide application for MOFs. In the case of MOF-derived materials, some concerning issues include the structural stability upon high-temperature treatment, long-term stability at a high current density, and the cost-effectiveness of large-scale synthesis.

Table 4B.5 compares the HER properties of some representative MOFs.

Tab. 4B.5: Comparison of HER performances of various MOFs.

Material	Substrate	Electrolyte	Overpotential (mV)		Tafel slope (mV/dec)	Year	Ref
			η_{10}	η_{100}			
NENU-500	GCE	0.5 M H$_2$SO$_4$	237	–	96	2015	[136]
UiO-66-NH$_2$-Mo-5	GCE	0.5 M H$_2$SO$_4$	200	340	59	2016	[146]
THTNi 2DSP	GCE	0.5 M H$_2$SO$_4$	333	–	81	2017	[138]
THTA-Co	GCE	0.5 M H$_2$SO$_4$	283	–	71	2016	[147]
CTGU-5	GCE	0.5 M H$_2$SO$_4$	388	–	125	2020	[148]
HUST-200	GCE	0.5 M H$_2$SO$_4$	131	–	51	2018	[149]

Tab. 4B.5 (continued)

Material	Substrate	Electrolyte	Overpotential (mV)		Tafel slope (mV/dec)	Year	Ref
			η_{10}	η_{100}			
AB/CTGU-9	GCE	1.0 M KOH	128	–	87	2018	[150]
NU-1000-Ni -S	FTO	0.1 M HCl	238	–	132	2015	[151]

NENU-500, $[TBA]_3[\varepsilon\text{-}PMo^V_8Mo^{VI}_4O_{36}(OH)_4\text{-}Zn_4][BTB]_{4/3}$ xGuest (BTB, benzene tribenzoate; TBA^+, tetrabutylammonium ion); THTA-M, metal dithiolene–diamine coordination; THT, 1,2,5,6,9,10-triphenylenehexathiol; 2DSPs, two-dimensional supramolecular polymers; CTGU-5, polymorphic Co-MOFs; HUST, two-dimensional iron-based metal-organic framework; HUST, Huazhong University of Science and Technology; AB, acetylene black; CTGU-9, $[Co_{1.5}(TTAB)_{0.5}(4,4'\text{-bipy})(H_2O)]$; NU-1000 is one member of a growing family of porous MOFs.

4B.3.6 Other metal compounds

Metal borides (e.g., MoB_2, NiB, CoB, and CoNiB) and borates (denoted as B_i, e.g., $Ni\text{-}B_i$, $Co\text{-}B_i$, and $Ni\text{-}Co\text{-}B_i$) [152, 153] have been subjected to an intensive investigation recently due to their excellent stability and unexpected HER performance (Fig. 4B.13).

As for amorphous metal borides such as cobalt boride, their excellent HER performance, according to Patel and coworkers [154], can be explained. The value of Pauling's electronegativity for B (2.01) is larger than that of Co (1.70). So, the expected electronic interaction of Co–B leads to the transfer of electrons from Co to B, making the metal site electron deficient. DFT studies on crystalline Co–B and Co_2B clusters further confirm this. For the crystalline metal borides, different conclusions might be drawn. For example, Fe-based borides exist in two crystalline forms of Fe_2B and FeB_2 [155]. According to the theoretical calculation, the top B site of the low index surface (001) of FeB_2 shows the lowest ΔG_H value, implying that it is the main site responsible for HER. The iron-rich Fe_2B shows much higher ΔG_H values on both (001) and (110) surfaces than in the case of FeB_2. These results suggest that the boron-rich FeB_2 is more likely to show better HER performance, which matches well with what has been found experimentally.

From the viewpoint of basic understanding, many aspects related to the catalytic mechanism of the metal boride/borate family are ambiguous. The metal boride system's second metalloid (P) might completely change the electrocatalysis mechanism. The change of active catalytic centers in such systems has been discussed in the literature, but no direct evidence has been provided. As mentioned above, in situ/operational testing combined with theoretical calculation should be used to understand the role of heteroatoms. The introduction of other nonmetals such as S,

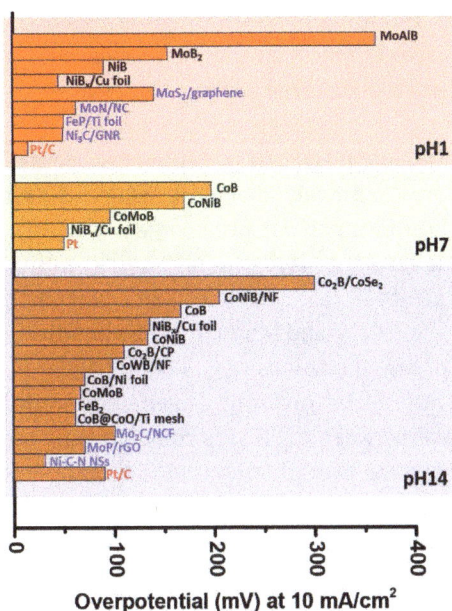

Fig. 4B.13: Comparison of HER performances of various metal boride electrocatalysts in different pH solutions, Pt, and other nonprecious metal electrocatalysts [153]. C, Vulcan carbon; NC, N-doped carbon; GNR, graphene nanoribbons; CP, carbon paper; NF, Ni foam; NCF, nanoparticles dispersed on carbon microflowers; NSs, nanosheets.

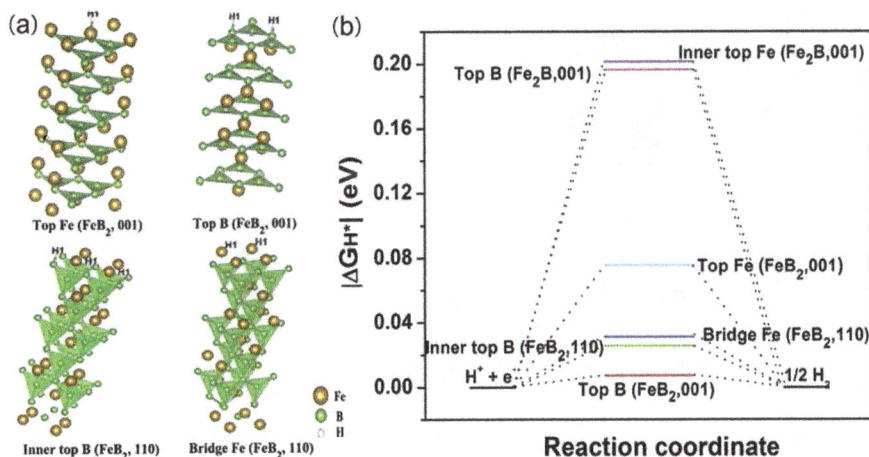

Fig. 4B.14: (a) There were four specific H* adsorption sites with relatively low ΔG_{H*} on the (001) and (110) surfaces of FeB_2. (b) Calculated free-energy diagram of HER over the low-index (001) and high-index (110) surfaces of FeB_2 and Fe_2B at the equilibrium potential [155].

C, and N to develop hybrid electrocatalysts and further study of their basic properties and electrochemical behavior has also been a topical area in the past decade.

The layered double hydroxide (LDH)-based materials have attracted increasing attention as the potential HER catalysts due to their unique layered structure with high specific surface area and unique electron distribution [156]. LDHs represent a class of ion lamellar crystals, which consist of mainly positively charged brucite-like host layers, and the interlayer anions between host layers to balance charge and solvent molecules (represented by the general formula $[M_{1-x}^{2+}M_x^{3+}(OH)_2]^{x+}[A_{x/n}]^{n-} \cdot mH_2O$, Fig. 4B.15). The LDH layers incorporate divalent M^{2+} (e.g., Fe^{2+}, Co^{2+}, Ni^{2+}, and Zn^{2+}) and trivalent M^{3+} metal cations (e.g., Al^{3+}, Fe^{3+} Cr^{3+}, and Mn^{3+}), forming positively charged layers. A^{n-} are mainly inorganic or organic anions (e.g., CO_3^{2-}, NO_3^-, Cl^-, SO_4^{2-}, and RCO_2^-), where x is the molar ratio of M^{2+}/M^{3+} and is generally in the range of 0.20–0.33, which occupy the space between the layers compensating for the positive charge and inducing stability in the overall structure [157]. Each hydroxyl group in the LDH layer faces the middle layer area and can form hydrogen bonds with interlayer anions and water molecules. The unique 2D lamellar structure endows LDHs with the following advantages: (1) ready tunability of metal cations in the host layers; (2) ease of change of interlayer anions and distance; and (3) ease of exfoliation of LDHs into ultrathin nanosheets.

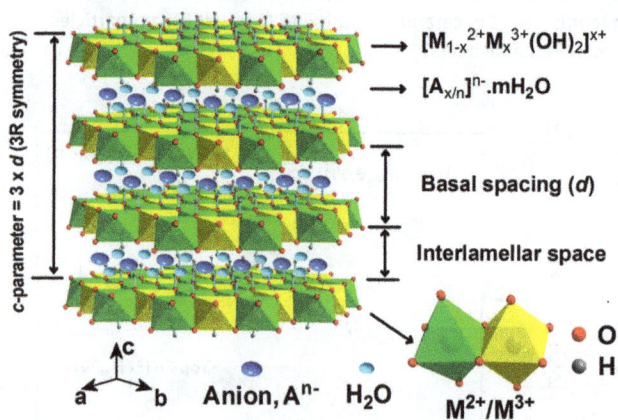

Fig. 4B.15: A schematic illustration of the chemical composition of LDH [158].

Nevertheless, the limited active sites and their poor intrinsic activity and low electronic conductivity impede the further improvement in the electrocatalytic activity of LDHs. To overcome these, some efficient strategies have been proposed, including (1) control of morphology and microstructure; (2) cationic doping; (3) tuning the anion and interlayer spacing; (4) exfoliation of LDHs and tuning electronic structure and introducing defects, and (5) combining LDHs with the conductive substrate to form hybrids/composites.

The LDH-derived materials, including metal hydroxides, oxyhydroxides, oxides, bimetallic nitrides, phosphides, sulfides, and selenides, have been extensively studied as HER electrocatalysts. They show excellent electrocatalytic activity and performance, which could be attributed to the significant advantages of LDHs themselves (e.g., adjustable composition, layered structure, and unique electronic structure) and their derivatives' unique physical and chemical properties, as well as the synergistic effects.

Carbon materials such as CNTs and graphene with large surface area and good conductivity are often used to improve the HER performance of LDHs, and many published results have verified the feasibility of this approach.

Several other strategies, such as exfoliation of LDHs, tuning the electronic structure, and introducing defects, can also improve the HER activity of LDHs. Bulk LDHs generally suffer from limited specific surface area and poor electrical conductivity, resulting in poor HER activity. Previous studies have shown that 2D nanomaterials with a single or few atomic layers can significantly increase the exposed active sites and greatly improve the physical and chemical properties. In addition, the exfoliation of bulk materials to thin-layered nanomaterials is always accompanied by the formation of numerous edges and corner sites, which will act as additional active sites. For example, single-layered NiFe LDHs, NiCo LDHs, and CoCo LDH nanosheets could be successfully prepared via the liquid exfoliation of bulk LDHs in the presence of degassed formamide [159]. According to Liu et al. [160], the exfoliated defect-rich CoFe LDH nanosheets exhibited satisfactory HER activity with 166 mV to drive the current density of 10 mA/cm^2, which was superior to most of the bulk CoFe LDHs.

The HER activity of NiFe-LDH in an alkaline medium is poor. However, the partial substitution of Fe atoms by Ru can greatly accelerate the hydrogen release kinetics, and the content of Ru plays a significant role in the HER performance [161]. With the optimal doping level (16 at.% Ru), the resultant NiFeRu-LDH only requires 29 mV to achieve the current density of 10 mA/cm^2 in a KOH solution, which is much lower than in the case of using a commercial Pt/C electrode (31 mV). Both experimental and theoretical results indicate that introducing Ru atoms into NiFe-LDH can effectively reduce the barrier of the Volmer reaction step, ultimately accelerating the HER kinetics. Apart from the strategies mentioned above, the derivatives of LDH-based materials, including metal oxyhydroxides, oxides, bimetal nitrides, phosphides, sulfides, and selenides, have been explored as HER electrocatalysts recently [162]. The LDH derivative electrocatalysts could inherit unique features of LDHs, such as the tunability of composition, layered structures, and unique electronic structure, generating synergistic effects on improving HER performance.

In addition to metal borides and LDHs, metal alkoxides [163], metal nitrides [164], and metal carbides [165] have been used as electrocatalysts for HER. The HER properties of some other representative types of metal compounds are listed in Tab. 4B.6.

Tab. 4B.6: Comparison of HER performances of various metal compounds.

Material	Substrate	Electrolyte	Overpotential (mV)		Tafel slope (mV/dec)	Year	Ref.
			η_{10}	η_{100}			
MoB	CPE	1.0 M KOH	~210	–	59	2012	[166]
Mo_2B_4	Carbon sheet	0.5 M H_2SO_4	310	–	80	2017	[167]
FeB_2 NPs	GCE	1.0 M KOH	61	~170	87.5	2017	[155]
NiB	Ni foam	1.0 M KOH	41.2	–	106.5	2020	[168]
VB_2	Cu sheet	0.5 M H_2SO_4	204	–	108.3	2020	[169]
Co-B NPs	Ni foil	1.0 M KOH	70	270	–	2018	[170]
MoAlB	Sn-plated Cu wire	0.5 M H_2SO_4	301	–	68	2017	[171]
WN NA	Carbon cloth	0.5 M H_2SO_4	198	340	92	2015	[172]
MoN NSs	GCE	0.5 M H_2SO_4	~220	–	90	2014	[173]
WN/Co	GCE	1.0 M KOH	76	~285	98	2018	[174]
NiMoN	Carbon cloth	1.0 M KOH	109	~165	95	2016	[175]
Mo_2N-Mo_2C/Gr	GCE	0.5 M H_2SO_4	157	320	55	2018	[176]
NiFeW-LDHs	Carbon paper	1.0 M KOH	114	–	112	2021	[177]
$Ni_{1-x}Fe_x$ LDHs	Ni foam	1.0 M KOH	169	–	31	2017	[178]
Cu/P-FeCo LDHs	Ni foam	1.0 M KOH	63	–	41.74	2021	[179]
B, P-NiVFe LDHs	Ni foam	1.0 M KOH	117	–	68	2021	[180]
WC_x NWs	Carbon cloth	0.5 M H_2SO_4	118	175	55	2017	[181]
Mo_2C NTs	GCE	0.1 M KOH	112	–	55	2015	[182]
Co_2C NPs	GCE	0.1 M KOH	181	–	89	2017	[183]
Fe_3C/GR	GCE	0.5 M H_2SO_4	49	–	46	2015	[184]
Fe-Ni_3C NSs	GCE	0.5 M H_2SO_4	178	–	36.5	2017	[185]
Co_3Mo_3C/CoNC	GCE	1.0 M KOH	68	178	63	2021	[186]

CPE, carbon paste electrode NS, nanosheets; NPs, nanoparticles; NWs, nanowires; NA, nanoarrays; GCE, glassy carbon electrode; NT, nanotubes.

4B.3.7 Metal-free materials

Although most metal (metal compounds)-based electrocatalysts are stable, some suffer from gradual oxidation and agglomeration upon long-term exposure to air. Furthermore, metal (metal compounds)-based electrocatalysts are often electrochemically unstable at high potentials in acidic media or easily reunite during the electrolysis process. As a new type of metal-free high-efficiency electrocatalyst, carbon materials have become a strong competitor of noble metal-based electrocatalysts in HER. Carbon materials are less expensive and have good electrical/thermal conductivity, excellent mechanical strength, and versatile morphologies, so they have been widely used in electrochemical energy storage and conversion systems. Carbon materials such as graphene, CNTs, carbon nanofibers, and other nanostructured carbon materials have been explored as cost-effective alternative catalysts for HER.

Theoretically, pristine carbon materials show limited HER activities since they are electrochemically inactive. For example, the activated CNTs need an overpotential up to 220 mV to afford a current density of 10 mA/cm^2, which is far from that required for practical applications [187]. The ease of changing the carbon structure and doping surface chemistry can lead to additional active sites, making carbon materials more feasible for HER electrocatalysis application.

Doping is an effective way to tune carbon's electronic and electrochemical properties. Due to the difference in electronegativity between carbon atoms and heteroatoms, the doping usually leads to redistribution of the charge density and spin density of carbon atoms in the crystal lattice, effectively regulating the electronic structure and increasing the adsorption of reactants at specific active sites. Due to the difference in atomic size between carbon and dopants, the local geometry around the heteroatoms might change, exerting a significant impact on the HER performances of the materials.

As illustrated in Fig. 4B.16, the absolute difference in electronegativity between heteroatoms and carbon atoms determines the nature of the charge density redistribution. For example, theoretical investigation indicates that carbon linked with nitrogen near the edge (pyridinic) significantly enhances the electrocatalytic activity [188]. Based on the electronegativity difference between heteroatom dopants and carbon atoms, the electronic effects of different heteroatom doping can be classified into charge-dominated and spin-dominated.

In general, two main approaches are used to introduce heteroatoms into the carbon framework: carbonization of heteroatom-containing compounds (also referred to as the in situ synthesis) and posttreatment of carbon materials with reactive heteroatom sources Fig. 4B.17 [190]. Taking N doping as an example, by direct annealing N-containing materials, such as biomass, polyaniline, polypyrrole, polyacrylonitrile, both N doping and the synthesis of carbon materials can occur simultaneously. Recently, other materials like MOFs and ionic liquids have also been used as precursors. The well-established synthesis of these materials has attracted

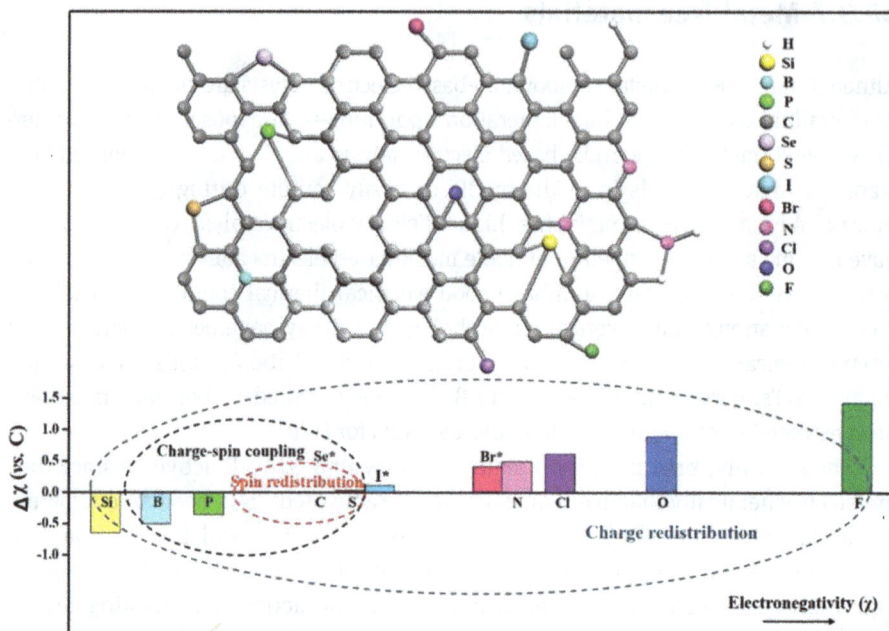

Fig. 4B.16: Illustration of carbon materials with different dopants. Upper panel: schematic diagram of carbon lattice structure doped by different heteroatoms. Lower panel: classification of the origins of doping effects as charge redistribution, spin redistribution, and charge–spin coupling [189].

increasing attention since N-doped carbon materials with various structures and doping contents can be prepared. In addition to the intrinsic properties of N-containing precursors, processing parameters such as carbonization temperature and time play a key role in regulating both N groups and their content. As found by Quílez-Bermejo et al. [191], with the increase of carbonization temperature from 800 to 900 °C, the quaternary N appears, and amines disappear due to their lower thermal stability. Once the temperature increases to 1,000 °C, the N–O species are formed, along with increased quaternary N and decreased pyridine N species. When the temperature increases to 1,200 °C, the pyridine N species almost disappear due to the conversion to quaternary N via the condensation reactions. As another general rule, the N content decreases with the increase of carbonization temperature.

As for the posttreatment for preparing N-doped carbon, materials are normally done via the direct reaction between presynthesized carbon materials and N-containing sources (e.g., urea, melamine, NH$_3$, ethylenediamine, and amines). The posttreatment usually requires a high temperature and a long reaction time to facilitate the incorporation of N into the carbon skeleton. Despite these drawbacks, the simplicity of postsynthesis doping routes to prepare N-doped carbons is the main reason behind their wide use.

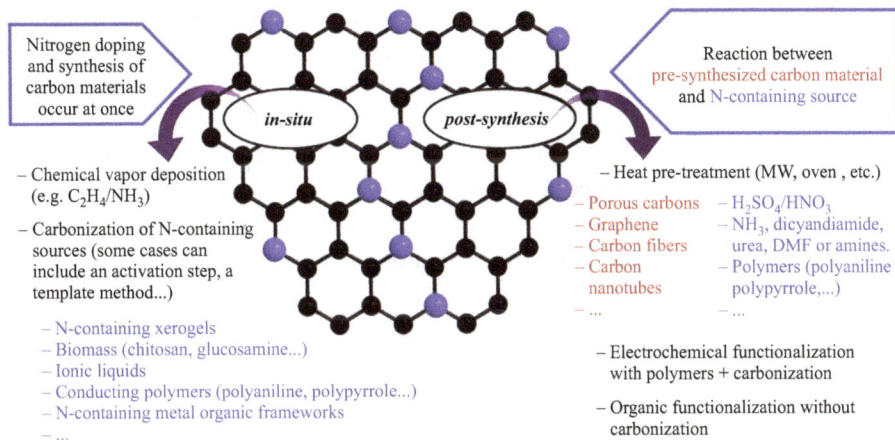

Nitrogen doping and synthesis of carbon materials occur at once

in-situ

Reaction between pre-synthesized carbon material and N-containing source

post-synthesis

– Chemical vapor deposition (e.g. C_2H_4/NH_3)

– Carbonization of N-containing sources (some cases can include an activation step, a template method...)

  – N-containing xerogels
  – Biomass (chitosan, glucosamine...)
  – Ionic liquids
  – Conducting polymers (polyaniline, polypyrrole...)
  – N-containing metal organic frameworks
  – ...

– Heat pre-treatment (MW, oven , etc.)

– Porous carbons	– H_2SO_4/HNO_3
– Graphene	– NH_3, dicyandiamide,
– Carbon fibers	urea, DMF or amines.
– Carbon	– Polymers (polyaniline
nanotubes	polypyrrole,...)
– ...	– ...

– Electrochemical functionalization with polymers + carbonization

– Organic functionalization without carbonization

Fig. 4B.17: Methods used to synthesize N-doped carbon materials [192].

Since the discovery of excellent ORR electrocatalytic behaviors of N-doped vertically aligned CNTs in 2009, various electroactive metal-free catalysts have been explored [193]. Recent studies have found that doping and co-doping carbon materials with heteroatoms such as B, P, and S are also effective in further improving the electrocatalytic activity of carbon catalysts. Due to the difference in electronegativity, doping with heteroatoms may induce charge transfer from neighbor carbon atoms, facilitating the chemisorption of reactants and the change of local density of states and electronic structure. The most investigated dopant is N, which is more electronegative than C and can remove charge density from adjacent carbon atoms, displaying the n-type doping behavior. More importantly, due to the similar bond length of C–C and C–N, incorporating N atoms into the carbon skeleton will cause minimal distortion of the carbon structure [194].

As determined using DFT calculations, the available spin states reveal that doping N atoms into graphene can induce asymmetrical charge distributions on the adjacent carbon atoms, leading to larger polarizations and stronger affinities toward H [195]. Moreover, doping with N atoms can also affect the energy levels of the valence orbitals in the graphene matrix, which is beneficial to the transfer of electrons from graphene to catalytically active sites, rapidly reducing the adsorbed H* species into H_2 [3]. Researchers have strived hard to find out what specific types of N (pyridinic, pyrrolic, graphitic, or oxidized) contribute to the improvement of electroactivity. However, no consensus has been reached [195, 196]. For example, the 3D N-doped, plasma-etched graphene exhibited a low overpotential of 128 mV to achieve a current density of 10 mA/cm^2 in an acidic electrolyte [197]. It also achieved a small Tafel slope of 66 mV/dec and a large exchange current density up to 0.11 mA/cm, which were comparable or superior to those in most cases of metal-free carbon catalysts.

Extensive research has been carried out to investigate the HER activities of carbon materials doped with elements that possess low (B) or similar (S) electronegativity [198]. For example, similar to N doping, doping of B atoms into the carbon skeleton can also improve the graphitization degree [199] and reduce the resistivity [200] (7.4×10^{-7}–7.7×10^{-6} Ω for B-doped CNTs, and 5.3×10^{-6}–1.9×10^{-5} Ω for CNTs). It was also demonstrated that doping with B could shift the Fermi level of the CNTs to the valence band, resulting in improved HER electrocatalytic activity [201]. Sathe et al. synthesized B-doped graphene via a facile wet-chemical method using borane tetrahydrofuran (BH_3-THF) as the borylation agent [202]. Its enhanced HER electrocatalytic activity concerning pristine graphene can be attributed to the introduction of B dopants, as the doping-induced surface-defect sites and lots of surface-active (electron-rich) reduction centers can reduce the conversion barriers from H^+ ions to H_2.

DFT calculations reveal that S doping could lead to a dramatic change in the electronic energy structures and significantly enhanced adsorption of H atoms on the carbon framework, showing a higher efficiency in enhancing the HER activity than in the case of N doping [203]. Zhou et al. [204] reported that the CVD-grown 3D S-doped graphene exhibited a low-onset overpotential of 237 mV with a small Tafel slope of 64 mV/dec in the acid electrolyte. In contrast, the pristine 3D graphene showed an onset overpotential of 515 mV and a Tafel slope of 225 mV/dec. Single-atom metal (e.g., Ni, Co)-doped carbon frameworks can also be used as electrocatalysts for HER. For example, Qiu et al. [205] prepared nanoporous graphene doped with single-atom nickel. Compared to conventional nickel-based catalysts and pure graphene, this material exhibits superior HER performance with a low overpotential of 50 mV at 10 mA/cm^2, and a small Tafel slope of 45 mV/dec in an acidic electrolyte, along with excellent cycling stability. Experimental and theoretical studies indicate that the unusual HER activity of this catalyst is attributed to the sp–d orbital charge transfer between the atoms of Ni dopants and the adjacent carbon atoms.

Despite the intensive studies on the synthesis and application of doped carbon materials as HER electrocatalysts, there is still an ongoing debate about the dopants' actual effects, chemical configuration, and defects on the electrocatalytic activity [190]. Along with the rapid developments in heteroatom doping of carbon structures, hybridization with other metals (metal compounds) has also been explored to increase the number of active sites to improve the performance of electrocatalysis.

N- and B-doped carbon materials have been proved to be promising HER catalysts for water splitting, and their performance can be further improved by co-doping with other heteroatoms, for example, S or P, because of the increased number of heteroatoms and the electronic interactions between different heteroatom dopants could bring about synergistic effects [206, 207]. The N/S co-doped nanoporous graphene prepared by CVD using pyridine and thiophene as the nitrogen and sulfur precursors shows an onset potential of −0.13 V and a Tafel slope of 80.5 mV/dec, which are better than those of the undoped and single N- or S-doped counterparts [208]. In DFT calculations, carbon defects alone in the graphene lattice are not catalytically active for HER. However, the coupling of S and N atoms with geometric defects gives synergistic effects on tuning the ΔG_{H*} and thus contributes to the higher HER activity. DTF calculates a highly positive ΔG_{H*} value of 1.85 eV is calculated by DTF, indicating its negligible HER activity [209]. N and/or P doping can reduce the ΔG_{H*} value to enhance the initial H* adsorption, in which the pyridinic N and P co-doping model shows the lowest $|\Delta G_{H*}|$ value of 0.08 eV, indicating its highest HER activity with the most favorable H* adsorption/desorption capability.

Although the strategies introduced above can increase the HER electrocatalytic activity either intrinsically or extrinsically, the overall performance is unsatisfactory compared to the precious metal-based catalysts. In electrocatalytic systems, conductive solid carbon material as a support for active sites is essentially required from both cost and performance viewpoints. Coupling nanostructured carbon materials with other active components such as metallic or nonmetallic moieties to form hybrids potentially can further improve the electrocatalytic activity. Such a strategy is developed based on the following: (1) increasing the surface area to increase the electroactive sites; (2) improving the electrical conductivity to facilitate electron/charge transfer; and (3) modifying the electronic structure/interaction to change the surface adsorption capability. These can be triggered individually or synergistically, leading to significantly enhanced HER activities of the formed hybrids.

Carbon supports can significantly increase the utilization of active sites by protecting the catalysts from agglomeration. Also, the electrical synergy generated between electrocatalytic centers and carbon surfaces may induce unique metal–support interactions, substantially impacting critical HER processes. Suliman et al. [210] investigated the impact of microstructural features of carbon support on the HER performance of FeP nanoparticles and found that higher specific and electrochemically active surface area was translated into a greater number of active sites or turnover frequency. However, analyses of Tafel slopes suggested that the HER mechanisms were insensitive to the carbon supports' surface area and electrical conductivity. For example, Mo_2C nanoparticles anchored on N-doped carbon nanosheets [211] exhibited promising HER activity in 1.0 M KOH electrolyte, with an onset potential of 0 V and low overpotential of 45 mV to achieve 10 mA/cm^2, which was comparable to the benchmark Pt/C.

Most of the current HER electrocatalysts are powder form, so conductive substrates such as glassy carbon, carbon papers, or metal foils (e.g., nickel foam, stainless steel mesh, and titanium plate) are inevitably used to support the electrocatalysts for electrochemical tests. In addition, other additives such as polymer binders might increase the mass and electron transport resistance and cause reduced active sites and attenuated electrocatalytic activity. As a result, binder-free 3D porous carbon architectures with interconnected porosity, high electrical conductivity, and mechanical stability can maximize the activity of the catalyst, addressing the above challenges.

Carbon nitride (C_3N_4), having a similar 2D crystal structure to that of graphene, is recognized as another promising low-cost metal-free electrocatalyst for HER despite its electrochemical inertness, which can be addressed by combining C_3N_4 with other conductive carbonaceous materials, doping with heteroatoms, and forming composites with other metal compounds. For example, thanks to the synergistic effects of chemical and electronic couplings on the enhancement of the proton adsorption/reduction kinetics, the C_3N_4/N-doped graphene composite showed the smallest $|\Delta G_{H*}|$ value of 0.19 eV in comparison with that of pure C_3N_4 (0.54 eV) or N-doped graphene (0.57 eV) [3]. Besides, it needed an overpotential of ~ 240 mV to achieve the current density of 10 mA/cm^2. Rao et al. [198] used 2D boron carbonitride (BCN) as highly active HER electrocatalysts, especially, the carbon-rich one (BC_7N_2) only required the onset potential of 56 mV and overpotential of 70 mV to achieve 10 mA/cm^2, close to the case of using commercial Pt. A recent theoretical study indicates that, due to the coexistence of unpaired electrons and empty states, unsaturated boron atoms along the hexagon holes act as the active sites, making 2D boron sheets (α and β_{12}) with ΔG_{H*} close to zero a promising candidate catalyst for HER [212].

The HER properties of some representative metal-free materials are listed in Tab. 4B.7.

Tab. 4B.7: Comparison of HER performances of various metal-free materials.

Material	Substrate	Electrolyte	Overpotential (mV)		Tafel slope (mV/dec)	Year	Ref
			η_{10}	η_{100}			
N, O, P-hollow carbon	GCE	1.0 M KOH	290	–	102	2019	[213]
B-GR	GCE	0.5 M H_2SO_4	~440		99	2014	[214]
TpPAM	GCE	0.5 M H_2SO_4	250	–	106	2017	[215]
C_{60}-SWCNTs	GCE	0.1 M KOH	380	–	120.8	2019	[216]
N, S-graphitic sheets	GCE	0.1 M KOH	310	–	112	2016	[217]
SiO_2/PPy NTs	GCE	1.0 M NaH_2PO_4/ Na_2HPO_4	~190	–	100.2	2017	[218]
N, P-porous carbon	GCE	0.5 M H_2SO_4	204	–	58.4	2015	[219]
g-C_3N_4 nanoribbon/GR	GCE	0.5 M H_2SO_4	70	–	54	2014	[220]
NG	GCE	0.5 M H_2SO_4	220	–	109	2014	[195]
NG	GCE	1.0 $HClO_4$	200	–	74	2013	[221]
N, S-carbon tubes	GCE	0.5 M H_2SO_4	76	–	126	2017	[222]

Tab. 4B.7 (continued)

Material	Substrate	Electrolyte	Overpotential (mV)		Tafel slope (mV/dec)	Year	Ref
			η_{10}	η_{100}			
N, P-carbon tubes	GCE	0.5 M H_2SO_4	89	–	122	2015	[223]
N, F, P-GR	GCE	0.1 M KOH	520	–	106.5	2016	[224]
g-C_3N_4/N-GR	GCE	0.5 M H_2SO_4	240	–	51.5	2014	[3]
BC_7N_2	GCE	0.5 M H_2SO_4	298	–	100	2016	[198]
ILS-CNTs	GCE	0.5 M H_2SO_4	135	~260	38	2021	[225]

TpPAM, porphyrin-based metal-free covalent organic polymer; PPy, polypyrrole; ILs, imidazolium ionic liquids.

4B.4 Conclusion

Production of H_2 via electrochemical water splitting plays a crucial role in the future application of renewable energy, and thus exploring inexpensive and high-efficiency electrocatalysts is currently a research hotspot. Pt is regarded as the ideal HER electrocatalyst, but its high cost makes it impractical. Moreover, since Pt must be employed along with the carbon support, the mechanical stability of the nanocomposite and the utilization of Pt nanoparticles are the critical factors that need to be considered carefully. Meanwhile, the research focus should be shifted to nonprecious electrocatalysts. Benefit from the considerable amount of work devoted to this field of research, the general characteristics of competitive electrocatalysts can be summarized as follows:

1) The importance of good carbon catalyst support with high surface area and robust structure is self-evident. The use of an appropriate carbon catalyst support provides a reliable conductive network for highly efficient reactions (not only a low overpotential) and improves the efficiency of HER electrocatalysts.
2) Of the electrocatalysts reported, molybdenum/tungsten sulfides, selenides, phosphides, and carbides show many advantages, warranting further investigations.
3) Optimal construction of materials architecture seems feasible to increase the active sites for the HER electrocatalysis. For instance, the exfoliation of layered materials to 2D materials effectively exposes more active sites, significantly enhancing the HER performance.
4) Defect engineering is an efficient way to redistribute the charge to activate the electrocatalytic sites. It can further improve the catalytic properties. It does not target the electroactive material but the carbon catalyst support. Doping can make the carbon electrochemically active to perform better in the charge transfer process while maintaining a better interaction with the catalyst. Nevertheless,

there is still a lack of clear understanding of the influence mechanisms in these systems (e.g., specific engineering of doping sites/types/locations on the HER performance), and further work in this area should be done in the future. A combination of experimental and theoretical studies will be the key to unraveling the electrocatalytic active centers in defect-engineered materials.

5) Developing in situ/operando characterization methodologies is vital to understanding the structure–composition–performance relationships, which will provide a new perspective for the rational design of highly efficient HER electrocatalysts.

6) Heterostructured HER catalysts with refined nanostructures and substantially exposed edges can provide sufficient adsorption sites for HER intermediates, enable fast mass diffusion, and improve the electrocatalyst's durability. This and other relevant subjects would be a fruitful area for future research.

References

[1] Larcher, D. & Tarascon, J. M. (2015). Nature Chemistry, 7(1), 19–29.
[2] Wang, S. & Wang, L. (2019). Tungsten, 1(1), 19–45.
[3] Zheng, Y., Jiao, Y., Zhu, Y., Li, L. H., Han, Y., Chen, Y., Du, A., Jaroniec, M. & Qiao, S. Z. (2014). Nature Communications, 5(1), 3783.
[4] Chen, S., Thind, S. S. & Chen, A. (2016). Electrochemistry Communications, 63, 10–17.
[5] You, B. & Sun, Y. (2018). Accounts of Chemical Research, 51(7), 1571–1580.
[6] Roger, I., Shipman, M. A. & Symes, M. D. (2017). Nature Reviews Chemistry, 1(1).
[7] Xu, Y., Wang, C., Huang, Y. & Fu, J. (2021). Nano Energy, 80, 105545.
[8] Trasatti, S. (1972). Journal of Electroanalytical Chemistry and Interfacial Electrochemistry, 39(1), 163–184.
[9] Zeradjanin, A. R., Grote, J.-P., Polymeros, G. & Mayrhofer, K. J. J. (2016). Electroanalysis, 28(10), 2256–2269.
[10] Cai, J., Javed, R., Ye, D., Zhao, H. & Zhang, J. (2020). Journal of Materials Chemistry, 8(43), 22467–22487.
[11] Tymoczko, J., Calle-Vallejo, F., Schuhmann, W. & Bandarenka, A. S. (2016). Nature Communications, 7(1), 10990.
[12] Eftekhari, A. (2004). Journal of the Electrochemical Society, 151(9), E291.
[13] Ehsan, M. A., Suliman, M. H., Rehman, A., Hakeem, A. S., Al Ghanim, A. & Qamar, M. (2020). International Journal of Hydrogen Energy, 45(30), 15076–15085.
[14] Liu, D., Li, L. & You, T. (2017). Journal of Colloid and Interface Science, 487, 330–335.
[15] Zhang, L., Roling, L. T., Wang, X., Vara, M., Chi, M., Liu, J., Choi, S.-I., Park, J., Herron, J. A., Xie, Z., Mavrikakis, M. & Xia, Y. (2015). Science, 349(6246), 412.
[16] Rajala, T., Kronberg, R., Backhouse, R., Buan, M. E. M., Tripathi, M., Zitolo, A., Jiang, H., Laasonen, K., Susi, T., Jaouen, F. & Kallio, T. (2020). Applied Catalysis. B, Environmental, 265, 118582.
[17] Bao, X., Gong, Y., Zheng, X., Chen, J., Mao, S. & Wang, Y. (2020). Journal of Energy Chemistry, 51, 272–279.
[18] Duan, S., Du, Z., Fan, H. & Wang, R. (2018). Nanomaterials, 8(11), 949.
[19] Shao, Q., Li, F., Chen, Y. & Huang, X. (2018). Advanced Materials Interfaces, 5(16), 1800486.

[20] Chen, C., Kang, Y., Huo, Z., Zhu, Z., Huang, W., Xin, H. L., Snyder, J. D., Li, D., Herron, J. A., Mavrikakis, M., Chi, M., More, K. L., Li, Y., Markovic, N. M., Somorjai, G. A., Yang, P. & Stamenkovic, V. R. (2014). Science, 343(6177), 1339.

[21] Xiao, W., Lei, W., Gong, M., Xin, H. L. & Wang, D. (2018). ACS Catalysis, 8(4), 3237–3256.

[22] Ma, Z., Tian, H., Meng, G., Peng, L., Chen, Y., Chen, C., Chang, Z., Cui, X., Wang, L., Jiang, W. & Shi, J. (2020). Science China Materials, 63(12), 2517–2529.

[23] Lin, L., Sun, Z., Yuan, M., He, J., Long, R., Li, H., Nan, C., Sun, G. & Ma, S. (2018). Journal of Materials Chemistry A, 6, 8068–8077.

[24] Cheng, N., Stambula, S., Wang, D., Banis, M. N., Liu, J., Riese, A., Xiao, B., Li, R., Sham, T.-K., Liu, L.-M., Botton, G. A. & Sun, X. (2016). Nature Communications, 7(1), 13638.

[25] Zhang, S., Li, J. & Wang, E. (2020). ChemElectroChem, 7(22), 4526–4534.

[26] Ngamlerdpokin, K. & Tantavichet, N. (2014). International Journal of Hydrogen Energy, 39(6), 2505–2515.

[27] Guo, H., Youliwasi, N., Zhao, L., Chai, Y. & Liu, C. (2018). Applied Surface Science, 435, 237–246.

[28] Yang, Y., Lin, Z., Gao, S., Su, J., Lun, Z., Xia, G., Chen, J., Zhang, R. & Chen, Q. (2017). ACS Catalysis, 7(1), 469–479.

[29] Yao, R.-Q., Zhou, Y.-T., Shi, H., Wan, W.-B., Zhang, Q.-H., Gu, L., Zhu, Y.-F., Wen, Z., Lang, X.-Y. & Jiang, Q. (2021). Advanced Functional Materials, 31(10), 2009613.

[30] Lu, Z., Zhu, W., Yu, X., Zhang, H., Li, Y., Sun, X., Wang, X., Wang, H., Wang, J., Luo, J., Lei, X. & Jiang, L. (2014). Advanced Materials, 26(17), 2683–2687.

[31] Shen, H., Thomas, T., Rasaki, S. A., Saad, A., Hu, C., Wang, J. & Yang, M. (2019). Electrochemical Energy Reviews, 2(2), 252–276.

[32] Zhou, W., Zhou, J., Zhou, Y., Lu, J., Zhou, K., Yang, L., Tang, Z., Li, L. & Chen, S. (2015). Chemistry of Materials, 27(6), 2026–2032.

[33] Mo, Q., Chen, N., Deng, M., Yang, L. & Gao, Q. (2017). ACS Applied Materials & Interfaces, 9(43), 37721–37730.

[34] Hu, C., Ma, Q., Hung, S.-F., Chen, Z.-N., Ou, D., Ren, B., Chen, H. M., Fu, G. & Zheng, N. (2017). Chem, 3(1), 122–133.

[35] Tavakkoli, M., Kallio, T., Reynaud, O., Nasibulin, A. G., Johans, C., Sainio, J., Jiang, H., Kauppinen, E. I. & Laasonen, K. (2015). Angewandte Chemie International Edition, 54(15), 4535–4538.

[36] Zou, X., Huang, X., Goswami, A., Silva, R., Sathe, B. R., Mikmeková, E. & Asefa, T. (2014). Angewandte Chemie International Edition, 126(17), 4461–4465.

[37] Fei, H., Yang, Y., Peng, Z., Ruan, G., Zhong, Q., Li, L., Samuel, E. L. G. & Tour, J. M. (2015). ACS Applied Materials & Interfaces, 7(15), 8083–8087.

[38] Chen, Z., Wu, R., Liu, Y., Ha, Y., Guo, Y., Sun, D., Liu, M. & Fang, F. (2018). Advanced Materials, 30(30).

[39] Zhang, Y., Ma, Y., Chen, -Y.-Y., Zhao, L., Huang, L.-B., Luo, H., Jiang, W.-J., Zhang, X., Niu, S., Gao, D., Bi, J., Fan, G. & Hu, J.-S. (2017). ACS Applied Materials & Interfaces, 3(42), 36857–36864.

[40] Qiu, H. J., Ito, Y., Cong, W., Tan, Y., Liu, P., Hirata, A., Fujita, T., Tang, Z. & Chen, M. (2015). Angewandte Chemie International Edition, 54(47), 14031–14035.

[41] Chen, W., Pei, J., He, C.-T., Wan, J., Ren, H., Zhu, Y., Wang, Y., Dong, J., Tian, S., Cheong, W.-C., Lu, S., Zheng, L., Zheng, X., Yan, W., Zhuang, Z., Chen, C., Peng, Q., Wang, D. & Li, Y. (2017). Angewandte Chemie International Edition, 56(50), 16086–16090.

[42] Fei, H., Dong, J., Arellano-Jiménez, M. J., Ye, G., Dong Kim, N., Samuel, E. L. G., Peng, Z., Zhu, Z., Qin, F., Bao, J., Yacaman, M. J., Ajayan, P. M., Chen, D. & Tour, J. M. (2015). Nature Communications, 6(1), 8668.

[43] Fang, S., Zhu, X., Liu, X., Gu, J., Liu, W., Wang, D., Zhang, W., Lin, Y., Lu, J., Wei, S., Li, Y. & Yao, T. (2020). Nature Communications, 11(1), 1029.

[44] Nsanzimana, J. M. V., Peng, Y., Miao, M., Reddu, V., Zhang, W., Wang, H., Xia, B. Y. & Wang, X. (2018). ACS Applied Nano Materials, 1(3), 1228–1235.

[45] Zhang, J., Wang, T., Liu, P., Liao, Z., Liu, S., Zhuang, X., Chen, M., Zschech, E. & Feng, X. (2017). Nature Communications, 8(1), 15437.

[46] Fang, M., Gao, W., Dong, G., Xia, Z., Yip, S., Qin, Y., Qu, Y. & Ho, J. C. (2016). Nano Energy, 27, 247–254.

[47] Shen, Y., Zhou, Y., Wang, D., Wu, X., Li, J. & Xi, J. (2018). Advanced Energy Materials, 8(2), 1701759.

[48] Deng, J., Ren, P., Deng, D. & Bao, X. (2015). Angewandte Chemie International Edition, 54(7), 2100–2104.

[49] Wang, P., Jiang, K., Wang, G., Yao, J. & Huang, X. (2016). Angewandte Chemie International Edition, 128(41), 13051–13055.

[50] Han, Z., Zhang, R.-L., Duan, -J.-J., Wang, A.-J., Zhang, Q.-L., Huang, H. & Feng, -J.-J. (2020). International Journal of Hydrogen Energy, 45(11), 6110–6119.

[51] Zhu, Y., Lin, Q., Zhong, Y., Tahini, H. A., Shao, Z. & Wang, H. (2020). Energy & Environmental Science, 13(10), 3361–3392.

[52] Zhang, W., Li, H., Firby, C. J., Al-Hussein, M. & Elezzabi, A. Y. (2019). ACS Applied Materials & Interfaces, 11(22), 20378–20385.

[53] Zhong, X., Sun, Y., Chen, X., Zhuang, G., Li, X. & Wang, J.-G. (2016). Advanced Functional Materials, 26(32), 5778–5786.

[54] Shi, Y., Zhou, Y., Yang, D.-R., Xu, W.-X., Wang, C., Wang, F.-B., Xu, -J.-J., Xia, X.-H. & Chen, H.-Y. (2017). Journal of the American Chemical Society, 139(43), 15479–15485.

[55] Jin, Y., Wang, H., Li, J., Yue, X., Han, Y., Shen, P. K. & Cui, Y. (2016). Advanced Materials, 28(19), 3785–3790.

[56] Fang, L., Jiang, Z., Xu, H., Liu, L., Guan, Y., Gu, X. & Wang, Y. (2018). Journal of Catalysis, 357, 238–246.

[57] Zheng, T., Sang, W., He, Z., Wei, Q., Chen, B., Li, H., Cao, C., Huang, R., Yan, X., Pan, B., Zhou, S. & Zeng, J. (2017). Nano Letters, 17(12), 7968–7973.

[58] Chen, J., Yu, D., Liao, W., Zheng, M., Xiao, L., Zhu, H., Zhang, M., Du, M. & Yao, J. (2016). ACS Applied Materials & Interfaces, 8(28), 18132–18139.

[59] Zhang, T., Wu, M.-Y., Yan, D.-Y., Mao, J., Liu, H., Hu, W.-B., Du, X.-W., Ling, T. & Qiao, S.-Z. (2018). Nano Energy, 43, 103–109.

[60] Zhao, Y., Chang, C., Teng, F., Zhao, Y., Chen, G., Shi, R., Waterhouse, G. I. N., Huang, W. & Zhang, T. (2017). Advanced Energy Materials, 7(18), 1700005.

[61] Liu, K., Zhang, W., Lei, F., Liang, L., Gu, B., Sun, Y., Ye, B., Ni, W. & Xie, Y. (2016). Nano Energy, 30, 810–817.

[62] Hu, Q., Wang, Z., Huang, X., Qin, Y., Yang, H., Ren, X., Zhang, Q., Liu, J. & He, C. (2020). Energy Environmental Science, 13(12), 5097–5103.

[63] Singh, K. P., Shin, C.-H., Lee, H.-Y., Razmjooei, F., Sinhamahapatra, A., Kang, J. & Yu, J.-S. (2020). ACS Applied Nano Materials, 3(4), 3634–3645.

[64] Maitra, S., Chakraborty, P. K., Mitra, R. & Nath, T. K. (2020). Current Applied Physics, 20(12), 1404–1415.

[65] Tang, Y. J., Gao, M. R., Liu, C. H., Li, S. L., Jiang, H. L., Lan, Y. Q., Han, M. & Yu, S. H. (2015). Angewandte Chemie International Edition, 127(44), 13120–13124.

[66] Gong, M., Zhou, W., Tsai, M.-C., Zhou, J., Guan, M., Lin, M.-C., Zhang, B., Hu, Y., Wang, D.-Y., Yang, J., Pennycook, S. J., Hwang, B.-J. & Dai, H. (2014). Nature Communications, 5(1), 4695.

[67] Wang, R., Han, J., Yang, B., Wang, X., Zhang, X. & Song, B. (2020). Chemistry – An Asian Journal, 15(23), 3961–3972.
[68] Choi, W., Choudhary, N., Han, G. H., Park, J., Akinwande, D. & Lee, Y. H. (2017). Materials Today, 20(3), 116–130.
[69] Tsai, C., Chan, K., Abild-Pedersen, F. & N?rskov, J. K. J. P. C. C. P (2014). 16(26), 13156–13164.
[70] Hinnemann, B., Moses, P. G., Bonde, J. & Kristina, P. (2005). Society, J. J. J. O. T. A. C.
[71] Lin, L., Sherrell, P., Liu, Y., Lei, W., Zhang, S., Zhang, H., Wallace, G. G. & Chen, J. (2020). Advanced Energy Materials, 10(16), 1903870.
[72] Yu, Y., Huang, S.-Y., Li, Y., Steinmann, S. N., Yang, W. & Cao, L. (2014). Nano Letters, 14(2), 553–558.
[73] Lv, R., Robinson, J. A., Schaak, R. E., Sun, D., Sun, Y., Mallouk, T. E. & Terrones, M. (2015). Accounts of Chemical Research, 48(1), 56–64.
[74] Shi, Y., Li, H. & Li, L.-J. (2015). Chemical Society Reviews, 44(9), 2744–2756.
[75] Lei, W., Xiao, J.-L., Liu, H.-P., Jia, Q.-L. & Zhang, H.-J. (2020). Tungsten, 2(3), 217–239.
[76] Shi, J., Ma, D., Han, G.-F., Zhang, Y., Ji, Q., Gao, T., Sun, J., Song, X., Li, C., Zhang, Y., Lang, X.-Y., Zhang, Y. & Liu, Z. (2014). ACS Nano, 8(10), 10196–10204.
[77] Wang, D.-Y., Gong, M., Chou, H.-L., Pan, C.-J., Chen, H.-A., Wu, Y., Lin, M.-C., Guan, M., Yang, J., Chen, C.-W., Wang, Y.-L., Hwang, B.-J., Chen, -C.-C. & Dai, H. (2015). Journal of the American Chemical Society, 137(4), 1587–1592.
[78] Kong, D., Wang, H., Lu, Z. & Cui, Y. (2014). Journal of the American Chemical Society, 136(13), 4897–4900.
[79] McCarthy, C. L., Downes, C. A., Schueller, E. C., Abuyen, K. & Brutchey, R. L. (2016). ACS Energy Letters, 1(3), 607–611.
[80] Sun, Y., Xu, K., Wei, Z., Li, H., Zhang, T., Li, X., Cai, W., Ma, J., Fan, H. J. & Li, Y. (2018). Advanced Materials, 30, 35. 1802121.
[81] Ma, X., Zhang, W., Deng, Y., Zhong, C., Hu, W. & Han, X. (2018). Nanoscale, 10(10), 4816–4824.
[82] Hu, T., Bian, K., Tai, G., Zeng, T., Wang, X., Huang, X., Xiong, K. & Zhu, K. (2016). The Journal of Physical Chemistry C, 120(45), 25843–25850.
[83] Cheng, L., Huang, W., Gong, Q., Liu, C., Liu, Z., Li, Y. & Dai, H. (2014). Angewandte Chemie International Edition, 126(30), 7994–7997.
[84] Kwon, H., Ji, B., Bae, D., Lee, J.-H., Park, H. J., Kim, D. H., Kim, Y.-M., Son, Y.-W., Yang, H. & Cho, S. (2020). Applied Surface Science, 515, 145972.
[85] Setayeshgar, S., Karimipour, M., Molaei, M., Moghadam, M. R. & Khazraei, S. (2020). International Journal of Hydrogen Energy, 45(11), 6090–6101.
[86] Sun, Y., Alimohammadi, F., Zhang, D. & Guo, G. (2017). Nano Letters, 17(3), 1963–1969.
[87] Jin, Q., Liu, N., Dai, C., Xu, R., Wu, B., Yu, G., Chen, B. & Du, Y. (2020). Advanced Energy Materials, 10(20), 2000291.
[88] Liu, S., Jiang, Y., Yang, M., Zhang, M., Guo, Q., Shen, W., He, R. & Li, M. (2019). Nanoscale, 11(16), 7959–7966.
[89] Pu, X., Qian, J., Li, J., Gao, D. & Zhang, R. (2021). FlatChem, 29, 100278.
[90] Kou, T., Smart, T., Yao, B., Chen, I., Thota, D., Ping, Y. & Li, Y. (2018). Advanced Energy Materials, 8(19), 1703538.
[91] Cui, Z., Ge, Y., Chu, H., Baines, R., Dong, P., Tang, J., Yang, Y., Ajayan, P. M., Ye, M. & Shen, J. (2017). Journal of Materials Chemistry A, 5(4), 1595–1602.
[92] Lonkar, S. P., Pillai, V. V. & Alhassan, S. M. (2020). International Journal of Hydrogen Energy, 45(17), 10475–10485.
[93] He, H.-Y., He, Z. & Shen, Q. (2020). Materials Science Engineering: B, 260, 114659.
[94] Du, H., Kong, R.-M., Guo, X., Qu, F. & Li, J. (2018). Nanoscale, 10(46), 21617–21624.

[95] Oyama, S. T., Gott, T., Zhao, H. & Lee, Y.-K. (2009). Catalysis Today, 143(1), 94–107.
[96] Voiry, D., Yang, J. & Chhowalla, M. (2016). Advanced Materials, 28(29), 6197–6206.
[97] Liu, P. & Rodriguez, J. A. (2005). Journal of the American Chemical Society, 127(42), 14871–14878.
[98] Wang, Y., Kong, B., Zhao, D., Wang, H. & Selomulya, C. (2017). Nano Today, 15, 26–55.
[99] Kibsgaard, J. & Jaramillo, T. F. (2014). Angewandte Chemie International Edition, 53(52), 14433–14437.
[100] Su, J., Zhou, J., Wang, L., Liu, C. & Chen, Y. (2017). Science Bulletin, 62(9), 633–644.
[101] Sumboja, A., An, T., Goh, H. Y., Lübke, M., Howard, D. P., Xu, Y., Handoko, A. D., Zong, Y. & Liu, Z. (2018). ACS Applied Materials & Interfaces, 10(18), 15673–15680.
[102] Jiang, P., Liu, Q. & Sun, X. (2014). Nanoscale, 6(22), 13440–13445.
[103] Adam, A., Suliman, M. H., Siddiqui, M. N., Yamani, Z. H., Merzougui, B. & Qamar, M. (2018). ACS Applied Materials & Interfaces, 10(35), 29407–29416.
[104] Lv, X., Hu, Z., Ren, J., Liu, Y., Wang, Z. & Yuan, Z.-Y. (2019). Inorganic Chemistry Frontiers, 6(1), 74–81.
[105] Liu, B., Wu, C., Chen, G., Chen, W., Peng, L., Yao, Y., Wei, Z., Zhu, H., Han, T., Tang, D. & Zhou, M. (2019). Journal of Power Sources, 429, 46–54.
[106] Regmi, Y. N., Roy, A., King, L. A., Cullen, D. A., Meyer, H. M., Goenaga, G. A., Zawodzinski, T. A., Labbé, N. & Chmely, S. C. (2017). Chemistry of Materials, 29(21), 9369–9377.
[107] Men, Y., Li, P., Zhou, J., Chen, S. & Luo, W. (2020). Cell Reports Physical Science, 1(8), 100136.
[108] Laursen, A. B., Patraju, K. R., Whitaker, M. J., Retuerto, M., Sarkar, T., Yao, N., Ramanujachary, K. V., Greenblatt, M. & Dismukes, G. C. (2015). Energy & Environmental Science, 8(3), 1027–1034.
[109] Popczun, E. J., McKone, J. R., Read, C. G., Biacchi, A. J., Wiltrout, A. M., Lewis, N. S. & Schaak, R. E. (2013). Journal of the American Chemical Society, 135(25), 9267–9270.
[110] Popczun, E. J., Read, C. G., Roske, C. W., Lewis, N. S. & Schaak, R. E. (2014). 53(21), 5427–5430.
[111] Tian, J., Liu, Q., Asiri, A. M. & Sun, X. (2014). Journal of the American Chemical Society, 136(21), 7587–7590.
[112] Yang, H., Zhang, Y., Hu, F. & Wang, Q. (2015). Nano Letters, 15(11), 7616–7620.
[113] Callejas, J. F., Read, C. G., Popczun, E. J., McEnaney, J. M. & Schaak, R. E. (2015). Chemistry of Materials, 27(10), 3769–3774.
[114] Wu, T., Pi, M., Wang, X., Guo, W., Zhang, D. & Chen, S. (2017). Journal of Alloys and Compounds, 729, 203–209.
[115] Guan, C., Xiao, W., Wu, H., Liu, X., Zang, W., Zhang, H., Ding, J., Feng, Y. P., Pennycook, S. J. & Wang, J. (2018). Nano Energy, 48, 73–80.
[116] Wang, T., Wu, L., Xu, X., Sun, Y., Wang, Y., Zhong, W. & Du, Y. (2017). Scientific Reports, 7(1), 11891.
[117] Tian, L., Yan, X. & Chen, X. (2016). ACS Catalysis, 6(8), 5441–5448.
[118] Jiang, P., Liu, Q., Liang, Y., Tian, J., Asiri, A. M. & Sun, X. (2014). 53(47), 12855–12859.
[119] Schipper, D. E., Zhao, Z., Thirumalai, H., Leitner, A. P., Donaldson, S. L., Kumar, A., Qin, F., Wang, Z., Grabow, L. C., Bao, J. & Whitmire, K. H. (2018). Chemistry of Materials, 30(10), 3588–3598.
[120] Tian, J., Liu, Q., Cheng, N., Asiri, A. M. & Sun, X. (2014). 53(36), 9577–9581.
[121] Xiao, P., Sk, M. A., Thia, L., Ge, X., Lim, R. J., Wang, J.-Y., Lim, K. H. & Wang, X. (2014). Energy Environmental Science, 7(8), 2624–2629.
[122] Xing, Z., Liu, Q., Asiri, A. M. & Sun, X. (2014). Advanced Materials, 26(32), 5702–5707.

[123] Zhang, X., Yu, X., Zhang, L., Zhou, F., Liang, Y. & Wang, R. (2018). Advanced Functional Materials, 28(16), 1706523.

[124] Zhang, G., Wang, G., Liu, Y., Liu, H., Qu, J. & Li, J. (2016). Journal of the American Chemical Society, 138(44), 14686–14693.

[125] McEnaney, J. M., Chance Crompton, J., Callejas, J. F., Popczun, E. J., Read, C. G., Lewis, N. S. & Schaak, R. E. (2014). Chemical Communications, 50(75), 11026–11028.

[126] Pi, M., Wu, T., Zhang, D., Chen, S. & Wang, S. (2016). Nanoscale, 8(47), 19779–19786.

[127] Tang, C., Gan, L., Zhang, R., Lu, W., Jiang, X., Asiri, A. M., Sun, X., Wang, J. & Chen, L. (2016). Nano Letters, 16(10), 6617–6621.

[128] Pei, Y., Zhang, H., Han, L., Huang, L., Dong, L., Jia, Q. & Zhang, S. (2020). Nanotechnology, 32(2), 024001.

[129] Pei, Y., Huang, L., Han, L., Zhang, H., Dong, L., Jia, Q. & Zhang, S. (2020). Green Energy & Environment.

[130] Wang, X.-D., Xu, Y.-F., Rao, H.-S., Xu, W.-J., Chen, H.-Y., Zhang, W.-X., Kuang, D.-B. & Su, C.-Y. (2016). Energy Environmental Science, 9(4), 1468–1475.

[131] Zhang, R., Wang, X., Yu, S., Wen, T., Zhu, X., Yang, F., Sun, X., Wang, X. & Hu, W. (2017). Advanced Materials, 29(9), 1605502.

[132] Lu, -S.-S., Zhang, L.-M., Dong, Y.-W., Zhang, J.-Q., Yan, X.-T., Sun, D.-F., Shang, X., Chi, J.-Q., Chai, Y.-M. & Dong, B. (2019). Journal of Materials Chemistry A, 7(284), 16859–16866.

[133] Silva, P., Vilela, S. M. F., Tomé, J. P. C. & Almeida Paz, F. A. (2015). Chemical Society Reviews, 44(19), 6774–6803.

[134] Tan, J.-B. & Li, G.-R. (2020). Journal of Materials Chemistry A, 8(29), 14326–14355.

[135] Huang, X., Yao, H., Cui, Y., Hao, W., Zhu, J., Xu, W. & Zhu, D. (2017). ACS Applied Materials & Interfaces, 9(46), 40752–40759.

[136] Qin, J.-S., Du, D.-Y., Guan, W., Bo, X.-J., Li, Y.-F., Guo, L.-P., Su, Z.-M., Wang, -Y.-Y., Lan, Y.-Q. & Zhou, H.-C. (2015). Journal of the American Chemical Society, 137(22), 7169–7177.

[137] Li, Y.-S., Yi, J.-W., Wei, J.-H., Wu, Y.-P., Li, B., Liu, S., Jiang, C., Yu, H.-G. & Li, D.-S. (2020). Journal of Solid State Chemistry, 281, 121052.

[138] Dong, R., Pfeffermann, M., Liang, H., Zheng, Z., Zhu, X., Zhang, J. & Feng, X. (2015). Angewandte Chemie International Edition, 54(41), 12058–12063.

[139] Wang, W., Xu, X., Zhou, W. & Shao, Z. (2017). Advanced Science, 4(4), 1600371.

[140] Zhao, R., Liang, Z., Zou, R. & Xu, Q. (2018). Joule, 2(11), 2235–2259.

[141] Wang, H.-F., Chen, L., Pang, H., Kaskel, S. & Xu, Q. (2020). Chemical Society Reviews, 49(5), 1414–1448.

[142] Xu, M., Han, L., Han, Y., Yu, Y., Zhai, J. & Dong, S. (2015). Journal of Materials Chemistry A, 3(43), 21471–21477.

[143] Huang, Z.-F., Song, J., Li, K., Tahir, M., Wang, Y.-T., Pan, L., Wang, L., Zhang, X. & Zou, -J.-J. (2016). Journal of the American Chemical Society, 138(4), 1359–1365.

[144] Yu, L., Wu, H. B. & Lou, X. W. D. (2017). Accounts of Chemical Research, 50(2), 293–301.

[145] Wang, S., Qin, J., Meng, T. & Cao, M. (2017). Nano Energy, 39, 626–638.

[146] Dai, X., Liu, M., Li, Z., Jin, A., Ma, Y., Huang, X., Sun, H., Wang, H. & Zhang, X. (2016). The Journal of Physical Chemistry C, 120(23), 12539–12548.

[147] Dong, R., Zheng, Z., Tranca, D. C., Zhang, J., Chandrasekhar, N., Liu, S., Zhuang, X., Seifert, G. & Feng, X. (2017). Chemistry-A European Journal, 23(10), 2255–2260.

[148] Wu, Y.-P., Zhou, W., Zhao, J., Dong, W.-W., Lan, Y.-Q., Li, D.-S., Sun, C. & Bu, X. (2017). Angewandte Chemie International Edition, 56(42), 13001–13005.

[149] Zhang, L., Li, S., Gómez-García, C. J., Ma, H., Zhang, C., Pang, H. & Li, B. (2018). ACS Applied Materials & Interfaces, 10(37), 31498–31504.

[150] Zhou, W., Wu, Y.-P., Wang, X., Tian, J.-W., Huang, -D.-D., Zhao, J., Lan, Y.-Q. & Li, D.-S. (2018). CrystEngComm, 20(33), 4804–4809.

[151] Hod, I., Deria, P., Bury, W., Mondloch, J. E., Kung, C.-W., So, M., Sampson, M. D., Peters, A. W., Kubiak, C. P., Farha, O. K. & Hupp, J. T. (2015). Nature Communications, 6(1), 8304.

[152] Wang, S., He, P., Xie, Z., Jia, L., He, M., Zhang, X., Dong, F., Liu, H., Zhang, Y. & Li, C. (2019). Electrochimica Acta, 296, 644–652.

[153] Gupta, S., Patel, M. K., Miotello, A. & Patel, N. (2020). Advanced Functional Materials, 30(1), 1906481.

[154] Gupta, S., Patel, N., Miotello, A. & Kothari, D. C. (2015). Journal of Power Sources, 279, 620–625.

[155] Li, H., Wen, P., Li, Q., Dun, C., Xing, J., Lu, C., Adhikari, S., Jiang, L., Carroll, D. L. & Geyer, S. M. (2017). Advanced Energy Materials, 7(17), 1700513.

[156] Yang, Z.-Z., Zhang, C., Zeng, G.-M., Tan, X.-F., Wang, H., Huang, D.-L., Yang, K.-H., Wei, J.-J., Ma, C. & Nie, K. (2020). Journal of Materials Chemistry A, 8(8), 4141–4173.

[157] Evans, D. G. & Slade, R. C. T. (2006). Structural aspects of layered double hydroxides. In: Layered Double Hydroxides Duan, X. & Evans, D. G. eds., Berlin, Heidelberg: Springer Berlin Heidelberg, 1–87.

[158] Li, T., Miras, H. N. & Song, Y.-F. (2017). Catalysts, 7(9), 260.

[159] Song, F. & Hu, X. (2014). Nature Communications, 5(1).

[160] Liu, P. F., Yang, S., Zhang, B. & Yang, H. G. (2016). ACS Applied Materials & Interfaces, 8(50), 34474–34481.

[161] Chen, G., Wang, T., Zhang, J., Liu, P., Sun, H., Zhuang, X., Chen, M. & Feng, X. (2018). Advanced Materials, 30(10), 1706279.

[162] Wang, Y., Yan, D., El Hankari, S., Zou, Y. & Wang, S. (2018). J. A. S., 5(8), 1800064.

[163] Liu, X., Gong, M., Deng, S., Zhao, T., Zhang, J. & Wang, D. (2020). Journal of Materials Chemistry A, 8(20), 10130–10149.

[164] Peng, X., Pi, C., Zhang, X., Li, S., Huo, K. & Chu, P. K. (2019). Sustainable Energy & Fuels, 3(2), 366–381.

[165] Chen, P., Ye, J., Wang, H., Ouyang, L. & Zhu, M. (2021). Journal of Alloys and Compounds, 883, 160833.

[166] Vrubel, H. & Hu, X. (2012). Angewandte Chemie International Edition, 51, 12703–12706.

[167] Park, H., Encinas, A., Scheifers, J. P., Zhang, Y. & Fokwa, B. P. T. (2017). Angewandte Chemie International Edition, 56(20), 5575–5578.

[168] Zhang, R., Liu, H., Wang, C., Wang, L., Yang, Y. & Guo, Y. (2020). ChemCatChem, 12(11), 3068–3075.

[169] Lee, E., Park, H., Joo, H. & Fokwa, B. P. T. (2020). Angewandte Chemie International Edition, 59(29), 11774–11778.

[170] Hao, W., Wu, R., Zhang, R., Ha, Y., Chen, Z., Wang, L., Yang, Y., Ma, X., Sun, D., Fang, F. & Guo, Y. (2018). Advanced Energy Materials, 8(26), 1801372.

[171] Alameda, L. T., Holder, C. F., Fenton, J. L. & Schaak, R. E. (2017). Chemistry of Materials, 29(21), 8953–8957.

[172] Shi, J., Pu, Z., Liu, Q., Asiri, A. M., Hu, J. & Sun, X. (2015). Electrochimica Acta, 154, 345–351.

[173] Xie, J., Li, S., Zhang, X., Zhang, J., Wang, R., Zhang, H., Pan, B. & Xie, Y. (2014). Chemical Science, 5(12), 4615–4620.

[174] Jin, H., Zhang, H., Chen, J., Mao, S., Jiang, Z. & Wang, Y. (2018). Journal of Materials Chemistry A, 6(23), 10967–10975.

[175] Zhang, Y., Ouyang, B., Xu, J., Chen, S., Rawat, R. S. & Fan, H. J. (2016). Advanced Energy Materials, 6(11), 1600221.

[176] Yan, H., Xie, Y., Jiao, Y., Wu, A., Tian, C., Zhang, X., Wang, L. & Fu, H. (2018). Advanced Materials, 30(2), 1704156.

[177] Ding, L., Li, K., Xie, Z., Yang, G., Yu, S., Wang, W., Cullen, D. A., Yu, H. & Zhang, F. (2021). Electrochimica Acta, 395, 139199.

[178] Liu, Q., Wang, H., Wang, X., Tong, R., Zhou, X., Peng, X., Wang, H., Tao, H. & Zhang, Z. (2017). International Journal of Hydrogen Energy, 42(8), 5560–5568.

[179] Feng, H., Yu, J., Tang, L., Wang, J., Dong, H., Ni, T., Tang, J., Tang, W., Zhu, X. & Liang, C. (2021). Applied Catalysis. B, Environmental, 297, 120478.

[180] Ma, X., Zhang, S., He, Y., He, T., Li, H., Zhang, Y. & Chen, J. (2021). Journal of Electroanalytical Chemistry, 886, 115107.

[181] Ren, B., Li, D., Jin, Q., Cui, H. & Wang, C. (2017). Journal of Materials Chemistry A, 5(25), 13196–13203.

[182] Ma, F.-X., Wu, H. B., Xia, B. Y., Xu, C.-Y. & Lou, X. W. (2015). Angewandte Chemie International Edition, 54(51), 15395–15399.

[183] Li, S., Yang, C., Yin, Z., Yang, H., Chen, Y., Lin, L., Li, M., Li, W., Hu, G. & Ma, D. (2017). Nano Research, 10(4), 1322–1328.

[184] Fan, X., Peng, Z., Ye, R., Zhou, H. & Guo, X. (2015). ACS Nano, 9(7), 7407–7418.

[185] Fan, H., Yu, H., Zhang, Y., Zheng, Y., Luo, Y., Dai, Z., Li, B., Zong, Y. & Yan, Q. (2017). Angewandte Chemie International Edition, 56(41), 12566–12570.

[186] Gao, S., Lin, L., Wang, H., Nie, P., Jian, J., Li, J. & Chang, L. (2021). Journal of Alloys and Compounds, 875, 160052.

[187] Cui, W., Liu, Q., Cheng, N., Asiri, A. M. & Sun, X. (2014). Chemical Communications, 50(66), 9340–9342.

[188] Liu, X. & Dai, L. (2016). Nature Reviews Materials, 1(11), 16064.

[189] Gao, K., Wang, B., Tao, L., Cunning, B. V., Zhang, Z., Wang, S., Ruoff, R. S. & Qu, L. (2019). Advanced Materials, 31(13), 1805121.

[190] Wang, J., Kong, H., Zhang, J., Hao, Y., Shao, Z. & Ciucci, F. (2021). Progress in Materials Science, 116, 100717.

[191] Quílez-Bermejo, J., Morallón, E. & Cazorla-Amorós, D. (2018). Chemical Communications, 54(35), 4441–4444.

[192] Salinas-Torres, D., Navlani-García, M., Mori, K., Kuwahara, Y. & Yamashita, H. (2019). Applied Catalysis. A, General, 571, 25–41.

[193] Gong, K., Du, F., Xia, Z., Durstock, M. & Dai, L. (2009). Science, 323(5915), 760.

[194] Shao, Y., Zhang, S., Engelhard, M. H., Li, G., Shao, G., Wang, Y., Liu, J., Aksay, I. A. & Lin, Y. (2010). Journal of Materials Chemistry, 20(35), 7491–7496.

[195] Huang, X., Zhao, Y., Ao, Z. & Wang, G. (2014). Scientific Reports, 4(1), 7557.

[196] Jiang, H., Gu, J., Zheng, X., Liu, M., Qiu, X., Wang, L., Li, W., Chen, Z., Ji, X. & Li, J. (2019). Energy & Environmental Science, 12(1), 322–333.

[197] Tian, Y., Ye, Y., Wang, X., Peng, S., Wei, Z., Zhang, X. & Liu, W. (2017). Applied Catalysis. A, General, 529, 127–133.

[198] Chhetri, M., Maitra, S., Chakraborty, H., Waghmare, U. V. & Rao, C. N. R. (2016). Energy & Environmental Science, 9(1), 95–101.

[199] Han, W., Bando, Y., Kurashima, K. & Sato, T. (1999). Chemical Physics Letters, 299(5), 368–373.

[200] Wei, B., Spolenak, R., Kohler-Redlich, P., Rühle, M. & Arzt, E. (1999). Applied Physics Letters, 74(21), 3149–3151.

[201] Mondal, K. C., Strydom, A. M., Erasmus, R. M., Keartland, J. M. & Coville, N. J. (2008). Materials Chemistry and Physics, 111(2), 386–390.

[202] Sathe, B. R., Zou, X. & Asefa, T. (2014). Catalysis Science & Technology, 4(7), 2023–2030.

[203] Zhou, Y., Leng, Y., Zhou, W., Huang, J., Zhao, M., Zhan, J., Feng, C., Tang, Z., Chen, S. & Liu, H. (2015). Nano Energy, 16, 357–366.
[204] Zhou, J., Qi, F., Chen, Y., Wang, Z., Zheng, B. & Wang, X. (2018). Journal of Materials Science, 53(10), 7767–7777.
[205] Qiu, H. J., Ito, Y., Cong, W., Tan, Y., Liu, P., Hirata, A., Fujita, T., Tang, Z. & Chen, M. (2015). J. A. C. I. E, 54(47), 14031–14035.
[206] Gong, X., Liu, S., Ouyang, C., Strasser, P. & Yang, R. (2015). ACS Catalysis, 5(2), 920–927.
[207] Choi, C. H., Chung, M. W., Park, S. H. & Woo, S. I. (2013). Physical Chemistry Chemical Physics, 15(6), 1802–1805.
[208] Ito, Y., Cong, W., Fujita, T., Tang, Z. & Chen, M. (2015). J. A. C, 127(7), 2159–2164.
[209] Zheng, Y., Jiao, Y., Li, L. H., Xing, T., Chen, Y., Jaroniec, M. & Qiao, S. Z. (2014). ACS Nano, 8(5), 5290–5296.
[210] Suliman, M. H., Adam, A., Siddiqui, M. N., Yamani, Z. H. & Qamar, M. (2019). Catalysis Science & Technology, 9(6), 1497–1503.
[211] Lu, C., Tranca, D., Zhang, J., Rodríguez Hernández, F. N., Su, Y., Zhuang, X., Zhang, F., Seifert, G. & Feng, X. (2017). ACS Nano, 11(4), 3933–3942.
[212] Liu, C., Dai, Z., Zhang, J., Jin, Y., Li, D. & Sun, C. (2018). The Journal of Physical Chemistry C, 122(33), 19051–19055.
[213] Huang, S., Meng, Y., Cao, Y., He, S., Li, X., Tong, S. & Wu, M. (2019). Applied Catalysis. B, Environmental, 248, 239–248.
[214] Sathe, B. R., Zou, X. & Asefa, T. (2014). Catalysis Science Technology, 4(7), 2023–2030.
[215] Patra, B. C., Khilari, S., Manna, R. N., Mondal, S., Pradhan, D., Pradhan, A. & Bhaumik, A. (2017). ACS Catalysis, 7(9), 6120–6127.
[216] Gao, R., Dai, Q., Du, F., Yan, D. & Dai, L. (2019). Journal of the American Chemical Society, 141(29), 11658–11666.
[217] Hu, C. & Dai, L. (2017). Advanced Materials, 29(9), 1604942.
[218] Feng, J. X., Xu, H., Ye, S. H., Ouyang, G., Tong, Y. X. & Li, G. R. (2017). Angewandte Chemie International Edition, 129(28), 8232–8236.
[219] Wei, L., Karahan, H. E., Goh, K., Jiang, W., Yu, D., Birer, Ö., Jiang, R. & Chen, Y. (2015). Journal of Materials Chemistry A, 3(14), 7210–7214.
[220] Zhao, Y., Zhao, F., Wang, X., Xu, C., Zhang, Z., Shi, G. & Qu, L. (2014). Angewandte Chemie International Edition, 53(50), 13934–13939.
[221] Sim, U., Yang, T.-Y., Moon, J., An, J., Hwang, J., Seo, J.-H., Lee, J., Kim, K. Y., Lee, J. & Han, S. (2013). Energy Environmental Science, 6(12), 3658–3664.
[222] Sun, T., Wu, Q., Jiang, Y., Zhang, Z., Du, L., Yang, L., Wang, X. & Hu, Z. (2016). Chemistry-A European Journal, 22(30), 10326–10329.
[223] Zhang, J., Qu, L., Shi, G., Liu, J., Chen, J. & Dai, L. (2016). Angewandte Chemie International Edition, 128(6), 2270–2274.
[224] Zhang, J. & Dai, L. (2016). Angewandte Chemie International Edition, 55(42), 13296–13300.
[225] Li, T., Chen, Y., Hu, W., Yuan, W., Zhao, Q., Yao, Y., Zhang, B., Qiu, C. & Li, C. M. (2021). Nanoscale, 13(8), 4444–4450.

Tanveer Ul Haq, Yousef Haik

Chapter 4C
Recent progress in cobalt-based nanosheets for electrochemical water oxidation

Abstract: Oxygen evolution reaction (OER) is a core reaction of electrochemical water splitting accountable for converting electricity into an ideal energy carrier, "hydrogen." This anodic reaction is kinetically sluggish due to multi-intermediates and the complex proton-coupled electron transfer process. The electrocatalyst with optimum surface and electronic structure reduces the kinetic energy barrier for challenging OER. Cobalt-based nanosheets have recently been recognized as highly efficient materials for OER due to their intrinsically active sites, rapid charge and mass transport, and fast reaction kinetics. This review summarizes the recent progress in cobalt-based nanosheets (oxide/hydroxide, boride, nitride, phosphide, and sulfide) for OER. The established mechanisms, structure design for high-output OER, and performance analysis of different Co-based nanosheets are discussed to promote the rational design of a highly efficient and cost-effective OER electrocatalyst.

Keywords: cobalt, nanosheets, electrochemical water splitting, OER, electronic and surface optimization

4C.1 Introduction

Electrochemical water splitting or water electrolysis to produce oxygen and hydrogen has been considered a possible way for sustainable and clean energy production and to mitigate the environmental threat related to the combustion of traditional fossil fuels [1]. The water-splitting process is highly energy-intensive and needs high input energy to overcome the reaction barriers [2]. The electrolysis process intrinsically depends on the performance of electrode materials, and thus the energy conversion

---

Acknowledgments: The authors would like to acknowledge the Department of Mechanical and Industrial Engineering, Texas A&M University-Kingsville.

Authors' contribution: T. Ul Haq developed the draft, and Y. Haik edited the manuscript and corresponded with the editors.

Tanveer Ul Haq, Sustainable Energy Engineering, Frank H. Dotterweich College of Engineering, Texas A&M University-Kingsville, TX 78363-8202, USA
Yousef Haik, Department of Mechanical and Industrial Engineering, Frank H. Dotterweich College of Engineering, Texas A&M University-Kingsville, TX 78363-8202, USA, e-mail: yousef.haik@tamuk.edu

https://doi.org/10.1515/9783110739879-012

efficiency is directly dependent on the improvement in the catalyst's reactivity. Like any electrochemical process, water splitting comprises two half-cell reactions: a cathodic response is known as hydrogen evolution reaction (HER) and an anodic reaction is termed oxygen evolution reaction (OER). Anodic OER is a multistep, multielectron process and impedes the overall electrochemical reaction's kinetics. The slow rate of OER is responsible for material degradation, efficiency loss, and retarding the installation of earthbound electrocatalyst in commercial electrolyzers. This needs the engineering of anode material to increase the intrinsic activity of electrocatalyst and degradation resistance for massive H_2 production from water [3].

The precious metals, that is, Ru and Ir oxides, possess superior water oxidation performance in the acidic electrolyte; however, their scarcity in nature limited its global commercialization for clean energy technologies. Recently, researchers have put severe efforts into rationally designing the transition metal (TM)-based catalyst with different electronic configurations, surfaces, and structures, as a promising alternative for the precious metal-based catalyst [4]. Various TM oxides and derivatives, for example, boride, nitride, sulfide, selenide, carbide, layered double hydroxide (LDH), and phosphide, have revealed unusual activity and stability of anode material in the water-splitting process [5]. Among TM, cobalt has emerged as an exciting catalyst to fulfill the primary demand of anode materials, that is, reveals high catalytic activity with the optimum exposed surface for the adsorption of OER intermediate, and shows outstanding durability at a high anodic potential [6, 7]. The practical application of these materials requires the maximum exposure of active sites that retain the surface reactivity during OER. It has been found that the two-dimensional (2D) nanostructure possesses unique physicochemical properties due to quantum size effect, high aspect ratio, maximum exposed edges, highly entropic facets, and uncoordinated surface. After graphene, the vast library of 2D materials such as graphitic carbon nitride (g-C_3N_4), boron nitride, metal dichalcogenides, LDH, TM carbide, black phosphorus, 2D covalent organic framework, metal-organic framework (MOF), 2D polymers, and 2D perovskite have been tested for water splitting [8]. Among the above-mentioned 2D materials, cobalt oxide and derivative sheets have fascinating properties and show high surface reactivity due to anisotropic structure.

Here, we summarize the recent progress in cobalt-based nanosheets as anode material for water splitting and systematically discuss the perceptions of the relationship between the structural and catalytic activity. Finally, the bottlenecks with personal outlooks are featured for developing OER catalysts to increase the kinetics of the electrochemical water splitting process to produce green and sustainable H_2 fuel.

4C.2 OER mechanism under alkaline conditions

In alkaline electrolytes, water is reduced to H_2 and OH^- ions following the reaction pathway

$$2H_2O + 2e^- \rightarrow H_2 + 2OH^-$$

while the following reaction occurs on the anode

$$2OH^- \rightarrow O_2 + 2H_2O + 4e^-$$

The OER is a multistep reaction, and each step accounts for a different type of voltage and ohmic loss. The various proposed mechanisms on TM surfaces for O–O coupling are as follows:

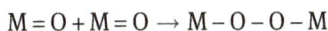

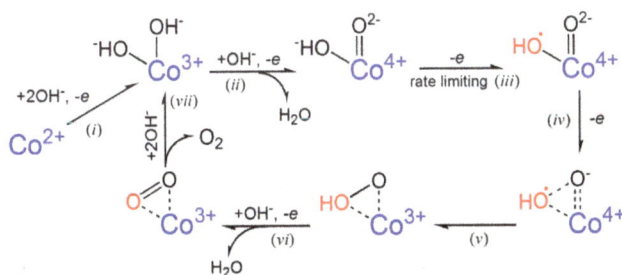

$$M + OH^- \rightarrow M-OH + e^-$$

$$M-OH + OH^- \rightarrow M=O + e^- + H_2O$$

$$M=O + OH^- \rightarrow M-OOH + e^-$$

$$M-OOH + OH^- \rightarrow M + H_2O + e^- + O_2$$

$$M=O + M=O \rightarrow M-O-O-M$$

$$M-O-O-M \rightarrow 2M + O_2$$

Fig. 4C.1: Oxygen evolution reaction mechanism on Co single site. Reproduced with permission from [9]. Copyright 2017, American Chemical Society.

In a single-site mechanism, one active metal site facilitates the four-concerted proton-coupled electron transfer steps and mediates the O–O bond formation through the nucleophilic attack of OH$^-$ ions from the electrolyte on the M=O intermediate which is considered a more energy-demanding step in OER (Fig. 4C.1) [9]. While in a dual-site mechanism, two active metal sites are involved in O–O coupling and support the oxidation reaction [10]. Recently, researchers investigated if the O–O coupling can proceed through a bifunctional mechanism to diminish the potential barrier of M-OOH bond formation. The overall activation energy might be reduced. In a bifunctional mechanism, a single-metal site mediates the O–O bond formation through the nucleophilic attack of hydroxyl ions where the adjacent acceptor site accepts the protons in the concerted reaction [11]:

$$M = O + OH^- + A \rightarrow M + A - H + O_2 + e^-$$

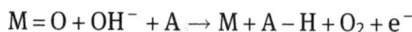

This route has been rarely explored for water oxidation and is known as a multisite-concerted electron–proton transfer. Besides the discrepancies in the O–O coupling route, it is widely recognized that the M–O bond in the intermediate plays a crucial role in regulating the M active center's activity. The M–O bond strength can be used as a descriptor for OER selectivity and activity; a too weak M–O bond precludes the intermediate binding. At the same time, a too strong bond decreases the desorption rate of gaseous products and reduces the contact portion between active sites and electrolyte molecules. The chemisorption energy or Gibbs free energy for each step in OER is expressed as follows:

$$\Delta G_1 = \Delta G(OH^*) - \Delta G(\ast) + k_b T \ln a_H^+ - eU$$

$$\Delta G_2 = \Delta G(O^*) - \Delta G(OH^*) + k_b T \ln a_H^+ - eU$$

$$\Delta G_1 = G(OOH^*) - \Delta G(O^*) + k_b T \ln a_H^+ - eU$$

$$\Delta G_1 = \Delta G(\ast) - \Delta G(OOH^*) + k_b T \ln a_H^+ - eU$$

Each step involved the removal of electron (oxidation process), and ΔG is the related Gibbs free energy for each step. While ΔG (*), ΔG (OH*), ΔG (O*), and ΔG (OOH*) are the chemisorption energy (Gibbs free energy) of the pristine active center and OH, O, and OOH intermediate adsorbed on the surface, respectively, eU is the shift in electron transfer under applied anodic potential, k_b is the Boltzmann constant (1.38×10^{-23} m^2 kg/s^2 K) and a_H^+ represents the proton activity, and T is the room temperature ($T = 298.15$ K) [10]. For an ideal catalyst, the potential barrier related to the chemisorption energy of each step is equally spaced and needs the same amount of energy in each step: $\Delta G_1 = \Delta G_2 = \Delta G_3 = \Delta G_4$. However, experimental findings demonstrated that the Gibbs free energy related to the peroxide formation (OOH) is higher than in the other step and supposed the following order: $\Delta G_3 > \Delta G_2 = \Delta G_1 > \Delta G_4$ [12]. It is found that the peroxide formation is the rate-determining step (RDS) due to the electron-dense nature of the MO surface that inhibits the attack of OH$^-$. The improvement in the chemisorption energy of this intermediate to reduce the potential barrier improved the performance and kinetics of the OER electrocatalyst.

4C.3 Morphological modulation

Morphology engineering is a promising approach to create the exposed active sites while controlling the catalyst surface texture to accelerate mass transfer and bubble detachment [13]. The structural modulation combined with the electronic optimization to enhance the intrinsic activity of active sites advances the catalytic efficiency and durability. For example, the morphology-controlled development of intermetallic nanostructure and its interaction with conductive current collectors have revealed appealing performance for water oxidation [14]. Characteristically, materials are classified dimensional-wise in three main categories: zero-dimensional (0D, the electrons are confined in all directions, e.g., nanoclusters or nanoparticles), one-dimensional (1D, the electrons are confined in two directions, e.g., nanofibers, nanowires, nanotubes, and nanorods), and 2D (the electrons are confined in one direction, e.g., nanoflakes, nanoplates, and nanosheets) [15]. The recent findings demonstrated that 2D nanostructures offer abundant active sites, reduce the charge transfer resistance (R_{ct}), and facilitate the mass transfer kinetics. The exposed enlarged surface area and surface defects make it an appealing candidate for electrochemical redox reactions. The surface engineering of 2D materials, for example, to create the defects and porosity, enhanced the electrode–electrolyte contact area and accelerated the bubbles detachment to evacuate the adsorption sites for the chain reaction [16]. Among 2D nanomaterials, TM nanosheets acquired unique physiochemical properties and structural characteristics, including high specific surface area, high aspect ratio, abundant least coordinated atoms, space and volume confinement, and higher electron and gas mobility [17, 18]. These surface features have provided them substantial benefits over other morphologies. The direct growth of metal nanosheets on conductive support creates a strong catalyst support interaction that increases the catalyst's electronic conduction and mechanical strength [19, 20].

Many synthesis routes have been employed to synthesize 2D metal nanosheets, including epitaxial growth, chemical synthesis, liquid exfoliation, and mechanical rolling. Molecular beam epitaxial growth has been considered a potential route for developing 2D nanosheets, but its large-scale fabrications are complicated due to special substrate requirements and other lavish synthetic conditions. Similarly, the liquid exfoliation process applies to those materials whose bulk crystal is layered. Chemical synthesis routes, including electrodeposition, anodization, and hydrothermal/solvothermal, are most widely employed to design metal nanosheets, producing a high percentage yield at a low cost [21].

4C.3.1 Co-based oxide nanosheets

The two most common phases of cobalt oxides (CoO) are the rock salt CoO phase (high intrinsic activity but easily be chemically transformed to inactive phase) and spinel cobalt oxide (Co_3O_4) phase (high chemical stability but has lower catalytic efficiency for water oxidation) [22]. It has been observed that the engineering of heterophase containing both active rock salt phase and stable spinel phase introduced unique features and increased the activity stability factor (ASF) [23]. The heterointerface facilitates the charge transfer and increases the active sites' reaction kinetics and durability at a high anodic potential [24]. Wang and coworkers investigated that the atomic arrangement of the heterointerface in 2D nanosheet pattern introduced the surface oxygen vacancies, and tetrahedral Co sites diffused to the interstitial octahedral sites offering flexible active sites to destabilize the water molecule [25]. Density functional theory (DFT) calculations demonstrated that the high catalytic efficiency is due to the optimum interaction of O 2p orbital with metal d states arranged in a 2D frame.

Similarly, Liu et al. have reported a facile and scalable method for developing a hybrid network of mesoporous nanosheets composed of Co_3O_4 and cobalt phosphate $(Co_3(PO_4)_2)$ [26]. The $Co_3(PO_4)_2$ received electrons from Co_3O_4 and arranged porous nanosheet structure with oxygen vacancies that reveal outstanding performance for water oxidation. The author described that the synthesis procedure initially decomposed guanosine 5-monophosphate to guanine, pentose, and phosphate ions in hydrothermal conditions:

$$C_{10}H_{12}O_8N_5P^{2-} + 2H_2O \rightarrow C_5H_{10}O_5 + C_5H_5N_5O + PO_4^{3-} + H^+$$

The carbon phase has been formed through dehydration of pentose, followed by carbonization of the organic residue:

$$C_5H_{10}O_5 \rightarrow 5C + 5H_2O$$

Then the hydrophilic surface of the carbon phase triggers the adsorption of Co^{2+} ions through electrostatic interactions or coordination processes. Meanwhile, the concentration of OH^- ions and CO_3^{2-} ions increases in the reaction due to the guanine hydrolysis that works as a pH buffer:

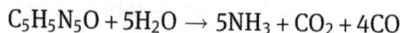

$$C_5H_5N_5O + 5H_2O \rightarrow 5NH_3 + CO_2 + 4CO$$

$$NH_3 + H_2O \rightarrow NH_4^+ + OH^-$$

$$CO_2 + H_2O \rightarrow CO_3^{2-} + 2H^+$$

Finally, the Co ions react with CO_3^{2-}, OH^-, and phosphate ions, yielding cobalt hydroxide and $Co_3(PO_4)_2$:

$$Co^{2+} + OH^- + 0.5\,CO_3^{2-} + 0.11\,H_2O \rightarrow Co(OH)_1(CO_3)_{0.5}0.11\,H_2O$$

$$Co^{2+} + 2PO_4^{3-} \rightarrow CO_3(PO_4)_2$$

The electronic interaction between $Co_3(PO_4)_2$ and Co_3O_4 has increased the acidity of Co_3O_4, which triggers the destabilization of a water molecule (Lewis's base) through acid–base interaction and decreases the potential kinetic barrier for O–H bond breakage. The electronic interaction at the 2D nanosheet interface reduces the charge transfer resistance and increases active sites' intrinsic activity. Recent findings revealed that the pure cobalt oxide has poor electronic conductivity, and a limited number of exposed active sites reduced its electrocatalytic performance. Therefore, the small grain size at the interface boundary with structural and morphological defects enhanced the exposed surface area, altered the electronic conduction of Co by modifying its d band center, and increased the electrochemical reaction kinetics. The structural and morphological optimization with enhanced oxygen vacancies optimizes the reaction center chemisorption energy, decreases the metal dissolution rate, and increases the ASF [27].

The porous nanosheet structure increases the mass transfer kinetics. Its strong linkage with the underlying support reduces the charge transfer resistance at the interface, facilitating water oxidation at low overpotential. Zhang et al. have optimized the hydrothermal process to control the structure and morphology of CoO [28]. The results demonstrated that the structural and morphological alterations substantially impact the overall performance. The ultrathin CoO nanosheets sustained exposed active sites at facets with $Co^{2+/Co3+}$ configuration, high surface oxygen vacancies, and significant amorphous portion trigger the formation of Co-OOH during OER act as an active phase for O–O coupling. The catalyst has preserved an active phase, and a porous structure that facilitates the mass diffusion of electrolytes stabilizes the OER intermediate and increases the dynamic removal of O_2 gas from the active center.

Researchers unveiled that different nanostructures, for example, nanosheets, mesoporous structure, nanowires, nanotubes, and nanocrystal, have abundant active sites and high electrochemical active surface area (ECSA) and promote water oxidation at the electrode surface. Nanosheets can easily chemically control the surface with particular facets with high electron mobility and benefit charge transfer [29]. Zhou and coworkers reported the two-step process for synthesizing ultrathin Co_3O_4 nanosheets comprising hydrothermal method followed by oxidation [30]. The results demonstrated that the peroxidation step introduced the irregular porous structure in the 2D nanosheet framework to increase the specific activity of the electrocatalyst. The surface oxygen vacancies in the nanosheet structure increased the Co 2p content, increased the active sites' adsorption property, and accelerated the water oxidation kinetics.

The morphology of electrocatalysts has a substantial impact on the reactivity and stability of the exposed active sites. Researchers found that the nanosheets have more exposed active edges with high specific surface area and higher mass activity [31]. Liu et al. had synthesized Mn-doped Co–OOH nanosheets with an average thickness of 1.3 nm. They demonstrated high efficiency for OER attributed to the great active sites with higher surface energy [32]. The bottleneck in 2D nanosheets is the stacked structure that limits mass transportation and impedes electron transfer kinetics at the interface boundary. Recent findings demonstrated that the nanopore's introduction substantially improves the mass transfer rate and increases the edges' surface reactivity. Similarly, it has been shown that holey nanosheets possess more reactive sites and mass transport penetration compared to porous nanosheets. Zhong et al. fabricated S-doped Ni–Fe–Co holey nanosheets using a sacrificial template procedure, revealing higher electronic conductivity and low potential barrier for O–O bond formation [33]. Chen and coworkers reported the facile cyanogen-$NaBH_4$

method for developing nickel foam (NF)-supported Fe-doped CO_3O_4 holey nanosheets (Fig. 4C.2) [34]. The growth mechanism demonstrated that chloride and cyano ligand create the porosity, while the 2D planer unit of divalent Co provides a self-platform for nanosheets' growth. The Fe-doped Co_3O_4 nanosheets revealed excellent performance as anode with lower overpotential, charge transfer resistance, and Tafel slope value endorsed the favorable kinetics of OER on the exposed active centers.

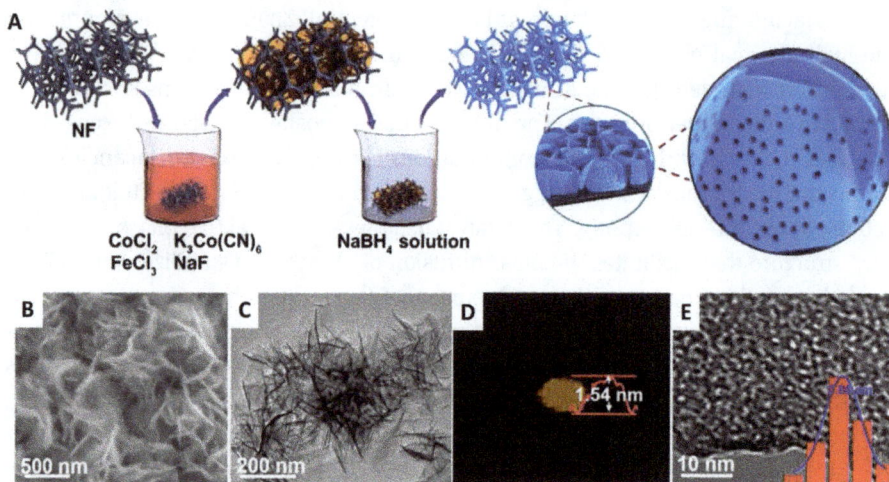

Fig. 4C.2: (A) Preparation procedure and (B) SEM, (C) TEM, (D) AFM, and (E) HRTEM images of Fe-doped Co_3O_4 holey nanosheets on nickel foam. Reproduced with permission from []. Copyright 2018, Elsevier.

The authors attributed the remarkable performance to the synergistic effect of highly conductive Ni foam as a current collector, microporous structure for effective mass transfer, and Fe-doped Co-active sites with optimum adsorption energy. A series of Ni–Co-based nanosheets have been prepared with different metal stoichiometry, chemical composition, and surface structures for OER. The results unveiled that Co content increases the Ni–Co oxide/nitride nanosheets' array activity. It is crucial for developing Ni-OOH nanosheets and needs a small overpotential of 247 mV to deliver a geometric activity of 10 mA/cm^2 with enhanced OER kinetics [35]. Qi and coworkers developed Ru-doped CoO hetero-interstructure on carbon cloth with hollow nanosheet morphology [36].

The nanosheets were generated by converting the Co-MOF via an ion-exchange process followed by pyrolysis in air. The optimized electronic structure for OER intermediate, composition, structural synergism, and intense interfacial contact triggers the charge and mass transfer kinetics at low overpotential. It has been investigated that surface oxygen vacancies modulate the electronic structure of 2D nanosheets, optimize the orbital energy of the active metal center, facilitate the OH^- attack, and desorption of oxygen molecules. Yao and coworkers modified the electronic structure and chemisorption energy of Fe-CoO nanosheets by creating the oxygen vacancies through S substitution [37]. The S atom stabilized the surface oxygen vacancies, created an active phase, and triggered the water dissociation. The DFT calculation study demonstrated that the water adsorption energy for Co sites near oxygen vacancies is preferable for water adsorption. The heteroatom doping modulated the electronic energy and adsorption property of active sites for water molecules to different extents depending on the orbital energy of heteroatoms. The Bader charge number study revealed that S is less electronegative than oxygen, reduced the charge transfer value, modulated the OH^* adsorption, stabilized the OOH^* intermediate, and triggered the OER.

Similarly, the Zhu group has synthesized Fe-CoO nanosheets using a simple wet chemistry method. The ultrathin nanosheets were treated with $NaBH_4$ and had abundant oxygen vacancies and a large specific surface area. The DFT calculation revealed that the O–O bond is formed through the electrolyte route (OH^*, O^*, OOH^*), where Co^{3+} close to the oxygen vacancies stabilizes the OER intermediates. The Co^{3+} oxidizes to Co^{4+}, creating an electrophilic center that facilitates the nucleophilic OH^- attack and OOH^* intermediate formation. Fe doping changes the reaction pathway, decreases the activation energy, and improves the OER kinetics. The authors attributed the activity enhancement to the decrease in Gibbs free energy difference on the Co site due to the electronic transfer from Co^{3+} sites to the areas with oxygen vacancies [38].

4C.3.2 Co-based boride nanosheets

Cobalt boride (CoB) is more catalytically reactive among all TM borides due to its low electronegativity and large atomic radii. The B insertion introduces lattice strain on Co's high entropic surface, decreasing the kinetics barrier for electron transfer during electrochemical redox reactions. It has also been investigated that amorphous Co has a more specific surface area, reconstruction capability, and more active sites than its crystalline counterpart. Researchers have synthesized CoB_x by simply soaking the cobalt oxide in an $NaBH_4$ solution. The pH of the solution decreases when cobalt oxide reacts with the BH_4^- and releases vigorous H^- ions. When sufficient cobalt oxide reduces, the pH increases swiftly as now the main reaction is the hydrolysis of $NaBH_4$. Researchers demonstrated that the B content in the vicinity of Co has a linear relation with the BH_4^- concentration.

The hydride ions (H^-) released from $NaBH_4$ reveal a strong reducing ability that can act as an oxygen scavenger to generate oxygen defects and modify the structure of Gd–Co bimetallic film:

$$4BH_4^- + 2Co^{2+} + 9H_2O \rightarrow Co_2B + 3B(OH)_3 + 12.5H_2$$

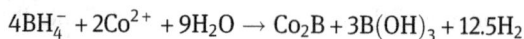

We have synthesized gold-supported gadolinium-doped CoB amorphous sheets (Gd–CoB@Au) with less than 5 nm average sheet thickness [39]. The Au support has the inherent ability to interact with the TM due to the relativistic effect and orbital confinement and facilitate the perfect/smooth thin-film formation. This has been observed that the smooth Au film assists the consistent growth of Gd–CoB to arise from Au and ensures a high amount of loading. The uniform distribution and horizontally aligned Au substrate sheet are clearer from the magnified scanning electronic microscopy)/transmission electron microscopy results, indicating the large contact areas between support and catalyst. Such support-dependent growth and alignment warranted the high interfacial contact and low R_{ct}, thereby strong interfacial collaboration for the concerted synergistic effect. The increased mass activity, turnover frequency (TOF), and low R_{ct} were attributed to the energetic transfer of electrons across the resistance-free interface for the dynamic release of the O_2 molecule as a final product. The amorphous CoB nanosheets supported the high current density (4,000 mA/cm^2), most likely due to the strong electronic wiring of Au with amorphous sheets. The Gd–CoO modification to Gd–CoB optimized the surface energy of Co that selectively binds with the OER intermediate. Compared with O, B is less oxophilic, which could optimally interact with the oxygen-containing intermediate and provide exposed sites for smooth adsorption/desorption. TM borides are highly corrosion-resistant, mechanically robust, electrically conductive, and considered promising electrocatalysts for water oxidation. Recent study unveiled that the electronic optimization of TM by introducing selective metals or heteroatoms ameliorates the catalytic performance. We have synthesized Au NC-decorated Gd–Co$_2$B nanoflakes embedded in titanium oxide (TiO$_2$) nanosheets, demonstrating the outstanding activity, selectivity, and durability in seawater electrolysis (Fig. 4C.3) [40]. It was found that the Au incorporation could modify the binding energy of oxygen-containing intermediates by perturbing the electronic properties of real active sites (Gd–CoB nanoflakes), directly participating in the reaction, and also providing bidentate sites for intermediates to increase the kinetics of OER [41, 42]. The role of gold, in particular, has been investigated for cobalt-boride due to its electron-withdrawing properties, which facilitate the positive shift in Co and serve as active species for the OER. The X-ray results demonstrated that the negative B.E energy shift in Au $4f$ spectrum endorsed the electronic shifting from the highest occupied molecular orbital (HOMO) of Co to the lowest unoccupied molecular orbital (LUMO) of Au due to the electron-withdrawing nature of Au. The positive chemical shift of 0.6 eV for B when compared with pure B was observed, revealing that the elemental B is chemically bonded with Co and not present in the free state [43]. Electrons from B are partially transferred to Co by occupying d-orbitals through hybridization of B $2p$ states, making boron electron-deficient and metal-enriched in electrons [43]. Usually, B insertion increases the amorphous content in the structure, enhances the ECSA and specific activity, and reduces the kinetic energy barrier, increasing TMs' catalytic efficiency.

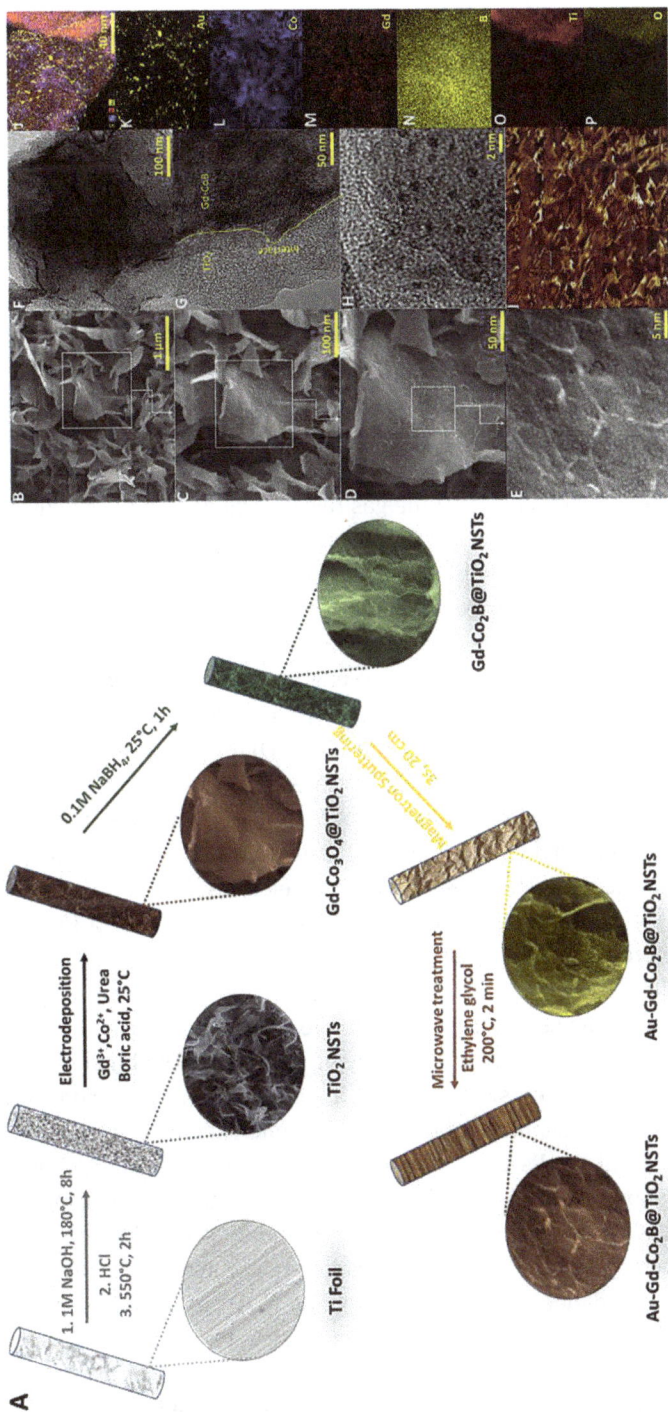

Fig. 4C.3: (A) Systematic illustration for the fabrication of Au-Gd-Co$_2$B@TiO$_2$ nanosheets. (B–E) SEM, (F, G) TEM, (H) HRTEM, (I) AFM images Au-Gd-Co$_2$B@TiO$_2$ nanosheets, (J) STEM, and (K–P) corresponding elemental mapping. Reproduced with permission from [40]. Copyright 2021, Elsevier.

Additionally, coordination of B with metals introduces structural defects, optimizes active sites' chemisorption energy, and improves the charge transfer rate. Lin and coworkers reported that the development of Co–Fe–B directly grown on the NF reveals strong catalyst support interaction and reduces the electron transfer resistance [44]. The rational design of nanosheets with less coordinated facets and edges provides monodentate and bidentate sites for intermediate adsorption and accelerates the O–O coupling. The B insertion provides open channels in the structure, facilitates the water diffusion, increases the gas release rate, and vacates the active sites for a chain reaction. The 2D nanostructures possess shortened charge transfer distance with abundant exposed active sites and large specific surface areas appealing for water oxidation. Its integration with carbon surface and electronic optimization through selective metal insertion boosted the catalytic performance multiple times. Zhang and coworkers prepared a Ni–Co–B composite with reduced graphene oxide [45]. The heterojunction between two sheet structures provides unique charge transfer sites, increases the water absorption coefficient, and enhances the gas release ability. The GO stabilizes the boride nanosheets and improves TM borides' chemical, structural, and mechanical resistance. The high electronic conductivity and abundant exposed active sites ensured the fast OER kinetics at metal boride nanosheets' surfaces. The theoretical calculations and experimental findings demonstrated that the peroxide formation on active metal sites is associated with the highest Gibbs free energy and is the potential determining step for OER. This high energy barrier is due to the electron-dense nature of the second existing intermediate on the active surface (M = O) that impedes the nucleophilic attack of OH$^-$ ions. The recent findings revealed that the presence of B in the vicinity of TM modifies the electronic structure of active sites, decreases the M–H bond, and reduces the kinetic and thermodynamic barrier for O–O coupling through M-OOH intermediate formation. The presence of B accelerates the O–O coupling through a single-site mechanism and avoided the double-site involvement in O–O bond formation; therefore, it has high resistance to metal dissolution. The monometallic boride has limited electronic conductivity that impedes electron transport across the interface. Electronic engineering to modulate the charge transfer kinetics is a feasible way to improve the catalytic efficiency of TM boride. The binary metal boride catalyst (Co–Fe–B) has been synthesized by a simple reduction method, and the detailed characterization demonstrated the electronic interaction between Fe and Co [46]. The electrochemical results show that Fe incorporation increases the TOF of metal boride sheets and promotes water oxidation kinetics. The stichometry of metals on the surface of boride sheets determines the overpotential needed for a specific geometric activity lowered for Co_2–Fe–B. The bimetallic boride revealed synergistic reactivity through improved electronic conductivity and OOH* intermediate stabilization on Co induced by Fe incorporation. Guo et al. developed an interface of $Co_2B@CoSe_2$ by growing the Co_2B nanostructure on $CoSe_2$ nanosheets [47]. The synergistic effect of interface with high reactive sites triggers the OER kinetics at low overpotential. The chemical or physical linkage of two heterointerfaces rescues charge transfer resistance, and the inert layer prevents the electrode–electrolyte contact. The efficient contact at the interface boundary increased the structural stability by reducing the metal dissolution rate and increasing the ASF. The catalyst demonstrated high structural and chemical strength and showed 100 % faradic efficiency for OER attributed to the unique structural features of nanosheets.

4C.3.3 Co-based nitride nanosheets

Recently, cobalt nitride (CoN) has been considered a promising catalyst for overall water splitting due to its high electronic conductivity, mechanical resistance, and quasi-metallic nature. Usually, CoN is derived from the corresponding oxides or hydroxides, possesses an oxidized surface with N coordination, and offers active redox reaction sites. Many groups have synthesized CoN and evaluated its electrochemical performance for OER under different conditions. For example, Co_4N had been synthesized and found that the surface oxidation enhanced the catalytic activity [48]. In oxidized CoN, the Co atom is in a 3+ valance state with six valence electrons in $3d$ orbital and followed the different electronic configurations depending on the energy gap between t_{2g} e_g, and electron filling pattern's high spin t_{2g}^4 e_g^2 or low spin t_{2g}^6 e_g^0 [49]. According to the Shao-Horn principle, TM's intrinsic activity for OER depends on the electronic population of e_g orbital [50]. It shows excellent activity with occupancy of 1.2, deviating from the electrons of Co^{3+} active sites in Co_4N. Liu et al. have optimized the configuration and electron filling in the e_g orbital of cobalt in the Co_4N structure by changing the N content in the network [51]. It has been found that reducing the N content from Co_4N to $Co_4N_{0.67}$ nanosheets increases the Co, e_g, electrons from 0.83 to 1.33.

The electronic population in the e_g orbital reached its optimum value (1.2) in the $Co_4N_{0.82}$ nanosheets. The $Co_4N_{0.82}$ nanosheets with optimum energy density demonstrated the highest OER activity and needed a small overpotential of 190 mV to deliver a current density of 10 mA/cm^2. The theoretical findings unveiled that the N content alteration tailored the electronic structure of Co_4N, reduced the potential energy barrier, and dramatically enhanced the OER performance. The Hubbard U-corrected DFT calculations have been performed for the perfect $Co_4N(011)$ surface, and N-removed Co_4N_x $(0 < x < 1)$ (011) surface (Fig. 4C.4) [51]. The partial density of Co $3d$ e_g orbitals revealed that N vacancies increased the electrons from 0.77 to 1.34. These theoretical calculations have well-matched with the experimental results of magnetization and ESR measurement and endorsed that the N removal increased the orbital population. The free energy profile for each coordinate depicted the interesting picture that e_g electronic configuration alteration changes the reaction mechanism. In perfect Co_4N structure, the O–O bond has been formed through the conventional electrolyte route (OH*-O*-OOH*-OO*-OH*). The N deficiency changed the single-site mechanism to a dual-site where adjacent Co

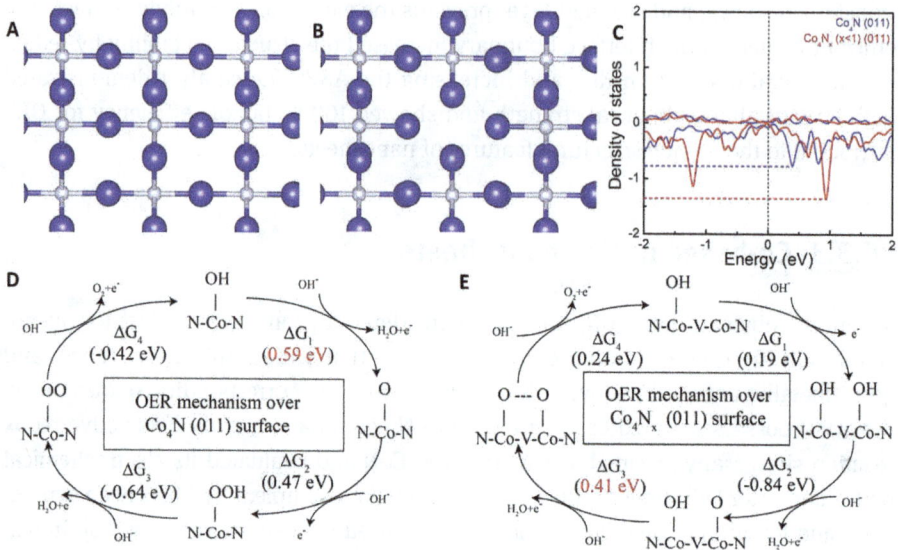

Fig. 4C.4: Visual depiction of (A) pristine Co₄N(011) surface and (B) active Co₄Nₓ OER site used for the DFT calculations. (C) Density of states of Co 3d eg in pristine Co₄N(011) and active Co₄Nₓ(011). (D) The change in Gibbs free energy at 1.23 V versus RHE for pristine Co₄N(011) surface and (E) active Co₄Nₓ. Reproduced with permission from [51]. Copyright 2021, Elsevier.

atom participated in the O–O coupling and followed the order of OH*-OH*OH*O*OH*-O-O*-OH*. The highest energy profile revealed that the second step (OH* to O*) was the RDS for the perfect structure with an energy barrier of 0.59 eV. The N vacancies reduced the energy barrier to 0.41 eV and changed the reaction pathway where O–O coupling is the RDS.

CoN nanosheets with porosity enhanced the OER reaction kinetics due to fast mass and electron transport rate. The ultrathin CoN nanosheets had been synthesized with 3D porosity, revealed high performance for water oxidation, needed low overpotential, further small charge transfer resistance, and Tafel slope value endorsed the kinetics favorability of catalyst for the oxidation reaction. The porous nanosheets sustained a high current density for a long time without structural collapse due to the structural flexibility of nanosheets [52]. Different morphologies and structures have different ECSA and roughness factors, the reactivity of active sites, electrolyte penetration, and bubble release ability, therefore, revealing different activities toward redox reactions. Synergistic electronic and structural modulation with exposed surface texture profoundly enhanced the OH^- adsorption, O–O coupling and O_2-release kinetics. Zheng group reported the direct growth of Co–Ni-based nanotubes on the current collector to decrease the charge transfer resistance [53]. The development of the active nanosheet phase revealed outstanding performance in water oxidation. The synthesis methods directly include the Cu_2O nanowire growth on Cu foil through an anodization approach. The Co–Ni–OH phase was grown on the nanowire structure through solution-phase cation exchange with intertwined tube structure attributed to the etching of actual Cu_2O structure. At the same time, the new active phase of Co–Ni–OH nanosheets had grown on the nanotube structure by gas-phase anion exchange.

The interface between the 1D nanowire and 2D nanosheets offers active sites with optimum chemisorption energy, low charge transfer resistance, and yield high TOF at low input voltage. Shi et al. have synthesized a heterostructure of Co–N/Mo_2N nanosheets and modulated the electronic structure by inserting the N-doped carbon via making a Mo-/Co–N–C bond, and impeccably supported over 3D nanoporous Cu structure revealed the high catalytic efficiency for water oxidation [54]. The heterointerface provides a dual site for both water OER and HER. The mechanistic study demonstrated that Mo_2N active sites increase the water polarizability and weaken the O–H bond while Co–N facilitated the adsorption of HER intermediates. The C–N network increases the charge and mass transport at the electrode–electrolyte junction. The nanoporous Cu framework enhanced the catalyst–support interaction, increased the charge transfer kinetics, reduced the metal dissolution rate, and increased the ASF of hybrid material. In TM nitride, the N atom is chemically bonded to the parent metal and remains the part of the inherent structure, and has an expanded lattice compared to parent metals. The N insertion enhanced the intermetal distance, decreased the interatomic interaction, contracted the TM d band, and increased the density of states near the Fermi level. These features increase the valance electrons and alter TMs' whole electronic structure. TM also has mixed ionic and covalent bonding and has unique absorption capabilities. These features increased the number of active sites with high intrinsic activity and enhanced the parent TM's structural, chemical, and mechanical resistance against transformation during electrocatalysis [55]. As discussed, multimetallic nitride revealed that the higher catalytic performance due to the synergetic electronic modulation between different metals, higher electron transfer coefficient, optimum chemisorption energy, and super hydrophilicity–aerophobicity triggers the OER reaction kinetics. Jiang and coworkers prepared Ni foam-supported Co–Mo nitride nanosheets via a scalable hydrothermal route followed by a nitridation process (Fig. 4C.5) [56]. The nanostructure has a robust chemical linkage with the underlying support and electronic interaction between two metals.

The N impurities in the vicinity of metals alter Mo and Co's electronic structure, increasing the water molecule polarizability and rate of dissociation. The 3D interconnected nanosheets have viable space for electrolyte interaction, improving OER intermediate adsorption and gas-release kinetics. The nanosheets with exposed active sites impeded the formation of a nonconducting phase on the catalyst's surface that hindered the electronic transfer at the interface.

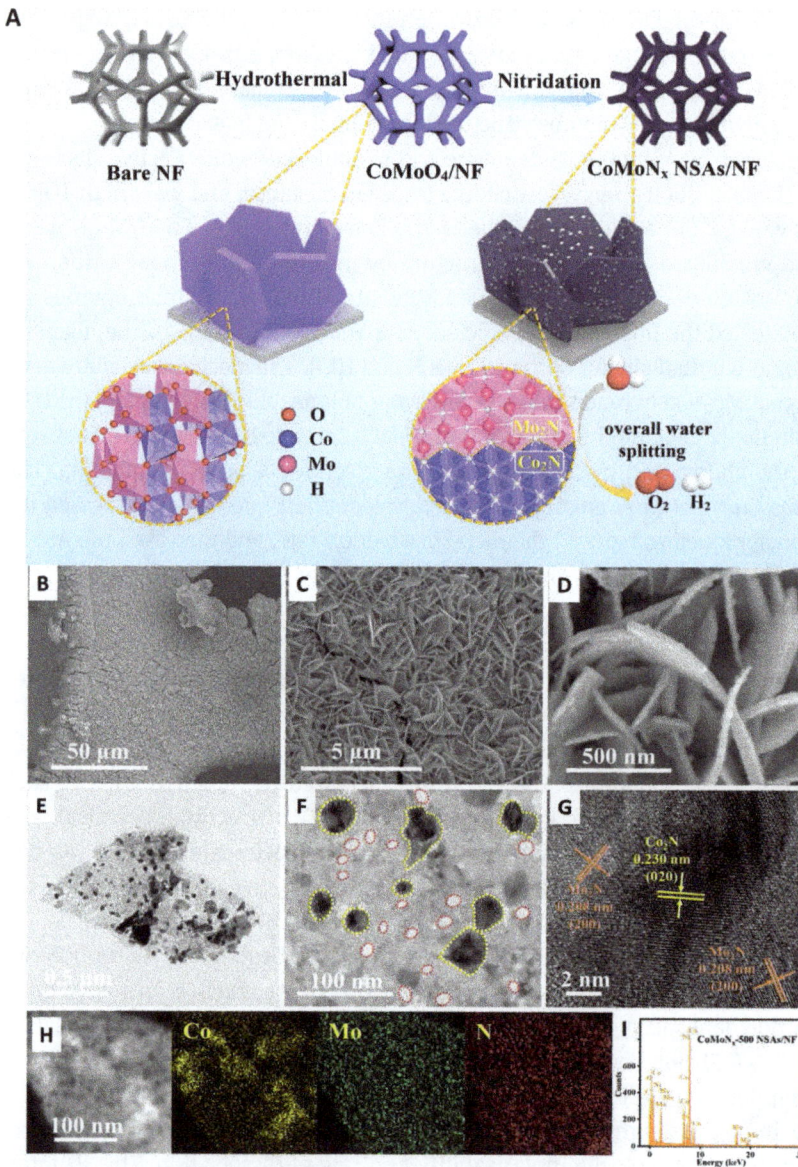

Fig. 4C.5: (A) Systematic representation of the formation of Ni foam-supported cobalt–molybdenum nitride nanosheet arrays (Co–Mo–N$_x$ NSA/NF), (B–D) SEM images, (E–F) TEM, (G) HRTEM, (H) elemental mapping, and (I) EDS spectrum of (Co–Mo–N$_x$ NSA/NF). Reproduced with permission from [56]. Copyright 2021, Elsevier.

4C.3.4 Co-based phosphide nanosheets

Metal phosphides are good electrical conductors and have high chemical and structural stability. The surface of metal phosphides is chemically oxidized to the metal oxide/hydroxide overlayer and offers a low resistive channel for electron transportation during a redox reaction. Recent findings demonstrated that P atoms chemically linked to metals oxidize to the corresponding phosphate and provide new active reaction sites. The phosphate content on the surface of metal phosphide alters its chelating mode and coordination number during the redox reaction and accelerates the OER kinetics. Cobalt phosphides (CoP) have structural and compositional advantages, inherent high intrinsic activity, and promising electrocatalysts for overall water splitting. Recently, researchers have established various synthesis routes for developing CoP comprising organic ligand-assisted synthesis, solid-phase reaction, electrodeposition, and hydrothermal synthesis. Generally, metal-organic decomposition yields CoP with 0D structure (NPS), whereas the electrodeposition method is beneficial for developing 3D design (e.g., nanoflowers). The solvothermal/hydrothermal route could produce the 2D structure (nanosheets), and by using these nanosheets as Co source in the phosphorization process via solid-phase reaction yields the CoP [57]. Recent findings have demonstrated that Co-P nanosheets growth on conducting current collector substantially enhances the water dissociation kinetics. The nanostructure coupling with the conducting substrate efficiently propagates the proton and electron transfer, exposing abundant reactive sites to increase OER performance. Yue et al. reported the growth of Co–P nanosheets on Cu nanowires by a simple electrodeposition method to enhance the exposed ECSA and increase active sites' intrinsic activity [58]. The experimental findings revealed that Co–P nanosheets coupled with Cu nanowires needed small overpotential to drive a high current density compared to another substrate for the same guest material. The structural and compositional characterization of the catalyst after long-term electrolysis endorsed the structure flexibility and stability in harsh conditions. These findings demonstrated that electronic modulation is crucial to increasing the intrinsic reactivity of each active site. Structural optimization also plays a pivotal role in the stability and kinetics of redox reactions. It has been investigated that CoP surface in situ chemically oxidized to Co–O/OH/OOH and is considered as actual active sites for the OER. Due to the nontoxicity, low cost, and synergistic electronic contribution between the actual CoP and in-situ-generated oxide/hydroxide/oxy-hydroxide layer, CoP has been considered a promising electrocatalyst for water oxidation. However, the bulk CoP needs more electronic and surface engineering optimization to expose particular active sites for OER with higher activity and stability. Che and coworkers reported the scalable strategy for developing CoP nanosheets and optimizing the electronic structure through Mn doping [59]. Experimental findings demonstrated that Mn substitution in the CoP nanosheets optimized the charge transfer kinetics, engineered the surface structure, and introduced lattice distortion that substantially enhanced the ECSA. The porous structure is beneficial for the favorable kinetics of mass transport and triggers the bubble detachment ability of active sites. The porous nanosheet structure through structural optimization and electronic engineering through Mn doping enhanced the intrinsic activity of active sites, and the reported catalyst needed small overpotential to deliver a high current density. The Tafel slope value unveiled the oxidation pathway and revealed that M-OH* intermediate formation is the RDS step with a high potential barrier. DFT simulation demonstrated that Mn doping modulates the electronic configuration of active O sites, reduces the potential energy barrier, changes the reaction pathway, and triggers the charge transfer at the interface boundary. Besides the electronic optimization, the nanosheet structure also facilitates the adsorption–desorption kinetics, destabilizes the water molecule, and enhances the dissociation kinetics at active sites. Theoretically speaking, the chemical linkage of the favorable atom with CoP altered the chemical environment. The coordination number of metal cations increased the intermetallic electronic interaction and improved the catalytic performance of CoP. Researchers found that the Fe interaction with the CoP substantially enhanced the activity of the catalyst by offering abundant active sites with suitable chemisorption energy to stabilize the OH* intermediate and trigger the OER kinetics. In addition to the electronic interaction through Fe doping, the geometry of the hybrid material has a considerable impact on the overall performance. The 2D nanosheets facilitate the electron and proton transfer and offer abundant active sites with coordination. Developing these nanosheets on the 3D hierarchical porous substrate with high electrical conductivity enhanced active sites' catalytic performance and structural and mechanical stability. The intense interaction between metal nanosheets with the underlying support prevents the catalyst exfoliation during the catalyst and retains the catalytic activity of active sites. Pang et al. reported that the freestanding electrode for water oxidation comprises Fe-doped Ni–Co–P nanosheets on carbon fiber paper [60]. The direct growth of nanosheets on carbon fiber paper through electrodeposition ensured the strong catalyst support interaction and demonstrated the small charge transfer resistance at the interface. The electrochemical characterization revealed that 2D nanosheets have abundant ECSA and deliver high current density at low overpotential. Fe doping enhanced the intrinsic activity of nanosheets, changed the reaction pathway, and considerably reduced the kinetics barrier. The Fe doping enhanced the selectivity of active sites for OER intermediate, and the hybrid material revealed 98 % faradic efficiency for OER. The catalyst sustained the high current density for continuous 35 h without considerable loss in the activity. These durability results showed that the structure has high stability under anodic bias and in an alkaline environment. Similarly, the Shen group reported the electronic modulation of TM phosphide to accelerate the OER performance of the catalyst [61]. The electronic structure of Co–P has been optimized by introducing Cu and stabilizing the active sites on carbon paper. Cu doping tailors the chemisorption energy of active sites, increases the electronic activity of the catalyst, and reduces the interface charge resistance. The nanosheet morphology further ensured the availability of abundant exposed active sites with facets and edges orientation, destabilized the water molecule, and preserved the functional areas from agglomeration. The hybrid material had been synthesized by hydrothermal method followed by phosphorization. The optimum surface ensures the intermediate adsorption and desorption without considerable voltage and ohmic losses and decreases the overall overpotential needed for a specific geometric activity. The detailed DFT calculation unveiled the influence of Cu doping in CoP nanosheets; Cu–Co–OOH and pristine Co–OOH surfaces were considered (Fig. 4C.6) [61]. The free energy profile demonstrated that the OH* to O* step is the RDS for pristine Co–OOH with the potential barrier of 2 eV. After Cu insertion in the CoP nanosheet structure, the energy barrier for this step remarkably reduces to 1.08 eV, implying that Cu substitution enhances the OER performance and changes the reaction pathway. These results validate that electronic modulation of CoP nanosheets through Cu doping reduces the energy barrier and facilitates the O–O coupling.

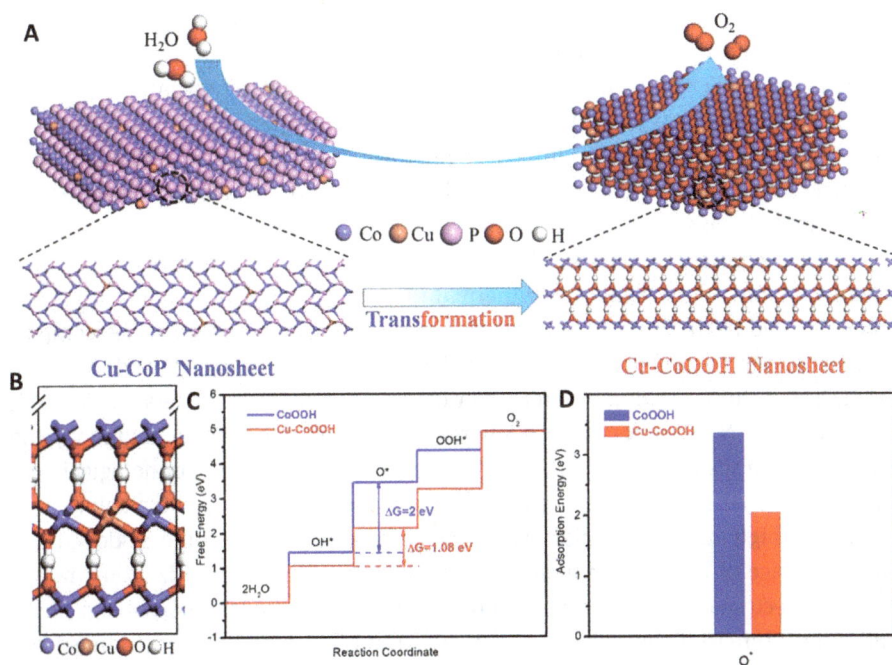

Fig. 4C.6: (A) Representation of the transformation of Cu-CoP to Cu-Co-OOH phase and the mechanism diagram of OER on the Cu-CoP surface (B) Atomic structure of Cu-Co-OOH. (C) Free energy profile at 0 V on the Cu-CoOOH and CoOOH surfaces. (D) Calculated O* adsorption energy for the Cu-Co-OOH and Co-OOH. Reproduced with permission from [61] Copyright 2019, Elsevier.

4C.3.5 Co-based sulfide nanosheets

Cobalt sulfide (CoS) has a low potential energy barrier for water adsorption and offers a fast electronic transportation rate to active sites. Recently, many struggles have been put to further facilitate the electrochemical performance of CoS by engineering the electronic structural and morphological modulation and integrating the active sites with various support. It has been investigated that cation or anion doping enhanced the water oxidation kinetics at the CoS surface and stabilized the functional areas from agglomeration. Anion doping vacated the orbitals effectively, increased the intermetallic strength, and promoted high current density without structural collapse. Cation doping can enhance active sites' intrinsic activity and decrease the potential energy barrier, reducing the required overpotential. The 2D structure enthralled unique physical and chemical properties. It improved the catalytic activity of CoS compared to other morphologies owing to the high surface-to-volume ratio, exposed facets and edges, sizeable electrode–electrolyte interface, and fast electron and ion diffusion path. Cao group integrated the ultrathin Co–S nanosheets using a microwave-assisted approach [62]. The catalyst revealed high catalytic activity, selectivity, and durability for the overall water splitting reaction. The electrochemical results and in situ X-rays photoelectron spectroscopy findings demonstrated that CoS chemically transformed to Co–O/OH/OOH depending on the applied potential and served as real active sites for OER.

He et al. reported that porous C–N sheets supported Ni–Co_2S_4 nanosheets synthesized by simple and scalable hydrothermal [63]. The morphological and electronic synergism had been observed at the interface boundary of nanosheets. The composite material demonstrated low charge transfer and diffusion resistance and increased the electrochemical performance for water oxidation. The in-situ-generated Ni–Co_2S_4 confined the C–N network created a low resistive interface for charge migration, and exposed active sites to stabilize the OER intermediate. Due to the morphological benefits, the catalyst revealed high catalytic activity and durability in the alkaline electrolyte at a high anodic potential. Wang and coworkers developed the Se-doped CoS on the surface of graphene nanofoam through an in situ recrystallization strategy [64]. The electronic modulation via Se doping in the structure and morphological optimization increased the electronic density around active sites and induced the metal–oxygen and selenate/selenite species formation to accelerate the OER kinetics (Fig. 4C.7) [64]. The 2D graphene foam with hierarchical porosity increased the durability of active sites and mass diffusion kinetics. The in-depth DFT simulations unveiled that (100) planes of Se-doped CoS were more favorable for water adsorption and considered real active sites. The Se introduction in the CoS nanosheets altered the electronic density around Co-active areas due to large ionic radius and lower electronegativity compared to S anion and increased the OER kinetics. The metallic Co–Se bond chemically oxidizes to oxide/hydroxide at the surface with exposed edges, improves the ECSA, and stabilizes the OER intermediate. The experimental results revealed that Se etching corroded the structure surface, increased the roughness factor, and enhanced the electrochemical response. The catalytic performance of CoS nanosheets can be enhanced by structural engineering of facets, defects, interfaces, and strains. The 3D structure turning to nanosheet morphology altered the electronic system due to the inherent large specific surface area and absence of

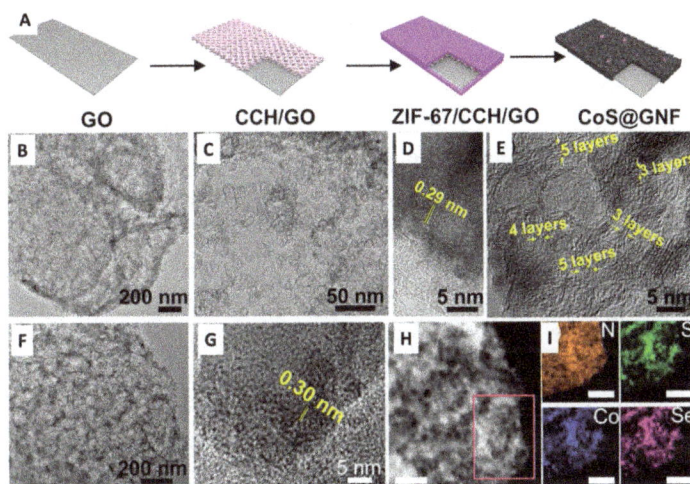

Fig. 4C.7: (A) Schematic sketch of the development of the 2D cobalt sulfide@graphene nanofoam (CoS@GNF). (B, C) SEM and (D, E) HRTEM images of the CoS@GNF. (F) TEM and (G) HRTEM images of the Co(S, Se)@GNF. (H) STEM and (I) EDS mapping images of the Co (S, Se)@GNF. Reproduced with permission from [64].

dimensions that created additional active sites and offered quick charge diffusion pathways. The interface boundary sandwiched between two nanosheets are considered an important region, and the presence of structural defects at the interface facilitates the adsorption and activation of OER intermediate and reduces the potential barrier for O–O coupling. The defect-rich interface provides abundant active sites to decrease the reaction activation energy. Recently, Li et al. reported the defected FeS_2/CoS_2 interface nanosheets for the overall water splitting [65]. The heterointerface demonstrated outstanding performance for water oxidation due to the interface nanosheets structure with abundant exposed active sites and needed a small overpotential of 302 mV to drive a large current density of 100 mA/cm^2. The heterostructure ensured the maximum FE and maintained steady behavior at a high current density.

4C.4 Catalyst design for high-output OER

The catalyst should simultaneously fulfill the following merits to make the energy-intensive water electrolysis process viable for H_2 fuel. (1) Facile, scalable, and low-cost synthetic procedure: A green synthesis route would be applied to develop large-scale electrocatalyst from abundant and low-cost raw materials without releasing threatened gases, for example, sulfide [66]. (2) Abundant active sites: Structural optimization such as porosity and amorphousness can create abundant active sites and increase the ECSA [67]. Nanosheets reveal a high density of grain boundaries and offer an active site for redox reactions. (3) Highly active sites: synergistic morphological and electronic modulation through cations, anion doping, and oxygen vacancies enhanced the intrinsic activity of each active site, optimized the chemisorption energy of reactive sites to stabilize and activate the OER intermediates, and reduced the kinetics and potential energy barrier [68]. (4) Efficient electron transport and strong catalyst support interaction increased the mechanical strength of hybrid material, decreased the charge transfer resistance, facilitated the redox reaction, and avoided the metal sintering or leaching from the underlying support [69]. (5) Efficient mass transport and gas releases: the porous structure with low-dimensional sites (nanosheets) offers a short mass diffusion pathway, increases the electrode–electrolyte contact, and secures efficient mass transfer with gas release ability [70].

4C.5 Conclusion

Industrial-scale hydrogen production from electrochemical water splitting needed a highly active, selective, and durable earth-abundant catalyst with a lower kinetics barrier for OER. A variety of TM-based catalysts have been designed proficiently for water splitting. Co-based catalysts with electronic and structural optimization have gained extensive attention for alkaline water oxidation due to their 3d orbital confinement with the O 2p orbital, electronic configurations, and unique physiochemical properties. The development of nanomaterials in nanosheet morphology has some special features. First, the atomic thickness and in-plane native or covalent bond in 2D order increased the structural flexibility and mechanical durability needed for water oxidation, especially at high anodic bias. Second, the electron confinement in 2D modulates the electronic properties and enhances the interaction with the incoming intermediates. Third, it offers highly exposed surface sites, and it is easy to change the structural properties of nanosheets by element doping strain or defect engineering or functionalization. Fourth, the large lateral size without atomic stacking ensured the high specific surface area. This review profiles the progress of cobalt-based nanosheets' (oxide/hydroxide, boride, nitride, phosphide, and sulfide) electrocatalyst in terms of electrocatalytic performance for OER. Co-based nanosheets' excellent electrochemical response can be assigned to the following factors: (1) the unique structure with abundant active sites, large specific surface area, and short diffusion pathways for mass transfer; (2) the enhanced synergy between catalyst and support with high electron transfer rate, and chemical and structural durability; and (3) the particular sites with chemisorption energy optimally adsorb the OER intermediates.

References

[1]   Munir, A., Joya, K. S., Ul Haq, T., Babar, N. U. A., Hussain, S. Z., Qurashi, A., Ullah, N. & Hussain, I. (2019). Metal nanoclusters: new paradigm in catalysis for water splitting, solar and chemical energy conversion. ChemSusChem, 12, 1517–1548. 10.1002/cssc.201802069.
[2]   Nasir, J. A., Munir, A., Ahmad, N., Ul Haq, T., Khan, Z. & Rehman, Z. (2021). Photocatalytic Z-Scheme overall water splitting: recent advances in theory and experiments. Advanced Materials, 10.1002/adma.202105195.
[3]   Munir, A., Haq, T. U., Qurashi, A., Rehman, H. U., Ul-Hamid, A. & Hussain, I. (2019). Ultrasmall Ni/NiO nanoclusters on thiol-functionalized and-exfoliated graphene oxide nanosheets for durable oxygen evolution reaction. ACS Applied Energy Materials, 2, 363–371. 10.1021/acsaem.8b01375.
[4]   Jiang, W. J., Tang, T., Tang, T., Zhang, Y., Hu, J. S. & Hu, J. S. (2020). Synergistic modulation of non-precious-metal electrocatalysts for advanced water splitting. Accounts of Chemical Research, 53, 1111–1123. 10.1021/acs.accounts.0c00127.
[5]   Suen, N. T., Hung, S. F., Quan, Q., Zhang, N., Xu, Y. J. & Chen, H. M. (2017). Electrocatalysis for the oxygen evolution reaction: recent development and future perspectives. Chemical Society Reviews, 46, 337–365. 10.1039/c6cs00328a.
[6]   Wang, J., Cui, W., Liu, Q., Xing, Z., Asiri, A. M. & Sun, X. (2016). Recent progress in cobalt-based heterogeneous catalysts for electrochemical water splitting. Advanced Materials, 10.1002/adma.201502696.
[7]   Abbas, M., Ul Haq, T., Arshad, S. N. & Zaheer, M. (2020). Fabrication of cobalt doped titania for enhanced oxygen evolution reaction. Molecular Catalysis, 488, 110894. 10.1016/j.mcat.2020.110894.
[8]   Dong, R., Zhang, T. & Feng, X. (2018). Interface-assisted synthesis of 2D materials: trend and challenges. Chemical Reviews, 10.1021/acs.chemrev.8b00056.
[9]   Wang, J., Ge, X., Liu, Z., Thia, L., Yan, Y., Xiao, W. & Wang, X. (2017). Heterogeneous electrocatalyst with molecular cobalt ions serving as the center of active sites. Journal of the American Chemical Society, 10.1021/jacs.6b10307.
[10]   Craig, M. J., Coulter, G., Dolan, E., Soriano-López, J., Mates-Torres, E., Schmitt, W. & García-Melchor, M. (2019). Universal scaling relations for the rational design of molecular water oxidation catalysts with near-zero overpotential. Nature Communications, 10.1038/s41467-019-12994-w.
[11]   Song, F., Busch, M. M., Lassalle-Kaiser, B., Hsu, C. S., Petkucheva, E., Bensimon, M., Chen, H. M., Corminboeuf, C. & Hu, X. (2019). An unconventional iron nickel catalyst for the oxygen evolution reaction. ACS Central Science, 5, 558–568. 10.1021/acscentsci.9b00053.
[12]   Dau, H., Limberg, C., Reier, T., Risch, M., Roggan, S. & Strasser, P. (2010). The mechanism of water oxidation: from electrolysis via homogeneous to biological catalysis. ChemCatChem, 2, 724–761. 10.1002/cctc.201000126.

[13] Zheng, D., Yu, L., Liu, W., Dai, X., Niu, X., Fu, W., Shi, W., Wu, F. & Cao, X. (2021). Structural advantages and enhancement strategies of heterostructure water-splitting electrocatalysts. Cell Reports Physical Science, 10.1016/j.xcrp.2021.100443.

[14] Haq, T. U., Haik, Y., Hussain, I., Rehman, H. U. & Al-Ansari, T. A. (2021). Gd-Doped Ni-oxychloride nanoclusters: new nanoscale electrocatalysts for high-performance water oxidation through surface and structural modification. ACS Applied Materials & Interfaces, 13, 468–479. 10.1021/acsami.0c17216.

[15] Saleh, T. A. (2020). Nanomaterials: classification, properties, environmental toxicities. Environmental Technology & Innovation, 10.1016/j.eti.2020.101067.

[16] Chen, Y., Yang, K., Jiang, B., Li, J., Zeng, M. & Fu, L. (2017). Emerging two-dimensional nanomaterials for electrochemical hydrogen evolution. Journal of Materials Chemistry A, 10.1039/c7ta00816c.

[17] Peng, J., Dong, W., Wang, Z., Meng, Y., Liu, W., Song, P. & Liu, Z. (2020). Recent advances in 2D transition metal compounds for electrocatalytic full water splitting in neutral media. Materials Today Advances, 10.1016/j.mtadv.2020.100081.

[18] Ul Haq, T., Bicer, Y., Munir, A., Mansour, S. A. & Haik, Y. (2020). Surface assembling of highly interconnected and vertically aligned porous nanosheets of Gd–CoB on TiO$_2$ nanoflowers for durable methanol oxidation reaction. ChemCatChem, 12, 3585–3597. 10.1002/cctc.202000392.

[19] Ul Haq, T. & Haik, Y. (2021). S doped Cu2O-CuO nanoneedles array: free standing oxygen evolution electrode with high efficiency and corrosion resistance for sea water splitting. Catalysis Today, 10.1016/j.cattod.2021.09.015.

[20] Arshad, F., Munir, A., Kashif, Q. Q., Ul Haq, T., Iqbal, J., Sher, F. & Hussain, I. (2020). Controlled development of higher-dimensional nanostructured copper oxide thin films as binder-free electrocatalysts for oxygen evolution reaction. International Journal of Hydrogen Energy, 45, 16583–16590. 10.1016/j.ijhydene.2020.04.152.

[21] Chen, Y., Fan, Z., Zhang, Z., Niu, W., Li, C., Yang, N., Chen, B. & Zhang, H. (2018). Two-Dimensional metal nanomaterials: synthesis, properties, and applications. Chemical Reviews, 10.1021/acs.chemrev.7b00727.

[22] Fu, G., Chen, Y., Cui, Z., Li, Y., Zhou, W., Xin, S., Tang, Y. & Goodenough, J. B. (2016). Novel hydrogel-derived bifunctional oxygen electrocatalyst for rechargeable air cathodes. Nano Letters, 16, 6516–6522. 10.1021/acs.nanolett.6b03133.

[23] Kim, Y., Lopes, P. P., Park, S., Lee, A., Lim, J., Lee, H., Back, S., Jung, Y., Danilovic, N., Stamenkovic, V., Erlebacher, J., Snyder, J. & Markovic, N. M. (n.d.). Oxygen evolution catalysts. Nature Communications, 1–8, 10.1038/s41467-017-01734-7.

[24] Ul Haq, T. & Haik, Y. (2021). Electrocatalysis for the water splitting: recent strategies for improving the performance of electrocatalyst. Advanced and Sustainable Energy, 10.1007/978-3-030-74406-9_11.

[25] Liu, Z., Xiao, Z., Luo, G., Chen, R., Dong, C. L., Chen, X., Cen, J., Yang, H., Wang, Y., Su, D., Li, Y. & Wang, S. (2019). Defects-induced in-plane heterophase in cobalt oxide nanosheets for oxygen evolution reaction. Small, 10.1002/smll.201904903.

[26] Liu, B., Peng, H. Q., Ho, C. N., Xue, H., Wu, S., Ng, T. W., Lee, C. S. & Zhang, W. (2017). Mesoporous nanosheet networked hybrids of cobalt oxide and cobalt phosphate for efficient electrochemical and photoelectrochemical oxygen evolution. Small, 10.1002/smll.201701875.

[27] Tao, H. B., Fang, L., Chen, J., Bin Yang, H., Gao, J., Miao, J., Chen, S. & Liu, B. (2016). Identification of surface reactivity descriptor for transition metal oxides in oxygen evolution reaction. Journal of the American Chemical Society, 138, 9978–9985. 10.1021/jacs.6b05398.

[28] Zhang, N., Wang, Y., Hao, Y. C., Ni, Y. M., Su, X., Yin, A. X. & Hu, C. W. (2018). Ultrathin cobalt oxide nanostructures with morphology-dependent electrocatalytic oxygen evolution activity. Nanoscale, 10.1039/c8nr05337e.

[29] Lai, C., Liu, X., Wang, Y., Cao, C., Yin, Y., Yang, H., Qi, X., Zhong, S., Hou, X. & Liang, T. (2020). Modulating ternary Mo–Ni–P by electronic reconfiguration and morphology engineering for boosting all-pH electrocatalytic overall water splitting. Electrochimica Acta, 10.1016/j.electacta.2019.135294.

[30] Zhang, S., Wei, N., Yao, Z., Zhao, X., Du, M. & Zhou, Q. (2021). Oxygen vacancy-based ultrathin Co3O4 nanosheets as a high-efficiency electrocatalyst for oxygen evolution reaction. International Journal of Hydrogen Energy, 10.1016/j.ijhydene.2020.11.072.

[31] Munir, A., Ul Haq, T., Hussain, I., Ullah, I., Hussain, S. Z., Qurashi, A., Iqbal, J., Rehman, A. & Hussain, I. (2020). Controlled assembly of Cu/Co-oxide beaded nanoclusters on thiolated graphene oxide nanosheets for high-performance oxygen evolution catalysts. Chemistry – A European Journal, 26, 11209–11219. 10.1002/chem.202000491.

[32] Huang, Y., Zhao, X., Tang, F., Zheng, X., Cheng, W., Che, W., Hu, F., Jiang, Y., Liu, Q. & Wei, S. (2018). Strongly electrophilic heteroatoms confined in atomic CoOOH nanosheets realizing efficient electrocatalytic water oxidation. Journal of Materials Chemistry A, 10.1039/c7ta09412d.

[33] Cao, L. M., Wang, J. W., Zhong, D. C. & Lu, T. B. (2018). Template-directed synthesis of sulfur-doped NiCoFe layered double hydroxide porous nanosheets with enhanced electrocatalytic activity for the oxygen evolution reaction. Journal of Materials Chemistry A, 10.1039/c7ta09734d.

[34] Li, Y., Li, F. M., Meng, X. Y., Wu, X. R., Li, S. N. & Chen, Y. (2018). Direct chemical synthesis of ultrathin holey iron-doped cobalt oxide nanosheets on nickel foam for oxygen evolution reaction. Nano Energy, 10.1016/j.nanoen.2018.10.032.

[35] Li, Y., Hu, L., Zheng, W., Peng, X., Liu, M., Chu, P. K. & Lee, L. Y. S. (2018). Ni/Co-based nanosheet arrays for efficient oxygen evolution reaction. Nano Energy, 10.1016/j.nanoen.2018.08.010.

[36] Wang, C. & Qi, L. (2020). Heterostructured inter-doped ruthenium–cobalt oxide hollow nanosheet arrays for highly efficient overall water splitting. Angewandte Chemie International Edition, 10.1002/anie.202005436.

[37] Zhuang, L., Jia, Y., Liu, H., Li, Z., Li, M., Zhang, L., Wang, X., Yang, D., Zhu, Z. & Yao, X. (2020). Sulfur-modified oxygen vacancies in iron-cobalt oxide nanosheets: enabling extremely high activity of the oxygen evolution reaction to achieve the industrial water splitting benchmark. Angewandte Chemie International Edition, 10.1002/anie.202006546.

[38] Hu, X., Zhang, S., Sun, J., Yu, L., Qian, X., Hu, R., Wang, Y., Zhao, H. & Zhu, J. (2019). 2D Fe-containing cobalt phosphide/cobalt oxide lateral heterostructure with enhanced activity for oxygen evolution reaction. Nano Energy, 10.1016/j.nanoen.2018.11.047.

[39] Ul Haq, T., Mansour, S. A., Munir, A. & Haik, Y. (2020). Gold-Supported Gadolinium Doped CoB Amorphous Sheet: a New Benchmark Electrocatalyst for Water Oxidation with High Turnover Frequency. Advanced Functional Materials, 1910309, 1–11. 10.1002/adfm.201910309.

[40] Ul Haq, T., Pasha, M., Tong, Y., Mansour, S. A. & Haik, Y. (2022). Au nanocluster coupling with Gd-Co2B nanoflakes embedded in reduced TiO2 nanosheets: seawater electrolysis at low cell voltage with high selectivity and corrosion resistance. Applied Catalysis B: Environmental, 10.1016/j.apcatb.2021.120836.

[41] Ng, J. W. D., García-Melchor, M., Bajdich, M., Chakthranont, P., Kirk, C., Vojvodic, A. & Jaramillo, T. F. (2016). Gold-supported cerium-doped NiOx catalysts for water oxidation. Nature Energy, 1, 1–8. 10.1038/nenergy.2016.53.

[42] Ma, T. Y., Dai, S., Jaroniec, M. & Qiao, S. Z. (2014). Metal-organic framework derived hybrid Co3O4-carbon porous nanowire arrays as reversible oxygen evolution electrodes. Journal of the American Chemical Society, 136, 13925–13931. 10.1021/ja5082553.

[43] Li, H., Yang, H. & Li, H. (2007). Highly active mesoporous Co-B amorphous alloy catalyst for cinnamaldehyde hydrogenation to cinnamyl alcohol. Journal of Catalysis, 10.1016/j.jcat.2007.07.022.

[44] Li, Y., Jiang, X., Tang, M., Zheng, Q., Huo, Y., Xie, F. & Lin, D. (2020). A high-performance oxygen evolution electrocatalyst based on partially amorphous bimetallic cobalt iron boride nanosheet. International Journal of Hydrogen Energy, 10.1016/j.ijhydene.2020.07.140.

[45] Sun, J., Zhang, W., Wang, S., Ren, Y., Liu, Q., Sun, Y., Tang, L., Guo, J. & Zhang, X. (2019). Ni-Co-B nanosheets coupled with reduced graphene oxide towards enhanced electrochemical oxygen evolution. Journal of Alloys and Compounds, 10.1016/j.jallcom.2018.10.296.

[46] Chen, H., Ouyang, S., Zhao, M., Li, Y. & Ye, J. (2017). Synergistic activity of Co and Fe in amorphous Cox-Fe-B catalyst for efficient oxygen evolution reaction. ACS Applied Materials & Interfaces, 10.1021/acsami.7b13939.

[47] Guo, Y., Yao, Z., Shang, C. & Wang, E. (2017). Amorphous Co2B grown on CoSe2 nanosheets as a hybrid catalyst for efficient overall water splitting in alkaline medium. ACS Applied Materials & Interfaces, 10.1021/acsami.7b10605.

[48] Chen, P., Xu, K., Fang, Z., Tong, Y., Wu, J., Lu, X., Peng, X., Ding, H., Wu, C. & Xie, Y. (2015). Metallic Co4N porous nanowire arrays activated by surface oxidation as electrocatalysts for the oxygen evolution reaction. Angewandte Chemie International Edition, 10.1002/anie.201506480.

[49] Zhou, S., Miao, X., Zhao, X., Ma, C., Qiu, Y., Hu, Z., Zhao, J., Shi, L. & Zeng, J. (2016). Engineering electrocatalytic activity in nanosized perovskite cobaltite through surface spin-state transition. Nature Communications, 7, 1–7. 10.1038/ncomms11510.

[50] Suntivich, J., May, K. J., Gasteiger, H. A., Marković, N. M., Gasteiger, H. A., Goodenough, J. B. & Shao-Horn, Y. (2009). A perovskite oxide optimized for oxygen evolution catalysis from molecular orbital principles. Science.

[51] Liu, H., Lei, J., Yang, S., Qin, F., Cui, L., Kong, Y., Zheng, X., Duan, T., Zhu, W. & He, R. (2021). Boosting the oxygen evolution activity over cobalt nitride nanosheets through optimizing the electronic configuration. Applied Catalysis B: Environmental, 10.1016/j.apcatb.2021.119894.

[52] Liu, C., Bai, G., Tong, X., Wang, Y., Lv, B., Yang, N. & Guo, X. Y. (2019). Mesoporous and ultrathin arrays of cobalt nitride nanosheets for electrocatalytic oxygen evolution. Electrochemistry Communications, 10.1016/j.elecom.2018.11.022.

[53] Li, S., Wang, Y., Peng, S., Zhang, L., Al-Enizi, A. M., Zhang, H., Sun, X. & Zheng, G. (2016). Co-Ni-based nanotubes/nanosheets as efficient water splitting electrocatalysts. Advanced Energy Materials, 10.1002/aenm.201501661.

[54] Shi, H., Dai, T. Y., Bin Wan, W., Wen, Z., Lang, X. Y. & Jiang, Q. (2021). Mo-/Co-N-C Hybrid Nanosheets Oriented on Hierarchical Nanoporous Cu as Versatile Electrocatalysts for Efficient Water Splitting. Advanced Functional Materials, 10.1002/adfm.202102285.

[55] Cheng, Z., Qi, W., Pang, C. H., Thomas, T., Wu, T., Liu, S. & Yang, M. (2021). Recent Advances in Transition Metal Nitride-Based Materials for Photocatalytic Applications. Advanced Functional Materials, 10.1002/adfm.202100553.

[56] Lu, Y., Li, Z., Xu, Y., Tang, L., Xu, S., Li, D., Zhu, J. & Jiang, D. (2021). Bimetallic Co-Mo nitride nanosheet arrays as high-performance bifunctional electrocatalysts for overall water splitting. Chemical Engineering Journal, 10.1016/j.cej.2021.128433.

[57] Li, Z., Feng, H., Song, M., He, C., Zhuang, W. & Tian, L. (2021). Advances in CoP electrocatalysts for water splitting. Materials Today Energy, 10.1016/j.mtener.2021.100698.

[58] Yue, Q., Gao, T., Yuan, H. & Xiao, D. (2021). An efficient way to improve water splitting electrocatalysis by electrodepositing cobalt phosphide nanosheets onto copper nanowires. International Journal of Hydrogen Energy, 10.1016/j.ijhydene.2021.03.113.

[59] Liu, Y., Ran, N., Ge, R., Liu, J., Li, W., Chen, Y., Feng, L. & Che, R. (2021). Porous Mn-doped cobalt phosphide nanosheets as highly active electrocatalysts for oxygen evolution reaction. Chemical Engineering Journal, 10.1016/j.cej.2021.131642.

[60] Pang, L., Liu, W., Zhao, X., Zhou, M., Qin, J. & Yang, J. (2020). Engineering electronic structures of nickel-cobalt phosphide via iron doping for efficient overall water splitting. ChemElectroChem, 10.1002/celc.202001390.

[61] Yan, L., Zhang, B., Zhu, J., Li, Y., Tsiakaras, P. & Kang Shen, P. (2020). Electronic modulation of cobalt phosphide nanosheet arrays via copper doping for highly efficient neutral-pH overall water splitting. Applied Catalysis B: Environmental, 10.1016/j.apcatb.2019.118555.

[62] Souleymen, R., Wang, Z., Qiao, C., Naveed, M. & Cao, C. (2018). Microwave-assisted synthesis of graphene-like cobalt sulfide freestanding sheets as an efficient bifunctional electrocatalyst for overall water splitting. Journal of Materials Chemistry A, 6, 7592–7607. 10.1039/c8ta01266k.

[63] He, J. Z., Niu, W. J., Wang, Y. P., Sun, Q. Q., Liu, M. J., Wang, K., Liu, W. W., Liu, M. C., Yu, F. C. & Chueh, Y. L. (2020). In-situ synthesis of hybrid nickel-cobalt sulfide/carbon-nitrogen nanosheet composites as highly efficient bifunctional oxygen electrocatalyst for rechargeable Zn-air batteries. Electrochimica Acta, 10.1016/j.electacta.2020.136968.

[64] Xie, C., Wang, Q., Xiao, C., Yang, L., Lan, M., Yang, S., Xiao, J., Xiao, F. & Wang, S. (2021). Two-dimensional metal-organic framework-derived selenium-doped cobalt sulfide@graphene nanofoam for oxygen electrocatalysis. Carbon N. Y, 10.1016/j.carbon.2021.03.054.

[65] Li, Y., Yin, J., An, L., Lu, M., Sun, K., Zhao, Y. Q., Gao, D., Cheng, F. & Xi, P. (2018). FeS2/CoS2 Interface Nanosheets as Efficient Bifunctional Electrocatalyst for Overall Water Splitting. Small, 10.1002/smll.201801070.

[66] Rong, H., Ji, S., Zhang, J., Wang, D. & Li, Y. (2020). Synthetic strategies of supported atomic clusters for heterogeneous catalysis. Nature Communications, 10.1038/s41467-020-19571-6.

[67] Anantharaj, S. & Noda, S. (2020). Amorphous catalysts and electrochemical water splitting: an untold story of harmony. Small, 16, 1–24. 10.1002/smll.201905779.

[68] Wang, X. H., Ling, Y., Wu, B., Li, B. L., Li, X. L., Lei, J. L., Li, N. B. & Luo, H. Q. (2021). Doping modification, defects construction, and surface engineering: design of cost-effective, high-performance electrocatalysts and their application in alkaline seawater splitting. Nano Energy, 10.1016/j.nanoen.2021.106160.

[69] Yu, L., Wu, L., Song, S., McElhenny, B., Zhang, F., Chen, S. & Ren, Z. (2020). Hydrogen generation from seawater electrolysis over a sandwich-like NiCoN|NixP|NiCoN microsheet array catalyst. ACS Energy Lett, 5, 2681–2689. 10.1021/acsenergylett.0c01244.

[70] Wang, Q., Zhang, Z., Cai, C., Wang, M., Zhao, Z. L., Li, M., Huang, X., Han, S., Zhou, H., Feng, Z., Li, L., Li, J., Xu, H., Francisco, J. S. & Gu, M. (2021). Single iridium atom doped Ni2P catalyst for optimal oxygen evolution. Journal of the American Chemical Society, 143, 13605–13615. 10.1021/jacs.1c04682.

Xuan Wang, Sajid Liu

Chapter 4D
Nanoapplication: carbon capture and conversions

4D.1 Carbon capture

Continuous industrialization has created an exceeding amount of energy demand, leading to a yearly release of gigatons of CO_2. The predominant energy resources remain from the combustion of fossil fuels, which accumulate excessive CO_2 concentrations in the atmosphere. The cumulative climate changes, such as global warming and hypercapnia, threaten the life of every living species. This environmental concern is so severe that a significant amount of research and development has boomed in carbon capture.

Different approaches have been considered to track the increasing amount of CO_2 levels. Industries have well established the use of solvent-based techniques, aqueous carbonate, and amine solution, for example, to capture CO_2 [1]. Those aqueous alkaline solutions could degrade over time, are corrosive, and require high-cost equipment and particular handling procedure in the regeneration [2]. Solid-based techniques for carbon capture were introduced to fast and facile CO_2 capture and possible conversion to useful products like methanol or methane. In particular, the study of porous materials has become a new hot topic. Their superior properties include the large surface area, flexibility for functionalization, and reusability. The long-term solution to mitigate the harmful impact of CO_2 is to use a sustainable energy source to transform CO_2 into valuable chemicals and chemical-related products.

This chapter focuses on applying nanomaterials in CO_2 storage, utilization, and sequestration. In the first part, we investigate the fundamentals of carbon capture and review some novel classes of nanomaterials as solid sorbents. In the second part, we highlight the most recent advances in CO_2 conversion with the assistance of two renewable energies – light and electricity.

4D.1.1 Nanomaterial and CO_2

Besides the direct capture from air, another three strategies are considered for carbon capture, as shown in Fig. 4D.1. The precombustion methods involve the water-gas shift

Xuan Wang, Texas A&M Higher Education Center at McAllen, McAllen, TX, USA,
e-mail: xuan.wang@science.tamu.edu
Sajid Liu, The Department of Chemistry, Texas A&M University-Kingsville, MSC161, 700
University Blvd., Kingsville, TX, USA

https://doi.org/10.1515/9783110739879-013

reaction, which increases the percentage of CO_2 in the stream at high pressure, which is more suitable for carbon capture. Therefore, this method has been widely implemented in integrated coal gasification combined cycle plants. Post-combustion relies on separating CO_2 and N_2 after fuel gas combustion, demanding high-performance absorbers. The third method requires separating O_2 from the air through cryogenic distillation and then using pure O_2 in the sequential combustion step to drive more extensive CO_2 contents in the output stream.

There are three sorption mechanisms: physisorption, chemisorption, and moisture-swing sorption [4]. Physisorption is based on weaker van der Waals or ion–dipole interactions between CO_2 and sorbents. Therefore, porous sorbents with a large surface area give rise to more interaction sites with gas, and the inclusion of charged particles could also induce a high affinity for CO_2. Since these interactions are weak, it is easier to regenerate the sorbent for future uses. On the other hand, chemisorption occurs when CO_2 reacts with sorbents to form a chemical bond. For example, oxides, hydroxides, and alkaline salts react with CO_2 to form carbonates [3, 5]. Besides, amine scrubbing is the most predominant technology in the industry. The interactions between CO_2 and primary, secondary, or tertiary amines or aqueous ammonia drive the carbon capture process. In those chemisorption processes, the chemical bonds are more robust than those of physisorption. Thus, chemical bonding demands much energy to recover the sorbents. In the amine-scrubbing method, the solution will be reheated to high temperatures with steam to recover both CO_2 and amines. One drawback of this process is the intensive demand for heat for water [3]. Another safety concern arises due to the corrosive nature of amine solutions. Besides, some amine solutions could be unstable at high temperatures, constrain the regeneration temperature, and cause loss from evaporation.

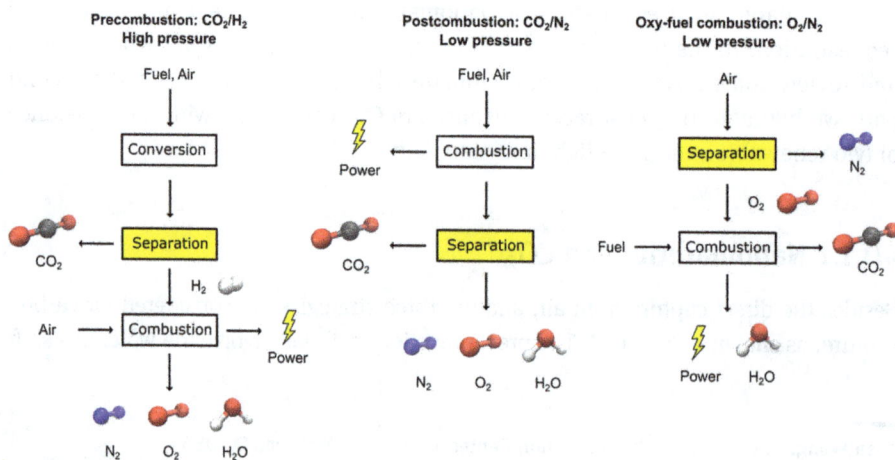

Fig. 4D.1: General scheme of CO_2 capture during precombustion (left), Post-combustion (middle), and oxy-fuel combustion (right). Adapted from ref. [3]. Copyright @ 2018 American Chemical Society.

Extensive research has focused on developing microcrystalline solid-based sorbents as the new class of sorbents demonstrated in Fig. 4D.2. Microporous porous materials include linear polymers, amorphous networks, network polymers, molecular solids, and crystalline frameworks. Crystalline frameworks and amorphous networks have been extensively studied among all the microporous materials. Crystalline frameworks, such as zeolites and zeolitic imidazolate frameworks, metal-organic frameworks (MOFs), and covalent organic frameworks, inherit well-defined pore structures. Pore engineering allows the pore size adjustment and the grafting CO_2-philic groups to improve the overall CO_2 capacity. In comparison, amorphous networks, such as activated carbon and carbon molecular sieves, do not contain well-organized pores and have also been explored. Studies on materials with crystalline and amorphous parts focus on metal oxides. The metal oxides capture CO_2 through chemical bonding, which could be energy-consuming to regenerate the sorbents. At the same time, the crystalline framework and carbon-based networks capture CO_2 predominantly through physical interactions; this makes it more feasible to regenerate the sorbents with reasonable amounts of energy [4].

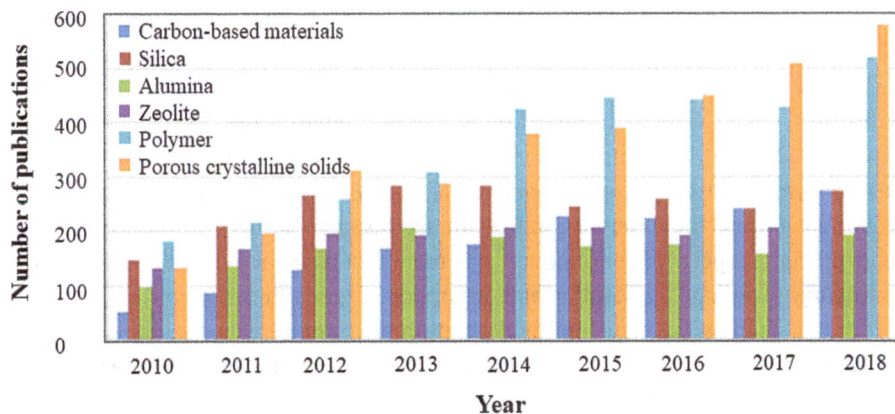

Fig. 4D.2: An increasing number of publications reporting various porous materials since 2010. Reproduced from ref. [6]. Copyright @American Chemical Society 2019.

The general procedure for CO_2 sorption with solid-based sorbents includes the following: (1) the solid sorbents adsorb on the surface through physical or chemical interactions; (2) the CO_2-rich materials get regenerated through temperature or pressure swing system while liberating CO_2. Urgent care of CO_2 emission is a global problem requiring new generations of regenerative sorbent. In practice, regenerable sorbents are desired to capture the massive CO_2 emissions. In this case, sorbent's regeneration energy in a desorber becomes critical [4]. The regeneration process typically involves the swing of sorbents and desorber. The swing operation could be driven by temperature, pressure, or both. In the pressure swing process, a vacuum is commonly used when there is a difference in partial pressures between the absorber and the desorber.

The CO_2 sorption works need to consider the maximum capacity, weight percent of uptake, the heat of adsorption, and selectivity of CO_2 over other species in the gas streams. Researchers use single-component gas adsorption isotherms at different temperatures to determine the isosteric heats of adsorption. This heat of adsorption could later be applied to the ideal adsorbed solution theory (IAST) to evaluate the gas selectivity of each sorbent under a mixed gas stream.

Another common method is breakthrough simulation. In temperature swing sorption, sorbents could favor CO_2 capture at one temperature, while a higher temperature enhances the release of CO_2. When a mixed gas stream goes through the packed beds filled with solid sorbents, the preferred gas stays in the sorbents while the less desired gas exits from the bed. A cycle of adsorption and desorption is achieved by lowering and raising the system temperature. Here, researchers calculated the dimensionless time, the ratio between the actual time to breakthrough divided by the contact time between gas and sorbents. The longer the dimensionless time is, the better selectivity the sorbent obsesses.

Current ongoing studies discovered a couple of checkpoints for sorbents' most desired features. Sorbents are expected to demonstrate high selectivity of CO_2 over other nontarget gases present in the emission stream. Therefore, the adsorption/absorption enthalpy of CO_2 capture should be lower than the standard value. The desired sorbent should display high maximum capacity, excellent chemical and thermal stability, low cost of synthesis, nontoxic and noncorrosive nature, and high reusability. Here, the working capacity, the difference in CO_2 uptake between sorbent and desorber, tells more practical aspects of performance. It is more cost-efficient to use well-established technology to operate the sorption and desorption of gas with the desired sorbents [4].

In this chapter, we focus on highlighting representative examples from zeolites, MOFs, carbon-based materials, porous coordination polymers, and other nanomaterials in carbon capture. The CO_2 emission by source and representative structures of porous materials are summarized in Fig. 4D.3. Those porous materials not only bear the intrinsic physisorption but also could be developed through two strategies to improve the CO_2 uptake. Efforts have been reported to increase the surface area of porous sorbents and increase CO_2 affinity through surface functionalization to sorb chemically. Many other studies have reported an exhaustive list of materials in each category [4, 7].

4D.1.2 Activated carbon

Activated carbon is known for its easy synthesis, low cost, and thermal stability. Various kinds of feedstock biomass could pyrolyze through physical and chemical activation methods to produce the activated carbon. The study reveals that the CO_2 capture of activated carbon is very sensitive to pressure change [8]. Activated carbon

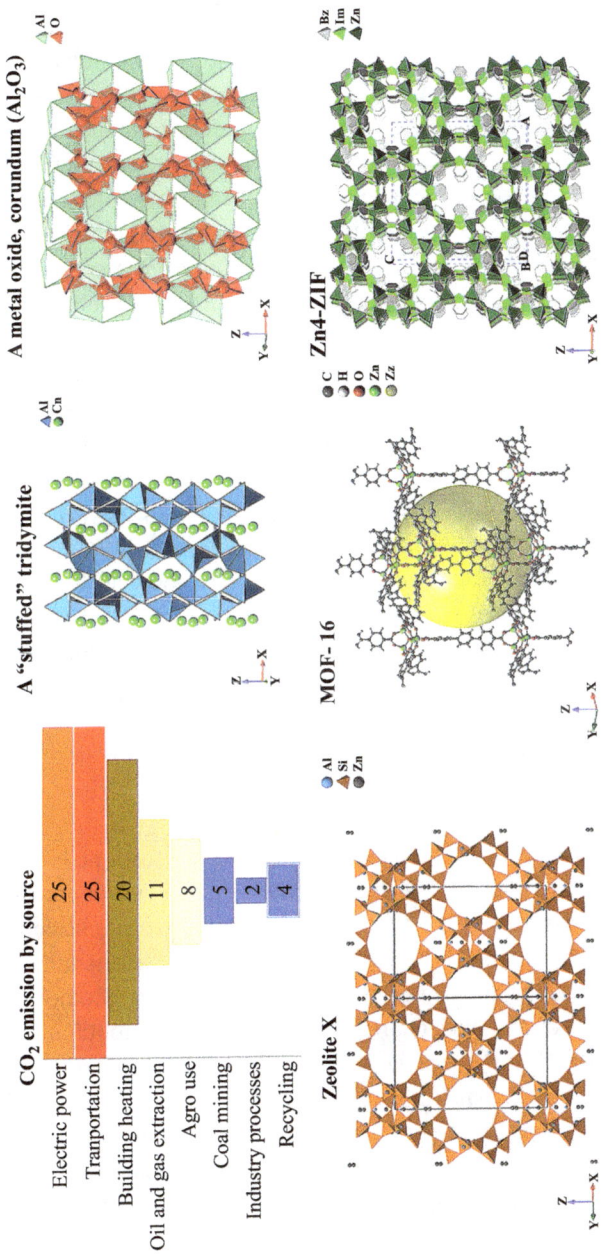

Fig. 4D.3: A summary of CO_2 emissions by sources and commonly used porous materials for CO_2 capture with high thermal stability and high uptake.

demonstrated a low adsorption capacity at low pressures. The capacity significantly increased at higher pressures. This pressure sensitivity allows the regeneration of activated carbon through pressure-swing sorption. Considering the acidic nature of CO_2, the decoration of basic functional groups on the carbon surface should enhance the interactions between CO_2 and activated carbon [9]. Though activated carbon has a relatively lower CO_2 capacity than other porous materials, the low heat of sorption makes it easier for regeneration. Due to the relatively low cost of materials, engineers used activated carbon to capture 65 tons of CO_2 per day from a coal-fired power plant through a vacuum swing and a two-stage process [10]. The total cost of the process per ton of CO_2 capture is lower than the result from economic analysis. Water exists as one by-product of any three combustion methods (Fig. 4D.1). Though there could be water separators before the regeneration stage, the existing stream may still be saturated with water, significantly affecting the reuse of any sorbents in the following cycles. Studies found that moisture is the most favorable gas adsorbed in activated carbon compared to CO_2, methane, CO, nitrogen, and hydrogen, which could adversely influence the efficiency of CO_2 capture [11]. Besides, different research groups have been working on improving the CO_2 capture and selectivity of activated carbon through modification of synthetic methods [12], and activation methods and activation agents [13]. Though those inexpensive carbon materials are robust, active carbon overall lacks the specificity toward CO_2.

4D.1.3 Zeolites

Zeolites, a class of aluminosilicates of metal ions, are common commercial materials known as molecular sieves as drying agents, detergent adsorbents, and water and air purifiers. The negatively charged framework requires cations, like sodium, lithium, calcium, and barium, for charge balance. The readily exchangeable cations induce more vital attractions to nonpolar CO_2 molecules, resulting in CO_2 capture. Studies on zeolites, like 13X, X, and Y, reveal that size, charge, density, cation distribution, humidity, and temperature could affect CO_2 uptake [14]. The dominant sorption is primarily between the cations and CO_2 and the affinity from hydrogen bonding of silanol groups on the surface. Typically, zeolites with larger pore diameters enclose higher CO_2 capacity. As a result, zeolite 13X displayed the largest capacity among those studies on pure zeolites [15]. In addition, increasing amounts of water content and higher temperatures prevent the adsorption of CO_2 [14b, 16]. We also need to consider the cost analysis for practical use. In Wiley's study, zeolites like 13X with a working capacity of 2.2 mol/kg and CO_2/N_2 selectivity of 54 demonstrated that $51 per ton of CO_2 was avoided through vacuum desorption [17].

Compared to other materials, solely zeolites yield neither good sorption capacities nor good selectivity over N_2. One way to remedy these problems is to modify the surface by introducing chemisorption, such as amine-grafting strategies (Fig. 4D.4).

Impregnation of monoethanolamine on the zeolite 13X increased the CO_2 capacity and improved the selectivity, particularly at a higher temperature [18]. Moreover, the Ahn group used a mesopore-generating agent to prepare mesoporous zeolite 13X and then impregnate the polyethyleneimine to enhance further the performance of the hybrid materials in both capture and selectivity [19]. Du and coworkers also reported similar efforts on mesoporous ZSM-5 with tetraethylenepentamine [20].

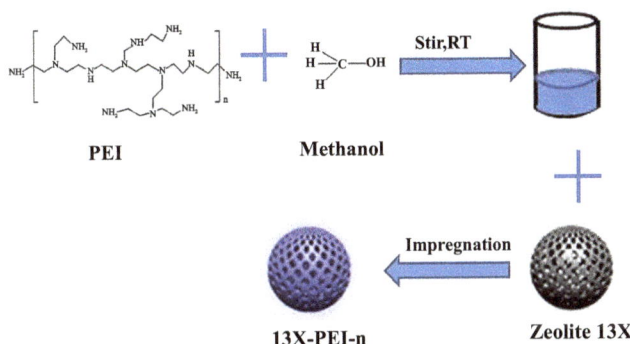

PEI   **Methanol**

13X-PEI-n   **Zeolite 13X**

Fig. 4D.4: Synthetic scheme of amine-modified zeolite 13X. Reproduced from ref. [21]. Copyright @ American Chemical Society 2019.

Like activated carbon, the CO_2 capacity of zeolite in the presence of water is also poor. Since the framework is highly charged, the electrostatic interactions make it en-ergy-deficient and time-intensive to regenerate to maintain the adsorption capacity.

4D.1.4 MOF series

MOF-74 is one of the novel frameworks, which resulted from divalent metal cations and H_4dobdc (1,4-dioxide-2,5-benzene dicarboxylic acid) [22]. Among all the iso-structural frameworks, the Mg-based framework features a high density of exposed metal ions (Fig. 4D.5), which show a high affinity to CO_2 [23]. Mg-MOF-74 absorbs 5.28 mmol/g CO_2 at 0.15 bar and 40 °C and a volumetric capacity of 4.83 mmol/cm^3. The isosteric heat of adsorption of MOF-74 is around 42 kJ/mol, and this value reveals that one CO_2 molecule is on every Mg cation. More CO_2 loading will significantly de-crease the heat of adsorption. In the natural purification studies, researchers found that Mg-MOF-74 is superior to zeolite 13X [24], evident by the high dynamic gravimet-ric capacity of CO_2 and faster CH_4 adsorption kinetics. In addition, Mg-MOF-24 also displayed mild regeneration condition with full activation at 80 °C [25]. The Glier group built an integrated model to mimic the application of porous sorbents in a su-percritical pulverized coal power plant [26]. They found that the net plant efficiency of MOF-based sorbents is the best compared to amine-based sorbents in both solid and liquid systems.

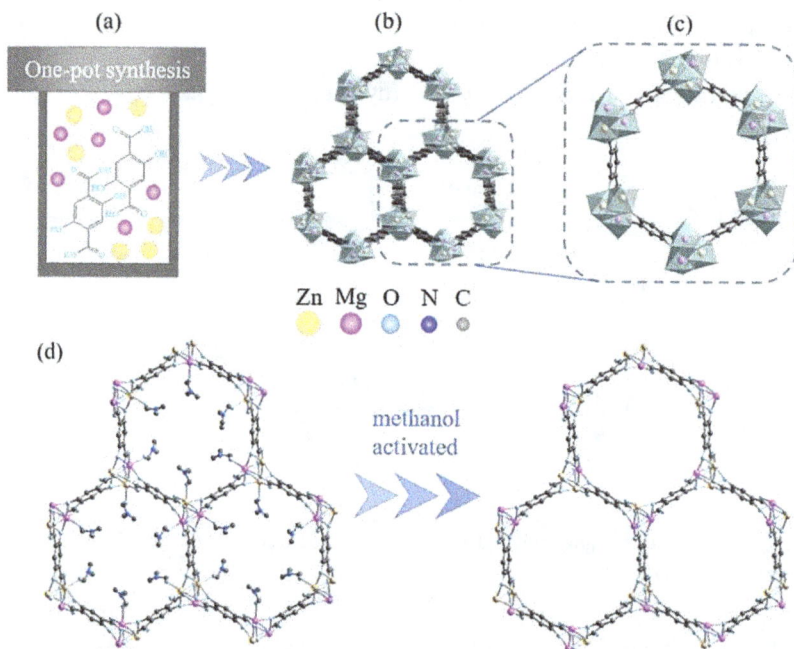

Fig. 4D.5: MOF-74 is constructed from H2dobdc and Mg or Zn with 1D channels enclosed with open metal sites. Reproduced from ref. [27]. Copyright @ American Chemical Society 2021.

Besides the unsaturated metal coordination sites, introducing polar functional groups in situ or post-synthetically could also attract more CO_2 into the frameworks. Later, Long and coworkers improved the CO_2 uptake by ligand extension and post-synthetical modification [28]. The isoreticular analog of Mg-MOF-74 obsesses a larger channel size, which allows the diamine functionalization on the open metal sites. This amine-functionalized Mg-MOF-74 demonstrates a large CO_2 capacity and a high selectivity in the dry air and flue gas with faster kinetics.

In another study, the Ahn group adjusted the ultrasonic power levels in the synthesis of MOF to control the catenation level of the framework [29]. The CO_2 capacity is increased in the catenated framework CuTATB-60 (TATB = 4,4′,4″-s-triazine-2,4,6-triyltribenzoate, and 60 is the power level) compared to the noncatenated framework.

One of the drawbacks of MOF sorbents is the metal–ligand interactions. Those co-ordination bonds are usually not strong enough, so most frameworks are sensitive to the moisture in the fuel stream. Many of the frameworks collapse in the presence of water. The construction of MOFs with large surface area or post-synthetic functionalization is a multistep synthesis, which is expensive and often requires a toxic solvent.

4D.1.5 Porous organic polymer (POP)

Porous organic polymers consist of covalent bonding between lightweight elements like C, N, and H. Like MOFs, POPs also demonstrate good flexibility for molecular design on pore sizes. Organic synthesis through Suzuki–Miyaura homo/coupling reaction, Friedel–Crafts arylation, Schiff-base reactions, Yamamoto-type Ullmann cross-coupling reaction, and Sonogashira–Hagihara reactions provide various opportunities for the adjustable pore sizes from simple phenyl rings to arenes. Moreover, the careful design of a rigid framework endorses the network with high surface area, low density, and good stability.

Similar to other porous species, POPs also have a large surface area. The porous polymer network (PPN-4) has a BET surface area of 6,461 m^2/g with CO_2 uptake of 2,121 mg/g (38.86 mmol/g) at 295 K and 50 bar [30]. Another class example, porous aromatic framework (PAF-1), displayed 29.55 mmol CO_2/g at 298 K and 40 bar, but 1.09 mmol CO_2/g below 1 bar [31]. It can be explained that there was too much dead volume in the framework at low pressure, and the low capacity at low pressure hinders the practical use of those materials.

High surface area is not the only key leading to a high CO_2 capacity of sorbent. Instead, introducing the polar functional groups on the surface will reduce the dead volume of large pore size and provide additional chemical sorption, ultimately increasing the packing efficiency. An excellent example is the modification of robust PAF-1 (also known as PPN-6). The treatment of PPN-6 through aromatic sulfonation and ensuing lithiation gave PPN-6-SO_3H, and the latter could be further neutralized to PPN-6-SO_3Li [32]. This modification reduced the BET surface from 4,023 to 1,254 to 1,186 m^2/g. Mesopores in the original framework PPN-6 mostly disappeared. However, the CO_2 uptake at 295 K and 1 bar increased from 1.16 to 3.60 mmol/g (PPN-6-SO_3H) to 3.70 mmol/g (PPN-6-SO_3Li). The heat of adsorption increased from 17 kJ/mol of PPN-6 to 35.7 kJ/mol of PPN-6-SO_3Li. Besides, the IAST calculation also indicated an excellent selectivity of CO_2 over N_2 for the functionalized PPN-6 at 1 bar. To further utilize the robust framework and improve the performance of this polymer, the same group of researchers tethered polyamines onto the framework [33]. Among all the amines, the introduction of diethylenediamine significantly increased the capacity to 4.3 mmol CO_2/g (15.8 wt%) at 295 K and 1 bar.

4D.2 Transformation of CO_2 to fine chemicals

As important as carbon capture, the storage and conversion of CO_2 need to be addressed in the whole carbon-neutral cycle. The captured greenhouse gas carbon dioxide (CO_2) has been stored underground or in the sea. The risks are leakage and change in the pH of water and earth, which potentially threaten the entire ecosystem.

Therefore, the utilization of CO_2 into useful products presents a feasible way to achieve a sustainable CO_2 value chain. We expect to convert the cheap and abundant CO_2 to liquid fuels and chemicals. We focus on applying porous materials to the two types of CO_2 conversion: photocatalytic and electrocatalytic reduction of CO_2 (Fig. 4D.6). One example is using solar-powered thin-film devices to facilitate a photoelectrochemical reaction. The cell is made of transition metal oxide films (e.g., TiO_2) to absorb the ultraviolet (UV), visible, and infrared (IR) portions. Using this technique, the conversion of CO_2 into fuels showed many benefits, such as efficient conversion, high efficiency to use sustainable energy, inexpensive operations, reduced emission, and solar energy and the power source.

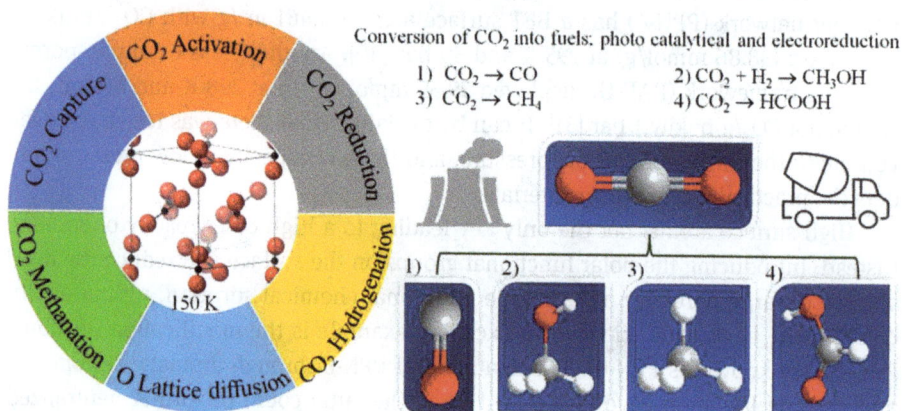

Fig. 4D.6: The CO_2 conversion into the fuel supplies using photocatalytic reduction and electroreduction.

4D.2.1 Photocatalytic reduction of CO_2

Mimicking photosynthesis, the photoreduction of CO_2 requires a photosensitizer, a photocatalyst, and the starting material CO_2. Porous materials provide a way to condense CO_2 in the network and facilitate active centers to use CO_2 as feedstocks.

4D.2.1.1 Zeolites

In most cases, all those porous materials potentially serve as the porous support or the host to build heterogeneous photocatalysts. However, the building blocks of zeolites are limited compared to those of MOFs and POPs. The charged nature of zeolites makes it an excellent platform to stabilize catalytic active radicals and ions and trap electrons [34]. Immobilization of photocatalytic centers into zeolites provides good charge transfer properties and ultimately facilitates CO_2 reduction. For example, the encapsulation of semiconductor polythiophene [35], titanium, and TiO_2 nanoparticles [36] in zeolites provides excellent electronic properties, consequently improving photoreduction. Another approach to enhance the hydrophilicity of zeolites is by introducing hydrophilic anions [37]. The framework presents various degrees of hydrophilic properties, which could attract water during CO_2 hydrogenation reactions and direct the final product based on preference.

4D.2.1.2 MOFs

There are three potential catalytic sites in MOFs: ligand, open metal sites, and defects on the metal cluster. In the early study of MOFs, many MOF-derived catalytic sites were reported on the linker part. Most photocatalysts are neither chemically inert nor thermally stable, so they will not survive the synthetic condition of MOF. Therefore, those catalysts must be introduced into the framework post-synthetically. The bipyridine-modified UiO-67 binds to $Ru(CO)_3Cl$ through the two nitrogen atoms on the framework (Fig. 4D.7) [38]. As a result, this MOF became photoactive to reduce CO_2 to CO with a turnover number of 5 in 6 h under UV light. This MOF system was grown onto Ag nanotubes to make a photocatalyst composite, which resulted in a sevenfold enhancement of CO_2-to-CO reduction under visible light [39]. Similar approaches based on ligand-to-metal charge transfer have been reported on many other MOF systems [40]. One of the best photocatalytic MOFs is NNU-28 [41]. The two building blocks of NNU-28, the Zr_6O cluster and the anthracene-based ligand, catalyzed the visible-light-driven CO_2 reduction to formate. Thus, NNU-28 displayed a TON of 18 over 10 h. More recent studies are exploring the porphyrinic MOFs on the conversion of CO_2. A comprehensive list of MOF-based photocatalysts can be found in these reviews [7g, 42].

4D.2.1.3 POPs

In nature, phthalocyanine- and porphyrin-based chemicals approve to facilitate photocatalytic CO_2 fixation. Researchers use electron-rich moieties to construct multiple POPs with microporosity and mesoporosity [43]. POPs with an inbuilt metal center become excellent heterogeneous catalysts for CO_2 conversion. In particular, the combination of Lewis acid metal center and nucleophile nitrogen-rich rings greatly facilitates the cycloaddition of CO_2 with a wide range of epoxides [44]. A Ru(III) complex was supported on a microporous POP in another study. The resultant composite presented a TON of 2,254 in the hydrogenation of CO_2 into formic acid in the presence of H_2 [45].

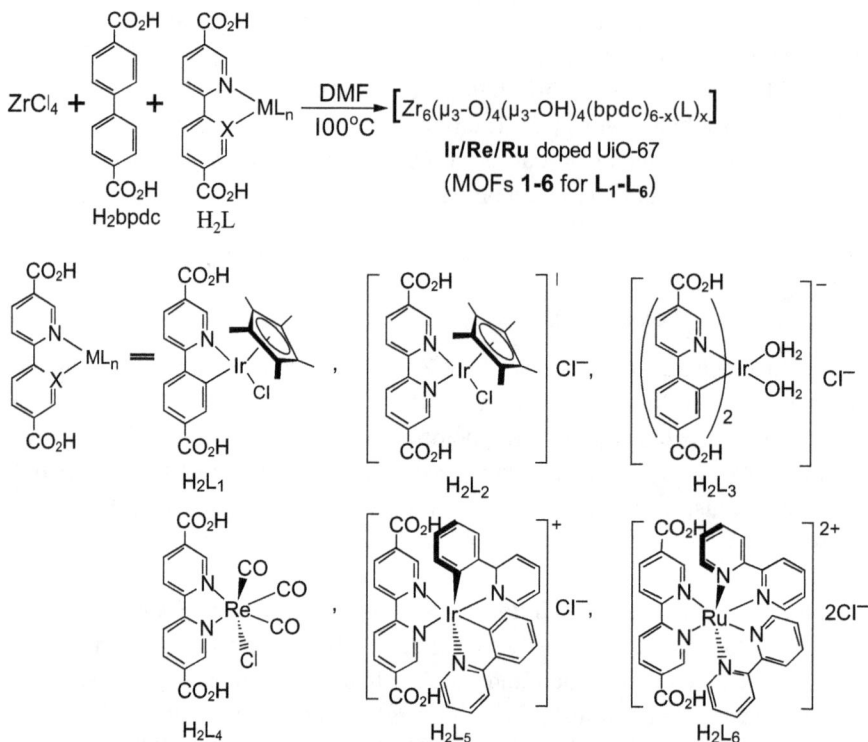

Fig. 4D.7: Synthetic preparation of doped UiO-67. Reproduced from ref. [38]. Copyright @ American Chemical Society 2011.

4D.2.2 Electrocatalytic reduction of CO_2

As most leading countries encourage the use of renewable energy resources, the study of electrical energy has surged recently due to the high environmental compatibility. In the electroreduction reaction, an overpotential is required to obtain the substantial radical intermediate. Here CO_2 reacts with water to produce CO, formate, HCHO, CH_3OH, and CH_4 [46]. Porous solid materials as electrocatalysts provide access to bulk materials so that the reduction could occur on the surface and in the pore. The physical separation prevents the deactivation of the active center.

4D.2.2.1 Zeolite

The structural nature makes zeolite with poor electronic conductivity, which hinders the further study of using zeolites solely as the electrocatalyst of CO_2 reduction. Zeolites have to hybridize with other active sites to catalyze the reduction reaction, such as transition metals. One Ni-zeolite catalyst improved the CO_2 conversion from 2 % to 14 %, which could be further enhanced by doping Ce to form CeO_2 in the framework [47]. The selectivity of the final product over CO_2 should be considered.

4D.2.2.2 MOFs

Along with the development of MOFs, the study of the electronic properties of those materials started on Cu-based MOFs [48] (Fig 4D.8a). In both works, the Cu-MOFs presented high selectivity for the conversion of CO_2. The catalysis started by reducing Cu (II) to Cu(I), then the electron is transferred to CO_2, so the faradic efficiency for both instances was low. To improve the efficiency of electroreduction, researchers further templated Cu-based MOF-74 to produce Cu nanoparticles [49] (Figure 4D.8b). Another approach is to use Cu-based porphyrinic MOF nanosheets [50]. The electron transfer occurred between the copper in the copper paddlewheel and the porphyrin–Cu(II) complex. As a result, this composite significantly enhanced the formate production with a faradic efficiency of 68.4 % and the acetate generation through C–C coupling with a faradic efficiency of 38.8–85.2 %.

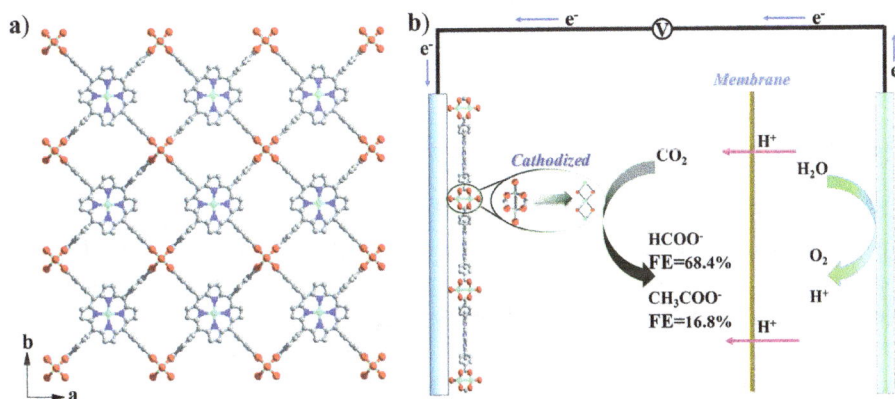

Fig. 4D.8: (a) Structure of MOF-based photocatalyst nanosheets. (b) The setup for the electrochemical transformation of CO_2 to formate and acetate with MOF-based nanosheets on the electrodes. Adapted with permission from ref. [49]. Copyright @2019 Royal Society of Chemistry.

Following this trend, many other porphyrinic MOFs have been reported as potential materials for CO_2 electrocatalysis. Porphyrinic Fe-MOF-525 was deposited onto the electrode, which concentrated the surface coverage of the metalloporphyrin catalysts [51]. As a result, the electroreduction of CO_2 showed the formation of CO and H_2 (1:1 ratio) with a faradic efficiency of 100 %. After increasing the current density, a TON of 1,520 over 3.2 h was obtained. Another parallel study used the Al-based Co-porphyrinic MOF to make a thin film [52]. The thin film on the electrodes allows mass and charge transfer to transform CO_2 into CO. The improvement was evident by a faradic efficiency of

76 % and a TON of 1,400 per site over 7 h. In both cases, the metal at the center of the porphyrin linker went through an oxidation state change.

4D.2.2.3 POPs

POPs with electron-donating N-sites are suitable for loading metallic and metal oxide nanoparticles, and the resulting system is a good instance to carry the electrochemical conversion of CO_2 [53]. It is found that both the mobile and fixed ions with the polymer affected the activation of CO_2. Similarly, pyrimidine-based POPs and graphene resulted in a new composite [54]. This system displayed electrochemical activity toward CO_2 reduction, which was not detected on individual components due to the low CO_2 capacity and poor conductivity. In another study, tetrazine-based POPs selectively transformed CO_2 to CO in water with a faradic efficiency of 82 % with an overpotential of 560 mV [55]. The Yoon group introduced Ir (III)-N-heterocyclic carbene to the tetrazine-based POPs [56]. The strong σ-donating and poor π-accepting nature of N-heterocyclic carbene contributed to an electron-rich environment for Ir (III), which ultimately led to outstanding electrocatalysis. The heterogenized complex converted CO_2 to formate with a TON for 16,000 h^{-1}.

4D.3 Conclusion and outlook

CO_2 is released in most industrial operations, and human beings have suffered from the consequence of dramatic climate changes. Current societal pressure for CO_2 fixation and conversion spurs more research opportunities to discover, develop, and deploy cost-efficient and environmental-friendly technologies. As commercial availability of carbon capture and conversion is limited, this chapter highlighted the current advance of solid-state sorbents in CO_2 storage and conversion. We mainly focus on porous materials – zeolites, MOFs, and porous organic polymers. The focus of this field is now transferring from the fundamental study of adsorption to practical considerations. We have witnessed many research studies on implementing those materials in real plants. There is a trend to study the synergic effect of porous materials and other catalytic species in CO_2 conversion. However, few of them have been tested on a production line on an industrial scale. Most of the studies have been published on the simulated flue gas, and a few studied the water effects in the adsorption. Considering that the real flue gas consists of water and other acidic gas, more studies should be encouraged to explore those effects. From the economic point of view, the expensive cost of MOFs and POPs is one of the leading reasons hindering their use in real situations. We expect more interdisciplinary collaboration across chemistry, physics, materials, process engineering, chemical engineering, environmental engineering, and politicians. A strategic partnership between all countries is strongly needed to take on this challenge, and it is a global problem.

References

[1] a. Rochelle, G. T. (2009). Science, 325, 1652–1654. b. Yu, W., Wang, T., Park, A.-H. A., Fang, M., Nanoscale 2019, 11, 17137–17156.

[2] a. Veawab, A., Tontiwachwuthikul, P. & Chakma, A. (1999). Industrial & Engineering Chemistry Research, 38, 3917–3924. b. Rao, A. B., Rubin, E. S., Environmental Science & Technology 2002, 36, 4467–4475.

[3] Kupgan, G., Abbott, L. J., Hart, K. E. & Colina, C. M. (2018). Chemical Reviews, 118, 5488–5538.

[4] Shi, X., Xiao, H., Azarabadi, H., Song, J., Wu, X., Chen, X. & Lackner, K. S. (2020). Angewandte Chemie International Edition, 59, 6984–7006.

[5] a. Nikulshina, V., Gálvez, M. E. & Steinfeld, A. (2007). Chemical Engineering Journal, 129, 75–83. b. Nikulshina, V., Steinfeld, A., Chemical Engineering Journal 2009, 155, 867–873; c. Zeman, F., Lackner, K. S., World Res. Rev. 2004, 16, 157–172; d. Zhao, S., Ma, L., Yang, J., Zheng, D., Liu, H., Yang, J., Energy & Fuels 2017, 31, 9824–9832.

[6] Pardakhti, M., Jafari, T., Tobin, Z., Dutta, B., Moharreri, E., Shemshaki, N. S., Suib, S. & Srivastava, R. (2019). ACS Applied Materials & Interfaces, 11, 34533–34559.

[7] a. Dirar, Q. H. & Loughlin, K. F. (2013). Adsorption, 19, 1149–1163. b. He, X., Energy, Sustainability and Society 2018 8, 34; c. Kumar, S., Srivastava, R., Koh, J., Journal of CO_2 Utilization 2020, 41, 101251; d. Megías-Sayago, C., Bingre, R., Huang, L., Lutzweiler, G., Wang, Q., Louis, B., Frontiers in Chemistry 2019, 7; e. Murge, P., Dinda, S., Roy, S., Langmuir 2019, 35, 14751–14760; f. Piscopo, C. G., Loebbecke, S., ChemPlusChem 2020, 85, 538–547; g Trickett, C. A., Helal, A., Al-Maythalony, B. A., Yamani, Z. H., Cordova, K. E., Yaghi, O. M., Nature Reviews Materials 2017, 2, 17045.

[8] Siriwardane, R. V., Shen, M.-S., Fisher, E. P. & Poston, J. A. (2001). Energy & Fuels, 15, 279–284.

[9] a. Pevida, C., Plaza, M. G., Arias, B., Fermoso, J., Rubiera, F. & Pis, J. J. (2008). Applied Surface Science, 254, 7165–7172. b. Plaza, M. G., Pevida, C., Arias, B., Casal, M. D., Martín, C. F., Fermoso, J., Rubiera, F., Pis, J. J., Journal of Environmental Engineering 2009, 135, 426–432.

[10] Chen, E. R., Katie, & Stillman, Z., Senior Design Reports (CBE) 2017, 89.

[11] Lopes, F. V. S., Grande, C. A., Ribeiro, A. M., Loureiro, J. M., Evaggelos, O., Nikolakis, V. & Rodrigues, A. E. (2009). Separation Science and Technology, 44, 1045–1073.

[12] Sibera, D., Narkiewicz, U., Kapica, J., Serafin, J., Michalkiewicz, B., Wróbel, R. J. & Morawski, A. W. (2019). Journal of Porous Materials, 26, 19–27.

[13] Wang, R., Wang, P., Yan, X., Lang, J., Peng, C. & Xue, Q. (2012). ACS Applied Materials & Interfaces, 4, 5800–5806.

[14] a. Bertsch, L. & Habgood, H. W. (1963). The Journal of Physical Chemistry, 67, 1621–1628. b. Brandani, F., Ruthven, D. M., Industrial & Engineering Chemistry Research 2004, 43, 8339–8344; c. Gallei, E., Stumpf, G., Journal of Colloid and Interface Science 1976, 55, 415–420; d. Ward, J. W., Habgood, H. W., Journal of Physical Chemistry 1966, 70, 1178–1182.

[15] Siriwardane, R. V., Shen, M. S. & Fisher, E. P. (2005). Energy & Fuels, 19, 1153–1159.

[16] Son, K. N., Cmarik, G. E., Knox, J. C., Weibel, J. A. & Garimella, S. V. (2018). Journal of Chemical and Engineering Data, 63, 1663–1674.

[17] Ho, M. T., Allinson, G. W. & Wiley, D. E. (2008). Industrial & Engineering Chemistry Research, 47, 4883–4890.

[18] Jadhav, P. D., Chatti, R. V., Biniwale, R. B., Labhsetwar, N. K., Devotta, S. & Rayalu, S. S. (2007). Energy & Fuels, 21, 3555–3559.

[19] Chen, C., Kim, -S.-S., Cho, W.-S. & Ahn, W.-S. (2015). Applied Surface Science, 332, 167–171.

[20] Wang, Y., Du, T., Song, Y., Che, S., Fang, X. & Zhou, L. (2017). Solid State Sciences, 73, 27–35.

[21] Karka, S., Kodukula, S., Nandury, S. V. & Pal, U. (2019). ACS Omega, 4, 16441–16449.

[22] Dietzel, P. D. C., Morita, Y., Blom, R. & Fjellvåg, H. (2005). Angewandte Chemie International Edition, 44, 6354–6358.

[23] Mason, J. A., Sumida, K., Herm, Z. R., Krishna, R. & Long, J. R. (2011). Energy & Environmental Science, 4, 3030–3040.

[24] Bao, Z., Yu, L., Ren, Q., Lu, X. & Deng, S. (2011). Journal of Colloid and Interface Science, 353, 549–556.

[25] Britt, D., Furukawa, H., Wang, B., Glover, T. G. & Yaghi, O. M., Proceedings of the National Academy of Sciences 2009, 106, 20637.

[26] Glier, J. C. & Rubin, E. S. (2013). Energy Procedia, 37, 65–72.

[27] Gao, Z., Liang, L., Zhang, X., Xu, P. & Sun, J. (2021). ACS Applied Materials & Interfaces, 13, 61334–61345.

[28] McDonald, T. M., Lee, W. R., Mason, J. A., Wiers, B. M., Hong, C. S. & Long, J. R. (2012). Journal of the American Chemical Society, 134, 7056–7065.

[29] Kim, J., Yang, S.-T., Choi, S. B., Sim, J., Kim, J. & Ahn, W.-S. (2011). Journal of Materials Chemistry, 21, 3070–3076.

[30] Yuan, D., Lu, W., Zhao, D. & Zhou, H.-C. (2011). Advanced Materials, 23, 3723–3725.

[31] Ben, T., Ren, H., Ma, S., Cao, D., Lan, J., Jing, X., Wang, W., Xu, J., Deng, F., Simmons, J. M., Qiu, S. & Zhu, G. (2009). Angewandte Chemie International Edition, 48, 9457–9460.

[32] Lu, W., Yuan, D., Sculley, J., Zhao, D., Krishna, R. & Zhou, H.-C. (2011). Journal of the American Chemical Society, 133, 18126–18129.

[33] Lu, W., Sculley, J. P., Yuan, D., Krishna, R., Wei, Z. & Zhou, H.-C. (2012). Angewandte Chemie International Edition, 51, 7480–7484.

[34] Sakar, M. & Do, T.-O. (2019). Chemistry of silica and zeolite-based materials, vol. 2 (pp. 89–103), Eds. Douhal, A. & Anpo, M., Elsevier.

[35] Kianička, J., Čík, G., Šeršeň, F., Špánik, I., Sokolík, R. & Filo, J. (2019). Molecules, 24, 992.

[36] Tong, Y., Chen, L., Ning, S., Tong, N., Zhang, Z., Lin, H., Li, F. & Wang, X. (2017). Applied Catalysis. B, Environmental, 203, 725–730.

[37] a. Hoeven, N., Mali, G., Mertens, M. & Cool, P. (2019). Microporous and Mesoporous Materials, 288, 109588. b. Ikeue, K., Yamashita, H., Anpo, M., Takewaki, T., The Journal of Physical Chemistry B 2001, 105, 8350–8355.

[38] Wang, C., Xie, Z., de Krafft, K. E. & Lin, W. (2011). Journal of the American Chemical Society, 133, 13445–13454.

[39] Choi, K. M., Kim, D., Rungtaweevoranit, B., Trickett, C. A., Barmanbek, J. T. D., Alshammari, A. S., Yang, P. & Yaghi, O. M. (2017). Journal of the American Chemical Society, 139, and 356–362.

[40] a. Fu, Y., Sun, D., Chen, Y., Huang, R., Ding, Z., Fu, X. & Li, Z. (2012). Angewandte Chemie International Edition, 51, 3364–3367. b. Li, L., Zhang, S., Xu, L., Wang, J., Shi, L.-X., Chen, Z.-N., Hong, M., Luo, J., Chemical Science 2014, 5, 3808–3813; c. Sun, D., Gao, Y., Fu, J., Zeng, X., Chen, Z., Li, Z., Chemical Communications 2015, 51, 2645–2648; d. Wang, S., Yao, W., Lin, J., Ding, Z., Wang, X., Angewandte Chemie International Edition 2014, 53, 1034–1038; e. Zhang, S., Li, L., Zhao, S., Sun, Z., Hong, M., Luo, J., Journal of Materials Chemistry A 2015,

3, 15764–15768; f. Fei, H., Sampson, M. D., Lee, Y., Kubiak, C. P., Cohen, S. M., Inorg Chem 2015, 54, 6821–6828.

[41] Chen, D., Xing, H., Wang, C. & Su, Z. (2016). Journal of Materials Chemistry A, 4, 2657–2662.

[42] Li, X. & Zhu, Q.-L. (2020). EnergyChem, 2, 100033.

[43] a. McKeown, N. B. & Budd, P. M. (2010). Macromolecules, 43, 5163–5176. b. McKeown, N. B., Journal of Materials Chemistry 2010, 20, 10588–10597.

[44] a. Kumar, S., Wani, M. Y., Arranja, C. T., Silva, J. D. A. E., Avula, B. & Sobral, A. J. F. N. (2015). Journal of Materials Chemistry A, 3, 19615–19637. b. Sheng, X., Guo, H., Qin, Y., Wang, X., Wang, F., RSC Advances 2015, 5, 31664–31669.

[45] Yang, -Z.-Z., Zhang, H., Yu, B., Zhao, Y., Ji, G. & Liu, Z. (2015). Chemical Communications, 51, 1271–1274.

[46] Zhang, L., Zhao, Z.-J. & Gong, J. (2017). Angewandte Chemie International Edition, 56, 11326–11353.

[47] Graça, I., González, L. V., Bacariza, M. C., Fernandes, A., Henriques, C., Lopes, J. M. & Ribeiro, M. F. (2014). Applied Catalysis. B, Environmental, 147, 101–110.

[48] a. Hinogami, R., Yotsuhashi, S., Deguchi, M., Zenitani, Y., Hashiba, H. & Yamada, Y. (2012). ECS Electrochemistry Letters, 1, H17–H19. b. Senthil Kumar, R., Senthil Kumar, S., Anbu Kulandainathan, M., Electrochemistry Communications 2012, 25, 70–73.

[49] Wu, J.-X., Hou, S.-Z., Zhang, X.-D., Xu, M., Yang, H.-F., Cao, P.-S. & Gu, Z.-Y. (2019). Chemical Science, 10, 2199–2205.

[50] Kim, M. K., Kim, H. J., Lim, H., Kwon, Y. & Jeong, H. M. (2019). Electrochimica Acta, 306, 28–34.

[51] Hod, I., Sampson, M. D., Deria, P., Kubiak, C. P., Farha, O. K. & Hupp, J. T. (2015). ACS Catalysis, 5, 6302–6309.

[52] Kornienko, N., Zhao, Y., Kley, C. S., Zhu, C., Kim, D., Lin, S., Chang, C. J., Yaghi, O. M. & Yang, P. (2015). Journal of the American Chemical Society, 137, 14129–14135.

[53] a. Aeshala, L. M., Uppaluri, R. G. & Verma, A. (2013). Journal of CO2 Utilization, 3–4, 49–55. b. Haikal, R. R., Soliman, A. B., Amin, M., Karakalos, S. G., Hassan, Y. S., Elmansi, A. M., Hafez, I. H., Berber, M. R., Hassanien, A., Alkordi, M. H., Applied Catalysis B: Environmental 2017, 207, 347–357.

[54] Soliman, A. B., Haikal, R. R., Hassan, Y. S. & Alkordi, M. H. (2016). Chemical Communications, 52, 12032–12035.

[55] Zhu, X., Tian, C., Wu, H., He, Y., He, L., Wang, H., Zhuang, X., Liu, H., Xia, C. & Dai, S. (2018). ACS Applied Materials & Interfaces, 10, 43588–43594.

[56] Gunasekar, G. H., Park, K., Ganesan, V., Lee, K., Kim, N.-K., Jung, K.-D. & Yoon, S. (2017). Chemistry of Materials, 29, 6740–6748.

Robert S. Luckett Ed.D., MSW

Postface: social impact, consequences, and results of nanotechnology

Abstract: Here, we conclude on ethical questions of "if a specific task can be accomplished" or "should it be?" Dr. Luckett argues about the ethical issues related to nanotechnology: analyze, discuss, and deploy ethical discussion into the interdisciplinary approach related to nanotechnology. Nanotechnology is viewed as an application of nanoscience. Here, chemical bonding, reactivity, stability, and function are related to the atomic topography, composition, and speciation of the atoms and molecules and their manipulation to form matter, that is, one-, two-, or three-dimensional atomic, molecular, or supramolecular forms, from which we derive the intrinsic functionalities. These are different from bulk materials due to the unavailability of the majority of the atoms on the surface or within the structure, as only peripheral or surface atoms engage in catalysis, binding, dissociation, or entrapment of guest molecules.

In this regard, nanoscience is the fundamental understanding of physical law and how these laws are manipulated to generate physical systems through a top-down or bottom-up synthesizes approach, which specifies morphology, dimensionality, shape, and composition at the nanometer-level ordering. The technology aspects leverage these fundamental rules for specific purposes, and in this realm, the scope of allowable, ethical, and democratic must play a leading role. Due to the dimensionality of nanomaterials, their usage and applications are broad, diverse, and widespread and impact all aspects of our lives, from communication and medicine to production of new catalysts for environmental remediation, energy production and storage, drug delivery, or fabrication of new alloys, sensors, dyes, paints, or bioimaging contrast agents.

The thesis here is that new materials, devices, or technologies must possess an inherent benefit or advantage of the current and, as such, benefit humankind. Idea acceptance implies a degree of public trust, which is achieved through public discourse, debate, and openness. Trust is based on the concept of reasonableness. Through transparency and openness, the public becomes enlightened on its merits through education and example, and osmosis leads to acceptance. The current climate of mistrust and pseudoscience may generate skepticism and a form of neo-Ludditism in passive resistance to consumerism and the increasing intrusion of digital devices such as surveillance cameras, laptops, and cellphones into the everyday sphere of operation.

Robert S. Luckett Ed.D., MSW, Department of Social Work, Texas A&M University-Kingsville, 700 University Blvd., MSC 123, Kingsville, Texas 78363, e-mail: Robert.luckett@tamuk.edu

https://doi.org/10.1515/9783110739879-014

The ethical questions that arise from the interdisciplinary nanotechnology have parallels in biotechnology and genetics or genetically modified crops. There was an ethical concern about the generation of self-replicating systems in these areas, which could modify their code and gain new functions. Certain nanosystems can self-replicate and potentially gain new catalytic functions, such as nanobots for tissue repair. The concern is similar to an earlier concern that genetically modified corn could endanger the monarch butterfly [1]. In the present context, it could be argued that the toxicity of nanoparticles is akin to lead, mercury, and asbestos [2]. If inhaled, nanotubes, such as long nanotubes, carbon nanotubes, buckminsterfullerene, and cube structures, may trigger an immune response and cause fiber-like damage to the lungs and other tissues similar to known scar-like damage from asbestos fibers [3]. What is known is that carbon nanotubes can aggregate to form bundles that induce interstitial inflammation and scarring [4]. Like gene therapy, which can restore loss of function or gain a new function, nanoparticles may treat disease, enhance the immune system, and vision beyond the visible spectrum or increase stamina [5], limiting such opportunities to elites who can afford them. Like human subject research, the ethical debate is if these threads should be universal or specific to human subjects related to the principles of respect for autonomy, the principles of beneficence, nonmaleficence, justice, and respect for integrity [6], where nanotechnology impinges on human wellness and the nature of humanity (propensity to become sick, mortal, loss of function due to age, limited lifespan). If nanotechnology coupled with gene insertion could abolish these traits and humans become sickness-free, immortal, never lost any functions (sight, hearing, ability to think better, augmented memory, and retain or gain athleticism regardless of chronological age), should these be available to all or only a select few who could afford these "treatments" [7]? Ethics are defined by human frailties and inevitabilities of old age, loss of function, and death. Along its current development arc, nanotechnology offers a transgression to overcome these human frailties [8]. Therefore, how relevant are the human-centric principles of autonomy, integrity, beneficence, nonmaleficence, and justice [9].

Since nanotechnology exists microscopically, manual labor has comparatively little to contribute [10], so what will be the overall human role in developing or maintaining it? Fear of technology has many angles, including concerns about job displacement, the environment, and robots taking over [11]. That could be an exaggeration, but if not a takeover, what about losing control? Humanity learned an international lesson from the impact of chlorofluorocarbons (CFCs) on the ozone layer (Solomon, 2004). Global awareness is the realization that large-scale industrial effects, which since the second world war appear to know no boundaries, have an associated cost. These costs adversely affect the environment and human or animal health, realizing that the polluters and everyone else share the same space. This has increased the tendency to question the merits of continued exploitation of natural resources. The result was a relatively quick, sharp decline in the use of freon by wealthier nations who were, and still are, most dependent on refrigeration. Ironically, freon, a CFC, was intended to protect the environment from other coolant threats in the 1930s. Alternative sources of refrigeration were developed to meet the demand [12]. However, one common replacement for CFCs, hydrofluorocarbons, is now recommended by environmentalists to be phased out for similar reasons [13]. Any benefits to profiteers for their role in saving the planet are discouraged by such setbacks.

Unfortunately, many damaging industrial practices are an even greater threat to the environment and seem to have no viable solution. The lack of trust is exemplified by nuclear crises such as Pennsylvania's Three Mile Island in 1979, Ukraine's Chernobyl in 1986, and Japan's Fukushima in 2011. It is not inappropriate to ask what the industry has learned since then. Are we now able to address these problems any better? Nuclear waste disposal has always been a risk: the possibility of widespread damage, possibly making the earth uninhabitable. In addition to greed and human error, there is also the concern of deliberate damage. Terrorists are constantly being monitored for interest in sabotage, and Ukraine suffered a Russian missile strike on their largest nuclear power plant in a deliberate act of war. If nanotechnology can bring new energy sources, can it also lead to safer environmental practices, enhanced security, and, in other ways, serve humankind?

What we learned from COVID-19

Reluctance to accept technology is nothing new. Self-serve gas pumps first appeared in the United States during the energy crisis of the 1970s, but this practice is now permanent. Many thought consumers would balk at this idea, but now it is a commonplace, overriding class, gender, and dry-cleaning expense. There are now very few people employed to fill gas tanks. No matter what job skills are required, promises of innovations provide more jobs than they displace and omit workers' critical details. For example, while it may be true that more jobs will come, they will go to those with qualifications that displaced workers may not have [10]. Some technology predicts this and avoids the need for labor from the start. Imagine what failure personal computers would have been if knowledge of Boolean language were required to operate them instead of the point-and-click world we now live in. Simplified computer operation has allowed warehouse work to grow. However, even still, few warehouse workers will be in a position to develop the skills to design, install, and operate the robotics waiting to replace them.

Given that nuclear power and freon were forced on humanity. At the same time, environmental impact was neglected, so it should not be a surprise that new technology will continue to be met with skepticism. While no reasonable person believes that COVID-19 was manufactured, then unleashed on the world, businesses have recognized opportunities to profit that would not otherwise have occurred. For example, long-neglected and staffed by an underpaid workforce, customer service had already brought the use of technology to pass tasks on to consumers. Many things that can be accomplished void of human contact by consumers have become the norm and accelerated during COVID-19 [14].

Future growth of technology's impact on self-service

Some retail outlets began installing self-serve checkout areas at the expense of space previously used for full-service checkout before COVID-19 and imposed it further during the outbreak, citing the advantages of reducing physical contact between staff and customers. Now that, self-checkout has become ubiquitous. It is likely to increase, given that full-service checkout lines were not well staffed before the pandemic, forcing people to use them instead. Fast food jobs have been among the most disrespected for decades. Are vending machines really what consumers want instead of dealing with human beings? If replacing workers takes hold, will other restaurants follow? Even restaurants that still employ people who rely on tips now incorporate some of the processes via computer: mini kiosks at tables that allow point-and-click ordering, paying, and tipping. Will all restaurants eliminate people? Will this be OK? What precedent does it set for other businesses? How does COVID-19-related unemployment factor into future customer experiences?

Technology, capitalism, and the future of working conditions

For those who did not return to low skills work after the pandemic eased, will concerns over the health impacts of low pay and high stress prompt them to learn new skills? Is it realistic to believe that those displaced from these jobs will find employment in technology that requires specialized skills at the least and credentials in higher education at the most? Will people have the time and money to retrain? Will consumers agree that self-service allows for convenience and speed and decide it outweighs the impersonality of scanning their purchases? Will they care, even if the reduction of workforces does not result in reduced labor costs but more profit instead of price cuts for consumers? Will the promise of new technologies be balanced with the pursuit of profit and fairer wages? If not, will there be enough people for profiteers to exploit to keep profits up without being fair [15]? Recent history tells us that big business will instead rely on Third World labor [15] and that attention to social justice from profiteers will be a struggle at best [16] and unusual in its success [17]. Indeed, social control continues to be the more common goal of the powerful, as witnessed by the Indian government's goal of using technology to assign unique identifying numbers to everyone possible [18].

The coal industry is old enough never to have broken the habit of high exploitation and low pay. Now that coal is being phased out internationally, it is likely too late for the better working conditions of newer jobs in energy, such as oil and natural gas, to catch up. In the United States, oil as a natural resource and First World workforce, coupled with the fast-growing auto industry, at least provided factories to work in instead of underground mines. Arguably, precious metals and diamond mining skipped complications of regulations altogether due to these resources in Third World countries [19], where more desperate people rushed to work without the luxury or regard for health concerns.

However, First World countries also continue to experience long-term effects of technology, especially in lower socioeconomic status (SES) communities [20]. Advocated for this to be explored with emphasis on the sequence of events, using the following three suggestions: (1) compare communities with and without toxic worksites, with a focus on SES, (2) explore documentation related to choosing toxic worksites, comparing changes in property values before and after the sites began operating, and (3) focus on what kind of pollutants are present, whether the waste is dumped nearby versus transported elsewhere, and if transported elsewhere, what the SES is of the area receiving the waste. Attorneys investigating cancer in and near a nuclear plant in South Carolina seek equal compensation for employees and residents living near the site [39], again reminding us that pollutants are difficult to contain.

Prioritizing low-paid workforces has not disappeared simply because we are in a period of high-tech innovation or because work now tends to be more closely related to that technology. Cell phone manufacturers in China have come under scrutiny for dangerous working conditions, but since the demand for cell phones continues, so do the dangerous working conditions [21]. In the United States, the military explores hearing loss associated with jet fuel exposure [22]. Free of most environmental restrictions, will the US military begin to lead the way in reducing emissions, or will the government hand off those responsibilities to infrastructure subcontractors of the Build Back Better program? Will grant-funded nanotech research and employment act more responsibly by comparison?

Alternative sources of fuel and the insatiable demand for energy

Fracking made its debut in Pennsylvania in 2004, targeting the natural gas of the Marcellus Shale reservoir. Being less of a polluter than oil, new access to natural gas generated excitement from big oil executives to consumers. The former hoped for greater profit, and the latter hoped for lower heating bills. The dream began a slow death when fracking's environmental impact became more widely known. The practice has been under fire ever since due to its overuse of water, another vital resource under environmental threat [23]. Cheaper is better in some ways, but if increased numbers of earthquakes are caused by fracking, as some suspect [24], then concerns over excessive water use may combine to limit further growth of fracking.

Although just recently making the rounds in news outlets, ethanol's environmental footprint has been known since at least 2005 to be worse than gasoline due to the amount of land required to produce the corn it requires. First presented as a gift to agriculture, its tiny contribution to the gasoline industry has brought only minor savings at the pump and a significant strain on resources required to process corn into the fuel supply [25].

Finally, a favorite of environmentalists, solar energy is not a perfect green option. While it may show the most promise of all alternative fuels in recent history, like any other fuel, construction, transportation, delivery, and waste associated with any energy source must be taken into account to assess whether or not it offsets its competitors. Much of the focus on solar technology is not on the sun itself but on producing batteries and heat storage captured from the sun. Solar energy remains promising, but it is not as simple as aiming panels into the sky and unplugging electric meters [26].

Making fuel from waste

Efforts to put waste to use as a fuel source are promising [27]. Currently, there are projects underway to use biodiesel wastewater [28], dairy wastewater [29], rice mill wastewater [30], seafood industrial wastewater [31], and urea [32]. When a big business claims to commit to reducing carbon emissions, such reductions only tend to be slow and minor, rather than working toward eliminating them [33].

Acceptance and growth of nanotechnology depend not only on economic considerations and ease of use but also on convincing governments and end-users that it is appropriate to replace existing sources. It needs to be reliable, affordable, safe for the environment, and credible in its protection claims from sabotage.

Conclusion

Nanotechnology is interdisciplinary that connects scientific principles (nanoscience) to practice (nanobots). However, from an ethical perspective, the relationship is not linear since nanomedicine, sensors, therapeutics, and gain of function transcends experimentation on human subjects because nanosystems can self-replicate and thus use noncarbon systems for testing. The procedural approach for human testing does not necessarily apply to nanotechnology due to the operational systems' nature and implementation. In general, most applications offer a clear benefit and, therefore, the practice of impartiality and rational decision making and the principle of nonmaleficence [34] and how far these ethical dimensions reach into personal, professional, and political realms, where a pregnant individual with a genetic may make a personal decision to terminate the pregnancy, which her doctors may agree, but is illegal under state law [35]. Here, political ethics is based on legislation, which may not factor in autonomy and justice. However, a policy design is to be supported by politically minded voters or opinion policymakers believe is correct, even when most citizens do not subscribe to the same opinion. Nanotechnology offers tremendous opportunities and is it correct for the general ethical questions to be posed around the relevance of autonomy, integrity, beneficence, nonmaleficence, and justice [36]. Since the outreach of nanotechnology and nanoscience spans all spheres of our lives, such as disease, communication, and restoring loss of function, many bioethical questions address the right to privacy, access, and autonomy issues cut across personal, professional, and political ethical beliefs. Nanotechnology will become more common, relevant, and ubiquitous [37]. We must address these questions to ender public confidence and trust in the science, its practitioners, and its regulators and do not fall into pseudoscience distractions and disagreements as was used with COVID-19 or the effect of power lines or aircraft contrails when the scientific benefits were clear cut [38]. However, the public perception was that these applications yielded cancer or limited autonomy.

References

[1]   Losey, J. E., Rayor, L. S. & Carter, M. E. (1999). Transgenic pollen harms monarch larvae. Nature, 399(6733), 214–214.
[2]   Nel, A., Xia, T., Madler, L. & Li, N. (2006). Toxic potential of materials at the nanolevel. Science, 311(5761), 622–627.
[3]   Donaldson, K., Stone, V., Tran, C. L., Kreyling, W. & Borm, P. J. (2004). Nanotoxicology. Occupational and Environmental Medicine, 61(9), 727–728.
[4]   Warheit, D. B., Laurence, B. R., Reed, K. L., Roach, D. H., Reynolds, G. A. & Webb, T. R. (2004). Comparative pulmonary toxicity assessment of single-wall carbon nanotubes in rats. Toxicological Sciences, 77(1), 117–125.
[5]   Walters, L., Palmer, J. G. & Palmer, J. G. (1997). The ethics of human gene therapy, USA: Oxford University Press.
[6]   Gillon, R. (1994). Medical ethics: four principles plus attention to scope. Bmj, 309(6948), 184.
[7]   Duprex, W. P., Fouchier, R. A., Imperiale, M. J., Lipsitch, M. & Relman, D. A. (2015). Gain-of-function experiments: time for a real debate. Nature Reviews. Microbiology, 13(1), 58–64.

[8]   Selgelid, M. J. (2016). Gain-of-function research: an ethical analysis. Science and Engineering Ethics, 22(4), 923–964.

[9]   Beauchamp, T. L. & Childress, J. F. (2001). Principles of biomedical ethics, USA: Oxford University Press.

[10]   Invernizzi, N. & Foladori, G. (2010). Nanotechnology Implications for Labor. Nanotechnology Law & Business, 7(1), 68–78.

[11]   Bainbridge, W. S. & Roco, M. C. (2006). Managing nano-bio-info-cogno innovations, Heidelberg: Springer.

[12]   Rusch, G. M. (2018). The development of environmentally acceptable fluorocarbons. Critical Reviews in Toxicology, 48(8), 615–665. https://0-doi-org.oasis.lib.tamuk.edu/10.1080/10408444.2018.1504276.

[13]   World leaders agree on a deal to cut climate-changing HFCs. (2016). *TCE: The Chemical Engineer, 905,* 10.

[14]   Li, M., Yin, D., Qiu, H. & Bai, B. (2022). Examining the effects of A.I. contactless services on customer psychological safety, perceived value, and hospitality service quality during the COVID-19 pandemic. Journal of Hospitality Marketing & Management, 31(1), 24–48. https://0-doi-org.oasis.lib.tamuk.edu/10.1080/19368623.2021.1934932.

[15]   Bainbridge, W. S. (2006). Ideological differences. Nature Nanotechnology, 1(3), 159. https://0-doi-org.oasis.lib.tamuk.edu/10.1038/nnano.2006.147.

[16]   Anwana, E. (2020). Social Justice, corporate social responsibility, and sustainable development in South Africa. Hervormde Teologiese Studies, 76(3), 1–10. https://0-doi-org.oasis.lib.tamuk.edu/10.4102/hts.v76i3.6095.

[17]   Alexander, J. K. (2020). An engineering career as industrial mission: jack Keiser in post-war Britain. History and Technology, 36(2), 263–291. https://0-doi-org.oasis.lib.tamuk.edu/10.1080/07341512.2020.1817407.

[18]   Krishna, S. (2021). Digital identity, datafication, and social justice: understanding Aadhaar use among informal workers in south India. Information Technology for Development, 27(1), 67–90. 10.1080/02681102.2020.1818544.

[19]   David, M., Wallkamm, M. & Bleicher, A. (2017). Resource Extraction Technologies: is a More Responsible Path of Development Possible?. Perspectives on Global Development & Technology, 16(4), 367–391. https://0-doi-org.oasis.lib.tamuk.edu/10.1163/15691497-12341440.

[20]   Mitchell, J. T., Thomas, D. S. K. & Cutter, S. L. (1999). Dumping in Dixie revisited: the evolution of environmental injustices in South Carolina. Social Science Quarterly (University of Texas Press), 80(2), 229.

[21]   Chan, J., Distelhorst, G., Kessler, D., Lee, J., Martin-Ortega, O., Pawlicki, P., Selden, M. & Selwyn, B. (2022). After the Foxconn Suicides in China: a Roundtable on Labor, the State and Civil Society in Global Electronics. Critical Sociology (Sage Publications, Ltd.), 48(2), 211–233. https://0-doi-org.oasis.lib.tamuk.edu/10.1177/089692052110134.

[22]   Morata, T. C., Hungerford, M. & Konrad-Martin, D. (2021). Potential Risks to Hearing Functions of Service Members From Exposure to Jet Fuels. American Journal of Audiology, 30, 922–927. https://0-doi-org.oasis.lib.tamuk.edu/10.1044/2021_AJA-20-00226.

[23]   Powers, M., Saberi, P., Pepino, R., Strupp, E., Bugos, E. & Cannuscio, C. (2015). Popular Epidemiology and "Fracking": citizens' Concerns Regarding the Economic, Environmental, Health and Social Impacts of Unconventional Natural Gas Drilling Operations. Journal of Community Health, 40(3), 534–541. https://0-doi-org.oasis.lib.tamuk.edu/10.1007/s10900-014-9968-x.

[24]   Skoumal, R. J., Ries, R., Brudzinski, M. R., Barbour, A. J. & Currie, B. S. (2018). Earthquakes Induced by Hydraulic Fracturing Are Pervasive in Oklahoma. Journal of Geophysical Research.

Solid Earth, 123(12), 10 918–10, 935 https://0-doi-org.oasis.lib.tamuk.edu/10.1029/
2018JB016790.

[25] Dias De Oliveira, M. E., Vaughan, B. E. & Rykiel, E. J. Jr. (2005). Ethanol as Fuel: energy,
Carbon Dioxide Balances, and Ecological Footprint. BioScience, 55(7), 593–602. https://0-doi
-org.oasis.lib.tamuk.edu/10.1641/0006-3568(2005)055[0593:EAFECD]2.0.CO;2.

[26] Mahmud, M. A. P., Huda, N., Farjana, S. H. & Lang, C. (2018). Environmental Impacts of Solar-
Photovoltaic and Solar-Thermal Systems with Life-Cycle Assessment. Energies (19961073),
11(9), 2346. https://0-doi-org.oasis.lib.tamuk.edu/10.3390/en11092346.

[27] Pekhota, A. N., Khroustalev, B. M., Phap, V. M., Romaniuk, V. N., Pekhota, E. A., Vostrova,
R. N. & Nguyen, T. N. (2021). Multicomponent Solid Fuel Production Technology Using Waste
Water. Energetika (0579-2983), 64(6), 525–537. https://0-doi-org.oasis.lib.tamuk.edu/10.
21122/1029-7448-2021-64-6-525-537.

[28] Yazici Guvenc, S. & Varank, G. (2020). Box-Behnken Design Optimization of Electro-Fenton/-
Persulfate Processes Following the Acidification for Tss Removal from Biodiesel Wastewater.
Sigma: Journal of Engineering & Natural Sciences / Mühendislikve Fen Bilimleri Dergisi,
38(4), 1767–1780.

[29] Senousy, H. H. & Ellatif, S. A. (2020). Mixotrophic Cultivation of Coccomyxa subellipsoidea
Microalga on Industrial Dairy Wastewater as an Innovative Method for Biodiesel Lipids
Production. Jordan Journal of Biological Sciences, 13(1), 47–54.

[30] Ramu, S. M., Dinesh, G. H., Thulasinathan, B., Thondi Rajan, A. S., Ponnuchamy, K.,
Pugazhendhi, A. & Alagarsamy, A. (2021). Dark fermentative biohydrogen production from
rice mill wastewater. International Journal of Energy Research, 45(12), 17233–17243.
https://0-doi-org.oasis.lib.tamuk.edu/10.1002/er.5829.

[31] Pugazhendi, A., Al, M. A. E., Jamal, M. T., Jeyakumar, R. B. & Palanisamy, K. (2020). Treatment
of seafood industrial wastewater coupled with electricity production using air cathode
microbial fuel cells under saline conditions. International Journal of Energy Research, 44(15),
12535–12545. https://0-doi-org.oasis.lib.tamuk.edu/10.1002/er.5774.

[32] Magotra, V. K., Kumar, S., Kang, T. W., Inamdar, A. I., Aqueel, A. T., Im, H., Ghodake, G.,
Shinde, S., Waghmode, D. P. & Jeon, H. C. (2020). Compost Soil Microbial Fuel Cell to
Generate power using Urea as Fuel. Scientific Reports, 10(1), 1–9. https://0-doi.org.oasis.lib.
tamuk.edu/10.1038/s41598-020-61038-7.

[33] Rice, J. L., Cohen, D. A., Long, J. & Jurjevich, J. R. (2020). Contradictions of the Climate-
Friendly City: new Perspectives on Eco-Gentrification and Housing Justice. International
Journal of Urban and Regional Research, 44(1), 145–165. https://0-doi-org.oasis.lib.tamuk.
edu/10.1111/1468-2427.12740.

[34] Rawls, A. (1971). Theories of social justice. https://polisci.wustl.edu/files/polisci/imce/l32_331.pdf

[35] Thompson, D. F. (1987). Political ethics and public office (pp 1–253, Editor: Dennis F.
Thompson), Harvard University Press, Cambridge, MA, USA.

[36] Powers, T. M. & Shah, S. I. (2013). Teaching a Course on Ethics in Nanoscience. Journal of
Nano Education, 5(2), 142–147.

[37] Pilarski, L. M., Mehta, M. D., Caulfield, T., Kaler, K. V. & Backhouse, C. J. (2004).
Microsystems and nanoscience for biomedical applications: a view to the future. Bulletin of
Science, Technology & Society, 24(1), 40–45.

[38] Angelos, P. (2020). Surgeons, ethics, and COVID-19: early lessons learned. Journal of the
American College of Surgeons, 230(6), 1119.

[39] Miller, P., Hill-Berry, N. P., Hylton-Fraser, K. & Powell, S. (2019). Social Justice Work as
Activism: the Work of Education Professionals in England and Jamaica. International Studies
in Educational Administration (Commonwealth Council for Educational Administration &
Management (CCEAM)), 47(1), 3–19.

Biography of the editors

Professor Xuan Wang is an instructional assistant professor at Texas A&M Higher Education at McAllen, where she holds appointments in the College of Science at Texas A&M University. Her research interests include synthesizing and developing metal-organic frameworks and porous coordination polymers for chemical sensing and their application in the environmental science field. Her academic training includes undergraduate study at Oklahoma Christian University (BS, 2011) and graduate study under the direction of Professor Hong-Cai Zhou at Texas A&M University (PhD, 2016). After graduation, she began an academic career as a visiting assistant professor at Colorado State University-Pueblo.

Professor Sajid Bashir received his PhD training in matrix-assisted laser desorption/ionization time-of-flight mass spectrometry from the University of Warwick (UK) in 2001 and previously graduated training in Fourier-transform ion cyclotron resonance mass spectrometry from the University of New York at Buffalo (USA). He was a postgraduate research associate at Cornell University (USA) in plant proteomics. Currently, Sajid is a full professor at Texas A&M University-Kingsville (TAMUK) and a past faculty fellow at the U.S. Air Force. He has directed and participated in more than 20 projects supported by the Welch Foundation, TAMUK, Texas Workforce Commission, and the U.S. National Institute of Health. He has coauthored more than 80 book chapters and peer-reviewed journal articles. He is a fellow of the Royal Society of Chemistry and holds Chartered Chemist and separately Chartered Scientists from the Science Council (UK). He is also the American Chemical Society Energy and Fuels Division Technical Secretary (2018–2022). During his service at TAMUK, he trained more than 3,000 students on both undergraduate and graduate levels. He created online courses and established safety training protocols in conjunction with risk management. Currently, he collaborates with local law enforcement as a consultant in forensic chemistry.

Dr. J. Louise Liu received her PhD from the University of Science and Technology Beijing in 2001 and completed her postdoctoral fellowship in the University of Calgary in 2004. She is the director of the Center for Teaching Effectiveness, tenured full professor at Texas A&M University (TAMU)-Kingsville in 2016 and affiliated with the TAMU Energy Institute. Her research covers nanostructured materials design, evaluation, and applications in alternative energy and biological science. She hosted 15 visiting scholars, taught more than 12,000 students; trained more than 200 undergraduates and 60 graduate students. She served as the National Science Foundation panelist and chair and reviewer of dozens of journals. Dr. Liu participated and directed in more than 40 sponsored projects, supported by the different funding agencies. She published more than 130 journal articles, books, and book chapters, and patents. Due to her leadership in chemical sciences and engineering, she was awarded the Chartered Scientist by the Science Council and Chartered Chemist by the Royal Society of Chemistry (RSC), and Fellow of the American Chemical Society (ACS). She was selected as the Distinguished Fellow in the ONR Summer Faculty Research Program, named as the 2021 Distinguished Women by IUPAC, elected as the fellow of the Linnean Society, and Fellow of RSC, in addition to the DEBI faculty fellow at the US Air Force, JSPS invitation fellow, and Israel summer

https://doi.org/10.1515/9783110739879-015

faculty fellow. She is a member of "Diversity, Equity, Inclusion, and Respect" experts panel and serves as a certified ACS Career Consultant, Councilor of Energy and Fuels and chair-elect of South Texas Chapter. She has been volunteering in the local Kleberg Law Enforcement on Forensics.

Special assistants to the editors

Mr. Sai Raghuveer Chava received his MS in chemistry from Texas A&M University-Kingsville in 2010. He was a graduate research assistant at NNTRC, Kingsville, who researched snake venoms. Currently, he is working as a scientist on vaccine research focusing on the analytical method development, formulation, and process-related problems using the analytical data. He has also worked on ophthalmic formulations, process characterization, method developments, and method validations. He coauthored five journal articles, a book chapter, and peer-reviewed journal articles. He is an elected member of the Royal Society of Chemistry and Sigma Xi.

Dr. Ashraf Abedin received his PhD from the Cain Department of Chemical Engineering at Louisiana State University under the supervision of Professor James J. Spivey in December 2021. He completed his bachelor's degree in 2016 from the Department of Chemical Engineering at Bangladesh University of Engineering and Technology. Toward the end of his PhD, Dr. Abedin worked as a Novelis Global Research and Technology Center researcher in Kennesaw, Georgia. He works at the National Energy Technology Laboratory, Morgantown, West Virginia, as a Leidos Research Scientist. His research focuses on catalytic converting fossil fuel to produce clean energy and value-added chemicals. He authored and coauthored numerous book chapters and peer-reviewed journal articles and worked as a guest editor and reviewer for prestigious scientific journals, including *Catalysis Today*.

Biography of the authors

Dr. Matthew Alexander
Associate Professor
Education
PhD in chemical engineering, 1990, Purdue University
MS in chemical engineering, 1986, Georgia Institute of Technology
BS in engineering science and chemistry, 1984, Trinity University

His research interests at Texas A&M University-Kingsville include renewable and biofuel production processes, sustainable energy production using chemical/biochemical processes, remediation of hazardous wastes in soil and groundwater, concentrating on bioremediation processes, industrial wastewater treatment, and chemical process simulation/modeling.

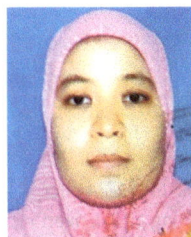

Dr. Telli Alia received her graduate's degree (2003), master's degree (2009), and doctor's degree (2017) in biochemistry from Kasdi Merbah University, Ouargla, Algeria. She is currently working as a lecturer at Kasdi Merbah University. Her research interests include an ethnobotanical survey of medicinal plants, antioxidant, antidiabetic and antimicrobial activities of herbal remedies, and the green synthesis of nanoparticles and their applications.

Bingbao Mei received his PhD in Shanghai Institute of Applied Physics, Chinese Academy of Sciences, in 2021. He got his BE degree from Nanjing University of Aeronautics and Astronautics in 2016. Currently, he is a research assistant at BL14W1 beamline in SSRF. His work focused on *operando* high-energy-resolution X-ray spectroscopy, including XES, RIXS, and HERFD-XAS, and its application in electrochemistry.

Dan Kong is an associate professor at the Department of Materials Science and Physics, China University of Mining and Technology, China. She obtained her PhD in chemical engineering in 2019 from the University College London, UK, and was then a postdoctoral researcher at Aalto University in Finland for one year. She completed her MSc (2014) from Zhejiang University and BSc (2011) from Wuhan University of Technology, China. Dan's research focuses on semiconductors for solar energy conversion, such as overall water splitting and CO_2 conversion. Dan's interests include efficient microwave-assisted synthetic techniques, photoelectrocatalysis, and mechanism investigation.

https://doi.org/10.1515/9783110739879-016

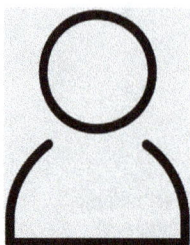

Sabrina DAREM received her master's degree in environmental science from the University of Ghardaia. She is currently a PhD student in the Department of Biology, Laboratory of Soils and Sustainable Development, University of Badji-Mokhtar, Annaba, Algeria. Her research interests include protecting and conserving soil in Algerian Saharan zones with a geostatistics modelization approach, medicinal plants, ecophysiology of plants, adaptation mechanism of plants to biotic and abiotic stress, and interaction of soil–plant–microorganisms.

Guanjie He is an associate professor in Materials Chemistry and Advanced Functional Materials Research Group Lead at the University of Lincoln and an Honorary Lecturer at University College London (UCL). Dr. He received his PhD in Chemistry Department, UCL, under his supervision. Dr. He's research focuses on electrochemical energy storage and conversion applications' materials, especially electrode materials and electrocatalysts in aqueous electrolyte systems.

Dr. Greeshma Gadikota

Research interests: Developing clean methods for producing clean energy carriers such as hydrogen while at the same time capturing and converting the carbon dioxide into a useful and environmentally harmless solid. One thread of her research involves using X-ray scattering, tomography, and spectroscopy techniques to observe chemical reactions as they happen, possibly allowing researchers to develop a method of producing hydrogen without creating greenhouse gases. More broadly, Gadikota would like to develop a method for observing the reaction environments for any solid–liquid–gas system. Gadikota is also studying the feasibility of pulling acid gases from dilute and heterogeneous gas streams and understanding how reactive fluids and solids behave in extreme subsurface and engineered environments.

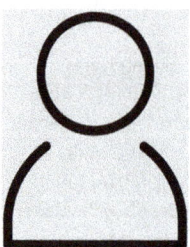

William Houf *and* **Pete Villarreal** are undergraduate students at Texas A&M University-Kingsville, who are taking a BS in chemical engineering with a minor in chemistry (William) and a BS in chemistry with a minor in biomedical science (Pete).

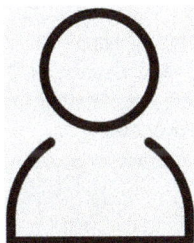

Hsuanyi Huang, Luis F Villanueva, Sushesh Palakurthi, and **Yesirat O Olaseni** were graduate students at Texas A&M University-Kingsville, who were graduated with an MS in chemistry and are working in the chemical industry and are pursuing a doctorate in pharmaceutical sciences (Sushesh).

Haijun Zhang received his PhD in physical chemistry in metallurgy from the University of Science and Technology in Beijing, China, in 1999. After 2 years working as a postdoctoral fellow at Functional Material Research Laboratory. He joined the High-Temperature Ceramics Institute, Zhengzhou University, in 2001 and was promoted to associate professor and then full professor there. In 2007, he joined Prof. Toshima's group at Tokyo University of Science Yamaguchi as a full-time researcher scientist. Currently, Prof. Zhang is a full professor at the State Key Laboratory of Refractories and Metallurgy, Wuhan University of Science and Technology (WUST, China). His research interests include refractories, high-temperature ceramics, and nanostructured materials. He has published more than 470 papers in refereed journals.

Zheng Jiang is a full professor at Shanghai Advanced Research Institute and Shanghai Institute of Applied Physics and beamline leader of X-ray absorption fine structure (BL14W1). After receiving his PhD from the University of Science and Technology of China in 2005, he then worked in Shanghai Synchrotron Radiation Facility (SSRF). In 2007 and 2013, he went to Australian Synchrotron and BESSY II HZB, respectively, as a senior visiting scholar. His main research interests focus on: (1) the design and construction of X-ray spectroscopy beamlines; (2) developing advanced synchrotron radiation X-ray spectroscopy methods, including high-energy-resolution XES, HERFD-XAFS, RIXS, and *operando*/in situ X-ray spectroscopy forming a research system for energy materials from a single atom to cluster.

Hao Zhang received his PhD in nuclear technology and application from the Shanghai Institute of Applied Physics, Chinese Academy of Sciences, in 2021. Currently, he is a lecturer at the Institute of Functional Nano & Soft Materials (FUNSOM), Soochow University. His research interests include applying synchrotron-based X-ray spectroscopy in various electrochemical reactions catalyzed by highly dispersed metal sites.

Ivan. P. Parkin is the dean of Mathematical and Physical Sciences (MAPS), Faculty at University College London. He received his PhD from Imperial College London, and his research focused on chemical vapor deposition, thin films for windows, superhydrophobic surfaces, inorganic materials for electrocatalysis, and zinc-ion batteries. Ivan produced over 600 publications in international journals, and all his works have over 39,000 citations. He has an h-index of 89 (Google scholar).

Jianhong-Jennifer Ren
Education
- PhD, Civil and Environmental Engineering, Northwestern University, 2003
- MS, Civil and Environmental Engineering, Drexel University, 1998
- BS, Environmental Engineering, Beijing Polytechnic University, Beijing, China, 1993

Her research interests at Texas A&M University-Kingsville are contaminant fate and transport; sediment transport; river system restoration and contaminant dynamics in streams, sediment beds, and estuaries; physicochemical particle–particle and particle–contaminant interactions.

Sohaib holds an MS in chemical engineering from the University of Illinois at Chicago and a BS in chemical engineering from Tikrit University in Iraq. Sohaib's research focuses on understanding confined fluids' properties and phase change and colloids' driven- and self-assembly at immiscible liquid interfaces. He utilizes computational tools, such as density functional theory and molecular dynamics simulations, and advanced experimental techniques, such as *in-operand* X-ray scattering, to probe the multiscale morphological, dynamical, and reactivity details of the systems of interest.

Hassnain Asgar is a PhD candidate in Prof. Gadikota's research group. He obtained his master's degree in engineering from Central Michigan University and his undergraduate degree in metallurgy and materials engineering from the University of the Punjab, Lahore. His research work in Prof. Gadikota's group involves understanding the influence of confinement geometry and chemistry on the organization of confined fluids for different energy and environmental applications.

Roli Mishra is working as an assistant professor of chemistry at the Institute of Advanced Research, Gandhinagar (India). She received her PhD (chemistry) in 2006 at the University of Allahabad (India) and did postdoctoral studies at the Indian Institute of Science, Bangalore (India); University of Minnesota, USA; Indian Institute of Technology, Delhi (India). She has worked broadly on peptide synthesis, developing a novel protecting group for nucleoside based on silyl, peptidomimetics, supramolecular chemistry, and thiazolium ionic liquids. Her current research focuses on the design, synthesis, and application of task-specific ionic liquids for organic synthesis, recognition of ions, stability of peptides/drugs, and fluorescence and optoelectronic properties.

Saurabh Vyas did his BSc and MSc in chemistry from the Institute of Advanced Research, Gandhinagar (India). He is currently pursuing his PhD under the supervision of Dr. Roli Mishra at the Institute of Advanced Research Gandhinagar. His research interests are in organic synthesis, synthesis of ionic liquids, and their application in organic transformation and optoelectronics based on complexity.

Shaowei Zhang received his PhD in materials science and engineering from Nagoya Institute of Technology, Japan, in 1996. Prof. Zhang is currently a professor and Royal Society Industry Fellow in the College of Engineering, Mathematics, and Physical Sciences at Exeter, U.K. Before moving to Exeter, he had worked for about 14 years in the Department of Materials Science and Engineering at the University of Sheffield as postdoctoral research associate, EPSRC advanced fellow, lecturer, and reader. His main research interests are in the processing, microstructures, and properties of structural and functional materials. Current research topics include low-temperature fabrication of nanomaterials, novel synthesis of 2D quantum dots, exploration of 2D nanocatalysts, and preparation of porous ceramics and ceramic matrix composites. Other works include the formation of functional coatings and the development of novel armor materials and ultrahigh-temperature ceramics.

With chemistry as his major subject, **Mr. Tanveer ul Haq** received his master's degree (2018) from the Lahore University of Management and Science (LUMS), Lahore. His master's thesis focused on developing a nanocluster for electrochemical water splitting. Since 2019, he has been a PhD student in the group of Prof. Yousef Haik. His current research interest includes the synthesis of self-supported nanoclusters (NCs) for overall water splitting and direct methanol fuel cell applications. He is the author of 11 publications in nanomaterial fields, which mainly focused on synthesizing ultra-small NCs for the electrochemical process. He has been awarded a merit scholarship (LUMS) and dean's honor award (LUMS).

Dr. Vivek Anand has been working as an assistant professor of chemistry at Chandigarh University, Department of Chemistry, since July 2021. He had two years of working experience as assistant professor at the Institute of Advanced Research, Gandhinagar (from August 2019 to July 2021). He received his PhD in physical and organic chemistry from the Department of Chemistry, IIT Madras, in July 2018. He has research expertise in the synthesis and characterization of conjugated organic monomers and polymers via various organic synthetic methodologies. He has applied Sonogashira coupling, Knoevenagel condensation, and Vilsmeyer–Heck reaction to vivid optoelectronic properties such as aggregation-induced emission, white light generation, and fluorescence sensing. To date, he has published eight research articles in international journals.

Wen Lei received his PhD in materials physics and chemistry from Huazhong University of Science and Technology, China, in 2018. He was a visiting student at the University of Waterloo to research in the field of plant genetics. Currently, he is an associate professor at Wuhan University of Science and Technology (WUST) and the "Macao Young Scholar" at the University of Macau. His research interests are materials and devices for electrochemical energy storage and conversion. So far, Dr. Lei has published more than 60 research papers, including *Energy Environ. Sci.*, *Mater. Today*, *Adv. Energy Mater.*, and *ACS Catal.* His papers have been cited more than 2,500 times.

Robert Luckett He is the assistant professor of social work at Texas A&M University-Kingsville.

Yeshu Tan received his master's degree from Soochow University in 2019. He is a PhD student in the Department of Chemistry, University College, London. He focused on the synthesis of nanoparticles and their application in electrocatalysis, including hydrogen evolution reaction and oxygen evolution reaction.

Dr. Zhaohui Wang received his PhD in electrical and computer engineering from the University of Arizona in 2011. He also received a Master of Science in bioengineering from the University of Toledo and Master of Science in mechanical engineering from the University of Arizona. From 2011, he continued postdoctoral training at the University of Pittsburgh Medical Center. In 2014, he was promoted to visiting assistant professor of Electrical Engineering and Computer Science at Texas A&M University-Kingsville. His current research includes medical imaging, bioinformatics, and nanomaterial.

His research interests at North Carolina Agricultural and Technical State University include data mining, embedded system design, biomedical physics, and nanomaterials.

Professor **Yousef Haik** currently serves as professor and department chair of the Mechanical and Industrial Engineering Department at Texas A & M University-Kingsville. His academic career includes appointments in engineering (mechanical, biomedical, and nanoengineering) and science (medicinal chemistry, physics, and nanoscience) programs. His academic and administrative career includes department chair, dean, and provost appointments. He is the inventor and coinventor of several US and international patents. He published and edited 9 books and technical proceedings, over 250 technical papers, and presented many papers at international conferences. He serves as editor-in-chief and on the editorial board for a dozen international journals. He serves on the review boards of International Research Foundations. He organized and chaired several international conferences and symposia. He is a member of a dozen professional associations. He is a fellow of the American Society of Mechanical Engineers and the National Academy for Innovators. His expertise includes nanoenergy, nanomaterials, biofuels, diagnostics, recycling, and environmental remediation. He earned his PhD from the Florida State University in Mechanical Engineering in 1997, an MS from the University of Iowa in 1994, and a BS in mechanical engineering from Jordan in 1986.

Author list

Ashraf Abedin
The Cain Department of Chemical Engineering
Louisiana State University
3307 Patrick F. Taylor Hall
Baton Rouge
LA 70803
USA
Email:mabedi1@lsu.edu

Matthew Alexander
The Department of Chemical Engineering
Texas A&M University-Kingsville
MSC 161, 700 University Blvd.
Kingsville, TX
USA

Telli Alia
Laboratoire de protection des écosystèmes
en zone aride and semi-aride
Université de KASDI Merbah, BP 511 la route
de Ghardaïa
Ouargla 30000
Algérie
Email: telli.alia@univ-ouargla.dz

Vivek Anand
Department of Chemistry, University Institute
of Science
Chandigarh University, Gharuan, Mohali
140413
Punjab
India
Email: vivekanandac88@gmail.com

Hassnain Asgar
Smith School of Chemical and Biological
Engineering
Cornell University
Ithaca, New York 14853
USA
Email: ha356@cornell.edu

Sai Raghuveer Chava
Department of Chemistry
Texas A&M University-Kingsville
MSC 161, 700 University Blvd.
Kingsville, TX 78363-8202
USA

Greeshma Gadikota
School of Civil and Environmental
Engineering
Cornell University
Ithaca
New York 14850
USA
Email: gg464@cornell.edu

Yousef Haik
Department of Mechanical and Industrial
Engineering
Frank H. Dotterweich College of Engineering
Texas A&M University-Kingsville
925 W. Avenue B
Kingsville, TX 78363-8202
USA
Email: yousef.haik@tamuk.edu

Guanjie He
Christopher Ingold Laboratory, Department of
Chemistry
University College London
20 Gordon Street, London WC1H 0AJ
UK
and
School of Chemistry
University of Lincoln
Joseph Banks Laboratories, Green Lane
Lincoln LN6 7DL
UK
Email: .he@ucl.ac.uk

https://doi.org/10.1515/9783110739879-017

William Houf
The Department of Chemistry
Texas A&M University-Kingsville
MSC 161, 700 University Blvd.
Kingsville, TX 78363
USA

Hsuan-Yi Huang
The Department of Chemistry
Texas A&M University-Kingsville
MSC 161, 700 University Blvd.
Kingsville, TX, USA

Zheng Jiang
Shanghai Synchrotron Radiation Facility
Zhangjiang Lab, Shanghai Advanced
Research Institute
Chinese Academy of Sciences
Shanghai 201210
PR China
E-mail: jiangzheng@sinap.ac.cn

Dan Kong
Department of Materials Science and Physics
China University of Mining and Technology
No. 1, Daxue Road, Xuzhou City
Jiangsu Province
PR China
Email: dan.kong.14@ucl.ac.uk

Wen Lei
The State Key Laboratory of Refractories and
Metallurgy
Wuhan University of Science and Technology
Wuhan 430081
PR China

Cuixia Li
School of Materials Science and Engineering
Lanzhou University of Technology
287 Langongping Rd
Qilihe District
Lanzhou, Gansu
PR China

Robert S. Luckett Ed.D.
Department of Social Work
Texas A&M University-Kingsville
MSC 123, 700 University Blvd.
Kingsville, TX 78363-8202
USA
Email: robert.luckett@tamuk.edu

Bingbao Me
Shanghai Synchrotron Radiation Facility
Zhangjiang Lab, Shanghai Advanced
Research Institute
Chinese Academy of Sciences
Shanghai 201210
PR China

Sohaib Mohammed
School of Civil and Environmental
Engineering
Cornell University
Ithaca, New York 14850
USA
Email: sm2793@cornell.edu

Yesirat O. Olaseni
The Department of Chemistry
Texas A&M University-Kingsville
MSC 161, 700 University Blvd.
Kingsville, TX
USA

Sushesh Srivatsa Palakurthi
The Department of Chemistry
Texas A&M University-Kingsville
MSC 161, 700 University Blvd.
Kingsville, TX
USA

Ivan P. Parkin
Christopher Ingold Laboratory
Department of Chemistry
University College London
20 Gordon Street
London WC1H 0AJ
UK
Email: i.p.parkin@ucl.ac.uk

Jianhong Ren
The Department of Environmental
Engineering
Texas A&M University-Kingsville
MSC 213, 917 W. Avenue B
Kingsville, TX 78363
USA

Darem Sabrine
Laboratoire de sol et développement durable
Université Badji Mokhtar
PB 12 Annaba 23000
Algérie

Yeshu Tan
Christopher Ingold Laboratory
Department of Chemistry
University College London
20 Gordon Street
London WC1H 0AJ
UK
Email:yeshu.tan.19@ucl.ac.uk

Luis Villanueva
The Department of Chemistry
Texas A&M University-Kingsville
MSC 161, 700 University Blvd.
Kingsville, TX 78363
USA

Pete Villarreal
The Department of Chemistry
Texas A&M University-Kingsville
MSC 161, 700 University Blvd.
Kingsville, TX
USA

Tanveer ul Haq
Sustainable Energy Engineering
Frank H. Dotterweich College of Engineering
Texas A&M University-Kingsville
Kingsville, TX 78363-8202
USA

Saurabh Vyas
Department of Engineering and Physical
Sciences
Institute of Advanced Research
Gandhinagar 382426
Gujarat
India
Email:saurabhvyas.phd2020@iar.ac.in

Zhaohui Wang
(Formerly at)
The Department of Electrical Engineering and
Computer Science
Texas A&M University-Kingsville
MSC 192, 700 University Blvd.
Kingsville, TX
USA
(Currently at)
North Carolina A&T State University
Department of Computer Systems Technology
Greensboro, NC 27411
USA

Haijun Zhang
The State Key Laboratory of Refractories and
Metallurgy
Wuhan University of Science and Technology
Wuhan 430081
PR China
Email: zhanghaijun@wust.edu.cn
Tel: +86-15697181003

Hao Zhang
Institute of Functional Nano & Soft Materials
(FUNSOM)
Jiangsu Key Laboratory for Carbon-Based
Functional Materials and Devices
Soochow University
199 Ren'ai Road
Suzhou 215123, Jiangsu
PR China

Shaowei Zhang
College of Engineering
Mathematics and Physical Sciences
University of Exeter
Exeter EX4 4QF
UK

Index

https://doi.org/10.1515/9783110739879-018

www.ingramcontent.com/pod-product-compliance
Lightning Source LLC
Chambersburg PA
CBHW060938210326

41598CB00031B/4668